Hans Hahn
Gesammelte Abhandlungen
Band 3

L. Schmetterer und K. Sigmund (Hrsg.)

SpringerWienNewYork

Hans Hahn
Collected Works
Volume 3

L. Schmetterer and K. Sigmund (eds.)

SpringerWienNewYork

em. Univ.-Prof. Dr. Leopold Schmetterer
Univ.-Prof. Dr. Karl Sigmund
Institut für Mathematik, Universität Wien
Strudlhofgasse 4, A-1090 Wien

Gedruckt mit Unterstützung des Fonds zur
Förderung der wissenschaftlichen Forschung

Satz: Vogel Medien GmbH, A-2100 Korneuburg
Druck: Novographic, Ing. W. Schmid, A-1238 Wien
Bindearbeiten: Fa. Papyrus, A-1100 Wien
Graphisches Konzept: Ecke Bonk
Gedruckt auf säurefreiem, chlorfrei gebleichtem Papier – TCF

ISBN 3-211-82781-1 Springer Wien New York

Vorwort

Die gesammelten mathematischen und philosophischen Werke von Hans Hahn erscheinen hier in einer dreibändigen Ausgabe. Sie enthält sämtliche Veröffentlichungen von Hahn, mit Ausnahme jener, die ursprünglich in Buchform erschienen – dazu gehören neben dem zweibändigen Werk über *Reelle Funktionen* auch die *Einführung in die Elemente der höheren Mathematik,* die er gemeinsam mit Heinrich Tietze schrieb, seine Anmerkungen zu Bolzanos *Paradoxien des Unendlichen* und mehrere Kapitel für E. Pascals *Repertorium der höheren Mathematik.* Nicht aufgenommen wurden auch die Buchbesprechungen von Hahn, bis auf seine Besprechung von Pringsheims *Vorlesungen über Zahlen- und Funktionslehre,* die einen eigenen Aufsatz über die Grundlagen des Zahlbegriffs darstellt.

Hahn war nicht nur einer der hervorragendsten Mathematiker dieses Jahrhunderts: Sein Einfluß auf die Philosophie war auch höchst bedeutsam. Das kommt in der Einleitung, die sein ehemaliger Schüler Sir Karl Popper für diese Gesamtausgabe geschrieben hat, deutlich zum Ausdruck. (Diese Einleitung ist der letzte Essay, den Sir Karl Popper verfaßte.)

Hahn schrieb ausschließlich auf deutsch. Wir haben seine Arbeiten in Teilgebiete zusammengefaßt (was auch auf andere Art geschehen hätte können) und ihnen jeweils einen englischsprachigen Kommentar vorangestellt. Diese Kommentare, die von hervorragenden Experten stammen, beschreiben Hahns Arbeiten und ihre Wirkung. Die Teilgebiete sind *Theorie der Kurven* (Kommentar von Hans Sagan), *Funktionalanalysis* (Harro Heuser), *Geordnete Gruppen* (Laszlo Fuchs), *Variationsrechnung* (Wilhelm Frank), *Reelle Funktionen* (David Preiss), *Strömungslehre* (Alfred Kluwick), *Maß- und Integrationstheorie* (Heinz Bauer), *Harmonische Analysis* (Jean-Pierre Kahane), *Funktionentheorie* (Ludger Kaup) und *Philosophie* (Christian Thiel). Die Einleitung von Sir Karl Popper und die Kurzbiographie Hahns sind in deutscher und englischer Sprache abgedruckt.

Wir danken den Professoren Olga Taussky-Todd, Georg Nöbeling, Leopold Vietoris, Fritz Haslinger, Harald Rindler und Walter Schachermayer für ihre Unterstützung und dem österreichischen Fonds zur Förderung der Wissenschaftlichen Forschung für Druckkostenbeiträge.

L. S.

K. S.

Preface

This three-volume edition contains the collected mathematical and philosophical writings of Hans Hahn. It covers all publications of Hahn, with the exception of those which appeared originally in book form – i. e. the two-volume treatise of real functions, a joint work (with Heinrich Tietze) on elementary mathematics, Hahn's annotated edition of Bolzano's *Paradoxien des Unendlichen* and several chapters of the *Repertorium der höheren Mathematik* edited by E. Pascal. We also did not include Hahn's book reviews, with the exception of his review of Pringsheim's *Vorlesungen über Zahlen- und Funktionslehre,* which is, in fact, a self-contained essay on the foundations of the concept of number.

Hahn was not only one of the leading mathematicians of the twentieth century: his influence in philosophy was also most remarkable. This is underscored by the introduction written by Hahn's former student, Sir Karl Popper, for this comprehensive edition. (Incidentally, this is the last essay that was written by Sir Karl.)

Hahn published exclusively in German. We have grouped his papers according to subject matter (a task with a non-unique solution); each group of papers is preceded by an English commentary resuming Hahn's results and discussing their place in the history of ideas. The subject areas are *curve theory* (commented by Hans Sagan), *functional analysis* (Harro Heuser), *ordered groups* (Laszlo Fuchs), *calculus of variations* (Wilhelm Frank), *real functions* (David Preiss), *hydrodynamics* (Alfred Kluwick), *measure and integration* (Heinz Bauer), *harmonic analysis* (Jean-Pierre Kahane), *complex functions* (Ludger Kaup) and *philosophical writings* (Christian Thiel). Sir Karl's introduction and the short biography of Hans Hahn are presented both in English and in German.

We are grateful to Professors Olga Taussky-Todd, Georg Nöbeling, Leopold Vietoris, Fritz Haslinger, Harald Rindler and Walter Schachermayer for their help, and to the Austrian Fonds zur Förderung der Wissenschaftlichen Forschung for financial support.

L. S.
K. S.

Inhaltsverzeichnis
Table of Contents

Hahn's work in complex analysis / Hahns Arbeit zur Funktionentheorie

Hahn's philosophical writings / Hahns philosophische Schriften

Comments on Hans Hahn's work in measure theory

Heinz Bauer
Erlangen

The contributions of Hahn to the theory of measure and integration belong to the period between 1914 and 1933.

The first two papers [H;1912] and [H;1915] discuss remarkable properties of the Riemann integral and compare the Lebesgue integral with an integral devised by Borel which is related to the Riemann integral. Hahn here adds to work by Lebesgue and Borel which is little known today.

The remaining five papers by Hahn on measure and integration were published between 1928 and 1933. They are contributions to the so-called abstract measure theory which was founded in 1915 by Fréchet [12], in immediate temporal and causal connection, incidentally, with the famous paper by Radon [26] from 1913. However, Hahn deals in most cases directly with signed abstract measures, i. e. with countably additive (or totally additive) set functions with values in $\bar{\mathbb{R}} = [-\infty, +\infty]$.

We shall devote one paragraph each to the two groups of papers mentioned in the previous sections. A third paragraph deals uniquely with the paper [H;1933]. It contains results on products of measures which today belong to the classical stock-in-trade of measure theory.

1. Contributions to the Riemann integral and its generalisations

1.1. The paper [H;1914] is based on the following setup: let $[a, b]$ be a compact interval in $\mathbb{R}$. For every partition $z = (x_0, \ldots, x_n)$ of $[a, b]$ by finitely many points $a = x_0 < x_1 < \ldots < x_n = b$, let δ_z denote the *mesh* of z, i. e. the maximum of the lengths $x_j - x_{j-1}$ of the subintervals $[x_{j-1}, x_j]$, for $j = 1, \ldots, n$. *A sequence* z_n of partitions of the interval $[a, b]$ is said to be *distinguished* if (δ_{z_n}) converges to 0. If one chooses for a given partition

$z = (x_0, \ldots, x_n)$ intermediate points $\xi_1, \ldots, \xi_n$ so that $\xi_j \in [x_{j-1}, x_j]$, this yields the usual Riemannian sum

$$S_z := S_z^{(\xi)} := \sum_{j=1}^{n} f(\xi_j)(x_j - x_{j-1}),$$

where we shall mostly use the first (incomplete) notation.

If f is bounded and integrable in the sense of Lebesgue (or L-integrable, for short), i. e. if f is integrable with respect to the Lebesgue measure (the completed measure of Lebesgue-Borel) on $[a, b]$, then not only the Lebesgue integral $\int_a^b f\,dx$, but every limit of a convergent sequence of Riemann sums (S_{z_n}) – based on a distinguished sequence of partitions and a selection of corresponding intermediate points – belongs to the interval $I_f := [\underline{\int}_a^b f\,dx, \overline{\int}_a^b f\,dx]$ having as end points the upper and lower Riemann integral. Even more holds: the interval I_f is the set of all such limits of convergent sequences (S_{z_n}), where it is always assumed that (z_n) is a distinguished sequence of partitions. In particular, the L-integral $\int_a^b f\,dx$ itself is the limit of a convergent sequence (S_{z_n}), with both the distinguished sequence of partitions and the corresponding intermediate points depending on the function f.

This simple result had already been pointed out in 1909 by Lebesgue [18], who had shown even more: the function f can be replaced by an arbitrary finite set $f_1, \ldots, f_k$ of L-integrable, *not necessarily bounded* functions on $[a, b]$. There still exists a distinguished sequence of partitions (z_n) (depending only on the $f_1, \ldots, f_k$) and, for each $i \in \{1, \ldots, k\}$ a corresponding sequence of intermediate points $(\xi^{(n,i)})$ with $\lim_{n\to\infty} S_{z_n}^{(\xi^{n,i})} = \int_a^b f_i dx$ for $i = 1, \ldots, k$.

Hahn's main result sharpens this statement in two respects. Firstly, the hitherto finite sequence of L-integrable functions is replaced by an arbitary sequence $\mathcal{F} = (f_i)_{i\in\mathbb{N}}$ of L-integrable functions. Secondly, a distinguished sequence of partitions (z_n) is given in advance. Then one can choose for each $i \in \mathbb{N}$ a sequence of intermediate points (ξ^n) (depending only on the partitions z_n and hence on $n \in \mathbb{N}$) such that

$$\lim_{n\to\infty} S_{z_n}^{(\xi^n)} = \int_a^b f_i dx \qquad \text{for all } i \in \mathbb{N}.$$

This remarkable result is based on the following argument which is of interest in its own right.

Let f be a bounded L-integrable function on $[a, b]$ taking its values in the compact interval $[m, M]$. Then there exists, for any two positive real numbers $\epsilon > 0$ and $\alpha > 0$, a continuous real function g on $[a, b]$ with values in $[m, M]$ sucht that

$$\lambda^1\{|f-g|>\varepsilon\}<\alpha \tag{1}$$

where λ^1 denotes Lebesgue measure on $[a, b]$. Hahn uses here the notion of *stochastic convergence* (but not today's terminology or a notation as in (1)). The statement (1) allows him to view the function f as the stochastic limit of a sequence (g_n) of continuous functions on $[a, b]$ with values in $[m, M]$ and to effectuate the passage to the limit

$$\lim_{n\to\infty}\int_a^b g_n dx = \int_a^b f dx. \tag{2}$$

In order to prove (1), Hahn uses a result by Vitali [33] from 1905. It is the result which contains „Lusin's Theorem" in its original form (cf. Bourbaki [8], p. 117, footnote 3). Lusin rediscovered the theorem in 1912 without being aware of Vitali's work. In its modern form (see [2], p. 187, and Bourbaki [8], p. 169–173), Lusin's theorem, together with Tietze's extension theorem (which appeared only in 1915) would have enabled Hahn to obtain statement (1) even for $\varepsilon = 0$. To validate the passage to the limit in (2), Hahn quotes the paper [5] by Borel. Today, the corresponding result is contained in a well-known convergence theorem linking stochastic convergence with uniform integrability (see [2], p. 142).

In the last part of his paper [H, 1914] Hahn improves upon a result of Zoard de Geöcze [34], p. 273. For this, one associates with every partition $z = (x_0, \ldots, x_n)$ of $[a, b]$ its so-called *quotient* defined by

$$Q_z := \min\left\{\frac{x_i - x_{i-1}}{x_j - x_{j-1}} : i, j = 1, \ldots, n, i \neq j\right\}$$

Obviously, Q_z is the largest real $Q > 0$ such that

$$Q \leq \frac{x_i - x_{i-1}}{x_j - x_{j-1}} \leq \frac{1}{Q} \qquad \text{for all } i, j = 1, \ldots, n, i \neq j.$$

In particular one has $0 < Q_z \leq 1$.

Hahn then shows: for every L-integrable real function f on $[a, b]$, there exists a distinguished sequence of partitions (z_n) of $[a, b]$ with the following properties:

(a) z_{n+1} is a refinement of z_n;

(b) $\lim_{n\to\infty} Q_{z_n} = 1$;

(c) $\lim_{n\to\infty} \sum_{j=1}^{j_n} \frac{f(x_j^{(n)}) + f(x_{j-1}^{(n)})}{2} (x_j^{(n)} - x_{j-1}^{(n)}) = \int_a^b f dx,$

where $z_n = (x_0^{(n)}, \ldots, x_{j_n}^{(n)})$.

In [34] this result is obtained under the additional assumption that the L-integrable function f is semi-continous. Furthermore, (b) is replaced by the weaker condition that $Q_{z_n} \geq \frac{1}{6}$ for all $n \in \mathbb{N}$. Hahn's result remains valid if one replaces the function f by a sequence $\mathcal{F}$ of L-integrable real functions on $[a, b]$. Together with (a) and (b), condition (c) then holds for all $f \in \mathcal{F}$. The proof can be carried out analogously to the afore-mentioned main result of the paper. This is pointed out at the end of [H; 1915].

1.2. The paper[H; 1915] appeared at a time when, on the one hand, Lebesgue's theory of integration was well on its way to success, but on the other hand several attempts were undertaken to modify Riemann's definition of the integral so that it would accommodate unbounded integrands and encompass conditionally convergent integrals. For the history of this development we refer to the article by Montel and Rosenthal in the Encyclopedia [23], especially from p. 1053 onward. Borel had announced such a generalisation of the Riemann integral in his two notes [3] and [4] from 1910, and gave a detailed presentation in his paper [5] from 1912.

In the paper to be discussed, this definition is slightly generalised and then compared with the integrals of Lebesgue and Riemann.

Let us first turn to the extension by Hahn and Borel of Riemann's definition. We consider a real-valued function f defined almost everywhere on a compact interval $[a, b]$ of the real line; hence we assume that there exists a Lebesgue null set $K \subset [a, b]$ such that f is defined on $[a, b]\setminus K$. The set K will be called a *set of singularities* if f is bounded on every closed subset F of $[a, b]$ which is disjoint from K. The following condition formulated by Hahn is equivalent to this: let (J_k) be a sequence of pairwise disjoint intervals $J_k \subset [a, b]$ which are open in $[a, b]$ and cover K. Then f is bounded on $[a, b] \setminus \bigcup_{k=1}^{\infty} J_k$. Obviously one can additionally assume without restricting generality that $J_k \cap K \neq \emptyset$ for all $k \in \mathbb{N}$. We shall say that (J_k) is a *reduced covering* of K by intervals .

Let $J := (J_k)$ be such a reduced covering of K by intervals. Furthermore, let $z = (x_0, \ldots, x_n)$ be a partition of $[a, b]$ such that the points x_i with $i \neq 0$ and $i \neq n$ do not belong to any of the intervals J_k. Let us put

$$S_z(J) := \sum_{j=1}^{n} f(\xi_j) h_j,$$

where we assume that each ξ_j belongs to $F_j := [x_{j-1}, x_j] \setminus \bigcup J_k$ and where

$$h_j := \lambda^1(F_j).$$

The quantity

$$\delta_z(J) := \max\{h_j : j = 1, \ldots, n\}$$

will be said (for our purpose) to be the *modified mesh* of z with respect to J. In Hahn's terminology, h_j is the length of $[x_{j-1}, x_j]$ reduced by the total length ‚of the parts of the intervals J which belong to the subinterval $[x_{j-1}, x_j]$'.

Hahn defines f to be Borel-integrable if the following two conditions hold:

1. The sequence of the sums $S_{z_n}(J)$ converges whenever (z_n) is a sequence of partitions of the type described just now, and the modified mesh $\delta_{z_n}(J)$ converges to 0.

2. The corresponding limits $\sigma(J)$ converge if the total length $l(J) := \lambda^1(\bigcup J_k)$ of the reduced coverings by intervals $J = (J_k)$ tends to zero.

Finally, the limit obtained in this way is denoted by

$$\text{(B)} \int_a^b f dx := \lim_{l(J) \to 0} \sigma(J) \tag{3}$$

and is said to be the *Borel integral* of f. (In the paper Borel [5] which is the starting point of Hahn's investigation, the situation is somewhat simpler: the set of singularities K is assumed to be countable.)

For a given reduced covering by intervals $J = (J_k)$ of K, we denote by L_J the closed set $[a, b] \setminus \bigcup_{k=1}^{\infty} J_k$. Therefore, the h_j defined above satisfies

$$h_j = \lambda^1([x_{j-1}, x_j] \cap L_J)$$

for any given partition $z = (x_0, \ldots, x_n)$ with $x_1, \ldots, x_n \notin K$. If the Borel integral exists, then the limit $\sigma(J)$ introduced above will also exist. It is said to be the Riemann integral with respect to the set L_J. If the set of singularities K is empty, then this integral coincides with the usual Riemann integral $\int_a^b f dx$ of the function f (which then is bounded and defined on the whole interval $[a, b]$).

In [H; 1915] it is shown that, for a bounded integrand f, the existence of the Borel integral implies that of the Lebesgue integral and that both yield the same value. The converse, however, does not hold. But Hahn's definition can be slightly modified (by means of the Riemann integral $\sigma(J)$ defined above) so that the resulting definition is equivalent to that of the Lebesgue integral for bounded integrands.

At the time Hahn's paper appeared, the case of unbounded integrands f was of particular interest, in view of the attempts to comprehend improper integrals. But Hahn notes that the definition of (Hahn-)Borel leads only to a very restricted set of interesting improper integrals. For this he shows that on the one hand, the existence of the Borel integral for f and (additionally) for its absolute value $|f|$ implies that the Lebesgue integral of f exists and yields the same value as the Borel integral. On the other hand, Hahn shows that if f is Borel integrable, then the Lebesgue integral of f over the closure $\bar{K}$ of the set of singularities always exists. Furthermore, the Borel integral turns out to be invariant with respect to alterations of the set of singularities, in the following sense: if the Borel integral of f exists with respect to two sets of singularities, then these two integrals coincide.

The paper of Hahn discussed here was published in 1915, during World War I, and therefore it appears to have remained unnoticed for a considerable time. Indeed, after the war, Lebesgue and Borel engaged in an uncommonly acrimonious dispute which can be retraced in the papers [19],[20] by Lebesgue and [6],[7] by Borel. In the first paper [19] Lebesgue writes (p.193): „M. Borel affirme que sa théorie est plus générale que la mienne, mais il n'en donne aucune preuve."

Both authors seem at the time to have been unaware of Hahn's paper. Otherwise, the dispute between Lebesgue and Borel could have been settled to a large extent, especially since Hahn writes in the introduction to [H;1915] that „the question concerning the range of Borel's definition is not yet properly clarified." In his paper Hahn provides the missing clarification.

The main purpose of Hahn's investigation was, as a sequel of Borel's work, to get a grip upon integrals which are not absolutely convergent. This was also the topic of an important survey lecture held by T. H. Hildebrand in april 1917 at a symposion in Chicago, whose printed version (see Hildebrand [17]) is still highly readable today. Hahn's paper is discussed in detail. In particular, Hahn's results are placed in the proper perspective of a general evolution leading from Jordan, Harnack and Moore via Lebesgue and Young, Borel and Hahn to Denjoy.

2. Abstract measures

2.1. The theory of abstract measures emerged – mainly under the influence of Fréchet[12] – soon after Radon's famous paper [26]. In the sixth chapter of his book „Theorie der reellen Funktionen", which appeared in 1921, Hahn deals with the foundations of this theory and proves (on p. 403) the statement, well-known today, that a σ-additive (= totally additive) set function φ on a σ-algebra (i. e. a signed measure) always admits a largest and a smallest value. In the very same step he identifies these two values with the „total mass" of the positive resp. negative part of φ. At the same stroke Hahn derives, as an immediate consequence, the decomposition theorem which today bears his name.

In his short note [H; 1928] Hahn emphasises that the signed measures considered in his book of 1921 may well have their values on the extended real line $[-\infty, +\infty]$, while the σ-additivity of φ implies that $-\infty$ and $+\infty$ cannot both occur as values of φ. In the note Hahn sketches a particularly simple proof for the existence of the largest and the smallest value of a signed measure. The proof is short and does not require the construction of the positive and negative parts of φ. This proof is presented in greater detail on p. 17 of the monograph [15] of Hahn and Rosenthal, which was published in 1948, fourteen years after Hahn's death. The theorem itself was re-discovered (without prior knowledge of Hahn's paper) by Franck [11] (cf. in this context Nikodym [24]). In volume 5 of the *Fundamenta Mathematicae,* immediately after the paper by Franck, one finds a further proof of the same theorem by Sierpinski [30]. But both proofs are considerably more complicated than the proof by Hahn.

2.2. The theorem on the existence of the largest and the smallest value of every signed measure φ is closely connected with the intermediate value theorem dating back to Fréchet [13] and Sierpinski [30]. According to this theorem φ admits every value between its extreme values, provided φ is ‚free of singularities'. (It is well known that this theorem has been extended in Lyapunov [22] to measures whose values are vectors in $\mathbb{R}^k$: the connectedness of the set of values is replaced by its convexity. This insight yielded the decisive impetus for the investigation of general vector-valued measures.)

The paper [H; 1928a] announces results that were proved in full detail in the paper [H; 1930] with the same title. This title does not mention the intermediate value theorem. Nevertheless, the latter gave probably the impetus for the results in both papers.

Hahn considers an infinite series $\sum_{\nu\in\mathbb{N}} a_\nu$ with elements $a_\nu > 0$ converging monotonically towards zero. Let $g \in]0, +\infty]$ be the sum of the series. It is shown first that there exists for every $x \in]0, g]$ a subseries $\sum_{i\in\mathbb{N}} a_{\nu_i}$ – i. e. with indices $\nu_i < \nu_{i+1}$ – with the property

$$x = \sum_{i\in\mathbb{N}} a_{\nu_i} \tag{4}$$

if and only if

$$\sum_{\nu>n} a_\nu \geq a_n \tag{5}$$

holds for all $n \in \mathbb{N}$.

The monotonicity of the sequence (a_ν) is only used in the proof of the necessity of condition (5). It follows that for a divergent series $\sum a_\nu$ with elements $a_\nu > 0$ decreasing monotonically to 0 the following holds: every real number $x > 0$ admits at least one representation in the form (4) by a suitably chosen subseries of the original series.

Using this tool, the intermediate value theorem can be proved quite easily. Indeed, let φ be a signed measure on a σ-algebra $\mathcal{A}$ in a set Ω such that φ is *free of singularities.* This means that $\mathcal{A}$ contains no set S which is a *singularity* in the following sense: $\varphi(S) \neq 0$ and for all $A \in \mathcal{A}$ with $A \subset S$ either $\varphi(A) = 0$ or $\varphi(A) = \varphi(S)$.

To avoid misunderstandings, let it be added that a singularity S is said to be an *atom* if, in addition, $|\varphi(S)| < +\infty$. Every atom, therefore, is a singularity, but the converse does not hold. Correspondingly, a signed measure which is free of atoms is free of singularities, but not the other way round. (If φ is a positive, inner regular Borel-measure on a locally compact space E, i. e. a Radon measure, then „free of atoms" is equivalent to „diffuse", i. e. to $\varphi(\{a\}) = 0$ for all a $\in E$, cf. Bourbaki [9], § 5, exercises 9, 10 and [1], chapter 6.5., p. 3.)

In the situation described above, the intermediate value theorem states that for a signed measure φ on the σ-algebra $\mathcal{A}$ which is free of singularities, the range $\varphi(\mathcal{A})$ is an interval in $\mathbb{R}$. (Since φ attains a largest and a smallest value, this interval must be closed, but it is a compact subset of $\mathbb{R}$ for real φ only.) In Hahn's paper the assumption is, in a formal sense, more general: in the terminology of Hahn's book [14] $\mathcal{A}$ is a ‚σ-field', i. e. a σ-ring in the terminology used today. Hence, in [H; 1928a] and [H; 1930] the intermediate value property is always formulated for the σ-algebra of all elements of a σ-ring which are contained in a given set of this σ-ring.

In order to prove the intermediate value theorem Hahn reduces the problem to the case $\varphi \geq 0$ of a (positive) measure on $\mathcal{A}$. He first shows that in the case $\varphi(\Omega) = +\infty$ the lack of singularities implies that the measure φ has to admit arbitrarily large real values. Hence one can assume $\varphi(\Omega) < +\infty$. But then it is possible to select a suitable sequence (A_n) of pairwise disjoint sets from $\mathcal{A}$ such that $\sum_{\nu>n} \varphi(A_\nu) \geq \varphi(A_n)$ holds for all $n \in \mathbb{N}$. Thus (5) is satisfied and the previously mentioned result concerning subseries yields (together with $\varphi(\emptyset) = 0$) that every number $x \in [0, \varphi(\Omega)]$ occurs as a value of φ.

The result on subseries is complemented, in [H; 1928a] and [H; 1930], by a necessary and sufficient condition for the property that, under the assumptions mentioned above ($a_n > 0$ decreasing monotonically to 0), every number x with $0 < x \leq g$ can be represented in a unique way as a subseries of thé given series, i. e. in the form

$$x = \sum_{i \in \mathbb{N}} a_{\nu_i}.$$

This result allows a remarkable application to *Cantor series* (cf. [25]), i. e. to series of the following form: there exists a sequence (n_k) of natural numbers such that

$$a_1 = \ldots = a_{n_1} = \frac{1}{n_1 + 1}, \quad a_{n_1+1} = \ldots = a_{n_1+n_2} = \frac{1}{(n_1+1)(n_2+1)}, \ldots$$

Such Cantor series turn out to be the only series $\sum_{n \in \mathbb{N}} a_n$ (again under the previous general assumptions) with the property that every number $x \in]0,1]$ can be represented by *exactly one* subseries of the original series. This result can be found in [H; 1930], but not in its announcement [H; 1928a].

2.3. The note [H; 1929] deals with the derivation of a little known, remarkably simple characterisation of the concepts of ‚integral' and ‚integrability' in the sense of Lebesgue's theory. The whole idea, however, is developed directly for abstract measures. The short note displays the ideas of the proof in the form of a sketch. In the same year (1929) a detailed elaboration of the sketch appears under the same title: [H; 1929a] is based on a lecture given by Hahn at the „57th Conference of German Philologists and School-men" in Salzburg.

In his paper Hahn starts out with a (positive) measure on a σ-ring (i. e. a σ-field in Hahn's terminology). In the more detailed version, even a signed measure is admitted. In the following we restrict ourselves – as we did before – to the case of a (positive) measure φ on a σ-algebra $\mathcal{A}$ in a given set Ω. Hahn assumes that φ is complete, but minor modifications allow to drop this requirement.

Let f be a function defined φ-almost everywhere on Ω and with range in $\bar{\mathbb{R}}$. Hahn defines f to be φ-integrable if there exists a signed measure λ on $\mathcal{A}$ which satisfies the following, unique condition:

If for a set $A \in \mathcal{A}$ there exist coefficients $\alpha, \beta \in \bar{\mathbb{R}}$ such that the double inequality

$$\alpha \leq f(x) \leq \beta \tag{6}$$

holds for all $x \in A$, then

$$\alpha\varphi(A) \leq \lambda(A) \leq \beta\varphi(A) \tag{7}$$

(with the usual convention that $\pm\infty \cdot 0 = 0$).

As Hahn shows, this implies that the signed measure λ is uniquely defined, that f is φ-integrable in the usual sense and that

$$\lambda(A) = \int_A f d\varphi \tag{8}$$

holds for all $A \in \mathcal{A}$. In other words, the signed measure $A \rightarrow \int_A f d\varphi$ on $\mathcal{A}$ (having density f with respect to φ) is singled out by the condition (6),(7). This characterises at the same stroke the φ-integrability of f.

Hahn emphasizes in [H; 1929a] that one can find in de la Valleé-Poussin [32] a hint towards this introduction of the concept of integral (p. 55), which however remains unproven there. In the book by Hahn and Rosenthal the concept of integral is introduced – even for signed measures – precisely according to Hahn's idea (cf. [15], p. 149 ff.).

3. Product measures

In [H; 1933] the construction of the product of two positive measures is dealt with in a comprehensive way. Hahn remarks in a footnote on the first page of his memoir that it is the elaboration of a set of lectures which he had delivered in Brünn (today Brno) in March 1933, and a few years earlier at the Viennese Mathematical Society. Hahn furthermore notes that the same topic had been treated in a completely independent way in Sak's book [27] which also appeared in 1933. A revised version of Hahn's paper can be found on the pages 223–245 of the book by Hahn and Rosenthal [15]. The traces which the ideas of Hahn and Saks have left in almost all text books on measure and integration cannot be overlooked.

Saks presents the theory of products of abstract measures in the form of a survey within a few pages (pp. 259–263) in the appendix of his book [27], up to the generalisations of the classical theorems of Fubini and Tonelli which date from the years 1907 and 1909 and relate to the theory of Lebesgue in its original sense. In contrast Hahn develops the theory in its full extension, but he formulates the theorem of Fubini and Tonelli only for indicator functions. His presentation concludes with this result, which coincides essentially with Hahn's *reduction theorem* at the end of his paper. From todays point of view it seems surprising that the names of Fubini and Tonelli are not even mentioned in Hahn's memoir. On the other hand, one finds at the end of the paper the editorial note: „Leonardo Tonelli – Direttore responsabile“.

Hahn's paper is not exactly easy to read, but it is full of important ideas which, at the time, were new. First Hahn deals with the process of completing a non-negative, σ-additive set function on a ring $\mathfrak{R}$ in a set Ω. By adjoining all subsets of φ-null sets from $\mathfrak{R}$, one first obtains the completion $(\Omega,\bar{\mathfrak{R}},\bar{\varphi})$ of φ. Here $\bar{\mathfrak{R}}$ is the smallest of all σ-rings $\tilde{\mathfrak{R}}$ in Ω which contain $\mathfrak{R}$ and for which there exists an extension $\tilde{\varphi}$ of φ to a non-negative, σ-additive set function with the additional property that $\tilde{\mathfrak{R}}$ is $\tilde{\varphi}$-complete, i. e. contains all subsets of $\tilde{\varphi}$-null sets. Hahn denotes $\bar{\mathfrak{R}}$ as the *complete hull* of $\mathfrak{R}$ (with respect to φ). Next, this complete hull is characterised in the manner which is well-known today: following Carathéodory [10], one associates to the given set function φ in the usual way an (outer) measure function φ^* defined on all subsets of Ω. Its restriction to the system of all subsets of Ω which are φ^*-measurable in the sense of Carathéodory is a complete measure.

At this point Hahn introduces the further assumption of the σ-*finiteness* of φ, i. e. the representability of Ω as union of a sequence of sets $A_n \in \mathfrak{R}$ with $\varphi(A_n) < +\infty$. He shows the fact, well-known today (cf. [2],[16]), that under the given asumptions the system of all φ^*-measurable subsets of Ω coincides with the complete hull $\bar{\mathfrak{R}}$ of $\mathfrak{R}$, and that $\bar{\varphi}$ is nothing else but the restriction of φ^* to $\bar{\mathfrak{R}}$. This is a result which erroneously is often associated with Carathéodory. Furthermore, Hahn considers the σ-ring $\mathcal{B}$ generated by the ring $\mathfrak{R}$ in Ω. Due to the σ-finiteness this is a σ-algebra, named by Hahn the system of Borel sets over $\mathfrak{R}$. $\bar{\mathfrak{R}}$ turns out to be the complete hull of the σ-algebra $\mathcal{B}$ of Borel sets, if the measure is chosen to be the restriction of $\bar{\varphi}$ to $\mathcal{B}$. This is illustrated by obtaining the Lebesgue measure λ^k, starting from the half-open rectangles parallel to the axes in $\mathbb{R}^k$. In this context the abstract Borel sets introduced above coincide with the usual Borel sets in $\mathbb{R}^k$.

If φ_1 and φ_2 are two non-negative σ-additive set functions defined on rings $\mathcal{R}_1$ resp. $\mathcal{R}_2$ contained in the sets Ω_1 resp. Ω_2, then the „product ring“ $\mathcal{R}$ on $\Omega_1 \times \Omega_2$ generated by the sets $A_1 \times A_2$ with $A_i \in \mathcal{R}_i$, $i = 1, 2$, supports exactly one σ-additive set function $\pi \geq 0$ with the property that $\pi(A_1 \times A_2) = \varphi_1(A_1)\varphi_2(A_2)$ holds for each of those sets $A_1 \times A_2$. The function π defines its complete hull and hence the complete measure $\bar{\pi}$. This is denoted as the *product measure* $\varphi_1 \times \varphi_2$ of φ_1 and φ_2. It is available without any finiteness assumptions concerning φ_1 and φ_2, and satisfies, according to its derivation,

$$(\varphi_1 \times \varphi_2)(A_1 \times A_2) = \varphi_1(A_1) \cdot \varphi_2(A_2) \tag{9}$$

for all $A_1 \times A_2 \in \mathcal{R}$.

If one asumes that φ_1 and φ_2 are *σ-finite,* then – according to the previously mentioned results – the restriction π_0 of the product measure on the σ-algebra of Borel sets over the product ring $\mathcal{R}$ is the unique measure ψ on $\mathcal{B}$ which coincides on $\mathcal{R}$ with π and hence also with $\varphi_1 \times \varphi_2$, i. e. satisfying condition (9) in the modified form

$$\psi(A_1 \times A_2) = \varphi_1(A_1) \cdot \varphi_2(A_2)$$

for all $A_1 \times A_2 \in \mathcal{R}$. The product measure $\varphi_1 \times \varphi_2$ itself is the completion of π_0.

A theorem of Tonelli-type for sets, i. e. – implicitly – for indicator functions holds for this product measure, again under the additional assumption of the σ-finiteness of φ_1 and φ_2. And only this situation is considered. In Hahn's paper, the relevant theorem remains without a name; but it contains, essentially, the statement of his reduction theorem placed at the end of the paper. For all sets $M \in \bar{\mathcal{R}}$ one has

$$(\varphi_1 \times \varphi_2)(M) = \int \varphi_2(M_{x_1}) d\varphi_1(x_1) \tag{10}$$

where M_{x_1} denotes the usual x_1-section of the set M and where, in addition, M_{x_1}, belongs up to a null set to the completion of $\mathcal{R}_2$ with respect to φ_2 and $x_1 \to \varphi_2(M_{x_1})$ is φ_1-measurable. An analogous statement holds if one interchanges x_1 and x_2:

$$(\varphi_1 \times \varphi_2)(M) = \int \varphi_1(M^{x_2}) d\varphi_2(x_2) \tag{11}$$

where the x_2-section M^{x_2} takes the place of the x_1-section. From this one derives – as is usual today (cf. for instance [2], p. 162–163) – the associativity of the product.

Apparently Hahn was the first to recognize the importance of the property which is subsumed under the notion of σ-finiteness and which was explicitly formulated by him. The notion received its name only much later (probably in the textbook by Halmos [16]). In the book by Hahn and Rosenthal [15] the notion is still used without bearing a name. In the corresponding section of Saks' monograph [27] the notion of σ-finiteness does not appear, although the properties (10) and (11) are precisely those emphasized in his book. It is only in the second (English) version of his book that Saks explicitly points out the importance of the σ-finiteness condition. On p. 87 he even gives an example (well-known today) which demonstrates the necessity of this condition for Hahn's investigations. Furthermore, Saks refers to Hahn's paper in the corresponding section ([28], p. 88). Incidentally, one also finds in Halmos [16], at the end of section 36 in exercise 8 a remark on Hahn's ideas, although Hahn's name remains unmentioned.

Finally, let us remark that the proof of the existence of products of σ-finite, abstract measures is nowadays usually carried through with the help of equations (10), (11). This has the additional advantage that it allows in a natural way to operate with arbitrary, not necessarily complete σ-algebras.

Thus Hahn's paper [H; 1933] has had a particularly decisive and lasting influence on the development of measure theory.

References

[1] Bauer H (1984) *Maße auf topologischen Räumen.* Fernuniversität Hagen

[2] Bauer H (1992) *Maß- und Integrationstheorie,* 2. Auflage. de Gruyter, Berlin New York

[3] Borel É (1910) Sur la définition de l'intégrale définie. C R Acad Sci 150: 375–377

[4] Borel É (1910) Sur une condition générale d'intégrabilité. C R Acad Sci 150: 508–511

[5] Borel É (1912) Le calcul des intégrales définies. J de math, 6ᵉ série, 8: 159–210

[6] Borel É (1919) Sur l'intégration des fonctions non bornées et sur les définitions constructives. Ann Éc Norm, 3ᵉ série, 36: 71–91

[7] Borel É (1920) A propos de la définition de l'intégrale définie. (Lettre à M. le Directeur des Annales Scientifiques de l'École Normale Supérieure). Ann Éc Norm, 3ᵉ série, 37: 461–462

[8] Bourbaki N (1956) *Intégration,* chap 5. Hermann, Paris

[9] Bourbaki N (1965) *Intégration,* chap 1–4. Hermann, Paris

[10] Carathéodory C (1914) Über das lineare Maß von Punktmengen – eine Verallgemeinerung des Längenbegriffs. Nachr K Ges d Wiss Göttingen, math-phys Kl: 404–426

[11] Franck R (1924) Sur une propriéte des fonctions additives d'ensemble. Fund Math 5: 252–261

[12] Fréchet M (1915) Sur l'intégrale d'une fonctionelle étendue à un ensemble abstrait. Bull Soc Math France 43: 249–267

[13] Fréchet M (1923) Des familles et fonctions additives d'ensembles abstraits. Fund Math 4: 329–365

[14] Hahn H (1921) *Theorie der reellen Funktionen.* Springer, Berlin

[15] Hahn H, Rosenthal A (1948) *Set Functions.* The University of New Mexico Press, Albuquerque NM

[16] Halmos PR (1950) *Measure Theory.* D. van Nostrand, New York London Toronto

[17] Hildebrandt TH (1918) On integrals related to and extensions of the Lebesgue integrals. Bull Am Math Soc 24: 113–144, 177–202

[18] Lebesgue H (1909) Sur les intégrales singulières. Ann Fac Sci Univ Toulouse (3) 1: 25–117

[19] Lebesgue H (1918) Remarques sur les théories de la mesure et de l'intégration. Ann Éc Norm, 3ᵉ série, 35: 191–250

[20] Lebesgue H (1920) Sur une définition due à M. Borel. Ann Éc Norm, 3ᵉ série, 37: 255–257

[21] Lusin N (1912) Sur les propriétés des fonctions mesurables. C R Acad Sci 154: 1688–1690

[22] Lyapunov A (1940) Sur les fonctions-vecteurs complètement additives. Bull Acad Sci URSS 4: 465–478

[23] Montel P, Rosenthal A (1924) Integration und Differentiation. Encyklop d Math Wiss II C 9b. Teubner, Leipzig

[24] Nikodym O (1930) Sur une généralisation des intégrales de M. J. Radon. Fund Math 15: 131–179, 358 (Errata et remarques)

[25] Perron O (1960) *Irrationalzahlen,* 4. Aufl. de Gruyter, Berlin

[26] Radon J (1913) Theorie und Anwendungen der absolut additiven Mengenfunktionen. Sitzungsber Kaiserl Akad Wiss Wien, Math-Naturw Kl 120: 1295–1438

[27] Saks St (1933) *Théorie de l'Intégrale.* Warszawa

[28] Saks St (1937) *Theory of the Integral.* Warszawa – Lwów

[29] Sierpinski W (1922) Sur des fonctions additives et continues. Fund Math 3: 240–246

[30] Sierpinski W (1924) Demonstration d'un théorème sur les fonctions additives d'ensemble. Fund Math 5: 262–264

[31] Tietze H (1915) Über Funktionen, die auf einer abgeschlossenen Menge stetig sind. J reine u angew Math 145: 9–14

[32] de la Vallée-Poussin C (1916) *Intégrales de Lebesgue, Fonctions d'Ensembles, Classes de Baire.* Gauthier-Villars, Paris

[33] Vitali G (1905) Una proprietà delle funzioni misurabili. R Inst Lombardo Rendiconti (2) 3: 599–603

[34] Zoard de Geöcze M (1911) Sur la fonction semi-continue. Bull Soc Math France 39: 256–295

Hahn's work in measure theory
Hahns Arbeiten zur Maßtheorie

Über Annäherung an Lebesgue'sche Integrale durch Riemann'sche Summen

von

Hans Hahn in Czernowitz.

(Vorgelegt in der Sitzung am 12. Februar 1914.)

Ist die Funktion $f(x)$ integrierbar im Sinne von Riemann im Intervalle $< a, b >$, so ist ihr Integral $\int_a^b f(x)dx$, der Riemann'schen Definition gemäß, Grenzwert von Summen folgender Gestalt: man nehme mit dem Intervalle $< a, b >$ eine Zerlegung D vor, durch Einschalten der Teilpunkte

$$a = x_0 < x_1 < \ldots < x_{\nu-1} < x_\nu = b$$

wähle in jedem Teilintervalle $< x_{i-1}, x_i >$ nach Belieben den Punkt ξ_i und bilde die Summe:

$$\sum_{i=1}^{\nu} f(\xi_i)(x_i - x_{i-1}); \tag{1}$$

das Integral ist der Grenzwert dieser Summen, wenn die Zerlegung D eine ausgezeichnete Zerlegungsfolge[1] durchläuft.

Jede im Riemann'schen Sinne integrierbare Funktion ist auch integrierbar im Sinne von Lebesgue, und zwar ergeben beide Integraldefinitionen denselben Wert. Ist hingegen $f(x)$ nur integrierbar im Sinne von Lebesgue, nicht aber im Sinne von

[1] Nach G. Kowalewski heißt eine Folge von Zerlegungen D_n des Intervalles $< a, b >$ eine ausgezeichnete Zerlegungsfolge, wenn das größte in D_n auftretende Teilintervall für $n = \infty$ den Grenzwert 0 hat.

Riemann, so liegt, wenn $f(x)$ geschränkt ist, das Integral $\int_a^b f dx$ zwischen oberem und unterem Riemann'schem Integral.

Da nun jeder beliebige zwischen oberem und unterem Riemann'schen Integral von f liegende Wert bei geeigneter Wahl einer ausgezeichneten Zerlegungsfolge und der Punkte ξ_i als Grenzwert Riemannscher Summen (1) darstellbar ist, so gilt dies auch vom Lebesgue'schen Integrale der geschränkten Funktion $f(x)$. H. Lebesgue hat, über diese evidente Bemerkung hinaus, bewiesen,[1] daß, wenn eine endliche Anzahl von Funktionen gegeben sind, die in seinem Sinne integrierbar sind,[2] eine ausgezeichnete Zerlegungsfolge und die Punkte ξ_i stets so gefunden werden können, daß für jede der gegebenen Funktionen die mit Hilfe dieser Zerlegungsfolge und dieser Punkte ξ_i gebildeten Riemann'schen Summen gegen das Integral der betreffenden Funktion konvergieren. Im folgenden wird dies Resultat nach zwei Richtungen hin wesentlich erweitert. Erstens wird gezeigt, daß die ausgezeichnete Zerlegungsfolge D_n noch beliebig vorgegeben werden kann, daß also durch geeignete Wahl der ξ_i allein bewirkt werden kann, daß die Riemannschen Summen (1) das Lebesgue'sche Integral zur Grenze haben. Zweitens aber wird gezeigt, daß, wieder bei vorgegebener ausgezeichneter Zerlegungsfolge D_n, die gleichzeitige Annäherung der Riemann'schen Summen an die Lebesgue'schen Integrale nicht nur für endlich viele, sondern auch für abzählbar unendlich viele Funktionen durch geeignete, für alle diese Funktionen gemeinsame Wahl der Punkte ξ_i erreicht werden kann.

Bekanntlich konvergieren für eine im Riemann'schen Sinne integrierbare Funktion nicht nur die Summen (1), sondern auch die Summen:

---

[1] Ann. de Toulouse (3), *1*, p. 30 ff.

[2] Wir nennen hier, wie im folgenden diejenigen Funktionen integrierbar im Sinne von Lebesgue, die Lebesgue selbst als »fonctions sommables« bezeichnet. Es sind dies alle (geschränkten wie ungeschränkten) Funktionen, denen durch das Verfahren von Lebesgue ein (absolut konvergentes) bestimmtes Integral zugeordnet ist.

$$\sum_{i=1}^{\nu} \frac{f(x_i)+f(x_{i-1})}{2}(x_i-x_{i-1}) \tag{2}$$

für jede ausgezeichnete Zerlegungsfolge gegen das Integral $\int_a^b f dx$. Mit diesen Summen (2) hat sich, für den Fall, daß f halbstetig ist, Zoard de Geöcze eingehend beschäftigt.[1] Er sagt von einer Zerlegung D, sie habe den Quotienten Q, wenn für irgend zwei ihrer Teilintervalle die Ungleichung gilt:

$$Q \leqq \frac{x_i - x_{i-1}}{x_j - x_{j-1}} \leqq \frac{1}{Q}$$

und beweist, daß für jede im Lebesgue'schen Sinne integrierbare halbstetige Funktion f das Integral $\int_a^b f dx$ als Grenzwert von Summen (2) dargestellt werden kann, mit Hilfe einer ausgezeichneten Folge von Zerlegungen D_n, deren Quotient $\geqq \frac{1}{6}$ ist, und wo D_{n+1} aus D_n durch Unterteilung entsteht. An Stelle dieses speziellen Resultates soll nun im folgenden das allgemeine gesetzt werden, daß eine solche Annäherung durch Summen (2) an $\int_a^b f dx$ für jede im Lebesgue'schen Sinne integrierbare Funktion möglich ist, und zwar mit Hilfe von ausgezeichneten Zerlegungsfolgen D_n, bei denen D_{n+1} aus D_n durch Unterteilung entsteht, und für deren Quotienten Q_n gilt: $\lim_{n=\infty} Q_n = 1$. Durch genau dieselben Methoden kann man übrigens beweisen, daß die Annäherung an $\int_a^b f dx$ durch die speziellen Riemann'schen Summen

$$\sum_{i=1}^{\nu} f(x_{i-1})(x_i - x_{i-1}) \quad \text{oder} \quad \sum_{i=1}^{\nu} f(x_i)(x_i - x_{i-1})$$

---

[1] Bull. soc. math., *39*, p. 256 ff.

möglich ist mit Hilfe ausgezeichneter Zerlegungsfolgen D_n, denen die eben angeführten Eigenschaften zukommen.[1]

§ 1.

Wir stellen zunächst einige bekannte Sätze zusammen, auf die wir uns stützen werden. Dabei sei ein für allemal bemerkt, daß, wo vom Inhalt einer Punktmenge die Rede ist, stets ihr Inhalt im Sinne von Lebesgue zu verstehen ist, und daß das Wort »integrierbar« stets bedeutet: »integrierbar im Sinne von Lebesgue«.

Bekanntlich heißt eine Funktion von erster Klasse, wenn sie Grenze von stetigen Funktionen ist, von zweiter Klasse, wenn sie Grenze von Funktionen erster Klasse ist, ohne selbst von erster Klasse zu sein. Die einseitigen oberen und unteren Ableitungen einer stetigen Funktion sind von höchstens zweiter (d. h. von zweiter oder erster) Klasse.

Eine im Intervall $< a, b >$ integrierbare[2] Funktion $f(x)$ unterscheidet sich von einer Funktion höchstens zweiter Klasse nur in den Punkten einer Menge des Inhaltes 0.

In der Tat, es ist das unbestimmte Integral

$$F(x) = \int_a^x f(x)dx$$

in $< a, b >$ stetig. Bekanntlich existiert die Ableitung $F'(x)$ und hat den Wert $f(x)$ überall in $< a, b >$, abgesehen von einer Punktmenge des Inhaltes 0. Also stimmt, abgesehen von einer solchen Punktmenge, $f(x)$ mit der rechten oberen Ableitung von $F(x)$ überein, und die Behauptung ist bewiesen.

Ist $f(x)$ integrierbar in $< a, b >$ und ist

$$m \leqq f(x) \leqq M,$$

so gibt es zu je zwei positiven Zahlen ε und α eine stetige Funktion $g(x)$, so daß

$$m \leqq g(x) \leqq M$$

---

[1] In anderer Richtung als die hier mitgeteilten Untersuchungen liegen die von W. H. Young, Phil. Trans. (A) *204* (1905) p. 221 und J. Pierpont, The theory of functions of real variables II (1912) p. 371, wo der Zusammenhang zwischen Lebesgue'schen Integralen und Riemann'schen Summen aus unendlich vielen Summanden behandelt wird.

[2] Es gilt dies übrigens für jede »meßbare« Funktion. G. Vitali Rend. Ist. Lomb. (2), *38*, p. 599.

und so, daß überall in $< a, b >$ abgesehen von einer Punktmenge des Inhaltes α die Ungleichung gilt:

$$|f(x)-g(x)| < \varepsilon.$$

Wir werden beim Beweise die Tatsache benutzen, daß, wenn man eine Funktion erster (beziehungsweise zweiter) Klasse überall dort, wo sie $> M$ ist, durch M ersetzt, überall dort, wo sie $< m$ ist, durch m, wieder eine Funktion erster (beziehungsweise höchstens zweiter) Klasse entsteht.

Um nun den ausgesprochenen Satz zu beweisen, ersetzen wir zunächst $f(x)$ durch eine Funktion höchstens zweiter Klasse $h(x)$, die sich von $f(x)$ nur in den Punkten einer Menge des Inhaltes 0 unterscheidet und von der nun immer angenommen werden kann, daß sie wie $f(x)$ der Ungleichung genügt:

$$m \leqq h(x) \leqq M.$$

Da $h(x)$ von höchstens zweiter Klasse ist, haben wir

$$h(x) = \lim_{\nu = \infty} h_\nu(x),$$

wo $h_\nu(x)$ von erster Klasse ist, und auf Grund der vorausgeschickten Bemerkung angenommen werden kann:

$$m \leqq h_\nu(x) \leqq M.$$

Da $h_\nu(x)$ von erster Klasse ist, so gilt weiter:

$$h_\nu(x) = \lim_{\mu = \infty} h_{\nu\mu}(x),$$

wo jedes $h_{\nu\mu}(x)$ stetig ist und wieder angenommen werden kann:

$$m \leqq h_{\nu\mu}(x) \leqq M.$$

Es genügt nun, um unsere Behauptung nachzuweisen, wenn gezeigt wird, daß die in ihr auftretende Eigenschaft bei Grenzübergang erhalten bleibt. Denn da diese Eigenschaft den stetigen Funktionen $h_{\nu\mu}(x)$ in trivialer Weise zukommt, so kommt sie dann auch den Grenzfunktionen $h_\nu(x)$ der $h_{\nu\mu}(x)$ zu und ebenso der Grenzfunktion $h(x)$ der $h_\nu(x)$. Kommt sie aber der Funktion $h(x)$ zu, so offenbar auch der von $h(x)$ nur in den

Punkten einer Menge des Inhaltes 0 verschiedenen Funktion $f(x)$. Es bleibt also nur folgendes zu beweisen:[1]

Genügen alle $f_\nu(x)$ der Ungleichung

$$m \leqq f_\nu(x) \leqq M$$

und gibt es zu je zwei positiven Zahlen ε und α eine der Ungleichung

$$m \leqq g_\nu(x) \leqq M$$

genügende stetige Funktion $g_\nu(x)$, für die, abgesehen von einer Punktmenge des Inhaltes α überall in $< a, b >$ die Ungleichung gilt:

$$|f_\nu(x) - g_\nu(x)| < \varepsilon,$$

so gilt für $f(x) = \lim f_\nu(x)$, dieselbe Eigenschaft, d. h. zu jedem ε und α gehört eine der Ungleichung

$$m \leqq g(x) \leqq M$$

genügende stetige Funktion $g(x)$, für die, abgesehen von einer Punktmenge des Inhaltes α überall in $< a, b >$ die Ungleichung gilt

$$|f(x) - g(x)| < \varepsilon.$$

Sei zum Beweise $\varepsilon_1, \varepsilon_2, \ldots, \varepsilon_\nu, \ldots$ eine Folge positiver Zahlen mit $\lim_{\nu=\infty} \varepsilon_\nu = 0$ und seien die positiven Zahlen η und $\alpha_1, \alpha_2, \ldots, \alpha_\nu, \ldots$ so gewählt, daß:

$$\eta + \sum_{\nu=1}^{\infty} \alpha_\nu < \alpha. \tag{3}$$

Nach Voraussetzung kann die stetige Funktion $g_\nu(x)$ so gefunden werden, daß

$$m \leqq g_\nu(x) \leqq M$$

und daß abgesehen von einer Menge $\mathfrak{A}_\nu$ des Inhaltes α_ν überall in $< a, b >$ die Ungleichung gilt:

$$|f_\nu(x) - g_\nu(x)| < \varepsilon_\nu.$$

---

[1] É. Borel, Journ. de math. (6), 8, p. 193.

Außerhalb der Vereinigungsmenge $\mathfrak{R}$ aller Mengen $\mathfrak{A}_\nu$ gilt also:

$$\lim_{\nu=\infty} g_\nu(x) = f(x),$$

und für den Inhalt ρ von $\mathfrak{R}$ gilt:

$$\rho \leqq \sum_{\nu=1}^{\infty} \alpha_\nu. \tag{4}$$

Bezeichnen wir nun mit $\mathfrak{B}_\nu$ die Menge aller jener Punkte von $< a, b >$, in denen für alle $i \geqq \nu$ die Ungleichung gilt:

$$|g_i(x) - f(x)| < \varepsilon,$$

so muß die Vereinigungsmenge aller Mengen $\mathfrak{B}_\nu$ alle nicht zu $\mathfrak{R}$ gehörigen Punkte von $< a, b >$ enthalten. Da $\mathfrak{B}_\nu$-Teil von $\mathfrak{B}_{\nu+1}$ ist, so hat man daher, wenn β_ν den Inhalt von $\mathfrak{B}_\nu$ bezeichnet,

$$\lim_{\nu=\infty} \beta_\nu \geqq b - a - \rho,$$

und für alle $\nu \geqq \nu_0$ wird, wenn ν_0 hinlänglich groß ist, die Ungleichung gelten:

$$b - a - \beta_\nu < \eta + \rho$$

und daher, bei Benutzung von (3) und (4):

$$b - a - \beta_\nu < \eta + \sum_{i=1}^{\infty} \alpha_i < \alpha,$$

d. h. die Menge der nicht zu $\mathfrak{B}_\nu$ gehörigen Punkte von $< a, b >$ hat einen Inhalt $< \alpha$. Da aber auf $\mathfrak{B}_\nu$ die Ungleichung gilt:

$$|g_\nu(x) - f(x)| < \varepsilon,$$

so kann jede der Funktionen $g_\nu(x)$ $(\nu \geqq \nu_0)$ für die Funktion $g(x)$ unserer Behauptung gewählt werden, und diese Behauptung ist bewiesen.

Endlich werden wir noch folgenden Satz benötigen:[1]

Ist $f(x)$ integrierbar, ist:

$$\lim_{\nu=\infty} \varepsilon_\nu = 0, \qquad \lim_{\nu=\infty} \alpha_\nu = 0,$$

---

[1] É. Borel, a. a. O., p. 195.

genügt die stetige Funktion $f_\nu(x)$ überall in $< a, b >$, abgesehen von einer Punktmenge $\mathfrak{A}_\nu$ des Inhaltes α_ν der Ungleichung

$$|f_\nu(x)-f(x)| < \varepsilon_\nu,$$

gibt es endlich eine Zahl F, so daß

$$|f(x)| \leqq F, \ |f_\nu(x)| \leqq F \quad (\nu = 1, 2, \ldots .),$$

so ist

$$\lim_{\nu=\infty} \int_a^b f_\nu(x)\,dx = \int_a^b f(x)\,dx.$$

In der Tat, bezeichnet $\mathfrak{B}_\nu$ die Komplementärmenge von $\mathfrak{A}_\nu$ bezüglich des Intervalles $< a, b >$ so hat man:

$$\int_a^b (f_\nu(x)-f(x))\,dx = \int_{\mathfrak{B}_\nu} (f_\nu(x)-f(x))\,dx + \int_{\mathfrak{A}_\nu} (f_\nu(x)-f(x))\,dx$$

und daher:

$$\left| \int_a^b f_\nu(x)\,dx - \int_a^b f(x)\,dx \right| \leqq \varepsilon_\nu(b-a) + 2F\alpha_\nu$$

und da

$$\lim_{\nu=\infty} (\varepsilon_\nu(b-a) + 2F\alpha_\nu) = 0$$

ist, ist die Behauptung erwiesen.

§ 2.

Wir kommen nun zu unserem eigentlichen Gegenstand und beweisen den Satz:

I. Ist f im Intervall $< a, b >$ integrierbar im Sinne von Lebesgue und ist irgend eine ausgezeichnete Zerlegungsfolge D_n des Intervalles $< a, b >$ gegeben so kann, wenn $\delta_1^{(n)}, \delta_2^{(n)}, \ldots, \delta_{\nu_n}^{(n)}$ die Teilintervalle der Zerlegung D_n sind, stets in $\delta_i^{(n)}$ der Punkt $\xi_i^{(n)}$ so gefunden werden, daß:

$$\lim_{n=\infty} \sum_{i=1}^{\nu_n} f(\xi_i^{(n)})\delta_i^{(n)} = \int_a^b f(x)\,dx.$$

Wir nehmen zunächst an, f sei in $< a, b >$ geschränkt.

Sei $\varepsilon_1, \varepsilon_2, \ldots, \varepsilon_m, \ldots$ irgend eine Folge positiver Zahlen mit $\lim_{m=\infty} \varepsilon_m = 0$ und seien die positiven Zahlen $\alpha_1, \alpha_2, \ldots, \alpha_m, \ldots$ so gewählt, daß $\sum_{i=1}^{\infty} \alpha_i$ konvergiert; wir setzen:

$$\rho_m = \sum_{i=m}^{\infty} \alpha_i$$

und haben $\lim_{m=\infty} \rho_m = 0$. Nun bezeichnen wir mit $f_m(x)$ eine zwischen oberer und unterer Grenze von $f(x)$ verbleibende stetige Funktion, die, abgesehen von einer Punktmenge $\mathfrak{A}_m$ des Inhaltes α_m in $< a, b >$ der Ungleichung genügt:

$$|f(x) - f_m(x)| < \varepsilon_m.$$

Wie in § 1 erinnert wurde, gibt es eine solche Funktion. Die Vereinigungsmenge der Mengen $\mathfrak{A}_m, \mathfrak{A}_{m+1}, \ldots, \mathfrak{A}_{m+i}, \ldots$ bezeichnen wir mit $\mathfrak{R}_m$; ihr Inhalt ist höchstens ρ_m und es ist $\mathfrak{R}_{m+1}$ Teil von $\mathfrak{R}_m$. Seien $\delta_i^{(n)}$ die Teilintervalle der Zerlegung D_n. Zu jedem dieser Intervalle $\delta_i^{(n)}$ gibt es einen kleinsten Wert m_i des Index m derart, daß $\delta_i^{(n)}$ nicht ganz in der Menge $\mathfrak{R}_{m_i}$ enthalten ist: in der Tat, sobald m so groß ist, daß ρ_m kleiner als die Länge des Intervalles $\delta_i^{(n)}$ geworden ist, was wegen $\lim_{m=\infty} \rho_m = 0$ sicher eintritt, enthält $\delta_i^{(n)}$ gewiß Punkte, die nicht zu $\mathfrak{R}_m$ gehören. In jedem Intervalle $\delta_i^{(n)}$ werde nun $\xi_i^{(n)}$ ganz beliebig außerhalb $\mathfrak{R}_{m_i}$ angenommen.

Ein Punkt $\xi_i^{(n)}$ gehört also nur dann zu $\mathfrak{R}_m$, wenn alle Punkte des betreffenden Teilintervalles $\delta_i^{(n)}$ zu $\mathfrak{R}_m$ gehören, woraus folgt: die Summe der Längen derjenigen Intervalle $\delta_i^{(n)}$, deren Punkt $\xi_i^{(n)}$ zu $\mathfrak{R}_m$ gehört, ist $\leqq \rho_m$. Und da $\mathfrak{A}_m$ Teil von $\mathfrak{R}_m$ ist, ist erst recht die Summe der Längen derjenigen Intervalle $\delta_i^{(n)}$ deren Punkt $\xi_i^{(n)}$ zu $\mathfrak{A}_m$ gehört, $\leqq \rho_m$.

Wir behaupten nun: ist $\eta > 0$ beliebig vorgegeben, so gibt es ein m_0 derart, daß für $m \geqq m_0$ und sämtliche Einteilungen D_n die Ungleichung gilt:

$$\left| \sum_{i=1}^{\nu_n} f_m(\xi_i^{(n)}) \delta_i^{(n)} - \sum_{i=1}^{\nu_n} f(\xi_i^{(n)}) \delta_i^{(n)} \right| < \eta. \tag{5}$$

In der Tat, sei irgend ein fester Wert von m gegeben, so bezeichnen wir diejenigen $\delta_i^{(n)}$, deren Punkt $\xi_i^{(n)}$ nicht zu $\mathfrak{A}_m$ gehört mit $\bar{\delta}_i^{(n)}$, die übrigen mit $\bar{\bar{\delta}}_i^{(n)}$, die in die Intervalle $\bar{\delta}_i^{(n)}$ fallenden Punkte $\xi_i^{(n)}$ mit $\bar{\xi}_i^{(n)}$, die in die Intervalle $\bar{\bar{\delta}}_i^{(n)}$ fallenden mit $\bar{\bar{\xi}}_i^{(n)}$.

Wie wir gesehen haben, ist:

$$\Sigma \bar{\bar{\delta}}_i^{(n)} \leqq \rho_m.$$

Für die in den $\bar{\delta}_i^{(n)}$ liegenden Punkte $\bar{\xi}_i^{(n)}$ gilt:

$$|f_m(\bar{\xi}_i^{(n)}) - f(\bar{\xi}_i^{(n)})| < \varepsilon_m.$$

Ist endlich F so groß gewählt, daß überall in $< a, b >$

$$|f(x)| \leqq F$$

und mithin auch

$$|f_m(x)| \leqq F,$$

so haben wir:

$$|\Sigma f_m(\bar{\xi}_i^{(n)}) \bar{\delta}_i^{(n)} - \Sigma f(\bar{\xi}_i^{(n)}) \bar{\delta}_i^{(n)}| < \varepsilon_m (b-a)$$

$$|\Sigma f_m(\bar{\bar{\xi}}_i^{(n)}) \bar{\bar{\delta}}_i^{(n)} - \Sigma f(\bar{\bar{\xi}}_i^{(n)}) \bar{\bar{\delta}}_i^{(n)}| \leqq 2 F \rho_m.$$

Da $\lim_{m=\infty} \varepsilon_m = 0$ und $\lim_{m=\infty} \rho_m = 0$ ist, kann m_0 so gewählt werden, daß für $m \geqq m_0$:

$$\varepsilon_m (b-a) + 2 F \rho_m < \eta$$

und somit gilt für $m \geqq m_0$ auch (5).

Da nach § 1:

$$\lim_{m=\infty} \int_a^b f_m \, dx = \int_a^b f \, dx$$

ist, kann m_0 auch so groß gewählt werden, daß für $m \geqq m_0$:

$$\left| \int_a^b f_m \, dx - \int_a^b f \, dx \right| < \eta \tag{6}$$

ist. Wir wählen irgendein $m \geqq m_0$ und können, da f_m stetig ist und mithin:

$$\int_a^b f_m \, dx = \lim_{n=\infty} \sum_{i=1}^{\nu_n} f_m(\xi_i^{(n)}) \delta_i^{(n)}$$

ist, n_0 so groß wählen, daß für $n \geqq n_0$:

$$\left| \sum_{i=1}^{\nu_n} f_m(\xi_i^{(n)}) \delta_i^{(n)} - \int_a^b f_m \, dx \right| < \eta. \tag{7}$$

Die drei Ungleichungen (5), (6), (7) zusammengenommen liefern für $n \geqq n_0$:

$$\left| \sum_{i=1}^{\nu_n} f(\xi_i^{(n)}) \delta_i^{(n)} - \int_a^b f \, dx \right| < 3\eta, \tag{8}$$

und da hierin $\eta > 0$ willkürlich war, ist dies die behauptete Gleichung von Satz I, der mithin für geschränktes f bewiesen ist.

Nun müssen wir uns noch von der Voraussetzung, daß f geschränkt sei, frei machen. Wir bezeichnen wieder mit ε_m eine Folge gegen 0 abnehmender positiver Zahlen und bezeichnen mit φ_k (k irgendeine natürliche Zahl) die Funktion, die mit f überall dort übereinstimmt, wo $|f| \leqq k$ ist und überall sonst den Wert 0 hat. Bekanntlich ist dann

$$\lim_{k=\infty} \int_a^b \varphi_k \, dx = \int_a^b f \, dx,$$

es kann also k_m so groß gewählt werden, daß:

$$\left| \int_a^b \varphi_{k_m} \, dx - \int_a^b f \, dx \right| < \varepsilon_m, \tag{9}$$

und es kann immer angenommen werden:

$$k_{m+1} > k_m.$$

Mit ρ_m bezeichnen wir eine Folge positiver Zahlen, wobei $\rho_{m+1} < \rho_m$ und:

$$\lim_{m=\infty} k_m \cdot \rho_m = 0$$

sei; es werde gesetzt:

$$\alpha_m = \rho_m - \rho_{m+1},$$

dann ist sicher $\alpha_m > 0$. Wir können nun nach § 1 eine stetige Funktion f_m, die wie φ_{k_m} der Ungleichung:

$$|f_m| \leqq k_m \tag{10}$$

genügt, so finden, daß überall in $< a, b >$, abgesehen von einer Punktmenge $\mathfrak{A}_m$ des Inhaltes α_m die Ungleichung:

$$|f_m - \varphi_{k_m}| < \varepsilon_m$$

besteht. Nun unterscheiden sich φ_{k_m} und f nur in der Menge $\mathfrak{K}_m$ jener Punkte, wo $|f| \geqq k_m$ ist, und bezeichnet man den Inhalt dieser Menge mit $\varkappa_m$, so ist bekanntlich

$$\lim_{m=\infty} k_m . \varkappa_m = 0.$$

Es gilt demnach die Ungleichung:

$$|f - f_m| < \varepsilon_m$$

überall in $< a, b >$ außerhalb der Vereinigungsmenge von $\mathfrak{A}_m$ und $\mathfrak{K}_m$, erst recht daher, wenn wie früher mit $\mathfrak{R}_m$ die Vereinigungsmenge von $\mathfrak{A}_m$, $\mathfrak{A}_{m+1}$, $\mathfrak{A}_{m+2}, \ldots$ verstanden wird, außerhalb der Vereinigungsmenge $\mathfrak{B}_m$ von $\mathfrak{R}_m$ und $\mathfrak{K}_m$. Für den Inhalt β_m von $\mathfrak{B}_m$ haben wir:

$$\beta_m \leqq \rho_m + \varkappa_m$$

und daher:

$$\lim_{m=\infty} k_m . \beta_m = 0. \tag{11}$$

Bemerken wir noch, daß $\mathfrak{B}_{m+1}$ Teil von $\mathfrak{B}_m$ ist.

Im Teilintervall $\delta_i^{(n)}$ werde nun der Punkt $\xi_i^{(n)}$ in folgender Weise gewählt. Es werde mit m_i der kleinste Wert von m bezeichnet derart, daß der Inhalt des nach $\delta_i^{(n)}$ fallenden Teiles von $\mathfrak{B}_m$ kleiner als $\frac{\delta_i^{(n)}}{2}$ ist. Ist $\mathfrak{D}_i^{(n)}$ die Menge der nicht zu $\mathfrak{B}_{m_i}$ gehörigen Punkte von $\vartheta_i^{(n)}$, so bezeichne man nun mit m'_i die kleinste natürliche Zahl m derart, daß in mindestens einem Punkt von $\mathfrak{D}_i^{(n)}$ die Ungleichung gilt:

$$|f| < m'_i,$$

und verstehe unter $\xi_i^{(n)}$ irgend einen Punkt von $\mathfrak{D}_i^{(n)}$, in dem diese Ungleichung gilt.

Wir behaupten wie früher: ist $\eta > 0$ beliebig vorgegeben, so gibt es ein m_0 derart, daß für $m \geqq m_0$ und sämtliche Einteilungen D_n die Ungleichung gilt:

$$\left| \sum_{i=1}^{\nu_n} f_m(\xi_i^{(n)})\, \delta_i^{(n)} - \sum_{i=1}^{\nu_n} f(\xi_i^{(n)})\, \delta_i^{(n)} \right| < \eta. \tag{12}$$

Ist in der Tat der Wert von m fest gegeben, so bezeichnen wir diejenigen $\delta_i^{(n)}$, deren Punkt $\xi_i^{(n)}$ nicht zu $\mathfrak{B}_m$ gehört, mit $\bar{\delta}_i^{(n)}$, die übrigen mit $\bar{\bar{\delta}}_i^{(n)}$ und haben wie früher:

$$|\Sigma f_m(\bar{\xi}_i^{(n)})\bar{\delta}_i^{(n)} - \Sigma f(\bar{\xi}_i^{(n)})\bar{\delta}_i^{(n)}| < \varepsilon_m . (b-a). \tag{13}$$

Was die Differenz

$$\bar{\bar{D}} = \Sigma f_m(\bar{\bar{\xi}}_i^{(n)})\bar{\bar{\delta}}_i^{(n)} - \Sigma f(\bar{\bar{\xi}}_i^{(n)})\bar{\bar{\delta}}_i^{(n)}$$

anlangt, so haben wir zunächst:

$$|\bar{\bar{D}}| \leqq \Sigma\, |f_m(\bar{\bar{\xi}}_i^{(n)})|\, \bar{\bar{\delta}}_i^{(n)} + \Sigma\, |f(\bar{\bar{\xi}}_i^{(n)})|\, \bar{\bar{\delta}}_i^{(n)}.$$

Berücksichtigen wir, daß nur diejenigen $\delta_i^{(n)}$ unter den $\bar{\bar{\delta}}_i^{(n)}$ auftreten, in denen $\mathfrak{B}_m$ einen Teil besitzt, dessen Inhalt $\geqq \dfrac{\delta_i^{(n)}}{2}$ ist, so ergibt sich sofort:

$$\Sigma \bar{\bar{\delta}}_i^{(n)} \leqq 2 . \beta_m \tag{14}$$

und hieraus im Verein mit (10):

$$|\bar{\bar{D}}| \leqq 2 k_m \beta_m + \Sigma\, |f(\bar{\bar{\xi}}_i^{(n)})|\, \bar{\bar{\delta}}_i^{(n)}. \tag{15}$$

Da einerseits, zufolge der Wahl der bei der Definition von $\bar{\bar{\xi}}_i^{(n)}$ auftretenden Zahl m_i' in allen Punkten von $\mathfrak{D}_i^{(n)}$ die Ungleichung:

$$|f| > |f(\bar{\bar{\xi}}_i^{(n)})| - 1$$

gilt, da andrerseits in jedem Intervall $\bar{\bar{\delta}}_i^{(n)}$ der Inhalt von $\mathfrak{D}_i^{(n)}$ größer als $\dfrac{\bar{\bar{\delta}}_i^{(n)}}{2}$ ist, so hat man sofort:

$$|f(\bar{\bar{\xi}}_i^{(n)})|\, \bar{\bar{\delta}}_i^{(n)} < 2 \int_{\bar{\bar{\delta}}_i^{(n)}} (|f| + 1)\, dx.$$

Wir haben also zufolge (15):

$$|\bar{\bar{D}}| < 2 k_m \beta_m + 2 \sum \int_{\bar{\bar{\delta}}_i^{(n)}} (|f| + 1) dx. \tag{16}$$

Da nun wegen (14) und $\lim\limits_{m=\infty} \beta_m = 0$ die Summe der Längen der Intervalle $\bar{\bar{\delta}}_i^{(n)}$ für $m = \infty$ den Grenzwert 0 hat, so kann man, nach einer bekannten Eigenschaft der Lebesgue'schen Integrale, m_0 so groß wählen, daß für $m \geqq m_0$

$$2 \sum \int_{\bar{\bar{\delta}}_i^{(n)}} (|f| + 1) dx < \frac{\eta}{2} \tag{17}$$

wird, ferner kann wegen $\lim\limits_{m=\infty} \varepsilon_m = 0$ und wegen (11) m_0 auch so groß gewählt werden, daß für $m \geqq m_0$:

$$\varepsilon_m . (b-a) + 2 k_m \beta_m < \frac{\eta}{2}$$

und dann hat man aus (13), (16) und (17) die behauptete Ungleichung (12) für $m \geqq m_0$. Nun ist, wie früher:

$$\lim_{n=\infty} \sum_{i=1}^{\nu_n} f_m(\xi_i^{(n)}) \delta_i^{(n)} = \int_a^b f_m\, dx. \tag{18}$$

Andrerseits, da in $< a, b >$ überall bis auf eine Menge des Inhalts α_m

$$|f_m - \varphi_{k_m}| < \varepsilon_m$$

war, überall in $< a, b >$ aber die beiden Ungleichungen

$$|f_m| \leqq k_m, \qquad |\varphi_{k_m}| \leqq k_m$$

gelten, so hat man:

$$\left| \int_a^b f_m dx - \int_a^b \varphi_{k_m} dx \right| < \varepsilon_m (b-a) + 2 k_m \alpha_m$$

und daher wegen (9):

$$\left| \int_a^b f_m dx - \int_a^b f dx \right| < \varepsilon_m (b-a+1) + 2 k_m \alpha_m.$$

Nun ist einerseits $\lim_{m=\infty} \varepsilon_m = 0$, andrerseits wegen $\alpha_m \leqq \beta_m$ und (11) auch $\lim_{m=\infty} k_m \alpha_m = 0$, daher:

$$\lim_{m=\infty} \int_a^b f_m dx = \int_a^b f dx. \tag{19}$$

Die Beziehungen (12), (18), (19) ergeben nun für nicht geschränktes integrables f unseren Satz genau so wie früher, im Falle, da f geschränkt angenommen war.

§ 3.

Wir behandeln nunmehr die gleichzeitige Annäherung an die Integrale abzählbar unendlich vieler integrabler Funktionen durch Riemann'sche Summen.

II. Sind die abzählbar unendlich vielen Funktionen $f^{(1)}, f^{(2)}, \ldots, f^{(h)}, \ldots$ in $<a, b>$ integrierbar im Sinne von Lebesgue und ist eine ausgezeichnete Zerlegungsfolge D_n des Intervalles $<a, b>$ gegeben, so kann, wenn $\delta_1^{(n)}, \delta_2^{(n)}, \ldots, \delta_{\nu_n}^{(n)}$ die Teilintervalle der Zerlegung D_n sind, stets in $\delta_i^{(n)}$ der Punkt $\xi_i^{(n)}$ so gefunden werden, daß für alle h die Beziehung gilt:

$$\lim_{n=\infty} \sum_{i=1}^{\nu_n} f^{(h)}(\xi_i^{(n)}) \delta_i^{(n)} = \int_a^b f^{(h)}(x) dx. \tag{20}$$

Wir nehmen die $f^{(h)}$ zunächst als gleichmäßig geschränkt an; dann gibt es eine Zahl F, so daß in $<a, b>$:

$$|f^{(h)}(x)| \leqq F \qquad (h = 1, 2, \ldots).$$

Sei wieder $\varepsilon_1, \varepsilon_2, \ldots, \varepsilon_m \ldots$ eine Folge positiver Zahlen mit $\lim_{m=\infty} \varepsilon_m = 0$, ferner $\alpha_m^{(h)}$ eine Doppelfolge positiver Zahlen mit konvergenter Doppelsumme $\sum_{h, m=1}^{\infty} \alpha_m^{(h)}$.

Wir setzen:

$$\beta_m = \sum_{h=1}^{\infty} \sum_{i=m}^{\infty} \alpha_i^{(h)}$$

und haben $\lim_{m=\infty} \rho_m = 0$. Die stetige Funktion $f_m^{(h)}(x)$ kann nach § 1 so gewählt werden, daß in $< a, b >$ überall außerhalb einer Menge $\mathfrak{A}_m^{(h)}$ des Inhaltes $\alpha_m^{(h)}$ die Ungleichung gilt:

$$|f^{(h)}(x) - f_m^{(h)}(x)| < \varepsilon_m$$

und daß in ganz $< a, b >$ die Ungleichung gilt:

$$|f_m^{(h)}(x)| \leqq F.$$

Nun bezeichnen wir die Vereinigungsmenge aller Mengen $\mathfrak{A}_i^{(h)} (h = 1, 2, \ldots; \; i = m, m+1, \ldots)$ mit $\mathfrak{R}_m$ und erkennen, daß der Inhalt von $\mathfrak{R}_m$ höchstens gleich ρ_m ist. Außerhalb der Menge $\mathfrak{R}_m$ gelten nun in $< a, b >$ die sämtlichen Ungleichungen

$$|f^{(h)}(x) - f_m^{(h)}(x)| < \varepsilon_m \quad (h = 1, 2 \ldots)$$

gleichzeitig. Die Menge $\mathfrak{R}_{m+1}$ ist Teil von $\mathfrak{R}_m$.

Die Punkte $\xi_i^{(n)}$ werden nun wörtlich so definiert wie in § 2 im Falle einer geschränkten Funktion f. Genau wie oben die Ungleichung (8), beweist man sodann: Ist $\eta > 0$ beliebig gegeben, so kann (bei gegebenem h) n_0 so gefunden werden, daß für $n \geqq n_0$ die Ungleichung gilt:

$$\left| \sum_{i=1}^{n_\nu} f^{(h)}(\xi_i^{(n)}) \, \delta_i^{(n)} - \int_a^b f^{(h)}(x) \, dx \right| < 3\eta.$$

Das ist aber gleichbedeutend mit der Behauptung von Satz II.

Wir beweisen nun Satz II ohne die Voraussetzung, daß alle f_h gleichmäßig geschränkt sind.

Da, wenn die Integrale $\int_a^b f^{(h)}(x) \, dx$ existieren, dasselbe von den Integralen $\int_a^b |f^{(h)}(x)| \, dx$ gilt, können die positiven Zahlen c_h so bestimmt werden, daß die unendliche Reihe

$$\sum_{h=1}^{\infty} c_h \int_a^b |f^{(h)}(x)| \, dx$$

konvergiert. Nach einem, von B. Levi herrührenden Satz, über Lebesgue'sche Integrale[1] ist aber dann:

$$\sum_{h=1}^{\infty} c_h \int_a^b |f^{(h)}(x)|\,dx = \int_a^b \sum_{h=1}^{\infty} c_h |f^{(h)}(x)|\,dx, \qquad (21)$$

wo die rechts unter dem Integralzeichen stehende unendliche Reihe überall, abgesehen von einer Menge des Inhaltes 0 konvergiert. Da es nun ganz gleichgültig ist, ob wir unseren Satz für die Funktionen $f^{(h)}(x)$ oder die Funktionen $c_h f^{(h)}(x)$ beweisen, können wir also von vorneherein annehmen, daß die Reihe

$$\sum_{h=1}^{\infty} |f^{(h)}(x)| = \mathfrak{F}(x) \qquad (22)$$

überall in $< a, b >$ bis auf eine Menge des Inhaltes 0 konvergiert, in deren Punkten etwa $\mathfrak{F}(x) = 0$ gesetzt werde und auch die Reihe

$$\sum_{h=1}^{\infty} \int_a^b |f^{(h)}(x)|\,dx = \int_a^b \mathfrak{F}(x)\,dx$$

konvergiert.

Wir bezeichnen wieder mit $\varepsilon_1, \varepsilon_2, \ldots, \varepsilon_m, \ldots$ eine Folge positiver Zahlen mit $\lim_{m=\infty} \varepsilon_m = 0$. Ferner werden wir mit $\varphi_k^{(h)}(x)$ diejenige Funktion bezeichnen, die überall dort mit $f^{(h)}(x)$ übereinstimmt, wo $|f^{(h)}| \leqq k$ und überall sonst den Wert 0 hat. Wir werden jede Funktion $f^{(h)}$ approximieren durch eine Folge von Funktionen $\varphi^{(h)}_{k_m^{(h)}}$, die kurz mit $\psi_m^{(h)}$ bezeichnet werden mögen.

Die Doppelfolge dieser Funktionen $\psi_m^{(h)}$ ordnen wir in eine einfache $\psi_1, \psi_2, \ldots, \psi_j, \ldots$ nach folgender Regel: die Funktionen $\psi_m^{(h)}$ werden geordnet nach steigender Indizessumme $h+m$; die Funktionen $\psi_m^{(h)}$ von gleicher Indizessumme $h+m$ werden untereinander nach steigenden unteren Indizes m geordnet. Dadurch ist gewährleistet, daß in der Folge der ψ_j jede Funktion $\psi_m^{(h)}$ allen anderen $\psi_m^{(h)}$ von gleichem oberen aber größerem unteren Index vorangeht.

[1] Siehe z. B. H. Lebesgue, Ann. Éc. Norm. (3), *27*, p. 380.

Jede der Funktionen ψ_j ist eine bestimmte Funktion $\psi_m^{(h)}$, und dies war eine Abkürzung für $\overset{(h)}{\varphi}_{k_m^{(h)}}$. Ist $\overset{(h)}{\varphi}_{k_m^{(h)}} = \psi_j$ so werden wir auch schreiben $k_m^{(h)} = k_j$. Wir haben noch die hierin auftretenden Größen $k_m^{(h)}$ zu definieren. Für alle hinlänglich großen $k_m^{(h)}$ gilt die Ungleichung:

$$\left| \int_a^b \overset{(h)}{\varphi}_{k_m^{(h)}} dx - \int_a^b f^{(h)} dx \right| < \varepsilon_m.$$

Zunächst werde jedenfalls $k_m^{(h)}$ so groß gewählt, daß diese Ungleichung gilt.

Sodann geben wir uns irgend eine konvergente Reihe positiver Zahlen $\sum_{j=1}^{\infty} \sigma_j$ vor, bezeichnen mit $\varkappa_m^{(h)}$ den Inhalt der Menge $\mathfrak{K}_m^{(h)}$ jener Punkte, in denen $|f^{(h)}| > k_m^{(h)}$ und schreiben auch $\mathfrak{K}_j$ statt $\mathfrak{K}_m^{(h)}$ und $\varkappa_j$ statt $\varkappa_m^{(h)}$, wenn $\overset{(h)}{\varphi}_{k_m^{(h)}} = \psi_j$ ist. Für alle hinlänglich großen k_j wird:

$$\varkappa_j < \sigma_j$$

sein; wir denken uns die $k_m^{(h)}$ auch so groß gewählt.

Sodann wählen wir $k_1^{(1)}$ $(= k_1)$ auch noch so groß, daß:

$$k_1^{(1)} \varkappa_1^{(1)} < \frac{\varepsilon_1}{2},$$

was immer möglich ist, weil

$$\lim_{k_1^{(1)} = +\infty} k_1^{(1)} \varkappa_1^{(1)} = 0$$

ist.

Die Größe $k_1^{(1)}$ wird keinen weiteren Bedingungen unterworfen. Nun wählen wir eine konvergente Reihe positiver Zahlen $\gamma_j^{(1)} (j = 2, 3, \ldots)$ so, daß ihre Summe der Ungleichung genügt:

$$k_1^{(1)} \sum_{j=2}^{\infty} \gamma_j^{(1)} < \frac{\varepsilon_1}{2},$$

und unterwerfen alle $k_m^{(h)}$ (außer $k_1^{(1)}$) der Bedingung: ist $\psi_m^{(h)} = \psi_j$, so soll $k_m^{(h)}$ so groß sein, daß $\varkappa_m^{(h)} < \gamma_j^{(1)}$. Dadurch ist jedenfalls erreicht, daß:

$$k_1^{(1)} \sum_{h,\, m=1}^{\infty} \varkappa_m^{(h)} = k_1 \sum_{j=1}^{\infty} \varkappa_j < \varepsilon_1$$

wird. Die Größe $k_1^{(2)} (= k_2)$ wird nun nur mehr der einen Bedingung unterworfen: sie sei so groß gewählt, daß:

$$k_1^{(2)} \varkappa_1^{(2)} < \frac{\varepsilon_2}{2}\,.$$

Nun werde wieder eine konvergente Reihe positiver Zahlen $\gamma_j^{(2)} (j = 3, 4, \ldots)$ so gewählt, daß:

$$k_1^{(2)} \sum_{j=3}^{\infty} \gamma_j^{(2)} = k_2 \sum_{j=3}^{\infty} \gamma_j^{(2)} < \frac{\varepsilon_2}{2}$$

und es werden alle $k_j (j > 2)$ der Bedingung unterworfen: k_j ist so groß zu wählen, daß $\varkappa_j < \gamma_j^{(2)}$. Dadurch ist erreicht, daß

$$k_2 \sum_{j=2}^{\infty} \varkappa_j < \varepsilon_2.$$

Man sieht, wie man, in dieser Weise weitergehend, durch geeignete Wahl der $k_m^{(h)}$ erreichen kann, daß für alle μ:

$$k_\mu \sum_{j=\mu}^{\infty} \varkappa_j < \varepsilon_\mu,$$

daß also:

$$\lim_{\mu=\infty} k_\mu \sum_{j=\mu}^{\infty} \varkappa_j = 0. \qquad (24)$$

Nachdem wir so über die $k_m^{(h)}$ verfügt haben und uns also die $\varphi_{k_m^{(h)}}^{(h)}$ festgelegt denken, verstehen wir unter $\sum_{j=1}^{\infty} \alpha_j$ eine Reihe positiver Zahlen, die so rasch konvergiert, daß:

$$\lim_{\mu=\infty} k_\mu \sum_{j=\mu}^{\infty} \alpha_j = 0. \tag{25}$$

Es gibt dann nach § 1 eine Funktion f_j, die überall in $< a, b >$, abgesehen von einer Punktmenge $\mathfrak{A}_j$ des Inhaltes α_j der Ungleichung genügt:

$$|f_j - \psi_j| < \varepsilon_j$$

und zwischen oberer und unterer Grenze von ψ_j verbleibt. Ist ψ_j die Funktion $\overset{(h)}{\varphi}_{k_m^{(h)}}$ so werde f_j auch mit $f_m^{(h)}$ bezeichnet, $\mathfrak{A}_j$ auch mit $\mathfrak{A}_m^{(h)}$. Da $\overset{(h)}{\varphi}_{k_m^{(h)}}$ mit $f^{(h)}$ überall übereinstimmt außer auf der Menge $\mathfrak{K}_m^{(h)}$, so gilt außerhalb der Vereinigungsmenge von $\mathfrak{A}_m^{(h)}$ und $\mathfrak{K}_m^{(h)}$ die Ungleichung:

$$|f_m^{(h)} - f^{(h)}| < \varepsilon_j,$$

während überall in $< a, b >$ die Ungleichung gilt:

$$|f_m^{(h)}| \leqq k_m^{(h)}.$$

Bezeichnen wir noch mit $\mathfrak{B}_\mu$ die Vereinigungsmenge der Mengen $\mathfrak{A}_j$ $(j = \mu, \mu+1, \ldots)$, $\mathfrak{K}_j (j = \mu, \mu+1, \ldots)$, mit β_μ den Inhalt von $\mathfrak{B}_\mu$, so ist:

$$\beta_\mu \leqq \sum_{j=\mu}^{\infty} \varkappa_j \dotplus \sum_{j=\mu}^{\infty} \alpha_j$$

und daher wegen (24) und (25) auch:

$$\lim_{\mu=\infty} k_\mu \beta_\mu = 0$$

und es ist $\mathfrak{B}_{\mu+1}$ Teil von $\mathfrak{B}_\mu$.

Wie im Falle einer Funktion $f(x)$ wählen wir nun wieder im Intervalle $\delta_i^{(n)}$ den Punkt $\xi_i^{(n)}$ in folgender Weise: Es werde mit μ_i der kleinste Wert von μ bezeichnet, für den der ins Intervall $\delta_i^{(n)}$ fallende Teil von $\mathfrak{B}_\mu$ einen Inhalt $< \frac{\delta_i^{(n)}}{2}$ hat. Ist $\mathfrak{D}_i^{(n)}$ die Menge der nicht zu $\mathfrak{B}_{\mu_i}$ gehörigen Punkte von $\delta_i^{(n)}$, so bezeichne μ', die kleinste natürliche Zahl μ derart, daß in mindestens einem Punkte von $\mathfrak{D}_i^{(n)}$ die durch (22) definierte Funktion $\mathfrak{F}(x)$ die Ungleichung erfüllt:

$$\mathfrak{F}(x) < \mu_i',$$

und es sei $\xi_i^{(n)}$ irgend ein Punkt von $\mathfrak{D}_i^{(n)}$, in dem diese Ungleichung gilt.

Wir können nun wieder beweisen: ist $\eta > 0$ beliebig vorgegeben, so gibt es ein m_0 derart, daß für $m \geqq m_0$ und sämtliche h sowie sämtliche Einteilungen D_n die Ungleichung gilt:

$$\left| \sum_{i=1}^{n_\nu} f_m^{(h)}(\xi_i^{(n)})\,\delta_i^{(n)} - \sum_{i=1}^{n_\nu} f^{(h)}(\xi_i^{(n)})\,\delta_i^{(n)} \right| < \eta. \tag{26}$$

Ist in der Tat der Wert von m fest gegeben, und ist $f_m^{(h)} = f_\mu$, so bezeichnen wir wieder diejenigen $\delta_i^{(n)}$, deren Punkt $\xi_i^{(n)}$ nicht zu $\mathfrak{B}_\mu$ gehört, mit $\bar{\delta}_i^{(n)}$, die übrigen mit $\bar{\bar{\delta}}_i^{(n)}$. Wir erhalten wie früher:

$$\left|\Sigma f_m^{(h)}(\bar{\xi}_i^{(n)})\,\bar{\delta}_i^{(n)} - \Sigma f^{(h)}(\bar{\xi}_i^{(n)})\,\bar{\delta}_i^{(n)}\right| < \varepsilon_\mu(b-a)$$

und, wenn

$$\bar{\bar{D}}_m^{(h)} = \Sigma f_m^{(h)}(\bar{\bar{\xi}}_i^{(n)})\,\bar{\bar{\delta}}_i^{(n)} - \Sigma f^{(h)}(\bar{\bar{\xi}}_i^{(n)})\,\bar{\bar{\delta}}_i^{(n)}$$

gesetzt wird, und die Ungleichung:

$$|f^{(h)}(x)| \leqq \mathfrak{F}(x)$$

berücksichtigt wird:

$$|\bar{\bar{D}}_m^{(h)}| < 2k_\mu\beta_\mu + 2\sum \int_{\bar{\bar{\delta}}_i^{(n)}} (\mathfrak{F}(x)+1)\,dx.$$

Schreiben wir endlich, wenn $f_m^{(h)} = f_\mu$ ist auch noch ausführlicher $\mu_m^{(h)}$ statt μ, so gehen bei festgehaltenem h die $\mu_m^{(h)}$ mit m gegen $+\infty$, d. h. ist μ_0 vorgegeben, so wird bei hinlänglich großem m für alle h gelten:

$$\mu_m^{(h)} \geqq \mu_0. \tag{27}$$

Wir werden also zunächst μ_0 so groß wählen, daß für alle $\mu \geqq \mu_0$:

$$\varepsilon_\mu(b-a)+2k_\mu\beta_\mu<\frac{\eta}{2}$$

wird und für jede Punktmenge $\mathfrak{D}$, deren Inhalt $<2\beta_\mu$ ist, die Ungleichung

$$2\int_{\mathfrak{D}}(\mathfrak{F}(x)+1)\,dx<\frac{\eta}{2} \tag{28}$$

gilt. Sodann werden wir m so groß wählen, daß für alle h Ungleichung (27) gilt.

Da stets

$$\Sigma\bar{\bar{\delta}}_i^{(n)}<2\beta_\mu$$

ist, gilt dann (28), und (26) ist bewiesen. Von (26) aus wird nun die Gültigkeit von Satz II genau so erschlossen, wie dies in § 2 für Satz I aus Ungleichung (12) geschah.

§ 4.

Wir wollen nun den Satz beweisen:

III. Ist $f(x)$ im Intervall $<a,\ b>$ integrierbar im Sinne von Lebesgue, und ist $q>0$ beliebig gegeben, so gibt es stets eine ausgezeichnete Folge von Zerlegungen D_n, deren jede einen Quotienten $>1-q$ hat, derart, daß, wenn $a=x_0^{(n)}<x_1^{(n)}<\ldots\ldots<x_{\nu_n-1}^{(n)}<x_{\nu_n}^{(n)}=b$ die Teilpunkte der Zerlegung D_n sind, die Beziehung gilt:

$$\lim_{n=\infty}\sum_{i=1}^{\nu_n}\frac{f(x_i^{(n)})+f(x_{i-1}^{(n)})}{2}\,(x_i^{(n)}-x_{i-1}^{(n)})=\int_a^b f(x)\,dx. \tag{29}$$

Wir führen den Beweis zunächst wieder für den Fall, daß f geschränkt ist in $<a,\ b>$; sei etwa:

$$|f|\leqq F. \tag{30}$$

Wir bezeichnen mit λ irgend eine feste Zahl, die der Bedingung genügt:

$$\lambda<\frac{1}{2}\cdot\frac{q}{2-q}\,.$$

Um die Zerlegung D_n zu finden, teilen wir zunächst das Intervall $< a, b >$ in n gleiche Teile durch die Punkte:

$$X_i^{(n)} = a + i\frac{(b-a)}{n} (i = 1, 2, \ldots, n-1)$$

und umgeben sodann jeden Punkt $X_i^{(n)}$ mit dem Intervalle:

$$d_i^{(n)} = < X_i^{(n)} - \lambda\frac{b-a}{n}, \quad X_i^{(n)} + \lambda\frac{b-a}{n} >. \tag{31}$$

Wählen wir nun den Punkt $x_i^{(n)} (i = 1, 2, \ldots, n-1)$ beliebig im Intervall $d_i^{(n)}$ und setzen $x_0^{(n)} = a$, $x_n^{(n)} = b$, so ist für irgend zwei $i, j (i, j = 1, 2, \ldots, n)$ der Quotient:

$$\frac{x_i^{(n)} - x_{i-1}^{(n)}}{x_j^{(n)} - x_{j-1}^{(n)}} > \frac{1-2\lambda}{1+2\lambda} > 1-q.$$

Wählen wir also $x_i^{(n)} (i = 1, 2, \ldots, n-1)$ als die Einteilungspunkte von D_n, so ist der Quotient jeder dieser Zerlegungen $> 1-q$.[1]

Nun bezeichnen wir mit $\varepsilon_1, \varepsilon_2, \ldots, \varepsilon_m \ldots$ eine Folge positiver Zahlen mit $\lim\limits_{m=\infty} \varepsilon_m = 0$. Ferner sei $\alpha_1, \alpha_2, \ldots, \alpha_m \ldots$ eine Folge positiver Zahlen derart, daß die Reihe $\sum\limits_{m=1}^{\infty} \alpha_m$ konvergiert; setzen wir

$$\rho_m = \sum_{i=m}^{\infty} \alpha_m,$$

so ist also $\lim\limits_{m=\infty} \rho_m = 0$.

Mit f_m bezeichnen wir eine der Ungleichung:

$$|f_m| \leqq F \tag{32}$$

---

1 Wir heben für spätere Anwendung hervor: durch D_n wird $< a, b >$ in n Teilintervalle zerlegt, für deren jedes die Ungleichung gilt:

$$\left| \frac{x_i^{(n)} - x_{i-1}^{(n)}}{\frac{b-a}{n}} - 1 \right| < 2\lambda.$$

Dabei konnte $\lambda > 0$ beliebig klein sein.

genügende Funktion, für die in $< a, b >$, abgesehen von einer Punktmenge $\mathfrak{A}_m$ des Inhaltes α_m die Ungleichung gilt:

$$|f_m - f| < \varepsilon_m.$$

Die Vereinigungsmenge von $\mathfrak{A}_m, \mathfrak{A}_{m+1}, \ldots$ bezeichnen wir mit $\mathfrak{R}_m$; ihr Inhalt ist $\leqq \rho_m$ und es ist $\mathfrak{R}_{m+1}$ Teil von $\mathfrak{R}_m$.

Wegen $\lim\limits_{m=\infty} \rho_m = 0$ muß es in jedem der Intervalle (31) eine erste Menge $\mathfrak{R}_m$ geben, die dieses Intervall nicht ausfüllt, der Punkt $x_i^{(n)}$ werde nun in $d_i^{(n)}$ außerhalb dieser Menge $\mathfrak{R}_m$, sonst aber beliebig gewählt. Dann können wir beweisen: Ist $\eta > 0$ beliebig gegeben, so gibt es ein m_0 so, daß für $m \geqq m_0$ und alle n die Ungleichung gilt:

$$\left| \sum_{i=1}^{n-1} (f^{(m)}(x_i^{(n)}) - f(x_i^{(n)})) \frac{x_{i+1}^{(n)} - x_{i-1}^{(n)}}{2} \right| < \eta. \tag{33}$$

Sei ein bestimmter Wert von m gegeben. Mit Σ' werde derjenige Teil der vorstehenden Summe $\sum\limits_{i=1}^{n-1}$ bezeichnet, der sich über diejenigen Indizes i erstreckt, für welche $x_i^{(n)}$ nicht zu $\mathfrak{R}_m$ gehört, mit Σ'' der übrig bleibende Teil der Summe. Der Punkt $x_i^{(n)}$ gehört dann und nur dann zu $\mathfrak{R}_m$, wenn $\mathfrak{R}_m$ das ganze Intervall $d_i^{(n)}$ ausfüllt. Da jedes Intervall $d_i^{(n)}$ die Länge $2\lambda \frac{b-a}{n}$ hat und $\mathfrak{R}_m$ höchstens den Inhalt ρ_m hat, ist die Anzahl N der Summanden in Σ'':

$$N \leqq \frac{\rho_m \cdot n}{2\lambda(b-a)}.$$

Da ferner:

$$x_{i+1}^{(n)} - x_i^{(n)} \leqq \frac{b-a}{n}(1+2\lambda) \tag{34}$$

ist, ist wegen (30) und (32) in Σ'' jeder Summand seinem absoluten Betrag nach:

$$\leqq 2F \cdot \frac{b-a}{n}(1+2\lambda)$$

und daher:

$$|\Sigma''| \leqq F \frac{1+2\lambda}{\lambda} \rho_m.$$

Andrerseits ist offenbar:

$$|\Sigma'| < \varepsilon_m(b-a);$$

wir brauchen also nur m_0 so groß zu wählen, daß für $m \geqq m_0$

$$\varepsilon_m(b-a) + F\frac{1+2\lambda}{\lambda}\rho_m < \eta,$$

und (33) ist bewiesen.

Nun wählen wir m_0 auch noch so groß, daß für $m \geqq m_0$:

$$\left|\int_a^b f_m dx - \int_a^b f dx\right| < \eta. \tag{35}$$

Sodann greifen wir irgend ein $m \geqq m_0$ heraus; da, wegen der Stetigkeit von f_m:

$$\lim_{n=\infty} \sum_{i=1}^{n-1} f^{(m)}(x_i^{(n)}) \frac{x_{i+1}^{(n)} - x_{i-1}^{(n)}}{2} = \int_a^b f_m dx$$

ist, kann n_0 so groß gewählt werden, daß für $n \geqq n_0$:

$$\left|\sum_{i=1}^{n-1} f^{(m)}(x_i^{(n)}) \frac{x_{i+1}^{(n)} - x_{i-1}^{(n)}}{2} - \int_a^b f_m dx\right| < \eta. \tag{36}$$

Wegen

$$\left|f(a).\frac{x_1^{(n)} - a}{2}\right| < \frac{F(1+\lambda)(b-a)}{2n};$$

$$\left|f(b)\frac{b - x_{n-1}^{(n)}}{2}\right| < \frac{F(1+\lambda)(b-a)}{2n}$$

kann n_0 auch so groß gewählt werden, daß für $n \geqq n_0$:

$$\left|\sum_{i=1}^{n} \frac{f^{(m)}(x_i^{(n)}) + f^{(m)}(x_{i-1}^{(n)})}{2}(x_i^{(n)} - x_{i-1}^{(n)}) - \sum_{i=1}^{n-1} f^{(m)}(x_i^{(n)}) \frac{x_{i+1}^{(n)} - x_{i-1}^{(n)}}{2}\right| < \eta. \tag{37}$$

Die Ungleichungen (33), (35), (36), (37) zusammengenommen, ergeben, daß für $n \geqq n_0$:

$$\left| \sum_{i=1}^{n} \frac{f(x_i^{(n)})+f(x_{i-1}^{(n)})}{2} (x_i^{(n)}-x_{i-1}^{(n)}) - \int_a^b f\,dx \right| < 4\eta$$

und das ist gleichbedeutend mit unserer Behauptung.

Der Übergang zu nicht geschränktem, aber integrablem f erfolgt ganz wie in § 2, nur daß an Stelle der Intervalle $\delta_i^{(n)}$ hier die Intervalle $d_i^{(n)}$ [siehe (31)] treten. Wenn also alle Buchstaben dieselbe Bedeutung haben, wie in § 2, wird der Punkt $x_i^{(n)}$ $(i = 1, 2, \ldots, n-1)$ in $d_i^{(n)}$ in folgender Weise gewählt: mit m_i wird der kleinste Wert von m bezeichnet, derart, daß der Inhalt des nach $d_i^{(n)}$ fallenden Teiles von $\mathfrak{B}_m$ einen Inhalt $< \frac{d_i^{(n)}}{2}$ hat. Ist $\mathfrak{D}_i^{(n)}$ die Menge der nicht zu $\mathfrak{B}_{m_i}$ gehörenden Punkte von $d_i^{(n)}$, so bezeichne wieder m_i' die kleinste natürliche Zahl m derart, daß in mindestens einem Punkte von $\mathfrak{D}_i^{(n)}$

$$|f| < m_i'$$

ist; für $x_i^{(n)}$ ist irgend ein Punkt von $\mathfrak{D}_i^{(n)}$ zu wählen, in dem diese Ungleichung gilt.

Wieder beweist man: ist $\eta > 0$ beliebig gegeben, so gibt es ein m_0 derart, daß für $m \geqq m_0$ und alle n die Ungleichung (33) gilt. Die in ihr auftretende Summe $\sum_{i=1}^{n-1}$ wird wieder zerlegt in zwei Teile Σ' und Σ'', der erste Teil, enthaltend alle Summanden, in denen $x_i^{(n)}$ nicht zu $\mathfrak{B}_m$ gehört, der zweite die übrigen. Wieder erhält man ohneweiteres:

$$|\Sigma'| < \varepsilon_m (b-a).$$

Der Punkt $x_i^{(n)}$ gehört nur dann zu $\mathfrak{B}_m$, wenn $\mathfrak{B}_m$ in $d_i^{(n)}$ einen Teil hat, dessen Inhalt $\geqq \frac{d_i^{(n)}}{2}$ ist. Da jedes $d_i^{(n)}$ die Länge $2\lambda \frac{b-a}{n}$ hat und β_m der Inhalt von $\mathfrak{B}_m$ ist, so ist die Anzahl N der Summanden in Σ'':

$$N \leqq \frac{\beta_m \cdot n}{\lambda(b-a)}. \tag{38}$$

Da nun

$$|\Sigma''| \leqq \Sigma'' |f^{(m)}(x_i^{(n)})| \frac{x_{i+1}^{(n)} - x_{i-1}^{(n)}}{2} + \Sigma'' |f(x_i^{(n)})| \frac{x_{i+1}^{(n)} - x_{i-1}^{(n)}}{2}$$

ist, so hat man zunächst, wegen (34):

$$|\Sigma''| \leqq \frac{1+2\lambda}{\lambda} k_m \beta_m + \Sigma'' |f(x_i^{(n)})| \frac{x_{i+1}^{(n)} - x_{i-1}^{(n)}}{2}. \tag{39}$$

Da ferner auf $\mathfrak{D}_i^{(n)}$:

$$|f| > |f(x_i^{(n)})| - 1$$

ist, da ferner der Inhalt von $\mathfrak{D}_i^{(n)}$ mindestens gleich $\frac{d_i^{(n)}}{2}$ ist, so hat man für jedes solche $d_i^{(n)}$:

$$|f(x_i^{(n)})| \frac{d_i^{(n)}}{2} \leqq \int_{\mathfrak{D}_i^{(n)}} (|f|+1)\,dx \leqq \int_{x_{i-1}^{(n)}}^{x_{i+1}^{(n)}} (|f|+1)\,dx$$

und da weiter:

$$x_{i+1}^{(n)} - x_{i-1}^{(n)} < \frac{1+2\lambda}{\lambda} d_i^{(n)}$$

ist, folgt daraus für die Summanden von Σ'':

$$|f(x_i^{(n)})| \frac{x_{i+1}^{(n)} - x_{i-1}^{(n)}}{2} < \frac{1+2\lambda}{\lambda} \int_{x_{i-1}^{(n)}}^{x_{i+1}^{(n)}} (|f|+1)\,dx,$$

und daher, zusammen mit (39):

$$|\Sigma''| < \frac{1+2\lambda}{\lambda} \Big(k_m \beta_m + \Sigma'' \int_{x_{i-1}^{(n)}}^{x_i^{(n)}} (|f|+1)\,dx + \\ + \Sigma'' \int_{x_i^{(n)}}^{x_{i+1}^{(n)}} (|f|+1)\,dx\Big).$$

Nun ist nach (38) die Anzahl der Summanden in jeder der rechts auftretenden Summen $\leqq \frac{\beta_m \cdot n}{\lambda(b-a)}$, jedes einzelne Integrationsintervall $< x_{i-1}^{(n)}, x_i^{(n)} >$ und $< x_i^{(n)}, x_{i+1}^{(n)} >$ nach (34), aber $\leqq \frac{b-a}{n}(1+2\lambda)$, die Summe der Integrationsintervalle der in jeder dieser Summen auftretenden Integrale daher $\leqq \frac{1+2\lambda}{\lambda}\beta_m$ und geht daher mit wachsendem m gegen 0.

Alle übrigen Schlüsse verlaufen nun, abgesehen von den schon im Fall eines geschränkten f angebrachten Modifikationen, genau so weiter wie in § 2 und brauchen nicht nochmals durchgeführt zu werden. Satz III ist damit allgemein bewiesen.

Wir können nun den bewiesenen Satz nach zwei Richtungen hin präzisieren; erstens können wir die in unserem Satz auftretende ausgezeichnete Zerlegungsfolge D_n ersetzen durch eine andere Δ_n bei der immer Δ_{n+1} aus Δ_n entsteht, indem alle Teilpunkte von Δ_n beibehalten und lediglich neue Teilpunkte hinzugefügt werden, so daß also Δ_{n+1} aus Δ_n durch Unterteilung entsteht; zweitens können wir erreichen, daß der Quotient der Zerlegung Δ_n mit wachsendem n geradezu den Grenzwert 1 hat. Wir wollen also zeigen:

IV. Ist $f(x)$ im Intervalle $\langle a, b \rangle$ integrierbar, so gibt es stets eine ausgezeichnete Folge von Zerlegungen Δ_n derart, daß: 1. Δ_{n+1} aus Δ_n durch Unterteilung entsteht, 2. der Quotient von Δ_n für $n=\infty$ den Grenzwert 1 hat, 3. wenn mit $a = x_0^{(n)} < x_1^{(n)} < \ldots\ldots x_{\nu_n-1}^{(n)} < x_{\nu_n}^{(n)} = b$ die Teilpunkte von Δ_n bezeichnet werden, die Beziehung gilt:

$$\lim_{n=\infty} \sum_{i=1}^{\nu_n} \frac{f(x_i^{(n)})+f(x_{i-1}^{(n)})}{2}(x_i^{(n)}-x_{i-1}^{(n)}) = \int_a^b f\,dx. \tag{40}$$

Wir wollen die abgekürzte Bezeichnung einführen:

$$\sum_{i=1}^{\nu_n} \frac{f(x_i^{(n)})+f(x_{i-1}^{(n)})}{2}(x_i^{(n)}-x_{i-1}^{(n)}) = S_{\Delta_n}. \tag{41}$$

Wir bezeichnen nun, um Satz IV zu beweisen, mit $\varepsilon_1, \varepsilon_2, \ldots \varepsilon_n \ldots$ eine Folge positiver Zahlen mit $\lim_{n=\infty} \varepsilon_n = 0$ mit $q_1. q_2, \ldots q_n \ldots$ ebenfalls eine Folge positiver Zahlen mit $\lim_{n=\infty} q_n = 0$. Angenommen, es sei die Zerlegung Δ_n, deren Quotient $> 1-q_n$

ist, gegeben, derart, daß, wenn ihre Einteilungspunkte mit $a=x_0^{(n)}, x_1^{(n)},\ldots, x_{\nu_n-1}^{(n)}, x_{\nu_n}^{(n)}=b$ bezeichnet werden, die Ungleichung gilt:

$$\left|S_{\Delta_n}-\int_a^b f\,dx\right|<\varepsilon_n.$$

Zufolge Satz III gibt es gewiß eine solche Zerlegung Δ_n.

Wir können nun, wenn $q'>0$ vorgegeben ist, ν_n natürliche Zahlen $m_1, m_2,\ldots, m_{\nu_n}$ so finden, daß:

$$\left|\frac{x_i^{(n)}-x_{i-1}^{(n)}}{x_j^{(n)}-x_{j-1}^{(n)}}:\frac{m_i}{m_j}-1\right|<q' \quad (i,j=1,2,\ldots,\nu_n). \tag{42}$$

Sodann können wir auf Grund des Beweises[1] von Satz IV (indem wir Satz IV statt auf das Intervall $<a, b>$ nun auf das Intervall $<x_{i-1}^{(n)}, x_i^{(n)}>$ anwenden) eine natürliche Zahl k so groß wählen, daß folgendes gilt: das Intervall $<x_{i-1}^{(n)}, x_i^{(n)}>$ wird durch Einschalten der Teilpunkte

$$x_{i-1}^{(n)}=x_0^{(n+1,i)}, x_1^{(n+1,i)},\ldots, x_{k.m_i-1}^{(n+1,i)}, x_{k.m_i}^{(n+1,i)}=x_i^{(n)}$$

so in $k.m_i$ Teilintervalle geteilt, daß (bei vorgegebenem $q''>0$):

$$\left|x_j^{(n+1,i)}-x_{j-1}^{(n+1,i)}:\frac{x_i^{(n)}-x_{i-1}^{(n)}}{k.m_i}-1\right|<q''\,(j=1,2,\ldots,k.m_i) \tag{43}$$

und daß (ν bedeutet die Anzahl der Teilintervalle der Zerlegung Δ_n):

$$\left|\sum_{j=1}^{k.m_i}\frac{f(x_j^{(n+1,i)})+f(x_{j-1}^{(n+1,i)})}{2}(x_j^{(n+1,i)}-x_{j-1}^{(n+1,i)}) -\int_{x_{i-1}^{(n)}}^{x_i^{(n)}} f\,dx\right|<\frac{\varepsilon_{n+1}}{\nu_n}.$$

Wählen wir dann für die Zerlegung Δ_{n+1} die durch die Gesamtheit der Punkte $x_j^{(n+1,i)}$ $(j=0,1,\ldots,k.m_i,\ i=1,2,\ldots,\nu_n)$ hervorgerufene Zerlegung, so entsteht Δ_{n+1} aus Δ_n durch Unterteilung; ferner ist:

---

1 Man beachte insbesondere die Anmerkung auf p. 735.

$$\left|S_{\Delta_{n+1}} - \int_a^b f\,dx\right| = \left|\sum_{i=1}^{\nu_n}\sum_{j=1}^{k.m_i} \frac{f(x_j^{(n+1,\,i)}) + f(x_{j-1}^{(n+1,\,i)})}{2}(x_j^{(n+1,\,i)} - x_{j-1}^{(n+1,\,i)}) - \int_a^b f\,dx\right| < \varepsilon_{n+1},$$

endlich ist, wegen (43), sobald q'' hinlänglich klein ($\mathfrak{d}_1$ und $\mathfrak{d}_2$ bedeuten Zahlen zwischen -1 und 1)

$$\left|\frac{x_{j_1}^{(n+1,\,i)} - x_{j_1-1}^{(n+1,\,i)}}{x_{j_2}^{(n+1,\,i)} - x_{j_2-1}^{(n+1,\,i)}} - 1\right| = \left|\frac{x_{j_1}^{(n+1,\,i)} - x_{j_1-1}^{(n+1,\,i)}}{\frac{x_i^{(n)} - x_{i-1}^{(n)}}{k.m_i}} : \frac{x_{j_2}^{(n+1,\,i)} - x_{j_2-1}^{(n+1,\,i)}}{\frac{x_i^{(n)} - x_{i-1}^{(n)}}{k.m_i}} - 1\right| = \left|\frac{1+\mathfrak{d}_1 q''}{1+\mathfrak{d}_2 q''} - 1\right| < q_{n+1} \qquad (44)$$

und, wenn wieder $\mathfrak{d}$, $\mathfrak{d}_1$, $\mathfrak{d}_2$ Zahlen zwischen -1 und $+1$ bedeuten, unter Berücksichtigung von (43) und (42):

$$\left|\frac{x_{j_1}^{(n+1,\,i_1)} - x_{j_1-1}^{(n+1,\,i_1)}}{x_{j_2}^{(n+1,\,i_2)} - x_{j_2-1}^{(n+1,\,i_2)}} - 1\right| = \left|\frac{\frac{x_{i_1}^{(n)} - x_{i_1-1}^{(n)}}{k.m_{i_1}}(1+\mathfrak{d}_1 q'')}{\frac{x_{i_2}^{(n)} - x_{i_2-1}^{(n)}}{k.m_{i_2}}(1+\mathfrak{d}_2 q'')} - 1\right| = \left|\frac{(1+\mathfrak{d} q')(1+\mathfrak{d}_1 q'')}{1+\mathfrak{d}_2 q''} - 1\right| < q_{n+1} \qquad (45)$$

für hinlänglich kleine q' und q''. Der Quotient der Einteilung Δ_{n+1} ist also, wie die Ungleichungen (44) und (45) besagen $> 1 - q_{n+1}$.

Es ist also aus jeder Zerlegung Δ_n, deren Quotient $> 1 - q_n$ ist und für die

$$\left|S_{\Delta_n} - \int_a^b f\,dx\right| < \varepsilon_n$$

gilt, eine aus ihr durch Unterteilung entstehende Zerlegung Δ_{n+1} hergeleitet, deren Quotient $> 1-q_{n+1}$ ist, und für die

$$\left| S_{\Delta_{n+1}} - \int_a^b f dx \right| < \varepsilon_{n+1}$$

ist. Damit ist die Existenz einer ausgezeichneten Zerlegungsfolge Δ_n nachgewiesen von folgenden Eigenschaften:

1. Es entsteht Δ_{n+1} aus Δ_n durch Unterteilung;
2. der Quotient von Δ_n ist $> 1-q$;
3. es ist $\lim\limits_{n=\infty} S_{\Delta_n} = \int_a^b f dx$, und Satz IV ist, da $\lim\limits_{n=\infty} q_n = 0$

ist, bewiesen.

Offenbar könnten die Resultate dieses Paragraphen, so wie es in § 3 mit den Resultaten von § 2 geschehen ist, dahin verallgemeinert werden, daß zu einem System abzählbar unendlich vieler Funktionen $f^{(1)}, f^{(2)}, \ldots, f^{(h)}, \ldots$ eine ausgezeichnete Folge von Zerlegung D_n, deren Quotienten für $n = \infty$ den Grenzwert 1 haben, so bestimmt wird, daß die zugehörigen Summen S_{D_n} für jede der Funktionen $f^{(h)}$ gegen $\int_a^b f^{(h)} dx$ konvergieren. Doch ist es wohl überflüssig, hierauf nochmals einzugehen.

Über eine Verallgemeinerung der Riemannschen Integraldefinition.

Von **Hans Hahn** in Czernowitz.

Es wurde vor kurzem von É. Borel eine Verallgemeinerung der Riemannschen Integraldefinition angegeben,[1]) die einerseits bei geschränktem Integranden in manchen Fällen zum Ziele führt, in denen die Riemannsche Definition selbst versagt (sie liefert dann denselben Integralwert, wie die Lebesguesche Definition), anderseits auch in Fällen von nicht geschränkten Integranden anwendbar bleibt, und dann, im Gegensatze zur Lebesgueschen Definition, auch auf bedingt konvergente Integrale führt, ähnlich wie Harnacks Definition der uneigentlichen Integrale, von der sie eine glückliche Verallgemeinerung darstellt. Da es mir scheint, daß in der vorliegenden Literatur, soweit sie mir bekannt wurde, die Frage nach der Tragweite dieser Borelschen Definition noch nicht hinlänglich geklärt ist, werden die folgenden Bemerkungen über diesen Gegenstand vielleicht nicht überflüssig sein.

§ 1.

Zunächst sei Borels Erweiterung der Riemannschen Integraldefinition, und zwar gleich in etwas verallgemeinerter Form[2]) wiedergegeben. Sie lautet:

Sei im Intervalle $< a, b >$ eine Funktion $f(x)$ und eine Punktmenge $\mathfrak{K}$ des Inhaltes 0 gegeben (sie werde kurz als die Singularitätenmenge bezeichnet) von folgender Eigenschaft: wird die Menge $\mathfrak{K}$ irgendwie in eine Menge sich nicht überdeckender Intervalle J eingeschlossen,[3]) so ist außerhalb der Intervalle J die Funktion $f(x)$ definiert und geschränkt. Bei jeder beliebigen Wahl

[1]) C. R. **150** (1910) S. 375, 508; Journal de math. (6) 8, (1912), S. 201.

[2]) Borel setzt die gleich einzuführende Singularitätenmenge als abzählbar voraus, während hier nur angenommen wird, sie habe den Inhalt 0. Der Begriff des Inhaltes ist stets im Sinne von Lebesgue zu verstehen.

[3]) Wir sagen, eine Menge $\mathfrak{K}$ sei in eine Intervallmenge J eingeschlossen, wenn jeder Punkt von $\mathfrak{K}$ im Innern eines Intervalles J liegt und jedes Intervall J mindestens einen Punkt von $\mathfrak{K}$ enthält. Die Intervalle J betrachten wir als sogenannte offene Intervalle, d. h. wir rechnen die Endpunkte nicht zum Intervalle.

1*

der die Menge $\mathfrak{K}$ einschließenden Intervallmenge J gelte folgendes: teilt man das Intervall $< a, b >$ durch Einschalten endlich vieler Zwischenpunkte

$$a = x_0 < x_1 < \ldots < x_{n-1} < x_n = b,$$

deren keiner einem Intervalle J angehöre, in n Teile und bezeichnet mit h_i die Länge des Teilintervalls $< x_{i-1}, x_i >$, vermindert um die Gesamtlänge[1]) der in dieses Teilintervall fallenden Stücke der Intervalle J, so möge, wie immer auch der Punkt ξ_i im Teilintervalle $< x_{i-1}, x_i >$ außerhalb der Intervalle J gewählt werden mag, die Summe

$$S(J) = \sum_{i=1}^{n} h_i f(\xi_i) \tag{1}$$

einen Grenzwert haben, wenn die Anzahl der eingeschalteten Teilpunkte x_i so über alle Grenzen wächst, daß die größte der Zahlen h_i gegen Null geht. Hat nun dieser Grenzwert selbst einen Grenzwert, wenn man die Gesamtlänge der die Singularitätenmenge einschließenden Intervalle J in beliebiger Weise gegen Null gehen läßt, so wird dieser letztere Grenzwert als das Integral $\int_a^b f(x)\,dx$ bezeichnet.

So definierte Integrale werden wir kurz als Integrale im Sinne von Borel bezeichnen und für sie, wo sie von anders definierten Integralen zu unterscheiden sind, schreiben $(B)\int_a^b f\,dx$. Ebenso wird das Zeichen $(R)\int$ Integrale im Sinne von Riemann, das Zeichen $(L)\int$ Integrale im Sinne von Lebesgue bedeuten.

Wir werden nun die eben mitgeteilte Borelsche Definition etwas näher betrachten. Sei J eine die Singularitätenmenge $\mathfrak{K}$ einschließende Intervallmenge, und sei p die bezüglich des Intervalles $< a, b >$ zur Intervallmenge J komplementäre Punktmenge, also jene in $< a, b >$ liegende abgeschlossene Punktmenge, deren punktfreie Intervalle die Intervalle J sind. Die in der Summe (1) auftretende Zahl h_i ist dann nichts anderes, als der Inhalt des ins Intervall $< x_{i-1}, x_i >$ fallenden Teiles von p.

[1]) Unter der Gesamtlänge einer Menge sich nicht überdeckender Intervalle verstehen wir die Summe der Längen aller dieser Intervalle. Zu jeder beliebigen Intervallmenge gibt es eine äquivalente (d. h. dieselben Punkte bedeckende) Menge sich nicht überdeckender Intervalle. Unter der Gesamtlänge einer beliebigen Intervallmenge werde die Gesamtlänge der äquivalenten Menge sich nicht überdeckender Intervalle verstanden.

Existiert, wie es in der Borelschen Definition verlangt wird, der Grenzwert von $S(J)$, wenn die größte der in $S(J)$ auftretenden Zahlen h_i gegen 0 geht, so wollen wir diesen Grenzwert als das über die Menge p erstreckte Riemannsche Integral von f bezeichnen:[1])

$$(R)\int_p f\,dx = \lim S(J). \tag{2}$$

Diese Definition ist gleichbedeutend mit folgender: ist $\eta > 0$ beliebig gegeben, so gibt es (bei festgehaltener Intervallmenge J) ein $\rho > 0$ derart, daß, sobald in $S(J)$ alle $h_i < \rho$ sind, die Ungleichung gilt:

$$\left| S(J) - (R)\int_p f\,dx \right| < \eta.$$

Existiert nun für jede die Singularitätenmenge $\mathfrak{K}$ einschließende Intervallmenge J das über die komplementäre Menge p erstreckte Integral $(R)\int_p f\,dx$, so ist das Borelsche Integral, wenn vorhanden, in folgender Weise charakterisiert: ist $\eta > 0$ beliebig gegeben, so existiert ein $\varepsilon > 0$ derart, daß, sobald die Gesamtlänge von J kleiner als ε ist, die Ungleichung gilt:

$$\left| (R)\int_p f\,dx - (B)\int_a^b f\,dx \right| < \eta.$$

Wir können leicht zeigen, daß so wie bei den gewöhnlichen Integralen im Sinne Riemanns auch für die Existenz von $(R)\int_p f\,dx$ notwendig[2]) ist, daß für jedes $k > 0$ die Menge $\mathfrak{M}$ aller jener Punkte von p, in denen der Unstetigkeitsgrad von f bezüglich p einen Wert $\geqq k$ hat, den Inhalt 0 habe. Bezeichnet in der Tat p_i den ins Intervall $< x_{i-1}, x_i >$ fallenden Teil von p und sind M_i und m_i obere und untere Grenze von f auf p_i, so muß im Falle der Existenz von $(R)\int_p f\,dx$ die Gleichung bestehen:

$$(R)\int_p f\,dx = \lim \sum_{i=1}^{n} h_i f(\xi_i) = \lim \sum_{i=1}^{n} h_i M_i = \lim \sum_{i=1}^{n} h_i m_i, \tag{3}$$

es muß also:

$$\lim \sum_{i=1}^{n} h_i (M_i - m_i) = 0 \tag{4}$$

---

[1]) Diese schon von W. H. Young (Phil. Trans. Royal Soc. Ser. A, 204 (1905), S. 221) diskutierte Definition deckt sich, wie man leicht sieht, mit der von J. Pierpont gegebenen: The theory of functions of real variables, Vol. I, S. 506 ff.

[2]) Die Bedingung ist auch hinreichend.

sein. Angenommen nun, es sei der Inhalt μ von $\mathfrak{M}$ nicht Null. Da für jede Menge p_i, die einen Punkt von $\mathfrak{M}$ enthält, $M_i - m_i \geqq k$ ist, hätte man also stets:

$$\sum_{i=1}^{n} h_i (M_i - m_i) \geqq \mu . k$$

im Gegensatze zu (4).

Es folgt daraus in bekannter Weise weiter, daß für die Existenz von $(R)\int_p f\,dx$ notwendig[1]) ist, daß die Menge aller Punkte von p, in denen f unstetig ist bezüglich p, den Inhalt 0 habe.

Daraus aber folgt leicht, daß wenn $(R)\int_p f\,dx$ existiert, f meßbar auf p ist: in der Tat, schließen wir die Menge der Unstetigkeitspunkte von f bezüglich p in eine Menge von Intervallen Δ ein, so ist f auf der Komplementärmenge $\mathfrak{R}$ der Δ bezüglich p stetig und daher meßbar. Läßt man die Intervallmenge Δ eine Folge $\Delta_1, \Delta_2, \ldots, \Delta_n, \ldots$ durchlaufen, so daß die Gesamtlänge von Δ_n gegen 0 geht, so ist, abgesehen von einer Menge des Inhaltes 0, die Menge p Vereinigungsmenge der Komplementärmengen $\mathfrak{R}_n$ von Δ_n bezüglich p. Da aber f auf jeder Menge $\mathfrak{R}_n$ meßbar ist, folgt hieraus, daß f auch auf p meßbar ist.

Beschränken wir uns nun in § 1 weiterhin auf geschränkte Funktionen f, so folgt also aus der Existenz von $(R)\int_p f\,dx$ auch die von $(L)\int_p f\,dx$, und zwar ist:

$$(R)\int_p f\,dx = (L)\int_p f\,dx.$$

In der Tat, wenden wir wieder die eben benützten Bezeichnungen p_i, M_i, m_i an, so ist:

$$(L)\int_p f\,dx = \sum_{i=1}^{n} (L)\int_{p_i} f\,dx$$

und somit:

$$\sum_{i=1}^{n} h_i m_i \leqq (L)\int_p f\,dx \leqq \sum_{i=1}^{n} h_i M_i,$$

es folgt also aus (3) die Behauptung.

[1]) Auch diese Bedingung ist hinreichend.

Daß auch bei geschränktem f die Borelsche Definition anwendbar bleibt in Fällen, in denen die Riemannsche Definition versagt, zeigt die Funktion f, die in allen irrationalen Punkten von $< a, b >$ den Wert 0, in allen rationalen den Wert 1 hat. Versteht man hier unter der Singularitätenmenge $\mathfrak{R}$ die Menge der rationalen Punkte, so ist bei beliebiger Wahl der Intervalle J auf der zu den Intervallen J komplementären abgeschlossenen Menge p durchweg $f = 0$, die Summen $S(J)$ haben also stets den Wert 0 und es ist somit nach der Borelschen Definition $\int_a^b f\,dx = 0$.

Weiter ist bei geschränktem f klar, daß aus der Existenz von $(B)\int_a^b f\,dx$ auch die Existenz von $(L)\int_a^b f\,dx$ und das Bestehen der Gleichung:

$$(B)\int_a^b f\,dx = (L)\int_a^b f\,dx \tag{5}$$

folgt.

In der Tat, bezeichne $\varepsilon_1, \varepsilon_2, \ldots, \varepsilon_\nu, \ldots$ eine Folge positiver Zahlen mit $\lim\limits_{\nu=\infty} \varepsilon_\nu = 0$. Zu jeder dieser Zahlen gibt es eine die Singularitätenmenge einschließende Intervallmenge J_ν, deren Gesamtlänge $< \varepsilon_\nu$ ist. Die zu J_ν komplementäre abgeschlossene Menge $p^{(\nu)}$ hat einen Inhalt $> b - a - \varepsilon_\nu$ und es ist auf ihr, wie wir schon gesehen haben, f meßbar. Das Intervall $< a, b >$ ist Vereinigungsmenge der Mengen $p^{(1)}, p^{(2)}, \ldots, p^{(\nu)}, \ldots$ und einer Menge des Inhaltes 0, woraus man sofort folgert, daß f in $< a, b >$ meßbar, und daher integrierbar im Sinne von Lebesgue ist. Nach einem bekannten Satze gilt nun aber, weil der Inhalt von $p^{(\nu)}$ gegen den des Intervalles $< a, b >$ konvergiert,

$$\lim_{\nu=\infty} (L)\int_{p^{(\nu)}} f\,dx = (L)\int_a^b f\,dx. \tag{6}$$

Nun ist aber, wie wir sahen:

$$(L)\int_{p^{(\nu)}} f\,dx = (R)\int_{p^{(\nu)}} f\,dx \tag{7}$$

und zufolge der Borelschen Definition ist:

$$(B)\int_a^b f\,dx = \lim_{\nu=\infty} (R)\int_{p^{(\nu)}} f\,dx. \tag{8}$$

Aus (6), (7) und (8) aber folgt sofort die behauptete Gleichung (5).

Gleichung (5) lehrt nun auch (zunächst bei geschränktem f) sofort, daß der Wert eines durch die Borelsche Definition gelieferten Integrales von der Wahl der Singularitätenmenge $\mathfrak{K}$ unabhängig ist in dem Sinne, daß wenn bei verschiedener Wahl der Singularitätenmenge der das Borelsche Integral definierende Grenzwert vorhanden ist, er für alle diese Wahlen der Singularitätenmenge denselben Wert hat.

Anderseits ist aber auch bei geschränkten Integranden die Borelsche Definition keineswegs mit der Lebesgueschen äquivalent: es kann im Sinne von Lebesgue ein Integral vorhanden sein, ohne daß auf Grund der Borelschen Definition ein Integral vorhanden ist. Wir zeigen dies an folgendem Beispiele:

Sei im Intervalle $< a, b >$ eine nirgends dichte perfekte Menge $\mathfrak{A}$ gegeben, deren Inhalt $\neq 0$ ist und es habe f in allen Punkten von $\mathfrak{A}$ den Wert 1, sonst den Wert 0. Dann existiert $(L)\int\limits_a^b f\,dx$ und ist gleich dem Inhalt von $\mathfrak{A}$. Wir wenden nun die Borelsche Definition auf f an. Da die Singularitätenmenge $\mathfrak{K}$ den Inhalt 0 haben muß, ist der Inhalt der Menge $\mathfrak{A}^*$ der nicht zu $\mathfrak{K}$ gehörigen Punkte von $\mathfrak{A}$ nicht 0; da aber der Inhalt einer Menge die obere Grenze der Inhalte der abgeschlossenen Teile dieser Menge ist, gibt es in $\mathfrak{A}^*$ einen abgeschlossenen Teil $\overline{\mathfrak{A}}$, dessen Inhalt $\neq 0$ ist. Seien (x'_ν, x''_ν) $(\nu = 1, 2, \ldots)$ die bezüglich des Intervalles $< a, b >$ zu $\overline{\mathfrak{A}}$ komplementären Intervalle (die punktfreien Intervalle von $\overline{\mathfrak{A}}$). Die Länge des Intervalles (x'_ν, x''_ν) werde abkürzend mit $2\delta_\nu$ bezeichnet: wir betrachten in (x'_ν, x''_ν) die Punkte

$$x'_{\nu,\mu} = x'_\nu + \frac{\delta_\nu}{\mu},\; x''_{\nu,\mu} = x''_\nu - \frac{\delta_\nu}{\mu} \quad (\mu = 1, 2, \ldots).$$

Da $\mathfrak{A}$ nirgends dicht ist, so ist der Inhalt des ins Intervall $(x'_{\nu,\mu+1}, x'_{\nu,\mu})$ oder $(x''_{\nu,\mu}, x''_{\nu,\mu+1})$ fallenden Teiles von $\mathfrak{A}$ kleiner als die Länge des betreffenden Intervalles und da $\mathfrak{K}$ den Inhalt 0 hat, so gibt es also in jedem Intervalle $(x'_{\nu,\mu+1}, x'_{\nu,\mu})$ einen Punkt $\xi'_{\nu,\mu}$, in jedem Intervalle $(x''_{\nu,\mu}, x''_{\nu,\mu+1})$ einen Punkt $\xi''_{\nu,\mu}$, der weder zu $\mathfrak{A}$ noch zu $\mathfrak{K}$ gehört. Fügen wir zu $\overline{\mathfrak{A}}$ noch die Menge aller Punkte $\xi'_{\nu,\mu}, \xi''_{\nu,\mu}$ $(\nu, \mu = 1, 2, \ldots)$ hinzu, so entsteht eine abgeschlossene Menge $\overline{\overline{\mathfrak{A}}}$, die mit $\mathfrak{K}$ keinen Punkt gemein hat. Da aber $\overline{\overline{\mathfrak{A}}}$ abgeschlossen ist, liegt nun jeder Punkt von $\mathfrak{K}$ im Inneren eines Intervalles, das keinen Punkt von $\overline{\overline{\mathfrak{A}}}$ enthält. Es kann also eine die Menge $\mathfrak{K}$ einschließende Intervallmenge J so gewählt werden, daß alle Punkte von $\overline{\overline{\mathfrak{A}}}$ außerhalb J liegen. Für die mit Bezug auf eine solche Intervallmenge J gebildeten Summen $S(J)$ kann aber der beim Borelschen Verfahren verlangte Grenzwert nicht existieren.

Denn bezeichnet p die zur Intervallmenge J komplementäre abgeschlossene Menge, so existiert, wie wir gesehen haben, dieser Grenzwert (das Riemannsche Integral über p) nur dann, wenn die Menge der Punkte, in denen der Unstetigkeitsgrad von f bezüglich p einen Wert $\geqq k$ hat, für jedes $k > 0$ vom Inhalte 0 ist. Hier nun ist $\overline{\overline{\mathfrak{A}}}$ Teil von p. Ferner war $\overline{\mathfrak{A}}$ Teil von $\overline{\overline{\mathfrak{A}}}$ und jeder Punkt von $\overline{\mathfrak{A}}$ ist Häufungspunkt von Punkten $\xi'_{\nu,\mu}$ oder $\xi''_{\nu,\mu}$. In allen Punkten von $\overline{\mathfrak{A}}$ nun ist $f = 1$, in allen Punkten $\xi'_{\nu,\mu}$ oder $\xi''_{\nu,\mu}$ aber ist $f = 0$. In jedem Punkt von $\overline{\mathfrak{A}}$ ist also der Unstetigkeitsgrad von f bezüglich p gleich 1, und da $\overline{\mathfrak{A}}$ nicht den Inhalt 0 hat, ist unsere Behauptung erwiesen.

Wir sehen also, daß die Borelsche und die Lebesguesche Integraldefinition auch bei geschränktem Integranden nicht äquivalent sind: existiert das Integral im Sinne von Borel, so auch im Sinne von Lebesgue und beide Definitionen ergeben denselben Wert. Doch kann sehr wohl ein Integral im Sinne von Lebesgue vorhanden sein, ohne daß ein Integral im Sinne von Borel vorhanden wäre.

Doch kann man, bei geschränktem Integranden, die Borelsche Definition so modifizieren, daß sie mit der Lebesgueschen völlig äquivalent wird. Diese modifizierte Definition lautet so:

Sei die Funktion f so beschaffen, daß, wie klein $\varepsilon > 0$ auch sein mag, es in $< a, b >$ eine abgeschlossene Menge p gibt, deren Inhalt $> b - a - \varepsilon$ ist und für die das Integral $(R)\int\limits_p f\,dx$ existiert. Es gibt dann eine Zahl J derart, daß, sobald ε hinlänglich klein ist, für jede abgeschlossene Menge p, deren Inhalt $> b - a - \varepsilon$ ist und auf der $(R)\int f\,dx$ existiert, die Ungleichung gilt:

$$\left|(R)\int\limits_p f\,dx - J\right| < \eta \quad (\eta > 0 \text{ beliebig gegeben});$$

diese Zahl J wird als das Integral $(B^*)\int\limits_a^b f\,dx$ bezeichnet.

In der Tat folgt hier, wo wir f als geschränkt vorausgesetzt haben (etwa $|f| < F$), von selbst, daß wenn es abgeschlossene Mengen p gibt, deren Inhalt beliebig nahe an $b - a$ liegt, und auf denen $(R)\int\limits_p f\,dx$ existiert, diese Integrale sich einem bestimmten Grenzwerte J nähern, wenn der Inhalt von p gegen $b - a$ geht. Denn ist sowohl der Inhalt von $\overline{p}$ als der von $\overline{\overline{p}}$ größer als $b - a - \varepsilon$, so hat die Menge der nicht zu $\overline{\overline{p}}$ gehörigen Punkte von $\overline{p}$, und ebenso die Menge der nicht zu $\overline{p}$ gehörigen Punkte von $\overline{\overline{p}}$ höchstens den Inhalt ε, es ist daher:

$$\left|(R)\int_{\overline{p}} f\,dx - (R)\int_{\overline{\overline{p}}} f\,dx\right| < 2\varepsilon F$$

und da F eine feste endliche Zahl ist, so ist dies gleichbedeutend mit der Existenz des behaupteten Grenzwertes J.

Daß weiter aus der Existenz von $(B^*)\int_a^b f\,dx$ die von $(L)\int_a^b f\,dx$ folgt, und daß diese beiden Werte gleich sind, zeigt man wie oben für die Integrale $(B)\int_a^b f\,dx$. Existiert aber umgekehrt $(L)\int_a^b f\,dx$, d. h. ist f meßbar, so gibt es bekanntlich bei beliebigem $\varepsilon > 0$ in $< a, b >$ abgeschlossene Mengen p, deren Inhalt $> b - a - \varepsilon$ ist, und bezüglich derer f stetig ist, sodaß also sicher $(R)\int_p f\,dx$ und somit nach dem Gesagten auch $(B^*)\int_a^b f\,dx$ existiert.

§ 2.

Wir wenden uns nun dem Falle eines nicht geschränkten Integranden zu. Da die Borelsche Definition die nach der Methode von Harnack definierten uneigentlichen Integrale mit umfaßt, die auch bedingt konvergent sein können, während alle Lebesgueschen Integrale absolut konvergieren, folgt sofort, daß nun auch ein Integral im Sinne von Borel existieren kann, ohne daß ein solches im Sinne von Lebesgue existiert. Hier gilt:

Existiert ein Integral im Sinne von Borel nicht nur für f, sondern auch für $|f|$, so existiert das Integral auch im Sinne von Lebesgue und hat denselben Wert.

In der Tat, existiere das Integral von f und $|f|$ im Sinne von Borel. Wie in § 1 folgt, daß f jedenfalls meßbar ist. Wir geben eine abnehmende Folge positiver Zahlen ε_ν mit $\lim_{\nu=\infty} \varepsilon_\nu = 0$ vor. Es sei J_ν eine $\mathfrak{R}$ einschließende Intervallmenge J_ν, deren Gesamtlänge $< \varepsilon_\nu$ sei; wir können annehmen, die Intervalle $J_{\nu+1}$ liegen im Innern der Intervalle J_ν. Endlich sei $p^{(\nu)}$ die zu J_ν bezüglich des Intervalles $< a, b >$ komplementäre abgeschlossene Menge. Wir haben schon gesehen, daß auf Grund der Borelschen Definition

$$(B)\int_a^b |f|\,dx = \lim_{\nu=\infty}(R)\int_{p^{(\nu)}} |f|\,dx = \lim_{\nu=\infty}(L)\int_{p^{(\nu)}} |f|\,dx \qquad (9)$$

ist. Sei nun $f_\nu = f$ in den Punkten von $p^{(\nu)}$ und außerhalb $p^{(\nu)}$ sei $f_\nu = 0$. Es ist dann $|f_{\nu+1}| \geqq |f_\nu|$ und abgesehen von einer Punktmenge des Inhaltes 0 ist $f = \lim_{\nu=\infty} f_\nu$. Die $|f_\nu|$ nähern sich also wachsend dem Werte $|f|$. Nun ist:

$$(L)\int_a^b |f_\nu|\,dx = (L)\int_{p^{(\nu)}} |f|\,dx.$$

nach (9) wachsen also die Integrale $(L)\int_a^b |f_\nu|\,dx$ nicht über alle Grenzen und nach einem bekannten Satze existiert also $(L)\int_a^b |f|\,dx$ und somit auch $(L)\int_a^b f\,dx$, und gleichfalls nach einem bekannten Satze, da der Inhalt von $p^{(\nu)}$ den von $< a, b >$ zur Grenze hat, ist

$$\lim_{\nu=\infty} (L)\int_{p^{(\nu)}} f\,dx = (L)\int_a^b f\,dx$$

das aber ist, da die linke Seite auch das Integral im Sinne von Borel darstellt, die Behauptung.

Nun aber wollen wir zeigen, daß die Borelsche Definition doch nur in sehr beschränktem Umfange zu bedingt konvergenten Integralen führt. Es gilt nämlich der Satz:

Existiert nach der Borelschen Definition das Integral von f, und wird mit $\mathfrak{Q}$ die Menge bezeichnet, die aus $\mathfrak{K}$ durch Hinzufügung aller Häufungspunkte entsteht, so existiert das Lebesguesche Integral von f (und somit auch von $|f|$) über die Menge $\mathfrak{Q}$.

In dem interessantesten Falle, daß $\mathfrak{K}$ in $< a, b >$ überall dicht angenommen werden muß, ergibt also die Borelsche Definition nur einen Spezialfall der Lebesgueschen. Wir gehen an den Beweis des aufgestellten Satzes.

Nehmen wir an, es existiere $(L)\int_{\mathfrak{Q}} f\,dx$ nicht, d. h. es habe $(L)\int_{\mathfrak{Q}} |f|\,dx$ den Wert $+\infty$. Setzen wir $g = f$ für $f \geqq 0$, $g = 0$ für $f < 0$; $h = 0$ für $f \geqq 0$, $h = -f$ für $f < 0$, so hat dann auch mindestens eines der Integrale $(L)\int_{\mathfrak{Q}} g\,dx$ und $(L)\int_{\mathfrak{Q}} h\,dx$ den Wert $+\infty$, nehmen wir etwa an, das erstere. Wie man durch das bekannte Zweiteilungsverfahren zeigt, existiert dann mindestens ein Punkt $\mathfrak{x}$ in $< a, b >$ derart, daß das Lebesguesche Integral von g erstreckt über den in irgend eine Umgebung von $\mathfrak{x}$ fallenden Teil von $\mathfrak{Q}$ stets den Wert $+\infty$ hat.

Sei ε irgend eine positive Zahl; wir werden zeigen, daß, wie groß auch A gewählt sei, es eine die Menge $\mathfrak{K}$ einschließende Intervallmenge J von einer Gesamtlänge $< \varepsilon$ gibt, derart, daß für ihre Komplementärmenge p bezüglich des Intervalles $< a, b >$ gilt:

$$(R)\int\limits_{p} f\,dx > A.$$

Damit ist dann nachgewiesen, daß ein Integral $\int\limits_a^b f\,dx$ im Sinne von Borel nicht existieren kann. Wir schließen den Punkt $\mathfrak{x}$ in ein Intervall δ ein, dessen Endpunkte nicht zu $\mathfrak{K}$ gehören und dessen Länge $< \frac{\varepsilon}{2}$ ist, und den außerhalb δ liegenden Teil von $\mathfrak{K}$ in eine Intervallmenge J', deren Gesamtlänge $< \frac{\varepsilon}{2}$ ist. Sei p' die zu den Intervallen J' und δ bezüglich $< a, b >$ komplementäre abgeschlossene Menge, so ist auf ihr f geschränkt und es hat $(R)\int\limits_{p'} f\,dx$ einen endlichen Wert.

Den ins Intervall δ fallenden Teil von $\mathfrak{K}$ schließen wir in eine Folge von Intervallmengen J''_ν ein ($\nu = 1, 2, \ldots$), so daß die Intervalle jeder dieser Mengen J''_ν sämtliche zu $\mathfrak{Q}$ bezüglich δ komplementären Intervalle bedecken, und daß die Gesamtlänge der Intervalle J''_ν für $\nu = \infty$ gegen die Gesamtlänge der zu $\mathfrak{Q}$ bezüglich δ komplementären Intervalle konvergiert; endlich mögen die Intervalle von $J''_{\nu+1}$ ganz im Innern der von J''_ν liegen. Für die Komplementärmengen p''_ν der J''_ν bezüglich δ gilt dann, daß p''_ν Teil von $p''_{\nu+1}$ ist, und daß die Vereinigungsmenge der p''_ν bis auf eine Menge des Inhaltes 0 identisch ist mit dem ins Intervall δ fallenden Teil von $\mathfrak{Q}$. Es folgt daraus, nach der Definition des Punktes $\mathfrak{x}$, unmittelbar, daß:

$$\lim_{\nu=\infty} (L)\int\limits_{p''_\nu} g\,dx = +\infty$$

ist. Auf jeder der Mengen p''_ν ist g geschränkt, etwa:

$$g < G_\nu \text{ auf } p''_\nu. \tag{10}$$

Sei $\mathfrak{H}$ die Menge aller jener nach δ fallenden Punkte von $\mathfrak{Q}$, in denen $h > 0$, (d. h. $f < 0$) ist. Wir schließen die Punkte von $\mathfrak{H}$

in eine Intervallmenge J_ν''' ein, deren Gesamtlänge sich vom Inhalte von $\mathfrak{H}$ um weniger als $\frac{1}{G_\nu}$ unterscheidet. Die ins Innere der Intervalle J_ν''' fallenden Punkte von p_ν'' gehören also, abgesehen von einer Punktmenge, deren Inhalt $< \frac{1}{G_\nu}$ ist, zu H, und da $g = 0$ ist in allen Punkten von $\mathfrak{H}$, ist also der Inhalt der ins Innere der Intervalle J_ν''' fallenden Teilmenge von p_ν'', auf der $g > 0$ ist, kleiner als $\frac{1}{G_\nu}$. Bezeichnen wir mit p_ν''' den Teil von p_ν'', der nicht im Innern der Intervalle J_ν''' liegt, so haben wir also wegen (10):

$$(L)\int_{p_\nu'''} g\,dx > (L)\int_{p_\nu''} g\,dx - 1$$

und daher auch:

$$\lim_{\nu=\infty} (L)\int_{p_\nu'''} g\,dx = +\infty.$$

Nach Konstruktion enthält jedes der Intervalle J' und J_ν'' einen Punkt von $\mathfrak{K}$; da aber jeder Punkt von $\mathfrak{Q}$ auch Punkt oder Häufungspunkt von $\mathfrak{K}$ ist, so enthält auch jedes Intervall J_ν''' einen Punkt von $\mathfrak{K}$. Die Intervallmenge J_ν, die alle Intervalle von J', J_ν'' und J_ν''' enthält, ist also eine die Menge $\mathfrak{K}$ einschließende Intervallmenge, deren Gesamtlänge offenbar $< \varepsilon$ ist. Ihre Komplementärmenge p_ν bezüglich $\langle a, b \rangle$ ist die Vereinigungsmenge der beiden Mengen p' und p_ν''', die höchstens die beiden Endpunkte des Intervalles δ gemein haben. Es ist also

$$(L)\int_{p_\nu} f\,dx = (L)\int_{p'} f\,dx + (L)\int_{p_\nu'''} f\,dx;$$

da p_ν''' keinen Punkt von $\mathfrak{H}$ enthält, ist weiter

$$(L)\int_{p_\nu'''} g\,dx = (L)\int_{p_\nu'''} f\,dx$$

und somit, da, wie schon erwähnt, $(L)\int_{p'} f\,dx$ endlich ist, und da wegen der vorausgesetzten Existenz von $(B)\int_a^b f\,dx$ auch $(R)\int_{p_\nu} f\,dx$ existiert und $= (L)\int_{p} f\,dx$ ist.

$$\lim_{\nu=\infty} (R)\int_{p_\nu} f\,dx = +\infty,$$

womit die Behauptung erwiesen ist.

Wir bezeichnen wieder die abgeschlossene Menge, die aus der Singularitätenmenge $\mathfrak{K}$ durch Hinzufügung aller Häufungspunkte entsteht, mit $\mathfrak{Q}$. Seien (a_ν, b_ν) die punktfreien Intervalle von $\mathfrak{Q}$. Existiert $(B)\int_a^b f\,dx$ nach der Borelschen Definition, so existiert offenbar für jedes im Innern von (a_ν, b_ν) gelegene Intervall (a'_ν, b'_ν) ein eigentliches Riemannsches Integral und es existiert

$$\lim_{a'_\nu=a_\nu,\ b'_\nu=b_\nu} (R)\int_{a'_\nu}^{b'_\nu} f\,dx.$$

Wir bezeichnen mit ω_ν die obere Grenze von $\left|\int_{a'_\nu}^{b'_\nu} f\,dx\right|$ für alle Teilintervalle $< a'_\nu, b'_\nu >$ von (a_ν, b_ν) und beweisen[1]):

Existiert das Integral $(B)\int_a^b f\,dx$, so konvergiert $\sum \omega_\nu$.

In der Tat, existiert dieses Integral, so gilt folgendes: zu jedem positiven η gehört ein positives ε so, daß wenn J' und J'' zwei $\mathfrak{K}$ einschließende Intervallmengen sind, deren jede eine Gesamtlänge $< \varepsilon$ hat, und p' und p'' die bezüglich $< a, b >$ zu J' und J'' komplementären abgeschlossenen Mengen bezeichnen, die Ungleichung besteht:

$$\left|\int_{p'} f\,dx - \int_{p''} f\,dx\right| < \eta.$$

Wir werden aber zeigen, daß im Falle der Divergenz von $\sum \omega_\nu$ bei beliebig gegebenem $\varepsilon > 0$ stets J' und J'' so gewählt werden können, daß:

$$\left|\int_{p'} f\,dx - \int_{p''} f\,dx\right| > 1.$$

Wir wählen zunächst ν_0 so groß, daß

$$\sum_{\nu=\nu_0}^{\infty} (b_\nu - a_\nu) < \varepsilon.$$

[1]) Für die Harnackschen Integrale wurde diese Tatsache zuerst bewiesen von E. H. Moore, Amer. Trans. 2, S. 324.

Im Innern von (a_ν, b_ν) kann $< a'_\nu, b'_\nu >$ so gefunden werden, daß

$$\left|\int_{a'_\nu}^{b'_\nu} f\,dx\right| > \omega_\nu - \frac{1}{2^\nu}.$$

Ferner gibt es wegen der Divergenz von $\Sigma\,\omega_\nu$ endlich viele $\nu > \nu_0$ (sie mögen mit $\nu_1, \nu_2, \ldots, \nu_p$ bezeichnet werden) derart, daß

$$\sum_{i=1}^{p} \omega_{\nu_i} > 2,$$

und daß $\int_{a'_{\nu_i}}^{b'_{\nu_i}} f\,dx$ für $i = 1, 2, \ldots, p$ einerlei Zeichen, z. B. das positive, hat. Nun können offenbar die zwei die Menge $\mathfrak{K}$ einschließenden Intervallmengen J' und J'' so gefunden werden, daß jede von beiden eine Gesamtlänge $< \varepsilon$ hat, und daß J' die Intervalle $< a_{\nu_i}, b_{\nu_i} >$ $(i = 1, 2, \ldots, p)$ vollständig bedeckt, während J'' von jedem dieser Intervalle gerade das Teilintervall $< a'_{\nu_i}, b'_{\nu_i} >$ freiläßt, während sonst J' und J'' völlig übereinstimmen. Dann ist

$$\int_{J''} f\,dx - \int_{J'} f\,dx = \sum_{i=1}^{p} \int_{a'_{\nu_i}}^{b'_{\nu_i}} f\,dx > \sum_{i=1}^{p} \left(\omega_{\nu_i} - \frac{1}{2^{\nu_i}}\right) > 1$$

wie behauptet. Damit ist für den Fall der Existenz von $(B)\int_a^b f\,dx$ die Konvergenz von $\Sigma\,\omega_\nu$ nachgewiesen, wie angekündigt.

Wir erkennen nun auch unmittelbar daß, wenn wieder $\mathfrak{Q}$ die Menge ist, die aus der Singularitätenmenge $\mathfrak{K}$ durch Hinzufügung aller Häufungspunkte entsteht und mit (a_ν, b_ν) die punktfreien Intervalle von $\mathfrak{Q}$ bezeichnet werden, für das Borelsche Integral die Gleichung gilt:

$$(B)\int_a^b f\,dx = \sum_\nu \int_{a_\nu}^{b_\nu} f\,dx + (L)\int_{\mathfrak{Q}} f\,dx. \tag{11}$$

Die in der Summe rechts auftretenden Integrale sind definiert durch

$$\int_{a_\nu}^{b_\nu} f\,dx = \lim_{h=+0,\,k=+0} (R)\int_{a_\nu+h}^{b_\nu-k} f\,dx$$

wo diese Grenzwerte, wie schon bemerkt, existieren; und die Summe aller dieser Integrale konvergiert absolut, da die oben mit $\sum\limits_{\nu} \omega_\nu$ bezeichnete Summe konvergiert. Um die Formel (11) nachzuweisen, gehen wir so vor: Ist $\varepsilon > 0$ beliebig gegeben, so schließen wir $\mathfrak{K}$ in eine Intervallmenge J ein, deren Gesamtlänge $< \varepsilon$ ist, und die von beiden Enden jedes Intervalles $< a_\nu, b_\nu >$ ein so kleines Stück abschneidet, daß für das übrigbleibende Intervall $< a'_\nu, b'_\nu >$ gilt:

$$\left| \int\limits_{a'_\nu}^{b'_\nu} f\,dx - \int\limits_{a_\nu}^{b_\nu} f\,dx \right| < \frac{\varepsilon}{2^\nu}. \tag{12}$$

Die zu J bezüglich $< a, b >$ komplementäre, abgeschlossene Menge p setzt sich zusammen aus den abzählbar unendlich vielen Intervallen $< a'_\nu, b'_\nu >$ und einer Menge p^*, die Teil von $\mathfrak{Q}$ ist; es ist daher:

$$(L)\int\limits_{p} f\,dx = \sum_{\nu} \int\limits_{a'_\nu}^{b'_\nu} f\,dx + (L)\int\limits_{p^*} f\,dx. \tag{13}$$

Geht ε gegen 0, so geht per definitionem die linke Seite dieser Gleichung in die linke Seite von (11) über, und beachtet man Ungleichung (12), sowie die Tatsache, daß p^* ein Teil von $\mathfrak{Q}$ ist, dessen Inhalt sich vom Inhalte von $\mathfrak{Q}$ um weniger als ε unterscheidet, so sieht man, da ja die Existenz von $(L)\int\limits_{\mathfrak{Q}} f\,dx$ bereits bewiesen ist, daß auch die rechte Seite von (13) in die rechte Seite von (11) übergeht, womit (11) bewiesen ist.

Wir können nun unschwer zeigen, daß der Wert eines Borelschen Integrales von der Wahl der Singularitätenmenge in folgendem Sinne unabhängig ist; existiert das Integral für zwei verschiedene Singularitätenmengen, so hat es in beiden Fällen denselben Wert. Es existiere also das Integral sowohl, wenn für die Singularitätenmenge die Menge $\mathfrak{K}$ als auch wenn für die Singularitätenmenge die Menge $\overline{\mathfrak{K}}$ gewählt wird, und habe die Werte B bezw. $\overline{B}$. Seien $\mathfrak{Q}$ und $\overline{\mathfrak{Q}}$ die Mengen, die aus $\mathfrak{K}$ und $\overline{\mathfrak{K}}$ durch Hinzufügung der Häufungspunkte entstehen, und seien (a_ν, b_ν) und $(\overline{a}_\nu, \overline{b}_\nu)$ die punktfreien Intervalle von $\mathfrak{Q}$ und $\overline{\mathfrak{Q}}$ in $< a, b >$. Es ist dann nach Formel (11):

$$B = \sum_{\nu} \int\limits_{a_\nu}^{b_\nu} f\,dx + (L)\int\limits_{\mathfrak{Q}} f\,dx.$$

Nach einer bekannten Eigenschaft der Lebesgueschen Integrale ist, wenn mit $\mathfrak{Q}_\nu$ der nach $(\overline{a}_\nu, \overline{b}_\nu)$ fallende Teil von $\mathfrak{Q}$ und mit $\mathfrak{D}$ der Durchschnitt von $\mathfrak{Q}$ und $\overline{\mathfrak{Q}}$ bezeichnet wird:

$$(L)\int_{\mathfrak{Q}} f\,dx = (L)\int_{\mathfrak{D}} f\,dx + \sum_\nu \int_{\mathfrak{Q}_\nu} f\,dx,$$

wo die rechts auftretende Reihe absolut konvergiert. Bezeichnen wir noch mit $\overline{\mathfrak{Q}}_\nu$ den nach (a_ν, b_ν) fallenden Teil von $\overline{\mathfrak{Q}}$, mit $(\overline{a}_{\nu,\mu}, \overline{b}_{\nu,\mu})$ $(\mu = 1, 2, \ldots)$ die punktfreien Intervalle von $\overline{\mathfrak{Q}}_\nu$ in (a_ν, b_ν), so ist

$$\int_{a_\nu}^{b_\nu} f\,dx = \sum_\mu \int_{\overline{a}_{\nu,\mu}}^{\overline{b}_{\nu,\mu}} f\,dx + (L)\int_{\overline{\mathfrak{Q}}_\nu} f\,dx; \tag{14}$$

dies folgt in der Tat unmittelbar daraus, daß f in jedem ganz im Inneren von (a_ν, b_ν) liegenden Intervalle (a'_ν, b'_ν) ein eigentliches Integral im Riemannschen Sinne besitzt; man hat nur die Formel (14) zunächst für das Intervall (a'_ν, b'_ν) anzusetzen und daraus für (a_ν, b_ν) durch Grenzübergang herzuleiten, wobei zu beachten ist, daß $(L)\int_{\overline{\mathfrak{Q}}_\nu} f\,dx$ sicher existiert, da $(L)\int_{\overline{\mathfrak{Q}}} f\,dx$ existiert, und daß $\sum_\mu \int_{\overline{a}_{\nu,\mu}}^{\overline{b}_{\nu,\mu}} f\,dx$ absolut konvergiert, da dies für $\sum_\nu \int_{\overline{a}_\nu}^{\overline{b}_\nu} f\,dx$ gilt, und jedes Intervall $(\overline{a}_{\nu,\mu}, \overline{b}_{\nu,\mu})$ — abgesehen höchstens von den beiden an die Endpunkte von (a_ν, b_ν) grenzenden — zugleich eines der Intervalle $(\overline{a}_\nu, \overline{b}_\nu)$ ist.

Wir haben also nun die Formel erhalten:

$$B = \sum_\nu \left(\sum_\mu \int_{\overline{a}_{\nu,\mu}}^{\overline{b}_{\nu,\mu}} f\,dx + (L)\int_{\overline{\mathfrak{Q}}_\nu} f\,dx \right) + \sum_\nu \int_{\mathfrak{Q}_\nu} f\,dx + \int_{\mathfrak{D}} f\,dx. \tag{15}$$

in der, wegen der Existenz von $(L)\int_{\overline{\mathfrak{Q}}} f\,dx$, die Summe der $(L)\int_{\overline{\mathfrak{Q}}_\nu} f\,dx$ absolut konvergiert. Wir können aber auch leicht die absolute Konvergenz der Doppelreihe $\sum_{\mu,\nu} \int_{\overline{a}_{\nu,\mu}}^{\overline{b}_{\nu,\mu}} f\,dx$ einsehen. Bezeichnen wir mit $(\bar{a}'_{\nu,\mu}, \bar{b}'_{\nu,\mu})$ ein beliebiges, ganz im Innern von $(\bar{a}_{\nu,\mu}, \bar{b}_{\nu,\mu})$ gelegenes Intervall, mit $\bar{\omega}_{\nu,\mu}$ die obere Grenze aller Werte von

$$\left| \int_{\bar{a}'_{\nu,\mu}}^{\bar{b}'_{\nu,\mu}} f\,dx \right|,$$

so genügt es, die Konvergenz von $\sum\limits_{\mu,\nu} \bar{\omega}_{\mu,\nu}$ nachzuweisen, was ganz ähnlich zu geschehen hat, wie oben die Konvergenz von $\sum \omega_\nu$ nachgewiesen wurde.

Die absolute Konvergenz aller in (15) auftretenden Reihen ermöglicht es nun, diese Gleichung auch in anderer Gestalt zu schreiben: Man bezeichne die punktfreien Intervalle von $\mathfrak{Q}_\nu$ in $(\bar{a}_\nu, \bar{b}_\nu)$ mit $(a_{\nu,\mu}, b_{\nu,\mu})$ $(\mu = 1, 2, \ldots)$ und beachte, daß die Gesamtheit aller Intervalle $(a_{\nu,\mu}, b_{\nu,\mu})$ $(\mu, \nu = 1, 2, \ldots)$ mit der aller Intervalle $(\bar{a}_{\nu,\mu}, \bar{b}_{\nu,\mu})$ $(\mu, \nu = 1, 2, \ldots)$ übereinstimmt, da jede dieser beiden Intervallmengen nichts anderes ist als die Menge aller punktfreien Intervalle der Vereinigungsmenge von $\mathfrak{Q}$ und $\bar{\mathfrak{Q}}$; man hat also:

$$B = \int_{\mathfrak{D}} f\,dx + \sum_\nu \int_{\mathfrak{Q}_\nu} f\,dx + \sum_\nu \left(\sum_\mu \int_{a_{\nu,\mu}}^{b_{\nu,\mu}} f\,dx + \int_{\mathfrak{Q}_\nu} f\,dx \right) =$$

$$= \int_{\bar{\mathfrak{Q}}} f\,dx + \sum_\nu \int_{\bar{a}_\nu}^{\bar{b}_\nu} f\,dx = \bar{B}$$

und die Behauptung ist erwiesen.

Das korr. Mitglied Hans Hahn übersendet die folgende von ihm verfaßte vorläufige Mitteilung: »Über additive Mengenfunktionen.«

Sei $\mathfrak{M}$ ein σ-Körper, φ eine auf $\mathfrak{M}$ definierte Mengenfunktion: sie heißt additiv, wenn für je zwei disjunkte Mengen A, B aus $\mathfrak{M}$ gilt $\varphi(A+B) = \varphi(A) + \varphi(B)$, sie heißt totaladditiv, wenn für jede Folge disjunkter Mengen A_n aus $\mathfrak{M}$ gilt

$$\varphi\left(\underset{n}{S} A_n\right) = \sum_n \varphi(A_n).$$

Dabei ist es keineswegs erforderlich, daß die Funktionswerte von φ alle endlich seien; gerade bei den bekanntesten Beispielen totaladditiver Mengenfunktionen (dem Lebesgue'schen Inhalt von Punktmengen) treten auch unendliche Funktionswerte auf. Damit die Definition der Additivität immer sinnvoll bleibe, ist nur erforderlich, daß φ nicht für zwei disjunkte Mengen aus $\mathfrak{M}$ die Werte $+\infty$ und $-\infty$ annehme, woraus sofort folgt, daß φ in $\mathfrak{M}$ überhaupt nicht unendliche Werte verschiedenen Zeichens annehmen darf.

Ich habe in meiner »Theorie der reellen Funktionen« (p. 403) gezeigt, daß es unter allen Werten, die die total additive

Funktion φ auf den Teilen einer Menge A aus $\mathfrak{M}$ annimmt, einen größten (und einen kleinsten) gibt. Mittlerweilen wurden zwei neue Beweise hiefür veröffentlicht.[1] Wenn ich nun auf meinen ursprünglichen Beweis zurückkomme, so geschieht es, weil dieser Beweis die Funktion φ nicht als endlich voraussetzen muß, und weil ich glaube, ihm nunmehr eine besonders einfache Gestalt geben zu können.

Gibt es unter den (zu $\mathfrak{M}$ gehörigen) Teilen von A einen A', für den $\varphi(A') = +\infty$, so ist die Behauptung trivial. Wir können also annehmen, für alle Teile A' von A sei $\varphi(A') < +\infty$.

Sei g die obere Schranke von φ für alle (zu $\mathfrak{M}$ gehörigen) Teile von A. Es gibt dann eine Folge A_n solcher Teile, so daß $\varphi(A_n) \to g$. Wir betrachten die n-Mengen $A_1, A_2, \ldots, A_n$ und ihre Komplemente $A - A_1$, $A - A_2, \ldots$, $A - A_n$.

Die Durchschnitte von je k $(0 \leqq k \leqq n)$ der Mengen $A_1, A_2, \ldots, A_n$ mit den Komplementen der übrigen $n - k$ bilden 2^n disjunkte Mengen B_i $(i = 1, 2, \ldots, 2^n)$, deren Summe A ist. Wir behalten von den Mengen B_i diejenigen bei, für die $\varphi(B_i) \geqq 0$, nennen ihre Summe C_n und setzen $C_n + C_{n+1} + \ldots = S_n$, dann ist

$$\varphi(S_n) \geqq \varphi(C_n) \geqq \varphi(A_n).$$

Setzen wir noch

$$S = S_1 . S_2 . \ldots . S_n . \ldots,$$

so ist S der sogenannte Limes superior der Mengenfolge $C_1, C_2, \ldots, C_n, \ldots$ und es gilt $\varphi(S_n) \to \varphi(S)$; wegen $\varphi(S_n) \geqq \varphi(A_n)$ gilt aber auch $\varphi(S_n) \to g$, d. h. es ist $\varphi(S) = g$. Damit ist die Behauptung bewiesen.

Das korr. Mitglied Hans Hahn übersendet ferner die folgende von ihm verfaßte vorläufige Mitteilung: »Über unendliche Reihen und totaladditive Mengenfunktionen«.

Sei $\sum_{\nu \geqq 1} a_\nu$ eine unendliche Reihe aus positiven, monoton gegen 0 konvergierenden Gliedern. Ist $\nu_1, \nu_2, \ldots, \nu_i, \ldots$ eine Folge wachsender natürlicher Zahlen, so bezeichnen wir die Reihe $\sum_{i \geqq 1} a_{\nu_i}$ als eine Teilreihe von $\sum_{\nu \geqq 1} a_\nu$. Die Summe der Reihe $\sum_{\nu \geqq 1} a_\nu$ bezeichnen wir mit g; falls die Reihe divergiert, ist $g = +\infty$ zu setzen. Man beweist ohne Schwierigkeit:

I. *Damit es zu jeder der Ungleichung* $0 < x \leqq g$ *genügenden Zahl* x *eine Teilreihe von* $\sum_{\nu \geqq 1} a_\nu$ *gebe, für die* $\sum_{i \geqq 1} a_{\nu_i} = x$ *ist, ist notwendig und hinreichend, daß für alle* $n \geqq 1$ *gelte:* $\sum_{\nu > n} a_\nu \geqq a_n$.

Die Bedingung ist hinreichend, auch wenn die a_ν nicht monoton abnehmen. Darin ist als Spezialfall enthalten:

II. *In jeder divergenten Reihe* $\sum_{\nu \geqq 1} b_\nu$ *aus positiven, gegen* 0 *konvergierenden Gliedern gibt es zu jeder Zahl* $x > 0$ *eine Teilreihe, so daß* $\sum_{i \geqq 1} b_{\nu_i} = x$.

Sei nun $\sum_{\nu \geqq 1} a_\nu \; (= g)$ eine Reihe aus positiven, monoton gegen 0 konvergierenden Gliedern, die der Bedingung: $\sum_{\nu > n} a_\nu \geqq a_n$ für alle n genügt; dann gilt:

III. *Damit es zu jeder der Ungleichung* $0 < x \leqq g$ *genügenden Zahl* x *nur eine einzige Teilreihe von* $\sum_{\nu \geqq 1} a_\nu$ *gebe, für die* $\sum_{i \geqq 1} a_{\nu_i} = x$ *ist, ist notwendig und hinreichend, daß für alle* n, *für die* $a_{n+1} < a_n$ *ist,* $\sum_{\nu > n} a_\nu = a_n$ *sei.*

Dabei gelten zwei Teilreihen $\sum_{i \geqq 1} a_{\nu_i}$ und $\sum_{i \geqq 1} a_{\mu_i}$ als dieselbe Teilreihe, wenn $a_{\nu_i} = a_{\mu_i}$ ist für alle i.

Man entnimmt daraus, daß die Cantor'schen Reihen[1] (die die Systembrüche einer Grundzahl g als Spezialfall enthalten) die

[1] Vgl. z. B. O. Perron, Irrationalzahlen, p. 111.

einzigen Reihen aus positiven Zahlen sind, deren sämtliche Teilreihen die sämtlichen der Ungleichung $0 < x \leqq 1$ genügenden Zahlen, und zwar jede solche Zahl x nur ein einziges Mal darstellen.

Aus Satz I gewinnt man einen sehr durchsichtigen Beweis für den von W. Sierpiński[2] und M. Fréchet[3] bewiesenen »Zwischenwertsatz« in der Theorie der totaladditiven Mengenfunktionen. Sei φ eine im σ-Körper $\mathfrak{M}$ totaladditive Mengenfunktion; alle Mengen, von denen im folgenden die Rede ist, mögen zu $\mathfrak{M}$ gehören. Eine Menge S heißt singulär für φ, wenn $\varphi(S) \neq 0$ ist, während für jeden Teil T von S entweder $\varphi(T) = 0$ oder $\varphi(T) = \varphi(S)$ ist. Gibt es keinen für φ singulären Teil von A, so heißt φ singularitätenfrei in A. Der Zwischenwertsatz besagt nun:

Ist die totaladditive Mengenfunktionen φ singularitätenfrei in A und sind A' und A'' Teile von A, so gibt es zu jeder zwischen $\varphi(A')$ und $\varphi(A'')$ liegenden Zahl x einen Teil B von A, für den $\varphi(B) = x$.

Man zeigt leicht, daß man sich auf den Fall beschränken kann, daß φ monoton wächst ($\varphi(M) \geqq 0$ für alle Teile M von A). Dann ist zu zeigen: ist $0 < x \leqq \varphi(A)$, so gibt es einen Teil B von A, so daß $\varphi(B) = x$. Ferner kann noch ohne weiteres $\varphi(A)$ als endlich angenommen werden. Wir bezeichnen mit M_1 alle diejenigen Teile von A, für die $0 < \varphi(M_1) \leqq \frac{1}{2}\varphi(A)$, und mit y_1 die obere Grenze aller Funktionswerte $\varphi(M_1)$; dann gibt es einen Teil A_1 von A, so daß $\frac{1}{2}y_1 < \varphi(A_1) \leqq y_1$. Sodann bezeichnen wir mit M_2 alle diejenigen Teile von $A - A_1$, für die $0 < \varphi(M_2) \leqq \frac{1}{2}\varphi(A - A_1)$, mit y_2 die obere Grenze aller $\varphi(M_2)$, mit A_2 einen Teil von $A - A_1$, so daß $\frac{1}{2}y_2 < \varphi(A_2) \leqq y_2$. Nun bezeichnen wir mit M_3 alle diejenigen Teile von $A - (A_1 + A_2)$ für die $0 < \varphi(M_3) \leqq \frac{1}{2}\varphi(A - (A_1 + A_2))$, mit y_3 die obere Grenze aller $\varphi(M_3)$, mit A_3 einen Teil von $A - (A_1 + A_2)$, so daß $\frac{1}{2}y_3 < \varphi(A_3) \leqq y_3$ usw. Man erhält so eine Folge zu je zweien fremder Mengen $A_1, A_2, \ldots, A_\nu, \ldots$ und zeigt leicht, daß $\varphi(A) = \sum_{\nu \geqq 1} \varphi(A_\nu)$ und $\sum_{\nu > n} \varphi(A_\nu) \geqq \varphi(A_n)$. Satz I ergibt nun: es gibt

[2] Fund. Math., *3* (1922), p. 240.
[3] Fund. Math., *4* (1923), p. 364.

zu jeder der Ungleichung $0 < x \leqq \varphi(A)$ genügenden Zahl x eine Teilreihe, so daß $\sum_{i \geqq 1} \varphi(A_{\nu_i}) = x$. Bezeichnet man noch mit B die Vereinigung von $A_{\nu_1}, A_{\nu_2}, \ldots, A_{\nu_i}, \ldots$, so ist also $\varphi(B) = x$, und der Zwischenwertsatz ist bewiesen.

Das korr. Mitglied Hans Hahn übersendet die folgende, von ihm verfaßte vorläufige Mitteilung: »Über den Integralbegriff.«

Es wurden verschiedentlich Systeme von Forderungen angegeben, durch die der Integralbegriff (im Sinne von Lebesgue) eindeutig festgelegt ist. Besonders einfach und naturgemäß scheint mir der folgende Vorgang. Sei E irgendeine Menge, $\mathfrak{K}$ ein aus Teilen von E bestehender σ-Körper, $\varphi(M)$ eine für alle $M \varepsilon \mathfrak{K}$ definierte und total-additive Mengenfunktion; es sei $E \varepsilon \mathfrak{K}$ und $\varphi(E)$ endlich, der Einfachheit halber nehmen wir an: $\varphi(M) \geqq 0$ für alle $M \varepsilon \mathfrak{K}$ (es hat keinerlei Schwierigkeit, sich von dieser Voraussetzung frei zu machen); wir nehmen ferner an, $\mathfrak{K}$ sei vollständig für φ, d. h. ist $M \varepsilon \mathfrak{K}$ und $\varphi(M) = 0$ (die Menge M heißt dann eine Nullmenge für φ), so gilt auch $M \varepsilon \mathfrak{K}$ für jeden Teil M' von M.

Eine auf E (abgesehen von einer Nullmenge für φ) definierte Funktion f heißt φ-meßbar, wenn für jedes reelle c die Menge $E[f > c]$ aller Punkte von E, in denen $f > c$ ist, zu $\mathfrak{K}$ gehört. Die Funktion f darf auch die Werte $+\infty$, $-\infty$ annehmen.

Wir definieren nun: die φ-meßbare Funktion f heißt φ-integrierbar auf E, wenn es eine für alle $M \varepsilon \mathfrak{K}$ definierte Mengenfunktion $\lambda(M)$ gibt, die folgende Forderungen erfüllt:

1. $\lambda(M)$ ist total-additiv in $\mathfrak{K}$;
2. ist $M \varepsilon \mathfrak{K}$ und $c' \leqq f \leqq c''$ auf M, so ist

$$c' \varphi(M) \leqq \lambda(M) \leqq c'' \varphi(M).$$

Dabei ist in dieser letzten Ungleichung $c\varphi(M) = 0$ zu setzen, wenn

$$c = \pm\infty, \quad \varphi(M) = 0.$$

Man beweist nun der Reihe nach:

I. Ist f φ-integrierbar auf E, so ist $\lambda(M)$ durch die Forderungen 1., 2. eindeutig bestimmt.

Man erkennt dies leicht durch Betrachtung der Zerlegung

$$(*)\qquad M = \mathop{\mathsf{S}}_{i=-\infty}^{+\infty} M_i + M_{+\infty} + M_{-\infty} + M^*,$$

wo

$$M_i = M[(i-1)\,\delta \leqq f < i\,\delta],$$
$$M_{+\infty} = M[f = +\infty],$$
$$M_{-\infty} = M[f = -\infty],$$

während M^* die Menge aller Punkte von M bedeutet, in denen f nicht definiert ist (δ bedeutet irgendeine positive Zahl); denn dann ist

$$(i-1)\,\delta\,\varphi(M_i) \leqq \lambda(M_i) < i\,\delta\,\varphi(M_i),$$
$$\lambda(M_{+\infty}) = +\infty \text{ oder } = 0,$$

je nachdem

$$\varphi(M_{+\infty}) > 0 \text{ oder } = 0,$$
$$\lambda(M_{-\infty}) = -\infty \text{ oder } = 0,$$

je nachdem

$$\varphi(M_{-\infty}) > 0 \text{ oder } = 0,$$

und wegen

$$\varphi(M^*) = 0 \text{ auch } \lambda(M^*) = 0.$$

Nachdem man erkannt hat, daß $\lambda(M)$ eindeutig bestimmt ist, setzt man:

$$\lambda(M) = (M)\int f\,d\,\varphi.$$

II. Ist f φ-integrierbar auf E, so auch auf jedem zu $\mathfrak{K}$ gehörigen Teile von E.

III. Ist f φ-integrierbar auf A und auf B, und sind

$$(A)\int f d\varphi \quad \text{und} \quad (B)\int f d\varphi$$

nicht unendlich verschiedenen Zeichens, so ist f auch φ-integrierbar auf $A+B$.

Denn setzt man

$$\lambda(M) = (AM)\int f d\varphi + (A-M)\int f d\varphi,$$

so genügt $\lambda(M)$ den Forderungen 1., 2. für alle zu $\mathfrak{K}$ gehörigen $M \subseteqq A+B$.

IV. Ist A eine Nullmenge für φ, so ist jede Funktion f φ-integrierbar auf A, und es ist

$$(A)\int f d\varphi = 0.$$

V. Ist f konstant auf M, so ist f auch φ-integrierbar auf M und

$$(M)\int f\,d\varphi = c\,\varphi(M).$$

Denn $\lambda(M) = c\,\varphi(M)$ genügt den Forderungen 1, 2.

VI. Ist f φ-meßbar auf M und nimmt f nur die endlich vielen (durchweg endlichen) Werte $c_1, c_2, \ldots, c_n$ an, so ist f auch φ-integrierbar auf M, und wenn $M[f = c_i] = M_i$ gesetzt wird, ist

$$(M)\int f\,d\varphi = c_1\,\varphi(M_1) + \ldots + c_n\,\varphi(M_n).$$

Die folgt aus III, V und Forderung 1.

VII. Sind f und g φ-integrierbar auf M und ist $f \leqq g$, so ist

$$(M)\int f\,d\varphi \leqq (M)\int g\,d\varphi.$$

Dies folgt durch Betrachtung der Zerlegung (*) von M.

VIII. Ist $f_1, f_2, \ldots, f_n, \ldots$ eine monoton wachsende Folge φ-integrierbarer Funktionen, und ist

$$(M)\int f_1\,d\varphi > -\infty,$$

so ist auch $f = \lim\limits_n f_n$ φ-integrierbar und

$$(M)\int f\,d\varphi = \lim_n\,(M)\int f_n\,d\varphi.$$

Zum Beweise zeigt man, daß die Mengenfunktion

$$\lambda(M) = \lim_n\,(M)\int f_n\,d\varphi$$

die Forderungen 1, 2 für $f = \lim\limits_n f_n$ erfüllt.

IX. Ist f φ-meßbar und $\geqq 0$ (oder $\leqq 0$) auf M, so ist f φ-integrierbar auf M.

Sei

$$f_n = \frac{i-1}{2^n},$$

wo

$$\frac{i-1}{2^n} \leqq f < \frac{i}{2^n} \quad (i = 1, 2, \ldots, n \,.\, 2^n),$$

und $f_n = n$, wo $f \geqq n$; dann ist f_n φ-integrierbar nach VI und da f_n monoton gegen f konvergiert, ist f φ-integrierbar nach VIII.

Wir setzen

$$M_+ = M[f \geqq 0], \quad M_- = M[f < 0].$$

X. Damit die φ-meßbare Funktion f φ-integrierbar sei auf M, ist notwendig und hinreichend, daß

$$(M_+)\int f\,d\varphi,\ (M_-)\int f\,d\varphi$$

nicht unendlich verschiedenen Zeichens seien.

Dies folgt aus IX und III.

XI. Ist f φ-integrierbar auf M und

$$(M)\int f\,d\varphi > -\infty \text{ (beziehungsweise } < +\infty),$$

ist ferner $g \geqq f$ (beziehungsweise $g \leqq f$), so ist auch g φ-integrierbar auf M.

Dies folgt aus X.

XII. Sind f_1 und f_2 φ-integrierbar auf M und sind

$$(M)\int f_1\,d\varphi,\ (M)\int f_2\,d\varphi$$

nicht unendlich verschiedenen Zeichens, so ist f_1+f_2 φ-integrierbar auf M und

$$(M)\int (f_1+f_2)\,d\varphi = (M)\int f_1\,d\varphi + (M)\int f_2\,d\varphi.$$

Man beweist dies, indem man vermöge der Zerlegung (*) von M zeigt, daß

$$\lambda(M) = (M)\int f_1\,d\varphi + (M)\int f_2\,d\varphi$$

die Forderungen 1, 2 für $f = f_1+f_2$ erfüllt.

Wir sagen: die Zahlen z_i $(i = 0, \pm 1, \pm 2, \ldots)$ bilden ein δ-Gitter $\{z_i\}$, wenn

$$z_i < z_{i+1} < z_i + \delta,\ \lim_{i\to+\infty} z_i = +\infty,\ \lim_{i\to-\infty} z_i = -\infty.$$

Wir machen wieder Gebrauch von der Zerlegung (*) von M und setzen

$$L(f,\{z_i\}) = \sum_{i=-\infty}^{+\infty} z_{i-1}\,\varphi(M_i) + (+\infty)\,\varphi(M_{+\infty}) + (-\infty)\,\varphi(M_{-\infty})$$

(falls diese Summe sinnvoll ist).

XIII. Ist f φ-integrierbar auf M, und ist $\{z_i\}$ ein δ-Gitter, so ist $L(f,\{z_i\})$ sinnvoll und

$$(M)\int f\,d\varphi - \delta\varphi(M) \leqq L(f,\{z_i\}) \leqq (M)\int f\,d\varphi.$$

Denn setzen wir $g = z_{i-1}$ auf M_i, $g = +\infty$ auf $M_{+\infty}$, $g = -\infty$ auf $M_{-\infty}$, so ist $f - \delta \leqq g \leqq f$, also ist g nach XI φ-integrierbar; wegen

$$(M)\int g\,d\varphi = L(f, \{z_i\})$$

ist also dieser Ausdruck sinnvoll und die behauptete Ungleichung folgt aus VII und XII.

Man erhält nun den Anschluß an Lebesgue's Integraldefinition, vermöge des aus XIII folgenden Satzes:

XIV. Ist f φ-integrierbar auf M, ist $\{z_i^n\}$ ein δ_n-Gitter und gilt $\delta_n \to 0$, so ist

$$(M)\int f\,d\varphi = \lim_n L(f, \{z_i^n\}).$$

Über den Integralbegriff.

Von Hans H a h n.

Der klassische Begriff des bestimmten Integrales hat zwei wesentliche Erweiterungen erfahren, die sich an die Namen S t i e l t j e s und L e b e s g u e knüpfen. Trotz ihrer außerordentlichen Bedeutung sind diese Erweiterungen des Integralbegriffes noch wenig in den Hochschulunterricht eingedrungen. Es mag daher nicht zwecklos sein, im folgenden einen besonders einfachen und naturgemäßen Zugang zur Theorie der Stieltjes-Lebesgue-Integrale mitzuteilen [1]).

Bekanntlich wird ein Mengensystem $\mathfrak{K}$ als σ-K ö r p e r bezeichnet, wenn 1. neben je abzählbar vielen in $\mathfrak{K}$ vorkommenden Mengen A_i auch deren Summe $\underset{i}{\mathbf{S}} A_i$ in $\mathfrak{K}$ vorkommt und 2. neben je zwei in $\mathfrak{K}$ vorkommenden Mengen A und B auch die Menge $A - B$ aller nicht zu B gehörenden Elemente von A in $\mathfrak{K}$ vorkommt. — Man zeigt leicht, daß dann auch neben je abzählbar vielen in $\mathfrak{K}$ vorkommenden Mengen A_i deren Durchschnitt $\underset{i}{\mathbf{D}} A_i$ in $\mathfrak{K}$ vorkommt.

Ist jeder Menge M aus $\mathfrak{K}$ eine reelle Zahl $\varphi(M)$ zugeordnet, so heißt $\varphi(M)$ eine in $\mathfrak{K}$ definierte M e n g e n f u n k t i o n; als Funktionswerte lassen wir dabei auch die Werte $+\infty$, $-\infty$ zu. Die Mengenfunktion φ heißt t o t a l - a d d i t i v, wenn für je abzählbar viele, zu je zweien fremde Mengen A_i aus $\mathfrak{K}$ gilt:

$$\varphi(\underset{i}{\mathbf{S}} A_i) = \sum_i \varphi(A_i).$$

Man erkennt leicht, daß unter den Werten einer total-additiven Mengenfunktion nicht unendliche verschiedenen Zeichens vorkommen können, da sonst für gewisse fremde Mengen A und B aus $\mathfrak{K}$ die Summe $\varphi(A) + \varphi(B)$ die sinnlose Form $+\infty + (-\infty)$ hätte und somit nicht $= \varphi(A + B)$ wäre. — Ist $\varphi(A)$ endlich, so

[1]) Nur bei C. d e l a V a l l é e P o u s s i n Intégrales de Lebesgue, Fonctions d'ensemble, Classes de Baire (Paris, 1916) S. 55 fand ich einen Hinweis auf diese Einführung des Integralbegriffes. — Den Gedankengang der folgenden Ausführungen habe ich schon veröffentlicht im Anzeiger der Wiener Akademie 1929, Nr. 2.

ist auch $\varphi(M)$ endlich für alle zu $\mathfrak{K}$ gehörigen Teile M von A. — Man zeigt leicht: gehören die M_n zu $\mathfrak{K}$, ist $M_n \leqq M_{n+1}$ und $M = \mathbf{S}_n M_n$, so ist: $\varphi(M) = \lim_n \varphi(M_n)$.

Die total-additive Mengenfunktion φ heißt monoton-wachsend, wenn $\varphi(M) \geqq 0$ ist für alle M aus $\mathfrak{K}$. Eine Menge aus $\mathfrak{K}$, für die $\varphi(M) = 0$ ist, nennen wir dann eine Nullmenge für φ. Der σ-Körper $\mathfrak{K}$ heißt vollständig für φ, wenn auch sämtliche Teile einer Nullmenge für φ zu $\mathfrak{K}$ gehören; natürlich ist dann jener Teil einer Nullmenge für φ selbst Nullmenge für φ. Durch Hinzufügen geeigneter Mengen kann $\mathfrak{K}$ stets zu einem für φ vollständigen σ-Körper ergänzt werden.

Sei nun φ eine im σ-Körper $\mathfrak{K}$ definierte, total-additive, monotonwachsende [2]) Mengenfunktion; $\mathfrak{K}$ sei vollständig für φ. Sei ferner A eine Menge aus $\mathfrak{K}$, für die $\varphi(A)$ endlich ist. Auf A sei eine Funktion definiert, die jedem Elemente x von A eine reelle Zahl $f(x)$ als Funktionswert zuordnet; auch die Werte $+\infty$, $-\infty$ sind als Funktionswerte zugelassen. Mit $A[f > c]$ bezeichnen wir die Menge aller Elemente x von A, in denen $f(x) > c$ ist, und analoge Bedeutung haben die Symbole $A[c' \leqq f < c'']$, $A[f = c]$ usf. Die Funktion f heißt φ-meßbar auf A, wenn für jedes c die Menge $A[f > c]$ zu $\mathfrak{K}$ gehört; bekanntlich gehören dann auch die Mengen $A[f \geqq c]$, $A[f < c]$, $A[f \leqq c]$, $A[c' \leqq f < c'']$, $A[f = c]$ usf. zu $\mathfrak{K}$. Ist f meßbar auf A, so auch auf jedem zu $\mathfrak{K}$ gehörigen Teile von A. Sind die Funktionen f_n (n = 1, 2, ...) φ-meßbar auf A und ist $f = \lim_n f_n$, so ist auch f φ-meßbar auf A.

Wir definieren nun: Die auf A φ-meßbare Funktion f heißt φ-integrierbar auf A, wenn es eine im σ-Körper $\mathfrak{K}'$ der zu $\mathfrak{K}$ gehörigen Teile M von A definierte Mengenfunktion $\lambda(M)$ gibt, die die beiden Forderungen erfüllt:

1. $\lambda(M)$ ist total-additiv in $\mathfrak{K}'$;
2. für jede Menge M aus $\mathfrak{K}'$, auf der $c' \leqq f \leqq c''$ ist, gilt:

$$c'\varphi(M) \leqq \lambda(M) \leqq c''\varphi(M). \tag{1}$$

Dabei ist in dieser letzten Ungleichung $c\varphi(M) = 0$ zu setzen, wenn $c = \pm\infty$, $\varphi(M) = 0$ ist; auch im folgenden halten wir an dieser

---

[2]) Diese Voraussetzung ist unwesentlich; wir machen sie nur, um die folgenden Entwicklungen etwas kürzer gestalten zu können.

Übereinkunft fest. Aus (1) folgt, indem wir $c' = -\infty$ bzw. $c'' = +\infty$ setzen: Ist $f \leqq c''$ auf M, so gilt $\lambda(M) \leqq c'' \varphi(M)$; ist $f \geqq c'$ auf M, so gilt $\lambda(M) \geqq c' \varphi(M)$.

I. Ist f φ-integrierbar auf A, so ist $\lambda(M)$ durch die Forderungen 1., 2. eindeutig bestimmt.

Wir setzen $M^* = M[f = +\infty]$, $M^{**} = M[f = -\infty]$. Ist $\varphi(M^*) > 0$, so folgt aus (1) für $c' = c'' = +\infty$: $\lambda(M^*) = +\infty$, also wegen $\lambda(M) = \lambda(M^*) + \lambda(M - M^*)$ auch $\lambda(M) = +\infty$, so daß in diesem Falle die Behauptung bewiesen ist; analog beweist man sie, wenn $\varphi(M^{**}) > 0$. Wir nehmen also an $\varphi(M^*) = 0$, $\varphi(M^{**}) = 0$. Seien nun λ_1, λ_2 zwei den Forderungen 1., 2. genügende Mengenfunktionen. Sei $\sigma > 0$ beliebig gewählt; wir setzen:

$$M_i = M[(i-1)\delta \leqq f < i\delta] \quad (i = 0, \pm 1, \pm 2, \ldots);$$

dann ist:

$$M = \mathop{\mathbf{S}}_{i=-\infty}^{+\infty} M_i + M^* + M^{**}, \tag{2}$$

also wegen der totalen Additivität von φ und wegen $\varphi(M^*) = 0$, $\varphi(M^{**}) = 0$:

$$\varphi(M) = \sum_{i=-\infty}^{+\infty} \varphi(M_i). \tag{3}$$

Aus (2) folgt wegen Forderung 1.:

$$\lambda_j(M) = \sum_{i=-\infty}^{+\infty} \lambda_j(M_i) + \lambda_j(M^*) + \lambda_j(M^{**}) \quad (j = 1, 2). \tag{4}$$

Aus $\varphi(M^*) = 0$, $\varphi(M^{**}) = 0$ folgt wegen (1):

$$\lambda_j(M^*) = 0, \quad \lambda_j(M^{**}) = 0; \tag{5}$$

ferner folgt aus (1):

$$(i-1)\,\delta\,\varphi(M_i) \leqq \lambda_j(M_i) \leqq i\,\delta\,\varphi(M_i),$$

somit:

$$\lambda_2(M_i) = \lambda_1(M_i) + \theta_i\,\delta\,\varphi(M_i) \quad (|\theta_i| \leqq 1);$$

aus (4) und (5) entnehmen wir also:

$$\lambda_2(M) = \lambda_1(M) + \delta . \sum_{i=-\infty}^{+\infty} \theta_i\,\varphi(M_i)$$

und somit wegen (3):

$$\lambda_2(M) = \lambda_1(M) + \delta . \theta\,\varphi(M) \quad (|\theta| \leqq 1);$$

da aber hierin $\delta > 0$ beliebig war, besagt dies $\lambda_2(M) = \lambda_1(M)$, und die Behauptung ist bewiesen.

13*

Nachdem wir erkannt haben, daß $\lambda(M)$ durch unsere Forderungen eindeutig bestimmt ist, bezeichnen wir den Funktionswert $\lambda(M)$ als das über M erstreckte φ-Integral von f, in Zeichen

$$\lambda(M) = (M) \int f \, d\varphi,$$

und unsere Forderungen besagen:

II. Ist f φ-integrierbar auf A, so ist das Integral $(M) \int f \, d\varphi$ eine im σ-Körper $\mathfrak{K}'$ der zu $\mathfrak{K}$ gehörigen Teile M von A total-additive Mengenfunktion.

III. Für jede Menge M aus $\mathfrak{K}'$, auf der $c' \leqq f \leqq c''$ ist, gilt:

$$c' \varphi(M) \leqq (M) \int f \, d\varphi \leqq c'' \varphi(M).$$

Dies ist der erste Mittelwertsatz der Integralrechnung. Wir haben also das Integral $(M) \int f \, d\varphi$ definiert durch die Forderung, es solle eine total-additive Mengenfunktion sein, für die der erste Mittelwertsatz gilt.

Sei nun B ein zu $\mathfrak{K}$ gehöriger Teil von A; dann bilden die zu $\mathfrak{K}$ gehörigen Teile M von B einen σ-Körper $\mathfrak{K}''$, und $(M) \int f \, d\varphi$ ist eine in $\mathfrak{K}''$ definierte Mengenfunktion, die in $\mathfrak{K}''$ den beiden an λ gestellten Forderungen 1., 2. genügt. Also:

IV. Ist f φ-integrierbar auf A, so auch auf jedem zu $\mathfrak{K}$ gehörigen Teile von A.

V. Gehören die Mengen A und B zu $\mathfrak{K}$, ist f φ-integrierbar auf A und auf B, und sind $(A) \int f \, d\varphi$ und $(B) \int f \, d\varphi$ nicht unendlich verschiedenen Zeichens, so ist f auch φ-integrierbar auf $A + B$.

Da f φ-meßbar auf A und auf B, so auch auf $A + B$. Für jeden zu $\mathfrak{K}$ gehörigen Teil M von $A + B$ setzen wir:

$$\lambda_1(M) = (MA) \int f \, d\varphi, \quad \lambda_2(M) = (M\text{-}A) \int f \, d\varphi;$$

dann sind $\lambda_1(M)$ und $\lambda_2(M)$ total-additiv im σ-Körper $\mathfrak{K}'$ der zu $\mathfrak{K}$ gehörigen Teile von $A + B$; und da λ_1, λ_2 nicht unendliche Werte verschiedenen Zeichens annehmen, ist auch $\lambda = \lambda_1 + \lambda_2$ total-additiv in $\mathfrak{K}'$. Aus III. folgt, daß λ auch der Forderung 2. genügt. Also genügt λ in $\mathfrak{K}'$ den beiden Forderungen 1., 2., d. h. f ist φ-integrierbar auf $A + B$.

VI. Ist A eine Nullmenge für φ, so ist jede Funktion f φ-integrierbar auf A, und es ist $(A) \int f \, d\varphi = 0$.

Da $\mathfrak{K}$ vollständig für φ, gehört jeder Teil von A zu $\mathfrak{K}$, und somit ist jede Funktion f φ-meßbar auf A. Setzen wir $\lambda(M) = 0$ für alle Teile M von A, so genügt λ den Forderungen 1., 2. im σ-Körper $\mathfrak{K}'$ aller Teile M von A.

VII. Gehört A zu $\mathfrak{K}$, und ist f auf A gleich der Konstanten c, so ist f φ-integrierbar auf A, und es ist:

$$(A)\int f\, d\varphi = c\,\varphi(A).$$

Denn die Mengenfunktion $\lambda(M) = c\,\varphi(M)$ genügt den Forderungen 1., 2. im σ-Körper $\mathfrak{K}'$ der zu $\mathfrak{K}$ gehörigen Teile von A.

Insbesondere sehen wir: Ist $f = +\infty$ (bzw. $= -\infty$) auf A, so ist $(A)\int f d\varphi = +\infty$ (bzw. $= -\infty$) oder $= 0$, je nachdem $\varphi(A) > 0$ oder $\varphi(A) = 0$.

VIII. Ist f φ-meßbar auf A und nimmt f nur die endlich vielen, durchwegs endlichen, Werte $c_1, c_2, \ldots, c_n$ an, so ist f auch φ-integrierbar auf A, und wenn $A[f = c_i] = A_i$ gesetzt wird, ist:

$$(A)\int f\, d\varphi = c_1\,\varphi(A_1) + c_2\,\varphi(A_2) + \ldots + c_n\,\varphi(A_n).$$

Dies folgt unmittelbar aus VII, V, II.

IX. Sind f und g φ-integrierbar auf A und ist $f \leqq g$, so ist auch:

$$(A)\int f\, d\varphi \leqq (A)\int g\, d\varphi. \tag{6}$$

Sei $\delta > 0$ beliebig gegeben; wir setzen

$$A_i = A\,[(i-1)\,\delta \leqq f < i\,\delta] \quad (i = 0, \pm 1, \pm 2, \ldots);$$
$$A^* = A\,[f = +\infty],\quad A^{**} = A\,[f = -\infty].$$

Ist $\varphi(A^{**}) > 0$, so ist nach VII: $(A^{**})\int f d\varphi = -\infty$, also, wegen $(A)\int f d\varphi = (A^{**})\int f d\varphi + (A - A^{**})\int f d\varphi$, auch $(A)\int f d\varphi = -\infty$, und (6) ist erfüllt. Wir nehmen also an $\varphi(A^{**}) = 0$. Nach VI ist dann:

$$(A^{**})\int f\, d\varphi = 0, \quad (A^{**})\int g\, d\varphi = 0.$$

Wegen:

$$A = \mathop{\mathbf{S}}_{i=-\infty}^{+\infty} A_i + A^* + A^{**} \tag{7}$$

ist also nach II:

$$(A)\int f\, d\varphi = \sum_{i=-\infty}^{+\infty} (A_i)\int f\, d\varphi + (A^*)\int f\, d\varphi; \tag{8}$$

$$(A)\int g\, d\varphi = \sum_{i=-\infty}^{+\infty} (A_i)\int g\, d\varphi + (A^*)\int g\, d\varphi.$$

Wegen $f \leqq g$ ist auf A^*: $f = g\,(= +\infty)$, also:

(9) $$(A^*) \int f\,d\,\varphi = (A^*) \int g\,d\,\varphi.$$

Aus III folgt (für $c' = -\infty$, $c'' = i\,\delta$):

$$(A_i) \int f\,d\,\varphi \leqq i\,\delta\,\varphi\,(A_i);$$

und wegen $f \leqq g$ folgt aus III (für $c' = (i-1)\,\delta$, $c'' = +\infty$):

$$(A_i) \int g\,d\,\varphi \geqq (i-1)\,\delta\,\varphi\,(A_i);$$

also ist:

(10) $$(A_i) \int g\,d\,\varphi \geqq (A_i) \int f\,d\,\varphi - \delta\,\varphi\,(A_i).$$

Aus (8), (9), (10) folgt:

(11) $$(A) \int g\,d\,\varphi \geqq (A) \int f\,d\,\varphi - \delta \sum_{i=-\infty}^{+\infty} \varphi\,(A_i)$$

Aus (7) folgt $\sum_{i=-\infty}^{+\infty} \varphi\,(A_i) \leqq \varphi(A)$; und da nach Annahme $\varphi\,(A)$ endlich ist, und da $\delta > 0$ beliebig war, folgt aus (11) die Behauptung (6).

X. Ist f_n ($n = 1, 2, \ldots$) eine monoton wachsende Folge auf A φ-integrierbarer Funktionen, und ist $(A) \int f_1\,d\,\varphi > -\infty$, so ist auch $f = \lim\limits_n f_n$ φ-integrierbar auf A und es ist:

$$(A) \int f\,d\,\varphi = \lim_n\, (A) \int f_n\,d\,\varphi.$$

Da die f_n φ-meßbar sind ist — wie schon erwähnt — auch f φ-meßbar. Sei M ein zu $\mathfrak{K}$ gehöriger Teil von A. Nach IX ist dann die Folge der Zahlen $(M) \int f_n\,d\,\varphi$ monoton wachsend; es existiert also der Grenzwert $\lambda\,(M) = \lim\limits_n\, (M) \int f_n\,d\,\varphi$. Wir haben zu zeigen, daß die so definierte Mengenfunktion $\lambda(M)$ im σ-Körper $\mathfrak{K}'$ der zu $\mathfrak{K}$ gehörigen Teile von A die Forderungen 1., 2. erfüllt. Wir zeigen zunächst, daß Forderung 1) erfüllt ist, d. h. daß λ total-additiv in $\mathfrak{K}'$ ist. Wir setzen für jede Menge M aus $\mathfrak{K}'$: $M\,[f_1 < 0] = M'$ und $\lambda^*(M) = (M') \int f_1\,d\,\varphi$; dann ist $-\infty < \lambda^*\,(M) \leqq 0$ und λ^* ist total-additiv in $\mathfrak{K}'$. Es genügt also, nachzuweisen, daß $\mu = \lambda - \lambda^*$ total-additiv in $\mathfrak{K}'$ ist. Setzen wir $\mu_n\,(M) = (M) \int f_n\,d\,\varphi - \lambda^*(M)$, so ist μ_n total-additiv und $\lim\limits_n \mu_n = \lambda - \lambda^* = \mu$. Nach IX ist $(M) \int f_n\,d\,\varphi \leqq (M) \int f_{n+1}\,d\,\varphi$, also auch $\mu_n \leqq \mu_{n+1}$; und nach Definition von λ^* ist $\mu_1\,(M) = (M-M') \int f_1\,d\,\varphi$, und da auf $M-M'$: $f_1 \geqq 0$ ist, so ist nach III: $\mu_1 \geqq 0$, also wegen $\mu_n \leq \mu_{n+1}$ auch $\mu_n \geq 0$ für alle n und mithin auch $\mu \geq 0$. Seien nun $M_1, M_2, \ldots, M_i, \ldots$ fremde Mengen aus $\mathfrak{K}'$ und $M = \mathbf{S}_i M_i$. Da μ_n total-additiv, ist

(12) $$\mu_n(M) = \sum_i \mu_n(M_i);$$

wegen $\mu_n \geq 0$ ist also $\mu_n(M_i) + \ldots + \mu_n(M_k) \leqq \mu_n(M)$, also durch Grenzübergang $n \to \infty$: $\mu(M_1) + \ldots + \mu(M_k) \leq \mu(M)$, und somit durch Grenzübergang $k \to \infty$:

(13) $$\sum_i \mu(M_i) \leq \mu(M).$$

Andererseits gibt es wegen $\mu_n \to \mu$ zu jedem $p < \mu(M)$ ein n, sodaß $\mu_n(M) > p$, also wegen (12) auch ein k, sodaß $\mu_n(M_1) + \ldots + \mu_n(M_k) > p$; durch $n \to \infty$ folgt daraus $\mu(M_1) + \ldots + \mu(M_k) > p$, wegen $\mu \geq 0$, also auch $\sum_i \mu(M_i) > p$; und da dies für jedes $p < \mu(M)$ gilt, ist:

(14) $$\sum_i \mu(M_i) \geq \mu(M);$$

aus (13) und (14) folgt $\mu(M) = \sum_i \mu(M_i)$, d. h. μ und somit auch λ ist total-additiv in $\mathfrak{K}'$, wie behauptet. — Nun zeigen wir, daß λ auch der Forderung 2. genügt. Sei M eine Menge aus $\mathfrak{K}'$ und sei $c' \leq f \leq c''$ auf M. Da die Folge der f_n monoton wächst, ist dann auch $f_n \leq c''$ und mithin $(M)\int f_n \, d\varphi \leq c'' \varphi(M)$, also durch Grenzübergang $n \to \infty$ auch:

(15) $$\lambda(M) \leq c'' \varphi(M).$$

Wir haben noch zu zeigen, daß auch

(16) $$\lambda(M) \geq c' \varphi(M).$$

Das ist trivial, wenn $c' = -\infty$; sei also $c' > -\infty$ und $p < c'$. Wir setzen $M[f_n \geq p] = M_n$; dann ist $M_n \leq M_{n+1}$ und wegen $f_n \to f$ ist $M = \mathbf{S}_n M_n$; wegen der totalen Additivität von φ und λ gilt also:

(17) $$\lim_n \varphi(M_n) = \varphi(M), \quad \lim_n \lambda(M_n) = \lambda(M).$$

Da $f_n \geq p$ auf M_n, ist $(M_n)\int f_n \, d\varphi \geq p\,\varphi(M_n)$, und da $(M_n)\int f_\nu \, d\varphi$ mit wachsendem ν monoton wachsend gegen $\lambda(M_n)$ konvergiert, ist auch $\lambda(M_n) \geq p\,\varphi(M_n)$, also wegen (17) auch $\lambda(M) \geq p\,\varphi(M)$, und da dies für jedes $p < c'$ gilt, ist (16) bewiesen.

XI. *Ist f φ-meßbar und ≥ 0 (oder ≤ 0) auf A, so ist f auch φ-integrierbar auf A.*

Sei $f_n = \frac{i-1}{2^n}$ auf $A\left[\frac{i-1}{2^n} \leqq f < \frac{i}{2^n}\right]$ $(i = 1, 2, \ldots)$ und $f_n = n$ auf $A[f_n \geqq n]$; dann ist f_n φ-integrierbar nach VIII, und da die Folge der f_n monoton wachsend gegen f konvergiert, ist nach X auch f φ-integrierbar auf A.

Ist f eine beliebige auf A φ-meßbare Funktion, so setzen wir:

$$A_+ = A\,[f \geqq 0], \quad A_- = A\,[f < 0];$$

dann ist nach X f φ-integrierbar auf A_+ und auf A_-, und es gilt:

XII. *Ist f φ-meßbar auf A, so ist, damit f φ-integrierbar sei auf A, notwendig und hinreichend, daß $(A_+)\int f d\varphi$ und $(A_-)\int f d\varphi$ nicht unendlich verschiedenen Zeichens seien.*

Notwendig: Dies folgt daraus, daß die total-additive Mengenfunktion $(M)\int f\, d\varphi$ nicht unendliche Werte verschiedenen Zeichens haben kann. *Hinreichend*: Dies folgt aus V.

XIII. *Ist f φ-integrierbar auf A und $(A)\int f d\varphi > -\infty$ (bzw. $< +\infty$), ist ferner g φ-meßbar auf A und $g \geqq f$ (bzw. $g \leqq f$), so ist auch g φ-integrierbar auf A.*

Wir setzen $A\,[g \geqq 0] = A_+$, $A\,[g < 0] = A_-$. Wegen $(A)\int f\,d\varphi > -\infty$, ist auch $(A_-)\int f\,d\varphi > -\infty$, also nach IX auch $(A_-)\int g\,d\varphi > -\infty$, also sind $(A_-)\int g\,d\varphi$ und $(A_+)\int g\,d\varphi$ nicht unendlich verschiedenen Zeichens, und die Behauptung folgt aus XII.

XIV. *Ist f φ-integrierbar auf A und c eine endliche Konstante, so ist auch $f + c$ φ-integrierbar auf A und*

$$(A)\int (f + c)\, d\varphi = (A)\int f\,d\varphi + c\,\varphi(A).$$

Da f φ-meßbar, ist offenbar auch $f + c$ φ-meßbar. Sei $\mathfrak{K}'$ der σ-Körper der zu $\mathfrak{K}$ gehörigen Teile M von A. Für jedes M aus $\mathfrak{K}'$ setzen wir $\lambda(M) = (M)\int f\,d\varphi + c\,\varphi(M)$ und haben zu zeigen, daß λ in $\mathfrak{K}'$ die Forderungen 1. und 2. für die Funktion $f + c$ erfüllt. Als Summe zweier total-additiver Mengenfunktionen ist auch λ total-additiv, Forderung 1. ist also erfüllt. Sei sodann $c' \leqq f + c \leqq c''$ auf M. Dann ist $c' - c \leqq f \leqq c'' - c$, also nach III:

$$(c' - c)\,\varphi(M) \leqq (M)\int f\,d\varphi \leqq (c'' - c)\,\varphi(M).$$

Durch Addition von $c\,\varphi(M)$ folgt daraus:

$$c'\,\varphi(M) \leqq \lambda(M) \leqq c''\,\varphi(M),$$

also ist auch Forderung 2. erfüllt.

Wir sagen: die Zahlen z_i ($i = 0, \pm 1, \pm 2, \ldots$) bilden ein δ-Gitter (z_i), wenn $z_i < z_{i+1} < z_i + \delta$ ($i = 0, \pm 1, \pm 2, \ldots$) und $\lim\limits_{i \to +\infty} z_i = +\infty$, $\lim\limits_{i \to -\infty} z_i = -\infty$.

Ist f φ-meßbar auf A und $[z_i]$ ein δ-Gitter, so setzen wir:

$$(18) \qquad \begin{aligned} A_i &= A[z_{i-1} \leqq f < z_i] \quad (i = 0, \pm 1, \pm 2, \ldots), \\ A^* &= A[f = +\infty], \quad A^{**} = A[f = -\infty]; \end{aligned}$$

dann ist:

$$(19) \qquad A = \mathop{\mathbf{S}}_{i=-\infty}^{+\infty} A_i + A^* + A^{**}.$$

Wir bilden:

$$L(f, [z_i]) = \sum_{i=-\infty}^{+\infty} z_{i-1}\, \varphi(A_i) + (+\infty)\, \varphi(A^*) + (-\infty)\, \varphi(A^{**}),$$

falls diese Summe sinnvoll ist, d. h. falls unter den vier Zahlen

$$\sum_{i=1}^{+\infty} z_{i-1}\varphi(A_i), \quad \sum_{i=-\infty}^{0} z_{i-1}\, \varphi(A_i), \quad (+\infty)\varphi(A^*), \quad (-\infty)\, \varphi(A^{**})$$

keine zwei unendlich verschiedenen Zeichens sind.

XV. Ist $[z_i]$ ein δ-Gitter, so ist, damit die auf A φ-meßbare Funktion f φ-integrierbar sei auf A, notwendig und hinreichend, daß $L(f, [z_i])$ sinnvoll sei.

Notwendig: Sei $g = z_{i-1}$ auf A_i, $g = +\infty$ auf A^*, $g = -\infty$ auf A^{**}, wo A_i, A^*, A^{**} die Bedeutung (18) haben; dann ist g φ-meßbar auf A und $f - \delta \leqq g \leqq f$. Sei f φ-integrierbar auf A; ist $(A) \int f\, d\varphi < +\infty$ so ist g wegen $g \leqq f$ nach XIII φ-integrierbar; ist hingegen $(A) \int f\, d\varphi = +\infty$, so ist nach XIV auch $(A) \int (f - \delta)\, d\varphi = +\infty$ also ist g wegen $g \geqq f - \delta$ nach XIII φ-integrierbar. Ist also f φ-integrierbar auf A, so auch g. Wegen (19) ist nach II: $(A) \int g\, d\varphi = \sum_{i=-\infty}^{+\infty} (A_i) \int g\, d\varphi + (A^*) \int g\, d\varphi + (A^{**}) \int g\, d\varphi = \sum_{i=-\infty}^{+\infty} z_{i-1}\varphi\, (A_i) + (+\infty)\, \varphi\, (A^*) + (-\infty)\, \varphi\, (A^{**})$ also:

$$(20) \qquad (A) \int g\, d\varphi = L(f, [z_i]);$$

also ist $L(f, [z_i])$ sinnvoll. Hinreichend: Es ist zu zeigen: ist f nicht φ-integrierbar auf A, so ist $L(f, [z_i])$ nicht sinnvoll. Wir setzen $A' = A[f \geqq z_0]$, $A'' = A[f < z_0]$. Nach XIII ist f φ-integrierbar auf A' und auf A''; wegen $A = A' + A''$ ist also nach V, weil f nicht φ-integrierbar auf A:

(21) $(A') \int f \, d\varphi = +\infty, \ (A'') \int f \, d\varphi = -\infty.$

Auch die vorhin eingeführte Funktion g ist nach XIII φ-integrierbar auf A' und auf A''; wegen $g \leqq f$, $g \geqq f - \delta$ folgt aus (21) auch:

(22) $(A') \int g \, d\varphi = +\infty, \ (A'') \int g \, d\varphi = -\infty.$

Nun ist aber:

$$(A') \int g \, d\varphi = \sum_{i=1}^{\infty} z_{i-1} \varphi(A_i) + (+\infty) \varphi(A'')$$

$$(A'') \int g \, d\varphi = \sum_{i=-\infty}^{0} z_{i-1} \varphi(A_i) + (-\infty) \varphi(A^{**}),$$

wegen (22) ist also $L(f, [z_i])$ nicht sinnvoll.

XVI. Ist f φ-integrierbar auf A und $[z_i]$ ein δ-Gitter, so ist:

(23) $(A) \int f \, d\varphi - \delta \varphi(A) \leqq L(f, [z_i]) \leqq (A) \int f \, d\varphi.$

Denn bedeutet g dieselbe Funktion, wie beim Beweise von XV, so ist, wie wir sahen, g φ-integrierbar, und es gilt (20). Aus $f - \delta \leqq g \leqq f$ folgt aber nach IX und XIV:

$$(A) \int f \, d\varphi - \delta \varphi(A) \leqq (A) \int g \, d\varphi \leqq (A) \int f \, d\varphi.$$

Aus (20) folgt also (23).

Aus XVI folgt unmittelbar:

XVII. Ist f φ-integrierbar auf A, ist $[z_i^n]$ ein δ_n-Gitter und gilt $\delta_n \to 0$, so gilt:

$$(A) \int f \, d\varphi = \lim_n L(f, [z_i^n]).$$

Damit ist die übliche Definitionsformel des Stieltjes-Lebesgue-Integrales gewonnen.

UEBER UNENDLICHE REIHEN UND ABSOLUT-ADDITIVE MENGENFUNKTIONEN

VON

HANS HAHN (IN WIEN)

[*Read August 19, 1928*]

Wir betrachten im folgenden unendliche Reihen aus positiven gegen 0 konvergierenden Gliedern a_ν :

(1) $$\lim_{\nu \to \infty} a_\nu = 0 ;$$

wir setzen :

(2) $$\sum_{\nu \geq 1} a_\nu = g ;$$

die Konvergenz der Reihe (2) wird nicht vorausgesetzt ; ist diese Reihe divergent, so ist $g = +\infty$ zu setzen. Ist $\nu_1, \nu_2, \nu_3, \ldots \nu_i, \ldots$ eine Folge wachsender natuerlicher Zahlen, so nennen wir die Reihe $\sum_{i \geq 1} a_{\nu_i}$ eine *Teilreihe* der Reihe (2). Wir behandeln nun die Frage, unter welchen Umstaenden *jede* der Ungleichung $0 < x \leq g$ genuegende Zahl x durch eine Teilreihe von (2) dargestellt werden kann, und unter welchen Umstaenden jedes solche x nur durch eine einzige Teilreihe von (2) dargestellt wird.

Die Beantwortung dieser Fragen ist keinesfalls schwierig ; wenn ich sie hier trotzdem mitteile, so geschieht es einerseits, weil ich diese einfachen Ueberlegungen in den ueblichen Lehrbuechern nicht gefunden habe, andrerseits, weil mir von hieraus der natuerlichste Weg zu einem tiefer liegenden Satze aus der Theorie der absolut-additiven Mengenfunktionen zu fuehren scheint, naemlich zu dem von W. Sierpinski (*Fund. math.* 3, 1922, p. 240) und Fréchet (*Fund. math.* 4, 1923, p. 364) bewiesenen "Zwischenwertsatze," dass unter einer gewissen einfachen Voraussetzung eine absolut-additive Mengenfunktion ϕ, die fuer zwei Teile A und B einer Menge M die Werte ϕ (A) und ϕ(B) annimmt, auch jeden zwischen ϕ (A) und ϕ (B) gelegenen Wert fuer mindestens einen Teil von M annimmt.

§ 1

Wir betrachten die Reihe (2) ; ihre Glieder a_ν seien positive Zahlen, die der Bedingung (1) genuegen. In ueblicher Weise bezeichnen wir mit s_n und r_n $(n \geq 1)$ die n-te Teilsumme und den Rest nach dem n-ten Gliede von (2) :

$$(3) \qquad s_n = \sum_{1 \leq \nu \leq n} a_\nu \; ; \qquad r_n = \sum_{\nu > n} a_\nu \; (n \geq 1) \; ;$$

fuer $n=0$ ergaenzen wir diese Definition durch :

$$(4) \qquad s_0 = 0.$$

Ebenso bezeichnen wir mit $s_{k,n}$, $r_{k,n}$ die Teilsummen und Reste von r_k, also :

$$(5) \qquad s_{k,n} = \sum_{k < \nu \leq k+n} = r_{k,n} \; , \qquad \sum_{\nu > k+n} a_\nu = r_{k+n} \; (n \geq 1)$$

$$(6) \qquad s_{k,0} = 0.$$

Wir beweisen nun den Satz :

I. *Sei (2) eine Reihe aus positiven, der Bedingung (1) genuegenden Gliedern und sei* $r_n \geq a_n$ für $n \geq 1$; *dann gibt es zu jeder, der Ungleichung* $0 < x \leq g$ *genuegenden Zahl* x *eine Teilreihe von (2), fuer die* $\sum_{i \geq 1} a_{\nu_i} = x$ *ist.*

Dies trifft sicher zu fuer $x=g$, da dann die Reihe (2) selbst die gewuenschte Teilreihe ist. Sei also $0 < x < g$. Nach (2) und (3) gilt : $s_n \longrightarrow g$; somit gibt es in der Folge s_1, s_2 ...ein erstes Glied s_{n_0}, das $\geq x$ ist. Dann ist :

$$(7) \qquad s_{n_0 - 1} < x \leq s_{n_0} \; (= s_{n_0 - 1} + a_{n_0}).$$

Da nach Annahme $r_{n_0} \geq a_{n_0}$, ist auch

$$s_{n_0 - 1} + r_{n_0} \geq x.$$

Gilt hierin das $=$ -Zeichen, so haben wir in $s_{n_0 - 1} + r_{n_0}$ eine Teilreihe von (2) vor uns, die $= x$ ist, und die Behauptung ist bewiesen. Sei also $s_{n_0 - 1} + r_{n_0} > x$, somit nach (7) :

$$s_{n_0 - 1} < x < s_{n_0 - 1} + r_{n_0}$$

Nach (3), (5) gilt $\lim_{n\to\infty} s_{n_0, n}=r_{n_0}$, es gibt also in der Folge $s_{n_0,1}, s_{n_0,2}, \ldots s_{n_0, n}$ ein erstes Glied s_{n_0, n_1}, fuer das $s_{n_0-1}+s_{n_0, n_1}\geq x$ ist. Dann ist:

$$(8) \qquad s_{n_0-1}+s_{n_0,n_1-1}<x\leq s_{n_0-1}+s_{n_0, n_1} \; (=s_{n_0-1} + s_{n_0, n_1-1}+a_{n_0+n_1}).$$

Da nach Annahme $r_{n_0+n_1}\geq a_{n_0+n_1}$, ist auch: $s_{n_0-1}+s_{n_0,n_1-1}+r_{n_0+n_1}\geq x$. Gilt hierin das $=$-Zeichen, so haben wir in $s_{n_0-1}+s_{n_0,n_1-1}+r_{n_0+n_1}$ ein Teilreihe von (2) vor uns, die $=x$ ist, und die Behauptung ist bewiesen. Sei also $s_{n_0-1}+s_{n_0, n_1-1}+r_{n_0+n_1}>x$, also nach (8)

$$s_{n_0-1}+s_{n_0, n_1-1}<x<s_{n_0-1}+s_{n_0, n_1-1}+r_{n_0+n_1}.$$

Wegen $\lim_{n\to\infty} s_{n_0+n_1, n}=r_{n_0+n_1}$ gibt es in der Folge $s_{n_0+n_1,1}, s_{n_0+n_1,2} \ldots s_{n_0+n_1,n}$ ein erstes Glied $s_{n_0+n_1, n_2}$ fuer das $s_{n_0-1}+s_{n_0, n_1-1}+s_{n_0+n_1, n_2}\geq x$ ist.

Dann ist:

$$s_{n_0-1}+s_{n_0, n_1-1}+s_{n_0+n_1, n_2-1} < x\leq s_{n_0-1}+s_{n_0, n_1-1}+s_{n_0+n_1, n_2-1}+a_{n_0+n_1+n_2},$$

und somit nach Annahme auch:

$$s_{n_0-1}+s_{n_0, n_1-1}+s_{n_0+n_1, n_2-1}+r_{n_0+n_1+n_2}\geq x.$$

Gilt hierin das $=$-Zeichen, so haben wir in

$$s_{n_0-1}+s_{n_0, n_1-1}+s_{n_0+n_1, n_2-1}+r_{n_0+n_1+n_2}$$

eine Teilreihe von (2) vor uns, die gleich x ist, und die Behauptung ist bewiesen. Andernfalls schliesst man in derselben Weise weiter. Kommt man so nach k Schritten zu einer Gleichung:

$$(9) \qquad s_{n_0-1}+s_{n_0, n_1-1}+s_{n_0+n_1, n_2-1}+\cdots + s_{n_0+n_1+\ldots+n_{k-1}, n_k-1}+r_{n_0+n_1+\ldots+n_k}=x$$

so ist die linke Seite eine Teilreihe von (2), die $=x$ ist, und die Behauptung ist bewiesen. Kommt man aber bei Fortsetzung des Verfahrens niemals auf eine Gleichung (9), so gilt fuer alle k:

$$s_{n_0-1}+s_{n_0,\,n_1-1}+\dots+s_{n_0+n_1+\dots+n_{k-1},\,n_k-1}<x$$

$$\leq s_{n_0-1}+s_{n_0,\,n_1-1}+\dots+s_{n_0+n_1+\dots n_{k-1},\,n_k-1}+a_{n_0+n_1+\dots+n_k}$$

und wegen (1) folgt daraus:

$$s_{n_0-1}+s_{n_0,\,n_1-1}+s_{n_0+n_1,\,n_2-1}+\dots+s_{n_0+n_1+\dots+n_{k-1},\,n_k-1}+\dots=x;$$

da hierin die linke Seite eine Teilreihe von (2) ist, ist Satz 1 bewiesen.
Als Korullar erhalten wir daraus:

Ist $\sum_{\nu\geq 1} a_\nu$ *eine divergente Reihe aus positiven, gegen o konvergierenden Gliedern, so gibt es in ihr zu jeder pesitiven Zahl* x *eine Teilreihe, fuer die* $\sum_{i\geq 1} a_{\nu_i}=x$ *ist.*

Denn da $r_n=+\infty$ ist fuer alle n, ist die Bedingung von Satz I erfuellt, und hier ist $g=+\infty$.

Wenn in der Reihe (2) die Glieder a_ν monoton abnehmen $(a_\nu\geq a_{\nu+1})$, so ist die in Satz I als hinreichend erwiesene Bedingung auch notwendig:

II. *Sei (2) eine Reihe aus positiven, monoton abnehmenden Gliedern: damit es zu jeder der Ungleichung* $0<x\leq g$ *genuegenden Zahl* x *eine Teilreihe von (2) gebe, fuer die* $\sum_{i\geq 1} a_{\nu_i}=x$ *ist, ist notwendig, dass fuer alle* $n\geq 1$ *gelte:* $r_n\geq a_n$.

Angenommen, es gebe ein $n^*\geq 1$, so dass $r_{n^*}<a_{n^*}$. Wegen

$$s_{n^*}=s_{n^*-1}+a_{n^*} \text{ ist dann } s_{n^*-1}+r_{n^*}<s_{n^*}.$$

Man waehle nun x gemaess der Ungleichung:

$$\text{(10)} \qquad s_{n^*-1}+r_{n^*}<x\leq s_{n^*}.$$

Sei sodann $\sum_{\geq 1} a_{\nu_i}$ eine beliebige Teilreihe von (2). Kommen $a_1, a_2, \dots a_{n^*}$ unter den a_{ν_i} vor, so ist $\sum_{i\geq 1} a_{\nu_i}>s_{n^*}$, also wegen (10) auch $\sum_{i\geq 1} a_{\nu_i}>x$.

Kommt hingegen eines der Glieder $a_1, a_2, \ldots a_{n^*}$ unter den a_{ν_i} nicht vor, so ist, da die a_ν monoton abnehmen und somit das fehlende Glied $\geq a_{n^*}$ ist: $\sum\limits_{i\geq 1} a_{\nu_i} \leq s_{n^*-1} + r_{n^*}$, also wegen (10) auch: $\sum\limits_{i\geq 1} a_{\nu_i} < x$. Jedenfalls ist also $\sum\limits_{i\geq 1} a_{\nu_i} \neq x$, es kann somit keine Teilreihe von (2) eine der Ungleichung (10) genuegende Zahl darstellen.

§ 2

Wir betrachten wieder die Reihe (2) aus positiven, monoton abnehmenden, der Bedingung (1) genuegenden Gliedern; wir nehmen an, die Bedingung $r_n \geq a_n$ von Satz I und II sei erfuellt und fragen: Unter welchen Umstaenden gibt es zu jeder der Ungleichung $0 < x \leq g$ genuegenden Zahl x nur eine einzige Teilreihe von (2), fuer die

$$\sum_{i\geq 1} a_{\nu_i} = x \text{ ist ?}$$

Dabei wollen wir, wenn unter den Gliedern a_ν einander Gleiche vorkommen, die beiden Teilreihen $\sum\limits_{i\geq 1} a_{\nu_i}$ und $\sum\limits_{i\geq 1} a_{\mu_i}$ als *dieselbe* Teilreihe von (2) betrachten, wenn $a_{\nu_i} = a_{\mu_i}$ ist fuer alle i:

III. *Sei (2) eine Reihe aus positiven, monoton abnehmenden, der Bedingung (1) genuegenden Gliedern, in der $r_n \geq a_n$ sei fuer alle $n \geq 1$; damit es zu jeder der Ungleichung $0 < x \geq g$ genuegenden Zahl x nur eine einzige Teilreihe von (2) gebe, fuer die $\sum\limits_{i\geq 1} a_{\nu_i} = x$ ist, ist notwendig und hinreichend, dass fuer alle, fuer die $a_{n+1} < a_n$ ist $r_n = a_n$ sei.*

Notwendig: Sei:

$$(11) \qquad a_{n^*+1} < a_{n^*},\ r_{n^*} > a_{n^*}.$$

Wir setzen,

$$(12) \qquad x^* = s_{n^*-1} + r_{n^*};$$

wegen $s_{n^*} = s_{n^*-1} + a_{n^*}$ folgt aus (11): $x^* > s_{n^*}$ und aus (12) folgt, bei Beachtung von (11):

$$x^* - s_{n^*} = s_{n^*-1} - s_{n^*} + r_{n^*} = r_{n^*} - a_{n^*} < r_{n^*} - a_{n^*+1} = r_{n^*+1}.$$

Wir haben also:

$$0<x^{*}-s_{n^{*}}<r_{n^{*}+1}\,.$$

Nach (3) ist aber $r_{n^{*}+1}=\sum_{\nu\geq 1} a_{n^{*}+1+\nu}$ eine unendliche Reihe aus positiven, monoton abnehmenden Gliedern, die der Bedingung (1) genuegen; und nach Satz I gibt es eine Teilreihe von $r_{n^{*}+1}$, sodass

$$(13)\qquad x^{*}-s_{n^{*}}=\sum_{i\geq 1} a_{n^{*}+1+\nu_i}$$

Nach (12) und (13) haben wir also fuer x^{*} die beiden Darstellungen:

$$(14)\qquad x^{*}=s_{n^{*}-1}+\sum_{\nu\geq 1} a_{n^{*}+\nu}\ ;\ x^{*}=s_{n^{*}}+\sum_{i\geq 1} a_{n^{*}+1+\nu_i}$$

In beiden Darstellungen ist die rechte Seite eine Teilreihe von (2) und zwar sind dies zwei verschiedene Teilreihen von (2); denn alle Glieder von $s_{n^{*}-1}$ kommen in beiden vor; auf diese $n^{*}-1$ gemeinsamen Glieder aber folgt in der ersten Teilreihe das Glied $a_{n^{*}+1}$, in der zweiten aber das Glied $a_{n^{*}}$, und nach (11) ist $a_{n^{*}}\neq a_{n^{*}+1}$. Es gibt also in (2) zwei verschiedene die Zahl x^{*} darstellende Teilreihen.

Hinreichend:

Seien $\sum_{i\geq 1} a_{\nu_i}$ und $\sum_{i\geq 1} a_{\mu_i}$ zwei verschiedene Teilreihen von (2) und es sei i^{*} der kleinste Index, fuer den $a_{\nu_{i^{*}}}\neq a_{\mu_{i^{*}}}$; wir koennen ohne weiteres annehmen, es sei $a_{\nu_{i^{*}}}<a_{\mu_{i^{*}}}$, also wegen der Monotonie der a_ν:

$$(15)\qquad \nu_{i^{*}}>\mu_{i^{*}}$$

Wir setzen:

$$\sum_{1\leq i<i^{*}} a_{\nu_i}=\sum_{1\leq i<i^{*}} a_{\mu_i}=\sigma$$

(wenn $i^{*}=1$ wird $\sigma=0$ gesetzt). Bezeichnen wir mit $a_{\mu^{*}}$ das letzte Glied von (2), das $=a_{\mu_{i^{*}}}$ ist, so ist wegen (15):

$$(16)\qquad \mu^{*}<\nu_{i^{*}}$$

und da wir die Bedingung von Satz III als erfuellt voraussetzen, ist:

$$(17)\qquad a_{\mu^{*}}=r_{\mu^{*}}$$

Ferner ist offenbar:

$$(18)\qquad \sum_{i\geq 1} a_{\nu_i} \leq \sigma + r_{\nu_{i^*}-1}\ ;\ \sum_{i\geq 1} a_{\mu_i} > \sigma + a_{\mu_{i^*}}\ (=\sigma + a_{\mu^*})$$

Wegen (16) ist hierin $\mu^* \leq \nu_{i^*-1}$ also bei Beachtung von (17):

$$r_{\nu_{i^*-1}} \leq r_{\mu^*} = a_{\mu^*}$$

so dass aus (18) folgt:

$$\sum_{i\geq 1} a_{\nu_i} < \sum_{i\geq 1} a_{\mu_i}\ .$$

Zwei verschiedene Teilreihen von (2) koennen also nicht dieselbe Zahl darstellen, und die Behauptung ist bewiesen.

Es ist nun leicht, alle moeglichen Reihen (2) aus positiven Gliedern anzugeben, durch deren saemtliche Teilreihen die saemtlichen der Unglei chung $0 < x \leq 1$ genuegenden Zahlen x dargestellt werden und zwar so, dass jeder solche x nur durch eine einzige Teilreihe dargestellt wird. Da es auf die Reihenfolge der Glieder nicht ankommt, koennen wir wieder die als monoton abnehmend voraussetzen. Jedenfalls muss in (2) $g=1$ sein; denn waere $g<1$, so waere keine der Ungleichung $g<x\leq 1$ genuegende Zahl x durch eine Teilreihe von (2) darzustellen, und waere $g>1$, so wuerde die Reihe (2) selbst, die ja auch eine ihrer Teilreihen ist, eine nicht der Ungleichung $0<x\leq 1$ genuegende Zahl darstellen. Die Reihe (2) muss also konvergent sein, so dass (I) von selbst erfuellt ist. Sei etwa $a_1 = a_2 = \ldots = a_{n_1} > a_{n_1+1}$. Dann muss nach III $r_{n_1} = a_{n_1}$ sein; aus:

$$\sum_{\nu\geq 1} a_\nu = a_1 + a_2 + \ldots + a_{n_1} + r_{n_1} = 1$$

folgt also:

$$a_1 = a_2 = \ldots = a_{n_1} = \frac{1}{n_1+1}\ ;\ r_{n_1} = \frac{1}{n_1+1}\ .$$

Sei sodann:

$$a_{n_1} = a_{n_1+2} = \ldots = a_{n_1+n_2} > a_{n_1+n_2+1}\ ;$$

nach III muss $r_{n_1+n_2} = a_{n_1+n_2}$ sein, und aus:

$$a_{n_1+1} + a_{n_1+2} + \ldots + a_{n_1+n_2} + r_{n_1+n_2} = r_{n_1} \left(= \frac{1}{n_1+1} \right)$$

folgt:

$$a_{n_1+1}+a_{n_1+2}+\ldots+a_{n_1+n_2}$$

$$=\frac{1}{(n_1+1)(n_2+1)};\ r_{n_1+n_2}=\frac{1}{(n_1+1)(n_2+1)}.$$

Indem man so fortschliesst, sieht man, dass die Reihe folgende Gestalt haben muss:

Es gibt eine Folge natuerlicher Zahlen $n_1, n_2, \ldots n_k$ so dass:

$$a_1=a_2=\ldots=a_{n_1}=\frac{1}{n_1+1};\ a_{n_1+1}=a_{n_1+2}=\ldots=a_{n_1+n_2}=\frac{1}{(n_1+1)(n_2+1)}.$$

Diese Reihen sind bekannt als *Cantorsche Reihen* (vergl. z.B.O. Perron, *Irrationalzahlen*, p. 111), und wir haben den Satz bewiesen:

IV. *Die Cantorschen Reihen sind die einzigen Reihen aus positiven Gliedern, deren saemtliche Teilreihen die saemtlichen der Ungleichung* $0<x\leq 1$ *genuegenden Zahlen so darstellen, dass jede solche Zahl nur durch eine einzige Teilreihe dargestellt wird.*

Die Systembrüche der Grundzahl $g(>1)$ sind in den Cantorschen Reihen als der Spezialfall $n_k=g-1$ $(k=1, 2, \ldots)$ enthalten.

§ 3

Wir verwenden nun Satz I zum Beweise des in der Einleitung erwaehnten Zwischenwertsatzes in der Theorie der absolut additiven Mengenfunktionen. Wegen aller im folgenden verwendeten Begriffe und Saetze aus der Theorie der absolut additiven Mengenfunktionen sei verwiesen auf H. Hahn, *Theorie der reellen Funktionen*, sechstes Kap.

Sei $\phi(M)$ eine im σ-Koerper M definierte absolut additive Mengenfunktion. Alle weiterhin auftretenden Mengen, gehoeren, auch wenn dies nicht ausdruecklich gesagt wird, zum σ-Koerper M. Eine Menge S heisst *singulaer* fuer ϕ, wenn $\phi(S)\neq 0$ ist und fuer jeden Teil J vom S entweder $\phi(J)=0$ oder $\phi(J)=\phi(S)$ ist. Besitzt die Menge A keinen fuer ϕ singulaeren Teil, so heisse ϕ *singularitaetenfrei* in **A**.

V. *Ist* $\phi(M)$ *singularitaetenfrei* in A, *so auch die Positivfunktion* $\pi(M)$ *und die Negativfunktion* $\nu(M)$ *von* ϕ. Wir beweisen dies etwa fuer $\pi(M)$. Angenommen, es gaebe in A einen fuer $\pi(M)$ singulaeren Teil P. Wir zerlegen (l.c. meine *Theorie*, p. 404, Satz IX.) P in zwei fremde Teile $P=P'+P''$, so dass $\pi(P')=\pi(P)$, $\pi(P'')=0$, $\nu(P')=0$, $\nu(P'')=\nu(P)$.

Fuer jeden Teil M von P′ gilt dann $\phi(M)=\pi(M)$. Da P singulaer fuer $\pi(M)$, ist $\pi(P)\neq 0$, also auch $\pi(P')\neq 0$, also auch $\phi(P')\neq 0$: und fuer jeden Teil M von P′ gilt $\pi(M)=0$ oder $\pi(M)=\pi(P)=\pi(P')$ also auch $\phi(M)=0$ oder $\phi(M)=\phi(P')$; d.h. P′ ist singulaer fuer ϕ. Gibt es also in A einen fuer π singulaeren Teil P, so gibt es in A auch einen fuer ϕ singulaeren Teil P′. Damit ist die Behauptung bewiesen.

Wir nennen die absolut additiv Mengenfunktion $\phi(M)$ monoton wachsend in A, wenn $\phi(M)\geq 0$ ist fuer alle Teile von A.

VI. *Ist ϕ in A monoton wachsend und singularitaetenfrei, und ist $\phi(A)>0$, so gibt es in A eine Folge von Teilen $B_1, B_2,\ldots B_{n_1},\ldots$ so dass $\phi(B_n)>0$ fuer alle n und $\lim_{n\to\infty}\phi(B_n)=0$.* Weil ϕ in A singularitaetenfrei, gibt es einen Teil A von A_1, so dass $0<\phi(A_1)<\phi(A)$, ebenso einen Teil A_2 von A, so dass $0<\phi(A_2)<\phi(A_1)$ usw. Man erhielt so eine Folge von Teilen A_1, $A_2,\ldots A_r$ von $A\ldots$ so dass $\phi(A_1)>\phi(A_2)>\ldots>\phi(A_n)>\ldots>0$.

Es existiert also $\lim_{n\to\infty}\phi(A_n)$. Setzen wir nun $A_n-A_{n+1}=B_n$, so ist $\phi(B_n)=\phi(A_n)-\phi(A_{n+1})$, und somit $\phi(B_n)>0$ und $\lim_{n\to\infty}\phi(B_n)=0$.

VII. *Ist ϕ in A monoton wachsend und singularitaetenfrei und ist $\phi(A)=+\infty$ so gibt es zu jeder noch so grossen Zahl z einen Teil B von A, so dass $z<\phi(B)<+\infty$.* Da ϕ singularitaetenfrei in A, gibt es jedenfalls Teile von M, fuer die $0<\phi(M)<+\infty$. Bilden wir fuer jeden solchen Teil von M den Funktionswert $\phi(M)$, so hat die Menge aller dieser $\phi(M)$ eine obere Grenze g. Unsere Behauptung ist gleichbedeutend mit: $g=+\infty$. Angenommen, es waere $g<+\infty$. Sicherlich gibt es in A eine Folge von Teilen $M_1, M_2,\ldots M_{n_1},\ldots$ so dass $\lim_{n\to\infty}\phi(M_n)=g$. Bezeichnen wir die Vereinigung von $M_1, M_n,\ldots M_n$, mit A_n so ist: $\phi(A_n)\leq\phi(M_1)+\phi(M_2)+\ldots\phi(M_n)$ also ist auch $\phi(A_n)$ endlich, mithin $\phi(A_n)\leq g$; und da $\phi(A_n)\geq\phi(M_n)$ folgt aus $\lim_{n\to\infty}\phi(M_n)=g$ auch $\lim_{n\to\infty}\phi(A_n)=g$. Sei nun B die Vereinigung von $A_1, A_2\cdots A_n,\ldots$ Da Die Mengenfolge $A_1, A_2,\cdots A_n,\ldots$ monoton waechst, ist dann bekanntlich $\phi(B)=\lim_{n\to\infty}\phi(A_n)$ also $\phi(B)=g$. Da $g<+\infty$ und $\phi(A)=+\infty$, Ist auch $\phi(A-B)=+\infty$. Da ϕ singularitaetenfrei in A gibt es in $A-B$ einen Teil ϕ, so dass $0<\phi(\phi)<+\infty$. Da B und ϕ Ist $\phi(B+\phi)=\phi(B)+\phi(\phi)$ $=g+\phi(\phi)$, also $g<\phi(B+\phi)<+\infty$; das aber widerspricht der Definition von g als der oberen Grenze aller fuer Teile M von A auftretenden endlichen Funktionswerte. Die Annahme $g<+\infty$ fuehrt also auf einen Widerspruch, und die Behauptung ist bewiesen.

Nach diesen Vorbereitungen kommen wir nun zum Beweise des Zwischenwertsatzes. Bekanntlich (l. c. meine *Theorie*, p. 401, Satz III, IV) gibt es unter allen Werten $\phi(M)$ die ϕ fuer Teile M von A annimmt, sowohl einen groessten als einen kleinsten und zwar sind dieser groesste und kleinste Wert gegeben durch π (A) und $-\nu(A)$. Der Zwischenwertsatz kann so ausgesprochen werden:

VIII. *Ist ϕ singularitaetenfrei in A, so gibt es zu jeder der Ungleichung $-\nu(A)\leq z\leq\pi(A)$ genuegenden Zahl z einen Teil M von A so dass $\phi(M)=z$.*

Beim Beweise koennen wir uns auf den Fall beschraenken, dass ϕ in A monoton wachsend ist. Denn nehmen wir an, der Satz gelte fuer monoton wachsende Mengenfunktionen; da $\pi(M)$ und ν (M) monoton wachsend und zufolge V in A singularitaetenfrei sind, kann dann der Satz auf π (M) und $\nu(M)$ angewendet werden; wir zerlegen nun A in zwei fremde Teile: $A=A'+A''$ so dass $\pi(A')=\pi(A)$, $\pi(A'')=0$, $\nu(A')=0$, $\nu(A'')=\nu(A)$ fuer jeden Teil M von A' ist dann π $(M)=\phi(M)$ und fuer jeden Teil M von A'' ist ν $(M)=-\phi(M)$; durch Anwendung unseres Satzes auf π und ν folgt dann: ist $0\leq z\leq\pi(A')$ $(=\pi(A))$ so gibt es einen Teil M von A', so dass $\pi(M)=z$ und mithin auch $\phi(M)=z$; ist $0\geq z\geq-\nu$ $(A'')(=-\nu(A))$ so gibt es einen Teil M von A'', so dass $-\nu(M)=z$ und mithin auch ϕ $(M)=z$; damit ist dann die Behauptung von Satz VIII fuer beliebiges ϕ bewiesen. Nehmen wir also nun $\phi(M)$ als monoton wachsendan, dann ist $\pi(M)=\phi(M)$, $\nu(M)=0$. Wir koennen weiter voraussetzen: $\phi(A)<+\infty$; denn nehmen wir an, der Satz sei bewiesen, wenn $\phi(A)$ endlich; ist nun ϕ $(A)=+\infty$, so gibt es nach Satz VII zu jedem $z\geq 0$ einen Teil B von A, so dass $\phi(B)<z$; nach Annahme kann nun der Satz statt auf A auf B angewendet werden, und ergibt die Existenz eines Teiles M von B, fuer den $\phi(M)=z$; da aber M auch Teil von A ist, ist damit auch die Behauptung fuer A bewiesen. Wir haben also nur mehr zu zeigen: Ist $\phi(M)$ in A monoton wachsend, und $\phi(A)<+\infty$ so gibt es zu jedem der Ungleichung $0\leq z\leq\phi(A)$ genuegenden Zahl z einen Teil M von A, so dass $\phi(M)=z$. Fuer $z=0$ ist die Behauptung trivial; denn fuer die leere Menge L gilt: $\phi(L)=0$. Wir nehmen also $z>0$ an. Wir betrachten alle diejenigen Teile M_1 von A, fuer die $0<\phi(M_1)\leq\frac{1}{2}\phi(A)$; nach VI gibt es solche Teile M_1; die obere Grenze der Werte $\phi(M_1)$. die auf diesen Teilen M_1 von A annimmt, bezeichnen wir mit g_1: dann ist

$$(19)\qquad 0<g_1\leq\tfrac{1}{2}\,\phi(A),$$

und es gibt gewiss einen Teil A_1 von A, so dass

$$(20)\qquad \tfrac{1}{2}\,g_1<\phi(A_1)\leq g_1$$

Aus (19) und (20) folgt $\phi(A_1)<\phi(A)$ also $\phi(A-A_1)>0$. Nun betrachten wir alle Teile M_2 von $A-A_1$, fuer die $0<\phi(M_2)\leq\frac{1}{2}\phi(A-A_1)$ die obige Grenze aller auf diesen Teilen M_2 auftretenden Funktionswerten $\phi(M_2)$ bezeichenen wir mit g_2 ; dann ist

$$(21) \qquad 0<g_2\leq\tfrac{1}{2}\phi(A-A_1)$$

und es gibt gewiss einen Teil A_2 von $A-A_1$ so dass:

$$(22) \qquad \tfrac{1}{2}g_2<\phi(A_2)\leq g_2$$

Aus (21) und (22) folgt $\phi(A_2)<\phi(A-A_1)$ also $\phi(A-(A_1+A_2))>0$.

Nun betrachten wir alle Teile M_3 von $A-(A_1+A_2)$, fuer die $0<\phi(M_3)\leq \frac{1}{2}\phi(A-(A_1+A_2))$; die obere Grenze aller auf diesen Teilen M_3 auftretenden Funktionswerten bezeichenen wir mit g_3 ; dann ist:

$$0<g_3\leq\tfrac{1}{2}\phi(A-(A_1+A_2))$$

und es gibt einen Teil A_3 von $A-(A_1+A_2)$, so dass:

$$\tfrac{1}{2}g_3<\phi(A_3)\leq g_3.$$

In dieser Weise schliessen wir fort. Wir erhalten so eine Folge von Teilen $A_1, A_2,\ldots A_n,\ldots$ von A und eine Folge von Zahlen $g_1, g_2,\ldots g_n,\cdots$ mit folgenden Eigenschaften:

(1) A_n ist Teil von $A-(A_1+A_2+\ldots+A_{n-1})$;

(2) $0<g_n\leq\frac{1}{2}\phi(A-(A_1+A_2+\ldots+A_{n-1}))$;

(3) $\frac{1}{2}g_n<\phi(A_n)\leq g_n$.

Wegen (1) sind die Mengen $A_1, A_2,\cdots A_n,\ldots$ zu je zweien fremd, also ist, wenn

$$(23) \qquad A_1+A_2+\cdots+A_n+\ldots=B \text{ gesetzt wird,}$$

$$(24) \qquad \sum_{n\geq1}\phi(A_n)=\phi(B)\leq\phi(A)$$

und mithin: $\lim\limits_{n\to\infty}\phi(A_n)=0$.

Aus (3) folgt daher auch:

$$(25) \qquad \lim_{n\to\infty} g_n=0$$

Wir behaupten nun, dass fuer die Menge (23) gilt:

(26) $$\phi(A-B)=0$$

Denn waere $\phi(A-B)>0$, so gaebe es nach VI einen Teil ϕ von $(A-B)$ so dass

(27) $$0<\phi(\phi)<\tfrac{1}{2}\,\phi\,(A-B).$$

Nach (23) ist ϕ Teil von $B-(A_1+A_2+\ldots+A_{n-1})$

also ist wegen (27): $g_n\geq\phi(\phi)$,

was wegen $\phi(\phi)>0$ mit (25) in Widerspruch steht. Damit ist (26) nachgewiesen. Aus (26) folgt nun $\phi(A)=\phi(B)$, also wegen (23):

(28) $$\phi(A)=\sum_{\nu\geq 1}\phi(A_\nu)$$

Da ferner $A-(A_1+A_2+\ldots+A_{n-1})=A_n+A_{n+1}+\ldots+(A-B)$ folgt aus (26) auch:

(29) $$\sum_{\nu\geq n}\phi(A_\nu)=\phi(A-(A_1+\cdots+A_{n-1})$$

Wegen (2) und (3) ist hierein

(30) $$\phi\,(A_n)\leq\tfrac{1}{2}\,\phi(A-(A_1+\ldots+A_{n-1}))$$

Aus (29) und (30) aber folgt:

$$\sum_{\nu\geq n}\phi\,(A_\nu)\geq\phi\,(A_n).$$

Die Reihe (24) erfuellt also die Bedingung von Satz I, und aus Satz I folgt: ist $0<z\leq\phi(A)$, so gibt es in (24) eine Teilreihe, so dass $\sum_{i\geq 1}\phi(A_{\nu_i})=z$. Bezeichnen wir also mit **M** die Vereinigung von A_{ν_1}, A_{ν_2},...so ist $\phi(M)=z$, und Satz VIII ist bewiesen.

ÜBER DIE MULTIPLIKATION TOTAL-ADDITIVER MENGENFUNKTIONEN

Von HANS HAHN (Wien).

Sind $\varphi(x)$ und $\psi(x)$ zwei im selben Mengensystem definierte, total-additive Mengenfunktionen, so ist auch $\varphi(x)+\psi(x)$ eine total-additive Mengenfunktion, nicht aber $\varphi(x)\cdot\psi(x)$. Wir wollen nun zeigen, wie auf andre Weise aus zwei total-additiven Mengenfunktionen $\varphi(x)$ und $\psi(x)$ eine gleichfalls total-additive Mengenfunktion hergeleitet werden kann, die man zweckmäßig als das *Produkt* $\varphi\times\psi$ bezeichnen kann. Bezeichnet man das n-dimensionale Lebesguesche Maß einer Punktmenge des n-dimensionalen Raumes mit μ_n, so ist im Sinne dieser Multiplikation: $\mu_{k+l}=\mu_k\times\mu_l$. Wir entwickeln den engen Zusammenhang dieser Multiplikation mit dem *Lebesgue-Stieltjesschen Integralbegriff* und zeigen insbesondere, daß das Reduktionstheorem mehrfacher Integrale nichts andres ist, als das assoziative Gesetz dieser Multiplikation ([1]).

§ 1. - Vollständige σ-Körper.

Sei E irgend eine Menge; wir bezeichnen sie als « Raum », ihre Elemente als « Punkte », ihre Teilmengen als « Punktmengen ». Ein System von Punktmengen, die zu je zweien keinen Punkt gemein haben, nennen wir *disjunkt.*

Sei $\mathfrak{M}$ irgend ein System von Punktmengen X des Raumes E und $\varphi(X)$ eine in $\mathfrak{M}$ definierte Mengenfunktion; sie heißt *additiv*, wenn für je zwei disjunkte Mengen A, B aus $\mathfrak{M}$, deren Summe $A+B$ gleichfalls zu $\mathfrak{M}$ gehört, gilt:

$$\varphi(A+B)=\varphi(A)+\varphi(B);$$

die in $\mathfrak{M}$ definierte und additive Mengenfunktion φ heißt *total-additiv*, wenn für jede Folge disjunkter Mengen $A_1, A_2, \ldots, A_n, \ldots$ aus $\mathfrak{M}$, deren Summe $\underset{n}{S} A_n$ gleichfalls zu $\mathfrak{M}$ gehört, gilt:

$$\varphi(\underset{n}{S} A_n)=\sum_n \varphi(A_n).$$

([1]) Die folgende Abhandlung ist eine Ausarbeitung von Vorträgen, die ich vor einigen Jahren in der Wiener Mathematischen Gesellschaft und im März 1933 an der Universität Brünn gehalten habe. Gänzlich unabhängig von mir hat Herr ST. SAKS denselben Gegenstand

Wir beschäftigen uns im Folgenden, der Einfachheit halber, nur mit *nicht-negativen* Mengenfunktionen; alle auftretenden Mengenfunktion sind also als nicht-negativ vorausgesetzt, ohne daß diese Voraussetzung jeweils ausdrücklich angegeben wird.

Ein Mengensystem $\mathfrak{M}$ heißt ein *Körper*, wenn aus $A \varepsilon \mathfrak{M}$ und $B \varepsilon \mathfrak{M}$ folgt: $A+B \varepsilon \mathfrak{M}$ und $A-B \varepsilon \mathfrak{M}$; bekanntlich (2) folgt dann von selbst auch $AB \varepsilon \mathfrak{M}$. Ist $\mathfrak{M}$ ein Körper, so kann jede Summe $\underset{n}{S} A_n$, deren Summanden A_n zu $\mathfrak{M}$ gehören, verwandelt werden in eine Summe $\underset{n}{S} B_n$ *disjunkter* Summanden B_n, die gleichfalls zu $\mathfrak{M}$ gehören, und Teile der entsprechenden Summanden A_n sind:

(1) $$\underset{n}{S} A_n = \underset{n}{S} B_n, \qquad B_n \leqq A_n, \qquad B_n \varepsilon \mathfrak{M};$$

man hat nur zu setzen $B_1 = A_1$, $B_n = A_n - (A_1 + \ldots + A_{n-1})$ $(n > 1)$.

Jeder nicht leere Körper enthält die leere Menge Λ. Ist φ additiv im Körper $\mathfrak{M}$ — und schließen wir den Fall $\varphi(X) = +\infty$ für alle $X \varepsilon \mathfrak{M}$ einfürallemal aus — so ist $\varphi(\Lambda) = 0$.

I. - *Ist die nichtnegative Mengenfunktion φ additiv im Körper $\mathfrak{M}$, so ist sie monoton wachsend, d. h. aus $A \varepsilon \mathfrak{M}$, $B \varepsilon \mathfrak{M}$, $A \leqq B$ folgt $\varphi(A) \leqq \varphi(B)$.*

Denn es ist $B = A + (B - A)$ eine Zerlegung von B in zwei disjunkte Summanden, und weil $\mathfrak{M}$ ein Körper, ist $B - A \varepsilon \mathfrak{M}$; also weil φ additiv und nicht-negativ:

$$\varphi(B) = \varphi(A) + \varphi(B - A) \geqq \varphi(A).$$

Man beweist leicht die Sätze (3):

II. - *Ist φ total-additiv im Körper $\mathfrak{M}$, ist $A = \underset{n}{S} A_n$, $A \varepsilon \mathfrak{M}$, $A_n \varepsilon \mathfrak{M}$, $A_n \leqq A_{n+1}$, so ist $\varphi(A) = \lim\limits_{n} \varphi(A_n)$.*

III. - *Ist φ total-additiv im Körper $\mathfrak{M}$, ist $A = \underset{n}{D} A_n$, $A \varepsilon \mathfrak{M}$, $A_n \varepsilon \mathfrak{M}$, $A_n \geqq A_{n+1}$, und sind die $\varphi(A_n)$ endlich, so ist $\varphi(A) = \lim\limits_{n} \varphi(A_n)$.*

Der Körper $\mathfrak{M}$ heißt ein *σ-Körper*, wenn aus $A_n \varepsilon \mathfrak{M}$ folgt: $\underset{n}{S} A_n \varepsilon \mathfrak{M}$. Bekanntlich (4) folgt dann aus $A_n \varepsilon \mathfrak{M}$ auch $\underset{n}{D} A_n \varepsilon \mathfrak{M}$.

Sei φ eine im Körper $\mathfrak{M}$ definierte, total-additive Mengenfunktion. Jede Menge $X \varepsilon \mathfrak{M}$ mit $\varphi(X) = 0$ nennen wir eine *Nullmenge für φ* (wobei der Zusatz « für φ »

---

in seinem während der Drucklegung dieser Abhandlung erschienenen Buche: *Théorie de l'intégrale*, Warszawa, 1933, p. 257 ff. behandelt.

(2) Vgl. z. B. H. HAHN: *Reelle Funktionen*, Leipzig, 1932, S. 12. Wir schließen uns in Terminologie und Bezeichnungsweise durchweg an dieses Buch (im Folgenden als R. F. zitiert).

(3) Vgl. z. B. H. HAHN: *Theorie der reellen Funktionen*, Berlin, 1921, S. 395 (im Folgenden als Th. R. F. zitiert).

(4) R. F. S. 16.

auch wegbleiben kann, wo kein Zweifel besteht). Die leere Menge Λ ist eine Nullmenge. Es gilt:

IV. - *Ist φ total-additiv im σ-Körper $\mathfrak{M}$, so ist das System $\mathfrak{N}$ aller Nullmengen für φ ein σ-Körper.*

Wir haben zu zeigen: ist $A \varepsilon \mathfrak{N}$, $B \varepsilon \mathfrak{N}$ so ist auch $A - B \varepsilon \mathfrak{N}$; ist $A_n \varepsilon \mathfrak{N}$, so ist auch $\underset{n}{S} A_n \varepsilon \mathfrak{N}$. Ist $A \varepsilon \mathfrak{N}$, $B \varepsilon \mathfrak{N}$, so ist auch $A \varepsilon \mathfrak{M}$, $B \varepsilon \mathfrak{M}$, also — weil $\mathfrak{M}$ ein Körper — auch $A - B \varepsilon \mathfrak{M}$; ferner ist $\varphi(A) = 0$, wegen $A - B \subseteqq A$ ist also nach I auch $\varphi(A - B) = 0$, also $A - B \varepsilon \mathfrak{N}$. Ist $A_n \varepsilon \mathfrak{N}$, also auch $A_n \varepsilon \mathfrak{M}$, so ist, weil $\mathfrak{M}$ ein σ-Körper, $\underset{n}{S} A_n \varepsilon \mathfrak{M}$; nach (1) ist $\underset{n}{S} A_n = \underset{n}{S} B_n$, wo die B_n disjunkte Mengen aus $\mathfrak{M}$ und $B_n \subseteqq A_n$; aus $\varphi(A_n) = 0$ folgt also nach I $\varphi(B_n) = 0$, also weil φ total-additiv: $\varphi(\underset{n}{S} A_n) = \sum_n \varphi(B_n) = 0$, also $\underset{n}{S} A_n \varepsilon \mathfrak{N}$.

Ist $\mathfrak{M}$ ein Körper und φ eine in $\mathfrak{M}$ definierte, total-additive Mengenfunktion, so heißt $\mathfrak{M}$ *vollständig für* φ, wenn jeder Teil einer Nullmenge für φ zu $\mathfrak{M}$ gehört. Aus I folgt sofort:

V. - *Ist der Körper $\mathfrak{M}$ vollständig für φ, so ist jeder Teil einer Nullmenge für φ auch eine Nullmenge für φ.*

Ist $\mathfrak{M}$ ein σ-Körper, φ eine in $\mathfrak{M}$ definierte, total-additive Mengenfunktion, so kann $\mathfrak{M}$, wie wir nun zeigen wollen, stets zu einem für φ vollständigen σ-Körper erweitert werden. Wir bezeichnen mit $\mathfrak{N}$ das System aller Nullmengen für φ, mit $\mathfrak{N}^*$ das System aller Mengen, die Teil einer Nullmenge für φ sind. Dann gilt:

VI. - *Aus $X \varepsilon \mathfrak{N}^*$ und $Y \subseteqq X$ folgt $Y \varepsilon \mathfrak{N}^*$.*

Nun zeigen wir:

VII. - *Das System $\mathfrak{N}^*$ ist ein σ-Körper.*

Wir haben zu zeigen: ist $A \varepsilon \mathfrak{N}^*$, $B \varepsilon \mathfrak{N}^*$, so ist auch $A - B \varepsilon \mathfrak{N}^*$; ist $A_n \varepsilon \mathfrak{N}^*$, so ist auch $\underset{n}{S} A_n \varepsilon \mathfrak{N}^*$. Wegen $A - B \subseteqq A$ folgt nach VI aus $A \varepsilon \mathfrak{N}^*$ auch $A - B \varepsilon \mathfrak{N}^*$. Ist $A_n \varepsilon \mathfrak{N}^*$, so gibt es ein $N_n \varepsilon \mathfrak{N}$, sodass $A_n \subseteqq N_n$; nach IV ist $\underset{n}{S} N_n \varepsilon \mathfrak{N}$; wegen $\underset{n}{S} A_n \subseteqq \underset{n}{S} N_n$ ist also $\underset{n}{S} A_n \varepsilon \mathfrak{N}^*$.

Wir bezeichnen nun mit $\mathfrak{M}_0$ das System aller Mengen X, die die Gestalt haben:

$$X = (A + N') - N'', \qquad (A \varepsilon \mathfrak{M},\ N' \varepsilon \mathfrak{N}^*,\ N'' \varepsilon \mathfrak{N}^*). \tag{2}$$

Da $A \varepsilon \mathfrak{N}^*$, ist $\mathfrak{M} \subseteqq \mathfrak{M}_0$; da $A \varepsilon \mathfrak{M}$, ist $\mathfrak{N}^* \subseteqq \mathfrak{M}_0$.

VIII. - *Das System $\mathfrak{M}_0$ ist ein σ-Körper.*

Sei $B_1 \varepsilon \mathfrak{M}_0$, $B_2 \varepsilon \mathfrak{M}_0$; dann ist $B_1 = (A_1 + N_1') - N_1''$, $B_2 = (A_2 + N_2') - N_2''$, wo $A_1 \varepsilon \mathfrak{M}$, $A_2 \varepsilon \mathfrak{M}$, $N_1' \varepsilon \mathfrak{N}^*$, $N_1'' \varepsilon \mathfrak{N}^*$, $N_2' \varepsilon \mathfrak{N}^*$, $N_2'' \varepsilon \mathfrak{N}^*$; man bestätigt leicht, dass $B_1 - B_2 = ((A_1 - A_2) + N') - N''$, wo $N' \subseteqq N_1' + N_2''$, $N'' \subseteqq N_1'' + N_2'$; da $\mathfrak{M}$ ein Körper, ist $A_1 - A_2 \varepsilon \mathfrak{M}$; da $\mathfrak{N}^*$ ein Körper, ist $N_1' + N_2'' \varepsilon \mathfrak{N}^*$, $N_1'' + N_2' \varepsilon \mathfrak{N}^*$, also nach VI auch $N' \varepsilon \mathfrak{N}^*$, $N'' \varepsilon \mathfrak{N}^*$; $B_1 - B_2$ hat also die Gestalt (2), d. h. es ist $B_1 - B_2 \varepsilon \mathfrak{M}_0$. - Sei $B_n \varepsilon \mathfrak{M}_0$; dann ist $B_n = (A_n + N_n') - N_n''$, wo $A_n \varepsilon \mathfrak{M}$, $N_n' \varepsilon \mathfrak{N}^*$,

$N_n'' \varepsilon \mathfrak{N}^*$; daraus folgt: $\underset{n}{S} B_n = (\underset{n}{S} A_n + N') - N''$, wo $N' \subseteqq \underset{n}{S} N_n'$, $N'' \subseteqq \underset{n}{S} N_n''$; da $\mathfrak{M}$ ein σ-Körper, ist $\underset{n}{S} A_n \varepsilon \mathfrak{M}$; da $\mathfrak{N}^*$ ein σ-Körper, ist $\underset{n}{S} N_n' \varepsilon \mathfrak{N}^*$, $\underset{n}{S} N_n'' \varepsilon \mathfrak{N}^*$, also nach VI auch $N' \varepsilon \mathfrak{N}^*$, $N'' \varepsilon \mathfrak{N}^*$; $\underset{n}{S} B_n$ hat also die Gestalt (2) d. h. $\underset{n}{S} B_n \varepsilon \mathfrak{M}_0$.

Nun wollen wir die Definition von φ auf $\mathfrak{M}_0$ erweitern. Sei $X \varepsilon \mathfrak{M}_0$; dann gilt für X die Darstellung (2); doch wird es für dieselbe Menge $X \varepsilon \mathfrak{M}_0$ verschiedene Darstellungen dieser Form geben; sei:

$$X = (A_1 + N_1') - N_1'', \qquad (A_1 \varepsilon \mathfrak{M},\ N_1' \varepsilon \mathfrak{N}^*,\ N_1'' \varepsilon \mathfrak{N}^*)$$

eine zweite solche Darstellung; wir behaupten, *dass dann* $\varphi(A_1) = \varphi(A)$ *ist.* In der Tat, offenbar ist $A - A_1 \subseteqq N_1' + N''$, $A_1 - A \subseteqq N' + N_1''$; da $\mathfrak{N}^*$ ein Körper, ist $N_1' + N'' \varepsilon \mathfrak{N}^*$, $N' + N_1'' \varepsilon \mathfrak{N}^*$; es gibt also eine Menge $M_1 \varepsilon \mathfrak{N}$ und eine Menge $M_2 \varepsilon \mathfrak{N}$, so dass $N_1' + N'' \subseteqq M_1$, $N' + N_1'' \subseteqq M_2$; dann ist auch $A - A_1 \subseteqq M_1$, $A_1 - A \subseteqq M_2$; da $\mathfrak{M}$ ein Körper, ist $A - A_1 \varepsilon \mathfrak{M}$, $A_1 - A \varepsilon \mathfrak{M}$; wegen $\varphi(M_1) = 0$, $\varphi(M_2) = 0$ ist nach I auch $\varphi(A - A_1) = 0$, $\varphi(A_1 - A) = 0$. Aus $A = AA_1 + (A - A_1)$, $A_1 = AA_1 + (A_1 - A)$ folgt also wegen der Additivität von φ weiter: $\varphi(A) = \varphi(AA_1)$, $\varphi(A_1) = \varphi(AA_1)$, mithin $\varphi(A) = \varphi(A_1)$, wie behauptet.

Ist $X \varepsilon \mathfrak{M}_0$, so hat also für alle Darstellungen der Menge X in der Gestalt (2) $\varphi(A)$ denselben Wert; wir können also $\varphi(X)$ für alle $X \varepsilon \mathfrak{M}_0$ definieren durch die Festsetzung:

$$(3) \quad \varphi(X) = \varphi(A), \quad \text{wenn} \quad X = (A + N') - N'', \qquad (A \varepsilon \mathfrak{M},\ N' \varepsilon \mathfrak{N}^*,\ N'' \varepsilon \mathfrak{N}^*).$$

Ist insbesondere $X \varepsilon \mathfrak{M}$, so kann man hierin $A = X$, $N' = \Lambda$, $N'' = \Lambda$ setzen, und der durch (3) gelieferte Funktionswert $\varphi(X)$ stimmt mit dem ursprünglich in $\mathfrak{M}$ gegebenen Funktionswerte $\varphi(X)$ überein. Für $X \varepsilon \mathfrak{N}^*$ erhält man aus (3) (indem man $A = \Lambda$, $N' = N$, $N'' = \Lambda$ setzt): $\varphi(X) = 0$.

IX. - *Die durch* (3) *in* $\mathfrak{M}_0$ *definierte Mengenfunktion* φ *ist total-additiv.*

Sei $X = S X_n$, wo die X_n disjunkte Mengen aus $\mathfrak{M}_0$ bedeuten; wir haben zu zeigen, dass:

$$(4) \qquad \varphi(X) = \sum_n \varphi(X_n).$$

Wegen $X_n \varepsilon \mathfrak{M}_0$ ist $X_n = (A_n + N_n') - N_n''$, wo $A_n \varepsilon \mathfrak{M}$, $N_n' \varepsilon \mathfrak{N}^*$, $N_n'' \varepsilon \mathfrak{N}^*$; dann ist, wie wir beim Beweise von VIII sahen: $X = (A + N') - N''$, wo $A = \underset{n}{S} A_n$ (also $A \varepsilon \mathfrak{M}$), $N' \varepsilon \mathfrak{N}^*$, $N'' \varepsilon \mathfrak{N}^*$; nach (3) ist also: $\varphi(X) = \varphi(A)$, und ebenso $\varphi(X_n) = \varphi(A_n)$, sodass sich (4) reduziert auf:

$$\varphi(A) = \sum_n \varphi(A_n).$$

Sei $A_1' = A_1$, $A_n' = A_n - (A_1 + \dots + A_{n-1})$ $(n > 1)$; dann ist auch $A = \underset{n}{S} A_n'$, $A_n' \varepsilon \mathfrak{M}$, und die A_n' sind disjunkt; da φ total-additiv in $\mathfrak{R}$, ist $\varphi(A)=\sum_n \varphi(A_n')$. Es ist also nur mehr zu zeigen: $\varphi(A_n')=\varphi(A_n)$. Nun ist: $A_n-A_n'=A_nA_1+\dots+A_nA_{n-1}$, und da die X_n disjunkt sind, ist $A_nA_i \subseteqq N_n''+N_i''$ $(i \neq n)$, also $A_n-A_n' \subseteqq N_1''+\dots+N_n''$, also $A_n-A_n' \varepsilon \mathfrak{N}^*$, also $\varphi(A_n-A_n')=0$; da $A_n-A_n' \varepsilon \mathfrak{R}$ und φ in $\mathfrak{R}$ additiv, ist also $\varphi(A_n)=\varphi(A_n')+\varphi(A_n-A_n')=\varphi(A_n')$, w. z. b. w.

X. - *Der σ-Körper* $\mathfrak{R}_0$ *ist vollständig für* φ.

Wir haben zu zeigen: Aus $X \varepsilon \mathfrak{R}_0$, $\varphi(X)=0$, $Y \subseteqq X$ folgt $Y \varepsilon \mathfrak{R}_0$. Aus (3) folgt: $X=(A+N')-N''$, wo $A \varepsilon \mathfrak{R}$, $N' \varepsilon \mathfrak{N}^*$, $N'' \varepsilon \mathfrak{N}^*$ und $\varphi(A)=0$; also ist $A \varepsilon \mathfrak{N}$, mithin auch $A \varepsilon \mathfrak{N}^*$; da $\mathfrak{N}^*$ ein Körper, ist auch $X \varepsilon \mathfrak{N}^*$, nach VI also auch $Y \varepsilon \mathfrak{N}^*$, und wegen $\mathfrak{N}^* \subseteqq \mathfrak{R}_0$ ist $Y \varepsilon \mathfrak{R}_0$.

Zusammenfassend haben wir gefunden:

XI. - *Ist* φ *total-additiv im σ-Körper* $\mathfrak{R}$, *so ist* $\mathfrak{R}_0$ *ein σ-Körper* $\geqq \mathfrak{R}$, *und die Definition von* φ *kann so auf* $\mathfrak{R}_0$ *erweitert werden, dass* φ *total-additiv in* $\mathfrak{R}_0$ *und* $\mathfrak{R}_0$ *vollständig für* φ.

Nun wollen wir noch zeigen, daß $\mathfrak{R}_0$ der kleinste $\mathfrak{R}$ umfassende, für φ vollständige Körper ist; präziser gesprochen:

XII. - *Ist* $\mathfrak{R}^*$ *ein Körper* $\geqq \mathfrak{R}$, *ist* ψ *eine in* $\mathfrak{R}^*$ *definierte total-additive Mengenfunktion, die in* $\mathfrak{R}$ *mit* φ *übereinstimmt, und ist* $\mathfrak{R}^*$ *vollständig für* ψ, *so ist* $\mathfrak{R}^* \geqq \mathfrak{R}_0$ *und* ψ *stimmt in* $\mathfrak{R}_0$ *mit* φ *überein.*

Sei $X \varepsilon \mathfrak{R}_0$, d. h. $X=(A+N')-N''$, wo $A \varepsilon \mathfrak{R}$, $N' \varepsilon \mathfrak{N}^*$, $N'' \varepsilon \mathfrak{N}^*$, und wegen VI ohneweiters angenommen werden kann, N' sei fremd zu A und $N'' \subseteqq A$. Nach Definition von $\mathfrak{N}^*$ gibt es Mengen A', A'' aus $\mathfrak{N}$, sodass $N' \subseteqq A'$, $N'' \subseteqq A''$. Da $\mathfrak{N} \subseteqq \mathfrak{R} \subseteqq \mathfrak{R}^*$, ist $A' \varepsilon \mathfrak{R}^*$, $A'' \varepsilon \mathfrak{R}^*$; aus $\varphi(A')=0$, $\varphi(A'')=0$ folgt wegen $A' \varepsilon \mathfrak{R}$, $A'' \varepsilon \mathfrak{R}$ nach Voraussetzung: $\psi(A')=0$, $\psi(A'')=0$, und weil $\mathfrak{R}^*$ vollständig für ψ, folgt daraus weiter $N' \varepsilon \mathfrak{R}^*$, $N'' \varepsilon \mathfrak{R}^*$; da wegen $\mathfrak{R} \subseteqq \mathfrak{R}^*$ auch $A \varepsilon \mathfrak{R}^*$, und da $\mathfrak{R}^*$ ein Körper, ist auch $X \varepsilon \mathfrak{R}^*$. Damit ist gezeigt, dass $\mathfrak{R}_0 \subseteqq \mathfrak{R}^*$. Nach I folgt aus $\psi(A')=0$, $\psi(A'')=0$ auch $\psi(N')=0$, $\psi(N'')=0$. Also ist $\psi(X)=\psi(A)+\psi(N')-\psi(N'')=\psi(A)$; wegen $A \varepsilon \mathfrak{R}$ ist also nach Voraussetzung $\psi(X)=\varphi(A)$, also nach (3) auch $\psi(X)=\varphi(X)$, wie behauptet.

§ 2. - Erweiterung eines Körpers zu einem vollständigen σ-Körper.

Nach C. Carathéodory nennen wir eine im System $\mathfrak{E}$ aller Punktmengen des Raumes E definierte Mengenfunktion ψ eine *Maßfunktion*, wenn sie folgende Eigenschaften hat:

1.) $\psi(A)=0$;

2.) Aus $A \subseteqq B$ folgt $\psi(A) \leqq \psi(B)$ (d. h. ψ ist monoton wachsend);

3.) Aus $A=\underset{n}{S} A_n$ folgt $\psi(A) \leqq \sum_n \psi(A_n)$.

Aus 1.) und 2.) folgt: $\psi(X) \geqq 0$ für alle $X \varepsilon \mathfrak{E}$.

Sei nun $\mathfrak{R}$ ein Körper $\subseteqq \mathfrak{E}$ und φ eine in $\mathfrak{R}$ definierte, *total-additive* Mengenfunktion. Wir wollen zeigen, dass φ zu einer in $\mathfrak{E}$ definierten Maßfunktion erweitert werden kann.

Wir bezeichnen mit $\mathfrak{R}_\sigma$ (bzw. $\mathfrak{R}_\delta$) das Mengensystem, das entsteht, indem man zu $\mathfrak{R}$ alle Mengen hinzufügt, die Summe (bzw. Durchschnitt) einer Mengenfolge aus $\mathfrak{R}$ sind, und mit $\mathfrak{R}_{\sigma\delta}$ (bzw. $\mathfrak{R}_{\delta\sigma}$) das Mengensystem, das entsteht, indem man zu $\mathfrak{R}_\sigma$ (bzw. zu $\mathfrak{R}_\delta$) alle Mengen hinzufügt, die Durchschnitt (bzw. Summe) einer Mengenfolge aus $\mathfrak{R}_\sigma$ (bzw. aus $\mathfrak{R}_\delta$) sind. Die Mengensysteme $\mathfrak{R}_\sigma$, $\mathfrak{R}_\delta$, $\mathfrak{R}_{\sigma\delta}$, $\mathfrak{R}_{\delta\sigma}$ sind dann i. a. keine Körper, wohl aber *Ringe*, d. h. gehören A und B zu einem dieser Systeme, so gehören auch $A+B$ und AB zu diesem System.

Wir erweitern zunächst die Definition von φ auf $\mathfrak{R}_\sigma$ durch die Festzetzung: ist $B\,\varepsilon\,\mathfrak{R}_\sigma-\mathfrak{R}$, so sei $\varphi(B)$ das Supremum der Funktionswerte $\varphi(X)$ für alle zu $\mathfrak{R}$ gehörigen (5) $X\subseteqq B$:

$$\varphi(B)=\sup\varphi(X),\qquad (X\,\varepsilon\,\mathfrak{R},\ X\subseteqq B). \tag{5}$$

Ist $B\,\varepsilon\,\mathfrak{R}$, so gilt, wie aus I unmittelbar folgt, gleichfalls (5), sodass φ durch (5) für alle $B\,\varepsilon\,\mathfrak{R}_\sigma$ gegeben ist.

XIII. - *Die in* $\mathfrak{R}_\sigma$ *definierte Mengenfunktion* φ *hat die Eigenschaften* 1), 2), 3).

Für 1.) und 2.) ist dies evident. Sei, um auch 3.) zu beweisen: $B_n\,\varepsilon\,\mathfrak{R}_\sigma$, $B=\underset{n}{S}B_n$, also auch $B\,\varepsilon\,\mathfrak{R}_\sigma$; dann ist $B_n=\underset{\nu}{S}B_{n\nu}$, wo $B_{n\nu}\,\varepsilon\,\mathfrak{R}$; also ist $B=\underset{n,\nu}{S}B_{n\nu}$. Nach (1) erhalten wir daraus eine Darstellung $B=\underset{i}{S}C_i$, wo die C_i disjunkte Mengen aus $\mathfrak{R}$ und jede Menge C_i Teil einer Menge $B_{n\nu}$, also auch Teil einer Menge B_n. Sei nun $X\,\varepsilon\,\mathfrak{R}$, $X\subseteqq B$; dann ist $X=\underset{i}{S}C_iX$, wo die C_iX disjunkte Mengen aus $\mathfrak{R}$; weil φ total-additiv in $\mathfrak{R}$, ist also $\varphi(X)=\sum_i\varphi(C_iX)$. Wegen $C_iX\subseteqq C_i$ ist jede Menge C_iX Teil einer Menge B_n; sind $X_{n1}, X_{n2},\ldots, X_{nj},\ldots$ die sämtlichen Mengen C_iX, die $\subseteqq B_n$ sind, so ist, da $X_{n1}+\ldots+X_{nj}\,\varepsilon\,\mathfrak{R}$ und $\varphi(X_{n1}+\ldots+X_{nj})=$ $=\varphi(X_{n1})+\ldots+\varphi(X_{nj})$, wegen (5): $\varphi(X_{n1})+\ldots+\varphi(X_{nj})\leqq\varphi(B_n)$, also auch $\sum_j\varphi(X_{nj})\leqq\varphi(B_n)$, also auch: $\varphi(X)=\sum_i\varphi(C_iX)\leqq\sum_n\varphi(B_n)$, also nach (5) auch $\varphi(B)\leqq\sum_n\varphi(B_n)$, w. z. b. w.

Nun erweitern wir die Definition von φ auf $\mathfrak{E}$ durch die Festsetzung: ist $A\,\varepsilon\,\mathfrak{E}-\mathfrak{R}_\sigma$ so sei $\varphi(A)$ das Infimum der Funktionswerte $\varphi(X)$ für alle zu $\mathfrak{R}_\sigma$ gehörigen (6) $X\supseteqq A$:

$$\varphi(A)=\inf\varphi(X),\qquad (X\,\varepsilon\,\mathfrak{R}_\sigma,\ X\supseteqq A). \tag{6}$$

(5) Zumindest die leere Menge Λ ist eine zu $\mathfrak{R}$ gehörige Menge $X\subseteqq B$.

(6) Gibt es kein zu $\mathfrak{R}_\sigma$ gehöriges $X\supseteqq A$, so setzen wir $\varphi(A)=+\infty$.

Ist $A \varepsilon \mathfrak{R}_\sigma$, so gilt, wie aus I unmittelbar folgt, gleichfalle (6), sodass φ durch (6) für alle $A \varepsilon \mathfrak{E}$ gegeben ist.

XIV. - *Die in $\mathfrak{E}$ definierte Mengenfunktion φ ist eine Maßfunktion.*

Dass φ die Eigenschaften 1.) und 2.) hat, ist evident. Sei, um auch 3.) zu beweisen: $A_n \varepsilon \mathfrak{E}$ und $A = \underset{n}{S} A_n$; wir haben zu zeigen:

(7) $$\varphi(A) \leqq \sum_n \varphi(A_n).$$

Das ist sicher richtig, wenn ein $\varphi(A_n) = +\infty$ ist; wir nehmen also an, alle $\varphi(A_n)$ seien endlich. Dann gibt es nach (6) zu jedem $\varepsilon > 0$ ein $B_n \varepsilon \mathfrak{R}_\sigma$, so dass $B_n \supseteqq A_n$ und $\varphi(B_n) < \varphi(A_n) + \frac{\varepsilon}{2^n}$. Setzen wir $B = \underset{n}{S} B_n$, so ist $B \varepsilon \mathfrak{R}_\sigma$, $B \supseteqq A$ und wegen XIII: $\varphi(B) \leqq \sum_n \varphi(B_n) < \sum_n \varphi(A_n) + \varepsilon$. Und da dies für jedes $\varepsilon > 0$ gilt, folgt (7) aus (6).

Ist φ eine im Körper $\mathfrak{R}$ definierte, total-additive Mengenfunktion, so nennen wir die durch (5) und (6) definierte Maßfunktion: *die zu φ gehörige Maßfunktion.*

Nach CARATHÉODORY nennen wir eine Menge $M \varepsilon \mathfrak{E}$ φ-messbar, wenn für *jede* Menge $A \varepsilon \mathfrak{E}$ gilt:

(8) $$\varphi(A) = \varphi(AM) + \varphi(A - M).$$

Das System aller φ-messbaren Mengen bildet einen σ-Körper, der insbesondere alle Mengen $N \varepsilon \mathfrak{E}$ enthält, für die $\varphi(N) = 0$ ist (die Nullmengen für φ). In diesem σ-Körper ist φ total-additiv (7). Wir bezeichnen den σ-Körper der φ-messbaren Mengen mit $\overline{\mathfrak{R}}$ und nennen ihn die *vollständige Hülle von $\mathfrak{R}$ für φ*. Dieser Name wird gerechtfertigt durch die folgenden Sätze.

XV. - *$\overline{\mathfrak{R}}$ ist vollständig für φ.*

Sei $N \varepsilon \overline{\mathfrak{R}}$ und $\varphi(N) = 0$; ist $X \subseteqq N$, so ist wegen Eigenschaft 2.) der Maßfunktionen auch $\varphi(X) = 0$, also ist, wie eben bemerkt, auch $X \varepsilon \overline{\mathfrak{R}}$.

Um zu zeigen, dass $\overline{\mathfrak{R}} \supseteqq \mathfrak{R}$, zeigen wir zunächst:

XVI. - *Die Mengenfunktion φ ist additiv in $\mathfrak{R}_\sigma$.*

Seien B', B'' disjunkte Mengen aus $\mathfrak{R}_\sigma$ und $B = B' + B''$; wir haben zu zeigen:

$$\varphi(B) = \varphi(B') + \varphi(B'').$$

Nach Eigenschaft 3.) der Maßfunktionen ist:

$$\varphi(B) \leqq \varphi(B') + \varphi(B'').$$

Es ist also nur mehr zu zeigen:

(9) $$\varphi(B) \geqq \varphi(B') + \varphi(B'').$$

---

(7) Vgl. z. B. Th. R. F. S. 424-430.

Das ist sicher richtig, wenn $\varphi(B)=+\infty$; wir nehmen also an, $\varphi(B)$ sei endlich; dann ist auch $\varphi(B')$ und $\varphi(B'')$ endlich. Nach (5) gibt es zu jedem $\varepsilon>0$ zu $\mathfrak{R}$ gehörige Teile X', X'' von B' bzw. B'', so dass: $\varphi(X')>\varphi(B')-\varepsilon$, $\varphi(X'')>\varphi(B'')-\varepsilon$. Dann ist $X'+X''\varepsilon\,\mathfrak{R}$, $X'+X''\subseteqq B$, und weil X', X'' disjunkt und φ in $\mathfrak{R}$ additiv:

$$\varphi(X'+X'')=\varphi(X')+\varphi(X'')>\varphi(B')+\varphi(B'')-2\varepsilon.$$

Nach (5) ist also $\varphi(B)>\varphi(B')+\varphi(B'')-2\varepsilon$, und da dies für jedes $\varepsilon>0$ gilt, ist (9) bewiesen.

XVII. - *Die Mengen aus $\mathfrak{R}$ sind φ-messbar, d. h. $\overline{\mathfrak{R}}\geqq\mathfrak{R}$.*

Wir haben zu zeigen, dass (8) für jede Menge $M\varepsilon\,\mathfrak{R}$ gilt. Da wegen Eigenschaft 3.) der Maßfunktionen jedenfalls

$$\varphi(A)\leqq\varphi(AM)+\varphi(A-M)$$

ist, ist nur zu zeigen:

(10) $$\varphi(A)\geqq\varphi(AM)+\varphi(A-M).$$

Das ist sicher richtig, wenn $\varphi(A)=+\infty$; wir nehmen also an, $\varphi(A)$ sei endlich; dann gibt es nach (6) zu jedem $\varepsilon>0$ ein $B\varepsilon\,\mathfrak{R}_\sigma$, so dass $B\geqq A$ und $\varphi(B)<\varphi(A)+\varepsilon$. Wegen $M\varepsilon\,\mathfrak{R}$, $B\varepsilon\,\mathfrak{R}_\sigma$, ist auch: $BM\varepsilon\,\mathfrak{R}_\sigma$, $B-M\varepsilon\,\mathfrak{R}_\sigma$. Aus $B=BM+(B-M)$ folgt also nach XVI:

$$\varphi(B)=\varphi(BM)+\varphi(B-M).$$

Wegen $AM\subseteqq BM$, $A-M\subseteqq B-M$ folgt daraus nach Eigenschaft 2.) der Maßfunktionen

$$\varphi(AM)+\varphi(A-M)\leqq\varphi(B)<\varphi(A)+\varepsilon,$$

und da dies für jedes $\varepsilon>0$ gilt, ist (10) bewiesen.

Wir fassen zusammen:

XVIII. - *Jede im Körper $\mathfrak{R}\subseteqq\mathfrak{E}$ definierte, total-additive Mengenfunktion φ kann erweitert werden zu einer in $\mathfrak{E}$ definierten Maßfunktion φ, die total-additiv ist in dem $\mathfrak{R}$ enthaltenden, für φ vollständigen σ-Körper $\overline{\mathfrak{R}}$.*

Seien $\mathfrak{R}$ und $\mathfrak{L}$ Körper aus $\mathfrak{E}$, und seien φ und ψ in $\mathfrak{R}$ bzw. $\mathfrak{L}$ definierte, total-additive Mengenfunktionen; wir bezeichnen auch die zu φ bzw. ψ gehörige Maßfunktion mit φ bzw. ψ; $\overline{\mathfrak{R}}$ sei die vollständige Hülle von $\mathfrak{R}$ für φ. Dann gilt:

XIX. - *Ist $\mathfrak{R}\subseteqq\mathfrak{L}\subseteqq\overline{\mathfrak{R}}$ und ist $\psi(X)=\varphi(X)$ für alle $X\varepsilon\,\mathfrak{L}$, so ist auch $\psi(A)=\varphi(A)$ für alle $A\,\varepsilon\,\mathfrak{E}$.*

Sei $B\varepsilon\,\mathfrak{L}_\sigma$; bei Beachtung von (1) kann B dargestellt werden in der Form: $B=\underset{n}{S}B_n$, wo die B_n disjunkte Mengen aus $\mathfrak{L}$; nach (5) ist $\psi(B)\geqq\psi(B_1+\ldots+B_n)=$ $=\psi(B_1)+\ldots+\psi(B_n)$, also $\psi(B)\geqq\sum\limits_n\psi(B_n)$; wegen Eigenschaft (3) der Maßfunktionen ist auch $\psi(B)\leqq\sum\limits_n\psi(B_n)$; also ist $\psi(B)=\sum\limits_n\psi(B_n)$, also wegen $B_n\,\varepsilon\,\mathfrak{L}$ auch $\psi(B)=\sum\limits_n\varphi(B_n)$; da $\mathfrak{L}\subseteqq\overline{\mathfrak{R}}$ und $\overline{\mathfrak{R}}$ ein σ-Körper, ist auch $\mathfrak{L}_\sigma\subseteqq\overline{\mathfrak{R}}$, also

$B_n \varepsilon \overline{\mathfrak{M}}$, $B \varepsilon \overline{\mathfrak{M}}$; und da φ nach XVIII total-additiv in $\overline{\mathfrak{M}}$, ist $\varphi(B)=\sum\limits_n \varphi(B_n)$, also $\psi(B)=\varphi(B)$. - Sei nun $A \varepsilon \mathfrak{E}$. Nach (6) ist $\varphi(A)=\inf \varphi(X)$ $(X \varepsilon \mathfrak{M}_\sigma, X \supseteqq A)$ $\psi(A)=\inf \psi(Y)$ $(Y \varepsilon \mathfrak{L}_\sigma, Y \supseteqq A)$; wegen $\mathfrak{M} \subseteqq \mathfrak{L}$ ist $\mathfrak{M}_\sigma \subseteqq \mathfrak{L}_\sigma$, also, wie eben bewiesen, $\psi(X)=\varphi(X)$ für alle $X \varepsilon \mathfrak{M}_\sigma$; alle $\varphi(X)$, deren Infimum $\varphi(A)$ ist, kommen also unter den $\psi(Y)$ vor, deren Infimum $\psi(A)$ ist; also ist $\psi(A) \leqq \varphi(A)$. Anderseits ist, wie schon gezeigt, $\psi(Y)=\varphi(Y)$ für alle $Y \varepsilon \mathfrak{L}_\sigma$; also wegen Eigenschaft 2.) der Maßfunktionen: $\varphi(A) \leqq \psi(Y)$ für alle $Y \varepsilon \mathfrak{L}_\sigma$, $Y \supseteqq A$; also auch $\varphi(A) \leqq \inf \psi(Y)$ $(Y \varepsilon \mathfrak{L}_\sigma, Y \supseteqq A)$, d. h. $\varphi(A) \leqq \psi(A)$. Mithin ist $\psi(A)=\varphi(A)$, wie behauptet.

Ist $\overline{\mathfrak{L}}$ die vollständige Hülle von $\mathfrak{L}$ für ψ, so folgt aus XIX:

XX. - *Ist* $\mathfrak{M} \subseteqq \mathfrak{L} \subseteqq \overline{\mathfrak{M}}$ *und ist* $\psi(X)=\varphi(X)$ *für alle* $X \varepsilon \mathfrak{L}$, *so ist* $\overline{\mathfrak{L}}=\overline{\mathfrak{M}}$.

Für den Rest dieses Paragrafen nehmen wir nun an, der Raum E habe die Gestalt:

(11) $$E=\underset{\nu}{S} E_\nu, \qquad E_\nu \varepsilon \mathfrak{M}, \qquad \varphi(E_\nu) \text{ endlich.}$$

Machen wir Gebrauch von der Umformung (1), so sehen wir, dass die E_ν ohneweiters als disjunkt angenommen werden können.

XXI. - *Gilt* (11), *so gibt es zu jeder Menge* $X \varepsilon \overline{\mathfrak{M}}$ *eine Menge* $A \varepsilon \mathfrak{M}_{\sigma\delta}$, *so dass* $A \supseteqq X$ *und* $\varphi(A-X)=0$.

Nach (11) ist $X=\underset{\nu}{S} XE_\nu$; da nach XVII: $\mathfrak{M} \subseteqq \overline{\mathfrak{M}}$ und $\overline{\mathfrak{M}}$ ein Körper, ist $XE_\nu \varepsilon \overline{\mathfrak{M}}$, und da $\varphi(E_\nu)$ endlich, ist auch $\varphi(XE_\nu)$ endlich. Nach (6) gibt es ein $A_{n\nu} \varepsilon \mathfrak{M}_\sigma$, so daß $A_{n\nu} \supseteqq XE_\nu$, $\varphi(A_{n\nu})<\varphi(XE_\nu)+\frac{1}{n \cdot 2^\nu}$; da $\mathfrak{M} \subseteqq \overline{\mathfrak{M}}$ und $\overline{\mathfrak{M}}$ ein σ-Körper, ist auch $\mathfrak{M}_\sigma \subseteqq \overline{\mathfrak{M}}$, also $A_{n\nu} \varepsilon \overline{\mathfrak{M}}$; da φ additiv in $\overline{\mathfrak{M}}$, ist $\varphi(A_{n\nu}-XE_\nu)=\varphi(A_{n\nu})-\varphi(XE_\nu)<\frac{1}{n \cdot 2^\nu}$. Setzen wir $A_n=\underset{\nu}{S} A_{n\nu}$, so ist $A_n \varepsilon \mathfrak{M}_\sigma$, $A_n \supseteqq X$, und wegen $A_n-X \subseteqq \underset{\nu}{S}(A_{n\nu}-XE_\nu)$ ist nach Eigenschaft (3) der Maßfunktionen $\varphi(A_n-X) \leqq \sum\limits_\nu \varphi(A_{n\nu}-XE_\nu)<\frac{1}{n}$. Setzen wir $A=\underset{n}{D} A_n$, so ist $A \varepsilon \mathfrak{M}_{\sigma\delta}$, $A \supseteqq X$, und wegen $A-X \subseteqq A_n-X$ ist $\varphi(A-X)<\frac{1}{n}$ für alle n, also $\varphi(A-X)=0$.

XXII. - *Gilt* (11), *so gibt es zu jeder Menge* $X \varepsilon \overline{\mathfrak{M}}$ *eine Menge* $C \varepsilon \mathfrak{M}_{\delta\sigma}$, *sodass* $C \subseteqq X$ *und* $\varphi(X-C)=0$.

Wir nehmen zunächst an, $\varphi(X)$ *sei endlich*. Dann gibt es nach (6) ein $A_n \varepsilon \mathfrak{M}_\sigma$, so dass $A_n \supseteqq X$, $\varphi(A_n)<\varphi(X)+\frac{1}{n}$, also $\varphi(A_n-X)<\frac{1}{n}$. Wieder nach (6) gibt es ein $B_n \varepsilon \mathfrak{M}_\sigma$, so dass $B_n \supseteqq A_n-X$, $\varphi(B_n)<\frac{1}{n}$. Dann ist $A_n-B_n \subseteqq X$ und:

(12) $$\varphi(A_n-B_n) \geqq \varphi(A_n)-\varphi(B_n)>\varphi(A_n)-\frac{1}{n}>\varphi(X)-\frac{1}{n}.$$

Da $A_n \varepsilon \mathfrak{M}_\sigma$, $B_n \varepsilon \mathfrak{M}_\sigma$, ist $A_n=\underset{i}{S} A_{ni}$ $(A_{ni} \varepsilon \mathfrak{M})$, $B_n=\underset{j}{S} B_{nj}$ $(B_{nj} \varepsilon \mathfrak{M})$, also $A_n-B_n=\underset{i}{S}(A_{ni}-B_n)=\underset{i}{S}\underset{j}{D}(A_{ni}-B_{nj})$; da $\mathfrak{M}$ ein Körper, ist hierin $A_{ni}-B_{nj} \varepsilon \mathfrak{M}$, also

$A_n - B_n \varepsilon \mathfrak{R}_{\delta\sigma}$. Setzen wir $A_n - B_n = C_n$, so haben wir aus (12): es gibt ein $C_n \varepsilon \mathfrak{R}_{\delta\sigma}$, so dass $C_n \subseteqq X$, $\varphi(C_n) > \varphi(X) - \frac{1}{n}$. Setzen wir $C = \underset{n}{S}\, C_n$, so ist auch $C \varepsilon \mathfrak{R}_{\delta\sigma}$, $C \subseteqq X$, und $\varphi(X) - \frac{1}{n} < \varphi(C) \leqq \varphi(X)$ für alle n, also $\varphi(C) = \varphi(X)$, also $\varphi(X - C) = 0$. - *Sei sodann* $\varphi(X) = +\infty$. Nach (11) ist $X = \underset{\nu}{S}\, XE_\nu$, wo $\varphi(XE_\nu)$ endlich; es gibt also, wie eben gezeigt, ein $C_\nu \varepsilon \mathfrak{R}_{\delta\sigma}$, so dass $C_\nu \subseteqq XE_\nu$, $\varphi(XE_\nu - C_\nu) = 0$. Setzen wir $C = \underset{\nu}{S}\, C_\nu$, so ist auch $C \varepsilon \mathfrak{R}_{\delta\sigma}$, $C \subseteqq X$, und wegen $X - C \subseteqq \underset{\nu}{S}(XE_\nu - C_\nu)$ ist $\varphi(X - C) \leqq \sum\limits_\nu \varphi(XE_\nu - C_\nu) = 0$, also $\varphi(X - C) = 0$.

Nun zeigen wir, immer unter der Voraussetzung (11), dass $\overline{\mathfrak{R}}$ *der kleinste* $\mathfrak{R}$ umfassende für φ vollständige σ-Körper ist; präziser gesprochen:

XXIII. - *Sei* $\mathfrak{S}$ *ein* σ*-Körper* $\subseteqq \mathfrak{E}$ *und* $\supseteqq \mathfrak{R}$; *sei* ψ *eine in* $\mathfrak{S}$ *definierte total-additive Mengenfunktion, sei* $\mathfrak{S}$ *vollständig für* ψ, *und sei* $\psi(X) = \varphi(X)$ *für alle* $X \varepsilon \mathfrak{R}$; *gilt dann* (11), *so ist* $\overline{\mathfrak{R}} \subseteqq \mathfrak{S}$ *und* $\psi(X) = \varphi(X)$ *für alle* $X \varepsilon \overline{\mathfrak{R}}$.

Aus $\mathfrak{R} \subseteqq \overline{\mathfrak{R}}$, $\mathfrak{R} \subseteqq \mathfrak{S}$ folgt, da $\overline{\mathfrak{R}}$ und $\mathfrak{S}$ σ-Körper sind:

$$\text{(13)} \qquad \mathfrak{R}_\sigma \subseteqq \overline{\mathfrak{R}}, \qquad \mathfrak{R}_{\sigma\delta} \subseteqq \overline{\mathfrak{R}}, \qquad \mathfrak{R}_\sigma \subseteqq \mathfrak{S}, \qquad \mathfrak{R}_{\sigma\delta} \subseteqq \mathfrak{S}.$$

Sei $X \varepsilon \mathfrak{R}_\sigma$; dann ist $X = \underset{n}{S}\, X_n$ mit $X_n \varepsilon \mathfrak{R}$, und indem man X_n ersetzt durch $X_1 + \ldots + X_n$, kann man annehmen $X_n \subseteqq X_{n+1}$. Da φ und ψ total-additiv in $\overline{\mathfrak{R}}$ bzw. $\mathfrak{S}$, folgt aus II: $\varphi(X) = \lim\limits_n \varphi(X_n)$, $\psi(X) = \lim\limits_n \psi(X_n)$; wegen $X_n \varepsilon \mathfrak{R}$ aber ist nach Voraussetzung $\varphi(X_n) = \psi(X_n)$, also ist $\psi(X) = \varphi(X)$ für alle $X \varepsilon \mathfrak{R}_\sigma$. - Sei sodann $X \varepsilon \mathfrak{R}_{\sigma\delta}$; dann ist $X = \underset{n}{D}\, X_n$ mit $X_n \varepsilon \mathfrak{R}_\sigma$, und indem man X_n ersetzt durch $X_1 X_2 \ldots X_n$, kann man annehmen $X_n \supseteqq X_{n+1}$. Nach (11) ist $X = \underset{\nu}{S}\, XE_\nu$, wo $XE_\nu \varepsilon \mathfrak{R}_{\sigma\delta}$ und die Summanden disjunkt angenommen werden können; wegen der totalen Additivität von φ und ψ ist also:

$$\text{(14)} \qquad \varphi(X) = \sum_\nu \varphi(XE_\nu), \qquad \psi(X) = \sum_\nu \psi(XE_\nu).$$

Hierin ist $XE_\nu = \underset{n}{D}\, X_n E_\nu$, wo $X_n E_\nu \varepsilon \mathfrak{R}_\sigma$, $X_n E_\nu \supseteqq X_{n+1} E_\nu$, und weil $\psi(E_\nu) = \varphi(E_\nu)$ endlich, sind auch $\psi(X_n E_\nu)$ und $\varphi(X_n E_\nu)$ endlich; nach III ist also $\varphi(XE_\nu) = \lim\limits_n \varphi(X_n E_\nu)$, $\psi(XE_\nu) = \lim\limits_n \psi(X_n E_\nu)$; da $X_n E_\nu \varepsilon \mathfrak{R}_\sigma$, ist, wie schon gezeigt, $\psi(X_n E_\nu) = \varphi(X_n E_\nu)$, also auch $\psi(XE_\nu) = \varphi(XE_\nu)$, also wegen (14) auch $\psi(X) = \varphi(X)$ für alle $X \varepsilon \mathfrak{R}_{\sigma\delta}$. - Sei sodann $X \varepsilon \overline{\mathfrak{R}}$ und $\varphi(X) = 0$; wir zeigen, dass dann auch $X \varepsilon \mathfrak{S}$ und $\psi(X) = 0$. Nach XXI gibt es ein $A \varepsilon \mathfrak{R}_{\sigma\delta}$, so dass $A \supseteqq X$ und $\varphi(A - X) = 0$, also auch $\varphi(A) = 0$. Wegen (13) ist $A \varepsilon \mathfrak{S}$, und wie eben gezeigt, ist auch $\psi(A) = 0$. Da $\mathfrak{S}$ vollständig für ψ, und da $X \subseteqq A$, ist also auch $X \varepsilon \mathfrak{S}$ und $\psi(X) = 0$. - Sei nun X eine beliebige Menge aus $\overline{\mathfrak{R}}$. Nach XXI gibt es ein $A \varepsilon \mathfrak{R}_{\sigma\delta}$, so dass

$A \supseteqq X$ und $\varphi(A-X)=0$; da nach (13) $A \varepsilon \overline{\mathfrak{R}}$, ist auch $A-X \varepsilon \overline{\mathfrak{R}}$, also, wie eben gezeigt, auch $A-X \varepsilon \mathfrak{S}$ und $\psi(A-X)=0$. Wegen $A \varepsilon \mathfrak{R}_{\sigma\delta}$ ist nach (13) $A \varepsilon \mathfrak{S}$ und, wie schon gezeigt, $\psi(A)=\varphi(A)$. Aus $X=A-(A-X)$ folgt nun wegen $A \varepsilon \mathfrak{S}$, $A-X \varepsilon \mathfrak{S}$ auch $X \varepsilon \mathfrak{S}$, und aus der Additivität von ψ folgt: $\psi(X)=\psi(A)-$ $-\psi(A-X)=\varphi(A)-\varphi(A-X)=\varphi(X)$, womit XXIII bewiesen ist.

Wir bezeichnen nun mit $\mathfrak{B}$ das System der Borelschen Mengen über $\mathfrak{R}$; da $\mathfrak{R}$ ein Körper, ist $\mathfrak{B}$ der kleinste σ-Körper über $\mathfrak{R}$ ([8]).

Da nach XVII $\overline{\mathfrak{R}}$ ein σ-Körper über $\mathfrak{R}$ ist, ist also $\mathfrak{B} \subseteq \overline{\mathfrak{R}}$; mithin ist φ total-additiv in $\mathfrak{B}$ und aus $\mathfrak{B}$ kann nach dem Verfahren von § 1 der kleinste für φ vollständige σ-Körper $\mathfrak{B}_0$ über $\mathfrak{B}$ gebildet werden. Wir zeigen:

XXIV. - *Gilt* (11) *so ist* $\overline{\mathfrak{R}}=\mathfrak{B}_0$.

Wie wir eben sahen, ist $\mathfrak{B} \subseteqq \overline{\mathfrak{R}}$. Da $\overline{\mathfrak{R}}$ nach XV ein für φ vollständiger σ-Körper und $\mathfrak{B}_0$ nach XII der kleinste für φ vollständige σ-Körper über $\mathfrak{B}$ ist, so ist $\mathfrak{B}_0 \subseteqq \overline{\mathfrak{R}}$. Da anderseits nach XXIII $\overline{\mathfrak{R}}$ der kleinste für φ vollständige σ-Körper über $\mathfrak{R}$ ist, muss auch $\overline{\mathfrak{R}} \subseteq \mathfrak{B}_0$ sein.

Wir können nun auch die Aussagen von Satz XIX und XX ein wenig verschärfen. Behalten wir die dort verwendeten Bezeichnungen bei, so erhalten wir:

XXV. - *Ist* $\mathfrak{R} \subseteqq \mathfrak{L} \subseteqq \overline{\mathfrak{R}}$, *ist* $\psi(X)=\varphi(X)$ *für alle* $X \varepsilon \mathfrak{R}$, *und gilt* (11), *so ist* $\psi(A)=\varphi(A)$ *für alle* $A \varepsilon \overline{\mathfrak{L}}$, *und mithin* $\overline{\mathfrak{L}}=\overline{\mathfrak{R}}$.

In der Tat, aus XXIII folgt für $\mathfrak{S}=\overline{\mathfrak{L}}$: es ist $\overline{\mathfrak{R}} \subseteq \overline{\mathfrak{L}}$ und $\psi(X)=\varphi(X)$ für alle $X \varepsilon \overline{\mathfrak{R}}$, also auch für alle $X \varepsilon \mathfrak{L}$. Die Behauptung folgt nun aus XIX und XX.

Die Theorie des k-dimensionalen Lebesgueschen Maßes ergibt sich als Spezialfall dieser Theorie. Man verstehe unter E den R_k (die Menge aller k-tupel $(x_1, x_2, \ldots, x_k)$ reeller Zahlen). Dem durch die Ungleichungen $a_i \leqq x_i < b_i$ $(i=1, 2, \ldots, k)$ gegebenen halboffenen Intervalle I des R_k ordne man den Funktionswert $\varphi(I)=(b_1-a_1)(b_2-a_2)\ldots(b_k-a_k)$ zu. Das System alle Punktmengen des R_k, die Summe endlich vieler solcher halboffener Intervalle sind, bildet einen Körper $\mathfrak{R}$, und zwar ist jede Menge $X \varepsilon \mathfrak{R}$ auch darstellbar in der Form $X=I_1+\ldots+I_n$, wo die I_i *disjunkte* halboffene Intervalle bedeuten. Setzt man $\varphi(X)=\varphi(I_1)+\ldots+$ $+\varphi(I_n)$, so ist φ total-additiv in $\mathfrak{R}$. Die zu φ gehörige Maßfunktion ist dann das k-dimensionale äussere Lebesguesche Maß, die vollständige Hülle $\overline{\mathfrak{R}}$ von $\mathfrak{R}$ ist das System der im Sinne von Lebesgue k-dimensional messbaren Mengen, und für alle $X \varepsilon \overline{\mathfrak{R}}$ ist $\varphi(X)$ das k-dimensionale Lebesguesche Maß von X.

§ 3. - Multiplikation additiver Mengenfunktionen.

Seien nun zwei Räume E', E'' gegeben. Als Produktraum $E' \times E''$ bezeichnen wir die Menge aller Paare (x', x'') mit $x' \varepsilon E'$, $x'' \varepsilon E''$. Ist $A' \subseteqq E'$, $A'' \subseteqq E''$, so be-

---

([8]) Siehe z. B. R. F. S. 258-262.

zeichnen wir als *Produktmenge* $A' \times A''$ die Menge aller Punkte $(x', x'') \varepsilon E' \times E''$ mit $x' \varepsilon A'$, $x'' \varepsilon A''$.

Sei $\mathfrak{R}'$ (bzw. $\mathfrak{R}''$) ein aus Punktmengen des Raumes E' (bzw. E'') bestehender Körper, und φ' (bzw. φ'') eine in $\mathfrak{R}'$ (bzw. $\mathfrak{R}''$) definierte additive Mengenfunktion. Mit $\mathfrak{P}$ bezeichnen wir das System aller Mengen der Gestalt $X' \times X''$ $(X' \varepsilon \mathfrak{R}', X'' \varepsilon \mathfrak{R}'')$. Wir definieren nun eine Mengenfunktion φ in $\mathfrak{P}$ durch die Festsetzung: ist $X = X' \times X''$ $(X' \varepsilon \mathfrak{R}', X'' \varepsilon \mathfrak{R}'')$, so sei:

$$\varphi(X) = \varphi'(X') \cdot \varphi''(X''), \tag{15}$$

wobei unter diesem Produkte der Wert 0 zu verstehen ist, wenn einer seiner Faktoren $=0$, der andere $=+\infty$ ist. Aus (15) folgt sofort: $\varphi(\Lambda) = 0$.

Ist $X = X_1 + \ldots + X_n$, wo $X_i = X_i' \times X''$, die X_i' disjunkte Mengen aus $\mathfrak{R}'$ und $X'' \varepsilon \mathfrak{R}''$ (oder $X_i = X' \times X_i''$, wo die X_i'' disjunkte Mengen aus $\mathfrak{R}''$ und $X' \varepsilon \mathfrak{R}'$), so ist:

$$\varphi(X) = \varphi(X_1) + \ldots + \varphi(X_n). \tag{16}$$

XXVI. - *Die Mengenfunktion* φ *ist additiv in* $\mathfrak{P}$.

Wir haben zu zeigen: Ist $P = P_1 + \ldots + P_n$, wo $P \varepsilon \mathfrak{P}$ und die P_i disjunkte Mengen aus $\mathfrak{P}$, so ist:

$$\varphi(P) = \varphi(P_1) + \ldots + \varphi(P_n). \tag{17}$$

Für $n=1$ ist (17) trivial; wir nehmen also an, die Behauptung gelte für weniger als n Summanden, und haben zu zeigen, dass sie dann auch für n Summanden gilt. Sei $P = P' \times P''$ $(P' \varepsilon \mathfrak{R}', P'' \varepsilon \mathfrak{R}'')$ und $P_i = P_i' \times P_i''$ $(P_i' \varepsilon \mathfrak{R}', P_i'' \varepsilon \mathfrak{R}'')$. Gilt $P_i'' = P''$ $(i = 1, 2, \ldots, n)$, so reduziert sich (17) auf (16). Wir nehmen also an, es sei etwa $P_n'' \subset P''$, d. h. $P'' - P_n'' \supset \Lambda$. Aus:

$$P = (P' \times P_n'') + (P' \times (P'' - P_n'')), \qquad P_i = (P_i' \times P_i'' P_n'') + (P_i' \times (P_i'' - P_n''))$$

folgt nach (16):

$$\begin{cases} \varphi(P) = \varphi(P' \times P_n'') + \varphi(P' \times (P'' - P_n'')), \\ \varphi(P_i) = \varphi(P_i' \times P_i'' P_n'') + \varphi(P_i' \times (P_i'' - P_n'')). \end{cases} \tag{18}$$

Wegen $P = \underset{i=1}{\overset{n}{S}} P_i = \underset{i=1}{\overset{n}{S}} (P_i' \times P_i'')$ ist offenbar:

$$P' \times P_n'' = \underset{i=1}{\overset{n}{S}} (P_i' \times P_i'' P_n''), \qquad P' \times (P'' - P_n'') = \underset{i=1}{\overset{n}{S}} (P_i' \times (P_i'' - P_n'')). \tag{19}$$

Wir zeigen, dass in jeder dieser beiden Formeln mindestens ein Summand der rechten Seite leer ist. Für die zweite Formel ist dies der n-te Summand; was die erste Formel anlangt, ist dies ebenfalls der n-te Summand, falls $P_n' = \Lambda$; ist hingegen $P_n' \supset \Lambda$, so gibt es wegen $P'' - P_n'' \supset \Lambda$ ein $a' \varepsilon P_n'$ und ein $a'' \varepsilon P'' - P_n''$; dann ist $(a', a'') \varepsilon P - P_n$, also gibt es genau ein $i < n$, so dass $(a', a'') \varepsilon P_i$; dann aber ist $P_i''P_n''=\Lambda$; denn wäre $b''\varepsilon P_i''P_n''$, so wäre wegen $a'\varepsilon P_n'$, $b''\varepsilon P_n''$ einerseits $(a', b'')\varepsilon P_n$, und wegen $(a', a'')\varepsilon P_i$ wäre $a'\varepsilon P_i'$, also wegen $b''\varepsilon P_i''$ wäre andererseits auch $(a', b'')\varepsilon P_i$, entgegen der Voraussetzung, dass die Mengen $P_1,\dots, P_n$ disjunkt sind. Da also $P_i''P_n''=\Lambda$, ist in der ersten Formel (19) der i-te Summand leer. Lassen wir in (19) die leeren Summanden weg, so enthalten also die rechten Seiten weniger als n Summanden; nach Annahme gilt also:

$$\varphi(P'\times P_n'')=\sum_{i=1}^{n}\varphi(P_i'\times P_i''P_n'');$$

$$\varphi(P'\times(P''-P_n''))=\sum_{i=1}^{n}\varphi(P_i'\times(P_i''-P_n'')),$$

(wo rechts die von den leeren Summanden in (19) herrührenden Glieder $=0$ sind). Durch Einsetzen in die erste Formel (18) ergibt sich:

$$\varphi(P)=\sum_{i=1}^{n}(\varphi(P_i'\times P_i''P_n'')+\varphi(P_i'\times(P_i''-P_n''))),$$

und wegen der zweiten Formel (18) ist das die zu beweisende Formel (17).

Das Mengensystem $\mathfrak{P}$ ist i. a. kein Körper; wohl aber folgt daraus, dass $\mathfrak{R}'$ und $\mathfrak{R}''$ Körper sind, nach einem bekannten Satze ([9]), dass das System $\mathfrak{R}$ aller Mengen, die Summe endlich vieler Mengen aus $\mathfrak{P}$ sind, *ein Körper ist, und dass jede Menge aus $\mathfrak{R}$ auch darstellbar ist als Summe endlich vieler disjunkter Mengen aus $\mathfrak{P}$*. Wir bezeichnen $\mathfrak{R}$ als den *Produktkörper* $\mathfrak{R}'\times\mathfrak{R}''$, und wollen die Definition der durch (15) nur in $\mathfrak{P}$ definierten Mengenfunktion φ auf $\mathfrak{R}$ erweitern. Zu dem Zweck zeigen wir vorerst:

XXVII. - *Sind $P_1,\dots, P_m$ disjunkte Mengen aus $\mathfrak{P}$ und $\overline{P}_1,\dots, \overline{P}_n$ disjunkte Mengen aus $\mathfrak{P}$, und ist $P_1+\dots+P_m=\overline{P}_1+\dots+\overline{P}_n$, so ist auch*:

$$\varphi(P_1)+\dots+\varphi(P_m)=\varphi(\overline{P}_1)+\dots+\varphi(\overline{P}_n). \tag{20}$$

Offenbar ist $\underset{i=1}{\overset{m}{S}}P_i=\underset{j=1}{\overset{n}{S}}\overline{P}_j=\underset{i=1}{\overset{m}{S}}\underset{j=1}{\overset{n}{S}}P_i\overline{P}_j$. Ist $P_i=P_i'\times P_i''$ ($P_i'\varepsilon\mathfrak{R}'$, $P_i''\varepsilon\mathfrak{R}''$) und $\overline{P}_j=\overline{P}_j'\times\overline{P}_j''$ ($\overline{P}_j'\varepsilon\mathfrak{R}'$, $\overline{P}_j''\varepsilon\mathfrak{R}''$), so ist $P_i\overline{P}_j=P_i'\overline{P}_j'\times P_i''\overline{P}_j''$; da $\mathfrak{R}'$ und $\mathfrak{R}''$ Körper, ist $P_i'\overline{P}_j'\varepsilon\mathfrak{R}'$, $P_i''\overline{P}_j''\varepsilon\mathfrak{R}''$, also $P_i\overline{P}_j\varepsilon\mathfrak{P}$; die $P_i\overline{P}_j$ sind also disjunkte Mengen aus $\mathfrak{P}$, und wegen $P_i=\underset{j=1}{\overset{n}{S}}P_i\overline{P}_j$, $\overline{P}_j=\underset{i=1}{\overset{m}{S}}P_i\overline{P}_j$, ist nach XXVI:

$$\varphi(P_i)=\sum_{j=1}^{n}\varphi(P_i\overline{P}_j),\qquad \varphi(\overline{P}_j)=\sum_{i=1}^{m}\varphi(P_i\overline{P}_j);$$

([9]) R. F. S. 14 (Satz 3.3.22).

also ist:

$$\sum_{i=1}^{m} \varphi(P_i) = \sum_{i=1}^{m} \sum_{j=1}^{n} \varphi(P_i \bar{P}_j), \qquad \sum_{j=1}^{n} \varphi(\bar{P}_j) = \sum_{j=1}^{n} \sum_{i=1}^{m} \varphi(P_i \bar{P}_j),$$

womit (20) bewiesen ist.

Sei nun X eine beliebige Menge aus $\mathfrak{R}$; dann ist X in der Form darstellbar $X = P_1 + \dots + P_n$, wo die P_i disjunkte Mengen aus $\mathfrak{P}$ sind. Wir setzen:

(21) $$\varphi(X) = \varphi(P_1) + \dots + \varphi(P_n),$$

und XXVII lehrt, dass der durch (21) definierte Wert $\varphi(X)$ unabhängig ist von der Art, wie X als Summe endlich vieler disjunkter Summanden aus $\mathfrak{P}$ dargestellt wird. Durch (21) ist nun φ in ganz $\mathfrak{R}$ definiert, und wir zeigen noch:

XXVIII. - *Die durch* (21) *in* $\mathfrak{R}$ *definierte Mengenfunktion* φ *ist additiv.*

Seien X und Y disjunkte Mengen aus $\mathfrak{R}$; dann ist $X = P_1 + \dots + P_m$, wo die P_i disjunkte Mengen aus $\mathfrak{P}$ bedeuten, und $Y = \mathfrak{Q}_1 + \dots + \mathfrak{Q}_n$, wo die $\mathfrak{Q}_i$ disjunkte Mengen aus $\mathfrak{P}$ bedeuten; also ist $X + Y = P_1 + \dots + P_m + \mathfrak{Q}_1 + \dots + \mathfrak{Q}_n$ eine Darstellung von $X + Y$ als Summe endlich vieler disjunkter Mengen aus $\mathfrak{P}$. Nach (21) ist:

$$\varphi(X) = \varphi(P_1) + \dots + \varphi(P_m), \qquad \varphi(Y) = \varphi(\mathfrak{Q}_1) + \dots + \varphi(\mathfrak{Q}_n)$$
$$\varphi(X+Y) = \varphi(P_1) + \dots + \varphi(P_m) + \varphi(\mathfrak{Q}_1) + \dots + \varphi(\mathfrak{Q}_n)$$

also $\varphi(X+Y) = \varphi(X) + \varphi(Y)$, w. z. b. w.

Wir bezeichnen die durch (15) und (21) in $\mathfrak{R} = \mathfrak{R}' \times \mathfrak{R}''$ definierte Mengenfunktion φ als *das Produkt* $\varphi' \times \varphi''$ der in $\mathfrak{R}'$ bzw. $\mathfrak{R}''$ definierten und additiven Mengenfunktionen φ' und φ''. Zusammenfassend haben wir dann:

XXIX. - *Ist* φ' *eine im Körper* $\mathfrak{R}'$ *definierte, additive Mengenfunktion und* φ'' *eine im Körper* $\mathfrak{R}''$ *definierte, additive Mengenfunktion, so ist* $\varphi' \times \varphi''$ *eine im Körper* $\mathfrak{R}' \times \mathfrak{R}''$ *definierte, additive Mengenfunktion.*

§ 4. - Multiplikation total-additiver Mengenfunktionen.

Verschärfen wir die Voraussetzung, φ' und φ'' seien in $\mathfrak{R}'$ bzw. $\mathfrak{R}''$ additiv, zur Voraussetzung, φ' und φ'' seien in $\mathfrak{R}'$ bzw. $\mathfrak{R}''$ *total-additiv*, so können wir auch zeigen, dass $\varphi' \times \varphi''$ in $\mathfrak{R} = \mathfrak{R}' \times \mathfrak{R}''$ total-additiv ist. Wir zeigen zunächst in Verschärfung von XXVI:

XXX. - *Ist* φ' *total-additiv in* $\mathfrak{R}'$ *und* φ'' *total-additiv in* $\mathfrak{R}''$, *so ist die durch* (15) *definierte Mengenfunktion* φ *total-additiv in* $\mathfrak{P}$.

Wir haben zu zeigen: Ist $P = \underset{i}{S} P_i$, wo $P \varepsilon \mathfrak{P}$ und die P_i disjunkte Mengen aus $\mathfrak{P}$ bedeuten, so ist:

(22) $$\varphi(P) = \sum_i \varphi(P_i).$$

Da $P \varepsilon \mathfrak{R}$, $P_1 + \ldots + P_n \varepsilon \mathfrak{R}$, und da $\mathfrak{R}$ ein Körper, ist auch $P-(P_1+\ldots+P_n) \varepsilon \mathfrak{R}$, und da jede Menge aus $\mathfrak{R}$ Summe endlich vieler disjunkter Mengen aus $\mathfrak{P}$ ist, so haben wir:

$$P = P_1 + \ldots + P_n + P_1^* + \ldots + P_m^*,$$

wo $P_1, \ldots, P_n, P_1^*, \ldots, P_m^*$ disjunkte Mengen aus $\mathfrak{P}$ sind. Nach XXVI ist also:

$$\varphi(P) = \varphi(P_1) + \ldots + \varphi(P_n) + \varphi(P_1^*) + \ldots + \varphi(P_m^*),$$

also $\varphi(P_1) + \ldots + \varphi(P_n) \leqq \varphi(P)$, mithin:

$$\sum_i \varphi(P_i) \leqq \varphi(P). \tag{23}$$

Es ist $P = P' \times P''$ ($P' \varepsilon \mathfrak{R}'$, $P'' \varepsilon \mathfrak{R}''$), $P_i = P_i' \times P_i''$ ($P_i' \varepsilon \mathfrak{R}'$, $P_i'' \varepsilon \mathfrak{R}''$). Wir bezeichnen mit $g_i(x')$ diejenige auf P' definierte Punkfunktion, die $= \varphi''(P_i'')$ ist für $x' \varepsilon P_i'$ und $=0$ für $x' \varepsilon P' - P_i'$, und setzen: $h_n(x') = g_1(x') + \ldots + g_n(x')$. Wir bilden die 2^n Mengen $D_{n1}, \ldots, D_{n2^n}$, die entstehen, indem man in $P_1' P_2' \ldots P_n'$ auf alle möglichen Weisen 0, 1, 2,..., n Faktoren P_i' ersetzt durch $P' - P_i'$. Dann ist $P' = \underset{j=1}{\overset{2^n}{S}} D_{nj}$, und da $\mathfrak{R}'$ ein Körper, ist $D_{nj} \varepsilon \mathfrak{R}'$. Die Funktion $h_n(x')$ ist konstant auf jeder der Mengen D_{nj}; bezeichnen wir ihren Wert auf D_{nj} mit [10] y_{nj}, so ist offenbar:

$$\varphi(P_1) + \ldots + \varphi(P_n) = \sum_{j=1}^{n} y_{nj} \varphi'(D_{nj}). \tag{24}$$

Bezeichnen wir nun für jedes $x' \varepsilon P'$ mit $\mathfrak{Q}_i(x')$ die Menge P_i'' oder die leere Menge, jenachdem $x' \varepsilon P_i'$ oder $x' \varepsilon P' - P_i'$, so folgt aus $P = \underset{i}{S} P_i$ sofort: $P'' = \underset{i}{S} \mathfrak{Q}_i(x')$, und da die Mengen P_i disjunkt sind, sind es (für jedes feste $x' \varepsilon P'$) auch die Mengen $\mathfrak{Q}_i(x')$; weil φ'' total-additiv in $\mathfrak{R}''$, ist also $\varphi''(P'') = \sum_i \varphi''(\mathfrak{Q}_i(x'))$ für jedes $x' \varepsilon P'$, und wegen der Definition von $g_i(x')$ ist dies gleichbedeutend mit: $\varphi''(P'') = \sum_i g_i(x')$, wofür wir auch schreiben können:

$$\varphi''(P'') = \lim_n h_n(x') \quad \text{für jedes} \quad x' \varepsilon P'. \tag{25}$$

Sei nun z eine beliebige Zahl $< \varphi''(P'')$. Bedeutet D_n die Summe aller D_{nj}, auf denen $h_n(x') > z$ ist, so ist $D_n \varepsilon \mathfrak{R}'$ und es folgt aus (24):

$$\varphi(P_1) + \ldots + \varphi(P_n) \geqq z \varphi'(D_n). \tag{26}$$

Wegen $z < \varphi''(P'')$ aber folgt aus (25): $P' = \underset{n}{S} D_n$, und da $D_n \subseteq D_{n+1}$, folgt aus

---

([10]) Ist $D_{nj} = A$, so setze man etwa $y_{nj} = 0$.

der totalen Additivität von φ' in $\mathfrak{R}'$ nach II: $\varphi'(P') = \lim\limits_{n} \varphi'(D_n)$. Also folgt aus (26):

$$\sum_i \varphi(P_i) \geqq z\varphi'(P'),$$

und da dies für jedes $z < \varphi''(P'')$ gilt:

$$\sum_i \varphi(P_i) \geqq \varphi'(P') \cdot \varphi''(P'') = \varphi(P).$$

Zusammen mit (23) ergibt das (22).

Nun erhalten wir in Verschärfung von XXIX:

XXXI. - *Ist φ' total-additiv in $\mathfrak{R}'$ und φ'' total-additiv in $\mathfrak{R}''$, so ist $\varphi' \times \varphi''$ total-additiv in $\mathfrak{R}' \times \mathfrak{R}''$.*

Wir setzen wieder $\varphi' \times \varphi'' = \varphi$, $\mathfrak{R}' \times \mathfrak{R}'' = \mathfrak{R}$, und haben zu zeigen: Ist $\mathfrak{A} = \underset{i}{S}\, \mathfrak{A}_i$, wo $\mathfrak{A} \varepsilon \mathfrak{R}$ und die $\mathfrak{A}_i$ disjunkte Mengen aus $\mathfrak{R}$ bedeuten, so ist:

$$\varphi(\mathfrak{A}) = \sum_i \varphi(\mathfrak{A}_i). \tag{27}$$

Wegen $\mathfrak{A} \varepsilon \mathfrak{R}$ ist $\mathfrak{A} = P_1 + \ldots + P_n$, wo die P_j disjunkte Mengen aus $\mathfrak{P}$ bedeuten; ebenso ist $\mathfrak{A}_i = P_{i1} + \ldots + P_{in_i}$, wo die P_{ij} disjunkte Mengen aus $\mathfrak{P}$ bedeuten. Nach (21) ist:

$$\varphi(\mathfrak{A}) = \sum_{j=1}^{n} \varphi(P_j). \tag{28}$$

Wegen $P_j \subseteq \mathfrak{A}$, $\mathfrak{A} = \underset{i}{S}\, \mathfrak{A}_i$ ist:

$$P_j = \underset{i}{S}\, P_j \mathfrak{A}_i = \underset{i}{S} \underset{k=1}{\overset{n_i}{S}} P_j P_{ik},$$

also nach XXX:

$$\varphi(P_j) = \sum_i \sum_{k=1}^{n_i} \varphi(P_j P_{ik}),$$

mithin nach (28):

$$\varphi(\mathfrak{A}) = \sum_{j=1}^{n} \sum_i \sum_{k=1}^{n_i} \varphi(P_j P_{ik}) = \sum_i \sum_{j=1}^{n} \sum_{k=1}^{n_i} \varphi(P_j P_{ik}). \tag{29}$$

Wegen $\mathfrak{A}_i \subseteqq \mathfrak{A} = \overset{n}{\underset{j=1}{S}}\, P_j$ ist:

$$\mathfrak{A}_i = \underset{j=1}{\overset{n}{S}}\, P_j \mathfrak{A}_i = \underset{j=1}{\overset{n}{S}} \underset{k=1}{\overset{n_i}{S}} P_j P_{ik},$$

also nach (21):

$$\varphi(\mathfrak{A}_i) = \sum_{j=1}^{n} \sum_{k=1}^{n_i} \varphi(P_j P_{ik}).$$

Setzt man dies in (29) ein, so erhält man (27).

Sei nun $\mathfrak{E}$ das System aller Teilmengen von $E' \times E''$. Wegen XXXI kann $\varphi' \times \varphi''$ nach XVIII erweitert werden zu einer in $\mathfrak{E}$ definierten Maßfunktion φ, die wir als *die zu $\varphi' \times \varphi''$ gehörige Maßfuntion* bezeichnen; sie ist, wenn $\overline{\mathfrak{R}}$

die vollständige Hülle für φ des Körpers $\mathfrak{M}=\mathfrak{M}'\times\mathfrak{M}''$ bezeichnet, total-additiv in $\overline{\mathfrak{M}}$. Setzen wir $\varphi=\varphi'\times\varphi''$ auch für alle $X\,\varepsilon\,\overline{\mathfrak{M}}$, so können wir XXIX verschärfen zu:

XXXII. - *Ist φ' eine im Körper $\mathfrak{M}'$ definierte, total-additive Mengenfunktion und φ'' eine im Körper $\mathfrak{M}''$ definierte, total-additive Mengenfunktion, und $\mathfrak{M}=\mathfrak{M}'\times\mathfrak{M}''$, so ist $\varphi'\times\varphi''$ eine in $\overline{\mathfrak{M}}$ definierte total-additive Mengenfunktion.*

Wir nehmen nun an, die Räume E', E'' haben die Gestalt:

$$(30)\qquad \begin{cases} E'=\underset{\mu}{S}E_\mu', & E_\mu'\,\varepsilon\,\mathfrak{M}', \quad \varphi'(E_\mu') \text{ endlich;} \\ E''=\underset{\nu}{S}E_\nu'', & E_\nu''\,\varepsilon\,\mathfrak{M}'', \quad \varphi''(E_\nu'') \text{ endlich.} \end{cases}$$

Dann ist $E'\times E''=\underset{\mu,\nu}{S}E_\mu'\times E_\nu'$, wo $E_\mu'\times E_\nu'\,\varepsilon\,\mathfrak{M}$ und $\varphi(E_\mu'\times E_\nu')=\varphi'(E_\mu')\cdot\varphi''(E_\nu'')$ endlich, und indem wir die Mengen $E_\mu'\times E_\nu''$ in eine Folge ordnen, sehen wir, dass für den Raum $E=E'\times E''$ (11) gilt.

Sei $\mathfrak{E}'$ das System aller Teilmengen von E', sei $\mathfrak{E}''$ das System aller Teilmengen von E'', und $\mathfrak{E}$ das System aller Teilmengen von $E'\times E''$; seien $\mathfrak{M}'$, $\mathfrak{L}'$ Körper aus $\mathfrak{E}'$, und $\mathfrak{M}''$, $\mathfrak{L}''$ Körper aus $\mathfrak{E}''$; seien φ' und ψ' in $\mathfrak{M}'$ bzw. $\mathfrak{L}'$ definierte, total-additiv Mengenfunktionen, φ'' und ψ'' in $\mathfrak{M}''$ bzw. $\mathfrak{L}''$ definierte, total-additive Mengenfunktionen; $\overline{\mathfrak{M}}'$ sei die vollständige Hülle von $\mathfrak{M}'$ für φ', und $\overline{\mathfrak{M}}''$ die vollständige Hülle von $\mathfrak{M}''$ für φ''. Seien φ und ψ die zu $\varphi'\times\varphi''$ bzw. $\psi'\times\psi''$ gehörigen Maßfunktionen, und seien $\overline{\mathfrak{M}}$ und $\overline{\mathfrak{L}}$ die vollständigen Hüllen für φ bzw. ψ von $\mathfrak{M}=\mathfrak{M}'\times\mathfrak{M}''$ und $\mathfrak{L}=\mathfrak{L}'\times\mathfrak{L}''$. Dann gilt:

XXXIII. - *Ist $\mathfrak{M}'\subseteqq\mathfrak{L}'\subseteqq\overline{\mathfrak{M}}'$, $\mathfrak{M}''\subseteqq\mathfrak{L}''\subseteqq\overline{\mathfrak{M}}''$, ist $\varphi'(X')=\psi'(X')$ für alle $X'\varepsilon\,\mathfrak{M}'$, $\varphi''(X'')=\psi''(X'')$ für alle $X''\varepsilon\,\mathfrak{M}''$, und gilt* (30), *so ist $\varphi(A)=\psi(A)$ für alle $A\,\varepsilon\,\mathfrak{E}$, und mithin $\overline{\mathfrak{M}}=\overline{\mathfrak{L}}$. Es stimmen also die Produkte $\varphi'\times\varphi''$ und $\psi'\times\psi''$ in ihrem gemeinsamen Definitionsbereiche $\overline{\mathfrak{M}}=\overline{\mathfrak{L}}$ überein.*

Offenbar ist $\mathfrak{M}\subseteqq\mathfrak{L}$ und $\psi(X)=\varphi(X)$ für alle $X\,\varepsilon\,\mathfrak{M}$; nach XXV ist also nur mehr zu zeigen: $\mathfrak{L}\subseteqq\overline{\mathfrak{M}}$. Sei $\mathfrak{Q}$ das System aller Mengen $X'\times X''$ mit $X'\varepsilon\,\overline{\mathfrak{M}}'$, $X''\varepsilon\,\overline{\mathfrak{M}}''$; da $\mathfrak{L}\subseteqq\overline{\mathfrak{M}}'\times\overline{\mathfrak{M}}''$, und jede Menge aus $\overline{\mathfrak{M}}'\times\overline{\mathfrak{M}}''$ Summe endlich vieler Mengen aus $\mathfrak{Q}$ ist, genügt es also zu zeigen, dass $\mathfrak{Q}\subseteqq\overline{\mathfrak{M}}$. Wir beweisen das schrittweise.

Hilfsatz 1. - Ist $B=B'\times B''$, wo $B'\varepsilon\,\mathfrak{M}_\sigma'$, $B''\varepsilon\,\mathfrak{M}_\sigma''$ so ist $B\,\varepsilon\,\overline{\mathfrak{M}}$ und $\varphi(B)=$ $=\varphi'(B')\cdot\varphi''(B'')$.

Wegen $B'\varepsilon\,\mathfrak{M}_\sigma'$, $B''\varepsilon\,\mathfrak{M}_\sigma''$ ist $B'=\underset{n}{S}B_n'$, $B''=\underset{n}{S}B_n''$, wo $B_n'\varepsilon\,\mathfrak{M}'$, $B_n''\varepsilon\,\mathfrak{M}''$, und es kann ohneweiters $B_n'\subseteqq B'_{n+1}$, $B_n''\subseteqq B''_{n+1}$ angenommen werden; dann ist $B_n'\times B_n''\varepsilon\,\mathfrak{M}$, also auch $B_n'\times B_n''\varepsilon\,\overline{\mathfrak{M}}$, und $B=\underset{n}{S}B_n'\times B_n''$, also $B\,\varepsilon\,\overline{\mathfrak{M}}_\sigma$; da aber $\overline{\mathfrak{M}}$ ein σ-Körper, ist $\overline{\mathfrak{M}}_\sigma=\overline{\mathfrak{M}}$, also $B\,\varepsilon\,\overline{\mathfrak{M}}$. Da $B=\underset{n}{S}B_n'\times B_n''$ und $B_n'\times B_n''\subseteqq$ $\subseteqq B'_{n+1}\times B''_{n+1}$, und da nach XXXII φ total-additiv in $\overline{\mathfrak{M}}$, ist nach II: $\varphi(B)=$

$=\lim\limits_n \varphi(B_n'\times B_n'')$. Hierin ist nach (15): $\varphi(B_n'\times B_n'')=\varphi'(B_n')\cdot\varphi''(B_n'')$. Da $\overline{\mathfrak{M}}'$ ein σ-Körper über $\mathfrak{M}'$, also $\mathfrak{M}_\sigma'\subseteq\overline{\mathfrak{M}}'$, und da nach XVIII φ' total-additiv in $\overline{\mathfrak{M}}'$, ist nach II $\lim\limits_n\varphi'(B_n')=\varphi'(B')$, und ebenso $\lim\limits_n\varphi''(B_n'')=\varphi''(B'')$. Also ist $\varphi(B)=\varphi'(B')\cdot\varphi''(B'')$. Damit ist Hilfsatz 1. bewiesen.

Hilfsatz 2. - Ist $A=A'\times A''$, wo $A'\varepsilon\mathfrak{M}'_{\sigma\delta}$, $A''\varepsilon\mathfrak{M}''_{\sigma\delta}$, so ist $A\,\varepsilon\,\overline{\mathfrak{M}}$.

Wegen $A'\varepsilon\mathfrak{M}'_{\sigma\delta}$, $A''\varepsilon\mathfrak{M}''_{\sigma\delta}$ ist $A'=\underset{n}{D}A_n'$, $A''=\underset{n}{D}A_n''$, wo $A_n'\varepsilon\mathfrak{M}_\sigma'$, $A_n''\varepsilon\mathfrak{M}_\sigma''$; dann ist $A'\times A''=\underset{n}{D}A_n'\times A_n''$; hierin ist nach Hilfsatz 1 $A_n'\times A_n''\varepsilon\overline{\mathfrak{M}}$, also ist $A\,\varepsilon\,\overline{\mathfrak{M}}_\delta$, und da $\overline{\mathfrak{M}}$ ein σ-Körper, ist auch $A\,\varepsilon\,\overline{\mathfrak{M}}$.

Hilfsatz 3. - *Ist $A=A'\times A''$, wo $A'\varepsilon\overline{\mathfrak{M}}'$, $\varphi'(A')=0$ und A'' eine beliebige Menge aus E'' (oder $A''\varepsilon\overline{\mathfrak{M}}''$, $\varphi''(A'')=0$ und A' eine beliebige Menge aus E'), so ist $A\,\varepsilon\,\overline{\mathfrak{M}}$ und $\varphi(A)=0$.*

Da $\overline{\mathfrak{M}}$ vollständig für φ, genügt es zu zeigen, dass $\varphi(A'\times E'')=0$. Da nach (30) $E''=\underset{\nu}{S}E_\nu''$, also $A'\times E''=\underset{\nu}{S}A'\times E_\nu''$, genügt es nach Eigenschaft 3.) der Maßfunktionen weiter, zu zeigen, dass $\varphi(A'\times E_\nu'')=0$. Da $\varphi'(A')=0$, gibt es nach (6) zu jedem $\varepsilon>0$ ein $B'\varepsilon\mathfrak{M}_\sigma'$, so dass $B'\supseteq A'$ und $\varphi'(B')<\varepsilon$. Dann ist nach Hilfsatz 1. $\varphi(B'\times E_\nu'')=\varphi'(B')\cdot\varphi''(E_\nu'')\leqq\varepsilon\varphi''(E_\nu'')$; nach Eigenschaft 2.) der Maßfunktionen ist also auch $\varphi(A'\times E_\nu'')\leqq\varepsilon\varphi''(E_\nu'')$, und da dies für jedes $\varepsilon>0$ gilt und $\varphi''(E_\nu'')$ endlich ist, ist $\varphi(A'\times E_\nu'')=0$, w. z. b. w.

Nun können wir XXXIII beweisen. Sei also $X\varepsilon\mathfrak{Q}$, d. h. $X=X'\times X''$, wo $X'\varepsilon\overline{\mathfrak{M}}'$, $X''\varepsilon\overline{\mathfrak{M}}''$. Nach XXI gibt es ein $A'\varepsilon\mathfrak{M}'_{\sigma\delta}$ und ein $A''\varepsilon\mathfrak{M}''_{\sigma\delta}$, so dass: $A'\supseteq X'$, $A''\supseteq X''$, $\varphi'(A'-X')=0$, $\varphi''(A''-X'')=0$. Nun ist:

$$X=A'\times A''-(((A'-X')\times A'')+(A'\times(A''-X'')));$$

nach Hilfsatz 2. ist hierin $A'\times A''\varepsilon\overline{\mathfrak{M}}$, nach Hilfsatz 3. ist $(A'-X')\times A''\varepsilon\overline{\mathfrak{M}}$, $A'\times(A''\times X'')\varepsilon\overline{\mathfrak{M}}$, also, da $\overline{\mathfrak{M}}$ ein Körper, auch $X\varepsilon\overline{\mathfrak{M}}$. Es ist also $\mathfrak{Q}\subseteq\overline{\mathfrak{M}}$, w.z.b.w.

Sei nun insbesondere E' der R_k, E'' der R_l und $\mathfrak{M}'$ das System aller Mengen des R_k, die Summe endlich vieler halboffener Intervalle $a_i'\leqq x_i'<b_i'$ $(i=1,2,\ldots,k)$ sind, $\mathfrak{M}''$ das System aller Mengen des R_l, die Summe endlich vieler halboffener Intervalle $a_i''\leqq x_i''<b_i''$ $(i=1,2,\ldots,l)$ sind. Für das durch $a_i'\leqq x_i'<b_i'$ $(i=1,2,\ldots,k)$ gegebene halboffene Intervall I' des R_k sei $\varphi'(I')=(b_1'-a_1')\ldots(b_k'-a_k')$, für das durch $a_i''\leqq x_i''<b_i''$ $(i=1,2,\ldots,l)$ gegebene halboffene Intervall I'' des R_l sei $\varphi''(I'')=(b_1''-a_1'')\ldots(b_l''-a_l'')$; ist $X'=I_1'+\ldots+I_m'$, $X''=I_1''+\ldots+I_n''$, wo die I_i' disjunkte halboffene Intervalle des R_k, die I_i'' disjunkte halboffene Intervalle des R_l sind, so sei $\varphi'(X')=\varphi'(I_1')+\ldots+\varphi'(I_m')$, $\varphi''(X'')=\varphi''(I_1'')+\ldots+\varphi''(I_n'')$. Dann ist offenbar $\mathfrak{M}'\times\mathfrak{M}''$ das System aller Mengen des R_{k+l}, die Summe endlich vieler hallboffener Intervalle $a_i\leqq x_i<b_i$ $(i=1,2,\ldots,k+l)$ sind. Die zu φ' bzw. φ'' gehörige Maßfunktion ist, wie wir in § 2 sahen, das k-dimensionale (bzw. l-dimensionale) äussere Lebesguesche Maß, und $\overline{\mathfrak{M}}'$ (bzw. $\overline{\mathfrak{M}}''$) ist das System aller

k-dimensional (bzw. l-dimensional) messbaren Mengen des R_k (bzw. R_l). Setzen wir $\varphi=\varphi'\times\varphi''$, so wird die zu φ gehörige Maßfunktion das $(k+l)$-dimensionale äussere Lebesguesche Maß, und $\overline{\mathfrak{R}}$ das System aller $(k+l)$-dimensional messbaren Mengen des R_{k+l}. Bezeichnen wir das r-dimensionale Lebesguesche Maß mit μ_r, so ist also $\mu_k\times\mu_l=\mu_{k+l}$. Satz XXXIII lehrt, dass es dabei gleichgiltig ist, ob wir von den Körpern $\mathfrak{R}'$ und $\mathfrak{R}''$ oder den Körpern $\overline{\mathfrak{R}}'$ und $\overline{\mathfrak{R}}''$ ausgehen.

§ 5. - Analytische Darstellung.

Sei wieder $\mathfrak{R}'$ ein aus Mengen des Raumes E' und $\mathfrak{R}''$ ein aus Mengen des Raumes E'' bestehender Körper, φ' eine in $\mathfrak{R}'$ und φ'' eine in $\mathfrak{R}''$ total-additive Mengenfunktion; $\overline{\mathfrak{R}}'$ sei die vollständige Hülle von $\mathfrak{R}'$ für φ'; dann kann φ' nach § 2 erweitert werden zu einer in $\overline{\mathfrak{R}}'$ total-addiven Mengenfunktion φ'. Die zu φ' bzw. φ'' gehörige Maßfunktion bezeichnen wir gleichfalls mit φ' bzw. φ''. Wir nehmen weiterhin an, es sei $E'\varepsilon\mathfrak{R}'$ und $\varphi'(E')$ endlich, und $E''=\underset{\nu}{S}E_\nu''$, wo $E_\nu''\varepsilon\mathfrak{R}''$ und $\varphi''(E_\nu'')$ endlich; dabei kann ohneweiteres $E_\nu''\subseteq E''_{\nu-1}$ angenommen werden.

Ist $M\subseteq E'\times E''$, so bezeichnen wir für jedes $x\,\varepsilon\,E'$ mit M_x die Menge aller $y\,\varepsilon\,E''$, für die $(x, y)\,\varepsilon\,M$; dann ist $\varphi''(M_x)$ eine in E' definierte Punktfunktion.

Eine in E' definierte Punktfunktion $f(x)$ heißt φ'-messbar, wenn für jede Zahl z die Menge $[f>z]$ aller Punkte x von E', in denen $f(x)>z$ ist zu $\overline{\mathfrak{R}}'$ gehört. Für jede nicht-negative, φ'-messbare Funktion f ist dann das über E' erstreckte Integral $(E')\int f d\varphi'$ definiert ([11]).

Setzen wir $\mathfrak{R}=\mathfrak{R}'\times\mathfrak{R}''$, $\varphi=\varphi'\times\varphi''$, so ist nach § 4 φ total-additiv in $\overline{\mathfrak{R}}$. Mit $\mathfrak{P}$ bezeichnen wir wieder das System aller Mengen P aus $E'\times E''$, die die Gestalt haben: $P=P'\times P''$, wo $P'\varepsilon\mathfrak{R}'$, $P''\varepsilon\mathfrak{R}''$.

XXXIV. - *Ist* $P\,\varepsilon\,\mathfrak{P}$, *so ist* $P_x\,\varepsilon\,\mathfrak{R}''$, $\varphi''(P_x)$ *ist* φ'-*messbar und* $\varphi(P)=$ $=(E')\int\varphi''(P_x)d\varphi'$.

In der Tat, für $x\,\varepsilon\,P'$ ist $P_x=P''$, für $x\,\varepsilon\,E'-P'$ ist $P_x=\Lambda$; also ist $P_x\,\varepsilon\,\mathfrak{R}''$, $\varphi''(P_x)=\varphi''(P'')$ für $x\,\varepsilon\,P'$ und $\varphi''(P_x)=0$ für $x\,\varepsilon\,E'-P'$. Die Funktion $\varphi''(P_x)$ ist also φ'-messbar und es ist:

$$(E')\int\varphi''(P_x)d\varphi'=\varphi'(P')\varphi''(P'')=\varphi(P).$$

XXXV. - *Ist* $M\,\varepsilon\,\mathfrak{R}$, *so ist* $M_x\,\varepsilon\,\mathfrak{R}''$, $\varphi''(M_x)$ *ist* φ'-*messbar und* $\varphi(M)=$ $=(E')\int\varphi''(M_x)d\varphi'$.

In der Tat, M ist darstellbar in der Form $M=P_1+\dots+P_n$, wo die P_i dis-

([11]) Vgl. z. B. Festschrift der 57. Vers. Deutscher Philologen u. Schulmänner, Salzburg, 1929, S. 193 ff.

junkte Mengen aus $\mathfrak{P}$ bedeuten. Dann ist $M_x = P_{1x} + \dots + P_{nx}$, wo die P_{ix} disjunkte Mengen aus $\mathfrak{M}''$ sind; mithin ist $M_x \,\varepsilon\, \mathfrak{M}''$, $\varphi''(M_x) = \varphi''(P_{1x}) + \dots + \varphi''(P_{nx})$; da $\varphi''(P_{ix})$ nach XXXIV φ'-messbar, so ist auch $\varphi''(M_x)$ φ'-messbar und:

$$(E')\int \varphi''(M_x)d\varphi' = (E')\int \varphi''(P_{1x})d\varphi' + \dots + (E')\int \varphi''(P_{nx})d\varphi' = \\ = \varphi(P_1) + \dots + \varphi(P_n) = \varphi(M).$$

XXXVI. - *Ist* $B \,\varepsilon\, \mathfrak{M}_\sigma$, *so ist* $B_x \,\varepsilon\, \mathfrak{M}''_\sigma$, $\varphi''(B_x)$ *ist* φ'-*messbar und* $\varphi(B) = (E')\int \varphi''(B_x)d\varphi'$.

In der Tat, B ist darstellbar in der Form $B = \underset{n}{S} B_n$, wo $B_n \,\varepsilon\, \mathfrak{M}$ und ohneweiters $B_n \subseteqq B_{n+1}$ angenommen werden kann; da $\mathfrak{M} \subseteqq \overline{\mathfrak{M}}$, $\mathfrak{M}_\sigma \subseteqq \overline{\mathfrak{M}}$, ist $B_n \,\varepsilon\, \overline{\mathfrak{M}}$, $B \,\varepsilon\, \overline{\mathfrak{M}}$, und da φ total-additiv in $\overline{\mathfrak{M}}$, ist nach II: $\varphi(B) = \lim\limits_n \varphi(B_n)$. Ferner ist $B_x = \underset{n}{S} B_{nx}$, $B_{nx} \subseteqq B_{n+1x}$; da nach XXXV $B_{nx} \,\varepsilon\, \mathfrak{M}''$, ist $B_x \,\varepsilon\, \mathfrak{M}''_\sigma$, also, da $\overline{\mathfrak{M}}''$ ein σ-Körper über $\mathfrak{M}''$, auch $B_x \,\varepsilon\, \overline{\mathfrak{M}}''$, und da φ'' total-additiv in $\overline{\mathfrak{M}}''$, ist nach II: $\varphi''(B_x) = \lim\limits_n \varphi''(B_{nx})$. Da nach XXV $\varphi''(B_{nx})$ φ'-messbar, ist auch $\varphi''(B_x)$ φ'-messbar, und da wegen $B_{nx} \subseteqq B_{n+1x}$ auch $\varphi''(B_{nx}) \leqq \varphi''(B_{n+1x})$ ist [12]:

$$(E')\int \varphi''(B_x)d\varphi' = \lim_n (E')\int \varphi''(B_{nx})d\varphi' = \lim_n \varphi(B_n) = \varphi(B).$$

XXXVII. - *Ist* $A \,\varepsilon\, \mathfrak{M}_{\sigma\delta}$, *so ist* $A_x \,\varepsilon\, \mathfrak{M}''_{\sigma\delta}$, $\varphi''(A_x)$ *ist* φ'-*messbar und* $\varphi(A) = (E')\int \varphi''(A_x)d\varphi'$.

In der Tat, es ist $A = \underset{n}{D} B_n$, wo $B_n \,\varepsilon\, \mathfrak{M}_\sigma$; also ist $A_x = \underset{n}{D} B_{nx}$, wo nach XXXVI $B_{nx} \,\varepsilon\, \mathfrak{M}''_\sigma$; also ist $A_x \,\varepsilon\, \mathfrak{M}''_{\sigma\delta}$. Wir setzen sodann $E' \times E''_\nu = E_\nu$; dann ist $E_\nu \,\varepsilon\, \mathfrak{M}$, $E_\nu \subseteqq E_{\nu+1}$ und $E' \times E'' = \underset{\nu}{S} E_\nu$. Setzen wir $A_\nu = E_\nu A$, so ist also $A_\nu \,\varepsilon\, \mathfrak{M}_{\sigma\delta}$, $A_\nu \subseteqq A_{\nu+1}$, $A = \underset{\nu}{S} A_\nu$ und nach II:

$$\varphi(A) = \lim_\nu \varphi(A_\nu). \tag{31}$$

Wegen $A_\nu \,\varepsilon\, \mathfrak{M}_{\sigma\delta}$ ist $A_\nu = \underset{n}{D} A_{\nu n}$, wo $A_{\nu n} \,\varepsilon\, \mathfrak{M}_\sigma$ und ohneweiters $A_{\nu n+1} \subseteqq A_{\nu n}$ angenommen werden kann; indem man noch $A_{\nu n}$ ersetzt durch $E_\nu A_{\nu n}$, kann auch $A_{\nu n} \subseteqq E_\nu$ angenommen werden; dann ist $\varphi(A_{\nu n})$ endlich und somit nach III:

$$\varphi(A_\nu) = \lim_n \varphi(A_{\nu n}). \tag{32}$$

Ferner ist $A_{\nu x} = \underset{n}{D} A_{\nu n x}$, $A_{\nu n+1 x} \subseteqq A_{\nu n x}$; da nach XXXVI $A_{\nu n x} \,\varepsilon\, \mathfrak{M}''_\sigma$, ist $A_{\nu x} \,\varepsilon\, \mathfrak{M}''_{\sigma\delta}$, also $A_{\nu x} \,\varepsilon\, \overline{\mathfrak{M}}''$; wegen $A_{\nu n} \subseteqq E_\nu$ ist $A_{\nu n x} \subseteqq E''_\nu$, also $\varphi''(A_{\nu n x})$ endlich, und da φ''

---

[12] A. a. O. S. 198.

total-additiv in $\overline{\mathfrak{R}}''$, ist nach III $\varphi''(A_{\nu x})=\lim_n \varphi''(A_{\nu n x})$. Da nach XXXVI $\varphi''(A_{\nu n x})$ φ'-messbar, ist also auch $\varphi''(A_{\nu x})$ φ'-messbar; ferner ist wegen $A_{\nu n+1 x}\subseteqq A_{\nu n x}$ auch $\varphi''(A_{\nu n+1 x})\leqq\varphi''(A_{\nu n x})$ und weil $\varphi''(A_{\nu n x})\leqq\varphi''(E_\nu'')$, ist $(E')\int\varphi''(A_{\nu n x})\,d\varphi'$ endlich; also ist [13] bei Beachtung von XXXVI:

$$(E')\int\varphi''(A_{\nu x})d\varphi'=\lim_n\,(E')\int\varphi''(A_{\nu n x})d\varphi'=\lim_n\,\varphi(A_{\nu n}),$$

also nach (32):

$$\varphi(A_\nu)=(E')\int\varphi''(A_{\nu x})d\varphi'. \tag{33}$$

Da $A=\underset{\nu}{S}A_\nu$, $A_\nu\subseteqq A_{\nu+1}$, ist $A_x=\underset{\nu}{S}A_{\nu x}$, $A_{\nu x}\subseteqq A_{\nu+1 x}$. Da $A_{\nu x}\,\varepsilon\,\overline{\mathfrak{R}}''$ und $\overline{\mathfrak{R}}''$ ein σ-Körper, ist auch $A_x\,\varepsilon\,\overline{\mathfrak{R}}''$ und nach II: $\varphi''(A_x)=\lim_\nu\varphi''(A_{\nu x})$. Also ist auch $\varphi''(A_x)$ φ'-messbar und wegen (33):

$$(E')\int\varphi''(A_x)d\varphi'=\lim_\nu\,(E')\int\varphi''(A_{\nu x})d\varphi'=\lim_\nu\,\varphi(A_\nu),$$

also nach (31): $\varphi(A)=(E')\int\varphi''(A_x)d\varphi'$.

Nennen wir wieder jede Menge $N'\subseteqq E'$ mit $\varphi'(N')=0$ eine Nullmenge für φ', so gilt:

XXXVIII. - *Ist $N\,\varepsilon\,\overline{\mathfrak{R}}$ und $\varphi(N)=0$, so ist $\varphi''(N_x)=0$ für alle $x\,\varepsilon\,E'$, abgesehen von einer Nullmenge für φ', mithin $N_x\,\varepsilon\,\overline{\mathfrak{R}}''$ für alle $x\,\varepsilon\,E'$, abgesehen von einer Nullmenge für φ'.*

Nach XXI gibt es eine Menge $A\,\varepsilon\,\mathfrak{R}_{\sigma\delta}$, so dass $A\supseteqq N$ und $\varphi(A-N)=0$; dann ist auch $\varphi(A)=0$, nach XXXVII ist also $(E')\int\varphi''(A_x)d\varphi'=0$, also ist $\varphi''(A_x)=0$, abgesehen von einer Nullmenge für φ'. Wegen $N\subseteqq A$ ist auch $N_x\subseteqq A_x$, also ist nach Eigenschaft 2.) der Maßfunktionen auch $\varphi''(N_x)=0$, abgesehen von einer Nullmenge für φ'.

XXXIX. - *Ist M eine beliebige Menge aus $\overline{\mathfrak{R}}$, so gilt für alle $x\,\varepsilon\,E'$, abgesehen von einer Nullmenge für φ': $M_x\,\varepsilon\,\overline{\mathfrak{R}}''$, es ist $\varphi''(M_x)$ φ'-messbar und $\varphi(M)=(E')\int\varphi''(M_x)d\varphi'$.*

Nach XXI gibt es eine Menge $A\,\varepsilon\,\mathfrak{R}_{\sigma\delta}$, so dass $A\supseteqq M$ und $\varphi(A-M)=0$, also $\varphi(A)=\varphi(M)$. Setzen wir $A-M=N$, so ist $A=M+N$, also $A_x=M_x+N_x$; hierin ist nach XXXVII $A_x\,\varepsilon\,\overline{\mathfrak{R}}''$ und nach XXXVIII $N_x\,\varepsilon\,\overline{\mathfrak{R}}''$ für alle $x\,\varepsilon\,E'$ abgesehen von einer Nullmenge für φ', also auch $M_x\,\varepsilon\,\overline{\mathfrak{R}}''$ für alle $x\,\varepsilon\,E'$ abgesehen von einer Nullmenge für φ'; und da nach XXXVIII $\varphi''(N_x)=0$, abgesehen von einer Nullmenge für φ', ist $\varphi''(A_x)=\varphi''(M_x)$, abgesehen von einer Nullmenge

(13) Vgl. Fußnote (12).

für φ'. Da nach XXXVII $\varphi''(A_x)$ φ'-messbar, ist auch $\varphi''(M_x)$ φ'-messbar, und es ist zufolge XXXVII:

$$(E')\int\varphi''(M_x)d\varphi'=(E')\int\varphi''(A_x)d\varphi'=\varphi(A);$$

da aber $\varphi(A)=\varphi(M)$ war, ist $\varphi(M)=(E')\int\varphi''(M_x)d\varphi'$, wie behauptet.

§ 6. - Das Reduktionstheorem.

Seien nun drei Räume gegeben: E_1, E_2, E_3; sein $\mathfrak{R}_1$, $\mathfrak{R}_2$, $\mathfrak{R}_3$ Körper aus Mengen des Raumes E_1, bzw. E_2, bzw. E_3, und seien φ_1, φ_2, φ_3 in $\mathfrak{R}_1$, bzw. $\mathfrak{R}_2$, bzw. $\mathfrak{R}_3$ total-additive Mengenfunktionen.

Wir bezeichnen mit $E_1\times E_2\times E_3$ die Menge aller Tripel (x, y, z), für die $x\,\varepsilon\,E_1$, $y\,\varepsilon\,E_2$, $z\,\varepsilon\,E_3$, und können schreiben: $E_1\times E_2\times E_3=(E_1\times E_2)\times E_3=E_1\times\times(E_2\times E_3)$. Ist $M_1\subseteq E_1$, $M_2\subseteq E_2$, $M_3\subseteq E_3$, so bezeichnen wir mit $M_1\times M_2\times M_3$ die Menge aller (x, y, z), für die $x\,\varepsilon\,M_1$, $y\,\varepsilon\,M_2$, $z\,\varepsilon\,M_3$, und können schreiben: $M_1\times M_2\times M_3=(M_1\times M_2)\times M_3=M_1\times(M_2\times M_3)$. Wir bezeichnen mit $\mathfrak{P}$ das System aller Mengen $M\subseteq E_1\times E_2\times E_3$, die die Gestalt haben $M_1\times M_2\times M_3$, wo $M_1\,\varepsilon\,\mathfrak{R}_1$, $M_2\,\varepsilon\,\mathfrak{R}_2$, $M_3\,\varepsilon\,\mathfrak{R}_3$; mit $\mathfrak{R}_1\times\mathfrak{R}_2\times\mathfrak{R}_3$ bezeichnen wir das System aller Mengen, die Summe endlich vieler Mengen aus $\mathfrak{P}$ sind; offenbar ist $\mathfrak{R}_1\times\mathfrak{R}_2\times\mathfrak{R}_3=(\mathfrak{R}_1\times\mathfrak{R}_2)\times\mathfrak{R}_3=\mathfrak{R}_1\times(\mathfrak{R}_2\times\mathfrak{R}_3)$; da $\mathfrak{R}_1\times\mathfrak{R}_2$ und $\mathfrak{R}_2\times\mathfrak{R}_3$ Körper, folgt daraus ([14]), dass auch $\mathfrak{R}_1\times\mathfrak{R}_2\times\mathfrak{R}_3$ ein Körper, und dass jede Menge aus $\mathfrak{R}_1\times\mathfrak{R}_2\times\mathfrak{R}_3$ Summe endlich vieler *disjunkter* Mengen aus $\mathfrak{P}$ ist.

Da $\varphi_1\times\varphi_2$ nach XXXI total-additiv in $\mathfrak{R}_1\times\mathfrak{R}_2$ und $\varphi_2\times\varphi_3$ total-additiv in $\mathfrak{R}_2\times\mathfrak{R}_3$, ist sowohl $(\varphi_1\times\varphi_2)\times\varphi_3$ als auch $\varphi_1\times(\varphi_2\times\varphi_3)$ nach XXXI total-additiv in $\mathfrak{R}_1\times\mathfrak{R}_2\times\mathfrak{R}_3$.

XL. - *In* $\mathfrak{R}_1\times\mathfrak{R}_2\times\mathfrak{R}_3$ *gilt*: $(\varphi_1\times\varphi_2)\times\varphi_3=\varphi_1\times(\varphi_2\times\varphi_3)$.

Wir setzen $\varphi_1\times\varphi_2=\varphi'$, $\varphi_2\times\varphi_3=\varphi''$, $\varphi'\times\varphi_3=\psi'$, $\varphi_1\times\varphi''=\psi''$. Ist $M\,\varepsilon\,\mathfrak{P}$, etwa $M=M_1\times M_2\times M_3$ mit $M_1\,\varepsilon\,\mathfrak{R}_1$, $M_2\,\varepsilon\,\mathfrak{R}_2$, $M_3\,\varepsilon\,\mathfrak{R}_3$, so ist nach (15):

$$\varphi'(M_1\times M_2)=\varphi_1(M_1)\varphi_2(M_2),\quad \varphi''(M_2\times M_3)=\varphi_2(M_2)\varphi_3(M_3),$$

also:

$$\psi'(M_1\times M_2\times M_3)=\varphi'(M_1\times M_2)\varphi_3(M_3)=\varphi_1(M_1)\varphi_2(M_2)\varphi_3(M_3)$$

und $\psi''(M_1\times M_2\times M_3)=\varphi_1(M_1)\varphi''(M_2\times M_3)=\varphi_1(M_1)\varphi_2(M_2)\varphi_3(M_3)$; wir haben also $\psi'(M)=\psi''(M)$ für alle $M\,\varepsilon\,\mathfrak{P}$. - Sei nun $X\,\varepsilon\,\mathfrak{R}_1\times\mathfrak{R}_2\times\mathfrak{R}_3$; dann ist $X=X_1+\ldots+X_n$, wo die X_i disjunkte Mengen aus $\mathfrak{P}$, und nach (21) ist:

$$\psi'(X)=\psi'(X_1)+\ldots+\psi'(X_n),\qquad \psi''(X)=\psi''(X_1)+\ldots+\psi''(X_n);$$

da, wie eben gezeigt: $\psi'(X_i)=\psi''(X_i)$, ist auch $\psi'(X)=\psi''(X)$, w. z. b. w.

([14]) Vgl. Fußnote ([9]).

Wir setzen nun $\mathfrak{M}_1 \times \mathfrak{M}_2 \times \mathfrak{M}_3 = \mathfrak{M}$ und können zufolge XL für alle Mengen aus $\mathfrak{M}$ statt $(\varphi_1 \times \varphi_2) \times \varphi_3$ und statt $\varphi_1 \times (\varphi_2 \times \varphi_3)$ einfach $\varphi_1 \times \varphi_2 \times \varphi_3$ schreiben; $\varphi_1 \times \varphi_2 \times \varphi_3$ ist total-additiv in $\mathfrak{M}$. Die zu $\varphi_1 \times \varphi_2 \times \varphi_3$ gehörige Maßfunktion φ ist dann gleichzeitig die zu $(\varphi_1 \times \varphi_2) \times \varphi_3$ und die zu $\varphi_1 \times (\varphi_2 \times \varphi_3)$ gehörige Maßfunktion; sie ist total-additiv in der für φ vollständigen Hülle $\overline{\mathfrak{M}}$ von $\mathfrak{M}$. Wir haben also nach XXXII:

XLI. - *Sowohl* $(\varphi_1 \times \varphi_2) \times \varphi_3$ *als* $\varphi_1 \times (\varphi_2 \times \varphi_3)$ *sind in* $\overline{\mathfrak{M}}$ *definierte, total-additive Mengenfunktionen, und in* $\overline{\mathfrak{M}}$ *gilt*: $(\varphi_1 \times \varphi_2) \times \varphi_3 = \varphi_1 \times (\varphi_2 \times \varphi_3)$.

Wir können also auch für alle Mengen aus $\overline{\mathfrak{M}}$ statt $(\varphi_1 \times \varphi_2) \times \varphi_3$ und statt $\varphi_1 \times (\varphi_2 \times \varphi_3)$ einfach $\varphi_1 \times \varphi_2 \times \varphi_3$ schreiben.

Wir nehmen nun an, es sei $E_1 \varepsilon \mathfrak{M}_1$, $\varphi_1(E_1)$ endlich, $E_2 \varepsilon \mathfrak{M}_2$, $\varphi_2(E_2)$ endlich und $E_3 = \underset{\nu}{S} E_{3\nu}$, wo $E_{3\nu} \varepsilon \mathfrak{M}_3$ und $\varphi_3(E_{3\nu})$ endlich; dann ist, wenn wieder $\varphi_1 \times \varphi_2 = \varphi'$, $\varphi_2 \times \varphi_3 = \varphi''$ gesetzt wird: $E_1 \times E_2 \varepsilon \mathfrak{M}_1 \times \mathfrak{M}_2$ und $\varphi'(E_1 \times E_2) = \varphi_1(E_1)\varphi_2(E_2)$ endlich, und $E_2 \times E_3 = \underset{\nu}{S} E_2 \times E_{3\nu}$, wo $E_2 \times E_{3\nu} \varepsilon \mathfrak{M}_2 \times \mathfrak{M}_3$ und $\varphi''(E_2 \times E_{3\nu}) = \varphi_2(E_2)\varphi_3(E_{3\nu})$ endlich.

Sei nun M eine Menge aus $E_1 \times E_2 \times E_3$; ist $x \varepsilon E_1$, so bezeichnen wir mit M_x die Menge aller $(y, z) \varepsilon E_2 \times E_3$, für die $(x, y, z) \varepsilon M$; ist $(x, y) \varepsilon E_1 \times E_2$, so bezeichnen wir mit M_{xy} die Menge aller $z \varepsilon E_3$, für die $(x, y, z) \varepsilon M$.

Sei nun insbesondere $M \varepsilon \overline{\mathfrak{M}}$. Setzen wir $\varphi = \varphi_1 \times \varphi_2 \times \varphi_3$, $\varphi' = \varphi_1 \times \varphi_2$, $\varphi'' = \varphi_2 \times \varphi_3$, so ist nach XXXIX wegen $\varphi = \varphi' \times \varphi_3$:

(34) $$\varphi(M) = (E_1 \times E_2)\int \varphi_3(M_{xy})d\varphi',$$

und wegen $\varphi = \varphi_1 \times \varphi''$:

(35) $$\varphi(M) = (E_1)\int \varphi''(M_x)d\varphi_1;$$

hierin ist nach XXXIX, wenn wir $\mathfrak{M}_2 \times \mathfrak{M}_3 = \mathfrak{M}''$ setzen: $M_x \varepsilon \overline{\mathfrak{M}}''$ für alle $x \varepsilon E_1$ abgesehen von einer Nullmenge für φ_1; also gilt nach XXXIX für alle $x \varepsilon E_1$, abgesehen von einer Nullmenge für φ_1:

$$\varphi''(M_x) = (E_2)\int \varphi_3(M_{xy})d\varphi_2;$$

durch Einsetzen in (35) erhalten wir also:

$$\varphi(M) = (E_1)\int \left((E_2)\int \varphi_3(M_{xy})d\varphi_2\right)d\varphi_1,$$

und durch Vergleich mit (34), indem wir für φ' wieder $\varphi_1 \times \varphi_2$ schreiben:

(36) $$(E_1 \times E_2)\int \varphi_3(M_{xy})d(\varphi_1 \times \varphi_2) = (E_1)\int \left((E_2)\int \varphi_3(M_{xy})d\varphi_2\right)d\varphi_1.$$

Wir sehen also, *dass das assoziative Gesetz der Multiplikation der total-additiven Mengenfunktionen sich analytisch durch das Reduktionstheorem der Doppelintegrale ausdrückt.*

Will man dieses Reduktionstheorem aus dem assoziativen Gesetz der Multiplikation herleiten, so hat man noch zu zeigen, dass für $\varphi_3(M_{xy})$ in (36) *jede beliebige* nicht negative $(\varphi_1 \times \varphi_2)$-messbare Funktion $f(x, y)$ auftraten kann (der Übergang zu Funktionen $f(x, y)$ beliebigen Zeichens erfolgt dann leicht in bekannter Weise).

Wir wählen zu dem Zwecke für E_3 den R_1 (die Menge aller reellen Zahlen), für $\mathfrak{M}_3$ das System der im Sinne von Lebesgue messbaren Mengen des R_1, für φ_3 das (eindimensionale) Lebesgue Maß μ_1; da dann $\mathfrak{M}_3$ ein für μ_1 vollständiger σ-Körper, ist $\overline{\mathfrak{M}}_3 = \mathfrak{M}_3$. Ist dann M die Menge aller (x, y, z) mit $x \varepsilon E_1$, $y \varepsilon E_2$, $z \varepsilon R_1$ für die $0 \leqq z < f(x, y)$, so ist $\varphi_3(M_{xy}) = f(x, y)$. Es ist also nur zu zeigen: *Ist $f(x, y)$ $(\varphi_1 \times \varphi_2)$-messbar, so ist $M \varepsilon \overline{\mathfrak{M}}$.*

Wir setzen $\mathfrak{M}_1 \times \mathfrak{M}_2 = \mathfrak{M}'$. Da $f(x, y)$ $(\varphi_1 \times \varphi_2)$-messbar, gehört für jedes z die Menge $[f > z]$ aller $(x, y) \varepsilon E_1 \times E_2$, in denen $f(x, y) > z$ ist, zu $\overline{\mathfrak{M}}'$. Seien $r_1, r_2, \ldots, r_n, \ldots$ die sämmtlichen positiven rationalen Zahlen und sei $M_n = [f > r_n]$; dann ist $M_n \varepsilon \overline{\mathfrak{M}}'$; bezeichnen wir noch mit I_n das durch $0 \leqq z \leqq r_n$ gegebene Intervall des R_1, so ist offenbar $M = \underset{n}{S} M_n \times I_n$. Da hierin $M_n \times I_n \varepsilon \overline{\mathfrak{M}}' \times \overline{\mathfrak{M}}_3$, und nach XXXIII $\overline{\mathfrak{M}}' \times \overline{\mathfrak{M}}_3 \subseteqq \overline{\mathfrak{M}}$, ist $M_n \times I_n \varepsilon \overline{\mathfrak{M}}$, und da $\overline{\mathfrak{M}}$ ein σ-Körper, ist auch $M \varepsilon \overline{\mathfrak{M}}$, w. z. b. w.

So erweist sich also das Reduktionstheorem der Lebesgue-Stieltjesschen Doppelintegrale als identisch mit dem assoziativen Gesetz der Multiplikation der total-additiven Mengenfunktionen.

LEONIDA TONELLI - *Direttore responsabile.* Finito di stampare il 30 settembre 1933 - XI.

Comments on Hans Hahn's work in Fourier analysis

Jean-Pierre Kahane

Orsay

Introduction

Let me summarize my feelings after reading the works of Hahn on Fourier series and related topics:

1. They are highly original, deep and far reaching

2. They are little known by people working in Fourier analysis

3. They are not the only works of Hahn used in Fourier analysis

4. Conversely, their interest and use is not limitated to Fourier analysis.

As a comment on points 2, 3, 4, let us have a look on a few books.

Zygmund 1935 gives two references to Hahn : *Ueber Fejér's Summierung der Fourierschen Reihe* (section V *Gesammelte Werke*) and *Ueber Folgen linearer Operationen* (section VI). Zygmund 1959, none.

Nina Bari quotes *Ueber die Menge der Konvergenzpunkte einer Funktionenfolge* (section IV), nothing in section V.

Kaczmarz-Steinhaus gives no reference at all.

Titchmarsh quotes the two main papers on Fourier integrals, and states two theorems of Hahn (pp. 14 and 20) in a restricted way.

On the other hand, Dunford and Schwartz give five references to Hahn and reserve a special treatment to the second paper on singular integrals *(Ueber die Darstellung gegebener Funktionen durch singuläre Integrale II)* when they study the sources of the principle of uniform boundedness, p. 80.

Now let us try to explore the works in section V, compare them with contemporary works, explain point 1 and reinforce point 4.

The works in section V can be divided into two groups:

– the first is related to singular integrals and orthogonal expansions,

– the second to Fejer's theorem and generalized harmonic analysis.

We shall divide our comments accordingly.

A. Singular integrals and orthogonal expansions

This group contains three articles, published in 1916 and 1918, that we shall denote by SI1, SI2, SI3.

SI1 and SI2 are two long and consecutive communications *Ueber die Darstellung gegebener Funktionen durch singuläre Integrale,* I, II (71 + 36 pages). SI3, *Einige Anwendungen der Theorie der singuläre Integrale,* continues SI2 and contains a correction to a theorem of SI2 that we shall discuss (23 pages).

The terms and notions in use at the time

In order to explain the content of these articles I shall use notions and symbols like L^1, L^2, L^∞, L^0, $\sigma(L^1, L^\infty)$. But it should be reminded that these did not exist at the time. Hahn worked on intervals (closed intervals are denoted by $< a, b >$, open intervals by (a, b)) or on the line $(-\infty, \infty)$, equipped with the Lebesgue measure. He considered Lebesgue integrable functions and very often used convergence in L^1; but there was no term for that, only formulas like

$$\lim_{n\,\infty} \int_a^b | g_n(x) - g(x) | \, dx = 0.$$

Convergence in L^2 already deserved a name (convergence in quadratic mean) because of orthogonal series. We shall discuss convergence in L^0 and in $\sigma(L^1, L^\infty)$ when explaining SI3. The duality between L^1 and L^∞ is used extensively in SI2 in relation with the Parseval formula, but L^∞ functions needed always a long presentation: bounded with the possible exception of a set of vanishing measure.

Orthogonal expansions, complete orthonormal systems (that we shall denote by CONS) were a familiar notion. Also, complete orthonormal system consisting of bounded (not uniformly bounded!) functions (CONS B), in order to define expansions of L^1-functions via formulas of the type $f_\nu = \int f\omega_\nu$.

Singular integrals appear as limits of integrals. Nowadays the paradigm for singular integrals is the Hilbert transform, and modern theory of singular integrals originates in the works of Calderón and Zygmund (see the book of E. M. Stein). In 1916 however singular integrals were essentially a way to represent a function:

$$f(x) = \lim_{n \to \infty} \int \varphi_n(\xi, x) f(\xi) \, d\xi. \qquad (1)$$

The theories of Fourier series and Fourier integrals gave a series of examples. The general problem of characterizing the sequences of kernels $\varphi_n(\xi, x)$ such that the above equality holds for a given class of functions f was initiated by Hobson in the *Proceedings of the London mathematical society* and by Lebesgue in *Annales de Toulouse* in 1909. The paper of Lebesgue proved very influential: it is quoted and used by Hahn at many places, from the beginning to the end of this series of papers on singular integrals, in particular, in the first sections of SI1 and SI3. Fifteen years later the trace of this paper is visible in the first chapter of Banach's book on linear operators.

The first important original idea of Hahn was to consider singular integrals for derivatives:

$$f^{(m)}(x) = \lim_{n \to \infty} \int \varphi_n(\xi, x) f(\xi)\, d\xi. \tag{2}$$

Results on (2) are related to (1) in two ways. First, they give new results on (1), when (2) is applied to the m-th iterated integral of f. Secondly, they give conditions for formal derivations of singular integrals of type (1) to hold.

Formulas (1) and (2) appear at the very beginning of SI1, together with an explanation on the meaning of $f^{(m)}(x)$. Usually $f^{(m)}$ is defined as the derivative of $f^{(m-1)}$. Hahn however did not want the definition of $f^{(m)}(x)$ to depend on the existence of $f^{(m-1)}$ in the neighbourhood of x. His first definition in SI1 is what we now call the Peano derivatives (this is Zygmund's terminology, without any explicit reference to Peano's works): whenever

$$f(x+h) = a_0 + a_1 h + \dots + a_m h^m + o(h^m) \qquad (h \to 0),$$

we write $f(x) = a_0, f'(x) = a_1, \dots f^{(m)}(x) = m! a_m$. Later on in SI1 he used also the „generalized derivatives", what we call the de la Vallée Poussin derivatives (actually introduced and used by de la Vallée Poussin in a beautiful paper published in 1908 in the Bulletin of the Royal Academy of Belgium, that Hahn knew very well and appreciated very much): the generalized derivatives of even order are defined by

$$\tfrac{1}{2}(f(x+h) + (f(x-h)) = a_0 + a_2 h^2 + \dots + a_{2p} h^{2p} + o(h^{2p})$$

and the generalized derivatives of odd order by

$$\tfrac{1}{2}(f(x+h) - f(x-h)) = a_1 h + a_3 h^3 + \dots + a_{2p+1} h^{2p+1} + o(h^{2p+1}).$$

Finally, in the last section of SI2, Hahn used a still less restrictive definition of $f^{(m)}(x)$, namely

$$f^{(m)}(x) = \lim_{h \to 0} \frac{\Delta_h^{m+1} F}{(2h)^{m+1}}(x),$$

where F denotes the indefinite integral of f, and Δ_h is the difference operator ($\Delta_h F(x) = F(x+h) - F(x-h)$). The relation between different kinds of high-order derivatives was studied later by Marcinkiewicz and Zygmund and the conclusion is, roughly speaking, that whenever the m-th derivative of a function f exists on a set E in a weak sense, it exists almost everywhere on E in a strong sense (see Zygmund 1959, chap. XI).

Charles de la Vallée Poussin had published the second edition of his *Cours d'analyse* in 1912. This also was an excellent reference for Hahn. The Lebesgue dominated convergence theorem was known in the stronger form given by Vitali (point convergence plus uniform integrability imply convergence of the integrals), though the concept of uniform integrability was expressed in a more complicated wording (uniform absolute continuity of the indefinite integral). The Fubini theorem on multiple integrals was also known and used. SI2 and SI3 depend on these ideas and tools.

SI1

Let me write the singular integral as $I_n(f, x)$:

$$I_n(f, x) = \int \varphi_n(\xi, x) f(\xi)\, d\xi.$$

The main purpose of SI1 is to establish pointwise theorems of the form

$$f(x) = \lim_{n \to \infty} I_n(f, x) \qquad (1)$$

when f belongs to a given class (L^1, L^2, L^∞ in particular), not only when f is continuous at x, but also under a weaker condition such as $f(x) = (F_{m+1})^{(m+1)}(x)$, F_{m+1} denoting the $m+1$-th iterated indefinite integral of f. As a help for this purpose and also for the sake of its own interest, other singular integral are introduced (see formula (2) above), giving theorems of the type

$$f^{(m)}(x) = \lim_{n \to \infty} \frac{d^m}{dx^m} I_n(f, x)$$

under specified conditions.

The classes are those considered by Lebesgue: $\mathcal{F}_1 = L^1, \mathcal{F}_2 = L^2, \mathcal{F}_3 = L^\infty$, $\mathcal{F}_4$ = class of functions having no discontinuity of the second kind ($f(x+0)\emptyset$) and $f(x-0)$ exist at each interior point x, as well as $f(a+0)$ and $f(b-0)$) . In § 1 Hahn states the theorems of Lebesgue which characterize the sequences of functions $\varphi_n(\xi)$ such that $I_n(f) = \int \varphi_n f = \int \varphi_n(\xi) d\xi$ converges to zero when $f \in \mathcal{F}_i (i = 1, 2, 3, 4)$. In Lebesgue's theorems functions are defined on a bounded interval. Hahn gives the extensions to functions defined on $(-\infty, \infty)$.

The hors d'œuvre continues in § 2 with a lemma inspired by Haar and Lebesgue: when the φ_n are not bounded in L^1, there exists a continuous function f, vanishing on a given finite set, such that $I_n(f)$ does not tend to zero.

The main course is given in § 3. Theorem VI, due to Lebesgue, characterizes the sequences of kernels $\varphi_n(\xi, x)$ for which (1) is true whenever $f \in \mathcal{F}_i$ and f is continuous at x. Theorem VII, due to Hahn, expresses the necessary and sufficient conditions on the kernels when (1) is replaced by (2): representation of the m-th derivative (in the Peano sense) by a singular integral.

Theorem VIII in § 4 deals with generalized derivatives (in the sense of de la Vallée Poussin), under a symmetry condition on the kernels around x depending on the parity of m.

Then § 5 considers the matter of the preceding paragraphs from the point of view of uniform convergence. Theorems IX, X, XI deal with the case of bounded intervals, theorems IX_a, X_a, XI_a with $(-\infty, \infty)$.

§ 6 shows the power of theorems VII and VIII in order to extend (1). The starting point is again a theorem of Lebesgue (th. XII) which gives sufficient conditions in order to have (1) at each point x where $f(x) = F'(x)$ (F being the indefinite integral), therefore, to have (1) almost everywhere. It appears as a corollary of theorem VII, and another statement (th. XIII) as corollary of VIII. Moreover theorems XII and XIII are just the beginning of a chain of sharper and sharper theorems obtained in the same way (§ 7).

§ 8 contains the applications of theorems VII and VIII to differentiation of singular integrals, with a particular emphasis on the case $\varphi_n(\xi, x) = \varphi_n(\xi - x)$, when $I_n(f) = \varphi_n * f$ (of course, the notation and even the notion of convolution did not exist); generalized derivatives appear when $\varphi_n(x) = \varphi_n(-x)$. Extensions to $(-\infty, \infty)$ and results on uniform convergence are given in § 9.

§ 10 to 17 apply the general theory to three important types of singular integrals $I_n(f) = \varphi_n * f$.

The Stieltjes type is defined by

$$\varphi_n(u) = c_n(\varphi(u))^{i_n}$$

where $u \in (-\ell,\ell)$, $\varphi(u) \geq 0$, $\varphi(0) = \lim_{u\to 0} \varphi(u) = 1$, $\sup_{|u|>\epsilon} \varphi(u) < 1$ for all $\epsilon > 0$, $i_n \to \infty$ and the c_n are chosen in such a way that $\int \varphi_n$ tends to 1. Hahn chose the name on the basis of a letter of Stieltjes to Hermite (see theorem XXI and the footnote); the particular cases $\ell = 1$, $\varphi(u) = 1 - u^2$ and $\ell = \pi$, $\varphi(u) = (\cos\frac{u}{2})^2$ (read π instead of 2π on top of p. 630) were studied by Landau and de la Vallée Poussin. Hahn considers also the case $\ell = \infty$. § 10 gives (1) $f(x) = \lim_n I_n(f)(x)$ when $f \in F_i$ and f is continuous at x, § 11 an estimate for c_n when

$$\varphi(u) = 1 - \alpha|u|^p + o(|u|^p) \qquad (u \to 0)$$

(the case $p = 2$ belongs to Stieltjes), § 12 the extension of (1) to points of discontinuity and conditions for (2) $f^{(i)}(x) = \lim_n (I_n(f))^i(x)$.

The Weierstass singular integral can be considered as the particular case $\ell = \infty$

$$\varphi(u) = e^{-u^2}, \qquad i_n = n, \qquad c_n = \sqrt{\tfrac{n}{\pi}}.$$

Hahn enlarges the frame in another way. The Weierstrass type (§ 13) is defined by

$$\varphi_k(u) = \tfrac{k}{\omega}\varphi(ku) \qquad (k \to \infty)$$

where $\varphi \in L^1(\mathbb{R})$ and $\int \varphi = \omega \neq 0$. Here (1) reads

$$f(x) = \lim_{k\to\infty} (f * \varphi_k)(x).$$

Assuming f continuous at x, (1) holds true whatever $f \in L^\infty$. It holds true whatever $f \in L^2$ if and only if

$$\int_{-\infty}^{-u} + \int_u^\infty \varphi^2 = O(\tfrac{1}{u}) \qquad (u \to \infty)$$

(theorem XXVI). It holds true whatever $f \in L^1$ if and only if

$$\varphi(u) = O(\tfrac{1}{|u|}) \qquad (|u| \to \infty)$$

(theorem XXVII). These remarkable results are completed in § 13 and 14 by considering discontinuity points and expressions for derivatives (formula (2) above).

Finally the Poisson type (§ 16) is defined by

$$\varphi_k(u) = c_k \frac{1}{1 + k\varphi(u)}$$

where

$$\varphi(u) = \alpha \, | \, u \, |^p + o(\, | \, u \, |^p) \qquad (u \to 0)$$

and $\inf_{|u|>\varepsilon} \varphi(u) > 0$ whatever $\varepsilon > 0$. The classical Poisson kernel corresponds to $\varphi(u) = 1 - \cos u$. Hahn gives estimates for the c_k and gives conditions for (1) or (2) as in the preceding cases (§ 17).

Let me add that SI1 is beautifully written, in a very precise way. Tough it is a very long paper, it is quite pleasant and easy to read.

SI2

SI2 is a continuation of SI1 and applies it to the theory of orthogonal expansions. However the point of view is quite different. Hahn is no more interested only in pointwise convergence of $I_n(f)$ to f, but in the L^1, $\sigma(L^\infty, L^1)$ or $\sigma(L^1, L^\infty)$ convergence, which play a key role when Hahn investigates the validity of the Parseval relation

$$\int fg = \sum f_\nu g_\nu$$

with $f \in L^1$ and $g \in L^\infty$. Parseval's relation proves a good way to go back to pointwise representations of functions, to which the end of the article is dedicated. Again, particular cases are studied at length: summation processes of Poisson, Fejér, de la Vallée Poussin.

In this review I shall insist on the beginning of the article (§ 1 to 4). Theorem II needs a special comment. § 3 and 4 can be viewed as an application of the principle of uniform boundedness; from a historical point of view however they are one of the sources of this principle.

In § 1 all kernels are defined on a rectangle $(a, b) \times (a', b')$ and usually $a \le a' < b \le b'$. The Fubini theorem gives a condition on the kernel φ for

$$I(f, x) = \int_a^b f(\xi, x) d\xi$$

to exist almost everywhere and to belong to $L^1(a', b')$ whenever $f \in L^1(a, b)$

(theorem I), resp. $f \in L^2$ resp. $f \in L^\infty$ (theorem I"). In theorem II we are given a sequence of kernels φ_n satisfying conditions

1) $\forall x, \int_a^b |\varphi_n(\xi, x)|\, d\xi < M$

2) $\forall \xi, \int_{a'}^{b'} |\varphi_n(\xi, x)|\, dx < M$

3) $\forall x, \forall a_x$, (subset of (a, b) with density 1 at x)

$$\lim_n \int_{a_x} \varphi_n(\xi, x)\, d\xi = 1$$

and Hahn states the conclusion: whenever $f \in L^1(a', b')$

$$\lim_{n\to\infty} \int_{a'}^{b'} |f - I_n(f)| = 0. \tag{7}$$

However there is a gap in the proof: the argument for formula (8) of the paper is not conclusive. This is pointed out in SI3, where a corrected version of theorem II is stated and proved: instead of (7), the conclusion is

$$\lim_{n\to\infty} \int_{a''}^{b''} |f - I_n(f)| = 0$$

whenever $a' < a'' < b'' < a'$. In order to have (7) Hahn gives an extra condition

3') $\lim_n \int |\varphi_n(\xi, x)|\, d\xi = 1$

Since theorem II is used through the whole article SI2 (in particular, in theorem VIII, § 5) it is appropriate to decide whether it is true or not. Let me check that it is true.

Under assumption 2) the mapping $f \to I_n(f)$ is bounded from L^1 to L^1 because

$$\begin{aligned}\int |I_n(f)| &= \int \left| \int f(\xi)\varphi_n(\xi, x)\, d\xi \right| dx \\ &\le \int\int |f(\xi)|\, |\varphi_n(\xi, x)|\, d\xi\, dx \\ &\le M \int |f|.\end{aligned}$$

Under assumption 3), applied when a_x is an interval, $I_n(f)(x)$ tends to $f(x)$ almost everywhere (namely, for $x \neq \alpha, \beta$) as $n \to \infty$ whenever $f = 1_{[\alpha,\beta]}$, the indicator function of an interval. Taking $f = 1_{|\alpha,\beta|}$ again, assumption 1) gives $|I_n(f)(x)| < M$. The Lebesgue dominated convergence applies, therefore (7) holds when $f = 1_{|\alpha,\beta|}$. Since linear combinations of such f are dense in L^1 and the mapping $f \to I_n(f)$ is bounded in L^1, (7) holds whenever $f \in L^1$.

This is much simpler than the tentative proof of Hahn, but relies on ideas familiar to us but not at the time: Banach spaces, bounded operators, density in function spaces.

§ 2 gives theorems similar to I and II when (a, b) and (a', b') are replaced by $(-\infty, \infty)$. Theorem IIa, whose conclusion is

$$\lim_{n\to\infty} \int_{-\infty}^{\infty} |f - I_n(f)| = 0,$$

uses 1), 2), 3) (mutatis mutandis) and an extra condition

$$4) \qquad \forall \xi, h > 0, \ \int_{-\infty}^{\xi-h} + \int_{\xi+h}^{\infty} |\varphi_n(\xi, x)|\, dx \underset{n\to\infty}{\longrightarrow} 0.$$

It applies to convolution transforms as the Weierstrass, Cauchy, Poisson transforms where all conditions are checked easily. Hahn remarks that condition 4) cannot be removed – a simple and interesting counter example is provided by $\varphi_n(\xi, x) = \varphi_n^*(\xi - x)$ and

$$\varphi_n^* = \tfrac{n}{2} 1_{\left|-\frac{1}{n}, \frac{1}{n}\right|} + 1_{|n, n+1|}.$$

§ 3 contains two general theorems on functions defined on intervals:

$$\overline{\lim} \int |\varphi_n| = \infty \Rightarrow \exists g \text{ continuous}, \overline{\lim} \int g\varphi_n = \infty$$

(theorem III) and

$$\exists \text{intervals } I_i, \text{ integers } n_i, \int_{I_i} \varphi_{n_i} = \infty \Rightarrow$$
$$\exists g \text{ absolutely continuous}, \overline{\lim}\, g\varphi_n = \infty$$

(theorem IV). These theorems implicitely involve 1) a principle of uniform boundedness; 2) the description of continuous linear functionals on the spaces of continuous resp. absolutely continuous functions.

In § 4 begins the application to orthonormal systems. Hahn always considers a complete orthonormal system (CONS) ω_ν, even when the completeness condition is not formulated explicitely, as later on in theorems XIII and XIV. When $f \in L^2$, the coefficients of f are the

$$f_\nu = \int f\omega_\nu.$$

When $f \in L^2$, $g \in L^2$, then

$$\int fg = \sum f_\nu g_\nu, \tag{P}$$

the Parseval relation. Hilbert called it the completeness relation (Vollständigkeitsrelation), and beforehand, at the beginning of the century, it was for Hürwitz the fundamental theorem on Fourier series (Fundamentalsatz der Fourier Reihen), when (ω_ν) is the trigonometric system. In this case and for L^2-functions, the theorem is due to Fatou, and the name of Parseval was used by Fatou, Lebesgue, de la Vallée Poussin (see the books of Lebesgue and de la Vallée Poussin). The name of Parseval relation in the general case of a CONS was given by Hahn. Anyway (P) plays a central role in SI2.

Hahn is interested in the question: when is (P) true under the assumptions $f \in L^1$, $g \in L^\infty$? In order to define f_ν an extra condition is needed on the ω_ν: each ω should be bounded (CONSB). This is assumed in the sequel. Given $g \in L^\infty$, (P) holds whatever $f \in L^1$ if and only if the L^∞-norms of the partial sums $\Sigma_1^n g_\nu\omega_\nu$ are bounded; that is theorem V. If we want to use a modern wording „(P) holds whatever $f \in L^1$" expresses that the series $\Sigma_1^\infty g_\nu\omega_\nu$ converges to g in $\sigma(L^\infty, L^1)$. Theorem VI gives a necessary and sufficient condition for (P) to hold whatever $f \in L^1$ and $g \in L^\infty$; it reads

$$\sup_n \operatorname{ess.sup}_x \int \Big|\sum_1^n \chi_\nu(\xi)\omega_\nu(x)\Big|\, d\xi < \infty. \tag{i}$$

Though the principle of uniform boundedness has many sources, Dunford and Schwartz mention this result in a particular way, indicating the very page where it appears (p. 678). Theorem VII is the analogue for $f \in L^1$ and $g \in V$ (functions with bounded variation); here the condition is

$$\sup_{n,\alpha,\beta} \operatorname{ess.sup}_x \Big|\int_\alpha^\beta \sum_1^n \omega_1(\xi)\omega_\nu(x)\, d\xi\Big| < 1. \tag{ii}$$

In the case of the trigonometric system the first condition fails and the second holds. To consider g in V is far from being artificial. It applies to the case when g is the indicator function of an interval, giving

$$\int_{\alpha}^{\beta} f = \sum_{1}^{\infty} f_{\nu} \int_{\alpha}^{\beta} \omega_{\nu}$$

and actually the Parseval relation for all $g \in V$ is equivalent to the above equality for all intervals (α, β).

§ 5 is another elaboration on the case $f \in L^1$, $g \in L^{\infty}$. Whenever theorem II applies, it provides a process of summation of the series $\Sigma f_{\nu} g_{\nu}$, such that the Parseval relation still holds (with a new meaning of Σ). This is the basis of a long series of theorems in § 5 and § 6, devoted to the series $\Sigma f_{\nu} \omega_{\nu}$, with applications to ordinary Fourier series and ordinary processes of summation. Hahn forgot to number theorems XVI and XVII (p. 33: Wir haben das Resultat; p. 34: So erhalten wir den Satz). The last theorem (it should be XIX) says that a generalized Riemann process of summation provides a L^1 approximation for periodic functions. Results of this kind, involving convolutions on the circle, can be obtained easily through vectorial integration, as explained in chapter I of Katznelson's book.

In brief, the considerable interest of SI2 relies not so much on the precise results on Fourier series, but on the conceptual tools introduced by Hahn in order to obtain these results.

SI3

This is true also for SI3. SI3 contains some excellent applications (Anwendungen) to orthogonal series, involving the above conditions (*i*) and (*ii*). However the central part (§ 3) is a study on L^0, L^1, and $\sigma(L^1, L^{\infty})$ convergence.

§ 1 contains theorems on punctual convergence (III, IV), simple corollaries of results of Lebesgue (I, II), but generalizing results of Haar (1910) and Helly (1912). They give necessary and sufficient conditions for

$$f(x) = \sum_{1}^{\infty} f_{\nu} \omega_{\nu}(x)$$

to hold at continuity points, when f is either bounded without discontinuity of the second kind (theorem III) or with bounded variation (theorem IV). III involves a condition (*i*) and (IV) a condition (*ii*).

§ 2 is inspired by theorem VI of SI2. Here (ω_{ν}) is a CONSB and (*i*) appears as a necessary and sufficient condition for the equality $f = \Sigma_1^{\infty} f_{\nu} \omega_{\nu}$ to hold in L^1 (theorem V).

Theorem VIII in § 4 shows that the same holds when convergence in L^1 is replaced by $\sigma(L^1, L^{\infty})$ convergence. As a consequence there exists

$f \in L^1(\mathrm{T})$ whose Fourier series diverges in $\sigma(L^1, L^\infty)$. This was the best result of this kind before the examples of Kolmogorov of almost everywhere (1923), then everywhere (1926) divergent Fourier series.

Convergence in L^0, L^1, $\sigma(L^1, L^\infty)$ appear everywhere through this article, but it is specifically the matter of § 3. Convergence in L^0 ($f_n \to 0$ if $\int 1 \wedge |f_n| \to 0$) is another name for convergence in measure, then called asymptotic convergence („die sogennante asymptotische Konvergenz"). Convergence in L^1 appears everywhere but it has no name. The $\sigma(L^1, L^\infty)$ convergence is called complete integral convergence („vollständig integrierbare Konvergenz"), because

$$g_n \xrightarrow{\sigma(L^1, L^\infty)} \forall a \int_a g_n \to 0$$

a denoting a measurable subset of the interval of definition. The uniform integrability of a sequence (g_n) in L^1 is defined and named through the indefinite integrals of the g_n and the notion of uniform absolute continuity. Theorem V is a „translation" of a theorem of Vitali (1907) where this notion is introduced: the uniform integrability of the g_n is a necessary and sufficient condition for $g_n \to g$ in L^0 to imply $g_n \to g$ in L^1. This is a far reaching extension of the Lebesgue dominated convergence theorem, of constant use nowadays in probability theory. Theorem VI is related to a sequence (g_n) in L^1 such that $g_n \to g$ in L^0; then

$$g_n \to g \text{ in } L^1 \Leftrightarrow g_n \to g \text{ in } \sigma(L^1, L^\infty).$$

A counter example shows that this fails if the condition $g_n \to g$ in L^0 is dropped. From the context and the notes it is clear that L^0-convergence was known as the weakest of the kinds of convergence under consideration; but it is mentioned explicitely for the almost everywhere punctual convergence and the L^2 convergence only.

§ 5 is devoted to theorem II of SI2, as we already said.

The contribution of Hahn to the theory of Banach spaces is well known, if only by the Hahn-Banach theorem. His contribution to functional analysis in SI2 and SI3 goes far beyond the frame of Banach spaces but there is no hint that Hahn was interested in 1916 in formalizing the Banach space structure of L^1, though $L^1(a,b)$ and $L^1(-\infty, \infty)$ were the main objects of his study. His main purpose was to represent $f(x)$, not f. This will appear, in a different form, in the second group of papers that we shall consider.

B. The Fejér theorem and generalized harmonic analysis

This group contains three articles and a short résumé:
- *Ueber Fejérs Summierung der Fourier Reihe (1916) (8 pages), that we shall denote by FS*
- *Ueber Fouriersche Reihen und Integrale (1924), 1 page: FR1*
- *Ueber die Methode der arithmetischen Mittel in der Theorie der verallgemeinerten Fourier'schen Integrale (1925), 22 pages: VFI1*
- *Ueber eine Verallgemeinerung der Fourier Integralformel (1926), 53 pages: VFI2.*

FS

FS is the only paper quoted by Zygmund 1935 in section V, actually the only paper devoted entirely to Fourier series. It consists in one theorem and a delicate construction. The result is striking even now. Hahn observes a slight difference between the Fejér summation process and the other most important processes of summation (Poisson, Riemann, de la Vallée Poussin). The last are known to converge to $f(x)$ whenever

$$\int_0^t (f(x+u)+f(x-u)-2f(x))du = o(t) \qquad (t \downarrow 0). \tag{1}$$

The first is known to converge to $f(x)$ when

$$\int_0^t |f(x+u)+f(x-u)-2f(x)|\,du = o(t) \qquad (t \downarrow 0). \tag{2}$$

a stronger condition. Is this difference just a lack of knowledge, or is it essential? Hahn proves that it is essential, by constructing an example of f which satisfies (1), with divergent Fejér sums at x.

Though the result and the example are striking, they express a rather simple fact: unlike the other kernels under consideration, the Fejér kernel inheritates something from the oscillatory character of the Dirichlet kernel. Nowadays, the Banach Steinhaus theorem provides a quick proof for the existence of continuous functions whose Fourier series diverge at a given point, say, 0. It applied also in the present case, and Hahn's result can be presented in the following way. Consider the linear forms

$$\sigma_n^*(f) = \frac{1}{\pi n}\int_{-\frac{\pi}{2}}^{\frac{\pi}{2}} f(2t)\,\frac{\sin^2 nt}{t^2}\,dt$$

(modified Fejér sums of f at 0) defined on the Banach space H which consists of functions $f \in L^1(-\pi, \pi)$ satisfying (1) with $x = 0$ and $f(0) = 0$, equipped with the norms

$$\|f\|_H = \int_{-\pi}^{\pi} |f| + \sup_{0<t<\pi} \frac{1}{t} \left| \int_0^t (f(u) + f(-u))\, du \right|.$$

We have to prove that they are not bounded. The trick is to consider

$$f(t) = f_{p,\alpha}(t) = (p\,|t| \cos p\,|t| + \sin p\,|t|)\,1_{[-\alpha,\alpha]}(t)$$

where $\alpha = \frac{2\pi q}{p}$, q integer $< \sqrt{p}$. Such functions belong to H and their H-norms are bounded. On the other hand, an integration by parts gives

$$\sigma_n^*(f) = \frac{1}{\pi} \int_0^{\alpha} t \sin 2pt \,\frac{\sin 2nt}{t^2}\, dt + O(t).$$

Choosing $p = n$ and $g \to \infty$ $(n \to \infty)$ shows that the linear forms $\sigma_n^*(f)$ are not bounded on H,what we had to prove.

Let us add a comment about the sums of de la Vallée Poussin. Nowadays that is the name for combinations of Fejér sums, like $2\sigma_{2n} - \sigma_n$ (see for example Katznelson's book). The proof easily shows that Hahn's result holds for these sums also. When Hahn evokes the de la Vallée Poussin sums at the beginning of his article, he means the convolutions by kernels $c_n \,(\cos\frac{t}{2})^{2n}$, a quite different notion.

Generalized harmonic analysis, Hahn and Wiener

FRI, VFI1 and VFI2 are the contributions of Hahn to what Norbert Wiener later called Generalized Harmonic Analysis.

FRI is a 14 lines long announcement of the matter of VFI2, published in 1926. VFI1, published in 1925, is a continuation of VFI2, but it is self contained and shorter, so that it can be adviced to read the papers in the chronological order FRI, VFI1, VFI2.

When VFI2 was in print Wiener published a paper *On the representations of functions by trigonometric integrals.* Hahn had an idea of Wiener's investigations through two short notes of the *Bulletin of the American Mathematical Society* issued in 1925. This was not enough to decide whether Wiener's methods had any relation with his: *„Ob seine Methoden mit den meinen in irgend einer Beziehung stehen, entzieht sich meiner Kenntnis"* (VFI2, p. 302). When Wiener's paper appeared Hahn added a footnote insisting on their importance: *„die sehr bedeutungsvollen Untersuchungen des Herrn Wiener sind mittlerweile ausführlich erschienen"*. A few years later, in 1930, Wiener published his extensive study on Generalized Harmonic Analysis, a real monument. Hahn never returned to the subject, though he intended to do so in 1925: *„An anderer Stelle hoffe ich, bald zeigen zu können, wie sich hier vorgebrachte Untersuchungen auf alle Funktionen ausdehnen lassen, die im Unendlichen beschränkt sind"* (VFI2, p. 301). Therefore it may be appropriate to consider the relations between Wiener's and Hahn's investigations, the question that Hahn left open in 1925. To begin with let us have a look on the *„bedeutungsvollen Untersuchungen des Herrn Wiener"*, his 1925 paper.

The first extensive article of Harold Bohr on almost periodic functions had just appeared in *Acta Mathematica* (1924). This is a common source for Wiener and Hahn. Wiener starts from the Plancherel formula (1910) for Fourier integrals, insists on the L^2 convergence to $\hat{f}(\lambda)$ of the integrals

$$\int_{-A}^{A} f(\xi) e^{i\lambda\xi} d\xi$$

as $A \to \infty$ when $f \in L^1 \cap L^2$, compares this L^2 convergence with the L^2 convergence of Fourier series of periodic functions, then to the L^2 aspect of Bohr's theory, i. e.

$$\lim_{N\to\infty} \lim_{T\to\infty} \frac{1}{2T} \int_{-T}^{T} |f(x) - \sum_{1}^{N} a_n e^{i\lambda_n x}|^2 dx = 0$$

when f is an almost periodic function with frequencies λ_n and coefficients a_n. He defines a L^2 analogue of bounded functions, namely the functions which belong locally to L^2 and satisfy

$$\sup_x \int_x^{x+1} |f|^2 < \infty,$$

that he names „nearly bounded". For a nearly bounded function f the integrals

$$\frac{1}{\pi} \int_{-T}^{T} f(\lambda) \frac{\sin \alpha\lambda}{\lambda} d\lambda$$

converge locally in L^2 as $T \to \infty$ to a limit $\gamma(\alpha)$ and the even part of f can be obtained through a L^2 summability process from a kind a Stieltjes integrals that Wiener denotes by

$$I_0^T \cos \alpha x d\gamma(\alpha).$$

The odd part can be obtained in a similar way (theorems I and II in Wiener's paper). Then Wiener develops a theory of nearly bounded almost periodic functions, which he calls pseudo-periodic. Actually Wiener's theory coincides with that of Stepanoff, then a student of Harald Bohr, and the new class of locally L^2 almost periodic functions was named after Stepanoff and not Wiener. The term of pseudo-periodic stayed vacant for another use (the preceding class plus the condition $\inf(\lambda_{n+1} - \lambda_n) > 0$) developed in the book of Paley and Wiener (1934).

The introduction and the bibliography in Wiener's article on *Generalized Harmonic Analysis* point out the sources and the other streams. The sources, around 1900, are celestial mechanics (Bohl, Esclangon, who anticipated a kind of almost periodic functions) and, more important, the periodogram of Schuster used in interference phenomena and hidden periodicities of terrestrial magnetism. Also Harald Bohr belongs to the sources. Hahn appears as a parallel stream: „the work of Hahn seems to have a much more definitely pure mathematics motivation“ and then Wiener moves towards Plancherel and Titschmarsh. However his appreciation of Hahn's work is clear in his preliminary comments on the bibliography. Only four names appear: Schuster, Bohr, Hahn, Fourier. Wiener distinguishes ten groups of papers, and the seventh is defined as

> „the Hahn direction of work, treating generalized harmonic analysis from the stand point of ordinary convergence, rather than from that of convergence in the mean“.

This short sentence answers Hahn's question on the relation between Wiener's and his investigations in a perfect way. On the other hand, Wiener proved able to develop generalized harmonic analysis in so many ways and to connect it with so many important questions that the decision of Hahn, to leave the subject, appears as very natural.

Let us turn to the Hahn direction of work.

FRI

FRI is an excellent overall description. For simplicity let us suppose that $f(x)$ is an even function, $-\infty < x < \infty$. Under very general assumptions it is possible to define

$$\Phi(\mu) = \int_{-\infty}^{\infty} f(x) \frac{\sin \mu x}{x} dx$$

and to write

$$f(x) = \frac{1}{\pi}\int_0^\infty \cos \mu x \, d\Phi(\mu). \tag{0}$$

This contains the Fourier reciprocity formula for Fourier integrals, as well as the Fourier series expansion for periodic functions.

VFI1

In this program there are two points. First, define the meaning of (0) when $\Phi(\mu)$ is a continuous function. Then, prove that $\Phi(\mu)$ is continuous under convenient conditions on *f*.

The first point is treated at the beginning of VFI1. It is a short exposition of the theory of integrals

$$\int_a^b f\, dg \tag{$*$}$$

where *g* is continuous and *f* continuous with bounded variation, a kind of dual version of what we now call the Stieltjes integral (but for Hahn (∗) was a Stieltjes integral also, cf. FRI). This was developed already in Hahn's article of 1922 *Über Folgen linearer Operationen* (here in section VI) and would deserve to be called the Hahn integral.

The second point is in VFI2 and we shall turn to it later on.

When *f* is periodic and the Fourier series of *f* converges to *f* at point *x*, (0) holds with these definitions (we always suppose for simplicity that *f* is an even function). This is mentioned in VFI1 in several places (pp. 449, 462). The main purpose of VFI1 is to obtain a representation formula, generalizing Fejer's formula, when *f* is supposed to be continuous at *x*, locally integrable, and bounded at infinity. This formula appears at the beginning of the article p. 449, then as (32) p. 462, the functions Φ_2 and Ψ_2 being defined by (11). Let us express it when *f* is even:

$$\begin{aligned} f(x) &= \lim_{\mu\to\infty} \frac{1}{\mu\pi}\int_0^\mu \left(\int_0^\lambda \cos \mu x \,\frac{d^2\Phi_2(\tau)}{d\tau}\right) d\lambda \\ \Phi_2(\mu) &= \int_{-\infty}^\infty f(x)\,\frac{1-\cos \mu x}{x^2}\, dx. \end{aligned} \tag{00}$$

In (00) the inner integral has to be defined and that is the core of the article. Another form of (00) is (35) p. 464, where the analogy with Fejer's theorem is more visible; it reads

$$f(x) = \frac{1}{\pi} \lim_{\mu \to \infty} \int_0^\mu \left(\int_0^\lambda \cos \tau x d\Phi(\tau) \right) d\lambda$$

Φ being the same as in (0).

The end of VFI1 (from p. 467 on) is devoted to almost periodic functions. § 6 considers an interesting generalization: if f is locally integrable and bounded at infinity, and if the limit

$$A(\mu) = \lim_{T \to \infty} \frac{1}{T} \int_t^{t+T} f(x) \cos \mu x \, dx$$

exists uniformly with respect to t, then

$$A(\mu) = \frac{1}{2\pi} \lim_{h \to 0} \frac{\Phi_2(\mu + h) - 2\Phi_2(\mu) + \Phi_2(\mu - h)}{h}.$$

VFI2

The purpose of VFI2 is to study continuity properties of $\Phi(\mu)$ (odd function) and the analogue even function $\Psi(\mu)$, and to establish (0) and the corresponding formula for odd functions f, under appropriate conditions.

Before stating the conditions it is necessary to understand the meaning of the integrals. When Hahn writes $\int_{-\infty}^{\infty}$ it means $\lim \int_{-p}^{q}$ as p and q tend to ∞ (footnote p. 302). When he writes $\int_0^\infty$ it may mean $\lim \int_h^\infty$ as h $\downarrow 0$ (p. 321). On bounded intervals $\int f(x)dx$ is the Lebesgue integral, except when specified otherwise. On bounded intervals (a, b), $0 < a < b$, $\int_a^b f\,dg$ is always a Hahn integral, therefore, g is continuous.

In order to have (0) Hahn always assume that a Fourier series expansion of f in a neighbourhood of x converges to $f(x)$ at point x (this does not depend on the neighbourhood nor on the chosen period of the Fourier series); that is condition A). Another condition is needed, if only to guarantee the continuity of function $\Phi(\mu)$. The paper is organized according to the choice of this other condition. Anyhow f is supposed locally integrable.

The first additional condition is the integrability of $|f(x)/x|$ at infinity, which allows to write

$$f(x) = \frac{1}{\pi} \lim_{\lambda \to \infty} \int_{-\infty}^{\infty} f(y) \frac{\sin \lambda (y - x)}{y - x} dy \tag{1}$$

immediately: that is condition B). Under conditions A) and B) (0) holds, as well as the analogue for odd functions, $\Phi(\mu)$ and $\Psi(\mu)$ being continuous functions (theorem I). When moreover Φ and Ψ are absolutely continuous

(0) provides the usual reciprocity formula on Fourier integrals (theorem II). If we replace B) by the assumption $f = gh$, where g is monotonic at infinity and satisfies B), and h is periodic, the conclusions of theorems I and II are valid (theorem IV).

An alternative additional condition is the convergence to zero and the monotonicity at infinity of $f(x)/x$: that is condition C). Condition C) is enough for (1) to hold and for $\Phi(\mu)$ and $\Psi(\mu)$ to exist. However it does not insure the existence of the integral in (0). An additional condition is needed in order to have the analogue of theorem I: boundedness at infinity for f proves sufficient, and Hahn provides a number of variations about the theme $f = gh$, h periodic as above, g satisfying a modified condition of the type C) (theorems V to VIII).

Then Hahn treats the case f periodic, where condition A) is sufficient for the representation (0), the ordinary representation as a Fourier series (theorem IX). Again there are variations when $f = gh$, g being either the function sign x or monotonic and converging to zero at infinity, h beeing either periodic or almost periodic of a certain type (theorems X to XIV).

Hahn concludes his article in introducing condition D): $f \in L^2$. The last three pages contain three theorems (XV to XVII) which establish (0) and different forms of the ponctual reciprocity formula under assumption 4). He was aware of the importance of Plancherel's works (it is the final quotation of the article) and there is a striking similarity between his and Wiener's formulas. But Hahn considered L^2 as the very end while Wiener introduced it at the very beginning. Hahn did not consider the L^2 convergence as Wiener did, nor Wiener the ponctual convergence as Hahn did.

The Hahn direction of work in general harmonic analysis had just a common motivation with that of Wiener: to give a common frame and common formulas for Fourier integrals and Fourier series, including some almost periodic functions. The point of view of Wiener, L^2 first, proved more efficient and became classical. The point of view of Hahn, punctual convergence first, was inspired by the classical analysis of his time, and looks exotic by now. This may explain why the constribution of Hahn to generalized harmonic analysis is not very well known.

Conclusion

The history of mathematics shows that dominant tendancies are changing along time. Often the most promizing new tendancies are rooted deeply in the past. Fortunately the past of mathematics is a mine of jewels. Among these jewels are the works of Hahn on Fourier series and integrals, waiting for visitors.

References

1) To Hahn's works

See list in section V

2) To books

Zygmund A (1935) Trigonometrical series. Monografje matematyczne V, Warsaw-Lwow
Zygmund A (1959) Trigonometric series, volumes I and II. Cambridge University Press
Bari N (1961) Trigonometričeskie ryadi (in russian). Moskow
Kaczmarz S, Steinhaus H (1951) Theorie der Orthogonalreihen. Monografje matematyczne VI, Warsaw
Titchmarsh EC (1937) Introduction to the theory of Fourier integrals. Oxford University Press
Dunford N, Schwartz JT (1958) Linear operators, part I. Interscience, New York
Stein EM (1970) Singular integrals and differentiability properties of functions. Princeton University Press
Vallée Poussin Ch de la (1912) Cours d'analyse infinitésimale, tome II, 2ème èdition. Louvain et Paris
Lebesgue H (1906) Leçons sur les séries trigonométriques. Gauthier-Villars, Paris
Paley REAC, Wiener N (1934) Fourier transforms in the complex domain. American Mathematical Society Colloquium Publications XIX, New York
Katznelson Y (1968) An introduction to harmonic analysis. Wiley, New York
Banach S (1932) Théorie des opérations linéaires. Monografje matematyczne I, Warsaw-Lwow

3) To articles

Hobson EW (1908) On a general convergence theorem, and the theory of the representation of a function by series of normal functions. Proc London Math Soc (2) 6: 349–395
Lebesgue H (1909) Sur les intégrales singulières (followed by Remarques sur un énoncé dû à Stieltjes). Ann Fac Sci Toulouse (3) 1: 25–117 (plus 119–128)
Vallée Poussin Ch de la (1908) Sur l'approximation des fonctions d'une variable réelle et de leurs dérivées par des polynômes et des suites limitées de Fourier. Bull Acad Sci, Acad Roy Belgique (4) 10: 193–254
Kolmogoroff AN (1923) Une série de Fourier-Lebesgue divergente presque partout. Fund Math 4: 324–328
Kolmogoroff AN (1926) Une série de Fourier divergente partout. C R Acad Sci Paris 183: 1327–1328
Wiener N (1925) On the representation of functions by trigonometric integrals. Math Z 24: 576–616
Wiener N (1930) Generalized harmonic analysis. Acta Math 55: 117–258
Bohr H (1924) Zur Theorie der fastperiodischen Funktionen. Acta Math 45: 29–127; 46 (1925): 101–214

Hahn's work in Fourier analysis
Hahns Arbeiten zur Fourieranalysis

ÜBER DIE
DARSTELLUNG GEGEBENER FUNKTIONEN DURCH SINGULÄRE INTEGRALE
1. MITTEILUNG

VON

HANS HAHN
(BONN)

VORGELEGT IN DER SITZUNG AM 30. MÄRZ 1916

Bekanntlich führen viele Probleme der Analysis auf die Frage, ob für einen gegebenen »Kern« $\varphi(\xi, x, n)$ und eine (unter gewissen Beschränkungen verschiedener Art) willkürliche Funktion $f(\xi)$ die Beziehung gilt:

$$\lim_{n=\infty} \int_a^b f(\xi)\,\varphi(\xi, x, n)\,d\xi = f(x). \tag{1}$$

Soll diese Formel insbesondere an allen Stetigkeitsstellen von $f(\xi)$ bestehen, und gehören zu den zugelassenen Funktionen f speziell auch diejenigen, die in einem Intervalle $= 1$, außerhalb dieses Intervalles $= 0$ sind, so muß der Kern $\varphi(\xi, x, n)$ folgende Eigenschaft haben: für jedes Intervall I, das den Punkt x im Inneren enthält, ist:

$$\lim_{n=\infty} \int_I \varphi(\xi, x, n)\,d\xi = 1,$$

während, wenn x außerhalb I liegt:

$$\lim_{n=\infty} \int_I \varphi(\xi, x, n)\,d\xi = 0$$

ist. Das in Formel (1) auftretende Integral heißt dann ein singuläres Integral mit der singulären Stelle x.

Eine umfassende Untersuchung solcher singulärer Integrale rührt von H. Lebesgue her.[1] An die Ergebnisse dieser Untersuchung wird hier angeknüpft. Wir stellen der von Lebesgue mit großer Vollständigkeit gelösten Frage, für welche Kerne die Beziehung (1) an allen Stetigkeitsstellen von $f(\xi)$ gilt, die Frage an die Seite, für welche Kerne die Beziehung:

$$\lim_{n=\infty} \int_a^b f(\xi)\,\varphi(\xi, x, n)\,d\xi = f^{(m)}(x) \tag{2}$$

überall dort gilt, wo die m-te Ableitung $f^{(m)}(x)$ von $f(x)$ existiert. Dabei legen wir dem Begriffe der m-ten Ableitung nicht die übliche Definition zugrunde: »$f^{(m)}(x)$ ist die erste Ableitung von $f^{(m-1)}(x)$«, sondern

[1] Annales de Toulouse, Serie 3, Bd. 1, p. 25 ff. — Kurz vorher hatte einige hieher gehörige Theoreme bewiesen E. W. Hobson Proc. of the London math. Soc. Serie 2, Bd. 6, p. 349.

wir definieren in weiter tragender Weise[1]: existieren an der Stelle x die Ableitungen bis zur $(m-1)$-ten Ordnung und gilt die Entwicklung:

$$f(x+h)=f(x)+\frac{h}{1!}f'(x)+\ldots+\frac{h^{m-1}}{(m-1)!}f^{(m-1)}(x)+a.h^m+\omega(h).h^m,$$

wo a eine Konstante und $\lim\limits_{h=0}\omega(h)=0$ ist, so wird die m-te Ableitung definiert durch $f^{(m)}(x)=m!\,.\,a$.

Aus der Beantwortung der Frage nach der Gültigkeit von (2) lassen sich einerseits nun Bedingungen herleiten, unter denen (1) nicht nur an allen Stetigkeitsstellen von $f(\xi)$ gilt, sondern überall dort, wo $f(\xi)$ m-te Ableitung jener Funktion ist, die aus $f(\xi)$ durch m hintereinander ausgeführte unbestimmte Integrationen entsteht[2]; andrerseits gewinnt man aus der Beantwortung unserer Frage Bedingungen, unter denen Beziehung (1) m-mal nach x differenziert werden darf, und zwar in der Weise, daß man auf der linken Seite unter dem Integralzeichen differenziert.

Nach diesen allgemeinen Untersuchungen werden nun drei Typen singulärer Integrale näher betrachtet, die ich als den Stieltjes'schen Typus, den Weierstrass'schen Typus und den Poisson'schen Typus bezeichne. Der erste dieser Typen führt nämlich in einem besonders einfachen Falle auf eine Formel, die in einem Briefe von Stieltjes an Hermite enthalten ist, und die mit Formeln von Laplace und Darboux in engem Zusammenhange steht.[3] Der zweite dieser Typen diente Weierstrass zu seinem berühmten Beweise, daß jede stetige Funktion unbeschränkt durch Polynome approximierbar ist. Der dritte endlich enthält als Spezialfall das bekannte Poisson'sche Integral, das die erste Randwertaufgabe der Potentialtheorie für den Kreis löst und auf die sogenannte Poisson'sche Summierung nicht konvergenter Fourier'scher Reihen führt. Ein noch einfacherer Fall dieses dritten Typus liefert eine Formel, die als das Analogon der Poisson'schen Summierung für das Fourier'sche Integraltheorem betrachtet werden kann, da sie aus der Poisson'schen Summierung der Fourier'schen Reihe durch denselben Grenzübergang hergeleitet werden könnte, durch den die Fourier'sche Reihe ins Fourier'sche Integral übergeht.

§ 1. Einige Sätze von Lebesgue und ihre Übertragung auf unendliche Intervalle.

Sei $\langle a, b\rangle$ ein endliches Intervall[4]: $a\leqq x\leqq b$. Wir bezeichnen im Folgenden: mit $\mathfrak{F}_1$ die Klasse aller in $\langle a, b\rangle$ integrierbaren[5] Funktionen; mit $\mathfrak{F}_2$ die Klasse aller in $\langle a, b\rangle$ meßbaren Funktionen, deren Quadrat in $\langle a, b\rangle$ integrierbar ist; mit $\mathfrak{F}_3$ die Klasse aller in $\langle a, b\rangle$ geschränkten, meßbaren Funktionen; mit $\mathfrak{F}_4$ die Klasse aller Funktionen, die in $\langle a, b\rangle$ nur Unstetigkeiten erster Art besitzen.[6]

Es bedeute $\varphi(\xi, n)$ eine für alle ξ von $\langle a, b\rangle$ und alle nicht negativen ganzzahligen n definierte, als Funktion von ξ in $\langle a, b\rangle$ integrierbare Funktion. Wir setzen (allemal wenn dieses Integral existiert):

$$I_n(f)=\int_a^b f(\xi)\,\varphi(\xi, n)\,d\xi.$$

[1] Diese Definition setzt nicht, wie die übliche, voraus, daß $f^{(m-1)}(\xi)$ in einer Umgebung der Stelle x existiert.

[2] Man beachte die zugrunde gelegte Definition der m-ten Ableitung. Bei der üblichen Definition würde der Satz für beliebiges m nicht mehr aussagen, als für $m=1$, für welchen Fall er sich (in etwas abweichender Form und mit anderem Beweise) schon bei Lebesgue findet: a. a. O. p. 80.

[3] Wir verweisen diesbezüglich auf eine kleine Abhandlung, die Lebesgue dieser Formel widmet: Ann. de Toul. Serie 3, Bd. 1, p. 119 ff.

[4] Es bedeutet stets $\langle\alpha, \beta\rangle$ ein Intervall mit Einschluß, (α, β) ein Intervall mit Ausschluß seiner Endpunkte.

[5] Das Wort »integrierbar« ist hier, wie im Folgenden, im Sinne der Lebesgue'schen Integration zu verstehen.

[6] Das heißt: jede Funktion von $\mathfrak{F}_4$ hat in einem inneren Punkt von $\langle a, b\rangle$ einen rechtsseitigen und einen linksseitigen, im Punkt a einen rechtsseitigen, im Punkt b einen linksseitigen endlichen Grenzwert.

Von H. Lebesgue wurden folgende Sätze bewiesen:[1]

I. Damit $\lim_{n=\infty} I_n(f) = 0$ sei für alle Funktionen von $\mathfrak{F}_1$, ist notwendig und hinreichend, daß $\varphi(\xi, n)$ den Bedingungen genügt:

1. Es gibt eine Konstante M, so daß, abgesehen von Nullmengen:[2]

$$|\varphi(\xi, n)| < M \qquad \text{für alle } n \text{ und alle } \xi \text{ von } < a, b >.$$

2. Für jedes Teilintervall[3] $< \alpha, \beta >$ von $< a, b >$ ist:

$$\lim_{n=\infty} \int_\alpha^\beta \varphi(\xi, n)\, d\xi = 0.$$

II. Damit $\lim_{n=\infty} I_n(f) = 0$ sei für alle Funktionen von $\mathfrak{F}_2$, ist notwendig und hinreichend, daß $\varphi(\xi, n)$ den Bedingungen genügt:

1. Es gibt eine Konstante M, so daß:

$$\int_a^b (\varphi(\xi, n))^2\, d\xi < M \quad \text{für alle } n.$$

2. Für jedes Teilintervall $< \alpha, \beta >$ von $< a, b >$ ist:

$$\lim_{n=\infty} \int_\alpha^\beta \varphi(\xi, n)\, d\xi = 0.$$

III. Damit $\lim_{n=\infty} I_n(f) = 0$ sei für alle Funktionen von $\mathfrak{F}_3$, ist notwendig und hinreichend, daß $\varphi(\xi, n)$ den Bedingungen genügt:

1. Zu jedem $\mu > 0$ gibt es ein $\lambda > 0$, so daß für jede Menge I sich nicht überdeckender, in $< a, b >$ gelegener Intervalle, deren Gesamtinhalt $\leqq \lambda$ ist, die Ungleichung besteht:

$$\int_I |\varphi(\xi, n)|\, d\xi < \mu \quad \text{für alle } n.$$

2. Für jedes Teilintervall $< \alpha, \beta >$ von $< a, b >$ ist:

$$\lim_{n=\infty} \int_\alpha^\beta \varphi(\xi, n)\, d\xi = 0.$$

IV. Damit $\lim_{n=\infty} I_n(f) = 0$ sei für alle Funktionen von $\mathfrak{F}_4$, ist notwendig und hinreichend, daß $\varphi(\xi, n)$ den Bedingungen genügt:

1. Es gibt eine Konstante M, so daß

$$\int_a^b |\varphi(\xi, n)|\, d\xi < M \quad \text{für alle } n.$$

2. Für jedes Teilintervall $< \alpha, \beta >$ von $< a, b >$ ist:

$$\lim_{n=\infty} \int_\alpha^\beta \varphi(\xi, n)\, d\xi = 0.$$

Diese Sätze können, mit geringfügigen Abänderungen, auch auf Intervalle übertragen werden, die sich ins Unendliche erstrecken. Es wird genügen, dies für das Intervall $(-\infty, +\infty)$ zu besprechen.

---

[1] Ann. de Toulouse, Serie 3, Bd. 1, p. 51 ff. Dabei ist bei Anschreiben von $\lim_{n=\infty} I_n(f)$ stets mitverstanden, daß $I_n(f)$ für alle n existiert.

[2] Unter einer »Nullmenge« wird eine Menge verstanden, die, im Sinne von Lebesgue, den Inhalt 0 hat.

[3] Zu den Teilintervallen $< \alpha, \beta >$ von $< a, b >$, nicht aber zu denen von (a, b), rechnen wir auch das Intervall $< a, b >$ selbst.

Man erhält die Sätze für dieses unendliche Intervall am einfachsten, indem man sie durch die Substitution:

$$u = \frac{e^{\xi}-1}{e^{\xi}+1}, \qquad \xi = \lg \frac{1+u}{1-u}$$

auf die Sätze für das endliche Intervall $< -1, 1 >$ zurückführt. Die Substitutionsformeln lauten:

$$\int_{-\infty}^{+\infty} F(\xi)\,d\xi = 2\int_{-1}^{1} F\left(\lg\frac{1+u}{1-u}\right)\frac{1}{1-u^2}\,du \tag{1}$$

$$\int_{-1}^{1} G(u)\,du = 2\int_{-\infty}^{+\infty} G\left(\frac{e^{\xi}-1}{e^{\xi}+1}\right)\frac{e^{\xi}}{(e^{\xi}+1)^2}\,d\xi. \tag{2}$$

Die Definition der Klassen $\mathfrak{F}_i$ bleibt für unendliche Intervalle dieselbe wie für endliche Intervalle. Dabei hat man, was die Definition von $\mathfrak{F}_1$ anlangt, zu beachten, daß wir eine Funktion $f(\xi)$ dann und nur dann als integrierbar in $(-\infty, +\infty)$ bezeichnen, wenn:

$$\lim_{x=+\infty} \int_{-x}^{x} |f(\xi)|\,d\xi$$

einen endlichen Wert hat, so daß auch im unendlichen Intervall jede integrierbare Funktion absolut integrierbar ist. — Was die Definition von $\mathfrak{F}_4$ anlangt, so verlangen wir, daß für die zu $\mathfrak{F}_4$ gehörigen Funktionen auch die beiden Grenzwerte:

$$\lim_{\xi=-\infty} f(\xi) \quad \text{und} \quad \lim_{\xi=+\infty} f(\xi)$$

existieren und endlich sein sollen.

Dies vorausgeschickt erhalten wir, wenn wir nun:

$$J_n(f) = \int_{-\infty}^{+\infty} f(\xi)\,\varphi(\xi, n)\,d\xi$$

setzten, folgende Sätze:

Ia. »Damit $\lim_{n=\infty} J_n(f) = 0$ sei für alle Funktionen von $\mathfrak{F}_1$, ist notwendig und hinreichend, daß $\varphi(\xi, n)$ den Bedingungen genügt:

1. Es gibt eine Konstante M, so daß abgesehen von Nullmengen:

$$|\varphi(\xi, n)| < M \quad \text{für alle } n \text{ und alle } \xi.$$

2. Für jedes endliche Intervall $< \alpha, \beta >$ ist:

$$\lim_{n=\infty} \int_{\alpha}^{\beta} \varphi(\xi, n)\,d\xi = 0.«$$

Die Bedingungen sind notwendig. Für 2. ersieht man dies, indem man für f die zu $\mathfrak{F}_1$ gehörige Funktion wählt, die in $< \alpha, \beta >$ gleich 1, sonst $= 0$ ist.

Angenommen es wäre 1. nicht erfüllt, so wäre für:

$$\psi(u, n) = \varphi\left(\lg\frac{1+u}{1-u}, n\right) \tag{3}$$

in $< -1, 1 >$ Bedingung 1. von Satz I nicht erfüllt; es gäbe also eine in $< -1, 1 >$ integrierbare Funktion $g(u)$, für die nicht:

$$\lim_{n=\infty} \int_{-1}^{1} g(u)\,\psi(u, n)\,du = 0 \tag{4}$$

ist. Nach Formel (2) ist die Funktion

$$f(\xi) = 2g\left(\frac{e^{\xi}-1}{e^{\xi}+1}\right)\cdot\frac{e^{\xi}}{(e^{\xi}+1)^2}$$

integrierbar in $(-\infty, +\infty)$. Und vermöge derselben Formel folgt aus dem Nichtbestehen von (4) auch das Nichtbestehen von:

$$\lim_{n=\infty}\int_{-\infty}^{+\infty} f(\xi)\,\varphi(\xi, n)\,d\xi = 0.$$

Damit sind die Bedingungen als notwendig erwiesen.

Die Bedingungen sind hinreichend. In der Tat, wegen unserer Bedingung 1. genügt der Kern (3) der Bedingung 1. von Satz I in $<-1, 1>$, und wegen unserer Bedingung 2. ist, nach (1), in jedem Teilintervalle $<\alpha, \beta>$ von $(-1, 1)$:

$$\lim_{n=\infty}\int_{\alpha}^{\beta}\psi(u, n)\frac{1}{1-u^2}\,du = 0.$$

Da $\dfrac{1}{1-u^2}$ in $<\alpha, \beta>$ geschränkt ist, genügt der Kern:

$$\psi(u, n)\cdot\frac{1}{1-u^2}$$

in jedem Teilintervalle $<\alpha, \beta>$ von $(-1, 1)$ beiden Bedingungen von Satz I, so daß wir nach Satz I haben:

$$\lim_{n=\infty}\int_{\alpha}^{\beta}\psi(u, n)\,du = \lim_{n=\infty}\int_{\alpha}^{\beta}(1-u^2)\cdot\psi(u, n)\frac{1}{1-u^2}\,du = 0.$$

Es ist nun leicht einzusehen, daß auch:

(5) $$\lim_{n=\infty}\int_{-1}^{1}\psi(u, n)\,du = 0.$$

In der Tat, da (abgesehen von Nullmengen):

$$|\psi(u, n)| < M \quad \text{für alle } n \text{ und alle } u \text{ von } (-1, 1),$$

gibt es zu jedem $\varepsilon > 0$ ein $\eta > 0$, so daß für alle n:

(6) $$\left|\int_{-1}^{-1+\eta}\psi(u, n)\,du\right| < \frac{\varepsilon}{4}; \quad \left|\int_{1-\eta}^{1}\psi(u, n)\,du\right| < \frac{\varepsilon}{4}.$$

Da, wie bereits gezeigt:

$$\lim_{n=\infty}\int_{-1+\eta}^{1-\eta}\psi(u, n)\,du = 0$$

ist, so haben wir:

$$\left|\int_{-1+\eta}^{1-\eta}\psi(u, n)\,du\right| < \frac{\varepsilon}{2} \quad \text{für alle } n \geqq n_0,$$

und somit, wegen (6):

$$\left|\int_{-1}^{1}\psi(u, n)\,du\right| < \varepsilon \quad \text{für alle } n \geqq n_0,$$

womit (5) bewiesen ist.

So sehen wir, daß die Beziehung

$$\lim_{n=\infty}\int_{\alpha}^{\beta}\psi(u, n)\,du = 0$$

nicht nur für alle Teilintervalle $<\alpha, \beta>$ von $(-1, 1)$, sondern auch für alle Teilintervalle von $<-1, 1>$ besteht. – Es genügt also der Kern $\psi(u, n)$ in $<-1, 1>$ beiden Bedingungen von Satz I.

Ist nun $f(\xi)$ integrierbar in $(-\infty, +\infty)$, so ist nach (1) die Funktion

$$g(u) = 2f\left(\lg \frac{1+u}{1-u}\right) \frac{1}{1-u^2}$$

integrierbar in $<-1, 1>$. Nach Satz I ist also:

$$\lim_{n=\infty} J_n(f) = \lim_{n=\infty} \int_{-1}^{1} g(u)\, \psi(u, n)\, du = 0.$$

Damit sind die Bedingungen 1. und 2. als hinreichend erwiesen.

Bemerken wir noch ausdrücklich, daß Bedingung 2. nur von **endlichen** Intervallen $<\alpha, \beta>$ handelt. Folgendes Beispiel zeigt, daß sie in Intervallen, die sich ins Unendliche erstrecken, nicht erfüllt zu sein braucht: es sei:

$$\varphi(\xi, n) = \begin{cases} 1 & \text{in } (n, n+1) \\ 0 & \text{außerhalb } (n, n+1). \end{cases}$$

Dann ist für jede in $(-\infty, +\infty)$ integrierbare Funktion:

$$\lim_{n=\infty} \int_{-\infty}^{+\infty} f(\xi)\, \varphi(\xi, n)\, d\xi = 0,$$

während wir für jedes n haben:

$$\int_{-\infty}^{+\infty} \varphi(\xi, n)\, d\xi = 1.$$

II *a*. »Damit $\lim_{n=\infty} J_n(f) = 0$ sei für alle Funktionen von $\mathfrak{F}_2$, ist notwendig und hinreichend, daß $\varphi(\xi, n)$ den Bedingungen genügt:

1. Es gibt eine Konstante M, so daß:

$$\int_{-\infty}^{+\infty} (\varphi(\xi, n))^2\, d\xi < M \quad \text{für alle } n.$$

2. Für jedes **endliche** Intervall $<\alpha, \beta>$ ist:

$$\lim_{n=\infty} \int_{\alpha}^{\beta} \varphi(\xi, n)\, d\xi = 0.«$$

Die Bedingungen sind **notwendig**. Für 2. ist dies wieder evident. Wäre 1. nicht erfüllt, so würde, wegen (1), der Kern:

$$\psi(u, n) = \sqrt{2}\, \varphi\left(\lg \frac{1+u}{1-u}, n\right) \frac{1}{\sqrt{1-u^2}} \tag{7}$$

in $<-1, 1>$ der Bedingung 1. von Satz II nicht genügen; es gäbe also eine in $<-1, 1>$ samt ihrem Quadrate integrierbare Funktion $g(u)$, für die **nicht**:

$$\lim_{n=\infty} \int_{-1}^{1} g(u)\, \psi(u, n)\, du = 0 \tag{8}$$

ist. Nach Formel (2) ist das Quadrat der Funktion:

$$f(\xi) = g\left(\frac{e^\xi - 1}{e^\xi + 1}\right) \cdot \frac{\sqrt{2e^\xi}}{e^\xi + 1}$$

in $(-\infty, +\infty)$ integrierbar. Da aber:

$$\int_{-1}^{1} g(u)\,\psi(u, n)\,du = \int_{-\infty}^{+\infty} f(\xi)\,\varphi(\xi, n)\,d\xi$$

ist, folgt aus dem Nichtbestehen von (8) auch das Nichtbestehen von $\lim\limits_{n=\infty} J_n(f) = 0$.

Die Bedingungen sind hinreichend. In der Tat, wegen Bedingung 1. und Formel (1) genügt Kern (7) der Bedingung 1. von Satz II. Ferner folgt aus unserer Bedingung 2., daß in jedem Teilintervalle $<\alpha, \beta>$ von $(-1, 1)$:

$$\lim_{n=\infty} 2\int_{\alpha}^{\beta} \varphi\left(\lg\frac{1+u}{1-u}\right)\frac{du}{1-u^2} = \lim_{n=\infty} \sqrt{2}\int_{\alpha}^{\beta} \psi(u, n)\frac{du}{\sqrt{1-u^2}} = 0$$

ist. Da $\dfrac{1}{1-u^2}$ in $<\alpha, \beta>$ geschränkt ist, folgt, daß zugleich mit dem Kerne $\psi(u, n)$ auch der Kern $\sqrt{2}\dfrac{\psi(u, n)}{\sqrt{1-u^2}}$ in $<\alpha, \beta>$ der Bedingung 1. von Satz II genügt. Er genügt also in $<\alpha, \beta>$ beiden Bedingungen von Satz II, so daß, nach Satz II:

$$\lim_{n=\infty}\int_{\alpha}^{\beta} \psi(u, n)\,du = \lim_{n=\infty}\int_{\alpha}^{\beta} \sqrt{\frac{1-u^2}{2}}\cdot\psi(u, n)\cdot\sqrt{\frac{2}{1-u^2}}\,du = 0$$

ist. Wie oben sehen wir, daß hieraus wieder Gleichung (5) folgt, nur haben wir uns beim Beweise von (6) diesmal darauf zu berufen, daß nach der Schwarz'schen Ungleichung:

$$\left|\int_{1-\eta}^{1} \psi(u, n)\,du\right| < \sqrt{\eta\cdot\int_{1-\eta}^{1} (\psi(u, n))^2\,du} < \sqrt{\eta}\cdot M$$

ist. — Wir sehen nun wieder, daß der Kern $\psi(u, n)$ in $<-1, 1>$ beiden Bedingungen von Satz II genügt.

Gehört nun $f(\xi)$ in $(-\infty, +\infty)$ zu $\mathfrak{F}_2$, so gehört die Funktion

$$g(u) = \sqrt{2} f\left(\lg\frac{1+u}{1-u}\right)\sqrt{\frac{1}{1-u^2}}$$

in $<-1, 1>$ zu $\mathfrak{F}_2$. Nach Satz II ist also:

$$\lim_{n=\infty} J_n(f) = \lim_{n=\infty}\int_{-1}^{1} g(u)\,\psi(u, n)\,du = 0,$$

womit bewiesen ist, daß die Bedingungen von Satz II *a* hinreichend sind. Wie bei Satz I *a* sieht man, daß Bedingung 2. auch hier nur für endliche Intervalle $<\alpha, \beta>$ erfüllt sein muß.

III *a*. »Damit $\lim\limits_{n=\infty} J_n(f) = 0$ sei für alle Funktionen von $\mathfrak{F}_2$, ist notwendig und hinreichend, daß $\varphi(\xi, n)$ den Bedingungen genügt:

1. Zu jedem $\mu > 0$ gibt es ein $\lambda > 0$, so daß für jede Menge I sich nicht überdeckender Intervalle, deren Gesamtinhalt $\leqq \lambda$ ist, die Ungleichung besteht:

$$\int_I |\varphi(\xi, n)|\,d\xi < \mu \quad \text{für alle } n.$$

Und zu jedem $\mu > 0$ gibt es ein A, so daß:

$$\int_{-\infty}^{-A} |\varphi(\xi, n)|\,d\xi < \mu; \quad \int_{A}^{+\infty} |\varphi(\xi, n)|\,d\xi < \mu \quad \text{für alle } n. \tag{9}$$

2. Für jedes endliche Intervall $<\alpha, \beta>$ ist:

$$\lim_{n=\infty}\int_{\alpha}^{\beta} \varphi(\xi, n)\,d\xi = 0.«$$

Die Bedingungen sind notwendig. Für 2. ist das trivial. Wäre 1. nicht erfüllt, so sieht man sofort, daß der Kern:

$$\psi(u, n) = 2\varphi\left(\lg \frac{1+u}{1-u}, n\right) \frac{1}{1-u^2} \tag{9}$$

im Intervalle $<-1, 1>$ Bedingung 1. von Satz III nicht erfüllen würde; es gäbe also eine in $<-1, 1>$ zu $\mathfrak{F}_3$ gehörige Funktion $g(u)$, für die nicht:

$$\lim_{n=\infty} \int_{-1}^{1} g(u)\, \psi(u, n)\, du = 0$$

wäre. Setzt man:

$$f(\xi) = g\left(\frac{e^\xi - 1}{e^\xi + 1}\right),$$

so gehört $f(\xi)$ in $(-\infty, +\infty)$ zu $\mathfrak{F}_3$ und es wird:

$$\int_{-1}^{1} g(u)\, \psi(u, n)\, du = \int_{-\infty}^{+\infty} f(\xi)\, \varphi(\xi, n)\, d\xi,$$

so daß auch nicht $\lim\limits_{n=\infty} J_n(f) = 0$ wäre.

Die Bedingungen sind hinreichend. In der Tat folgt aus unserer Bedingung 1., daß der Kern (9) in $<-1, 1>$ der Bedingung 1. von Satz III genügt. Ferner folgt aus unserer Bedingung 2., daß für jedes Teilintervall $<\alpha, \beta>$ von $(-1, 1)$:

$$\lim_{n=\infty} \int_{\alpha}^{\beta} \psi(u, n)\, du = 0$$

ist. Wie bisher entnimmt man hieraus, daß auch:

$$\lim_{n=\infty} \int_{-1}^{1} \psi(u, n)\, du = 0$$

ist, nur hat man sich diesmal beim Beweise von (6) auf den zweiten Teil unserer Bedingung 1. zu berufen. — Es genügt also $\psi(u, n)$ in $<-1, 1>$ beiden Bedingungen von Satz III.

Gehört nun $f(\xi)$ in $(-\infty, +\infty)$ zu $\mathfrak{F}_3$, so gehört

$$g(u) = f\left(\lg \frac{1+u}{1-u}\right)$$

in $<-1, 1>$ zu $\mathfrak{F}_3$. Somit ist:

$$\lim_{n=\infty} J_n(f) = \lim_{n=\infty} \int_{-1}^{1} g(u)\, \psi(u, n)\, du = 0$$

und Satz III *a* ist bewiesen.

Da die Funktion $f(\xi) = 1$ (oder $f(\xi) = 1$ für $\xi \geqq 0$, $f(\xi) = 0$ für $\xi < 0$) zu $\mathfrak{F}_3$ gehört, ist es hier notwendig, daß auch für unendliche Intervalle Bedingung 2. erfüllt sei:

$$\lim_{n=\infty} \int_{-\infty}^{+\infty} \varphi(\xi, n)\, d\xi = 0; \quad \lim_{n=\infty} \int_{0}^{+\infty} \varphi(\xi, n)\, d\xi = 0.$$

Doch braucht dies in Bedingung 2. nicht ausdrücklich aufgenommen zu werden, da es vermöge Bedingung 1. aus der Gültigkeit von 2. für endliche Intervalle von selbst folgt.

IV *a*. Damit $\lim\limits_{n=\infty} J_n(f) = 0$ sei für alle Funktionen von $\mathfrak{F}_4$, ist notwendig und hinreichend, daß $\varphi(\xi, n)$ den Bedingungen genügt:

1. Es gibt eine Konstante M, so daß:

$$\int_{-\infty}^{+\infty} |\varphi(\xi, n)| \, d\xi < M \quad \text{für alle } n.$$

2. Für jedes endliche Intervall $< \alpha, \beta >$ ist

$$\lim_{n=\infty} \int_{\alpha}^{\beta} \varphi(\xi, n) \, d\xi = 0,$$

und es ist:

$$\lim_{n=\infty} \int_{-\infty}^{0} \varphi(\xi, n) \, d\xi = 0; \quad \lim_{n=\infty} \int_{0}^{\infty} \varphi(\xi, n) \, d\xi = 0.$$

Der Beweis wird geführt wie für Satz III *a*. Nur folgt diesmal die Tatsache, daß der Kern (9) in $< -1, 1 >$ der Bedingung 2. von Satz IV genügt, unmittelbar aus unserer Bedingung 2.

Hier wäre es nicht hinreichend, Bedingung 2. bloß für endliche Intervalle auszusprechen, wie wieder das schon einmal verwendete Beispiel zeigt:

$$\varphi(\xi, n) = \begin{cases} 1 & \text{in } (n, n+1) \\ 0 & \text{außerhalb } (n, n+1). \end{cases}$$

Es genügt dieses φ der Bedingung 1., sowie der Bedingung 2. für endliche Intervalle, während wenn wir $f = 1$ wählen, wir für jedes n haben:

$$J_n(1) = 1.$$

§ 2. Ein Hilfssatz von Haar und Lebesgue.

Wir kehren wieder zur Betrachtung endlicher Intervalle zurück, und beweisen folgenden Satz, den wir weiterhin verwenden werden: [1]

V. Ist die Menge der Zahlen

$$\int_a^b |\varphi(\xi, n)| \, d\xi$$

($n = 1, 2, \ldots$) nicht geschränkt, so gibt es eine in $< a, b >$ stetige, in endlich vielen vorgegebenen Punkten $x_1, x_2, \ldots, x_p$ von $< a, b >$ verschwindende Funktion $f(x)$, für die nicht $\lim_{n=\infty} I_n(f) = 0$ ist.

Wir erinnern daran, daß (bei gegebenem n) zu jedem $\sigma > 0$ ein $\tau_n > 0$ gehört, derart, daß für jede in $< a, b >$ gelegene meßbare Menge $\mathfrak{A}$, deren Inhalt $< \tau_n$ ist, die Ungleichung gilt:

$$\int_{\mathfrak{A}} |\varphi(\xi, n)| \, d\xi < \sigma. \tag{1}$$

Bezeichnen wir mit $h_n(\xi)$ die Funktion, die $= 1$ ist, wo $\varphi(\xi, n) \geqq 0$, und $= -1$, wo $\varphi(\xi, n) < 0$, so gibt es nach Voraussetzung eine (wachsende) Indizesfolge n_i mit $\lim_{i=\infty} n_i = \infty$, so daß:

$$\lim_{i=\infty} \int_a^b |\varphi(\xi, n_i)| \, d\xi = \lim_{i=\infty} \int_a^b h_{n_i}(\xi) \, \varphi(\xi, n_i) \, d\xi = +\infty. \tag{2}$$

Wir zeigen zunächst, daß es auch eine Folge in $< a, b >$ stetiger Funktionen $g_n(\xi)$ gibt, die, ebenso wie die $h_n(\xi)$ der Ungleichung genügen:

$$|g_n(\xi)| \leqq 1, \tag{3}$$

---

[1] H. Lebesgue, a. a. O., p. 61. A. Haar, Math. Ann. 69, p. 335. Der im Text gegebene Beweis unterscheidet sich nicht wesentlich vom Lebesgue'schen Beweise.

die sämtlich in $x_1, x_2, \ldots, x_p$ verschwinden, und für die gleichfalls:

$$\lim_{i=\infty} \int_a^b g_{n_i}(\xi)\, \varphi(\xi, n_i)\, d\xi = +\infty. \tag{4}$$

Sei also $\sigma > 0$ beliebig gegeben. Wir schließen die Menge $\mathfrak{M}_n$ aller Punkte von $< a, b >$, in denen $\varphi(\xi, n) \geqq 0$ ist, in eine abzählbare Menge I_n sich nicht überdeckender Intervalle ein, deren Gesamtlänge den Inhalt von $\mathfrak{M}_n$ um weniger als $\frac{\tau_n}{2}$ übersteigt.[1] Von den unendlich vielen Intervallen von I_n behalten wir eine endliche Menge J_n bei, deren Inhalt hinter dem von I_n um weniger als $\frac{\tau_n}{2}$ zurückbleibt.

Bezeichnen wir nun mit $h_n^*(\xi)$ die Funktion, die $= 1$ ist in den Punkten von J_n, sonst $= -1$; dann unterscheiden sich h_n und h_n^* voneinander nur in den Punkten einer Menge, deren Inhalt $\leqq \tau_n$ ist, und zwar ist in den Punkten dieser Menge:

$$|h_n(\xi) - h_n^*(\xi)| = 2.$$

Es ist also zufolge (1):

$$\left| \int_a^b h_n(\xi)\, \varphi(\xi, n)\, d\xi - \int_a^b h_n^*(\xi)\, \varphi(\xi, n)\, d\xi \right| < 2\sigma. \tag{5}$$

Die Funktion $h_n^*(\xi)$ ist auch noch unstetig, hat aber nur mehr endlich viele Unstetigkeitspunkte $\xi_1^{(n)}, \xi_2^{(n)}, \ldots, \xi_q^{(n)}$. Wir umgeben jeden derselben mit einem Intervalle $< \xi_i^{(n)} - h_i^{(n)}, \xi_i^{(n)} + h_i^{(n)} >$, wo die $h_i^{(n)} (> 0)$ so klein gewählt seien, daß alle diese Intervalle in (a, b) liegen, keine zwei einen Punkt gemein haben und:

$$2\,(h_1^{(n)} + h_2^{(n)} + \ldots + h_q^{(n)}) < \tau_n \tag{6}$$

ist. Nun definieren wir eine Funktion $g_n^*(\xi)$ durch die Vorschrift: es ist $g_n^*(\xi) = h_n^*(\xi)$ außerhalb der Intervalle $(\xi_i^{(n)} - h_i^{(n)}, \xi_i^{(n)} + h_i^{(n)})$; in jedem dieser Intervalle ist $g_n^*(\xi)$ gleich derjenigen linearen Funktion, die in den beiden Endpunkten des Intervalles mit $h_n^*(\xi)$ übereinstimmt. Dann unterscheiden sich g_n^* und h_n^* voneinander nur in den Punkten einer Menge, deren Inhalt $< \tau_n$ ist (wegen (6)) und in den Punkten dieser Menge ist:

$$|h_n^*(\xi) - g_n^*(\xi)| \leqq 2,$$

so daß, wegen (1):

$$\left| \int_a^b h_n^*(\xi)\, \varphi(\xi, n)\, d\xi - \int_a^b g_n^*(\xi)\, \varphi(\xi, n)\, d\xi \right| < 2\sigma. \tag{7}$$

Die Funktion $g_n^*(\xi)$ ist bereits stetig, genügt der Ungleichung $|g_n^*(\xi)| \leqq 1$, erfüllt aber noch nicht die Bedingung, in $x_1, x_2, \ldots, x_p$ zu verschwinden.

Um auch das zu erreichen, legen wir um jeden dieser Punkte ein Intervall[2] $< x_i - k_i^{(n)}, x_i + k_i^{(n)} >$, wo die $k_i^{(n)} (> 0)$ so klein gewählt seien, daß alle diese Intervalle in $< a, b >$ liegen, keine zwei einen Punkt gemein haben und:

$$2\,(k_1^{(n)} + k_2^{(n)} + \ldots + k_p^{(n)}) < \tau_n$$

ist. Sodann definieren wir die Funktion $g_n(\xi)$ durch die Vorschrift: es ist $g_n(\xi) = g_n^*(\xi)$ außerhalb der Intervalle $(x_i - k_i^{(n)}, x_i + k_i^{(n)})$; in jedem der Intervalle $< x_i - k_i^{(n)}, x_i >$, beziehungsweise $< x_i, x_i + k_i^{(n)} >$ ist $g_n(\xi)$ gleich derjenigen linearen Funktion, die in x_i verschwindet und in $x_i - k_i^{(n)}$, beziehungsweise $x_i + k_i^{(n)}$ mit $g_n^*(\xi)$ übereinstimmt. Dann unterscheiden sich g_n und g_n^* nur in den Punkten einer Menge, deren Inhalt $> \tau_n$ ist, und in den Punkten dieser Menge ist:

$$|g_n(\xi) - g_n^*(\xi)| < 2,$$

[1] Dabei bedeutet τ_n die zufolge der eingangs gemachten Bemerkung der Größe σ zugeordnete Größe.

[2] Fällt der Punkt x_i mit a oder mit b zusammen, hat es statt dessen zu heißen:

$$< a, a + k_i^{(n)} > \text{ oder } < b - k_i^{(n)}, b >.$$

so daß, wegen (1):

$$\left|\int_a^b g_n^*(\xi)\,\varphi(\xi, n)\,d\xi - \int_a^b g_n(\xi)\,\varphi(\xi, n)\,d\xi\right| < 2\sigma. \tag{8}$$

Die Funktionen $g_n(\xi)$ sind nun stetig, genügen der Ungleichung (3) und verschwinden in $x_1\, x_2, \ldots, x_p$. Die Ungleichungen (5), (7) und (8) zusammengenommen ergeben:

$$\left|\int_a^b h_n(\xi)\,\varphi(\xi, n)\,d\xi - \int_a^b g_n(\xi)\,\varphi(\xi, n)\,d\xi\right| < 6\sigma,$$

und wegen (2) ist daher auch (4) bewiesen.

Wir setzen nun:

$$\int_a^b |\varphi(\xi, n)|\,d\xi = L_n. \tag{9}$$

Beim weiteren Beweise von Satz V können wir annehmen, es sei für alle i:

$$\lim_{n=\infty} I_n(g_i) = 0. \tag{10}$$

Wäre nämlich dies für ein i nicht der Fall, so wäre ja, indem wir $f = g_i$ nehmen, die Behauptung schon bewiesen.

Wir können dann aus der Folge der Indizes n_i eine wachsende Teilfolge n_{i_ν} (mit $n_{i_1} = n_1$ beginnend) so herausgreifen, daß folgende drei Eigenschaften bestehen:

1. Für $n \geqq n_{i_{\nu+1}}$ ist:

$$\left|I_n\left(g_{n_{i_1}} + \frac{1}{2\,L_{n_{i_1}}} g_{n_{i_2}} + \ldots + \frac{1}{2^{\nu-1} \cdot L_{n_{i_{\nu-1}}}} g_{n_{i_\nu}}\right)\right| < 1,$$

2. $$I_{n_{i_{\nu+1}}}\left(g_{n_{i_{\nu+1}}}\right) \geqq 2^{\nu+1} L_{n_{i_\nu}},$$

3. $$L_{n_{i_{\nu+1}}} > L_{n_{i_\nu}};$$

und zwar läßt sich dies für Eigenschaft 1. wegen (10), für Eigenschaft 2. wegen (4), für Eigenschaft 3. wegen (2) erreichen.

Schreiben wir nun der Kürze halber:

$$g_{n_{i_\nu}} = g^{(\nu)}\;;\quad L_{n_{i_\nu}} = L^{(\nu)}\;;\quad I_{n_{i_\nu}} = I^{(\nu)},$$

so haben wir:

$$\left|I^{(n)}\left((g^{(1)} + \frac{1}{2\,L^{(1)}} g^{(2)} + \ldots + \frac{1}{2^{\nu-1}\,L^{(\nu-1)}} g^{(\nu)}\right)\right| < 1 \quad \text{für } n \geqq \nu + 1. \tag{11}$$

$$I^{(\nu+1)}\left(g^{(\nu+1)}\right) \geqq 2^{\nu+1} L^{(\nu)}. \tag{12}$$

$$L^{(\nu+1)} > L^{(\nu)}. \tag{13}$$

Wir setzen:

$$f = g^{(1)} + \frac{1}{2\,L^{(1)}} g^{(2)} + \ldots + \frac{1}{2^{\nu-1}\,L^{(\nu-1)}} g^{(\nu)} + \ldots; \tag{14}$$

dann ist, wegen (3), diese Reihe gleichmäßig konvergent, und somit f eine stetige Funktion, und es verschwindet f in den vorgeschriebenen Punkten $x_1, x_2, \ldots, x_p$.

Wir bilden:

$$I^{(\nu)}(f) = I^{(\nu)}\left(g^{(1)} + \frac{1}{2\,L^{(1)}} g^{(2)} + \ldots + \frac{1}{2^{\nu-2}\,L^{(\nu-2)}} g^{(\nu-1)}\right) + I^{(\nu)}\left(\frac{1}{2^{\nu-1}\,L^{(\nu-1)}} g^{(\nu)}\right) + I^{(\nu)}\left(\frac{1}{2^{\nu}\,L^{(\nu)}} g^{(\nu+1)} + \ldots\right).$$

Hierin ist, wegen (11):

$$\left|I^{(\nu)}\left(g^{(1)} + \frac{1}{2\,L^{(1)}} g^{(2)} + \ldots + \frac{1}{2^{\nu-2}\,L^{(\nu-2)}} g^{(\nu-1)}\right)\right| < 1 \tag{15}$$

und wegen (12):

$$I^{(\nu)}\left(\frac{1}{2^{\nu-1}\,L^{(\nu-1)}} g^{(\nu)}\right) \geqq 2. \tag{16}$$

Endlich ist, wegen der gleichmäßigen Konvergenz von (14):

$$I^{(\nu)}\left(\frac{1}{2^{\nu}\,L^{(\nu)}} g^{(\nu+1)} + \frac{1}{2^{\nu+1}\,L^{(\nu+1)}} g^{(\nu+2)} + \ldots\right) = \frac{1}{2^{\nu}\,L^{(\nu)}} I^{(\nu)}(g^{(\nu+1)}) + \frac{1}{2^{\nu+1}\,L^{(\nu+1)}} I^{(\nu)}(g^{(\nu+2)}) + \ldots,$$

und somit, wegen (3), (9) und (13):

$$\left|I^{(\nu)}\left(\frac{1}{2^{\nu}\,L^{(\nu)}} g^{(\nu+1)} + \frac{1}{2^{\nu+1}\,L^{(\nu+1)}} g^{(\nu+2)} + \ldots\right)\right| \leqq$$

$$\frac{1}{2^{\nu}\,L^{(\nu)}} \cdot L^{(\nu)} + \frac{1}{2^{\nu+1}\,L^{(\nu+1)}} L^{(\nu)} + \ldots \leqq \frac{1}{2} \quad (\text{für } \nu > 1). \tag{17}$$

Die Ungleichungen (15), (16), (17) ergeben nun zusammen:

$$I^{(\nu)}(f) = I_{n_{i_\nu}}(f) \geqq \frac{1}{2}$$

und da $\lim\limits_{\nu=\infty} n_{i_\nu} = \infty$ war, so ist nicht $\lim\limits_{n=\infty} I_n(f) = 0$, womit Satz V erwiesen ist.

§ 3. Darstellung der Ableitungen einer gegebenen Funktion.

Wir nehmen nun an, es sei $\varphi(\xi, x, n)$ für jedes nicht negative ganzzahlige n und für alle x des endlichen Intervalles (a, b) in $< a, b >$ als integrierbare Funktion von ξ gegeben; wir bezeichnen $\varphi(\xi, x, n)$ mit einem der Theorie der Integralgleichungen entlehnten Ausdruck als Kern und setzen:

$$I_n(f, x) = \int_a^b f(\xi)\, \varphi(\xi, x, n)\, d\xi.$$

Es wurde von H. Lebesgue folgender Satz bewiesen:[1]

VI. »Damit für alle der Klasse $\mathfrak{F}_i$ angehörigen Funktionen f, die im gegebenen Punkte x von (a, b) stetig sind, die Gleichung gelte:

$$f(x) = \lim_{n=\infty} I_n(f, x),$$

ist notwendig und hinreichend, daß der Kern $\varphi(\xi, x, n)$ folgenden Bedingungen genügt:

1. In jedem den Punkt x nicht enthaltenden Teilintervalle $< \alpha, \beta >$ von $< a, b >$ ist Bedingung 1. desjenigen der Sätze I bis IV erfüllt, der sich auf die Klasse $\mathfrak{F}_i$ bezieht.

2. Für jedes den Punkt x nicht enthaltende Teilintervall $< \alpha, \beta >$ von $< a, b >$ ist:

$$\lim_{n=\infty} \int_\alpha^\beta \varphi(\xi, x, n)\, d\xi = 0.$$

3. Es gibt eine Konstante N, so daß

$$\int_a^b |\varphi(\xi, x, n)|\, d\xi < N \quad \text{für alle } n.$$

[1] A. a. O., p. 69 ff.

4. Es ist:

$$\lim_{n=\infty} \int_a^b \varphi(\xi, x, n)\, d\xi = 1.$$

Wir wollen nun im Folgenden den Satz beweisen:

VII. Damit für alle der Klasse $\mathfrak{F}_i$ angehörigen Funktionen f, die im gegebenen Punkte x von (a, b) endliche Ableitungen der m ersten Ordnungen haben, die Gleichung gelte:

(1) $$f^{(m)}(x) = \lim_{n=\infty} I_n(f, x),$$

ist notwendig und hinreichend, daß der Kern $\varphi(\xi, x, n)$ folgenden Bedingungen genüge:

1. In jedem den Punkt x nicht enthaltenden Teilintervalle $<\alpha, \beta>$ von $<a, b>$ ist Bedingung 1. desjenigen der Sätze I bis IV erfüllt, der sich auf die Klasse $\mathfrak{F}_i$ bezieht.[1]

2. Für jedes den Punkt x nicht enthaltende Teilintervall $<\alpha, \beta>$ von $<a, b>$ ist:

$$\lim_{n=\infty} \int_\alpha^\beta \varphi(\xi, x, n)\, d\xi = 0.$$

3. Es gibt eine Konstante N, so daß

(2) $$\int_a^b |(\xi - x)^m \varphi(\xi, x, n)|\, d\xi < N \quad \text{für alle } n.$$

4. Es ist:

(3) $$\lim_{n=\infty} \int_a^b (\xi - x)^i \varphi(\xi, x, n)\, d\xi = 0 \quad (i = 0\ 1, 2 \ldots, m-1),$$

(4) $$\lim_{n=\infty} \int_a^b (\xi - x)^m \varphi(\xi, x, n)\, d\xi = m!$$

Die Bedingungen sind hinreichend: In der Tat, wenn $f(\xi)$ im Punkte x endliche Ableitungen der ersten m Ordnungen hat, so können wir schreiben:

(5) $$f(\xi) = f(x) + \sum_{i=1}^{m} \frac{(\xi - x)^i}{i!} f^{(i)}(x) + (\xi - x)^m \cdot \omega(\xi),$$

wo:

(6) $$\lim_{\xi = x} \omega(\xi) = 0$$

ist; setzen wir also $\omega(x) = 0$, so ist die Funktion $\omega(\xi)$ im Punkte x stetig.

Wegen (5) ist nun:

$$I_n(f, x) = f(x) \int_a^b \varphi(\xi, x, n)\, d\xi + \sum_{i=1}^{m} \frac{f^{(i)}(x)}{i!} \int_a^b (\xi - x)^i \varphi(\xi, x, n)\, d\xi + \int_a^b \omega(\xi)(\xi - x)^m \varphi(\xi, x, n)\, d\xi.$$

Wegen (3) und (4) erhält man daraus:

$$\lim_{n=\infty} \left(I_n(f, x) - \int_a^b \omega(\xi)(\xi - x)^m \varphi(\xi, x, n)\, d\xi \right) = f^{(m)}(x),$$

so daß zum Beweise von (1) nur mehr zu zeigen ist, daß:

(7) $$\lim_{n=\infty} \int_a^b \omega(\xi)(\xi - x)^m \varphi(\xi, x, n)\, d\xi = 0 \quad \text{ist.}$$

---

[1] Die in dieser Bedingung auftretende Konstante M (beziehungsweise bei der Klasse $\mathfrak{F}_3$ die Konstante λ) braucht keineswegs für alle diese Intervalle dieselbe zu sein.

Sei, um das nachzuweisen, $\varepsilon > 0$ beliebig gegeben. Es gibt wegen (6) ein $h > 0$, derart, daß:

$$|\omega(\xi)| < \varepsilon \quad \text{in} \quad \langle x-h, x+h \rangle.$$

Wegen (2) ist dann:

$$(8) \qquad \left| \int_{x-h}^{x+h} \omega(\xi)(\xi-x)^m \varphi(\xi, x, n)\, d\xi \right| < \varepsilon \cdot N \quad \text{für alle } n.$$

Zufolge (5) gehört $(\xi-x)^m \cdot \omega(\xi)$ zugleich mit $f(\xi)$ zur Klasse $\mathfrak{F}_i$. Wegen Bedingung 1. und 2. hat man daher, unter Berufung auf denjenigen der Sätze I bis IV, der sich auf die Klasse $\mathfrak{F}_i$ bezieht:

$$\lim_{n=\infty} \int_a^{x-h} \omega(\xi)(\xi-x)^m \varphi(\xi, x, n)\, d\xi = 0; \quad \lim_{n=\infty} \int_{x+h}^b \omega(\xi)(\xi-x)^m \varphi(\xi, x, n)\, d\xi = 0,$$

und somit gibt es ein n_0, so daß für $n \geqq n_0$:

$$(9) \qquad \left| \int_a^{x-h} \omega(\xi)(\xi-x)^m \varphi(\xi, x, n)\, d\xi \right| < \varepsilon; \quad \left| \int_{x+h}^b \omega(\xi)(\xi-x)^m \varphi(\xi, x, n)\, d\xi \right| < \varepsilon.$$

Die Ungleichungen (8) (9) zusammengenommen ergeben:

$$\left| \int_a^b \omega(\xi)(\xi-x)^m \varphi(\xi, x, n)\, d\xi \right| < (N+2) \cdot \varepsilon \quad \text{für } n \geqq n_0,$$

und da ε beliebig war, ist damit (7) und somit auch (1) bewiesen.

Die Bedingungen sind notwendig: Wäre in der Tat 1. oder 2. für ein den Punkt x nicht enthaltendes Teilintervall $\langle \alpha, \beta \rangle$ von $\langle a, b \rangle$ nicht erfüllt, so gäbe es, nach demjenigen der Sätze I bis IV, der sich auf die Klasse $\mathfrak{F}_i$ bezieht, in $\langle \alpha, \beta \rangle$ eine der Klasse $\mathfrak{F}_i$ angehörige Funktion $g(\xi)$, für die nicht:

$$\lim_{n=\infty} \int_\alpha^\beta g(\xi) \varphi(\xi, x, n)\, d\xi = 0$$

gilt. Definieren wir dann $f(\xi)$ durch die Vorschrift:

$$f(\xi) = g(\xi) \text{ in } \langle \alpha, \beta \rangle, \quad f(\xi) = 0 \text{ außerhalb } \langle \alpha, \beta \rangle,$$

so gehört $f(\xi)$ auch in $\langle a, b \rangle$ zur Klasse $\mathfrak{F}_i$ und es ist $f^{(m)}(x) = 0$, während die Gleichung:

$$\lim_{n=\infty} I_n(f, x) \left(= \lim_{n=\infty} \int_\alpha^\beta g(\xi) \varphi(\xi, x, n)\, d\xi \right) = 0$$

nicht gilt.

Angenommen es sei Bedingung 3. nicht erfüllt. Dann gibt es, nach Satz V, eine in $\langle a, b \rangle$ stetige für $\xi = x$ verschwindende Funktion $\omega(\xi)$, für die nicht:

$$\lim_{n=\infty} \int_a^b \omega(\xi)(\xi-x)^m \varphi(\xi, x, n)\, d\xi = 0$$

ist. Wir setzen:

$$f(\xi) = (\xi-x)^m \omega(\xi).$$

Dann besitzt $f(\xi)$ im Punkte x eine m-te Ableitung und es ist:

$$f^{(m)}(x) = 0,$$

während die Gleichung:

$$\lim_{n=\infty} I_n(f, x) = 0$$

nicht gilt.

Bedingung 4. ergibt sich als notwendig, indem man für $f(\xi)$ die Funktionen wählt:

$$f(\xi) = (\xi-x)^i \quad (i = 0, 1, \ldots, m).$$

Die Sätze VI und VII können ohne weiteres auf unendliche Intervalle ausgedehnt werden. Wir setzen:

$$J_n(f, x) = \int_{-\infty}^{+\infty} f(\xi)\, \varphi(\xi, x, n)\, d\xi,$$

bezeichnen, für $\varepsilon > 0$, mit $\psi(\xi, x, n, \varepsilon)$ die Funktion von ξ, die in $< x-\varepsilon, x+\varepsilon >$ den Wert 0 hat, sonst mit $\varphi(\xi, x, n)$ übereinstimmt, und haben:

VI *a.* »Damit für alle in $(-\infty, +\infty)$ der Klasse $\mathfrak{F}_i$ angehörigen Funktionen f, die im gegebenen Punkte x stetig sind, die Gleichung gelte:

$$f(x) = \lim_{n=\infty} J_n(f, x),$$

ist notwendig und hinreichend, daß $\varphi(\xi, x, n)$ folgenden Bedingungen genügt:

1. und 2. Für jedes $\varepsilon > 0$ genügt der Kern $\psi(\xi, x, n, \varepsilon)$ den Bedingungen 1. und 2. desjenigen der Sätze I *a* bis IV *a*, der sich auf die Klasse $\mathfrak{F}_i$ bezieht.

3. Zu jedem $\varepsilon > 0$ gibt es ein N, so daß

$$\int_{x-\varepsilon}^{x+\varepsilon} |\varphi(\xi, x, n)|\, d\xi < N \quad \text{für alle } n.$$

4. Für jedes $\varepsilon > 0$ ist:

$$\lim_{n=\infty} \int_{x-\varepsilon}^{x+\varepsilon} \varphi(\xi, x, n)\, d\xi = 1.\text{«}$$

VII *a.* »Damit für alle in $(-\infty, +\infty)$ der Klasse $\mathfrak{F}_i$ angehörigen Funktionen, die im gegebenen Punkte x von (a, b) endliche Ableitungen der m ersten Ordnungen haben, die Gleichung gelte:

$$f^{(m)}(x) = \lim_{n=\infty} J_n(f, x),$$

ist notwendig und hinreichend, daß $\varphi(\xi, x, n)$ folgenden Bedingungen genügt:

1. und 2. wie in Satz VI *a*.

3. Zu jedem $\varepsilon > 0$ gibt es ein N, so daß:

$$\int_{x-\varepsilon}^{x+\varepsilon} |(\xi - x)^m \varphi(\xi, x, n)|\, d\xi < N \quad \text{für alle } n.$$

4. Für jedes $\varepsilon > 0$ ist:

$$\lim_{n=\infty} \int_{x-\varepsilon}^{x+\varepsilon} (\xi - x)^i \varphi(\xi, x, n)\, d\xi = 0 \quad (i = 0, 1, \ldots m-1),$$

$$\lim_{n=\infty} \int_{x-\varepsilon}^{x+\varepsilon} (\xi - x)^m \varphi(\xi, x, n)\, d\xi = m!\text{«}$$

Zum Beweise von VI *a* und VII *a* hat man nur zu bemerken, daß wegen der Bedingungen 1. und 2. nach dem in Betracht kommenden der Sätze I *a* bis IV *a* für jedes $\varepsilon > 0$:

$$\lim_{n=\infty} \left\{ J_n(f, x) - \int_{x-\varepsilon}^{x+\varepsilon} f(\xi)\, \varphi(\xi, x, n)\, d\xi \right\} = \lim_{n=\infty} \int_{-\infty}^{+\infty} f(\xi)\, \psi(\xi, x, n, \varepsilon)\, d\xi = 0$$

ist, wodurch die Betrachtung von $\lim_{n=\infty} J_n(f, x)$ auf die von

$$\lim_{n=\infty} \int_{x-\varepsilon}^{x+\varepsilon} f(\xi)\, \varphi(\xi, x, n)\, d\xi$$

zurückgeführt ist. Auf dieses Integral aber sind die Sätze VI und VII anwendbar.

§ 4. Darstellung der verallgemeinerten Ableitungen.

Das Resultat von § 3 läßt sich noch ein wenig verallgemeinern, wenn wir annehmen, daß der Kern in einer Umgebung der Stelle x, etwa für $|t| \leqq k$, einer der zwei Bedingungen genügt:

(1) $$\varphi(x+t, x, n) = \varphi(x-t, x, n),$$

(2) $$\varphi(x+t, x, n) = -\varphi(x-t, x, n).$$

Den Ausführungen von § 3 lag folgende Definition der m-ten Ableitung einer Funktion zugrunde:

Sei $f(u)$ eine in der Umgebung der Stelle u_0 definierte Funktion, die in u_0 Ableitungen der $m-1$ ersten Ordnungen besitzt:

$$f'(u_0),\quad f''(u_0), \ldots, f^{(m-1)}(u_0).$$

Gibt es dann eine endliche Zahl a, so daß in einer Umgebung von u die Entwicklung gilt:

$$f(u)-f(u_0) - \sum_{i=1}^{m-1} \frac{(u-u_0)^i}{i!} f^{(i)}(u_0) = a(u-u_0)^m + \omega(u)\cdot(u-u_0)^m,$$

wo:

$$\lim_{u=u_0} \omega(u) = 0$$

ist, so wird:

$$m!\, a = f^{(m)}(u_0)$$

gesetzt und als die m-te Ableitung von $f(u)$ an der Stelle u_0 bezeichnet.

Diese Definition wurde von Ch. J. de la Vallée-Poussin in folgender Weise verallgemeinert: [1]

Sei $f(u)$ eine in der Umgebung der Stelle u_0 definierte Funktion, und es gelte für alle hinlänglich kleinen $|t|$ ($\neq 0$) eine der beiden Entwicklungen:

(3) $$\frac{f(u_0+t) + f(u_0-t)}{2} = a_0 + \sum_{i=1}^{m} a_{2i} \frac{t^{2i}}{(2i)!} + \omega(t)\, t^{2m}$$

(4) $$\frac{f(u_0+t) - f(u_0-t)}{2} = \sum_{i=0}^{m} a_{2i+1} \frac{t^{2i+1}}{(2i+1)!} + \omega(t)\, t^{2m+1},$$

worin:

(5) $$\lim_{t=0} \omega(t) = 0$$

sei. Es heißen dann, wenn (3) gilt, die Koeffizienten a_i ($i = 0, 2, \ldots, 2m$) [2], und wenn (4) gilt, die Koeffizienten a_i ($i = 1, 3, \ldots, 2m+1$) die verallgemeinerten Ableitungen i-ter Ordnung von $f(u)$ an der Stelle u_0.

Vergleicht man diese Definition mit der oben angeführten Definition der i-ten Ableitungen, so erkennt man sofort: wo die i-te Ableitung ($i > 0$) von $f(u)$ existiert, existiert auch die verallgemeinerte i-te Ableitung und ist gleich der i-ten Ableitung.

Wir werden die verallgemeinerte i-te Ableitung von $f(u)$ mit $\dot{f}^{(i)}(u)$ bezeichnen.

Wir haben dann folgende Sätze:

---

[1] Acad. Bruxelles, Bulletin, Classe des Sciences 1908, p. 214.

[2] Wie man sieht, ist eine »0-te Ableitung« von f an der Stelle u_0 sicher vorhanden, wenn f dort einen rechts- und einen linksseitigen Grenzwert $f(u_0+0)$, beziehungsweise $f(u_0-0)$ besitzt. Und zwar ist die 0-te Ableitung dann nichts anderes als das arithmetische Mittel $\frac{1}{2}(f(u_0+0) + f(u_0-0))$ dieser beiden einseitigen Grenzwerte.

VIII. Es genüge für ein gegebenes x von (a, b) und für alle $|t| \leqq k$ der Kern $\varphi(\xi, x, n)$ der Relation (1). Damit für alle der Klasse $\mathfrak{F}_i$ angehörigen Funktionen, die im Punkte x eine verallgemeinerte Ableitung $\overset{*}{f}{}^{(2m)}(x)$ besitzen, die Gleichung gelte:

$$\overset{*}{f}{}^{(2m)}(x) = \lim_{n=\infty} I_n(f, x), \tag{6}$$

ist notwendig und hinreichend, daß noch folgende Bedingungen erfüllt seien:[1]

1. und 2. Es sind die Bedingungen 1. und 2. von Satz VI erfüllt.

3. Es gibt eine Konstante N, so daß

$$\int_a^b |(\xi - x)^{2m} \varphi(\xi, x, n)| \, d\xi < N \quad \text{für alle } n.$$

4. Es ist:

$$\lim_{n=\infty} \int_a^b (\xi - x)^i \varphi(\xi, x, n) \, d\xi = 0 \quad (i = 0, 2, \ldots 2m-2), \tag{7}$$

$$\lim_{n=\infty} \int_a^b (\xi - x)^{2m} \varphi(\xi, x, n) \, d\xi = (2m)! \tag{8}$$

Es genüge für ein gegebenes x von (a, b) und alle $|t| \leqq k$ der Kern $\varphi(\xi, x, n)$ der Relation (2). Damit für alle der Klasse $\mathfrak{F}_i$ angehörigen Funktionen, die im Punkte x eine verallgemeinerte Ableitung $\overset{*}{f}{}^{(2m+1)}(x)$ besitzen, die Gleichung gelte:

$$\overset{*}{f}{}^{(2m+1)}(x) = \lim_{n=\infty} I_n(f, x), \tag{6a}$$

ist notwendig und hinreichend, daß noch folgende Bedingungen erfüllt seien:

1. und 2. Es sind die Bedingungen 1. und 2. von Satz VI erfüllt.

3. Es gibt eine Konstante N, so daß:

$$\int_a^b |(\xi - x)^{2m+1} \varphi(\xi, x, n)| \, d\xi < N \quad \text{für alle } n.$$

4. Es ist:

$$\lim_{n=\infty} \int_a^b (\xi - x)^i \varphi(\xi, x, n) \, d\xi = 0 \quad (i = 1, 3, \ldots, 2m-1), \tag{7a}$$

$$\lim_{n=\infty} \int_a^b (\xi - x)^{2m+1} \varphi(\xi, x, n) \, d\xi = (2m+1)! \tag{8a}$$

Wir beweisen die erste Hälfte dieses Satzes. Es handelt sich nur darum, daß die Bedingungen hinreichend sind; denn der Beweis, daß sie notwendig sind, ist derselbe wie in § 3.

Um nachzuweisen, daß unter den gemachten Voraussetzungen Gleichung (6) besteht, schreiben wir:

$$I_n(f, x) = \int_a^{x-k} f(\xi) \varphi(\xi, x, n) \, d\xi + \int_{x+k}^b f(\xi) \varphi(\xi, x, n) \, d\xi + \int_{x-k}^{x+k} f(\xi) \varphi(\xi, x, n) \, d\xi.$$

Da die Bedingungen 1. und 2. erfüllt sind, ist hierin, auf Grund desjenigen der Sätze I bis IV, der sich auf die Klasse $\mathfrak{F}_i$ bezieht:

$$\lim_{n=\infty} \int_a^{x-k} f(\xi) \varphi(\xi, x, n) \, d\xi = 0 \; ; \; \lim_{n=\infty} \int_{x+k}^b f(\xi) \varphi(\xi, x, n) \, d\xi = 0.$$

---

[1] Dies gilt auch für $m = 0$. Die Gleichungen (7) fallen dann weg, und in (8) steht auf der rechten Seite: 1.

Um also (6) zu beweisen, haben wir nur nachzuweisen:

$$\overset{*}{f}{}^{(2m)}(x) = \lim_{n=\infty} \int_{x-k}^{x+k} f(\xi)\, \varphi(\xi, x, n)\, d\xi. \tag{9}$$

Da aber für $|t| \leqq k$ Beziehung (1) gilt, können wir schreiben:

$$\int_{x-k}^{x+k} f(\xi)\, \varphi(\xi, x, n)\, d\xi = \int_0^k \{f(x+t) + f(x-t)\}\, \varphi(x+t, x, n)\, dt.$$

Setzen wir hierin die Entwicklung (3) ein ($u_0 = x$), so haben wir weiter:

$$\int_{x-k}^{x+k} f(\xi)\, \varphi(\xi, x, n)\, d\xi = 2a_0 \int_0^k \varphi(x+t, x, n)\, dt + 2 \sum_{i=1}^{m} \frac{a_{2i}}{(2i)!} \int_0^k t^{2i} \varphi(x+t, x, n)\, dt +$$

$$+2 \int_0^k \omega(t)\, t^{2m} \varphi(x+t, x, n)\, dt = a_0 \int_{x-k}^{x+k} \varphi(\xi, x, n)\, d\xi + \sum_{i=1}^{m} \frac{a_{2i}}{(2i)!} \int_{x-k}^{x+k} (\xi-x)^{2i} \varphi(\xi, x, n)\, d\xi + \tag{10}$$

$$+ \int_{x-k}^{x+k} \omega(\xi-x)(\xi-x)^{2m} \varphi(\xi, x, n)\, d\xi,$$

wobei berücksichtigt ist, daß zufolge (3) $\omega(t)$ eine gerade Funktion von t ist.

Nun ist, wegen der Sätze von § 1:

$$\lim_{n=\infty} \int_a^{x-k} (\xi-x)^{2i} \varphi(\xi, x, n)\, d\xi = 0\,; \quad \lim_{n=\infty} \int_{x+k}^{b} (\xi-x)^{2i} \varphi(\xi, x, n)\, d\xi = 0 \quad (i = 0, 1, \ldots, m),$$

also, wegen (7) und (8):

$$\lim_{n=\infty} \int_{x-k}^{x+k} (\xi-x)^{2i} \varphi(\xi, x, n)\, d\xi = 0 \quad (i = 0, 1, \ldots, m-1)$$

$$\lim_{n=\infty} \int_{x-k}^{x+k} (\xi-x)^{2m} \varphi(\xi, x, n)\, d\xi = (2m)!$$

Berücksichtigt man dies, sowie die Tatsache, daß $a_{2m} = \overset{*}{f}{}^{(2m)}(x)$ ist, so sieht man aus (10), daß (9) und damit auch (6) bewiesen sein wird, wenn:

$$\lim_{n=\infty} \int_{x-k}^{x+k} \omega(\xi-x)(\xi-x)^{2m} \varphi(\xi, x, n)\, d\xi = 0$$

nachgewiesen sein wird. Dies aber beweist man, da (5) gilt, genau so, wie Gleichung (7) in § 3, was nicht noch einmal durchgeführt werde.

Damit ist die erste Hälfte des ausgesprochenen Satzes bewiesen. Der Beweis der zweiten Hälfte ist analog, nur hat man sich, statt auf (1) und (3), auf (2) und (4) zu stützen.

VIII a. »Man erhält aus Satz VIII einen für das Intervall $(-\infty, +\infty)$ gültigen Satz, indem man in Bedingung 1. und 2. von Satz VIII statt »Satz VI« setzt: »Satz VI a« und in Bedingung 3. und 4. von Satz VIII die Integrationsgrenzen a und b ersetzt durch $x - \varepsilon$ und $x + \varepsilon$ ($\varepsilon > 0$).«

§ 5. Bedingungen für gleichmäßige Konvergenz.

Wir wollen nun Bedingungen dafür aufstellen, daß die Konvergenz von $I_n(f, x)$ gegen $f^{(m)}(x)$ eine gleichmäßige sei.

Es hänge in den Sätzen von § 1 die Funktion $\varphi(\xi, n)$ noch ab von einem Parameter α und werde deshalb geschrieben $\varphi(\xi, \alpha, n)$. Und zwar sei φ für alle einer Menge $\mathfrak{A}$ angehörigen Werte des Parameters α als in $< a, b >$ meßbare Funktion von ξ gegeben. Wir setzen:

$$I_n(f, \alpha) = \int_a^b f(\xi)\, \varphi(\xi, \alpha, n)\, d\xi.$$

IX. Damit für jede Funktion f der Klasse $\mathfrak{F}_i$ die Beziehung:

$$\lim_{n=\infty} I_n(f, \alpha) = 0 \tag{0}$$

gleichmäßig für alle α von $\mathfrak{A}$ gelte, ist notwendig und hinreichend, daß φ folgenden Bedingungen genügt:[1]

1 *a*. Für jedes einzelne α genügt φ der Bedingung 1. desjenigen der Sätze I bis IV, der sich auf die Klasse $\mathfrak{F}_i$ bezieht.[2]

1 *b*. Für die Klasse $\mathfrak{F}_1$: Es gibt einen Index n_0 und eine Konstante M, so daß, abgesehen von Nullmengen:

$$|\varphi(\xi, \alpha, n)| < M \quad \text{für alle } \xi \text{ von } \langle a, b \rangle, \text{ alle } n \geqq n_0 \text{ und alle } \alpha \text{ von } \mathfrak{A}.$$

Für die Klasse $\mathfrak{F}_2$: Es gibt einen Index n_0 und eine Konstante M, so daß:

$$\int_a^b (\varphi(\xi, \alpha, n))^2 \, d\xi < M \quad \text{für alle } n \geqq n_0 \text{ und alle } \alpha \text{ von } \mathfrak{A}.$$

Für die Klasse $\mathfrak{F}_3$: Zu jedem $\mu > 0$ gehört ein $\lambda > 0$ und ein Index n_0, so daß für jede Menge I sich nicht überdeckender Teilintervalle von $\langle a, b \rangle$, deren Gesamtlänge $< \lambda$ ist, für alle $n \geqq n_0$ und alle α von $\mathfrak{A}$:

$$\int_I |\varphi(\xi, \alpha, n)| \, d\xi < \mu.$$

Für die Klasse $\mathfrak{F}_4$: Es gibt einen Index n_0 und eine Konstante M, so daß:

$$\int_a^b |\varphi(\xi, \alpha, n)| \, d\xi < M \quad \text{für alle } n \geqq n_0 \text{ und alle } \alpha \text{ von } \mathfrak{A}.$$

2. Für jedes Teilintervall $\langle \alpha, \beta \rangle$ von $\langle a, b \rangle$ gilt die Beziehung:

$$\lim_{n=\infty} \int_\alpha^\beta \varphi(\xi, \alpha, n) \, d\xi = 0$$

gleichmäßig für alle α von $\mathfrak{A}$.

Die Bedingungen sind notwendig. Dies ist trivial für 1 *a* und 2. Wäre 1 *b* nicht erfüllt, so gäbe es eine Folge von zu $\mathfrak{A}$ gehörigen Werten α_i und von Indizes n_i mit $\lim\limits_{i=\infty} n_i = \infty$, so daß für $\bar{\varphi}(\xi, i) = \varphi(\xi, \alpha_i, n_i)$ Bedingung 1. des in Frage kommenden der Sätze I bis IV nicht erfüllt wäre. Es gäbe also in $\mathfrak{F}_i$ eine Funktion f, für die nicht

$$\lim_{i=\infty} \int_a^b f(\xi) \, \bar{\varphi}(\xi, i) \, d\xi = 0$$

wäre, so daß für dieses f Beziehung (0) nicht gleichmäßig für alle α von $\mathfrak{A}$ gelten würde.

Die Bedingungen sind hinreichend. Denn sind sie erfüllt, so genügt für jede Folge α_i aus $\mathfrak{A}$ und jede Indizesfolge n_i mit $\lim\limits_{i=\infty} n_i = \infty$ die Funktion $\bar{\varphi}(\xi, i) = \varphi(\xi, \alpha_i, n_i)$ den Bedingungen 1. und 2. des in Frage kommenden der Sätze I bis IV, so daß (für alle solchen Folgen α_i und n_i): $\lim\limits_{i=\infty} I_{n_i}(f, \alpha_i) = 0$ ist, was gleichbedeutend ist mit der gleichmäßigen Konvergenz von (0).

[1] Vgl. H. Lebesgue, a. a. O., p. 68.

[2] Dabei kann die in dieser Bedingung auftretende Konstante M (beziehungsweise, bei der Klasse $\mathfrak{F}_3$, die Konstante λ) noch von α abhängen.

Sei wie bisher der Kern $\varphi(\xi, x, n)$ für jedes x von (a, b) als eine in $< a, b >$ integrierbare Funktion von ξ gegeben. Sei ε irgend eine positive Zahl. Wir leiten, wie schon einmal, aus dem Kerne $\varphi(\xi, x, n)$ einen Kern $\psi(\xi, x, n, \varepsilon)$ her durch folgende Vorschrift:

$$\psi(\xi, x, n, \varepsilon) = \begin{cases} 0 & \text{in } < x-\varepsilon, x+\varepsilon > \\ \varphi(\xi, x, n) & \text{in } < a, b > \text{ außerhalb } < x-\varepsilon, x+\varepsilon >. \end{cases}$$

Dann gilt folgender Satz:[1]

X. Für den Kern $\varphi(\xi, x, n)$ seien die nachstehenden Bedingungen erfüllt:

1. und 2. Für jedes gegebene $\varepsilon > 0$ genügt der dem Kerne $\varphi(\xi, x, n)$ zugeordnete Kern $\psi(\xi, x, n, \varepsilon)$ im Intervalle $< a, b >$ den Bedingungen 1 a, 1 b, 2. des Satzes IX für die Klasse $\mathfrak{F}_i$, wenn unter α die Veränderliche x, unter $\mathfrak{A}$ ein Teilintervall $< a', b' >$ von (a, b) verstanden wird.

3. Zu jedem (hinlänglich kleinen) $\varepsilon > 0$ gehört ein N, so daß:

$$\int_{x-\varepsilon}^{x+\varepsilon} |\varphi(\xi, x, n)| \, d\xi < N \quad \text{für alle } n \text{ und alle } x \text{ von } < a', b' >.$$

4. Für jedes (hinlänglich kleine) $\varepsilon > 0$ gilt die Beziehung:

$$\lim_{n=\infty} \int_{x-\varepsilon}^{x+\varepsilon} \varphi(\xi, x, n) \, d\xi = 1$$

gleichmäßig für alle x von $< a', b' >$.

Dann gilt für jede der Klasse $\mathfrak{F}_i$ angehörige Funktion f, die in jedem Punkte von $< a', b' >$ stetig ist[2], die Beziehung:

$$f(x) = \lim_{n=\infty} I_n(f, x)$$

gleichmäßig für alle x von $< a', b' >$.

Wir geben ein $\eta > 0$ beliebig vor, und zeigen zunächst: es gibt ein $\varepsilon > 0$, so daß:

$$|f(\xi) - f(x)| < \eta \text{ für alle } x \text{ von } < a', b' > \text{ und alle } \xi \text{ von } < x-\varepsilon, x+\varepsilon >.$$

In der Tat, da f stetig ist in $< a', b' >$, gibt es, zufolge des Satzes von der gleichmäßigen Stetigkeit, ein $\varepsilon_1 > 0$, so daß:

$$|f(\xi) - f(x)| < \eta$$

für alle x von $< a', b' >$ und alle gleichfalls zu $< a', b' >$ gehörigen ξ von $< x-\varepsilon_1, x+\varepsilon_1 >$.

Da aber f auch stetig ist in a' und in b', so gibt es ein $\varepsilon_2 > 0$, so daß:

$$|f(\xi) - f(a')| < \frac{\eta}{2} \text{ für alle } \xi \text{ von } < a'-\varepsilon_2, a'+\varepsilon_2 >,$$

$$|f(\xi) - f(b')| < \frac{\eta}{2} \text{ für alle } \xi \text{ von } < b'-\varepsilon_2, b'+\varepsilon_2 >,$$

und somit:

$$|f(\xi) - f(x)| < \eta \text{ für alle } x \text{ von } < a', a'+\varepsilon_2 > \text{ und alle } \xi \text{ von } < a'-\varepsilon_2, a' >,$$

$$|f(\xi) - f(x)| < \eta \text{ für alle } x \text{ von } < b'-\varepsilon_2, b' > \text{ und alle } \xi \text{ von } < b', b'+\varepsilon_2 >.$$

Wie man sieht, hat man für das ε unserer Behauptung lediglich die kleinere der beiden Größen ε_1 und ε_2 zu nehmen, und diese Behauptung ist bewiesen.

---

[1] Vgl. H. Lebesgue, a. a. O., p. 73.

[2] Es muß also f auch in a' und b' stetig (nicht etwa bloß rechtsseitig, beziehungsweise linksseitig stetig) sein.

Sei also zum vorgegebenen η das ε in dieser Weise bestimmt. Wir wählen es obendrein so klein, daß $< a' - \varepsilon, b' + \varepsilon >$ in $< a, b >$ liegt. Da der mit diesem ε gebildete Kern $\psi(\xi, x, n, \varepsilon)$ in $< a, b >$ den Voraussetzungen von Satz IX genügt, [1] so sehen wir: es gibt ein n_0, so daß:

(1) $$\left|\int_a^b f(\xi)\,\psi(\xi, x, n, \varepsilon)\,d\xi\right| < \eta \quad \text{für } n \geqq n_0 \text{ und alle } x \text{ von } < a', b' >.$$

Nun kann geschrieben werden:

$$I_n(f, x) = f(x)\int_{x-\varepsilon}^{x+\varepsilon} \varphi(\xi, x, n)\,d\xi + \int_{x-\varepsilon}^{x+\varepsilon} (f(\xi) - f(x))\,\varphi(\xi, x, n)\,d\xi + \int_a^b f(\xi)\,\psi(\xi, x, n, \varepsilon)\,d\xi.$$

Weil $f(x)$ in $< a', b' >$, zufolge der Stetigkeit, geschränkt ist, kann, nach Bedingung 4. n_0 auch so groß gewählt werden, daß:

$$\left|f(x)\int_{x-\varepsilon}^{x+\varepsilon} \varphi(\xi, x, n)\,d\xi - f(x)\right| < \eta \quad \text{für } n \geqq n_0 \text{ und alle } x \text{ von } < a', b' >.$$

Wegen Bedingung 3. und unserer Wahl von ε haben wir weiter:

$$\left|\int_{x-\varepsilon}^{x+\varepsilon} (f(\xi) - f(x))\,\varphi(\xi, x, n)\,d\xi\right| < \eta \cdot N \quad \text{für alle } n \text{ und alle } x \text{ von } < a', b' >,$$

und die beiden letzten Ungleichungen, zusammen mit (1) ergeben sofort:

$$|I_n(f, x) - f(x)| < (2 + N) \cdot \eta \quad \text{für alle } n \geqq n_0 \text{ und alle } x \text{ von } < a', b' >.$$

Da η beliebig war, ist damit Satz X bewiesen.

XI. Für den Kern $\varphi(\xi, x, n)$ seien die nachstehenden Bedingungen erfüllt:

1. und 2. Der dem Kerne $\varphi(\xi, x, n)$ zugeordnete Kern $\psi(\xi, x, n, \varepsilon)$ genügt den Bedingungen 1. und 2. von Satz X.

3. Zu jedem (hinlänglich kleinen) $\varepsilon > 0$ gehört ein N, so daß:

$$\int_{x-\varepsilon}^{x+\varepsilon} |(\xi - x)^m\,\varphi(\xi, x, n)|\,d\xi < N \quad \text{für alle } n \text{ und alle } x \text{ von } < a', b' >.$$

4. Für jedes (hinlänglich kleine) $\varepsilon > 0$ gelten die Beziehungen:

$$\lim_{n=\infty} \int_{x-\varepsilon}^{x+\varepsilon} (\xi - x)^i\,\varphi(\xi, x, n)\,d\xi = 0; \quad (i = 0, 1, \ldots, m-1)$$

$$\lim_{n=\infty} \int_{x-\varepsilon}^{x+\varepsilon} (\xi - x)^m\,\varphi(\xi, x, n)\,d\xi = m!$$

gleichmäßig für alle x von $< a', b' >$.

Dann gilt für jede der Klasse $\mathfrak{F}_i$ angehörige Funktion f, die im Teilintervall $< a', b' >$ von (a, b) m-mal stetig differenzierbar [2] ist, die Beziehung:

$$f^{(m)}(x) = \lim_{n=\infty} \int_a^b f(\xi)\,\varphi(\xi, x, n)\,d\xi$$

gleichmäßig für alle x von $< a', b' >$.

---

[1] Dabei ist unter α die Veränderliche x, unter $\mathfrak{A}$ das Intervall $< a', b' >$ zu verstehen.

[2] Die Funktion f heißt m-mal stetig differenzierbar in $< \alpha, \beta >$, wenn in jedem Punkte x von $< \alpha, \beta >$ die m-te Ableitung $f^{(m)}(x)$ existiert, und $f^{(m)}(x)$ in jedem Punkte von (α, β) stetig, in α rechtsseitig, in β linksseitig stetig ist.

Wir setzen zum Beweise,[1] wenn x ein Wert von $< a', b' >$ ist:

$$R_m(\xi, x) = f(x) + \sum_{i=1}^{m} \frac{(\xi - x)^i}{i!} f^{(m)}(x)$$

und:

$$\omega(\xi, x) = \frac{f(\xi) - R_m(\xi, x)}{(\xi - x)^m} \qquad (\xi \neq x).$$

Wir behaupten: ist $\eta > 0$ beliebig gegeben, so gibt es ein $\varepsilon > 0$, so daß:

(2) $$|\omega(\xi, x)| < \eta \quad \text{für alle } x \text{ von } < a', b' > \text{ und alle } \xi \text{ von } < x - \varepsilon, x + \varepsilon >.$$

Liegt nicht nur x, sondern auch ξ in $< a', b' >$, so haben wir:

$$\omega(\xi, x) = \frac{1}{m!} \{ f^{(m)}(x + \vartheta(\xi - x)) - f^{(m)}(x) \} \quad (0 < \vartheta < 1).$$

Weil aber $f^{(m)}(\xi)$ als in $< a', b' >$ stetig vorausgesetzt wurde, gibt es ein $\varepsilon_1 > 0$, so daß:

$$|f^{(m)}(x') - f^{(m)}(x)| < \eta$$

für alle x und x' von $< a', b' >$, für die $|x - x'| \leqq \varepsilon_1$. Wir haben also bereits:

$$|\omega(\xi, x)| < \eta$$

für alle x von $< a', b' >$ und alle gleichfalls zu $< a', b' >$ gehörigen ξ von $< x - \varepsilon_1, x + \varepsilon_1 >$.

Da $f(\xi)$ in a' und in b' eine m-te Ableitung hat, so gibt es ein $\varepsilon_2 > 0$, so daß:

(3) $$|\omega(\xi, a')| < \frac{\eta}{2} \text{ für alle } \xi \text{ von } < a' - \varepsilon_2, a' + \varepsilon_2 >,$$

(3') $$|\omega(\xi, b')| < \frac{\eta}{2} \text{ für alle } \xi \text{ von } < b' - \varepsilon_2, b' + \varepsilon_2 >.$$

Wir zeigen weiter, daß es ein $\varepsilon_3 > 0$ gibt, so daß:

(3a) $$|R_m(\xi, x) - R_m(\xi, a')| < \frac{\eta}{2}(x - \xi)^m$$

für alle ξ von $< a' - \varepsilon_3, a' >$ und alle x von $< a', a' + \varepsilon_3 >$; und ebenso:

(3'a) $$|R_m(\xi, x) - R_m(\xi, b')| < \frac{\eta}{2}(\xi - x)^m$$

für alle ξ von $< b', b' + \varepsilon_3 >$ und alle x von $< b' - \varepsilon_3, b' >$.

Sei in der Tat $g(\xi)$ die Funktion, die in $< a', b' >$ übereinstimmt mit $f(\xi)$, in $< a, a' >$ mit $R_m(\xi, a')$, und in $< b', b >$ mit $R_m(\xi, b')$. Dann ist $g(\xi)$ in ganz $< a, b >$ m-mal stetig differenzierbar, so daß wir nun für jedes x von $< a', b' >$ und jedes ξ von $< a, b >$ haben:

$$g(\xi) - R_m(\xi, x) = \frac{(\xi - x)^m}{m!} (g^{(m)}(x + \vartheta(\xi - x)) - g^{(m)}(x)) \quad (0 < \vartheta < 1).$$

Weil $g^{(m)}$ in ganz $< a, b >$ stetig ist, gibt es ein $\varepsilon_4 > 0$, so daß:

$$|g^{(m)}(x') - g^{(m)}(x)| < \frac{\eta}{4}$$

[1] Der Beweis wird viel einfacher, wenn man sich darauf beschränkt zu beweisen, daß die Konvergenz von $I_n(f, x)$ gegen $f^{(m)}(x)$ gleichmäßig ist in jedem Teilintervalle $< a'', b'' >$ von (a', b').

für alle x und x' von $< a, b >$, für die $|x-x'| \leqq \varepsilon_4$ ist. – Wir setzen $\varepsilon_3 = \frac{\varepsilon_4}{2}$. Liegt dann x in $< a', a'+\varepsilon_3 >$ und ξ in $< a'-\varepsilon_3, a' >$, so haben wir:

$$|R_m(\xi, a') - g(\xi)| < \frac{\eta}{4}(a'-\xi)^m,$$

$$|R_m(\xi, x) - g(\xi)| < \frac{\eta}{4}(x-\xi)^m,$$

woraus sich ergibt:

$$|R_m(\xi, x) - R_m(\xi, a')| < \frac{\eta}{2}(x-\xi)^m.$$

Damit ist (3 a) bewiesen, und ebenso beweist man (3' a).

Liegt nun wieder x in $< a', a' + \varepsilon_3 >$ und ξ sowohl in $< a' - \varepsilon_2, a' >$ als in $< a' - \varepsilon_3, a' >$, so gelten (3) und (3 a) gleichzeitig und wir haben, wenn ϑ eine Zahl zwischen -1 und 1 bezeichnet:

$$\omega(\xi, x) = \frac{f(\xi) - R_m(\xi, x)}{(\xi - x)^m} = \frac{f(\xi) - R_m(\xi, a') - \vartheta \cdot \frac{\eta}{2}(x-\xi)^m}{(\xi - x)^m} = \omega(\xi, a') \cdot \frac{(\xi - a')^m}{(\xi - x)^m} \pm \vartheta \cdot \frac{\eta}{2}$$

und somit, wegen (3):

$$|\omega(\xi, x)| < \eta.$$

Ebenso beweist man die Gültigkeit dieser Ungleichung, wenn x in $< b' - \varepsilon_3, b' >$ und ξ sowohl in $< b', b' + \varepsilon_2 >$ als in $< b', b' + \varepsilon_3 >$ liegt.

Diese Ungleichung ist also jetzt bewiesen:

1. Wenn x in $< a', b' >$ und ξ sowohl in $< a', b' >$ als auch in $< x - \varepsilon_1, x + \varepsilon_1 >$; 2. wenn x in $< a', a' + \varepsilon_3 >$ und ξ sowohl in $< a' - \varepsilon_2, a' >$ als in $< a' - \varepsilon_3, a' >$; 3. wenn x in $< b' - \varepsilon_3, b' >$ und ξ sowohl in $< b', b' + \varepsilon_2 >$ als in $< b', b' + \varepsilon_3 >$ liegt. Versteht man also unter ε die kleinste der drei Zahlen $\varepsilon_1, \varepsilon_2, \varepsilon_3$, so gilt die Ungleichung für alle x von $< a, b >$ und alle ξ von $< x - \varepsilon, x + \varepsilon >$: das aber war die Behauptung (2).

Wir bilden nun mit diesem ε den Kern $\psi(\xi, x, n, \varepsilon)$ und sehen wieder durch Berufung auf Satz IX: es gibt einen Index n_0, so daß:

$$\left|\int_a^b f(\xi)\,\psi(\xi, x, n, \varepsilon)\,d\xi\right| < \eta \quad \text{für alle } n \geqq n_0 \text{ und alle } x \text{ von } < a'\ b' >. \tag{4}$$

Für jedes x von $< a', b' >$ und jedes ξ von $< a' - \varepsilon, b' + \varepsilon >$ gilt die Entwicklung:

$$f(\xi) = f(x) + \sum_{i=1}^{m} \frac{(\xi - x)^i}{i!} f^{(i)}(x) + (\xi - x)^m\,\omega(\xi, x).$$

Wir haben also:

$$\int_{x-\varepsilon}^{x+\varepsilon} f(\xi)\,\varphi(\xi, x, n)\,d\xi = f(x)\int_{x-\varepsilon}^{x+\varepsilon} \varphi(\xi, x, n)\,d\xi$$

$$+ \sum_{i=1}^{m} \frac{f^{(i)}(x)}{i!} \int_{x-\varepsilon}^{x+\varepsilon} (\xi - x)^i\,\varphi(\xi, x, n)\,d\xi + \int_{x-\varepsilon}^{x+\varepsilon} \omega(\xi, x)\,(\xi - x)^m\,\varphi(\xi, x, n)\,d\xi. \tag{5}$$

Wegen Bedingung 4. kann n_0 so groß gewählt werden, daß:

$$\left|\left\{f(x)\int_{x-\varepsilon}^{x+\varepsilon} \varphi(\xi, x, n)\,d\xi + \sum_{i=1}^{m} \frac{f^{(i)}(x)}{i!} \int_{x-\varepsilon}^{x+\varepsilon} (\xi - x)^i\,\varphi(\xi, x, n)\,d\xi\right\} - f^{(m)}(x)\right| < \eta \tag{6}$$

für $n \geqq n_0$ und alle x von $< a', b' >$.

Wegen (2) und Bedingung 3. haben wir:

$$(7) \qquad \left| \int_{x-\varepsilon}^{x+\varepsilon} \omega(\xi, x)\,(\xi - x)^m\, \varphi(\xi, x, n)\, d\xi \right| < \eta \cdot N$$

für alle n und alle x von $< a', b' >$.

Wir haben also aus (5), (6) und (7):

$$(8) \qquad \left| \int_{x-\varepsilon}^{x+\varepsilon} f(\xi)\, \varphi(\xi, x, n)\, d\xi - f^{(m)}(x) \right| < \eta\,(1+N)$$

für alle $n \geqq n_0$ und alle x von $< a', b' >$.

Beachten wir endlich noch, daß:

$$\int_a^b f(\xi)\, \psi(\xi, x, n, \varepsilon)\, d\xi = \int_a^{x-\varepsilon} f(\xi)\, \varphi(\xi, x, n)\, d\xi + \int_{x+\varepsilon}^b f(\xi)\, \varphi(\xi, x, n)\, d\xi,$$

so sehen wir, daß aus (4) und (8) folgt:

$$|I_n(f, x) - f^{(m)}(x)| < \eta\,(2+N)$$

für alle $n \geqq n_0$ und alle x von $< a', b' >$. Da aber η beliebig war, so ist damit Satz XI bewiesen.

Die Übertragung der Sätze X und XI auf unendliche Intervalle bietet keine Schwierigkeit. Zunächst tritt an Stelle von Satz IX folgender Satz:

IX *a.* »Damit für jede Funktion f von $\mathfrak{F}_i$ die Beziehung:

$$\lim_{n=\infty} \int_{-\infty}^{+\infty} f(\xi)\, \varphi(\xi, \alpha, n)\, d\xi = 0$$

gleichmäßig für alle α von $\mathfrak{A}$ gelte, ist notwendig und hinreichend, daß φ folgenden Bedingungen genügt:

1 *a.* Für jedes einzelne α von $\mathfrak{A}$ genügt φ der Bedingung 1. desjenigen der Sätze I *a* bis IV *a*, der sich auf die Klasse $\mathfrak{F}_i$ bezieht.

1 *b.* Für die Klasse $\mathfrak{F}_1$: Es gibt einen Index n_0 und eine Konstante M, so daß, abgesehen von Nullmengen:

$$|\varphi(\xi, \alpha, n)| < M \quad \text{für alle } \xi, \text{ alle } n \geqq n_0 \text{ und alle } \alpha \text{ von } \mathfrak{A}.$$

Für die Klasse $\mathfrak{F}_2$: Es gibt einen Index n_0 und eine Konstante M, so daß:

$$\int_{-\infty}^{+\infty} (\varphi(\xi, \alpha, n))^2\, d\xi < M \quad \text{für alle } n \geqq n_0 \text{ und alle } \alpha \text{ von } \mathfrak{A}.$$

Für die Klasse $\mathfrak{F}_3$: Zu jedem $\mu > 0$ gehört ein $\lambda > 0$ und ein Index n_0, so daß für jede Menge I sich nicht überdeckender Intervalle, deren Gesamtlänge $< \lambda$ ist, für alle $n \geqq n_0$ und alle α von $\mathfrak{A}$:

$$\int_I |\varphi(\xi, \alpha, n)|\, d\xi < \mu;$$

und zu jedem $\mu > 0$ gehört ein A und ein Index n_0, so daß:

$$\int_{-\infty}^{-A} |\varphi(\xi, \alpha, n)|\, d\xi < \mu; \quad \int_A^{+\infty} |\varphi(\xi, \alpha, n)|\, d\xi < \mu \quad \text{für alle } n \geqq n_0 \text{ und alle } \alpha \text{ von } \mathfrak{A}.$$

Für die Klasse $\mathfrak{F}_4$: Es gibt einen Index n_0 und eine Konstante M, so daß:

$$\int_{-\infty}^{+\infty} |\varphi(\xi, \alpha, n)|\, d\xi < M \quad \text{für alle } n \geqq n_0 \text{ und alle } \alpha \text{ von } \mathfrak{A}.$$

2. Für jedes endliche Intervall $<\alpha, \beta>$ gilt die Beziehung:

$$\lim_{n=\infty} \int_{\alpha}^{\beta} \varphi(\xi, \alpha, n)\, d\xi = 0$$

gleichmäßig für alle α von $\mathfrak{A}$. Im Falle der Klasse $\mathfrak{F}_4$ gilt dies auch für die Beziehungen:

$$\lim_{n=\infty} \int_{-\infty}^{0} \varphi(\xi, \alpha, n)\, d\xi = 0; \quad \lim_{n=\infty} \int_{0}^{+\infty} \varphi(\xi, \alpha, n)\, d\xi = 0.\text{«}$$

Unter Berufung auf diesen Satz IX *a* erkennt man dann sofort die Richtigkeit der beiden Sätze:

X *a*. »Für den Kern $\varphi(\xi, x, n)$ seien die nachstehenden Bedingungen erfüllt:

1. und 2. Für jedes $\varepsilon > 0$ genügt der dem Kerne $\varphi(\xi, x, n)$ zugeordnete Kern $\psi(\xi, x, n, \varepsilon)$ den Bedingungen 1 *a*, 1 *b* und 2. von Satz IX *a* für die Klasse $\mathfrak{F}_i$, wenn unter α die Veränderliche x, unter $\mathfrak{A}$ das Intervall $<a', b'>$ verstanden wird.

3. und 4. Es ist Bedingung 3. und 4. von Satz X erfüllt.

Dann gilt für jede der Klasse $\mathfrak{F}_i$ angehörige Funktion f, die in jedem Punkte von $<a', b'>$ stetig ist, die Beziehung:

$$f(x) = \lim_{n=\infty} \int_{-\infty}^{+\infty} f(\xi)\, \varphi(\xi, x, n)\, d\xi$$

gleichmäßig für alle x von $<a', b'>$.«

XI *a*. »Genügt der Kern $\varphi(\xi, x, n)$ den Bedingungen 1. und 2. von Satz X *a* und den Bedingungen 3. und 4. von Satz XI, so gilt für jede der Klasse $\mathfrak{F}_i$ angehörige Funktion f, die in $<a', b'>$ m-mal stetig differenzierbar ist, die Beziehung:

$$f^{(m)}(x) = \lim_{n=\infty} \int_{-\infty}^{+\infty} f(\xi)\, \varphi(\xi, x, n)\, d\xi$$

gleichmäßig für alle x von $<a', b'>$.«.

§ 6. Darstellung von $f(x)$ in gewissen Unstetigkeitspunkten.

Ein einfaches Korollar von Satz VII ist folgende, von H. Lebesgue direkt bewiesene Tatsache,[1] die, unter spezielleren Voraussetzungen über den Kern $\varphi(\xi, x, n)$, eine wesentliche Verschärfung von Satz VI enthält: Sei x ein fest gegebener Punkt von (a, b); dann gilt der Satz:

XII. Der Kern $\varphi(\xi, x, n)$ sei (als Funktion von ξ betrachtet) absolut stetig in einer Umgebung $<x-h, x+h>$ des Punktes x von (a, b), in der außerdem für jedes $\xi \neq x$ die Beziehung gelte:

$$\lim_{n=\infty} \varphi(\xi, x, n) = 0. \tag{1}$$

Ferner genüge der Kern $\varphi(\xi, x, n)$ noch folgenden Bedingungen:

1. und 2. Es genügt $\varphi(\xi, x, n)$ den Bedingungen 1. und 2. von Satz VI.

3. Zu jedem hinlänglich kleinen $h > 0$ gibt es ein N, so daß:

$$\int_{x-h}^{x+h} \left| (\xi - x) \cdot \frac{\partial}{\partial \xi} \varphi(\xi, x, n) \right| d\xi < N \quad \text{für alle } n.$$

4. Es ist:

$$\lim_{n=\infty} \int_{a}^{b} \varphi(\xi, x, n)\, d\xi = 1.$$

[1] A. a. O. p. 80.

Dann gilt für jede der Klasse $\mathfrak{F}_i$ angehörige Funktion f, die im Punkte x die erste Ableitung ihres unbestimmten Integrales ist, die Gleichung:

$$f(x) = \lim_{n=\infty} I_n(f, x). \tag{2}$$

Wir schicken die Bemerkung voraus, daß die in Bedingung 3. auftretende Ableitung $\frac{\partial}{\partial \xi} \varphi(\xi, x, n)$ in einer Umgebung $< x-h, x+h >$ von x überall, abgesehen von einer Nullmenge, als endliche Zahl existiert: dies folgt, nach einem bekannten Satze, daraus, daß f in einer solchen Umgebung als absolut stetig vorausgesetzt wurde.

Die Bedeutung des zu beweisenden Satzes liegt darin, daß, wenn seine Bedingungen in allen Punkten x von (a, b) erfüllt sind, die Gleichung $f(x) = \lim_{n=\infty} I_n(f, x)$ für jede Funktion f von $\mathfrak{F}_i$ überall in (a, b) gilt, abgesehen von einer Nullmenge. denn es ist jede Funktion von $\mathfrak{F}_i$ überall, abgesehen von einer Nullmenge, Ableitung ihres unbestimmten Integrales.

Um nun Satz XII aus Satz VII herzuleiten, setzen wir:

$$F_1(\xi) = \int_a^{\xi} f(t)\, dt.$$

Wir wählen ein k (> 0) so klein, daß in $< x-k, x+k >$ die über φ gemachten Voraussetzungen gelten und insbesondere auch Bedingung 3. angewendet werden kann.

Durch partielle Integration, die wegen der absoluten Stetigkeit von φ angewendet werden darf, erhalten wir:

$$\int_{x-k}^{x+k} f(\xi)\, \varphi(\xi, x, n)\, d\xi = F_1(\xi)\, \varphi(\xi, x, n)\Big]_{x-k}^{x+k} - \int_{x-k}^{x+k} F_1(\xi) \frac{\partial}{\partial \xi} \varphi(\xi, x, n)\, d\xi.$$

Nun ist, wegen Bedingung 1. und 2.:

$$\lim_{n=\infty} \left\{ I_n(f, x) - \int_{x-k}^{x+k} f(\xi)\, \varphi(\xi, x, n)\, d\xi \right\} = 0.$$

Ferner ist, wegen (1):

$$\lim_{n=\infty} F_1(\xi)\, \varphi(\xi, x, n)\Big]_{x-k}^{x+k} = 0.$$

Es wird also, um (2) nachzuweisen, genügen zu zeigen, daß:

$$\lim_{n=\infty} \int_{x-k}^{x+k} F_1(\xi) \left\{ -\frac{\partial}{\partial \xi} \varphi(\xi, x, n) \right\} d\xi = f(x) \tag{2a}$$

ist. Da wir voraussetzen, daß im Punkte x die Funktion f Ableitung ihres unbestimmten Integrales sei, das heißt, daß:

$$f(x) = F_1'(x)$$

sei, da weiter $F_1(\xi)$ stetig ist und mithin zur Klasse $\mathfrak{F}_4$ gehört, so wird $(2a)$ bewiesen sein, wenn wir zeigen, daß der Kern:

$$\Phi(\xi, x, n) = -\frac{\partial}{\partial \xi} \varphi(\xi, x, n)$$

im Intervalle $< x-k, x+k >$ allen Voraussetzungen von Satz VII für die Klasse $\mathfrak{F}_4$ und für $m = 1$ genügt.

Bedingung 2. von VII ist erfüllt wegen:

$$\int_\alpha^\beta \frac{\partial}{\partial \xi} \varphi(\xi, x, n)\, d\xi = \varphi(\xi, x, n)\Big]_\alpha^\beta \tag{3}$$

und wegen (1). — Bedingung 3. von VII (für das Intervall $< x-k, x+k >$) ist identisch mit unserer Bedingung 3. — Um nachzuweisen, daß Bedingung 1. von VII erfüllt ist, sei $< \alpha, \beta >$ ein beliebiges, den Punkt x nicht enthaltendes Teilintervall von $< x-k, x+k >$. Es liege etwa in $(x, x+k)$. Wir haben zu zeigen, daß es ein M gibt, so daß:

$$\int_\alpha^\beta \left| \frac{\partial}{\partial \xi} \varphi(\xi, x, n) \right| d\xi < M \quad \text{für alle } n. \tag{4}$$

Schreiben wir:

$$\frac{\partial}{\partial \xi} \varphi(\xi, x, n) = \frac{1}{\xi - x} \cdot \left\{ (\xi - x) \frac{\partial}{\partial \xi} \varphi(\xi, x, n) \right\},$$

so haben wir:

$$\int_\alpha^\beta \left| \frac{\partial}{\partial \xi} \varphi(\xi, x, n) \right| d\xi \leqq \frac{1}{\alpha - x} \int_\alpha^\beta \left| (\xi - x) \cdot \frac{\partial}{\partial \xi} \varphi(\xi, x, n) \right| d\xi,$$

und somit nach Bedingung 3.

$$\int_\alpha^\beta \left| \frac{\partial}{\partial \xi} \varphi(\xi, x, n) \right| d\xi < \frac{N}{\alpha - x},$$

womit (4) erwiesen ist.

Was endlich Bedingung 4. von Satz VII anlangt, so sind die beiden Gleichungen zu beweisen:

$$\lim_{n=\infty} \int_{x-k}^{x+k} \frac{\partial}{\partial \xi} \varphi(\xi, x, n)\, d\xi = 0,$$

$$\lim_{n=\infty} \int_{x-k}^{x+k} (\xi - x) \frac{\partial}{\partial \xi} \varphi(\xi, x, n)\, d\xi = -1.$$

Die erste folgt wegen (1) unmittelbar aus (3), angewendet auf $< x-k, x+k >$; die zweite folgt vermöge:

$$\int_{x-k}^{x+k} \varphi(\xi, x, n)\, d\xi = (\xi - x)\, \varphi(\xi, x, n)\Big]_{x-k}^{x+k} - \int_{x-k}^{x+k} (\xi - x) \frac{\partial}{\partial \xi} \varphi(\xi, x, n)\, d\xi$$

unter Berücksichtigung von (1) unmittelbar aus unseren Bedingungen 2. und 4.

Damit ist Satz XII nachgewiesen. Es sei, ohne Beweis, noch Folgendes bemerkt: Hält man an den Voraussetzungen von Satz XII, mit Ausnahme von Bedingung 3., fest, so ist, damit (2) für alle Funktionen von $\mathfrak{F}_1$ gelte, die im Punkte x Ableitungen ihres unbestimmten Integrales sind, n o t w e n d i g, daß die Bedingung erfüllt sei:

3 *a*. Zu jedem hinlänglich kleinen $h > 0$ gibt es ein N, so daß

$$\left| \int_\alpha^\beta (\xi - x) \frac{\partial}{\partial \xi} \varphi(\xi, x, n)\, d\xi \right| < N$$

für alle n und alle Teilintervalle $< \alpha, \beta >$ von $< x-h, x+h >$.

Satz XII läßt sich sofort noch etwas verallgemeinern:

XIII. Gelten alle Voraussetzungen und Bedingungen von Satz XII und ist noch für alle hinlänglich kleinen $|t|$ die Beziehung

$$\varphi(x+t, x, n) = \varphi(x-t, x, n) \tag{5}$$

erfüllt, so gilt (2) für jede Funktion f, die im Punkte x verallgemeinerte erste Ableitung ihres unbestimmten Integrales ist.

In der Tat, aus (5) folgt:

$$\frac{\partial}{\partial \xi} \varphi(x+t, x, n) = -\frac{\partial}{\partial \xi} \varphi(x-t, x, n),$$

so daß statt Satz VII Satz VIII verwendet werden kann.

Die Bedingung, daß f verallgemeinerte erste Ableitung seines unbestimmten Integrales sei, kann noch etwas umgeformt werden. Es ist nämlich nach § 4 die verallgemeinerte erste Ableitung von $F(\xi)$ im Punkte x nichts anderes als die erste Ableitung nach t für $t = 0$ von

$$\frac{F(x+t) - F(x-t)}{2}.$$

Ist $F_1(\xi)$ unbestimmtes Integral von $f(\xi)$, so wird:

$$F_1(x+t) - F_1(x-t) = \int_{-t}^{t} f(x+u)\, du = \int_0^t (f(x+u) + f(x-u))\, du.$$

Es wird also $f(\xi)$ im Punkte x verallgemeinerte erste Ableitung seines unbestimmten Integrales sein, wenn:

$$\lim_{t=0} \frac{1}{2t} \int_{-t}^{t} f(x+u)\, du = f(x)$$

ist, oder, noch anders formuliert, wenn:

$$\lim_{t=0} \frac{1}{t} \int_0^t (f(x+t) + f(x-t) - 2f(x))\, dt = 0$$

ist. In dieser letzteren Form findet sich die Bedingung bei H. Lebesgue.

§ 7. Darstellung von $f(x)$ in allgemeineren Unstetigkeitspunkten.

Die Sätze XII und XIII sind die einfachsten in einer Kette analoger, immer schärferer Sätze, die nun ausgesprochen und bewiesen werden sollen:

XIV. Der Kern $\varphi(\xi, x, n)$ besitze in einer Umgebung $\langle x-h, x+h \rangle$ des Punktes x von (a, b) eine absolut stetige $(m-1)$-te Ableitung[1] $\frac{\partial^{m-1}}{\partial \xi^{m-1}} \varphi(\xi, x, n)$, und es gelten in dieser Umgebung für $\xi \neq x$ die Beziehungen:

$$(1) \qquad \lim_{n=\infty} \varphi(\xi, x, n) = 0, \quad \lim_{n=\infty} \frac{\partial^i}{\partial \xi^i} \varphi(\xi, x, n) = 0 \quad (i = 1, 2 \ldots, m-1).$$

Ferner genüge der Kern $\varphi(\xi, x, n)$ noch folgenden Bedingungen:

1. und 2. Es genügt $\varphi(\xi, x, n)$ den Bedingungen 1. und 2. von Satz VI.

3. Zu jedem hinlänglich kleinen $h > 0$ gibt es ein N, so daß:

$$\int_{x-h}^{x+h} \left| (\xi - x)^m \frac{\partial^m}{\partial \xi^m} \varphi(\xi, x, n) \right| d\xi < N \quad \text{für alle } n.$$

4. Es ist:

$$\lim_{n=\infty} \int_a^b \varphi(\xi, x, n)\, d\xi = 1.$$

[1] Daraus folgt, daß $\frac{\partial^m}{\partial \xi^m} \varphi(\xi, x, n)$ in $\langle x-h, x+h \rangle$ überall, abgesehen von einer Nullmenge, als endliche Zahl existiert.

Dann gilt für jede der Klasse $\mathfrak{F}_i$ angehörige Funktion f, die im Punkte x m-te Ableitung ihres m-fach iterierten unbestimmten Integrales ist, die Gleichung:

$$f(x) = \lim_{n=\infty} I_n(f, x). \tag{2}$$

Dabei ist unter dem m-fach iterierten unbestimmten Integrale von f die Funktion F_m verstanden, die definiert ist durch:

$$F_1(\xi) = \int_a^\xi f(t)\,dt; \quad F_{i+1}(\xi) = \int_a^\xi F_i(t)\,dt.$$

Zum Beweise von Satz XIV wählen wir nun ein k (> 0) so klein, daß in der Umgebung $< x-k, x+k >$ von x alle über $\varphi(\xi, x, n)$ gemachten Voraussetzungen gelten, insbesondere auch Bedingung 3. angewendet werden kann.

Durch m-malige partielle Integration (was wegen der absoluten Stetigkeit von $\frac{\partial^{m-1}\varphi}{\partial\xi^{m-1}}$ und der daraus folgenden absoluten Stetigkeit von $\frac{\partial^i\varphi}{\partial\xi^i}$ ($i < m-1$) und von φ zulässig ist) erhalten wir:

$$\int_{x-k}^{x+k} f(\xi)\,\varphi(\xi, x, n)\,d\xi = F_1(\xi)\,\varphi(\xi, x, n)\Big]_{x-k}^{x+k} +$$

$$+ \sum_{i=1}^{m-1} (-1)^i F_{i+1}(\xi) \frac{\partial^i}{\partial\xi^i}\varphi(\xi, x, n)\Big]_{x-k}^{x+k} + (-1)^m \int_{x-k}^{x+k} F_m(\xi) \frac{\partial^m}{\partial\xi^m}\varphi(\xi, x, n)\,d\xi.$$

Wegen (1) verschwinden beim Grenzübergange die rechts außerhalb des Integralzeichens stehenden Glieder. Wegen der Bedingungen 1. und 2. haben wir:

$$\lim_{n=\infty}\left\{\int_a^b f(\xi)\,\varphi(\xi, x, n)\,d\xi - \int_{x-k}^{x+k} f(\xi)\,\varphi(\xi, x, n)\,d\xi\right\} = 0.$$

Es wird also, um (2) nachzuweisen, genügen zu zeigen, daß:

$$\lim_{n=\infty}\int_{x-k}^{x+k} F_m(\xi)\left\{(-1)^m \frac{\partial^m}{\partial\xi^m}\varphi(\xi, x, n)\right\}d\xi = f(x) \tag{2a}$$

ist. Da wir voraussetzen, daß im Punkte x die Funktion f die m-te Ableitung von F_m sei, da weiter F_m zur Klasse $\mathfrak{F}_4$ gehört, so wird (2 a) bewiesen sein, wenn wir zeigen, daß der Kern $(-1)^m \frac{\partial^m}{\partial\xi^m}\varphi(\xi, x, n)$ im Intervalle $< x-k, x+k >$ allen Voraussetzungen von Satz VII für die Klasse $\mathfrak{F}_4$ genügt.

Für Bedingung 2. von VII folgt dies wieder unmittelbar aus (1) und der Relation:

$$\int_\alpha^\beta \frac{\partial^m}{\partial\xi^m}\varphi(\xi, x, n)\,d\xi = \frac{\partial^{m-1}}{\partial\xi^{m-1}}\varphi(\xi, x, n)\Big]_\alpha^\beta.$$

Bedingung 3. von VII (für das Intervall $< x-k, x+k >$) ist identisch mit unserer Bedingung 3. Für Bedingung 1. von VII wird es ganz analog wie in § 6 aus unserer Bedingung 3. hergeleitet. Was endlich Bedingung 4. von VII anlangt, so sind die Gleichungen zu beweisen:

$$\lim_{n=\infty}\int_{x-k}^{x+k} (\xi - x)^i \frac{\partial^m}{\partial\xi^m}\varphi(\xi, x, n)\,d\xi = 0 \quad (i = 0, 1, \ldots, m-1),$$

$$\lim_{n=\infty}\int_{x-k}^{x+k} (\xi - x)^m \frac{\partial^m}{\partial\xi^m}\varphi(\xi, x, n)\,d\xi = (-1)^m m!$$

Sie folgen vermöge der Gleichungen (1) und unserer Bedingungen 2. und 4. aus der für $i < m$ gültigen Formel:[1]

$$\int_{x-k}^{x+k} (\xi - x)^i \frac{\partial^m}{\partial \xi^m} \varphi(\xi, x, n)\, d\xi = \sum_{r=0}^{i} (-1)^r\, i\,(i-1)\ldots(i-r+1)\,(\xi - x)^{i-r} \frac{\partial^{m-r-1}}{\partial \xi^{m-r-1}} \varphi(\xi, x, n) \Bigg|_{x-k}^{x+k}$$

und der Formel:

$$\int_{x-k}^{x+k} (\xi - x)^m \frac{\partial^m}{\partial \xi^m} \varphi(\xi, x, n)\, d\xi = \sum_{r=0}^{m-1} (-1)^r\, m\,(m-1)\ldots(m-r+1)\,(\xi - x)^{m-r} \frac{\partial^{m-r-1}}{\partial \xi^{m-r-1}} \varphi(\xi, x, n) \Bigg]_{x-k}^{x+k} +$$

$$+ (-1)^m \cdot m! \int_{x-k}^{x+k} \varphi(\xi, x, n)\, d\xi.$$

Unser Satz ist damit bewiesen.

Auch hier können wir ihn sofort noch etwas verallgemeinern:

XV. Gilt, außer den Voraussetzungen und Bedingungen von Satz XIV noch für alle hinlänglich kleinen $|t|$ die Beziehung:

$$\varphi(x+t, x, n) = \varphi(x-t, x, n), \tag{3}$$

so gilt (2) für jede Funktion f, die im Punkte x verallgemeinerte m-te Ableitung ihres m-fach iterierten unbestimmten Integrales ist.

In der Tat haben wir aus (3):

$$\frac{\partial^m}{\partial \xi^m} \varphi(x+t, x, n) = (-1)^m \frac{\partial^m}{\partial \xi^m} \varphi(x-t, x, n),$$

so daß Satz VIII angewendet werden kann.

Auch Satz XIV und XV können ohneweiters auf unendliche Intervalle ausgedehnt werden:

XIV a und XV a: »Man erhält aus den Sätzen XIV und XV Sätze für das Intervall $(-\infty, +\infty)$, indem man in den Bedingungen 1. und 2. statt »Satz VI« setzt »Satz VI a« und in Bedingung 4. die Integrationsgrenzen a und b ersetzt durch $x-h$ und $x+h$ $(h > 0)$.«

§ 8. Differenziation singulärer Integrale.

Die Sätze VII und VIII gestatten es auch, Theoreme über die Differenziation von Darstellungen gegebener Funktionen durch singuläre Integrale herzuleiten.

Nehmen wir an, es genüge $\varphi(\xi, x, n)$ für alle x von (a, b) den Bedingungen von Satz VI. Insbesondere ist dann für jedes Teilintervall $< \alpha, \beta >$ von $< a, b >$, das den Punkt x nicht enthält:

$$\lim_{n=\infty} \int_\alpha^\beta \varphi(\xi, x, n)\, d\xi = 0, \tag{1}$$

während für das Intervall $< a, b >$ gilt:

$$\lim_{n=\infty} \int_a^b \varphi(\xi, x, n)\, d\xi = 1. \tag{2}$$

Ist $f(x)$ stetig in allen Punkten von (a, b), so haben wir in ganz (a, b):

$$f(x) = \lim_{n=\infty} \int_a^b f(\xi)\, \varphi(\xi, x, n)\, d\xi. \tag{3}$$

[1] Es ist in dieser Formel $\frac{\partial^0 \varphi}{\partial \xi^0}$ durch φ zu ersetzen und $i\,(i-1)\ldots(i-r+1)$ für $r = 0$ zu ersetzen durch 1.

Differenzieren wir Formel (1), die nichts anderes ist als Formel (3) für diejenige Funktion f, die $= 1$ ist in $< \alpha, \beta >$ und $= 0$ außerhalb $< \alpha, \beta >$, m-mal nach x und nehmen an, was natürlich durchaus nicht immer der Fall ist, es könne die Differenziation unter dem Limes- und Integralzeichen ausgeführt werden, so erhalten wir:

$$\lim_{n=\infty} \int_\alpha^\beta \frac{\partial^m}{\partial x^m} \varphi(\xi, x, n)\, d\xi = 0,$$

das heißt es genügt der Kern $\frac{\partial^m}{\partial x^m} \varphi(\xi, x, n)$ der Bedingung 2. von Satz VII.

Schreiben wir neben Formel (2), die nichts anderes ist als (3) für $f = 1$, noch die aus (3) für $f = \xi^i$ $(i = 1, 2, \ldots, m)$ entstehenden Formeln auf:

$$\lim_{n=\infty} \int_a^b \xi^i \varphi(\xi, x, n)\, d\xi = x^i \qquad (i = 1, 2, \ldots, m). \tag{4}$$

Differenzieren wir die Formeln (2) und (4) m-mal nach x und nehmen wir wieder an, es könne die Differenziation unter dem Limes- und Integralzeichen ausgeführt werden, so erhalten wir:

$$\lim_{n=\infty} \int_a^b \xi^i \frac{\partial^m}{\partial x^m} \varphi(\xi, x, n)\, d\xi = 0 \qquad (i = 0, 1, \ldots, m-1),$$

$$\lim_{n=\infty} \int_a^b \xi^m \frac{\partial^m}{\partial x^m} \varphi(\xi, x, n)\, d\xi = m!$$

Hieraus folgt unmittelbar, daß der Kern $\frac{\partial^m}{\partial x^m} \varphi(\xi, x, n)$ der Bedingung 4. von Satz VII genügt. Nehmen wir also an, er genüge auch noch den Bedingungen 1. und 3. von Satz VII, welch letztere wir nun aufschreiben als Bedingung:

3 *a*. Es gibt ein N, so daß:

$$\int_a^b \left| (\xi - x)^m \frac{\partial^m}{\partial x^m} \varphi(\xi, x, n) \right| d\xi < N \quad \text{für alle } n,$$

so kann Satz VII angewendet werden, und wir haben das Resultat:

Genügt sowohl der Kern $\varphi(\xi, x, n)$ als auch der Kern $\frac{\partial^m}{\partial x^m} \varphi(\xi, x, n)$ der Bedingung 1. von Satz VI, genügt der Kern $\varphi(\xi, x, n)$ ferner den Bedingungen 2., 3. und 4. von Satz VI und unserer Bedingung 3 *a*, und darf Formel (3) für $f = \xi^i$ $(i = 0, 1, \ldots, m)$ sowie für die Funktionen f, die in einem den Punkt x nicht enthaltenden Teilintervalle $< \alpha, \beta >$ von $< a, b >$ den Wert 1, sonst den Wert 0 haben, m-mal unter dem Limes- und Integralzeichen nach x differenziert werden, so gilt dies für jedes f von $\mathfrak{F}_i$, das im Punkte x eine m-te Ableitung besitzt.

Ein befriedigenderes Resultat erhalten wir, wenn wir uns auf den (von H. Lebesgue ausschließlich betrachteten) Fall beschränken, daß der Kern die Form hat: $\varphi(\xi - x, n)$. Wir können dann den Satz aussprechen:

XVI. Sei $\varphi(u, n)$ eine im Intervalle $(-l, l)$ gegebene Funktion, die eine in jedem Teilintervalle $< \alpha, \beta >$ von $(-l, l)$ absolut stetige Ableitung $(m-1)$-ter Ordnung[1] besitzt.[2]

Ferner sei für jedes $u \neq 0$ von $(-l, l)$:

$$\lim_{n=\infty} \varphi(u, n) = 0; \quad \lim_{n=\infty} \varphi^{(i)}(u, n) = 0 \qquad (i = 1, 2, \ldots, m-1). \tag{5}$$

---

[1] Es ist (im Falle $m = 1$) unter der Ableitung 0-ter Ordnung hier wie im Folgenden die Funktion $\varphi(u, n)$ selbst zu verstehen.

[2] Daraus folgt, daß in $(-l, l)$ auch die m-te Ableitung $\varphi^{(m)}(u)$ überall, abgesehen von einer Nullmenge, als endliche Zahl existiert.

Damit für jede im Intervalle $< a, a+l >$ zur Klasse $\mathfrak{F}_i$ gehörige Funktion, die im beliebigen Punkte x von $(a, a+l)$ endliche Ableitungen der m ersten Ordnungen besitzt, die Formeln gelten:

(6) $$f(x) = \lim_{n=\infty} \int_a^{a+l} f(\xi)\, \varphi(\xi - x, n)\, d\xi,$$

(7) $$f^{(i)}(x) = \lim_{n=\infty} (-1)^i \int_a^{a+l} f(\xi)\, \varphi^{(i)}(\xi - x, n)\, d\xi \qquad (i = 1, 2, \ldots, m),$$

ist notwendig und hinreichend, daß $\varphi(u, n)$ folgenden Bedingungen genügt:

1. In jedem Teilintervalle $< \alpha, \beta >$ von $(-l, l)$, das den Punkt 0 nicht enthält, genügt $\varphi^{(m)}(u, n)$ der Bedingung 1. desjenigen der Sätze I bis IV, der sich auf die Klasse $\mathfrak{F}_i$ bezieht.[1]

2. Für jedes Teilintervall $< \alpha, \beta >$ von $(-l, l)$, das den Punkt 0 nicht enthält, ist:

$$\lim_{n=\infty} \int_\alpha^\beta \varphi(u, n)\, du = 0.$$

3. Es gibt ein N und ein $h > 0$, so daß:

(8) $$\int_{-h}^{h} |u^m \varphi^{(m)}(u, n)|\, du < N \quad \text{für alle } n.$$

4. Es gibt ein $h > 0$, für das:

(9) $$\lim_{n=\infty} \int_{-h}^{h} \varphi(u, n)\, du = 1.$$

Die Bedingungen sind hinreichend. Bemerken wir zunächst, daß man durch partielle Integration erhält:[2]

$$\int_0^h |u^{m-1} \varphi^{(m-1)}(u, n)|\, du = \frac{u^m}{m} |\varphi^{(m-1)}(u, n)| \Big]_0^h - \frac{1}{m} \int_0^h u^m \cdot \operatorname{sgn} \varphi^{(m-1)}(u, n)\, \varphi^{(m)}(u, n)\, du.$$

Es gibt also, wegen (8) und (5), ein M, so daß:

$$\int_0^h |u^{m-1} \varphi^{(m-1)}(u, n)|\, du < M \quad \text{für alle } n.$$

Ebenso sieht man, daß es ein M gibt, so daß:

$$\int_{-h}^0 |u^{m-1} \varphi^{(m-1)}(u, n)|\, du < M \quad \text{für alle } n.$$

Wir sehen also schließlich, daß aus (8) folgt: es kann N so groß angenommen werden, daß:

(10) $$\int_{-h}^{h} |u^{m-1} \varphi^{(m-1)}(u, n)|\, du < N \quad \text{für alle } n.$$

Ebenso wie (10) aus (8) hergeleitet wurde, kann aus (10) hergeleitet werden: es kann N auch so groß angenommen werden, daß:

$$\int_{-h}^{h} |u^{m-2} \varphi^{(m-2)}(u, n)|\, du < N \quad \text{für alle } n,$$

[1] Daraus folgt, daß dasselbe auch für $\varphi(u, n)$ und $\varphi^{(i)}(u, n)$ $(i = 1, 2, \ldots, m-1)$ gilt.

[2] In der folgenden Gleichung bedeutet $\operatorname{sgn} \varphi^{(m-1)}$ das Vorzeichen von $\varphi^{(m-1)}$, wo $\varphi^{(m-1)} \neq 0$ ist, und den Wert 0, wo $\varphi^{(m-1)} = 0$ ist.

und indem man so weiter schließt, sieht man, daß aus Bedingung 3. und (5) folgt: es gibt ein N, so daß:

(11) $$\int_{-h}^{h} |u^i \varphi^{(i)}(u, n)| du < N \quad \text{für alle } n \quad (i = 1, 2, \ldots, m)$$

(12) $$\int_{-h}^{h} |\varphi(u, n)| \, du < N \quad \text{für alle } n.$$

Ferner erhält man durch partielle Integration:[1]

$$\int_{-h}^{h} u^j \varphi^{(i)}(u, n) \, du = \sum_{r=0}^{j} (-1)^r j(j-1)\ldots(j-r+1)\, u^{j-r} \varphi^{(i-r-1)}(u, n) \Big]_{-h}^{h} \qquad (0 < j < i \leqq m)$$

$$\int_{-h}^{h} u^i \varphi^{(i)}(u, n) \, du = \sum_{r=0}^{i-1} (-1)^r \cdot i(i-1)\ldots(i-r+1)\, u^{i-r} \varphi^{(i-r-1)}(u, n) \Big]_{-h}^{h} +$$

$$+ (-1)^i \cdot i! \int_{-h}^{h} \varphi(u, n) \, du \qquad (0 < i \leqq m),$$

so daß wir wegen (9) und (5) die Formeln haben:

(13) $$\lim_{n=\infty} \int_{-h}^{h} u^j \varphi^{(i)}(u, n) \, du = 0 \qquad (0 < j < i \leqq m)$$

(14) $$\lim_{n=\infty} \int_{-h}^{h} u^i \varphi^{(i)}(u, n) \, du = (-1)^i \cdot i! \qquad (0 < i \leqq m).$$

Endlich folgt aus:

$$\int_{\alpha}^{\beta} \varphi^{(i)}(u, n) \, du = \varphi^{(i-1)}(u, n) \Big]_{\alpha}^{\beta}$$

wegen (5) unmittelbar für jedes den Punkt 0 nicht enthaltende Intervall $<\alpha, \beta>$, sowie für jedes Intervall $<-h, h>$:

(15) $$\lim_{n=\infty} \int_{\alpha}^{\beta} \varphi^{(i)}(u, n) \, du = 0 \qquad (i = 1, 2, \ldots, m).$$

Bedingung 1., 2., 4. unseres Satzes zusammen mit Ungleichung (12) besagt nun aber: liegt x in $(a, a + l)$, so genügt der Kern $\varphi(\xi, x, n) = \varphi(\xi - x, n)$ im Intervalle $<a, a + l>$ der Veränderlichen ξ allen Bedingungen von Satz VI. Ferner besagen die Bedingungen 1. und 3. unseres Satzes zusammen mit den Ungleichungen (11) und den Gleichungen (13), (14), (15): der Kern $\varphi(\xi, x, n) = (-i)^i \cdot \varphi^{(i)}(\xi - x, n)$ $(i = 1, 2 \ldots, m)$ genügt allen Bedingungen von Satz VII für $m = i$. Damit aber sind unsere Bedingungen als hinreichend erwiesen.

Die Bedingungen sind notwendig. Angenommen in der Tat, es wäre Bedingung 1. für ein den Nullpunkt nicht enthaltendes Teilintervall $<\alpha, \beta>$ von $(-l, l)$ nicht erfüllt. Da die Länge von $<\alpha, \beta>$ gewiß $< l$ ist, gäbe es in $(a, a + l)$ einen Punkt x, so daß $<\alpha, \beta>$ in $<a - x, a - x + l>$ enthalten wäre. Dann aber würde der Kern $\varphi^{(m)}(\xi - x, n)$ im Intervalle $<a, a + l>$ der Veränderlichen ξ nicht der Bedingung 1. von Satz VII genügen, so daß (7) für $i = m$ nicht für alle Funktionen von $\mathfrak{F}_i$ gelten könnte. Angenommen, Bedingung 2. wäre für ein den Nullpunkt nicht enthaltendes Teilintervall $<\alpha, \beta>$ von $(-l, l)$ nicht erfüllt, so sieht man ebenso, daß für ein gewisses x von $(a, a + l)$ der Kern $\varphi(\xi - x, n)$ in $<a, a + l>$ der Bedingung 2. von Satz VI nicht genügen würde, so daß (6) nicht für alle Funktionen von $\mathfrak{F}_i$ gelten könnte. Wäre Bedingung 3. nicht erfüllt, so könnte $\varphi^{(m)}(\xi - x, n)$ für kein x von $(a, a + l)$ der Bedingung 3. von Satz VII genügen. Bedingung 4. ergibt sich als notwendig durch Betrachtung der Funktion, die in $<x - h, x + h>$ den Wert 1, außerhalb $<x - h, x + h>$ den Wert 0 hat.

[1] In den folgenden Formeln ist φ^0 durch φ und $j \cdot (j-1) \ldots (j-r+1)$ für $r = 0$ durch 1 zu ersetzen.

Satz XVI wird ergänzt durch:

XVII. Genügt die Funktion $\varphi(u, n)$ außer den Voraussetzungen und Bedingungen von Satz XVI noch für alle hinlänglich kleinen $|u|$ der Beziehung:

$$\varphi(u, n) = \varphi(-u, n),$$

so gilt Satz XVI auch noch, wenn man darin die Ableitung $f^{(i)}(x)$ durch die verallgemeinerte i-te Ableitung $\overset{*}{f}{}^{(i)}(x)$ ersetzt.

Der Beweis ist derselbe, wie für Satz XVI, nur hat man sich, statt auf Satz VII, diesmal auf Satz VIII zu berufen.

Auch die Sätze XVI und XVII können sofort auf das Intervall $(-\infty, +\infty)$ übertragen werden. Wir führen zu diesem Zwecke wieder die Funktion $\psi(u, n, h)$ ein, die aus $\varphi(u, n)$ dadurch entsteht, daß man die Funktionswerte im Intervalle $< -h, h >$ durch 0 ersetzt. Die i-te Ableitung von $\psi(u, n, h)$ sei $\psi^{(i)}(u, n, h)$.

XVI*a*. »Sei $\varphi(u, n)$ eine für alle reellen u gegebene Funktion, die eine in jedem endlichen Intervalle $< \alpha, \beta >$ absolut stetige $(m-1)$-te Ableitung besitzt. Ferner sei für jedes $u \neq 0$:

$$\lim_{n=\infty} \varphi(u, n) = 0; \quad \lim_{n=\infty} \varphi^{(i)}(u, n) = 0 \quad (i = 1, 2, \ldots, m-1).$$

Damit für jede in $(-\infty, +\infty)$ zu $\mathfrak{F}_i$ gehörige Funktion, die im beliebigen Punkte x endliche Ableitungen der m ersten Ordnungen besitzt, die Gleichungen gelten:

$$f(x) = \lim_{n=\infty} \int_{-\infty}^{+\infty} f(\xi)\, \varphi(\xi - x, n)\, d\xi, \tag{6a}$$

$$f^{(i)}(x) = \lim_{n=\infty} (-1)^i \int_{-\infty}^{+\infty} f(\xi)\, \varphi^{(i)}(\xi - x, n)\, d\xi \quad (i = 1, 2, \ldots, m), \tag{7a}$$

ist notwendig und hinreichend, daß $\varphi(u, n)$ folgenden Bedingungen genügt:

1. Für jedes $h > 0$ genügen $\psi(u, n, h)$ und $\psi^{(i)}(u, n, h)$ $(i = 1, 2, \ldots, m)$ der Bedingung 1. desjenigen der Sätze I*a* bis IV*a*, der sich auf die Klasse $\mathfrak{F}_i$ bezieht.

2. In jedem endlichen, den Nullpunkt nicht enthaltenden Intervalle $< \alpha, \beta >$ ist:

$$\lim_{n=\infty} \int_{\alpha}^{\beta} \varphi(u, n)\, du = 0,$$

und wenn es sich um $\mathfrak{F}_4$ handelt, ist außerdem für jedes $h > 0$:

$$\lim_{n=\infty} \int_{-\infty}^{-h} \varphi(u, n)\, du = 0; \quad \lim_{n=\infty} \int_{h}^{+\infty} \varphi(u, n)\, du = 0,$$

$$\lim_{n=\infty} \int_{-\infty}^{-h} \varphi^{(i)}(u, n)\, du = 0; \quad \lim_{n=\infty} \int_{h}^{+\infty} \varphi^{(i)}(u, n)\, du = 0 \quad (i = 1, 2, \ldots, m).$$

3. und 4. Es sind die Bedingungen 3. und 4. von Satz XVI erfüllt.«

XVII*a*. »Ist außerdem für alle hinlänglich kleinen $|u|$:

$$\varphi(u, n) = \varphi(-u, n),$$

so kann in XVI*a* die Ableitung $f^{(i)}(x)$ auch durch die verallgemeinerte i-te Ableitung $\overset{*}{f}{}^{(i)}(x)$ ersetzt werden.«

§ 9. Bedingungen für gleichmäßige Konvergenz.

Wir geben nun Bedingungen an, unter denen die Konvergenz in den Formeln (6) und (7) von § 8 ein gleichmäßige ist.

Sei $\varphi(u, n)$ für alle nicht negativen ganzzahligen n als Funktion von u gegeben in $(-l, l)$. Ist $h > 0$ irgendwie gegeben, so bezeichnen wir wieder mit $\psi(u, n, h)$ die Funktion, die in $< -h, h >$ gleich 0, sonst gleich $\varphi(u, n)$ ist.

XVIII. Sei $\varphi(u, n)$ eine in $(-l, l)$ gegebene, in jedem den Nullpunkt nicht enthaltenden Teilintervalle $< \alpha, \beta >$ von $(-l, l)$ für jedes einzelne n geschränkte Funktion, für die die Beziehung:

$$\lim_{n=\infty} \varphi(u, n) = 0 \tag{1}$$

gleichmäßig in jedem den Nullpunkt nicht enthaltenden Teilintervalle $< \alpha, \beta >$ von $(-l, l)$ gilt und die außerdem folgenden Bedingungen genügt:[1]

3. Es gibt ein $h > 0$ und ein N, so daß:

$$\int_{-h}^{h} |\varphi(u, n)| \, du < N \quad \text{für alle } n.$$

4. Es gibt ein $h > 0$, so daß:

$$\lim_{n=\infty} \int_{-h}^{h} \varphi(u, n) \, du = 1.$$

Dann gilt für jede in $< a, a + l >$ zu $\mathfrak{F}_1$ gehörige[2] Funktion f die Formel:

$$f(x) = \lim_{n=\infty} \int_{a}^{a+l} f(\xi) \, \varphi(\xi - x, n) \, d\xi$$

in jedem Punkte x von $(a, a + l)$, in dem f stetig ist; sie gilt gleichmäßig in jedem Teilintervall $< a', b' >$ von $(a, a + l)$, in dessen sämtlichen Punkten f stetig ist.

Zum Beweise genügt es zu zeigen, daß der Kern

$$\varphi(\xi, x, n) = \varphi(\xi - x, n)$$

den Bedingungen von Satz X für die Klasse $\mathfrak{F}_1$ genügt.

Sei, um dies für Bedingung 1. von X nachzuweisen, $h > 0$ beliebig gegeben und x ein Punkt von $< a', b' >$. Die Werte von $\psi(\xi, x, n, h) = \psi(\xi - x, n, h)$ sind 0 in $< x - h, x + h >$ und stimmen in $< a, x - h >$ und $< x + h, a + l >$ überein mit denen von $\varphi(u, n)$ in $< a - x, -h >$ und $< h, a + l - x >$. Diese beiden letzteren Intervalle aber liegen für jedes x von $< a', b' >$ in den Intervallen $< a - b', -h >$ und $< h, a + l - a' >$, die ihrerseits den Nullpunkt nicht enthaltende Teilintervalle von $(-l, l)$ sind.

Wegen der gleichmäßigen Konvergenz von (1) gibt es also ein n_0, so daß:

$$|\varphi(u, n)| < 1 \quad \text{für } n \geqq n_0$$

für alle u dieser beiden Intervalle. Da aber nach Voraussetzung $\varphi(u, n)$ für jedes einzelne n in diesen Intervallen geschränkt ist, gibt es ein M $(\geqq 1)$, so daß:

$$|\varphi(u, n)| < M \quad (n = 1, 2 \ldots, n_0)$$

für alle u dieser beiden Intervalle. Wir sehen also, es ist:

$$|\psi(\xi, x, n, h)| < M \quad \text{für alle } n, \text{ alle } \xi \text{ von } < a, a + l >, \text{ alle } x \text{ von } < a', b' >.$$

Damit ist Bedingung 1. von X[3] nachgewiesen.

---

[1] Wir bezeichnen diese Bedingungen mit 3. und 4. wegen der Analogie mit unseren bisherigen Sätzen.

[2] Und mithin auch für jede zu $\mathfrak{F}_2$, $\mathfrak{F}_3$, $\mathfrak{F}_4$ gehörige Funktion.

[3] Das heißt Bedingung 1 *a* und 1 *b* von Satz IX.

Sei, um Bedingung 2. nachzuweisen, $<\alpha, \beta>$ ein beliebiges Teilintervall von $<a, a+l>$. Es ist:

$$\int_{\alpha}^{\beta} \psi(\xi, x, n, h)\, d\xi = \int_{\alpha-x}^{\beta-x} \psi(u, n, h)\, du.$$

Für alle x von $<a', b'>$ liegt das Integrationsintervall $<\alpha - x, \beta - x>$ in $<\alpha - b', \beta - a'>$, das seinerseits in $(-l, l)$ liegt. Es ist also:

$$\left|\int_{\alpha}^{\beta} \psi(\xi, x, n, h)\, d\xi\right| \leqq \int_{\alpha-b'}^{\beta-a'} |\psi(u, n, h)|\, du. \tag{2}$$

Nun ist aber, wegen der gleichmäßigen Konvergenz von (1):

$$\lim_{n=\infty} \psi(u, n, h) = 0$$

gleichmäßig in ganz $<\alpha - b', \beta - a'>$ und somit:

$$\lim_{n=\infty} \int_{\alpha-b'}^{\beta-a'} |\psi(u, n, h)|\, du = 0.$$

Da dieser Ausdruck von x nicht abhängt, ist zufolge (2) Bedingung 2. von X bewiesen.

Daß Bedingung 3. und 4. von X erfüllt sind, folgt wegen:

$$\int_{x-h}^{x+h} |\varphi(\xi, x, n)|\, d\xi = \int_{-h}^{h} |\varphi(u, n)|\, du; \quad \int_{x-h}^{x+h} \varphi(\xi, x, n)\, d\xi = \int_{-h}^{h} \varphi(u, n)\, du$$

(wo die rechten Seiten von x nicht abhängen) sofort aus den Bedingungen 3. und 4. unseres Satzes.

XIX. Sei $\varphi(u, n)$ eine in $(-l, l)$ gegebene Funktion, die eine in jedem Teilintervalle $<\alpha, \beta>$ von $(-l, l)$ absolut stetige $(m-1)$-te Ableitung besitzt.[1] Ferner möge die Beziehung:

$$\lim_{n=\infty} \varphi^{(m-1)}(u, n) = 0 \tag{3}$$

gleichmäßig in jedem den Nullpunkt nicht enthaltenden Teilintervalle $<\alpha, \beta>$ von $(-l, l)$ gelten. Sind dann noch die Bedingungen 1., 2., 3. und 4. von Satz XVI erfüllt, so gelten die Gleichungen (6) und (7) von Satz XVI gleichmäßig in jedem Teilintervalle $<a', b'>$ von $(a, a+l)$, in dem die zu $\mathfrak{F}_i$ gehörige Funktion f m-mal stetig differenzierbar ist.

Korollar zu Satz XIX. Existiert die Ableitung $\varphi^{(m)}(u, n)$ für jedes $u \neq 0$ von $(-l, l)$, ist sie für jedes einzelne n geschränkt in jedem den Nullpunkt nicht enthaltenden Teilintervalle $<\alpha, \beta>$ von $(-l, l)$ und ist:

$$\lim_{n=\infty} \varphi^{(m)}(u, n) = 0$$

gleichmäßig in jedem der genannten Teilintervalle $<\alpha, \beta>$, so ist in Satz XIX die die Beziehung (3) betreffende Voraussetzung sowie Bedingung 1. von Satz XVI für die Klasse $\mathfrak{F}_1$ von selbst erfüllt.

Wir beweisen Satz XIX. Was die Gleichung (6) von XVI anlangt, braucht nur gezeigt zu werden, daß $\varphi(u, n)$ den Voraussetzungen und Bedingungen von Satz XVIII genügt: daß in jedem den Nullpunkt nicht enthaltenden Teilintervalle $<\alpha, \beta>$ von $(-l, l)$ $\varphi(u, n)$ geschränkt ist für jedes einzelne n, folgt unmittelbar aus unseren Voraussetzungen über $\varphi^{(m-1)}(u, n)$. Was die in Satz XVIII vorausgesetzte Beziehung (1) anlangt, so folgt aus dem gleichmäßigen Bestehen von (3), daß in jedem den Nullpunkt nicht enthaltenden Teilintervalle $<\alpha, \beta>$ von $(-l, l)$ eine Beziehung:

$$\lim_{n=\infty} \{\varphi(u, n) + c_0(n) + c_1(n)\, u + \ldots + c_{m-2}(n)\, u^{m-2}\} = 0$$

[1] Für $m = 1$ ist darunter die Funktion $\varphi(u, n)$ selbst zu verstehen.

gleichmäßig gilt, was mit Bedingung 2. von Satz XVI nur dann verträglich ist, wenn:

$$\lim_{n=\infty} c_0(n) = \lim_{n=\infty} c_1(n) = \ldots = \lim_{n=\infty} c_{m-2}(n) = 0$$

ist; damit ist das gleichmäßige Bestehen von (1) nachgewiesen. — Bedingung 3. von XVIII ist erfüllt, wegen (12) von § 8, und Bedingung 4. von XVIII ist identisch mit 4. von XVI.

Was die Gleichungen (7) von XVI anlangt, braucht nur gezeigt zu werden, daß die Bedingungen von Satz XI erfüllt sind, wenn unter dem Kerne $\varphi(\xi, x, n)$ von Satz XI verstanden wird der Kern $(-1)^i \varphi^{(i)}(\xi - x, n)$ $(i = 1, 2, \ldots, m)$.

Um zu zeigen, daß Bedingung 1. von Satz XI erfüllt ist, bemerken wir: durchläuft in $\psi^{(i)}(\xi - x, n, h)$ die Veränderliche ξ das Intervall $< a, a + l >$, so durchläuft $\xi - x$ das Intervall $< a - x, a + l - x >$, und da $\psi^{(i)}(\xi - x, n, h)$ in $< x - h, x + h >$ den Wert 0 hat, kommt es nur an auf die Intervalle $< a - x, -h >$ und $< h, a + l - x >$, die, solange x in $< a', b' >$ liegt, ganz in den Intervallen $< a - b', -h >$ und $< h, a + l - a' >$ enthalten sind, die wieder ihrerseits Teilintervalle von $(-l, l)$ sind, die den Nullpunkt nicht enthalten. Es folgt also Bedingung 1. von XI für $i = m$ unmittelbar aus Bedingung 1. von XVI. — Für $i = m - 1$ beachte man, daß nach Voraussetzung $\varphi^{(m-1)}(u, n)$ in den Intervallen $< a - b', -h >$ und $< h, a + l - a' >$ absolut stetig und somit geschränkt ist, woraus, zusammen mit der gleichfalls vorausgesetzten gleichmäßigen Konvergenz von (3), wie wir schon beim Beweise von Satz XVIII gesehen haben, das Bestehen der Bedingung 1. von XI für die Klasse $\mathfrak{F}_1$, und damit auch für die Klassen $\mathfrak{F}_2$, $\mathfrak{F}_3$, $\mathfrak{F}_4$ leicht folgt. — Dasselbe gilt für $i = 1, 2, \ldots, m - 2$, da die Funktionen $\varphi^{(i)}(u, n)$ $(i = 1, 2, \ldots, m - 2)$ sicherlich gleichfalls absolut stetig sind und auch für sie die Beziehung:

$$\lim_{n=\infty} \varphi^{(i)}(u, n) = 0 \qquad (i = 1, 2, \ldots, m - 2) \tag{4}$$

gleichmäßig in jedem den Nullpunkt nicht enthaltenden Teilintervalle $< \alpha, \beta >$ von $(-l, l)$ gelten muß, was aus dem gleichmäßigen Bestehen dieser Beziehung für $i = m - 1$ und Bedingung 2. von XVI leicht folgt, wie wir gerade vorhin für $\varphi(u, n)$ gesehen haben.

Bedingung 2. von XI verlangt, daß in jedem Teilintervalle $< \alpha, \beta >$ von $< a, a + l >$ die Beziehung:

$$\lim_{n=\infty} \int_\alpha^\beta \psi^{(i)}(\xi - x, n, h)\, d\xi = 0 \quad (i = 1, 2, \ldots, m) \tag{5}$$

gleichmäßig für alle x von $< a', b' >$ gelte. Nun ist:[1]

$$\int_\alpha^\beta \psi^{(i)}(\xi - x, n, h)\, d\xi = \psi^{(i-1)}(u, n, h)\Big]_{\alpha - x}^{\beta - x} + \varphi^{(i-1)}(-h, n) - \varphi^{(i-1)}(h, n),$$

wenn (α, β) die beiden Punkte $x - h$ und $x + h$ enthält; ist einer dieser Punkte nicht in (α, β) enthalten, so ist diese Formel durch Weglassen eines oder beider Summanden $\varphi^{(i-1)}(-h, n)$, $\varphi^{(i-1)}(h, n)$ zu modifizieren. Ferner ist $\psi^{(i-1)}(u, n, h) = 0$ oder $= \varphi^{(i-1)}(u, n)$, je nachdem $|u| \leqq h$ oder $|u| > h$ ist. Solange nun x in $< a', b' >$ liegt, liegen $\alpha - x$ und $\beta - x$ in $< \alpha - b', \beta - a' >$, welches Intervall seinerseits in $(-l, l)$ liegt. Das gleichmäßige Bestehen von (5) folgt nun unmittelbar aus dem gleichmäßigen Bestehen von (3) und dem daraus folgenden gleichmäßigen Bestehen von (4) in den außerhalb $(-h, h)$ liegenden Teilen von $< \alpha - b', \beta - a' >$.

Endlich sind die Bedingungen 3. und 4. von XI erfüllt, wie die Formeln (11), (13), (14) und (15) von § 8 zeigen.

Es erübrigt noch, die Sätze XVIII und XIX für unendliche Intervalle auszusprechen.

XVIII *a.* »Sei $\varphi(u, n)$ eine für alle reellen u gegebene Funktion, für die die Beziehung:

$$\lim_{n=\infty} \varphi(u, n) = 0 \tag{6}$$

[1] Hierin ist $\varphi^{(0)}$ durch φ und $\psi^{(0)}$ durch ψ zu ersetzen.

gleichmäßig in jedem endlichen den Nullpunkt nicht enthaltenden Intervalle gilt. Damit für jede in $(-\infty, +\infty)$ zur Klasse $\mathfrak{F}_i$ gehörige Funktion in jedem Punkte x, in dem sie stetig ist, die Beziehung gelte:

$$(7) \qquad f(x) = \lim_{n=\infty} \int_{-\infty}^{+\infty} f(\xi)\, \varphi(\xi - x, n)\, d\xi,$$

ist notwendig und hinreichend, daß $\varphi(u, n)$ folgenden Bedingungen genügt:

1. Für jedes $h > 0$ genügt $\psi(u, n, h)$ der Bedingung 1. desjenigen der Sätze Ia bis IVa, der sich auf die Klasse $\mathfrak{F}_i$ bezieht.

2. Im Falle der Klasse $\mathfrak{F}_4$ ist für jedes $h > 0$:

$$\lim_{n=\infty} \int_{-\infty}^{-h} \varphi(u, n)\, du = 0; \qquad \lim_{n=\infty} \int_{h}^{+\infty} \varphi(u, n)\, du = 0.$$

3. und 4. Es sind die Bedingungen 3. und 4. von Satz XVIII erfüllt.

Unter diesen Voraussetzungen und Bedingungen ist die Konvergenz in (7) gleichmäßig in jedem endlichen Intervalle $< a', b' >$, in dessen sämtlichen Punkten f stetig ist.«

Der Beweis wird geführt durch Berufung auf die Sätze VIa und Xa. Der einzige gegenüber dem Beweise von Satz XVIII neu hinzukommende Punkt ist der, im Falle der Klasse $\mathfrak{F}_4$ nun zu führende Beweis, daß die Beziehungen:

$$(8) \qquad \lim_{n=\infty} \int_{-\infty}^{0} \psi(\xi - x, n, h)\, d\xi = 0; \quad \lim_{n=\infty} \int_{0}^{+\infty} \psi(\xi - x, n, h)\, d\xi = 0$$

gleichmäßig für alle x von $< a', b' >$ gelten. Um dies etwa für die erste dieser Beziehungen zu zeigen, bemerken wir, daß:

$$\int_{-\infty}^{0} \psi(\xi - x, n, h)\, d\xi = \int_{-\infty}^{-x} \psi(u, n, h)\, du$$

ist. Wird $A > h$ und $> b'$ gewählt, so haben wir also:

$$\int_{-\infty}^{0} \psi(\xi - x, n, h)\, d\xi = \int_{-\infty}^{-A} \varphi(u, n)\, du + \int_{-A}^{-x} \psi(u, n, h)\, du.$$

Hierin hängt der erste Summand der rechten Seite von x nicht ab, und es ist wegen Bedingung 2.

$$\lim_{n=\infty} \int_{-\infty}^{-A} \varphi(u, n)\, du = 0.$$

Solange nun x in $< a', b' >$ liegt, liegt $< -A, -x >$ in $< -A, -a' >$, und es ist daher:

$$\left| \int_{-A}^{-x} \psi(u, n, h)\, du \right| \leq \int_{-A}^{-a'} |\psi(u, n, h)|\, du.$$

Aus der vorausgesetzten gleichmäßigen Konvergenz von (6) folgt nun aber:

$$\lim_{n=\infty} \int_{-A}^{-a'} |\psi(u, n, h)|\, du = 0,$$

und da dieser Ausdruck von x nicht abhängt, ist damit die gleichmäßige Konvergenz von (8) nachgewiesen.

XIXa. »Sei $\varphi(u, n)$ eine für alle reellen u gegebene Funktion, die eine in jedem endlichen Intervalle absolut stetige $(m-1)$-te Ableitung besitzt, für die die Beziehung:

$$\lim_{n=\infty} \varphi^{(m-1)}(u, n) = 0$$

gleichmäßig in jedem endlichen, den Nullpunkt nicht enthaltenden Intervalle $< \alpha, \beta >$ gilt. Damit für jede in $(-\infty, +\infty)$ zur Klasse $\mathfrak{F}_i$ gehörige Funktion f, die im Punkte x endliche Ableitungen der m ersten Ordnungen besitzt, die Beziehungen gelten:

(9) $$f(x) = \lim_{n=\infty} \int_{-\infty}^{+\infty} f(\xi)\, \varphi\,(\xi - x, n)\, d\xi,$$

(10) $$f^{(i)}(x) = \lim_{n=\infty} (-1)^i \int_{-\infty}^{+\infty} f(\xi)\, \varphi^{(i)}(\xi - x, n)\, d\xi \quad (i = 1, 2, \ldots, m)$$

ist notwendig und hinreichend, daß $\varphi(u, n)$ den Bedingungen 1., 2., 3., 4. von Satz XVI*a* genügt.

Unter diesen Voraussetzungen und Bedingungen ist die Konvergenz in (9) und (10) gleichmäßig in jedem endlichen Intervalle $< a', b' >$, in dem f m-mal stetig differenzierbar ist.«

§ 10. Singuläre Integrale vom Stieltjes'schen Typus.

Wir wenden uns zum Studium eines besonders einfachen Spezialfalles.[1]

XX. Sei $\varphi(u)$ eine in $(-l, l)$ definierte, nicht negative Funktion. Im Punkte $u = 0$ sei sie stetig, und es sei $\varphi(0) = 1$, während in jedem den Nullpunkt nicht enthaltenden Teilintervalle $< \alpha, \beta >$ von $(-l, l)$ die obere Grenze von $\varphi(u)$ kleiner als 1 sei. Es sei $i_1, i_2, \ldots, i_n, \ldots$ eine wachsende Folge positiver Zahlen mit $\lim_{n=\infty} i_n = +\infty$, und es sei γ eine beliebige positive Zahl $< l$. Ist dann $c_1, c_2, \ldots, c_n, \ldots$ eine Folge von Zahlen, für die:

$$\lim_{n=\infty} c_n \cdot \int_{-\gamma}^{\gamma} (\varphi(u))^{i_n}\, du = 1$$

ist, so gilt für jede in $< a, a + l >$ zur Klasse $\mathfrak{F}_1$ gehörige Funktion[2] f, die im Punkte x von $(a, a + l)$ stetig ist, die Beziehung:

$$f(x) = \lim_{n=\infty} c_n \int_a^{a+l} f(\xi)\, (\varphi(\xi - x))^{i_n}\, d\xi.$$

Diese Beziehung gilt gleichmäßig in jedem Teilintervalle $< a', b' >$ von $(a, a + l)$, in dessen sämtlichen Punkten f stetig ist.

Zum Beweise berufen wir uns auf Satz XVIII, indem wir setzen:

(1) $$\varphi(u, n) = c_n \cdot (\varphi(u))^{i_n}.$$

Dann ist zunächst klar, daß $\varphi(u, n)$ für jedes einzelne u in $(-l, l)$ geschränkt ist.

Wir setzen zur Abkürzung:

(2) $$k_n = \int_{-\gamma}^{\gamma} (\varphi(u))^{i_n}\, du,$$

und haben somit:

(3) $$\lim_{n=\infty} c_n \cdot k_n = 1.$$

Sei ein beliebiges, den Nullpunkt nicht enthaltendes Teilintervall $< \alpha, \beta >$ von $(-l, l)$ gegeben. Weil die obere Grenze von $\varphi(u)$ in $< \alpha, \beta >$ kleiner als 1 ist, gibt es ein $\eta > 0$, so daß:

(4) $$0 \leqq \varphi(u) < 1 - \eta \quad \text{in } < \alpha, \beta >.$$

---

[1] Vgl. H. Lebesgue, a. a. O., p. 95.

[2] Und mithin erst recht für jede zu einer der Klassen $\mathfrak{F}_2$, $\mathfrak{F}_3$, $\mathfrak{F}_4$ gehörige Funktion.

Wegen der Stetigkeit von $\varphi(u)$ im Nullpunkte und wegen $\varphi(0) = 1$ gibt es ein $\delta > 0$, so daß:

$$\varphi(u) > 1 - \frac{\eta}{2} \quad \text{in} < -\delta, \delta >.$$

Nehmen wir dieses δ kleiner an als das γ in (2), so haben wir:

$$k_n > 2\delta \cdot \left(1 - \frac{\eta}{2}\right)^{i_n},$$

und infolgedessen nach (3) für alle hinlänglich großen n:

$$c_n < \frac{1}{\delta} \frac{1}{\left(1 - \frac{\eta}{2}\right)^{i_n}}. \tag{5}$$

Nach (1) und (4) ist also in $< \alpha, \beta >$ für alle hinlänglich großen n:

$$|\varphi(u, n)| < \frac{1}{\delta} \left(\frac{1-\eta}{1 - \frac{\eta}{2}} \right)^{i_n}.$$

Da hierin $\frac{1-\eta}{1 - \frac{\eta}{2}} < 1$ ist, so ist damit gezeigt, daß:

$$\lim_{n=\infty} \varphi(u, n) = 0$$

gleichmäßig in $< \alpha, \beta >$ gilt. Die in Satz XVIII zunächst über $\varphi(u, n)$ gemachten Voraussetzungen sind also hier erfüllt.

Daß auch die Bedingungen 3. und 4. von XVIII erfüllt sind, erkennt man auf den ersten Blick, da wegen (3) und wegen $\varphi(u) \geqq 0$:

$$\lim_{n=\infty} \int_{-\gamma}^{\gamma} |\varphi(u, n)|\, du = \lim_{n=\infty} \int_{-\gamma}^{\gamma} \varphi(u, n)\, du = 1$$

ist. Satz XX ist damit bewiesen.

XX*a*. Sei $\varphi(u)$ eine für alle reellen u definierte, nicht negative Funktion. Im Punkte $u = 0$ sei sie stetig, und es sei $\varphi(0) = 1$, während in jedem, den Nullpunkt nicht enthaltenden endlichen Intervalle $< \alpha, \beta >$ die obere Grenze von $\varphi(u)$ kleiner als 1 sei. Es sei γ eine beliebige positive Zahl, und i_n und c_n mögen dieselbe Bedeutung haben wie in Satz XX. Damit für jede in $(-\infty, +\infty)$ zur Klasse $\mathfrak{F}_1$ gehörige Funktion, die im Punkte x stetig ist, die Beziehung gelte:

$$f(x) = \lim_{n=\infty} c_n \int_{-\infty}^{+\infty} f(\xi) (\varphi(\xi - x))^{i_n} d\xi, \tag{6}$$

ist notwendig und hinreichend, daß $\varphi(u)$ sich höchstens in einer Nullmenge unterscheide von einer Funktion $\varphi^*(u)$, für die:

$$\overline{\lim_{u=-\infty}} \varphi^*(u) < 1; \quad \overline{\lim_{u=+\infty}} \varphi^*(u) < 1. \tag{7}$$

Ist auch diese Bedingung erfüllt, so gilt (6) gleichmäßig in jedem endlichen Intervalle $< a', b' >$, in dessen sämtlichen Punkten f stetig ist.

Die Bedingung (7) ist hinreichend: dies wird ebenso bewiesen wie Satz XX, nur hat man sich diesmal auf Satz XVIII *a* zu berufen und zu beachten, daß, abgesehen von Nullmengen, eine Ungleichung (4) nun auch (für $h > 0$) in jedem Intervalle $(-\infty, -h >$ und $< h, +\infty)$ gilt.

Die Bedingung (7) ist notwendig. In der Tat, es ist für jedes $\gamma > 0$:

(8) $$\lim_{n=\infty} \int_{-\gamma}^{\gamma} (\varphi(u))^{in}\, du = 0.$$

Denn ist $\varepsilon > 0$ und $< \gamma$ beliebig gegeben, so gibt es ein $\eta > 0$, so daß:

$$0 \leqq \varphi(u) < 1 - \eta \quad \text{in} < -\gamma, -\frac{\varepsilon}{4} > \text{und in} < \frac{\varepsilon}{4}, \gamma >.$$

Wegen $\varphi(u) \leqq 1$ ist also, indem man die Zerlegung:

$$\int_{-\gamma}^{\gamma} = \int_{-\gamma}^{-\frac{\varepsilon}{4}} + \int_{-\frac{\varepsilon}{4}}^{\frac{\varepsilon}{4}} + \int_{\frac{\varepsilon}{4}}^{\gamma}$$

anwendet:

$$0 \leqq \int_{-\gamma}^{\gamma} (\varphi(u))^{in}\, du \leqq 2 \cdot (1-\eta)^{in} \cdot \gamma + \frac{\varepsilon}{2},$$

und daher für alle hinlänglich großen n:

$$0 \leqq \int_{-\gamma}^{\gamma} (\varphi(u))^{in}\, du < \varepsilon,$$

womit (8) bewiesen ist.

Aus (8) aber folgt wegen (3):

(9) $$\lim_{n=\infty} c_n = +\infty.$$

Wäre nun Bedingung (7) nicht erfüllt, so gäbe es (außerhalb eines gegebenen endlichen Intervalles) eine Menge, die keine Nullmenge ist, und in deren Punkten $(\varphi(u))^{in} > \frac{1}{2}$ und somit $\varphi(u, n) > \frac{c_n}{2}$ wäre. Wegen (9) wäre also Bedingung 1. von Satz XVIII *a* für die Klasse $\mathfrak{F}_1$ nicht erfüllt.

XX *b*. Genügt die Funktion $\varphi(u)$ allen Voraussetzungen von Satz XX *a* und ist:

$$\overline{\lim_{u=-\infty}}\, \varphi(u) < 1; \quad \overline{\lim_{u=+\infty}}\, \varphi(u) < 1,$$

so gilt Formel (6) auch für jede, in $(-\infty, +\infty)$ zu einer der Klassen $\mathfrak{F}_2, \mathfrak{F}_3, \mathfrak{F}_4$ gehörige Funktion, die im Punkte x stetig ist, vorausgesetzt, daß im Falle der Klasse $\mathfrak{F}_2$ das Integral

$$\int_{-\infty}^{+\infty} (\varphi(u))^{2i_1}\, du,$$

im Falle der Klassen $\mathfrak{F}_3$ und $\mathfrak{F}_4$ das Integral

$$\int_{-\infty}^{+\infty} (\varphi(u))^{i_1}\, du$$

existiert. Und zwar gilt dann die Beziehung (6) gleichmäßig in jedem endlichen Intervalle $< a', b' >$, in dessen sämtlichen Punkten $f(x)$ stetig ist.

Zum Beweise bemerken wir, daß Ungleichung (5) offenbar auch so ausgesprochen werden kann: zu jedem $\zeta > 1$ gibt es ein n_0, so daß:

(10) $$c_n < \zeta^{in} \quad \text{für } n \geqq n_0.$$

Für den Fall der Klasse $\mathfrak{F}_2$ wird XX *b* bewiesen sein, wenn wir zeigen: für jedes $h > 0$ ist:

(11) $$\lim_{n=\infty} c_n^2 \int_{-\infty}^{-h} (\varphi(u))^{2in}\, du = 0; \quad \lim_{n=\infty} c_n^2 \int_{h}^{+\infty} (\varphi(u))^{2in}\, du = 0.$$

Beweisen wir dies etwa für die zweite dieser Formeln. Nach Voraussetzung gibt es ein $\vartheta < 1$, so daß:

$$0 \leqq \varphi(u) < \vartheta\,(< 1) \quad \text{in} < h, +\infty).$$

Bezeichnen wir den Inhalt der Menge aller Punkte von $< h, +\infty)$, in denen:

$$\frac{\vartheta}{\nu+1} \leqq \varphi(u) < \frac{\vartheta}{\nu}$$

ist mit e_ν, so haben wir:

$$\int_h^{+\infty} (\varphi(u))^{2i_1}\,du \geqq \sum_{\nu=1}^{\infty} \left(\frac{\vartheta}{\nu+1}\right)^{2i_1} e_\nu,$$

wo die rechts stehende Reihe wegen der vorausgesetzten Konvergenz des links stehenden Integrales konvergent ist. Es folgt hieraus sofort die Konvergenz der Reihen:

$$\sum_{\nu=1}^{\infty} \left(\frac{\vartheta}{\nu}\right)^{2i_n} e_\nu,$$

so daß die Ungleichung aufgeschrieben werden kann:

$$\int_h^{+\infty} (\varphi(u))^{2i_n}\,du \leqq \sum_{\nu=1}^{\infty} \left(\frac{\vartheta}{\nu}\right)^{2i_n} e_\nu\,.$$

Setzen wir noch:

$$s = \sum_{\nu=1}^{\infty} \frac{e_\nu}{\nu^{2i_1}},$$

so haben wir:

$$\int_h^{+\infty} (\varphi(u))^{2i_n}\,du \leqq s \cdot \vartheta^{2i_n},$$

und somit wegen (10):

(12) $$c_n^2 \int_h^{+\infty} (\varphi(u))^{2i_n}\,du \leqq s \cdot (\zeta \cdot \vartheta)^{2i_n} \quad \text{für } n \geqq n_0\,.$$

Hierin war ζ irgend eine Zahl > 1. Wählen wir sie gemäß:

$$1 < \zeta < \frac{1}{\vartheta},$$

so folgt nun aus (12) unmittelbar die zweite Gleichung (11). Ebenso beweist man die erste.

Im Falle der Klasse $\mathfrak{F}_3$ genügt es, folgende zwei Tatsachen zu beweisen:

a) Ist ein endliches, den Nullpunkt nicht enthaltendes Intervall $< \alpha, \beta >$ gegeben, so gehört zu jedem $\varepsilon > 0$ ein $\lambda > 0$, so daß für jede Menge I sich nicht überdeckender Intervalle von $< \alpha, \beta >$ deren Gesamtlänge $< \lambda$ ist:

(13) $$c_n \int_I (\varphi(u))^{i_n}\,du < \varepsilon \quad \text{für alle } n.$$

b) Zu jedem $\varepsilon > 0$ gehört ein A, so daß:

(13a) $$c_n \int_{-\infty}^{-A} (\varphi(u))^{i_n}\,du < \varepsilon; \quad c_n \int_A^{+\infty} (\varphi(u))^{i_n}\,du < \varepsilon \quad \text{für alle } n.$$

Um die Tatsache *a)* zu beweisen, zeigt man zuerst, wie beim Beweise von Satz XX, daß die Beziehung:

$$\lim_{n=\infty} c_n (\varphi(u))^{i_n} = 0$$

gleichmäßig in $<\alpha, \beta>$ gilt, woraus die Gleichung folgt:

$$\lim_{n=\infty} c_n \int_\alpha^\beta (\varphi(u))^{i_n}\, du = 0. \tag{14}$$

Ist sodann $\varepsilon > 0$ beliebig gegeben, so kann man n_0 so groß wählen, daß:

$$c_n \int_\alpha^\beta (\varphi(u))^{i_n}\, du < \varepsilon \quad \text{für } n > n_0. \tag{15}$$

Nun kann man, zufolge einer bekannten Eigenschaft der Lebesgue'schen Integrale für jedes einzelne n zu dem gegebenen ε ein λ_n so bestimmen, daß für dieses n und für $\lambda = \lambda_n$ Ungleichung (13) gilt. Wählt man nun für λ die kleinste der Zahlen $\lambda_1, \lambda_2, \ldots, \lambda_{n_0}$, so gilt nun, bei Berücksichtigung von (15), Ungleichung (13) für alle n und die Tatsache *a)* ist bewiesen.

Um die Tatsache *b)* zu beweisen, zeigt man zuerst, daß:

$$\lim_{n=\infty} c_n \int_1^{+\infty} (\varphi(u))^{i_n}\, du = 0, \quad \lim_{n=\infty} c_n \int_{-\infty}^{-1} (\varphi(u))^{i_n}\, du = 0, \tag{16}$$

was ebenso bewiesen wird wie (11). Sodann wählt man n_0 so groß, daß:

$$c_n \int_1^{+\infty} (\varphi(u))^{i_n}\, du < \varepsilon \quad \text{für } n > n_0. \tag{17}$$

Für jedes einzelne n aber gibt es, da jedes Integral:

$$\int_1^{+\infty} (\varphi(u))^{i_n}\, du$$

konvergent ist, ein A_n, so daß:

$$c_n \int_{A_n}^{+\infty} (\varphi(u))^{i_n}\, du < \varepsilon. \tag{18}$$

Wählt man also A größer als die Zahlen $1, A_1, A_2, \ldots, A_{n_0}$, so gilt, wegen (17) und (18), die zweite Ungleichung (13 *a*) für alle n. Analog zeigt man, daß die erste Ungleichung (13 *a*) gilt; und Tatsache *b)* ist bewiesen.

Daß Satz XX *b* auch für die Klasse $\mathfrak{F}_4$ richtig ist, folgt aus (14) und (16).

§ 11. Die verallgemeinerte Stieltjes'sche Formel.

Es genüge wieder $\varphi(u)$ den Voraussetzungen von Satz XX (beziehungsweise XX *a* oder XX *b*). Wir wählen für die in diesen Sätzen auftretende Zahlenfolge $i_1, i_2, \ldots, i_n, \ldots$ nun der Einfachheit halber die Folge $1, 2, \ldots, n, \ldots$ Ersetzt man in (2) von § 10 die Zahl γ durch eine andere positive Zahl γ' von $(-l, l)$ (beziehungsweise eine beliebige andere positive Zahl γ') und setzt:

$$k_n = \int_{-\gamma}^{\gamma} (\varphi(u))^n\, du; \quad k'_n = \int_{-\gamma'}^{\gamma'} (\varphi(u))^n\, du, \tag{1}$$

so gelingt es leicht, die Differenz $k_n - k'_n$ abzuschätzen. Sei etwa $\gamma' < \gamma$. Dann hat man:

$$k_n - k'_n = \int_{-\gamma}^{-\gamma'} (\varphi(u))^n\, du + \int_{\gamma'}^{\gamma} (\varphi(u))^n\, du.$$

Da in $<-\gamma, -\gamma'>$ sowie in $<\gamma', \gamma>$ die obere Grenze von $\varphi(u)$ kleiner als 1 ist, gibt es ein $\eta > 0$, so daß in diesen Intervallen:

$$0 \leqq \varphi(u) < 1 - \eta,$$

und somit:

$$k_n - k'_n < 2(\gamma - \gamma')(1-\eta)^n.$$

Wir haben also allgemein: es gibt ein positives $\vartheta < 1$, so daß für alle hinlänglich großen n:

(2) $$|k_n - k'_n| < \vartheta^n \quad (0 < \vartheta < 1).$$

Diese einfache Bemerkung ermöglicht es, unter weiteren spezialisierenden Annahmen über das Verhalten von $\varphi(u)$ in der Umgebung des Nullpunktes, eine asymptotische Auswertung von k_n vorzunehmen und dadurch einfache Ausdrücke für die Konstanten c_n zu ermitteln.

Es habe $\varphi(u)$ außer den schon bisher geforderten Eigenschaften die Gestalt:

(3) $$\varphi(u) = 1 - \alpha\,|u|^p + \omega(u)\cdot|u|^p,$$

worin:

(4) $$\alpha > 0; \quad p > 0; \quad \lim_{u=0} \omega(u) = 0$$

sei. Wir können dann, wenn ein $h > 0$ beliebig gegeben ist, γ' so klein wählen, daß in $< -\gamma', \gamma' >$ die Ungleichung gilt:

(5) $$1 - (\alpha + h)\,|u|^p \leqq \varphi(u) \leqq 1 - (\alpha - h)\cdot|u|^p,$$

und somit auch:

(6) $$2\int_0^{\gamma'} \{1 - (\alpha + h)|u|^p\}^n du \leqq \int_{-\gamma'}^{\gamma'} (\varphi(u))^n du \leqq 2\int_0^{\gamma'} \{1 - (\alpha - h)|u|^p\}^n du.$$

Dies führt uns auf die Aufgabe, eine asymptotische Auswertung des Integrales:

(7) $$\varkappa_n(\beta, \sigma, p, q) = \int_0^{\sigma} u^q (1 - \beta u^p)^n du \qquad (\beta > 0, \sigma > 0, p > 0, q > 0)$$

vorzunehmen. Nehmen wir die Substitution:

$$v = \beta\cdot u^p \qquad \tau = \beta\cdot\sigma^p$$

vor, so erhalten wir:

$$\varkappa_n(\beta, \sigma, p, q) = \frac{1}{p}\cdot\beta^{-\frac{q+1}{p}} \int_0^{\tau} v^{\frac{q+1}{p}-1} (1-v)^n dv.$$

Wir wollen noch annehmen, es sei:

(7a) $$\tau = \beta\cdot\sigma^p < 1.$$

Dann haben wir offenbar die Ungleichung:

(8) $$\left|\varkappa_n(\beta, \sigma, p, q) - \frac{1}{p}\beta^{-\frac{q+1}{p}} \int_0^1 v^{\frac{q+1}{p}-1}(1-v)^n dv\right| < \frac{1}{p}\beta^{-\frac{q+1}{p}}(1-\tau)^{n+1}\cdot\tau^{-1}.$$

Nun ist bekanntlich:

$$\int_0^1 v^{\frac{q+1}{p}-1}(1-v)^n dv = \frac{\Gamma\left(\frac{q+1}{p}\right)\Gamma(n+1)}{\Gamma\left(\frac{q+1}{p} + n + 1\right)},$$

und somit nach der Stirling'schen Formel:

$$\int_0^1 v^{\frac{q+1}{p}-1}(1-v)^n dv = \Gamma\left(\frac{q+1}{p}\right) n^{-\frac{q+1}{p}} A(n) \quad (\lim_{n=\infty} A(n) = 1).$$

Man erhält daher aus (8) wegen:

$$\lim_{n=\infty} n^{\frac{q+1}{p}} (1-\tau)^{n+1} = 0$$

die Beziehung:

(9) $$\lim_{n=\infty} n^{\frac{q+1}{p}} \varkappa_n(\beta, \sigma, p, q) = \frac{1}{p} \beta^{-\frac{q+1}{p}} \Gamma\left(\frac{q+1}{p}\right).$$

Sie gilt, wegen (7 *a*), für $\beta \cdot \sigma^p < 1$.

Sei nun ein $h > 0$ beliebig gegeben; es kann dann γ' so klein gewählt werden, daß in $< -\gamma', \gamma' >$ die Ungleichungen (5) gelten. Wir haben dann nach (6):

$$2\varkappa_n(\alpha+h, \gamma', p, 0) \leqq \int_{-\gamma'}^{\gamma'} (\varphi(u))^n \, du \leqq 2\varkappa_n(\alpha-h, \gamma', p, 0).$$

Wählen wir γ' auch so klein, daß:

$$(\alpha+h)\cdot\gamma'^p < 1,$$

so können wir (9) anwenden und haben, wenn wir wieder von der Bezeichnungsweise (1) Gebrauch machen:

Ist $\eta > 0$ beliebig gegeben, so ist für alle hinlänglich großen n:

$$\frac{2}{p}(\alpha+h)^{-\frac{1}{p}} \Gamma\left(\frac{1}{p}\right) - \eta \leqq n^{\frac{1}{p}} k'_n \leqq \frac{2}{p}(\alpha-h)^{-\frac{1}{p}} \Gamma\left(\frac{1}{p}\right) + \eta,$$

und somit, bei Benützung von (2), wieder für alle hinlänglich großen n:

$$\frac{2}{p}(\alpha+h)^{-\frac{1}{p}} \Gamma\left(\frac{1}{p}\right) - n^{\frac{1}{p}} \vartheta^n - \eta \leqq n^{\frac{1}{p}} k_n \leqq \frac{2}{p}(\alpha-h)^{-\frac{1}{p}} \Gamma\left(\frac{1}{p}\right) + n^{\frac{1}{p}} \vartheta^n + \eta.$$

Da hierin h und η beliebig waren und $0 < \vartheta < 1$ ist, ist das gleichbedeutend mit:

$$\lim_{n=\infty} n^{\frac{1}{p}} k_n = \frac{2}{p} \alpha^{-\frac{1}{p}} \Gamma\left(\frac{1}{p}\right).$$

Zufolge von Gleichung (3) in § 10 können wir also setzen:

$$c_n = \frac{p}{2} \alpha^{\frac{1}{p}} \frac{1}{\Gamma\left(\frac{1}{p}\right)} n^{\frac{1}{p}}.$$

Wir haben damit den Satz:

XXI. Sei $\varphi(u)$ eine in $(-l, l)$ gegebene, nicht negative Funktion der Form (3), (4), deren obere Grenze in jedem den Nullpunkt nicht enthaltenden Teilintervalle $<\alpha, \beta>$ von $(-l, l)$ kleiner als 1 ist. Dann gilt für jede in $<a, a+l>$ zur Klasse $\mathfrak{F}_1$ gehörige Funktion, die im Punkte x von $(a, a+l)$ stetig ist, die Formel:

(10) $$f(x) = \frac{p \cdot \alpha^{\frac{1}{p}}}{2 \cdot \Gamma\left(\frac{1}{p}\right)} \lim_{n=\infty} n^{\frac{1}{p}} \int_a^{a+l} f(\xi)(\varphi(\xi - x))^n \, d\xi.$$

Diese Beziehung gilt gleichmäßig in jedem Teilintervalle $<a', b'>$ von $(a, a+l)$, in dessen sämtlichen Punkten f stetig ist.

Formel (10) wurde für den Fall $p = 2$ aufgestellt von Stieltjes.[1] Man erhält wichtige Spezialfälle der Stieltjes'schen Formel, indem man für $(-l, l)$ das Intervall $(-1, 1)$ nimmt und setzt:[2]

$$\varphi(u) = 1 - u^2,$$

[1] Correspondance de Hermite et de Stieltjes, Bd. II, p. 185. Näheres hierüber: H. Lebesgue, a. a. O., p. 119.

[2] E. Landau, Rend. Pal., Bd. 25, p. 337; Ch. J. de la Vallée-Poussin, Acad. Bruxelles, Bull. Classe des Sciences 1908, p. 193.

oder indem man für $(-l, l)$ das Intervall $(-2\pi, 2\pi)$ nimmt und setzt:[1]

$$\varphi(u) = \left(\cos\frac{u}{2}\right)^2.$$

XXI *a.* Sei $\varphi(u)$ eine für alle reellen u gegebene nicht negative Funktion der Form (3), (4), deren obere Grenze in jedem Intervalle $(-\infty, -h>$ und $<h, +\infty)$ $(h > 0)$ kleiner als 1 ist. Dann gilt für jede in $(-\infty, +\infty)$ zur Klasse $\mathfrak{F}_1$ gehörige Funktion, die im Punkte x stetig ist, die Formel:

$$f(x) = \frac{p \cdot \alpha^{\frac{1}{p}}}{2\Gamma\left(\frac{1}{p}\right)} \lim_{n=\infty} n^{\frac{1}{p}} \int_{-\infty}^{+\infty} f(\xi)\,(\varphi(\xi - x))^n\, d\xi. \tag{11}$$

Diese Beziehung gilt gleichmäßig in jedem endlichen Intervalle $<a', b'>$, in dessen sämtlichen Punkten f stetig ist. — Existiert das Integral

$$\int_{-\infty}^{+\infty} (\varphi(u))^2\, du, \text{ beziehungsweise } \int_{-\infty}^{+\infty} \varphi(u)\, du,$$

so gilt dies auch für die in $(-\infty, +\infty)$ zu $\mathfrak{F}_2$, beziehungsweise zu $\mathfrak{F}_3$, gehörigen Funktionen f.

Einen bekannten Spezialfall[2] erhält man, indem man wählt:

$$\varphi(u) = e^{-u^2}.$$

§ 12. Konvergenz an Unstetigkeitsstellen und Differenziation.

Wir wollen nun auf den in den §§ 10 und 11 behandelten Typus singulärer Integrale Satz XII und XIII anwenden.

XXII. Es genüge $\varphi(u)$ außer den Voraussetzungen von Satz XXI noch folgenden Bedingungen: $\varphi(u)$ ist absolut stetig[3] in einer Umgebung des Nullpunktes und es genügt $\varphi'(u)$ in dieser Umgebung (abgesehen von einer Nullmenge) einer Ungleichung:

$$|\varphi'(u)| < A \cdot |u|^{p-1} \quad (A \text{ eine Konstante}). \tag{1}$$

Dann gilt Gleichung (10) von § 11 in jedem Punkte x von $(a, a+l)$, in dem die zur Klasse $\mathfrak{F}_1$ gehörige Funktion f Ableitung ihres unbestimmten Integrales ist. — Ist außerdem in einer Umgebung des Nullpunktes $\varphi(-u) = \varphi(u)$, so gilt (10) von § 11 in jedem Punkte x von $(a, a+l)$, in dem f verallgemeinerte erste Ableitung seines unbestimmten Integrales ist.

Wir haben uns zu überzeugen, daß der Kern:

$$\varphi(\xi, x, n) = \frac{p \cdot \alpha^{\frac{1}{p}}}{2\Gamma\left(\frac{1}{p}\right)} n^{\frac{1}{p}} (\varphi(\xi - x))^n \tag{2}$$

allen Voraussetzungen und Bedingungen von Satz XII genügt. Die absolute Stetigkeit wurde ausdrücklich vorausgesetzt. Daß Beziehung (1) von § 6 erfüllt ist, haben wir schon in § 10 gesehen. Ebenso wissen wir, daß die Bedingungen 1., 2. und 4. von Satz XII für die Klasse $\mathfrak{F}_1$ erfüllt sind. Es bleibt also nur Bedingung 3. von XII nachzuweisen; das heißt, daß es ein $h > 0$ und ein N gibt, so daß:

$$\int_{x-h}^{x+h} \left|(\xi - x)\frac{\partial}{\partial \xi}\varphi(\xi, x, n)\right| d\xi < N \quad \text{für alle } n. \tag{3}$$

[1] Ch. J. de la Vallée Poussin, a. a. O., p. 227.

[2] Weierstraß, Werke, Bd. III, p. 1.

[3] Es existiert also $\varphi'(u)$, abgesehen von einer Nullmenge.

Dazu genügt es nachzuweisen, daß es ein $h > 0$ und ein N gibt, so daß:

(4) $$n^{\frac{1}{p}+1}\int_{-h}^{h}|u\cdot\varphi'(u)\,(\varphi(u))^{n-1}|\,du < N \quad \text{für alle } n.$$

Nun haben wir aber, wenn γ hinlänglich klein gewählt wird, in $<-\gamma, \gamma>$, wegen (3) und (4) von § 11, für ein $\beta < \alpha$:

(5) $$0 < \varphi(u) < 1 - \beta|u|^p,$$

und somit bei Berücksichtigung von (1):

$$\int_{-\gamma}^{\gamma}|u\,\varphi'(u)\,(\varphi(u))^{n-1}|\,du < 2A\int_{0}^{\gamma}u^p\,(1-\beta\,u^p)^{n-1}\,du.$$

Dieses letztere Integral aber ist nach der Bezeichnungsweise (7) von § 11 nichts anderes als $\varkappa_{n-1}(\beta, \gamma, p, p)$. Wählen wir noch γ so klein, daß $\beta\cdot\gamma^p < 1$, so hat, nach (9) von § 11, $n^{\frac{p+1}{p}}\cdot\varkappa_{n-1}(\beta, \gamma, p, p)$ für $n = \infty$ einen endlichen Grenzwert. Damit ist also (4) und gleichzeitig (3) für $h = \gamma$ nachgewiesen und unser Satz ist bewiesen.

Der Spezialfall dieses Satzes $\varphi(u) = 1 - u^2$ wurde von Fr. Riesz,[1] der Fall $\varphi(u) = \left(\cos\frac{u}{2}\right)^2$ von H. Lebesgue[2] bewiesen. Weitergehende Resultate erhält man durch Anwendung von Satz XIV und XV:

XXIII. Es genüge $\varphi(u)$ außer den Voraussetzungen von Satz XXI (beziehungsweise XXI*a*) noch folgenden Bedingungen: $\varphi(u)$ besitzt in einer Umgebung des Nullpunktes eine absolut stetige $(m-1)$-te Ableitung, und es sind in dieser Umgebung, abgesehen von einer Nullmenge, die Ungleichungen erfüllt:

(6) $$|\varphi^{(i)}(u)| < A\,|u|^{p-i} \quad (i = 1, 2, \ldots, m).$$

Dann gilt Gleichung (10) (beziehungsweise (11)) von § 11 in jedem Punkte x von $(a, a+l)$ (beziehungsweise in jedem Punkte x), in dem f m-te Ableitung seines m-fach iterierten unbestimmten Integrales ist. — Ist außerdem in einer Umgebung des Nullpunktes $\varphi(-u) = \varphi(u)$, so gilt (10) (beziehungsweise (11)) von § 11 in jedem Punkte x von $(a, a+l)$ (beziehungsweise in jedem Punkte x) in dem f verallgemeinerte m-te Ableitung seines m-fach iterierten unbestimmten Integrales ist.

Wir haben nachzuweisen, daß der Kern (2) allen Bedingungen von Satz XIV genügt. Die absolute Stetigkeit der $(m-1)$-ten Ableitung wurde ausdrücklich vorausgesetzt. Um einzusehen, daß die Beziehungen (1) von § 7 gelten, hat man nur zu beachten, daß die i-te Ableitung von $(\varphi(u))^n$ $(i = 1, 2, \ldots, m-1)$ eine Summe aus einer endlichen (von n unabhängigen) Anzahl von Summanden der Form:

$$P(n)\,(\varphi(u))^{n-j}\,(\varphi'(u))^{j_1}\cdot(\varphi''(u))^{j_2}\ldots(\varphi^{(i)}(u))^{j_i}$$

ist, wo $P(n)$ ein Polynom in n ist und die Exponenten $j, j_1, \ldots, j_i$ von n unabhängig sind. Da $|\varphi(u)| < 1$ ist für $u \neq 0$, haben für $u \neq 0$ alle diese Ausdrücke auch nach Multiplikation mit $n^{\frac{1}{p}}$ für $n = \infty$ den Grenzwert 0.

Daß die Bedingungen 1., 2. und 4. von Satz XIV erfüllt sind, ist uns schon bekannt. Es bleibt nur noch nachzuweisen, daß auch Bedingung 3. erfüllt ist. Dazu genügt es wieder nachzuweisen, daß für ein hinlänglich kleines γ eine Ungleichung besteht:

$$n^{\frac{1}{p}}\int_{0}^{\gamma}u^m\left|\frac{d^m}{du^m}(\varphi(u))^n\right|du < N \quad \text{für alle } n.$$

[1] Jahresber. Math. Ver., Bd. 17, p. 196.

[2] A. a. O., p. 100.

Nun ist, abgesehen von einer Nullmenge, $\frac{d^m}{du^m}(\varphi(u))^n$ eine Summe aus einer endlichen (von n unabhängigen) Anzahl von Summanden der Form:

$$P(n)\,(\varphi(u))^{n-j}\,(\varphi'(u))^{j_1}\cdot(\varphi''(u))^{j_2}\ldots(\varphi^{(m)}(u))^{j_m}.$$

Berücksichtigen wir Ungleichung (5) und (6), so wird es also genügen nachzuweisen, daß für jeden einzelnen dieser Summanden eine Ungleichung gilt:

$$(7)\qquad n^{\frac{1}{p}}\,P(n)\int_0^{\gamma} u^{m+j_1(p-1)+j_2(p-2)+j_m(p-m)}\,(1-\beta u^p)^{n-j}\,du < N \quad (\text{für alle } n).$$

Dieses Integral ist nichts anderes als:

$$\varkappa_{n-j}(\beta,\gamma,p,m+j_1(p-1)+j_2(p-2)+\ldots+j_m(p-m)).$$

Bezeichnen wir noch mit k den Grad des Polynoms $P(n)$, so verhält sich also nach Formel (9) von § 11 der Ausdruck (7) für unendlich wachsendes n wie:

$$n^{k-\frac{1}{p}(m+j_1(p-1)+j_2(p-2)+\ldots+j_m(p-m))},$$

und unsere Behauptung wird bewiesen sein, wenn wir die Gleichheiten beweisen:

$$(8)\qquad \begin{aligned} j_1+2j_2+\ldots+mj_m &= m,\\ j_1+j_2+\ldots+j_m &= k. \end{aligned}$$

Im Falle $m=1$ sind diese Gleichheiten richtig. Denn die erste Ableitung von $(\varphi(u))^n$ besteht nur aus dem einen Gliede:

$$n\cdot(\varphi(u))^{n-1}\cdot\varphi'(u),$$

und es ist also $k=1, j_1=1$. Nehmen wir also an, die Gleichheiten (8) seien richtig für $m=i$, und zeigen wir, daß sie dann auch richtig sind für $m=i+1$.

Differenziert man das Glied:

$$(9)\qquad P(n)\,(\varphi(u))^{n-j}\,(\varphi'(u))^{j_1}\,(\varphi''(u))^{j_2}\ldots(\varphi^{(i)}(u))^{j_i}$$

von $\frac{d^i}{du^i}(\varphi(u))^n$ nach der Regel für die Differenziation eines Produktes, so erhält man $i+1$ Glieder von $\frac{d^{i+1}}{du^{i+1}}(\varphi(u))^n$. Die in diesen Gliedern auftretenden Zahlen $k, j_1, j_2, \ldots, j_{i+1}$ gehen aus den entsprechenden Zahlen von (9) in folgender Weise hervor: Differenziert man in (9) den Faktor $(\varphi(u))^{n-j}$ und läßt die übrigen ungeändert, so vermehrt sich in den als richtig vorausgesetzten Gleichheiten:

$$(10)\qquad \begin{aligned} j_1+2j_2+\ldots+ij_i &= i,\\ j_1+j_2+\ldots+j_i &= k, \end{aligned}$$

j_1 um $1, j_2, \ldots, j_i$ bleiben ungeändert, die Gradzahl k vermehrt sich um 1. Gleichzeitig geht i in $i+1$ über. Die Gleichheiten bleiben dabei bestehen. — Differenziert man in (9) den Faktor $(\varphi^{(h)}(u))^{j_h}$ $(h<i,\ j_h>0)$ und läßt die andern ungeändert, so vermindert sich j_h um 1 und es vermehrt sich j_{h+1} um 1, die übrigen j und k bleiben ungeändert, i vermehrt sich um 1; die Gleichheiten bleiben dabei bestehen. Differenziert man den Faktor $(\varphi^{(i)}(u))^{j_i}$ und läßt die anderen ungeändert, so vermindert sich j_i um 1, die anderen j und k bleiben ungeändert, aus i wird $i+1$ und es tritt in (10) auf der linken Seite in der ersten Gleichheit der Summand $i+1$, in der zweiten der Summand 1 hinzu. Die Gleichheiten bleiben dabei bestehen. Damit ist (8) nachgewiesen und somit auch Satz XXIII bewiesen.

Nehmen wir nun an, es sei $\varphi^{(m-1)}(u)$ nicht nur in einer Umgebung des Nullpunktes, sondern in jedem Teilintervall $<\alpha,\beta>$ von $(-l, l)$ absolut stetig, so sind nun offenbar alle Bedingungen von Satz XIX erfüllt, so daß wir den Satz aussprechen können:

XXIV. Es genüge $\varphi(u)$ außer den Voraussetzungen von Satz XXI noch folgenden Bedingungen: $\varphi(u)$ besitzt eine in jedem Teilintervalle $<\alpha, \beta>$ von $(-l, l)$ absolut stetige $(m-1)$-te Ableitung und es sind in jedem solchen Teilintervalle (abgesehen von einer Nullmenge) Ungleichungen der Gestalt:

$$|\varphi^{(i)}(u)| < A|u|^{p-i} \quad (i = 1, 2 \ldots, m) \tag{11}$$

erfüllt. Dann gilt in jedem Punkte x von $(a, a+l)$, in dem die in $<a, a+l>$ zur Klasse $\mathfrak{F}_1$ gehörige Funktion f eine endliche Ableitung m-ter Ordnung besitzt, die Formel:

$$f^{(i)}(x) = \frac{p\,\alpha^{\frac{1}{p}}}{2\Gamma\left(\frac{1}{p}\right)} \lim_{n=\infty} n^{\frac{1}{p}} \int_a^{a+l} f(\xi) \frac{\partial^i}{\partial x^i} (\varphi(\xi - x))^n \, d\xi \quad (i = 1, 2, \ldots, m), \tag{12}$$

und zwar gilt diese Beziehung gleichmäßig in jedem Teilintervalle $<a', b'>$ von $(a, a+l)$, in dem $f(x)$ i-mal stetig differenzierbar ist. – Genügt $\varphi(u)$ in einer Umgebung des Nullpunktes auch der Relation $\varphi(-u) = \varphi(u)$, so gilt (12) auch für die verallgemeinerte i-te Ableitung $\mathring{f}^{(i)}(x)$.

Ist speziell p eine gerade natürliche Zahl, so kann es vorkommen, daß alle Ableitungen von $\varphi(u)$ existieren und absolut stetig sind. Gilt dann in jedem Teilintervalle $<\alpha, \beta>$ von $(-l, l)$ für jedes i eine Ungleichung der Form:

$$|\varphi^{(i)}(u)| < A_i \, |u|^{p-i},$$

so gilt (12) für jedes i. Die Spezialfälle unseres Satzes: $\varphi(u) = 1 - u^2$ und $\varphi(u) = \left(\cos \frac{u}{2}\right)$ wurden von Ch. J. de la Vallée-Poussin bewiesen.[1]

Für unendliche Intervalle haben wir (unter Berufung auf Satz XIX *a*):

XXIV *a*. »Es genüge $\varphi(u)$ den Voraussetzungen von Satz XXI *a* für die Klasse $\mathfrak{F}_i$. Ferner besitze $\varphi(u)$ eine in jedem endlichen Intervalle absolut stetige $(m-1)$-te Ableitung; im Falle der Klasse $\mathfrak{F}_1$ sei ferner $\varphi^{(i)}(u)$ $(i = 1, 2, \ldots, m)$ geschränkt für alle u; im Falle der Klasse $\mathfrak{F}_2$ sei für alle hinlänglich großen $|u|$ (abgesehen von Nullmengen):

$$|\varphi^{(i)}(u)| < u^{-\left(\frac{1}{2}+\alpha\right)} \quad (\alpha > 0); \tag{13}$$

im Falle der Klasse $\mathfrak{F}_3$ sei für alle hinlänglich großen $|u|$ (abgesehen von Nullmengen):

$$|\varphi^{(i)}(u)| < u^{-(1+\alpha)}.$$

Endlich mögen in jedem endlichen Intervalle Ungleichungen der Form (11) erfüllt sein. Dann gilt für jede in $(-\infty, +\infty)$ zu $\mathfrak{F}_i$ gehörige Funktion, die im Punkte x eine endliche Ableitung i-ter Ordnung besitzt, die Formel:

$$f^{(i)}(x) = \frac{p\,\alpha^{\frac{1}{p}}}{2\Gamma\left(\frac{1}{p}\right)} \lim_{n=\infty} n^{\frac{1}{p}} \int_{-\infty}^{+\infty} f(\xi) \frac{\partial^i}{\partial x^i} (\varphi(\xi - x))^n d\xi \quad (i = 1, 2 \ldots, m). \tag{14}$$

Diese Beziehung gilt gleichmäßig in jedem endlichen Intervalle $<a', b'>$, in dem f i-mal stetig differenzierbar ist. – Genügt $\varphi(u)$ in einer Umgebung des Nullpunktes der Relation $\varphi(u) = \varphi(-u)$, so gilt (14) auch für die verallgemeinerte i-te Ableitung $\mathring{f}^{(i)}(x)$.«

[1] A. a. O., p. 204 ff. u. p. 238 ff. Die Resultate von de la Vallée-Poussin sagen auch in diesen Spezialfällen insofern etwas weniger aus als die des Textes, als sie die gleichmäßige Konvergenz von (12) nur behaupten für jedes Intervall $<a'', b''>$, das ganz in einem Intervalle (a', b') liegt, in dem f i-mal stetig differenzierbar ist.

Um dies zu beweisen, hat man nur noch zu zeigen, daß Bedingung 1. von Satz XIX a [1] erfüllt ist, was man unmittelbar erkennt, indem man beachtet, daß $\frac{d^i}{du^i}(\varphi(u))^n$ sich aus einer endlichen Anzahl Glieder der Form (9) zusammensetzt. Sei der Beweis etwa für den Fall der Klasse $\mathfrak{F}_2$ angedeutet: es ist, wenn ϑ eine Zahl zwischen 0 und 1 bedeutet:

$$\int_h^{+\infty} \{P(n)(\varphi(u))^{n-j}(\varphi'(u))^{j_1}(\varphi''(u))^{j_2}\ldots(\varphi^{(i)}(u))^{j_i}\}^2\,du \leqq$$

$$\leqq (P(n))^2\,\vartheta^{2(n-j)}\int_h^{+\infty}(\varphi'(u))^{2j_1}(\varphi''(u))^{2j_2}\ldots(\varphi^{(i)}(u))^{2j_i}\,du.$$

Das rechts auftretende Integral hat wegen (13) einen endlichen Wert und es ist wegen $0 < \vartheta < 1$:

$$\lim_{n=\infty} n^{\frac{2}{p}}(P(n))^2\,\vartheta^{2(n-j)} = 0.$$

So erkennt man, daß die Integrale:

$$n^{\frac{2}{p}}\int_h^{+\infty}\left\{\frac{d^i}{du^i}(\varphi(u))^n\right\}^2 du$$

geschränkt sind für alle n, wodurch Bedingung 1. von Satz XIX a (das ist Bedingung 1. von Satz XVI a) verifiziert erscheint. Ähnlich argumentiert man in den anderen Fällen.

§ 13. Singuläre Integrale vom Weierstraß'schen Typus.

Im Weierstraß'schen Falle:

$$\varphi(u) = e^{-u^2}$$

sind alle Voraussetzungen der Sätze XXI a, XXIII, XXIV a erfüllt, und zwar für jede unserer Klassen $\mathfrak{F}_i$. Die Sätze XXI a und XXIII lehren uns also, da bekanntlich:

$$\Gamma\left(\frac{1}{2}\right) = \sqrt{\pi}$$

ist, daß für jede in $(-\infty, +\infty)$ zu einer der Klassen $\mathfrak{F}_1$, $\mathfrak{F}_2$, $\mathfrak{F}_3$ gehörige Funktion die Formel:

$$f(x) = \lim_{n=\infty}\sqrt{\frac{n}{\pi}}\int_{-\infty}^{+\infty} f(\xi)\,e^{-n(\xi-x)^2}\,d\xi$$

in jedem Punkte gilt, in dem f (für irgend ein m) verallgemeinerte m-te Ableitung seines m-fach iterierten Integrales ist, und daß sie gleichmäßig in jedem endlichen Intervalle $\langle a', b'\rangle$ gilt, in dessen sämtlichen Punkten f stetig ist. — Satz XXIV a lehrt sodann, daß (für jedes i) die Formel:

$$\mathring{f}^{(i)}(x) = \lim_{n=\infty}\sqrt{\frac{n}{\pi}}\int_{-\infty}^{+\infty} f(\xi)\,\frac{d^i}{dx^i}\,e^{-n(\xi-x)^2}\,d\xi$$

in jedem Punkte gilt, in dem f eine verallgemeinerte endliche Ableitung i-ter Ordnung besitzt, und daß sie gleichmäßig in jedem endlichen Intervalle $\langle a', b'\rangle$ gilt, in dem f i-mal stetig differenzierbar ist.

So wie diese Resultate als Spezialfälle der in den letzten Paragraphen durchgeführten Erörterungen aufgefaßt werden können, kann man sie auch als Spezialfälle eines anderen Typus singulärer Integrale auffassen, den wir als den Weierstraß'schen Typus bezeichnen wollen, und der in mancher Hinsicht

[1] Das heißt Bedingung 1. von Satz XVI a.

einfacher ist, als der in den letzten Paragraphen betrachtete Typus. Es ist für das Folgende bequemer, statt wie bisher Integrale der Form:

$$I_n(f,x)=\int_{-\infty}^{+\infty} f(\xi)\,\varphi(\xi,x,n)\,d\xi$$

zu betrachten, nunmehr Integrale der Form:

$$I(f,k,x)=\int_{-\infty}^{+\infty} f(\xi)\,\varphi(\xi,x,k)\,d\xi$$

zu betrachten, wo k alle Werte $\geqq 1$ durchläuft.

XXV. Sei $\varphi(u)$ eine für alle reellen u definierte Funktion, für die das verallgemeinerte Lebesgue'sche Integral:

(1) $$\lim_{k=+\infty}\int_{-k}^{k}\varphi(u)\,du=\int_{-\infty}^{+\infty}\varphi(u)\,du\ (=\omega)$$

existiert und einen von Null verschiedenen Wert ω besitzt.

Damit für jede in $(-\infty, +\infty)$ zur Klasse $\mathfrak{F}_3$ gehörige Funktion f, die im Punkte x stetig ist, die Formel gelte:[1]

(2) $$f(x)=\lim_{k=+\infty}\frac{k}{\omega}\int_{-\infty}^{+\infty} f(\xi)\,\varphi(k(\xi-x))\,d\xi,$$

ist notwendig und hinreichend, daß das Integral:

(3) $$\int_{-\infty}^{+\infty}|\varphi(u)|\,du$$

einen endlichen Wert habe. Ist dies der Fall, so gilt (2) gleichmäßig in jedem endlichen Intervalle $<a', b'>$, in dessen sämtlichen Punkten f stetig ist.

Die Bedingung ist hinreichend (auch für gleichmäßige Konvergenz). Es genügt zu zeigen, daß für jede Folge positiver Werte k_n mit $\lim\limits_{n=\infty} k_n=+\infty$ der Kern:

$$\varphi(\xi,x,n)=\frac{k_n}{\omega}\,\varphi(k_n(\xi-x))$$

den Bedingungen von Satz X a für die Klasse $\mathfrak{F}_3$ genügt.

Um Bedingung 1. von Satz X a als erfüllt nachzuweisen, zeigen wir zunächst, daß für jedes $\varepsilon>0$:

(4) $$\lim_{n=\infty}\int_{-\infty}^{+\infty}|\psi(\xi,x,n,\varepsilon)|\,d\xi=0$$

gleichmäßig für alle x von $<a', b'>$ ist. In der Tat, es ist:

$$\int_{-\infty}^{+\infty}|\psi(\xi,x,n,\varepsilon)|\,d\xi=\frac{k_n}{\omega}\left\{\int_{-\infty}^{-\varepsilon}|\varphi(k_n u)|\,du+\int_{\varepsilon}^{+\infty}|\varphi(k_n u)|\,du\right\}=$$
$$=\frac{1}{\omega}\left\{\int_{-\infty}^{-k_n\varepsilon}|\varphi(v)|\,dv+\int_{k_n\varepsilon}^{+\infty}|\varphi(v)|\,dv\right\},$$

wodurch wegen der über das Integral (3) gemachten Voraussetzung und weil der letzte Ausdruck von x nicht abhängt, die gleichmäßige Konvergenz von (4) bewiesen ist.

Um nun zu beweisen, daß Bedingung 1. von X a für die Klasse $\mathfrak{F}_3$ erfüllt ist, zeigen wir zunächst: zu jedem $\mu>0$ gibt es ein $\lambda>0$, so daß für jede Menge I sich nicht überdeckender Intervalle, deren Gesamtlänge $<\lambda$ ist, sowie für alle x von $<a', b'>$:

$$\int_I|\psi(\xi,x,n,\varepsilon)|\,d\xi<\mu \quad \text{für alle } n$$

[1] Darunter ist hier wie im Folgenden auch mitverstanden, daß das in dieser Formel auftretende Integral für jedes $k\geqq 1$ existiert.

ist. Wir wählen zu dem Zwecke n_0 so groß, daß für $n > n_0$ und alle x von $< a', b' >$:

$$\int_{-\infty}^{+\infty} |\psi(\xi, x, n, \varepsilon)|\, d\xi < \mu,$$

was wegen (4) möglich ist; sodann bestimmen wir für jedes einzelne n ein λ_n, so daß für jede Menge $\mathfrak{A}$, deren Inhalt $< \lambda_n$ ist:

$$\frac{k_n}{\omega} \int_{\mathfrak{A}} |\varphi(k_n u)|\, du < \mu$$

ist, was wegen eines bekannten Satzes über Lebesgue'sche Integrale sicher möglich ist. Wir haben nun für das gewünschte λ lediglich die kleinste der Zahlen $\lambda_1, \lambda_2, \ldots, \lambda_{n_0}$ zu wählen.

Wir zeigen weiter: zu jedem $\eta > 0$ gibt es ein A, so daß:

(5) $$\int_{-\infty}^{-A} |\psi(\xi, x, n, \varepsilon)|\, d\xi < \eta; \quad \int_{A}^{+\infty} |\psi(\xi, x, n, \varepsilon)|\, d\xi < \eta$$

für alle n und alle x von $< a', b' >$. Wir wählen wieder n_0 so groß, daß:

$$\int_{-\infty}^{+\infty} |\psi(\xi, x, n, \varepsilon)|\, d\xi < \eta$$

für $n > n_0$ und alle x von $< a', b' >$. Sodann beachten wir, daß:

$$\int_{A}^{+\infty} |\psi(\xi, x, n, \varepsilon)|\, d\xi \leqq \frac{k_n}{\omega} \int_{A-x}^{+\infty} |\varphi(k_n u)|\, du,$$

und daher weiter für alle x von $< a', b' >$:

$$\int_{A}^{+\infty} |\psi(\xi, x, n, \varepsilon)|\, d\xi \leqq \frac{k_n}{\omega} \int_{A-b'}^{+\infty} |\varphi(k_n u)|\, du.$$

Nun kann aber zu jedem einzelnen n ein A_n so gewählt werden, daß:

$$\frac{k_n}{\omega} \int_{A_n-b'}^{+\infty} |\varphi(k_n u)|\, du < \eta.$$

Wir haben daher, um die zweite Ungleichung (5) zu befriedigen, für A lediglich die größte der Zahlen $A_1, A_2, \ldots, A_{n_0}$ zu wählen. Analog befriedigt man die erste Ungleichung (5). – Damit ist Bedingung 1. von X a als erfüllt nachgewiesen.

Daß Bedingung 2. von X a erfüllt ist, folgt unmittelbar aus dem gleichmäßigen Bestehen von (4) für alle x von $< a', b' >$.

Bedingung 3. von X a ist erfüllt, denn es ist:

$$\frac{k_n}{\omega} \int_{-\infty}^{+\infty} |\varphi(k_n(\xi - x))|\, d\xi = \frac{1}{\omega} \int_{-\infty}^{+\infty} |\varphi(u)|\, du,$$

wo die rechte Seite von n und x nicht abhängt und nach Voraussetzung endlich ist.

Bedingung 4. von X a ist erfüllt. Denn wegen der gleichmäßigen Konvergenz von (4) für alle x von $< a', b' >$ ist diese Bedingung nun gleichbedeutend mit folgender: es ist:

$$\lim_{n=\infty} \frac{k_n}{\omega} \int_{-\infty}^{+\infty} \varphi(k_n(\xi - x))\, d\xi = 1$$

gleichmäßig für alle x von $< a', b' >$.

Das aber ist der Fall, wegen der Gleichungen:

$$k_n \int_{-\infty}^{+\infty} \varphi(k_n(\xi - x))\, d\xi = \int_{-\infty}^{+\infty} \varphi(u)\, du = \omega.$$

Die Bedingung ist notwendig. Dies erkennt man durch Berufung auf Satz VI *a*. In der Tat, es ist:

$$\int_{x-\varepsilon}^{x+\varepsilon} |\varphi(\xi, x, n)| \, d\xi = \frac{k_n}{\omega} \int_{x-\varepsilon}^{x+\varepsilon} |\varphi(k_n(\xi - x))| \, d\xi = \frac{1}{\omega} \int_{-k_n\varepsilon}^{k_n\varepsilon} |\varphi(u)| \, du.$$

Hätte nun das Integral (3) nicht einen endlichen Wert, so wäre:

$$\lim_{n=\infty} \int_{-k_n\varepsilon}^{k_n\varepsilon} |\varphi(u)| \, du = +\infty,$$

und es wäre somit Bedingung 3. von VI *a* nicht erfüllt.

Man erkennt ohneweiters:

XXV *a*. »Genügt $\varphi(u)$ den zu Beginn von Satz XXV angeführten Voraussetzungen, so ist die das Integral (3) betreffende Bedingung auch notwendig und hinreichend dafür, daß in jedem Punkte x des beliebigen Intervalles (a, b), der ein Stetigkeitspunkt für die in $< a, b >$ zur Klasse $\mathfrak{F}_3$ gehörige Funktion f ist, die Formel gelte:

$$f(x) = \lim_{k=+\infty} \frac{k}{\omega} \int_a^b f(\xi) \, \varphi(k(\xi - x)) \, d\xi. \tag{6}$$

Ist auch Bedingung (3) erfüllt, so gilt (6) gleichmäßig in jedem Teilintervalle $< a', b' >$ von (a, b), in dessen sämtlichen Punkten f stetig ist.«

Wir wenden uns nunmehr zum Studium von Formel (2) für Funktionen der Klassen $\mathfrak{F}_1$ und $\mathfrak{F}_2$. Da für diese Klassen Bedingung 3. von VI *a* dieselbe ist wie für die Klasse $\mathfrak{F}_3$, so sehen wir aus dem zuletzt geführten Beweise, daß Formel (2) jedenfalls nur dann für alle Funktionen, die in $(-\infty, +\infty)$ zu $\mathfrak{F}_1$ oder zu $\mathfrak{F}_2$ gehören, gelten kann, wenn die das Integral (3) betreffende Bedingung von Satz XXV erfüllt ist. Wir setzen also diese Bedingung von jetzt an als erfüllt voraus.

XXVI. Sei $\varphi(u)$ eine für alle reellen u definierte Funktion, für die das Integral (3) einen endlichen Wert hat, und für die der Wert ω des Integrales (1) nicht verschwindet. Damit Beziehung (2) für jede im Punkte x stetige Funktion gelte, die in $(-\infty, +\infty)$ zur Klasse $\mathfrak{F}_2$ gehört, ist notwendig und hinreichend, daß zu jedem $h > 0$ ein A gehört, so daß:

$$\int_{-\infty}^{-u} (\varphi(v))^2 \, dv < \frac{A}{u}; \quad \int_u^{+\infty} (\varphi(v))^2 \, dv < \frac{A}{u} \quad \text{für } u \geqq h. \tag{7}$$

Ist diese Bedingung erfüllt, so gilt (2) gleichmäßig in jedem endlichen Intervalle $< a', b' >$, in dessen sämtlichen Punkten f stetig ist.

Die Bedingung ist hinreichend. Um Bedingung 1. von X *a* für die Klasse $\mathfrak{F}_2$ als erfüllt nachzuweisen, genügt es zu zeigen, daß es zu jedem $h > 0$ ein M gibt, so daß:

$$\int_{-\infty}^{+\infty} (\psi(\xi, x, k, h))^2 \, d\xi < M \quad \text{für alle } k \geqq 1 \text{ und alle } x \text{ von } < a', b' >. \tag{8}$$

Nun ist:

$$\int_{-\infty}^{+\infty} (\psi(\xi, x, k, h))^2 \, d\xi = \frac{k^2}{\omega^2} \left\{ \int_{-\infty}^{-h} (\varphi(ku))^2 \, du + \int_h^{+\infty} (\varphi(ku))^2 \, du \right\} =$$

$$= \frac{k}{\omega^2} \left\{ \int_{-\infty}^{-k.h} (\varphi(u))^2 \, du + \int_{k.h}^{+\infty} (\varphi(u))^2 \, du \right\}.$$

Also haben wir bei Benützung von (7):

$$\int_{-\infty}^{+\infty} (\psi(\xi, x, k, h))^2 \, d\xi < \frac{2}{\omega^2} \cdot \frac{A}{h},$$

wodurch (8) bewiesen ist.

Daß Bedingung 2., 3. und 4. von X a erfüllt sind, haben wir schon beim Beweise von Satz XXV gesehen.

Die Bedingung ist notwendig. Wäre sie nicht erfüllt, so gäbe es ein $h > 0$ und eine Folge von Zahlen A_n mit:

(9) $$\lim_{n=\infty} A_n = +\infty,$$

sowie eine Folge von Zahlen u_n, alle $\geqq h$ (oder alle $\leqq -h$), für die:

$$\int_{u_n}^{+\infty} (\varphi(v))^2\, dv = \frac{A_n}{u_n} \quad \left(\text{oder} \int_{-\infty}^{u_n} (\varphi(v))^2\, dv = \frac{A_n}{|u_n|}\right)$$

wäre. Sei etwa das erstere der Fall. Wir setzen: $u_n = k_n \cdot h$ (dann ist $k_n \geqq 1$) und haben:

$$k_n \int_{u_n}^{+\infty} (\varphi(v))^2\, dv = k_n^2 \int_{h}^{+\infty} (\varphi(k_n u))^2\, du = \frac{A_n}{h}.$$

Wegen (9) wäre also für den Kern:

$$\varphi(\xi, x, n) = \frac{k_n}{\omega} \varphi(k_n(\xi - x))$$

gewiss Bedingung 1. von Satz VI a für die Klasse $\mathfrak{F}_2$ nicht erfüllt. Damit ist unsere Behauptung erwiesen.

XXVI a. »Genügt $\varphi(u)$ allen zu Beginn von Satz XXVI angeführten Voraussetzungen, so ist notwendig und hinreichend dafür, daß in jedem Punkte x des beliebigen Intervalles (a, b), der ein Stetigkeitspunkt für die in $<a, b>$ zu $\mathfrak{F}_2$ gehörige Funktion f ist, die Formel (6) gelte, daß φ auch der Bedingung (7) genügt. Ist auch Bedingung (7) erfüllt, so gilt (6) gleichmäßig in jedem Teilintervalle (a', b') von $<a, b>$, in dessen sämtlichen Punkten f stetig ist.«

Die Bedingung ist hinreichend. Das folgt unmittelbar aus dem Beweise von Satz XXVI.

Die Bedingung ist notwendig. Um dies einzusehen, wird es genügen, statt der Notwendigkeit von (7) die Notwendigkeit der folgenden Bedingung nachzuweisen: Zu jedem $h > 0$ und $\lambda > 1$ gibt es ein A', so daß:

(10) $$\int_{-\lambda u}^{-u} (\varphi(u))^2\, du < \frac{A'}{u}; \quad \int_{u}^{\lambda u} (\varphi(u))^2\, du < \frac{A'}{u} \quad \text{für } u \geqq h.$$

In der Tat, ist Bedingung (10) erfüllt, so ist auch Bedingung (7) erfüllt, denn es ist (für $u \geqq h$ und $\lambda > 1$):

$$\int_{u}^{+\infty} (\varphi(u))^2\, du = \int_{u}^{\lambda u} (\varphi(u))^2\, du + \int_{\lambda u}^{\lambda^2 u} (\varphi(u))^2\, du + \ldots + \int_{\lambda^n u}^{\lambda^{n+1} u} (\varphi(u))^2\, du + \ldots$$

und somit wegen (10):

$$\int_{u}^{+\infty} (\varphi(u))^2\, du < \frac{A'}{u}\left(1 + \frac{1}{\lambda} + \ldots + \frac{1}{\lambda^n} + \ldots\right) = \frac{A'\lambda}{\lambda - 1} \cdot \frac{1}{u};$$

das heißt: es ist auch Bedingung (7) erfüllt.

Angenommen nun, es wäre Bedingung (10) nicht erfüllt; dann gäbe es ein $h > 0$ und ein $\lambda > 1$, so daß für eine Folge in $<h, +\infty)$ oder in $(-\infty, -h>$ gelegener u_n (wir nehmen etwa das erstere an):

(11) $$\int_{u_n}^{\lambda u_n} (\varphi(u))^2\, du = \frac{A_n}{u_n} \quad \left(\lim_{n=\infty} A_n = +\infty\right)$$

wäre. Sei x ein Punkt von (a, b). Wir wählen ein $\varepsilon > 0$, so daß:

$$\lambda\varepsilon \leqq b - x \quad \text{und} \quad \varepsilon \leqq h,$$

und setzen:

$$u_n = k_n \varepsilon.$$

Dann wird $k_n \geqq 1$ und nach (11):

$$k_n^2 \int_{x+\varepsilon}^{b} \{\varphi(k_n(\xi - x))\}^2 d\xi \geqq k_n \int_{k_n\varepsilon}^{k_n\lambda.\varepsilon} (\varphi(u))^2 du = \frac{A_n}{\varepsilon},$$

so daß für den Kern:

$$\varphi(\xi, x, n) = k_n \varphi(k_n(\xi - x))$$

Bedingung 1. von Satz VI nicht erfüllt wäre.

XXVII. Sei $\varphi(u)$ eine für alle reellen u definierte Funktion, für die das Integral (3) einen endlichen Wert hat und für die der Wert ω des Integrales (1) nicht verschwindet. Damit Beziehung (2) für jede im Punkte x stetige Funktion gelte, die in $(-\infty, +\infty)$ zur Klasse $\mathfrak{F}_1$ gehört, ist notwendig und hinreichend, daß zu jedem $h > 0$ ein A gehört, so daß (abgesehen von Nullmengen):

$$(12) \qquad |\varphi(u)| < \frac{A}{|u|} \quad \text{für } |u| \geqq h.$$

Ist diese Bedingung erfüllt, so gilt (2) gleichmäßig in jedem endlichen Intervalle $< a', b' >$, in dessen sämtlichen Punkten f stetig ist.

Die Bedingung ist hinreichend. Es handelt sich nur darum, nachzuweisen, daß Bedingung 1. von Satz X a für die Klasse $\mathfrak{F}_1$ erfüllt ist; denn für Bedingung 2., 3. und 4. ist es uns schon bekannt.

Wir haben zu zeigen: ist $h > 0$ beliebig gegeben, so gibt es ein M, so daß (abgesehen von Nullmengen):

$$(13) \qquad \frac{k}{\omega} |\varphi(ku)| < M \quad \text{für } |u| \geqq h \text{ und } k \geqq 1.$$

Nun folgt aus (12) (immer abgesehen von Nullmengen):

$$k|\varphi(ku)| < \frac{A}{|u|} \quad \text{für } |u| \geqq h \text{ und } k \geqq 1,$$

mithin auch:

$$k|\varphi(ku)| < \frac{A}{h} \quad \text{für } |u| \geqq h \text{ und } k \geqq 1,$$

womit (13) bewiesen ist.

Die Bedingung ist notwendig. Wäre sie nicht erfüllt, so hieße das: es gibt ein $h > 0$ und eine, sei es in $< h, +\infty)$, sei es in $(-\infty, -h >$ gelegene Folge von Mengen $\mathfrak{M}_1, \mathfrak{M}_2, \ldots, \mathfrak{M}_n \ldots$, deren jede einen von 0 verschiedenen Inhalt hat, und derart, daß in den Punkten von $\mathfrak{M}_n$:

$$|\varphi(u)| > \frac{n}{|u|}$$

ist. Nehmen wir etwa an, diese Mengen liegen in $< h, +\infty)$. Jedenfalls gibt es dann auch eine Folge von Punkten u_n $(\geqq h)$, so daß der in $< u_n, u_n + 1 >$ liegende Teil $\bar{\mathfrak{M}}_n$ von $\mathfrak{M}_n$ nicht den Inhalt 0 hat. Setzen wir:

$$u_n = k_n \cdot h,$$

so ist $k_n \geqq 1$, und es wird in den Punkten der Menge $\mathfrak{M}_n^*$, die aus $\overline{\mathfrak{M}_n}$ durch Ähnlichkeitstransformation im Verhältnisse $k_n : 1$ hervorgeht:

$$k_n \varphi(k_n u) > \frac{n}{u} \geqq \frac{n}{\frac{1}{k_n}(u_n+1)} = \frac{n}{h+\frac{1}{k_n}} \geqq \frac{n}{h+1}.$$

Und da die Menge $\mathfrak{M}_n^*$ in $< h, +\infty)$ liegt, wäre also Bedingung 1. von Satz VI *a* für die Klasse $\mathfrak{F}_1$ nicht erfüllt.

XXVII *a*. »Genügt $\varphi(u)$ allen zu Beginn von Satz XXVII aufgezählten Bedingungen, so ist, damit in jedem Punkte x des beliebigen Intervalles (a, b), der für die in $< a, b >$ zur Klasse $\mathfrak{F}_1$ gehörige Funktion f ein Stetigkeitspunkt ist, Formel (6) gelte, notwendig und hinreichend, daß $\varphi(u)$ auch der Bedingung (12) genüge. Ist auch Bedingung (12) erfüllt, so gilt (6) gleichmäßig in jedem Teilintervalle $< a', b' >$ von (a, b), in dessen sämtlichen Punkten f stetig ist.«

Eines Beweises bedarf hier nur die Behauptung, daß die Bedingung (12) notwendig ist. Wäre sie nicht erfüllt, so gäbe es wieder eine Folge, etwa in $< h, +\infty)$ gelegener Mengen $\mathfrak{M}_1, \mathfrak{M}_2, \ldots, \mathfrak{M}_n, \ldots$ mit von 0 verschiedenem Inhalte, so daß in den Punkten von $\mathfrak{M}_n$:

$$|\varphi(u)| > \frac{n}{u}. \tag{14}$$

Es gäbe also auch eine in $< h, +\infty)$ gelegene Punktfolge $u_1, u_2, \ldots, u_n, \ldots$ und in $< u_n, u_n + \frac{1}{n} >$ gelegene Mengen $\overline{\mathfrak{M}_n}$ mit von 0 verschiedenem Inhalte, auf denen (14) gilt.

Sei x ein Punkt von (a, b). Wir wählen ein $\varepsilon > 0$ gemäß:

$$\varepsilon \leqq h, \quad \varepsilon < b - x$$

und setzen:

$$k_n = \frac{u_n}{\varepsilon}; \quad u = k_n(\xi - x).$$

Gehört dann $k_n(\xi - x)$ zur Menge $\overline{\mathfrak{M}_n}$, das heißt gehört ξ zur Menge $\mathfrak{M}_n^*$, die aus $\overline{\mathfrak{M}_n}$ durch die lineare Transformation $u = k_n(\xi - x)$ entsteht, so haben wir wegen (14):

$$k_n |\varphi(k_n(\xi - x))| > \frac{n}{\xi - x}.$$

Die Menge $\mathfrak{M}_n^*$ aber liegt im Intervalle $< x + \varepsilon, x + \varepsilon + \frac{1}{n k_n} >$ der Veränderlichen ξ. Wegen $k_n \geqq 1$ liegen diese Intervalle für hinlänglich großes n in $< x + \varepsilon, b >$, so daß für den Kern:

$$\varphi(\xi, x, n) = k_n \varphi(k_n(\xi - x))$$

sicherlich Bedingung 1. von Satz VI nicht erfüllt ist.

§ 14. Konvergenz an Unstetigkeitsstellen.

Besonders einfach gestaltet sich für den jetzt betrachteten Typus singulärer Integrale die Anwendung der Sätze von §§ 7, 8 und 9, wenn wir bemerken, daß hier die Bedingung:

$$\lim_{k=+\infty} \varphi(\xi, x, k) = 0 \quad \text{für } \xi \neq x$$

gleichbedeutend ist mit:

$$\lim_{u=-\infty} u \cdot \varphi(u) = 0; \quad \lim_{u=+\infty} u \cdot \varphi(u) = 0, \tag{1}$$

und daß wegen:

$$\frac{\partial^i}{\partial \xi^i} \varphi(\xi, x, k) = (-1)^i \frac{\partial^i}{\partial x^i} \varphi(\xi, x, k) = k^{i+1} \varphi^{(i)}(k(\xi - x))$$

jede der Beziehungen:

$$\lim_{k=+\infty} \frac{\partial^i}{\partial \xi^i} \varphi(\xi, x, k) = 0; \quad \lim_{k=+\infty} \frac{\partial^i}{\partial x^i} \varphi(\xi, x, k) = 0 \quad \text{für } \xi \neq x$$

gleichbedeutend wird mit:

(1 *a*) $$\lim_{u=-\infty} u^{i+1} \varphi^{(i)}(u) = 0; \quad \lim_{u=+\infty} u^{i+1} \varphi^{(i)}(u) = 0.$$

Durch Anwendung von Satz XIV *a* und XV *a* finden wir also:

XXVIII. Sei $\varphi(u)$ eine für alle reellen u definierte Funktion, die eine in jedem endlichen Intervalle absolut stetige $(m-1)$-te Ableitung besitzt, und es sei:

(2) $$\lim_{u=-\infty} u^m \varphi^{(m-1)}(u) = 0; \quad \lim_{u=+\infty} u^m \varphi^{(m-1)}(u) = 0.$$

Ferner existiere das verallgemeinerte Integral:

(3) $$\lim_{A=+\infty} \int_{-A}^{A} \varphi(u)\, du = \int_{-\infty}^{+\infty} \varphi(u)\, du = \omega,$$

und es sei sein Wert $\omega \neq 0$. Endlich existiere das Integral:

(4) $$\int_{-\infty}^{+\infty} |u^m \varphi^{(m)}(u)|\, du.$$

Dann gilt für jede in $(-\infty, +\infty)$ zu einer der Klassen $\mathfrak{F}_1, \mathfrak{F}_2, \mathfrak{F}_3$ gehörige Funktion f die Beziehung:

(5) $$f(x) = \lim_{k=+\infty} \frac{k}{\omega} \int_{-\infty}^{+\infty} f(\xi)\, \varphi(k(\xi - x))\, d\xi$$

in jedem Punkte x, in dem f m-te Ableitung seines m-fach iterierten unbestimmten Integrales ist. Ist $\varphi(u)$ eine gerade Funktion, so gilt (5) in jedem Punkte x, in dem f verallgemeinerte m-te Ableitung seines m-fach iterierten unbestimmten Integrales ist.

Zum Beweise bemerken wir, daß gleichzeitig mit $\varphi^{(m-1)}$ auch $\varphi^{(m-2)}, \ldots, \varphi'$ und φ in jedem endlichen Intervalle absolut stetig sind. Daraus und aus der Existenz des verallgemeinerten Integrales (3) folgt ohneweiters: es gibt Punktfolgen $u_n, \bar{u}_n, u_n^{(i)}, \bar{u}_n^{(i)}$ mit:

(6) $$\lim_{n=\infty} u_n = +\infty, \quad \lim_{n=\infty} \bar{u}_n = -\infty, \quad \lim_{n=\infty} u_n^{(i)} = +\infty, \quad \lim_{n=\infty} \bar{u}_n^{(i)} = -\infty,$$

für die:

(7) $$\lim_{n=\infty} \varphi(u_n) = 0, \quad \lim_{n=\infty} \varphi(\bar{u}_n) = 0, \quad \lim_{n=\infty} \varphi^{(i)}(u_n^{(i)}) = 0, \quad \lim_{n=\infty} \varphi^{(i)}(\bar{u}_n^{(i)}) = 0 \quad (i = 1, 2, \ldots, m-1).$$

Aus (2) folgt: ist $\varepsilon > 0$ beliebig gegeben, so gibt es ein A, so daß:

$$|\varphi^{(m-1)}(u)| < \frac{\varepsilon}{u^m} \quad \text{für } u \geqq A.$$

Und daraus folgt durch Integration (für $u \geqq A$ und hinlänglich großes n):

$$|\varphi^{(m-2)}(u) - \varphi^{(m-2)}(u_n^{(m-2)})| \leqq \frac{\varepsilon}{m-1} \left| \frac{1}{u^{m-1}} - \frac{1}{(u_n^{(m-2)})^{m-1}} \right|.$$

Aus (6) und (7) folgt nun:

$$|\varphi^{(m-2)}(u)| \leqq \frac{\varepsilon}{m-1} \frac{1}{u^{m-1}} \quad \text{für } u \geqq A,$$

oder, was dasselbe heißt:

(8) $$\lim_{u=+\infty} u^{m-1} \varphi^{(m-2)}(u) = 0.$$

Ebenso beweist man:

(8a) $$\lim_{u=-\infty} u^{m-1} \varphi^{(m-2)}(u) = 0.$$

So wie (8) und (8a) aus (2) hergeleitet wurden, zeigt man, indem man in derselben Weise weiterschließt: es bestehen die Relationen:

(9) $$\lim_{u=-\infty} u\,\varphi(u) = 0 \quad \lim_{u=+\infty} u\,\varphi(u) = 0$$

$$\lim_{u=-\infty} u^{i+1} \varphi^{(i)}(u) = 0 \quad \lim_{u=+\infty} u^{i+1} \varphi^{(i)}(u) = 0 \quad (i = 1, 2, \ldots, m-1).$$

Wir zeigen sodann, daß aus unseren Voraussetzungen folgt, daß $\varphi(u)$ der Bedingung (3) von Satz XXV, der Bedingung (7) von Satz XXVI und der Bedingung (12) von Satz XXVII genügt.

Durch partielle Integration ergibt sich:

$$\int_0^u |v^{m-1} \varphi^{(m-1)}(v)|\,dv = \frac{v^m}{m} |\varphi^{(m-1)}(v)|\Big]_0^u - \frac{1}{m}\int_0^u v^m \, sgn.\,\varphi^{(m-1)}(v)\cdot\varphi^{(m)}(v)\,dv.$$

Hierin haben wir, wegen (2):

$$\lim_{u=+\infty} \frac{v^m}{m} |\varphi^{(m-1)}(v)|\Big]_0^u = 0,$$

der Subtrahend hat, wegen der vorausgesetzten Existenz des Integrales (4) einen endlichen Grenzwert für $u = +\infty$, so daß die Existenz des Integrales:

$$\int_0^{\infty} |u^{m-1} \varphi^{(m-1)}(u)|\,du$$

bewiesen ist. Ebenso beweist man die Existenz von:

$$\int_{-\infty}^0 |u^{m-1} \varphi^{(m-1)}(u)|\,du,$$

so daß aus der vorausgesetzten Existenz von (4) die Existenz von:

$$\int_{-\infty}^{+\infty} |u^{m-1} \varphi^{(m-1)}(u)|\,du$$

folgt. Indem man (unter Benützung von (9)) so weiter schließt, beweist man der Reihe nach die Existenz von:

(10) $$\int_{-\infty}^{+\infty} |u^i \varphi^{(i)}(u)|\,du \quad (i = m-1, m-2, \ldots, 1) \quad \text{und} \int_{-\infty}^{+\infty} |\varphi(u)|\,du.$$

Damit ist Bedingung (3) von Satz XXV erwiesen.

Wie aus der Stetigkeit von $\varphi(u)$ und den beiden ersten Relationen (9) unmittelbar folgt, gibt es eine Konstante B, so daß für alle u:

(11) $$|\varphi(u)| < \frac{B}{|u|}.$$

Also haben wir für alle $u \geqq h$:

$$\int_{-\infty}^{-u} (\varphi(v))^2 \, dv < \frac{B^2}{u}; \quad \int_{u}^{+\infty} (\varphi(v))^2 \, dv < \frac{B^2}{u},$$

womit Bedingung (7) von Satz XXVI nachgewiesen ist. — Und durch Ungleichung (11) ist gleichzeitig Bedingung (12) von Satz XXVII als erfüllt nachgewiesen.

Daraus folgt aber, wie die Beweise von XXV, XXVI und XXVII gezeigt haben, das Bestehen der Bedingungen 1., 2. und 4. von Satz XIV *a* für jede der Klassen $\mathfrak{F}_1, \mathfrak{F}_2, \mathfrak{F}_3$. — Um einzusehen, daß auch Bedingung 3. von Satz XIV *a* erfüllt ist, schreiben wir:

$$\int_{x-h}^{x+h} \left| (\xi - x)^m \frac{\partial^m}{\partial \xi^m} \varphi(\xi, x, u) \right| d\xi = \int_{-h}^{h} k^{m+1} |u^m \varphi^{(m)}(ku)| \, du = \int_{-k.h}^{k.h} |u^m \varphi^{(m)}(u)| \, du,$$

und dieser Ausdruck liegt, wegen der vorausgesetzten Existenz des Integrales (4), tatsächlich für alle $k \geqq 1$ unter einer endlichen Schranke N.

Damit ist Satz XXVIII bewiesen. Man beweist ebenso:

XXVIII *a*. »Genügt $\varphi(u)$ den Voraussetzungen von Satz XXVIII, so gilt für jede in $< a, b >$ zur Klasse $\mathfrak{F}_1$ gehörige Funktion f die Beziehung

(12)
$$f(x) = \lim_{k=+\infty} \frac{k}{\omega} \int_a^b f(\xi) \, \varphi(k(\xi - x)) \, d\xi$$

in jedem Punkte x von (a, b), in dem f m-te Ableitung seines m-fach iterierten unbestimmten Integrales ist. — Ist $\varphi(u)$ eine gerade Funktion, so gilt (12) in jedem Punkte x von (a, b), in dem f verallgemeinerte m-te Ableitung seines m-fach iterierten unbestimmten Integrales ist.«

§ 15. Differenziation der Integrale des Weierstraß'schen Typus.

Durch Berufung auf die Sätze XVI *a*, XVII *a*, XIX *a* finden wir:

XXIX. Es genüge $\varphi(u)$ allen Voraussetzungen von Satz XXVIII. Damit in jedem Punkte x, in dem die in $(-\infty, +\infty)$ zur Klasse $\mathfrak{F}_3$ gehörige Funktion f eine endliche Ableitung m-ter Ordnung besitzt, die Beziehungen gelten:

(1)
$$f(x) = \lim_{k=+\infty} \frac{k}{\omega} \int_{-\infty}^{+\infty} f(\xi) \, \varphi(k(\xi - x)) \, d\xi,$$

(2)
$$f^{(i)}(x) = \lim_{k=+\infty} (-1)^i \frac{k^{i+1}}{\omega} \int_{-\infty}^{+\infty} f(\xi) \, \varphi^{(i)}(k(\xi - x)) \, d\xi \quad (i = 1, 2, \ldots, m),$$

ist notwendig und hinreichend, daß das Integral (4) von § 14 existiere. Ist dies der Fall, so gelten die Beziehungen (1), (2) gleichmäßig in jedem endlichen Intervalle, in dem f m-mal stetig differenzierbar ist. Ist $\varphi(u)$ eine gerade Funktion, so kann die Ableitung $f^{(i)}(x)$ durch die verallgemeinerte Ableitung $\overset{*}{f}{}^{(i)}(x)$ ersetzt werden.

Die Bedingung ist hinreichend (auch für die gleichmäßige Konvergenz). Es genügt zu beweisen, daß für den Kern

$$\varphi(u, k) = \frac{k}{\omega} \varphi(ku)$$

alle Voraussetzungen und Bedingungen von Satz XIX *a* erfüllt sind; zunächst gilt tatsächlich die Beziehung:

(3)
$$\lim_{k=+\infty} \varphi^{(m-1)}(u, k) = \lim_{k=+\infty} \frac{k^m}{\omega} \varphi^{(m-1)}(k \cdot u) = 0$$

gleichmäßig in jedem endlichen, den Nullpunkt nicht enthaltenden Intervalle $<\alpha, \beta>$, wie unmittelbar aus (2) von § 14 folgt; denn sei etwa:

$$0<\alpha<\beta.$$

Wegen (2) von § 14 ist, bei beliebig gegebenem $\varepsilon>0$:

$$|u^m \varphi^{(m-1)}(u)|<\varepsilon \quad \text{für } u \geqq A.$$

Daher weiter für alle u von $<\alpha, \beta>$:

$$|k^m \varphi^{(m-1)}(uk)|<\frac{\varepsilon}{\alpha^m} \quad \text{für } k \geqq \frac{A}{\alpha},$$

womit das gleichmäßige Bestehen von (3) in $<\alpha, \beta>$ bewiesen ist.

Bedingung 1. von XIX a für die Klasse $\mathfrak{F}_3$ ist erfüllt, wenn für jedes $h>0$ die Kerne $\psi(u, k, h)$ und $\psi^{(i)}(u, k, h)$ $(i=1,2,\ldots, m)$ der Bedingung 1. von Satz IIIa genügen. Für den Kern $\psi(u, k, h)$ haben wir schon beim Beweise von Satz XXV gezeigt, daß dies tatsächlich aus der Existenz des Integrales

$$\int_{-\infty}^{+\infty}|\varphi(u)|\, du$$

folgt. Zeigen wir es nun für den Kern $\psi^{(i)}(u, k, h)$.

Es wird wieder genügen zu zeigen, daß:

$$\lim_{k=+\infty} \int_{-\infty}^{+\infty}|\psi^{(i)}(u, k, h)|\, du=0 \tag{4}$$

ist, da daraus das Bestehen der Bedingung 1. von III a für den Kern $\psi^{(i)}(u, k, h)$ ebenso gefolgert werden kann, wie beim Beweise von Satz XXV das Bestehen dieser Bedingung für den Kern $\psi(\xi, x, n, \varepsilon)$ aus der dortigen Relation (4) gefolgert wurde.

Um (4) zu beweisen, beachten wir, daß:

$$\int_{-\infty}^{+\infty}|\psi^{(i)}(u, k, h)|\, du=\frac{k^{i+1}}{\omega}\left\{\int_{-\infty}^{-h}|\varphi^{(i)}(ku)|\, du+\int_{h}^{+\infty}|\varphi^{(i)}(ku)|\, du\right\}= \tag{5}$$

$$=\frac{k^i}{\omega}\left\{\int_{-\infty}^{-kh}|\varphi^{(i)}(u)|\, du+\int_{kh}^{+\infty}|\varphi^{(i)}(u)|\, du\right\}.$$

Wegen der in § 14 bewiesenen Existenz des Integrales:

$$\int_{-\infty}^{+\infty}|u^i \varphi^{(i)}(u)|\, du$$

gibt es zu einem beliebig vorgegebenen $\varepsilon>0$ ein A, so daß:

$$\int_{-\infty}^{-u}|v^i \varphi^{(i)}(v)|\, dv<\varepsilon; \quad \int_{u}^{+\infty}|v^i \varphi^{(i)}(v)|\, dv<\varepsilon \quad \text{für } u \geqq A,$$

und somit auch:

$$\int_{-\infty}^{-u}|\varphi^{(i)}(v)|\, dv<\frac{\varepsilon}{u^i}; \quad \int_{u}^{+\infty}|\varphi^{(i)}(v)|\, dv<\frac{\varepsilon}{u^i} \quad \text{für } u \geqq A,$$

woraus man sofort entnimmt:

$$k^i \int_{-\infty}^{-kh}|\varphi^{(i)}(u)|\, du<\frac{\varepsilon}{h^i}; \quad k^i \int_{kh}^{+\infty}|\varphi^{(i)}(u)|\, du<\frac{\varepsilon}{h^i} \quad \text{für } k \geqq \frac{A}{h}.$$

Damit ist, im Hinblick auf (5), Beziehung (4) dargetan und mithin bewiesen, daß Bedingung 1. von Satz XIX a erfüllt ist.

Daß Bedingung 3. von XIX *a* erfüllt ist, entnimmt man daraus, daß:

$$\int_{-\infty}^{+\infty} |u^m \varphi^{(m)}(u,k)|\, du = \frac{k^{m+1}}{\omega}\int_{-\infty}^{+\infty} |u^m \varphi^{(m)}(ku)|\, du = \frac{1}{\omega}\int_{-\infty}^{+\infty} |u^m \varphi^{(m)}(u)|\, du \tag{6}$$

ist. Daß Bedingung 4. von XIX *a* erfüllt ist, ist evident.

Gleichung (6) lehrt aber auch, daß die Bedingung von Satz XXIX notwendig ist,[1] womit dieser Satz völlig bewiesen ist.

Ebenso beweist man:

XXIX *a*. »Es genüge $\varphi(u)$ allen Voraussetzungen von Satz XXVIII. Damit in jedem Punkte x des beliebigen Intervalles (a, b), in dem die in $< a, b >$ zur Klasse $\mathfrak{F}_3$ gehörige Funktion f eine endliche Ableitung m-ter Ordnung besitzt, die Beziehungen gelten:

$$f(x) = \lim_{k=+\infty} \frac{k}{\omega}\int_a^b f(\xi)\, \varphi(k(\xi - x))\, d\xi, \tag{7}$$

$$f^{(i)}(x) = \lim_{k=+\infty} (-1)^i \frac{k^{i+1}}{\omega}\int_a^b f(\xi)\, \varphi^{(i)}(k(\xi - x))\, d\xi \quad (i = 1, 2, \ldots, m), \tag{8}$$

ist notwendig und hinreichend, daß das Integral (4) von § 14 existiere. — Ist dies der Fall, so gelten die Beziehungen (7), (8) gleichmäßig in jedem Teilintervalle $< a', b' >$ von (a, b), in dem f m-mal stetig differenzierbar ist. — Ist $\varphi(u)$ eine gerade Funktion, so kann die Ableitung $f^{(i)}(x)$ durch die verallgemeinerte Ableitung $\overset{*}{f}{}^{(i)}(x)$ ersetzt werden.«

Wir wenden uns nunmehr der Frage nach der Gültigkeit der Formeln (1), (2) für Funktionen der Klassen $\mathfrak{F}_1$ und $\mathfrak{F}_2$ zu.

XXX. Es genüge $\varphi(u)$ allen Voraussetzungen von Satz XXVIII. Damit in jedem Punkte x, in dem die in $(-\infty, +\infty)$ zur Klasse $\mathfrak{F}_2$ gehörige Funktion f eine endliche Ableitung m-ter Ordnung besitzt, die Beziehungen (1), (2) bestehen, ist notwendig und hinreichend, daß das Integral (4) von § 14 existiere, und daß folgende Bedingung erfüllt sei: zu jedem $h > 0$ gehört ein A, so daß für $u \geqq h$:

$$\int_{-\infty}^{-u} (\varphi^{(m)}(v))^2\, dv < \frac{A}{u^{2m+1}}; \quad \int_u^{+\infty} (\varphi^{(m)}(v))^2\, dv < \frac{A}{u^{2m+1}}. \tag{9}$$

Sind diese Bedingungen erfüllt, so gelten die Beziehungen (1), (2) gleichmäßig in jedem endlichen Intervalle $< a', b' >$, in dem f m-mal stetig differenzierbar ist. — Ist $\varphi(u)$ eine gerade Funktion, so kann die Ableitung $f^{(i)}(x)$ ersetzt werden durch die verallgemeinerte Ableitung $\overset{*}{f}{}^{(i)}(x)$.

Die Bedingungen sind hinreichend. Es braucht nur mehr gezeigt zu werden, daß Bedingung 1. von Satz XIX *a* für die Klasse $\mathfrak{F}_2$ erfüllt ist; es genügt also zu zeigen: zu jedem $h > 0$ gibt es ein M, so daß für alle $k \geqq 1$:

$$\int_{-\infty}^{+\infty} (\psi(u, k, h))^2\, du < M, \tag{10}$$

$$\int_{-\infty}^{+\infty} (\psi^{(i)}(u, k, h))^2\, du < M \quad (i = 1, 2, \ldots, m). \tag{11}$$

Wir schreiben:

$$\int_{-\infty}^{+\infty} (\psi^{(i)}(u, k, h))^2\, du = \frac{k^{2(i+1)}}{\omega^2}\left\{\int_{-\infty}^{-h} (\varphi^{(i)}(ku))^2\, du + \int_h^{+\infty} (\varphi^{(i)}(ku))^2\, du\right\} = \tag{12}$$

$$= \frac{k^{2i+1}}{\omega^2}\left\{\int_{-\infty}^{-kh} (\varphi^{(i)}(u))^2\, du + \int_{kh}^{+\infty} (\varphi^{(i)}(u))^2\, du\right\}.$$

[1] Und zwar nicht nur, wenn f zu $\mathfrak{F}_3$ gehört, sondern auch, wenn f zu $\mathfrak{F}_1$ oder $\mathfrak{F}_2$ gehört.

Benützt man (9), so hat man daher weiter:

$$\int_{-\infty}^{+\infty} (\psi^{(m)}(u, k, h))^2 \, du \leqq \frac{2}{\omega^2} \cdot \frac{A}{h^{2m+1}},$$

womit (11) für $i = m$ bewiesen ist. Für $i = 1, 2, \ldots, m-1$, sowie für (10) folgt dies wieder aus (12), beziehungsweise der analogen Formel für $\psi(u, k, h)$, wenn man bemerkt, daß es wegen der Stetigkeit von $\varphi, \varphi', \ldots, \varphi^{(m-1)}$ und wegen (9) von § 14 eine Konstante B gibt, so daß für alle u die Ungleichungen bestehen:

$$|\varphi(u)| < \frac{B}{|u|}, \quad \varphi^{(i)}(u) < \frac{B}{|u|^{i+1}} \quad (i = 1, 2, \ldots, m-1).$$

Die Bedingungen sind notwendig. Für die das Integral (4) von § 14 betreffenden Bedingung haben wir das schon oben bemerkt.[1] Was (9) anlangt, so nehmen wir an, diese Bedingung sei nicht erfüllt.

Es gibt dann einen Wert $h > 0$ und eine sei es in $< h, +\infty)$, sei es in $(-\infty, -h >$ liegende Punktfolge $u_1, u_2, \ldots, u_n, \ldots$, für die die entsprechende der beiden Beziehungen:

$$\int_{u_n}^{+\infty} (\varphi^{(m)}(u))^2 \, du = \frac{A_n}{u_n^{2m+1}}; \quad \int_{-\infty}^{u_n} (\varphi^{(m)}(u))^2 \, du = \frac{A_n}{|u_n|^{2m+1}}$$

mit:

$$\lim_{n=\infty} A_n = +\infty$$

gilt. Nehmen wir etwa ersteren Fall an und setzen:

$$u_n = k_n \cdot h, \tag{13}$$

so haben wir $k_n \geqq 1$ und:

$$\int_{u_n}^{+\infty} (\varphi^{(m)}(u))^2 \, du = k_n \int_{\frac{u_n}{k_n}}^{+\infty} (\varphi^{(m)}(k_n u))^2 \, du = \frac{A_n}{u_n^{2m+1}},$$

oder wegen (13):

$$k_n^{2m+2} \int_h^{+\infty} (\varphi^{(m)}(k_n u))^2 \, du = \frac{A_n}{h^{2m+1}},$$

es wäre also für $\psi^{(m)}(u, k_n, h)$ Bedingung 1. von Satz XVI *a* für die Klasse $\mathfrak{F}_2$ nicht erfüllt.

Damit ist Satz XXX nachgewiesen. In analoger Weise (man vergleiche den Beweis von Satz XXVI *a*) zeigt man:

XXX *a*. »Es genüge $\varphi(u)$ allen Voraussetzungen von Satz XXVIII. Damit in jedem Punkte x des beliebigen Intervalles (a, b), in dem die in $< a, b >$ zur Klasse $\mathfrak{F}_2$ gehörige Funktion f eine endliche Ableitung m-ter Ordnung besitzt, die Beziehungen (7), (8) gelten, ist notwendig und hinreichend, daß das Integral (4) von § 14 existiere und daß folgende Bedingung erfüllt sei: zu jedem $h > 0$ gehört ein A, so daß für $u \geqq h$:

$$\int_{-\infty}^{-u} (\varphi^{(m)}(u))^2 \, du < \frac{A}{u^{2m+1}}; \quad \int_{u}^{+\infty} (\varphi^{(m)}(u))^2 \, du < \frac{A}{u^{2m+1}}. \tag{14}$$

Sind diese Bedingungen erfüllt, so gelten die Beziehungen (7), (8) gleichmäßig in jedem Teilintervalle $< a', b' >$ von (a, b), in dem f m-mal stetig differenzierbar ist. — Ist $\varphi(u)$ eine gerade Funktion, so kann die Ableitung $f^{(i)}(x)$ ersetzt werden durch die verallgemeinerte Ableitung $\dot{f}^{(i)}(x)$.«

[1] Anmerkung auf p. 61 [645].

XXXI. In Satz XXX kann die Klasse $\mathfrak{F}_2$ ersetzt werden durch die Klasse $\mathfrak{F}_1$, wenn Bedingung (9) ersetzt wird durch die Bedingung: zu jedem $h > 0$ gibt es ein A, so daß (abgesehen von Nullmengen):

(15) $$|\varphi^{(m)}(u)| < \frac{A}{|u|^{m+1}} \quad \text{für } |u| \geqq h.$$

Bedingung (15) ist hinreichend. Es genügt, gemäß Satz XIX *a* für die Klasse $\mathfrak{F}_1$, nachzuweisen: Zu jedem $h > 0$ gibt es ein M, so daß (abgesehen von Nullmengen) für alle $k \geqq 1$:

$$|\psi(u, k, h)| < M; \quad |\psi^{(i)}(u, k, h)| < M \quad (i = 1, 2, \ldots, m).$$

Aus (15) und den Beziehungen (9) von § 14 folgt, daß es zu jedem $h > 0$ ein B gibt, so daß für $|u| \geqq h$:

$$|\varphi(u)| < \frac{B}{|u|}; \quad |\varphi^{(i)}(u)| < \frac{B}{|u|^{i+1}} \quad (i = 1, 2, \ldots, m).$$

Infolgedessen ist für $|u| \geqq h$ und $k \geqq 1$:

$$\left|\frac{k}{\omega}\varphi(ku)\right| < \frac{B}{\omega \cdot h}; \quad \left|\frac{k^{i+1}}{\omega}\varphi^{(i)}(ku)\right| < \frac{B}{\omega \cdot h^{i+1}} \quad (i = 1, 2, \ldots, m),$$

womit Bedingung 1. von XIX *a* für $\mathfrak{F}_1$ nachgewiesen ist.

Bedingung (15) ist notwendig. Wäre sie nicht erfüllt, so hieße das: es gibt ein $h > 0$ und eine sei es in $< h, +\infty)$, sei es in $(-\infty, -h >$ gelegene Folge von Mengen $\mathfrak{M}_1, \mathfrak{M}_2, \ldots, \mathfrak{M}_n, \ldots$ mit von 0 verschiedenem Inhalte, derart, daß in den Punkten von $\mathfrak{M}_n$:

$$|\varphi^{(m)}(u)| > \frac{n}{|u|^{m+1}}$$

ist. Nehmen wir etwa an, diese Mengen liegen in $< h, +\infty)$. Es gibt dann auch eine Folge von Punkten u_n ($\geqq h$), so daß der in $< u_n, u_n + 1 >$ liegende Teil $\overline{\mathfrak{m}}_n$ von $\mathfrak{M}_n$ nicht den Inhalt 0 hat. Setzen wir:

$$u_n = k_n \cdot h,$$

so ist $k_n \geqq 1$, und es wird in den Punkten der Menge $\mathfrak{M}_n^*$, die aus $\overline{\mathfrak{M}}_n$ durch Ähnlichkeitstransformation im Verhältnisse $k_n : 1$ entsteht:

$$k_n^{m+1}\varphi^{(m)}(k_n u) > \frac{n}{u^{m+1}} \geqq \frac{n}{\left(\frac{1}{k_n}(u_n + 1)\right)^{m+1}} = \frac{n}{\left(h + \frac{1}{k_n}\right)^{m+1}} \geqq \frac{n}{(h+1)^{m+1}},$$

so daß also Bedingung 1. von Satz XVI für die Klasse $\mathfrak{F}_1$ nicht erfüllt wäre.

Analog beweist man (vergleiche den Beweis von Satz XXVII *a*):

XXXI *a*. »In Satz XXX *a* kann die Klasse $\mathfrak{F}_2$ ersetzt werden durch die Klasse $\mathfrak{F}_1$, wenn Bedingung (14) ersetzt wird durch Bedingung (15).«

§ 16. Singuläre Integrale vom Poisson'schen Typus.

Wir betrachten nun einen Typus singulärer Integrale, den wir, weil in ihm das aus der Potentialtheorie bekannte Poisson'sche Integral als Spezialfall enthalten ist, als den Poisson'schen Typus bezeichnen wollen.

Sei $\varphi(u)$ eine in $(-l, l)$ definierte, im Nullpunkt verschwindende Funktion der Gestalt:

(1) $$\varphi(u) = \alpha |u|^p + \omega(u) \cdot |u|^p,$$

worin:

(2) $$\alpha > 0; \quad p > 1; \quad \lim_{u=0} \omega(u) = 0.$$

In jedem den Nullpunkt nicht enthaltenden Teilintervalle $< \alpha, \beta >$ von $(-l, l)$ sei die untere Grenze von $\varphi(u)$ eine positive Zahl.

Sei $0 < \gamma < l$. Wir wollen für $k > 0$ eine Funktion $c(k)$ so bestimmen, daß:

(3) $$\lim_{k=+\infty} c(k) \int_{-\gamma}^{\gamma} \frac{du}{1 + k\,\varphi(u)} = 1$$

wird. — Sei $h > 0$ beliebig gegeben. Wir können dann, zufolge (1) und (2), $\gamma' > 0$ so klein wählen, daß in $< -\gamma', \gamma' >$:

(4) $$(\alpha - h)\,|u|^p \leqq \varphi(u) \leqq (\alpha + h)\,|u|^p,$$

und somit auch:

(5) $$2\int_0^{\gamma'} \frac{du}{1 + k\cdot(\alpha + h)\,u^p} \leqq \int_{-\gamma'}^{\gamma'} \frac{du}{1 + k\cdot\varphi(u)} \leqq 2\int_0^{\gamma'} \frac{du}{1 + k\cdot(\alpha - h)\,u^p}$$

Dies führt uns auf die Aufgabe, eine asymptotische Auswertung des Integrales:

$$I(k, \beta, \sigma, p) = \int_0^{\sigma} \frac{du}{1 + k\cdot\beta u^p}$$

vorzunehmen. Führen wir die Substitution:

$$v = k^{\frac{1}{p}} \cdot \beta^{\frac{1}{p}} \cdot u \qquad \tau = \beta^{\frac{1}{p}} \cdot \sigma$$

aus, so erhalten wir:

$$I(k, \beta, \sigma, p) = k^{-\frac{1}{p}}\, \beta^{-\frac{1}{p}} \int_0^{k^{\frac{1}{p}}\cdot\tau} \frac{dv}{1 + v^p}.$$

Setzen wir noch:

(6) $$\Omega(p) = \int_0^{+\infty} \frac{du}{1 + u^p},$$

so haben wir also:

(7) $$\lim_{k=+\infty} k^{\frac{1}{p}}\, I(k, \beta, \sigma, p) = \beta^{-\frac{1}{p}}\, \Omega(p).$$

Nun kann (5) auch geschrieben werden:

$$2\,k^{\frac{1}{p}}\, I(k, \alpha + h, \gamma', p) \leqq k^{\frac{1}{p}} \int_{-\gamma'}^{\gamma'} \frac{du}{1 + k\,\varphi(u)} \leqq 2\,k^{\frac{1}{p}}\, I(k, \alpha - h, \gamma', p).$$

Wegen (7) ist also, wenn $\eta > 0$ beliebig gegeben wird, für alle $k \geqq k_0$:

$$2\,(\alpha + h)^{-\frac{1}{p}}\, \Omega(p) - \eta \leqq k^{\frac{1}{p}} \int_{-\gamma'}^{\gamma'} \frac{du}{1 + k\cdot\varphi(u)} \leqq 2\,(\alpha - h)^{-\frac{1}{p}}\, \Omega(p) + \eta.$$

Sei nun γ eine beliebige Zahl aus $(0, l)$ und es sei γ' $(< \gamma)$ entsprechend (4) gewählt. Da die untere Grenze von $\varphi(u)$ in $< \gamma', \gamma >$ sowie in $< -\gamma, -\gamma' >$ positiv (etwa $= \vartheta > 0$) ist, haben wir:

$$\left| \int_{-\gamma}^{\gamma} \frac{du}{1 + k\,\varphi(u)} - \int_{-\gamma'}^{\gamma'} \frac{du}{1 + k\,\varphi(u)} \right| \leqq \frac{2\,(\gamma - \gamma')}{1 + k\vartheta}.$$

Wir haben also endlich für alle hinlänglich großen k:

$$2\,(\alpha + h)^{-\frac{1}{p}}\,\Omega(p) - k^{\frac{1}{p}}\,\frac{2\gamma}{1 + k\vartheta} - \eta \leqq k^{\frac{1}{p}} \int_{-\gamma}^{\gamma} \frac{du}{1 + k\cdot\varphi(u)} \leqq 2\,(\alpha - h)^{-\frac{1}{p}}\,\Omega(p) + k^{\frac{1}{p}}\,\frac{2\gamma}{1 + k\cdot\vartheta} + \eta,$$

und da hierin h und η beliebig waren, so ist das gleichbedeutend mit:

$$\lim_{k=+\infty} k^{\frac{1}{p}} \int_{-\gamma}^{\gamma} \frac{du}{1+k\cdot\varphi(u)} = 2\cdot\alpha^{-\frac{1}{p}}\,\Omega(p).$$

Wir sehen also, daß wir in (3) wählen können:

$$c(k) = \frac{1}{2}\,\alpha^{\frac{1}{p}}\,\frac{1}{\Omega(p)}\cdot k^{\frac{1}{p}}.$$

Wir können nun den Satz beweisen:

XXXII. Sei $\varphi(u)$ eine in $(-l, l)$ gegebene Funktion der Form (1), (2), deren untere Grenze in jedem den Nullpunkt nicht enthaltenden Teilintervalle $<\alpha, \beta>$ von $(-l, l)$ positiv ist. Mit $\Omega(p)$ werde der Wert (6) bezeichnet. Dann gilt für jede in $<a, a+l>$ zur Klasse $\mathfrak{F}_1$ gehörige Funktion f, die im Punkte x von $(a, a+l)$ stetig ist, die Formel:

(8)
$$f(x) = \frac{\alpha^{\frac{1}{p}}}{2\,\Omega(p)}\cdot \lim_{k=+\infty} k^{\frac{1}{p}} \int_a^{a+l} f(\xi)\,\frac{d\xi}{1+k\varphi(\xi-x)}\,.$$

Diese Formel gilt gleichmäßig in jedem Teilintervalle $<a', b'>$ von $(a, a+l)$, in dessen sämtlichen Punkten f stetig ist.

Zum Beweise haben wir uns nur auf Satz XVIII zu berufen. Ist $<\alpha, \beta>$ ein den Nullpunkt nicht enthaltendes Teilintervall von $(-l, l)$, so haben wir nach Voraussetzung in $<\alpha, \beta>$:

$$\varphi(u) > \vartheta > 0$$

und somit, wenn:

$$\varphi(u, k) = \frac{\alpha^{\frac{1}{p}}}{2\,\Omega(p)}\cdot k^{\frac{1}{p}}\,\frac{1}{1+k\varphi(u)}$$

gesetzt wird:

(9)
$$0 < \varphi(u, k) < \frac{\alpha^{\frac{1}{p}}}{2\,\Omega(p)}\,\frac{k^{\frac{1}{p}}}{1+k\vartheta},$$

so daß:

$$\lim_{k=+\infty} \varphi(u, k) = 0$$

gleichmäßig in $<\alpha, \beta>$ ist, wie Voraussetzung (1) von Satz XVIII verlangt.

Daß Bedingung 3. und 4. von XVIII erfüllt sind, folgt aus der wegen (3) gültigen Beziehung:

$$\lim_{k=+\infty} \int_{-\gamma}^{\gamma} \varphi(u, k)\,du = \lim_{k=+\infty} \int_{-\gamma}^{\gamma} |\varphi(u, k)|\,du = 1.$$

Einen bekannten Spezialfall erhält man, indem man für $(-l, l)$ das Intervall $(-2\pi, 2\pi)$ wählt und:

$$\varphi(u) = 1 - \cos u$$

setzt. Man erhält so:

$$f(x) = \frac{1}{\pi\cdot\sqrt{2}} \lim_{k=+\infty} k^{\frac{1}{2}} \int_a^{a+2\pi} f(\xi)\,\frac{d\xi}{1+k(1-\cos(\xi-x))}$$

Setzt man hierin:

$$k = \frac{2r}{(1-r)^2} \qquad (0 < r < 1)$$

und multipliziert unter dem Limeszeichen mit dem Faktor:

$$\frac{1+r}{2\sqrt{r}}$$

dessen $\lim\limits_{r=1}$ gleich 1 ist, so erhält man die bekannte Poisson'sche Formel:

$$f(x) = \frac{1}{2\pi} \lim_{r=1-0} \int_0^{2\pi} f(\xi) \frac{1-r^2}{1-2r\cos(\xi-x)+r^2}\, d\xi.$$

Die Übertragung von Satz XXXII auf unendliche Intervalle gelingt leicht durch Berufung auf Satz XVIII *a*:

XXXII *a*. »Sei $\varphi(u)$ eine für alle reellen u definierte Funktion der Gestalt (1), (2), deren untere Grenze in jedem endlichen, den Nullpunkt nicht enthaltenden Intervalle positiv ist. Mit $\Omega(p)$ werde der Wert (6) bezeichnet. Es gilt dann die Formel:

$$f(x) = \frac{\alpha^{\frac{1}{p}}}{2\Omega(p)} \lim_{k=+\infty} k^{\frac{1}{p}} \int_{-\infty}^{+\infty} f(\xi) \frac{d\xi}{1+k\varphi(\xi-x)} \tag{10}$$

in jedem Punkte x, in dem f stetig ist: *a)* für alle Funktionen, die in $(-\infty, +\infty)$ zur Klasse $\mathfrak{F}_1$ gehören wenn:

$$\lim_{u=-\infty} \varphi(u) > 0; \quad \lim_{u=+\infty} \varphi(u) > 0. \tag{11}$$

b) für alle Funktionen, die in $(-\infty, +\infty)$ zur Klasse $\mathfrak{F}_2$ gehören, wenn (11) erfüllt ist und:

$$\int_{-\infty}^{+\infty} \frac{du}{(1+\varphi(u))^2} \tag{12}$$

existiert; *c)* für alle Funktionen, die in $(-\infty, +\infty)$ zur Klasse $\mathfrak{F}_3$ gehören, wenn (11) erfüllt ist und:

$$\int_{-\infty}^{+\infty} \frac{du}{1+\varphi(u)} \tag{13}$$

existiert. In allen diesen Fällen gilt (10) gleichmäßig in jedem endlichen Intervalle $< a', b' >$, in dessen sämtlichen Punkten f stetig ist.«

In der Tat, ist Bedingung (11) erfüllt, so gibt es zu jedem $h > 0$ ein $\vartheta > 0$, so daß in $(-\infty, -h >$ und in $< h, +\infty)$:

$$\varphi(u) > \vartheta.$$

In diesen Intervallen ist daher Ungleichung (9) erfüllt, und somit $|\varphi(u, k)|$ geschränkt für alle k. Damit ist Bedingung 1. von XVIII *a* für die Klasse $\mathfrak{F}_1$ nachgewiesen.

Ferner hat man in den genannten Intervallen:

$$\frac{1+\varphi(u)}{1+k\varphi(u)} < \frac{1+\varphi(u)}{k\varphi(u)} < \frac{1}{k}\,\frac{\vartheta+1}{\vartheta}.$$

Existiert also das Integral (12), so hat man:

$$\int_h^{+\infty} (\varphi(u, k))^2\, du = \frac{\alpha^{\frac{2}{p}}}{4(\Omega(p))^2} k^{\frac{2}{p}} \int_h^{+\infty} \frac{du}{(1+k\varphi(u))^2} \leqq$$

$$\leqq \frac{\alpha^{\frac{2}{p}}}{4(\Omega(p))^2} k^{2\left(\frac{1}{p}-1\right)} \left(\frac{\vartheta+1}{\vartheta}\right)^2 \int_h^{+\infty} \frac{du}{(1+\varphi(u))^2}.$$

Es sind also gewiß die Integrale:

$$\int_h^{+\infty} (\varphi(u, k))^2\, du \quad \text{und} \quad \int_{-\infty}^{-h} (\varphi(u, k))^2\, du$$

geschränkt für alle $k \geqq 1$. Damit ist Bedingung 1. von XVIII *a* für die Klasse $\mathfrak{F}_2$ nachgewiesen.

Ebenso beweist man, wenn das Integral (13) existiert, die Ungleichung:

$$\int_h^{+\infty} \varphi(u, k)\, du \leqq \frac{\alpha^{\frac{1}{p}}}{2\,\Omega(p)} k^{\frac{1}{p}-1} \frac{\vartheta+1}{\vartheta} \int_h^{+\infty} \frac{du}{1+\varphi(u)},$$

woraus man entnimmt:

$$\lim_{k=+\infty} \int_h^{+\infty} \varphi(u, k)\, du = 0; \quad \lim_{k=+\infty} \int_{-\infty}^{-h} \varphi(u, k)\, du = 0.$$

Daraus aber kann man, wie wir nun schon wiederholt gesehen haben, weiter schließen, daß für jede Folge von Zahlen k_n mit $\lim_{n=\infty} k_n = +\infty$ der zum Kerne:

$$\varphi(u, n) = \frac{\alpha^{\frac{1}{p}}}{2\Omega(p)} k_n^{\frac{1}{p}} \frac{1}{1+k_n\varphi(u)}$$

gehörige Kern $\psi(u, n, h)$ für jedes $h > 0$ in $(-\infty, +\infty)$ der Bedingung 1. von Satz III *a* genügt, so daß Bedingung 1. von XVIII für die Klasse $\mathfrak{F}_3$ nachgewiesen ist. Damit ist Satz XXXII *a* bewiesen.

Als einfachsten Spezialfall erwähnen wir:[1]

$$\varphi(u) = u^2.$$

Man erhält so, indem man noch

$$\rho^2 = \frac{1}{k}$$

setzt, die in jedem Stetigkeitspunkte einer Funktion *f*, die in $(-\infty, +\infty)$ einer der drei Klassen $\mathfrak{F}_1$, $\mathfrak{F}_2$, $\mathfrak{F}_3$ angehört, gültige Formel:

(14) $$f(x) = \frac{1}{\pi} \lim_{\rho=+0} \int_{-\infty}^{+\infty} f(\xi) \frac{\rho}{\rho^2+(\xi-x)^2}\, d\xi.$$

§ 17. Konvergenz an Unstetigkeitsstellen und Differenziation.

Wir wenden auch auf unseren jetzigen Typus singulärer Integrale Satz XIV und XV, beziehungsweise XIV *a* und XV *a* an. Wir erhalten:

XXXIII. Es genüge $\varphi(u)$ außer den Voraussetzungen von Satz XXXII (beziehungsweise den Voraussetzungen von Satz XXXII*a* für die Klasse $\mathfrak{F}_i$) noch folgenden Bedingungen: $\varphi(u)$ besitzt in einer Umgebung des Nullpunktes eine absolut stetige $(m-1)$-te Ableitung und es sind in dieser Umgebung, abgesehen von einer Nullmenge, die Ungleichungen erfüllt:

(1) $$|\varphi^{(i)}(u)| < A\,|u|^{p-i} \quad (i = 1, 2, \ldots, m);$$

dann gilt Gleichung (8) (beziehungsweise (10)), von § 16 in jedem Punkte x von $(a, a+l)$ (beziehungsweise in jedem Punkte x), in dem die zur Klasse $\mathfrak{F}_1$ (beziehungsweise zur Klasse $\mathfrak{F}_i$) gehörige Funktion f m-te Ableitung ihres m-fach iterierten unbestimmten Integrales ist. — Ist außerdem in einer Umgebung des Nullpunktes: $\varphi(-u) = \varphi(u)$, so gilt (8) (beziehungsweise (10)) von § 16 in jedem Punkte x von $(a, a+l)$ (beziehungsweise in jedem Punkte x), in dem f verallgemeinerte m-te Ableitung seines m-fach itererierten unbestimmten Integrales ist.

Wir haben nachzuweisen, daß der Kern:

(2) $$\varphi(\xi, x, k) = \frac{\alpha^{\frac{1}{p}}}{2\,\Omega(p)} k^{\frac{1}{p}} \frac{1}{1+k\,\varphi(\xi-x)}$$

[1] Es ist dies gleichzeitig ein Spezialfall des Weierstraß'schen Typus.

allen Voraussetzungen und Bedingungen von Satz XIV genügt. Beachten wir zu dem Zwecke, daß die i-te Ableitung von $\frac{1}{1+k\varphi(u)}$ eine Summe aus einer endlichen Anzahl von Summanden der Form:

$$\frac{k^j (\varphi'(u))^{j_1} (\varphi''(u))^{j_2} \ldots (\varphi^{(i)}(u))^{j_i}}{(1+k\varphi(u))^{j+1}} \tag{3}$$

ist, wo für die Exponenten die Gleichheiten gelten:

$$j_1 + 2j_2 + \ldots + ij_i = i \tag{4}$$

$$j_1 + j_2 + \ldots + j_i = j. \tag{5}$$

In der Tat gilt dies für $i = 1$:

$$\frac{d}{du}\frac{1}{1+k\varphi(u)} = -\frac{k\cdot\varphi'(u)}{(1+k\varphi(u))^2},$$

und kann allgemein durch vollständige Induktion bewiesen werden, die genau so verläuft, wie beim analogen Beweise in § 12.

Zunächst sehen wir, daß in einer Umgebung der Stelle x für $\xi \neq x$ die Beziehung (1) von § 7:

$$\lim_{k=+\infty} \frac{\partial^i}{\partial \xi^i} \varphi(\xi, x, k) = 0 \quad (i = 1, 2, \ldots, m-1) \tag{6}$$

besteht: in der Tat, zufolge (2) und (3) enthält jeder Summand von $\frac{\partial^i}{\partial \xi^i} \varphi(\xi, x, k)$ im Zähler den Faktor $k^{j+\frac{1}{p}}$, während der Nenner groß wird wie $(\varphi(u)\,k)^{j+1}$. Da $\varphi(u) \neq 0$ und $\frac{1}{p} < 1$ ist, folgt unmittelbar (6).

Es bleibt noch nachzuweisen, daß Bedingung 3. von Satz XIV erfüllt ist, oder, was dasselbe ist, daß für ein hinlänglich kleines $\gamma > 0$ eine Ungleichung besteht:

$$k^{\frac{1}{p}} \int_{-\gamma}^{\gamma} u^m \left| \frac{d^m}{du^m} \frac{1}{1+k\varphi(u)} \right| du < N \quad \text{für alle } k \geqq 1\,. \tag{7}$$

Nun besteht

$$\frac{d^m}{du^m} \frac{1}{1+k\varphi(u)}$$

aus einer endlichen Anzahl Summanden der Form (3). Wegen (1) von § 16 ist hierin, wenn $h > 0$ beliebig und dazu $\gamma > 0$ hinlänglich klein gewählt wurde:

$$1 + k\varphi(u) > 1 + k(\alpha - h)|u|^p,$$

und somit ist, wegen der Ungleichungen (1), das Integral in (7) kleiner als eine endliche Anzahl Summanden der Form (C bedeutet eine Konstante):

$$C k^{\frac{1}{p}+j} \int_0^{\gamma} \frac{u^{m+j_1(p-1)+j_2(p-2)+\ldots+j_m(p-m)}}{(1+k(\alpha-h)\cdot u^p)^{j+1}} du;$$

indem wir die Substitution $k^{\frac{1}{p}}.u = v$ vornehmen und (4) und (5) beachten (für $i = m$), sehen wir, daß hierin alle Faktoren k sich vollständig wegheben und dieser Ausdruck übergeht in:

$$C \int_0^{k^{\frac{1}{p}}\cdot\gamma} \frac{v^{pj}}{(1+(\alpha-h)\,v^p)^{j+1}} dv;$$

wegen $p > 1$ aber existiert das Integral:

$$\int_0^{+\infty} \frac{v^{pj}}{(1+(\alpha-h)\,v^p)^{j+1}} dv,$$

womit (7) nachgewiesen ist. Der Beweis von Satz XXXIII ist damit beendet.

Für das Poisson'sche Integral wurde der einfachste Fall dieses Satzes ($m = 1$) zuerst von P. Fatou[1] bewiesen.

Durch Anwendung von Satz XIX erhalten wir den Satz:

XXXIV. Es genüge $\varphi(u)$ außer den Voraussetzungen von Satz XXXII nach folgenden Bedingungen: $\varphi(u)$ besitzt eine in jedem Teilintervalle $< \alpha, \beta >$ von $(-l, l)$ absolut stetige $(m-1)$-te Ableitung und es sind in jedem solchen Teilintervalle (abgesehen von einer Nullmenge) Ungleichungen der Gestalt:

(8) $$|\varphi^{(i)}(u)| < A\,|u|^{p-i} \qquad (i = 1, 2, \ldots, m)$$

erfüllt. Dann gilt in jedem Punkte x von $(a, a+l)$, in dem die in $< a, a+l >$ zur Klasse $\mathfrak{F}_1$ gehörige Funktion f eine endliche Ableitung m-ter Ordnung besitzt, die Formel:

(9) $$f^{(i)}(x) = \frac{\alpha^{\frac{1}{p}}}{2\,\Omega(p)} \lim_{k=+\infty} k^{\frac{1}{p}} \int_a^{a+l} f(\xi)\, \frac{\partial^i}{\partial x^i} \frac{1}{1+k\,\varphi(\xi-x)}\, d\xi \qquad (i = 1, 2, \ldots, m).$$

Diese Beziehung gilt gleichmäßig in jedem Teilintervalle $< a', b' >$ von $(a, a+l)$, in dem f m-mal stetig differenzierbar ist. — Genügt $\varphi(u)$ in einer Umgebung des Nullpunktes der Beziehung $\varphi(-u) = \varphi(u)$, so gilt (9) auch für die verallgemeinerte i-te Ableitung $\mathring{f}^{(i)}(x)$.

Beim Beweise hat man nur zu beachten, daß in jedem den Nullpunkt nicht enthaltenden Teilintervalle $< \alpha, \beta >$ von $(-l, l)$ die Beziehung:

$$\lim_{k=+\infty} k^{\frac{1}{p}} \frac{d^{m-1}}{du^{m-1}} \frac{1}{1+k\,\varphi(u)} = 0$$

gleichmäßig gilt. In der Tat gilt nach den in Satz XXXII über $\varphi(u)$ gemachten Voraussetzungen in $< \alpha, \beta >$ eine Ungleichung:

$$\varphi(u) > \vartheta > 0,$$

woraus nach der oben besprochenen Form von

$$\frac{d^{m-1}}{du^{m-1}} \frac{1}{1+k\,\varphi(u)}$$

die Behauptung unmittelbar folgt. Damit ist die Voraussetzung (3) vom Satz XIX erwiesen. Daß die übrigen Bedingungen dieses Satzes gelten ist evident.

Im speziellen Falle des Poisson'schen Integrales wurden die Behauptungen unseres Satzes bewiesen von Ch. J. de la Vallée-Poussin.[2]

Für unendliche Intervalle erhalten wir durch Berufung auf Satz XIX a:

XXXIVa. »Es genüge $\varphi(u)$ den zu Beginn von Satz XXXIIa gemachten Voraussetzungen.

Ferner besitze $\varphi(u)$ eine in jedem endlichen Intervalle absolut stetige $(m-1)$ te Ableitung und es seien in $(-\infty, +\infty)$ Ungleichungen der Form:

(10) $$\varphi(u) > \beta\,|u|^p \qquad (\beta > 0),$$

(11) $$\varphi^{(i)}(u) < A\,|u|^{p-i} \qquad (i = 1, 2, \ldots, m)$$

erfüllt. Dann gilt in jedem Punkte x, in dem die in $(-\infty, +\infty)$ zu einer der Klassen $\mathfrak{F}_1, \mathfrak{F}_2, \mathfrak{F}_3$ gehörige Funktion f eine endliche Ableitung m-ter Ordnung besitzt, die Formel:

---

[1] Acta Math., Bd. 30, p. 345, 373. Vgl. auch H. Lebesgue a. a. O., p. 87, und W. Groß, Sitzungsber. Akad. Wien. Abt. IIa, Bd. 124, p. 1020.

[2] Acad. Bruxelles, Bull. Classe des Sciences 1908, p. 245.

(12)
$$f^{(i)}(x) = \frac{\alpha^{\frac{1}{p}}}{2\,\Omega(p)} \lim_{k=+\infty} k^{\frac{1}{p}} \int_{-\infty}^{+\infty} f(\xi)\, \frac{\partial^i}{\partial x^i}\, \frac{1}{1+k\,\varphi(\xi-x)}\, d\xi \quad (i=1,2,\ldots,m).$$

Sie gilt gleichmäßig in jedem endlichen Intervalle $< a', b' >$, in dem f m-mal stetig differenzierbar ist. — Genügt $\varphi(u)$ in einer Umgebung des Nullpunktes der Beziehung $\varphi(-u) = \varphi(u)$, so gilt (12) auch für die verallgemeinerte i-te Ableitung $\mathring{f}^{(i)}(x)$.«

Es genügt, nach dem schon Bewiesenen, zu zeigen, daß Bedingung 1. von Satz XIXa erfüllt ist.

Zeigen wir dies zunächst für die Klasse $\mathfrak{F}_1$. Wir setzen:

$$\psi(u, k, h) = \begin{cases} \dfrac{\alpha^{\frac{1}{p}}}{2\,\Omega(p)} \cdot k^{\frac{1}{p}}\, \dfrac{1}{1+k\,\varphi(u)} & \text{außerhalb } < -h, h > \\ 0 & \text{in } < -h, h >. \end{cases}$$

und haben nach (10):

$$|\psi(u, k, h)| < \frac{\alpha^{\frac{1}{p}}}{2\,\Omega(p)} \cdot \frac{k^{\frac{1}{p}}}{1+k\beta\,.\,h^p},$$

also ist $|\psi(u, k, h)|$ geschränkt für alle u und alle k.

Ferner haben wir, unter Benützung von (3), (4), (5), (10) und (11), wenn durch das Symol Σ eine Summation über eine endliche Zahl von Summanden angedeutet wird:

(13)
$$|\psi^{(i)}(u, k, h)| < \sum C \frac{\alpha^{\frac{1}{p}}}{2\,\Omega(p)} \cdot k^{\frac{1}{p}}\, \frac{k^j\, |u|^{pj-i}}{(1+\beta\cdot k\, |u|^p)^{j+1}} \quad (i=1, 2, \ldots, m),$$

und indem wir $k\,|u|^p = v$ schreiben, haben wir außerhalb $< -h, h >$:

$$|\psi^{(i)}(u, k, h)| < \sum C \frac{\alpha^{\frac{1}{p}}}{2\,\Omega(p)}\, h^{-(i+1)}\, \frac{v^{j+\frac{1}{p}}}{(1+\beta\, v)^{j+1}}\,.$$

Es ist also, da $\frac{1}{p} < 1$ ist, auch $\psi^{(i)}(u, k, h)$ $(i = 1, 2, \ldots, m)$ geschränkt für alle u und alle k. — Damit ist Bedingung 1. von XIXa für die Klasse $\mathfrak{F}_1$ nachgewiesen.

Wir gehen sogleich über zur Klasse $\mathfrak{F}_3$, da für die Klasse $\mathfrak{F}_2$ der Nachweis ganz analog ist: Aus (13) entnehmen wir sofort:

$$\int_{-\infty}^{+\infty} |\psi^{(i)}(u, k, h)|\, du < \sum C \frac{\alpha^{\frac{1}{p}}}{\Omega(p)}\, k^{\frac{1}{p}+j} \int_h^{+\infty} \frac{u^{pj-i}}{(1+\beta\cdot k\cdot u^p)^{j+1}}\, du,$$

oder vermöge der Substitution: $v = k^{\frac{1}{p}} \cdot u$

(14)
$$\int_{-\infty}^{+\infty} |\psi^{(i)}(u, k, h)|\, du < \sum C \frac{\alpha^{\frac{1}{p}}}{\Omega(p)}\, k^{\frac{i}{p}} \int_{k^{\frac{1}{p}}.h}^{+\infty} \frac{v^{pj-i}}{(1+\beta\, v^p)^{j+1}}\, dv.$$

Nun haben wir:

$$\frac{v^{pj-i}}{(1+\beta\, v^p)^{j+1}} < \frac{1}{\beta^{j+1}}\, v^{-(i+p)};$$

infolgedessen:

$$\int_{k^{\frac{1}{p}}.h}^{+\infty} \frac{v^{pj-i}}{(1+\beta\, v^p)^{j+1}}\, dv < \frac{1}{\beta^{j+1}}\, \frac{1}{i+p-1} \cdot h^{-(i+p-1)}\, k^{-1-\frac{i-1}{p}},$$

und, indem man dies in (14) einführt und berücksichtigt, daß $p > 1$ ist:

$$\int_{-\infty}^{+\infty} |\psi^{(i)}(u, k, h)| \, du < c \cdot k^{-\eta} \qquad (\eta > 0).$$

Wir haben also:

$$\lim_{k=+\infty} \int_{-\infty}^{+\infty} |\psi^{(i)}(u, k, h)| \, du = 0,$$

woraus zusammen mit der ähnlich zu bestätigenden Beziehung:

$$\lim_{k=+\infty} \int_{-\infty}^{+\infty} |\psi(u, k, h)| \, du = 0$$

nach einem schon wiederholt durchgeführten Schlusse folgt, daß Bedingung 1. von XIX*a* für die Klasse $\mathfrak{F}_3$ erfüllt ist.

Wenden wir die Sätze XXXIII und XXXIV *a* auf den Spezialfall $\varphi(u) = u^2$ an, so sehen wir: Formel (14) von § 16 gilt für jede in $(-\infty, +\infty)$ einer der Klassen $\mathfrak{F}_1$, $\mathfrak{F}_2$, $\mathfrak{F}_3$ angehörende Funktion f, in jedem Punkte, in dem f verallgemeinerte m-te Ableitung seines m-fach iterierten unbestimmten Integrales ist; und in jedem Punkte, in dem f eine verallgemeinerte m-te Ableitung $\overset{*}{f}^{(m)}(x)$ besitzt, wird sie (für jedes m) dargestellt durch:

$$\overset{*}{f}^{(m)}(x) = \frac{1}{\pi} \lim_{\rho=+0} \rho \cdot \int_{-\infty}^{+\infty} f(\xi) \cdot \frac{\partial^m}{\partial x^m} \cdot \frac{1}{\rho^2 + (\xi - x)^2} \, d\xi. \tag{15}$$

Nun ist bekanntlich:

$$\frac{\rho}{\rho^2 + u^2} = \int_0^{+\infty} e^{-\rho\lambda} \cos \lambda u \, d\lambda; \quad \frac{d^m}{du^m} \frac{\rho}{\rho^2 + u^2} = \int_0^{+\infty} e^{-\rho\lambda} \frac{d^m}{du^m} \cos \lambda u \, d\lambda \qquad (\rho > 0).$$

Wir können Formel (14) von § 16 also auch so schreiben:

$$f(x) = \frac{1}{\pi} \lim_{\rho=+0} \int_{-\infty}^{+\infty} f(\xi) \left\{ \int_0^{+\infty} e^{-\rho\lambda} \cos \lambda\xi \cdot \cos \lambda x \, d\lambda + \int_0^{+\infty} e^{-\rho\lambda} \sin \lambda\xi \cdot \sin \lambda x \, d\lambda \right\} d\xi. \tag{16}$$

Sie gilt in dieser Form für jede Funktion von $\mathfrak{F}_1$, $\mathfrak{F}_2$ oder $\mathfrak{F}_3$ in jedem Punkte, in dem f verallgemeinerte m-te Ableitung seines m-fach iterierten unbestimmten Integrales ist. Die Formel (15) besagt, daß in jedem Punkte, in dem f eine verallgemeinerte m-te Ableitung besitzt, diese aus (16) durch m unter den Integralzeichen auszuführende Differenziationen nach x gewonnen wird.

Gehört f speziell zu $\mathfrak{F}_1$, so existieren die Ausdrücke:

$$A(\lambda) = \frac{1}{\pi} \int_{-\infty}^{+\infty} f(\xi) \cos \lambda\xi \, d\xi; \quad B(\lambda) = \frac{1}{\pi} \int_{-\infty}^{+\infty} f(\xi) \sin \lambda\xi \, d\xi.$$

Dann kann (16) auch so geschrieben werden:

$$f(x) = \lim_{\rho=+0} \int_0^{+\infty} e^{-\rho\lambda} (A(\lambda) \cos \lambda x + B(\lambda) \sin \lambda x) \, d\lambda, \tag{17}$$

und stellt für das Fourier'sche Integraltheorem das Analogon zur Poisson'schen Summierung der Fourier'schen Reihe dar. Formel (17) gilt für jede in $(-\infty, +\infty)$ absolut integrierbare Funktion f an jeder Stelle, wo sie verallgemeinerte m-te Ableitung ihres m-fach iterierten unbestimmten Integrales ist. Wo eine verallgemeinerte m-te Ableitung von f existiert, wird sie aus (17) gewonnen, indem man unter dem Integralzeichen m-mal nach x differenziert.

ÜBER DIE DARSTELLUNG GEGEBENER FUNKTIONEN DURCH SINGULÄRE INTEGRALE

2. MITTEILUNG

VON

HANS HAHN

(BONN)

VORGELEGT IN DER SITZUNG AM 11. MAI 1916

In dieser zweiten Mitteilung setze ich meine Untersuchungen über singuläre Integrale fort, und wende sodann die erhaltenen Resultate auf die Theorie der orthogonalen Funktionensysteme an. Es wird zunächst als Maß der Approximation des singulären Integrales:

$$I_n(f, x) = \int_a^b f(\xi)\, \varphi(\xi, x, n)\, d\xi$$

an die darzustellende Funktion $f(x)$ das Integral über den absoluten Betrag des Fehlers:

$$R_n = \int_a^b |f(x) - I_n(f, x)|\, dx$$

betrachtet. Unter sehr allgemeinen Bedingungen gelingt der Nachweis, daß:

$$\lim_{n=\infty} R_n = 0$$

ist. Von diesem Satze werden Anwendungen auf einige Spezialfälle, insbesondere auf die üblichen Summationsverfahren der Fourier'schen Reihe gemacht.

Sodann wird die Giltigkeit des sogenannten Parseval'schen Theoremes näher untersucht, und zwar sogleich für beliebige orthogonale Funktionensysteme. Es handelt sich um die Gleichung:

$$\int_a^b f(x)\, g(x)\, dx = \sum_{\nu=1}^{\infty} f_\nu g_\nu,$$

wo f_ν, g_ν die »Fourier'schen Konstanten« von $f(x)$ und $g(x)$ in Bezug auf ein vollständiges, normiertes Orthogonalsystem bedeuten. Es ist bekannt, daß diese Gleichung gilt, wenn die Quadrate von $f(x)$ und $g(x)$ integrierbar sind. Ein anderer typischer Fall, indem das Integral des Produktes $f(x) \cdot g(x)$ sicher existiert, ist der, daß von den beiden Funktionen $f(x)$ und $g(x)$ die eine integrierbar, die andere geschränkt ist. Es gelingt eine sehr einfache, notwendige und hinreichende Bedingung anzugeben, der ein Orthogonalsystem genügen muß, damit auch in diesem Falle die fragliche Formel gelte. Eine ebenso einfache Bedingung ist notwendig und hinreichend dafür, daß die Formel immer gelte, wenn von den zwei Funktionen $f(x)$ und $g(x)$ die eine integrierbar, die andere von geschränkter Variation ist. Im Falle der trigonometrischen Reihen ist die erste dieser Bedingungen nicht erfüllt, wohl aber die zweite. Sodann werden Summationsverfahren für die Reihe:

$$\sum_{\nu=1}^{\infty} f_\nu g_\nu$$

entwickelt, die stets gegen:

$$\int_a^b f(x)\, g(x)\, dx$$

konvergieren, wenn von den zwei Funktionen $f(x)$ und $g(x)$ die eine integrierbar, die andere geschränkt ist. Im Falle der trigonometrischen Reihen entspricht jedem der bekannten Summationsverfahren auch ein analoges Summationsverfahren für die Reihe:

$$\sum_{\nu=1}^{\infty} f_\nu, g_\nu .$$

Wendet man auf das singuläre Integral $I_n(f, x)$ die Parseval'sche Formel an, so erhält man Summationsverfahren für die Reihenentwicklung von f nach den Funktionen eines Orthogonalsystemes. Versteht man unter $I_n(f, x)$ ein Integral, das gegen die m-te Ableitung von $f(x)$ konvergiert (ich habe solche Integrale in meiner 1. Mitteilung eingehend behandelt), so erhält man Prozesse, die aus der Reihenentwicklung von $f(x)$ nach den Funktionen des Orthogonalsystemes Ausdrücke herleiten, die gegen $f^{(m)}(x)$ konvergieren überall, wo diese Ableitung existiert. Die Anwendung auf trigonometrische Reihen ergibt neben den bekannten Summationsverfahren bei Benützung der wohl einfachstmöglichen Kerne eine Kette sich immer verschärfender Verfahren, deren zwei erste Glieder die sogenannten Riemann'schen Summationsverfahren sind; ferner erhält man Summationsverfahren für die durch m-maliges gliedweises Differenzieren der Fourier'schen Reihe gebildeten Reihen, die überall gegen $f^{(m)}(x)$ konvergieren, wo diese Ableitung existiert.

Es hätte wohl keine Schwierigkeiten, all dies auch für kontinuierliche Orthogonalschaaren durchzuführen, deren wichtigstes Beispiel die Theorie der Fourier'schen Integrale bildet. Ich habe mich auf eine Bemerkung aus der Theorie der Fourier'schen Integrale beschränkt, die das Annalogon des Poisson'schen Summationsverfahrens, das ich in meiner ersten Mitteilung angegeben habe, sowie das Analogon des Summationsverfahrens von de la Vallée-Poussin betrifft.

---

§ 1.

Unter Benützung eines Gedankens von W. H. Young[1] beweisen wir den Satz:

I. Sei $\varphi(\xi, x)$ eine im Rechtecke $a \leqq \xi \leqq b$, $a' \leqq x \leqq b'$ meßbare Funktion der zwei Veränderlichen x, ξ. Als Funktion von x sei sie für alle Werte ξ aus $< a, b >$, abgesehen von einer Nullmenge, integrierbar[2] in $< a', b' >$, und es gebe ein M, so dass, wieder abgesehen von einer Nullmenge:

$$\int_{a'}^{b'} |\varphi(\xi, x)|\, dx < M \quad \text{für alle } \xi \text{ von } < a, b >. \tag{1}$$

---

[1] Comptes rendus, Bd. 155 (1912), p. 30.

[2] Das Wort »integrierbar« wird stets im Sinne der Lebesgue'schen Integration gebraucht.

Ist dann $f(\xi)$ eine in $< a, b >$ integrierbare Funktion, so existiert das Integral:

(2) $$I(f, x) = \int_a^b f(\xi)\, \varphi(\xi, x)\, d\xi$$

für alle x von $< a', b' >$ abgesehen von einer Nullmenge, und ist eine in $< a', b' >$ integrierbare Funktion von x.

Wir beweisen den Satz zunächst unter der Annahme;

(3) $$\varphi(\xi, x) \geqq 0; \quad f(\xi) \geqq 0.$$

Bezeichnen wir mit $\Phi_\nu(\xi, x)$ die Funktion, die aus $f(\xi) \cdot \varphi(\xi, x)$ entsteht, indem man alle Werte dieses Produktes, die $> \nu$ sind, durch ν ersetzt, so existiert das Doppelintegral:

$$\int_a^b \int_{a'}^{b'} \Phi_\nu(\xi, x)\, d\xi\, dx$$

und man hat bekanntlich:[1]

(4) $$\int_a^b \int_{a'}^{b'} \Phi_\nu(\xi, x)\, d\xi\, dx = \int_a^b \left(\int_{a'}^{b'} \Phi_\nu(\xi, x)\, dx \right) d\xi.$$

Wegen Voraussetzung (1) existiert nun das Integral:

$$\int_a^b f(\xi) \left(\int_{a'}^{b'} \varphi(\xi, x)\, dx \right) d\xi = \int_a^b \left(\lim_{\nu=\infty} \int_{a'}^{b'} \Phi_\nu(\xi, x)\, dx \right) d\xi,$$

und nach einem bekannten Satze hat man daher weiter, da $\Phi_\nu(\xi, x)$ mit ν monoton wächst:

$$\int_a^b f(\xi) \left(\int_{a'}^{b'} \varphi(\xi, x)\, dx \right) d\xi = \lim_{\nu=\infty} \int_a^b \left(\int_{a'}^{b'} \Phi_\nu(\xi, x)\, dx \right) d\xi;$$

wegen (4) aber heißt das nichts anderes als: es existiert das Doppelintegral:

(5) $$\int_a^b \int_{a'}^{b'} f(\xi)\, \varphi(\xi, x)\, d\xi\, dx.$$

Nach dem schon vorhin benutzten Satze von Fubini existiert dann auch, abgesehen von einer Nullmenge, das Integral:

$$\int_a^b f(\xi)\, \varphi(\xi, x)\, d\xi,$$

stellt eine in $< a', b' >$ integrierbare Funktion von x dar, und es ist:

(6) $$\int_{a'}^{b'} \left(\int_a^b f(\xi)\, \varphi(\xi, x)\, d\xi \right) dx = \int_a^b \int_{a'}^{b'} f(\xi)\, \varphi(\xi, x)\, d\xi\, dx.$$

Damit ist Satz I nachgewiesen unter der Annahme (3), und es ist obendrein der Wert des Integrales

$$\int_{a'}^{b'} I(f, x)\, dx$$

durch ein Doppelintegral ausgedrückt.

Den allgemeinen Fall führt man auf diesen speziellen zurück, indem man setzt:

$$\varphi(\xi, x) = \varphi_1(\xi, x) - \varphi_2(\xi, x) \qquad (\varphi_1(\xi, x) \geqq 0; \ \varphi_2(\xi, x) \geqq 0),$$
$$f(\xi) = f_1(\xi) - f_2(\xi) \qquad (f_1(\xi) \geqq 0; \ f_2(\xi) \geqq 0).$$

---

[1] Siehe etwa Ch. J. de la Vallée Poussin Cours d'analyse infinitésimale, II (2. éd), p. 122 (»Théorème de M. Fubini«).

Unter Berufung auf (6) ergänzen wir die Aussage von Satz I durch:

I'. Unter den Voraussetzungen von Satz I existiert das Doppelintegral (5) und der Wert von:

$$\int_{a'}^{b'} I(f, x)\, dx$$

ist gleich diesem Doppelintegrale.

Eine weitere Ergänzung von I liefert der ebenso zu beweisende Satz:

I''. Ersetzt man in Satz I die Voraussetzung (1) durch die Voraussetzung, es sei das Quadrat von:

$$\Phi(\xi) = \int_{a'}^{b'} |\varphi(\xi, x)|\, dx$$

integrierbar in $< a, b >$, so gilt die Behauptung von Satz I für jede Funktion f, deren Quadrat in $< a, b >$ integrierbar ist. – Ersetzt man in Satz I die Voraussetzung (1) durch die Voraussetzung, es sei $\Phi(\xi)$ integrierbar in $< a, b >$, so gilt die Behauptung von Satz I für jede in $< a, b >$ geschränkte, meßbare Funktion f.

Wir denken uns nun den Kern $\varphi(\xi, x)$ noch abhängig von der natürlichen Zahl n, und setzen, wo dieses Integral existiert:

$$I_n(f, x) = \int_a^b f(\xi)\, \varphi(\xi, x, n)\, d\xi;$$

dann können wir folgenden Satz beweisen:

II. Sei $< a', b' >$ ein Teilintervall von $< a, b >$,[1] und es sei $\varphi(\xi, x, n)$ meßbar im Rechtecke $a \leqq \xi \leqq b$, $a' \leqq x \leqq b'$. Abgesehen von einer Nullmenge sei $\varphi(\xi, x, n)$ für jedes x von $< a', b' >$ nach ξ integrierbar in $< a, b >$. Ferner genüge $\varphi(\xi, x, n)$ folgenden Bedingungen:

1. Es gibt ein M, so daß für alle x von $< a', b' >$, abgesehen von einer Nullmenge, und für alle n:

$$\int_a^b |\varphi(\xi, x, n)|\, d\xi < M.$$

2. Es gibt ein M, so daß für alle ξ von $< a, b >$, abgesehen von einer Nullmenge, und für alle n:

$$\int_{a'}^{b'} |\varphi(\xi, x, n)|\, dx < M.$$

3. Abgesehen von einer Nullmenge, gelte für alle x von $< a', b' >$ folgendes: für jede im Intervalle $< a, b >$ der Veränderlichen ξ gelegene Punktmenge $\mathfrak{A}_x$, die im Punkte x die Dichte[2] 1 hat, sei:

$$\lim_{n=\infty} \int_{\mathfrak{A}_x} \varphi(\xi, x, n)\, d\xi = 1.$$

---

[1] Es kann natürlich $< a', b' >$ auch mit $< a, b >$ identisch sein.

[2] Eine Punktmenge $\mathfrak{A}$ hat im Punkte x die Dichte p, wenn für den Inhalt $i(h)$ ihres ins Intervall $< x-h, x+h >$ fallenden Teiles die Beziehung gilt:

$$\lim_{h=0} \frac{1}{2h} \cdot i(h) = p.$$

Dann gilt für jede in $<a, b>$ integrierbare Funktion $f(\xi)$ die Beziehung:

$$\lim_{n=\infty} \int_{a'}^{b'} |f(x) - I_n(f, x)|\, dx = 0. \tag{7}$$

Es sei zunächst folgendes bemerkt: nach Satz I angewendet auf den Spezialfall $f = 1$, existiert das in Bedingung 2 auftretende Integral:

$$\int_{a'}^{b'} |\varphi(\xi, x, n)|\, dx$$

für alle ξ von $<a, b>$, abgesehen von einer Nullmenge. Ebenso lehrt Satz I die Existenz von $I_n(f, x)$ für alle x von $<a', b'>$, abgesehen von einer Nullmenge.

Aus Bedingung 3 folgt, daß für jede im Intervalle $<a, b>$ der Veränderlichen ξ gelegene Punktmenge $\mathfrak{B}_x$, die im Punkte x von $<a', b'>$ die Dichte 0 hat:

$$\lim_{n=\infty} \int_{\mathfrak{B}_x} |\varphi(\xi, x, n)|\, d\xi = 0 \tag{8}$$

ist. In der Tat sei $\mathfrak{B}_x^{(1)}$ der Teil von $\mathfrak{B}_x$, auf dem $\varphi(\xi, x, n) \geqq 0$, und $\mathfrak{B}_x^{(2)}$ der Rest von $\mathfrak{B}_x$. Sind dann $\mathfrak{A}_x^{(1)}$ und $\mathfrak{A}_x^{(2)}$ die Komplemente von $\mathfrak{B}_x^{(1)}$ und $\mathfrak{B}_x^{(2)}$ bezüglich des Intervalles $<a, b>$, so haben $\mathfrak{A}_x^{(1)}$ und $\mathfrak{A}_x^{(2)}$ im Punkte x die Dichte 1. Es ist also nach Bedingung 3:

$$\lim_{n=\infty} \int_a^b \varphi(\xi, x, n)\, d\xi = 1; \quad \lim_{n=\infty} \int_{\mathfrak{A}_x^{(1)}} \varphi(\xi, x, n)\, d\xi = 1; \quad \lim_{n=\infty} \int_{\mathfrak{A}_x^{(2)}} \varphi(\xi, x, n)\, d\xi = 1,$$

woraus sofort:

$$\lim_{n=\infty} \int_{\mathfrak{B}_x^{(1)}} \varphi(\xi, x, n)\, d\xi = 0; \quad \lim_{n=\infty} \int_{\mathfrak{B}_x^{(2)}} \varphi(\xi, x, n)\, d\xi = 0,$$

und somit auch (8) folgt.

Sei nun $\varepsilon > 0$ beliebig gegeben. Wir bezeichnen mit $\mathfrak{A}_x$ die Menge aller Punkte ξ von $<a, b>$, in denen:

$$|f(\xi) - f(x)| \leqq \varepsilon \tag{9}$$

ist, mit $\mathfrak{B}_x$ ihr Komplement bezüglich $<a, b>$. Nach einem bekannten Satze[1] hat dann für jedes x von $<a', b'>$, abgesehen von einer Nullmenge, $\mathfrak{A}_x$ im Punkte x die Dichte 1, und daher $\mathfrak{B}_x$ im Punkte x die Dichte 0.

Wir schreiben:

$$I_n(f, x) - f(x) = \int_{\mathfrak{A}_x} (f(\xi) - f(x))\, \varphi(\xi, x, n)\, d\xi + f(x) \left\{ \int_{\mathfrak{A}_x} \varphi(\xi, x, n)\, d\xi - 1 \right\} + \\ + \int_{\mathfrak{B}_x} f(\xi)\, \varphi(\xi, x, n)\, d\xi \tag{10}$$

---

[1] »Abgesehen von einer Nullmenge ist, wenn $f(\xi)$ integrierbar ist, $|f(\xi) - c|$ überall Ableitung seines unbestimmten Integrales für alle Werte von c« (de la Vallée-Poussin, Cours d'analyse II, 2-éd, p. 115). Es ist also, wieder abgesehen von einer Nullmenge:

$$\lim_{h=0} \frac{1}{2h} \int_{-h}^{h} |f(x+t) - f(x)|\, dt = 0.$$

Ist also $i(h, \varepsilon)$ der Inhalt der Menge aller jener Punkte von $<x-h, x+h>$, in denen $|f(\xi) - f(x)| > \varepsilon$ ist, so haben wir:

$$\lim_{h=0} \frac{1}{2h}\, i(h, \varepsilon) = 0.$$

Das aber ist die Behauptung des Textes.

Wegen (9) ist:

$$\left|\int_{\mathfrak{A}_x} (f(\xi)-f(x))\,\varphi(\xi, x, n)\,d\xi\right| \leqq \varepsilon \int_a^b |\varphi(\xi, x, n)|\,d\xi,$$

und mithin, wegen Bedingung 1:

(11) $$\int_{a'}^{b'} \left|\int_{\mathfrak{A}_x} (f(\xi)-f(x))\,\varphi(\xi, x, n)\,d\xi\right| dx \leqq \varepsilon M (b'-a').$$

Nach Bedingung 3 ist, abgesehen von einer Nullmenge, für alle x von $<a', b'>$:

$$\lim_{n=\infty} \left\{\int_{\mathfrak{A}_x} \varphi(\xi, x, n)\,d\xi - 1\right\} = 0,$$

und nach Bedingung 1 ist, wieder abgesehen von einer Nullmenge:

$$\left|\int_{\mathfrak{A}_x} \varphi(\xi, x, n)\,d\xi - 1\right| < M+1.$$

Daraus folgt nach einem bekannten Satze:

(12) $$\lim_{n=\infty} \int_{a'}^{b'} \left| f(x) \left\{\int_{\mathfrak{A}_x} \varphi(\xi, x, n)\,d\xi - 1\right\}\right| dx = 0.$$

Sei endlich $\bar{\varphi}(\xi, x, n)$ für jedes einzelne x in den Punkten von $\mathfrak{B}_x$ gleich $|\varphi(\xi, x, n)|$, sonst gleich 0. Dann ist, nach (8), abgesehen von einer Nullmenge, für alle x von $<a', b'>$:

(13) $$\lim_{n=\infty} \int_a^b \bar{\varphi}(\xi, x, n)\,d\xi = 0.$$

Ferner ist:

(14) $$\left|\int_{\mathfrak{B}_x} f(\xi)\,\varphi(\xi, x, n)\,d\xi\right| \leqq \int_a^b |f(\xi)| \cdot \bar{\varphi}(\xi, x, n)\,d\xi.$$

Nach Satz I' haben wir:

(15) $$\int_{a'}^{b'} \left(\int_a^b |f(\xi)|\,\bar{\varphi}(\xi, x, n)\,d\xi\right) dx = \int_a^b |f(\xi)| \left(\int_{a'}^{b'} \bar{\varphi}(\xi, x, n)\,dx\right) d\xi.$$

Nach Bedingung 2 ist, abgesehen von einer Nullmenge, für alle ξ von $<a, b>$ und alle n:

(16) $$0 \leqq \int_{a'}^{b'} \bar{\varphi}(\xi, x, n)\,dx \leqq M.$$

Wegen Bedingung 1 ist, abgesehen von einer Nullmenge, für alle x von $<a', b'>$ und alle n:

$$0 \leqq \int_a^b \bar{\varphi}(\xi, x, n)\,d\xi < M,$$

und somit wegen (13):

$$\lim_{n=\infty} \int_{a'}^{b'} \left(\int_a^b \bar{\varphi}(\xi, x, n)\,d\xi\right) dx = 0.$$

Da aber nach Satz I':

$$\int_{a'}^{b'} \left(\int_b^b \bar{\varphi}(\xi, x, n)\,d\xi\right) dx = \int_a^b \int_{a'}^{b'} \bar{\varphi}(\xi, x, n)\,d\xi\,dx$$

ist, so haben wir:

$$\lim_{n=\infty}\int_a^b\int_{a'}^{b'}\overline{\varphi}(\xi, x, n)\, d\xi\, dx = 0,$$

und somit, da $\overline{\varphi} \geqq 0$ ist, auch für jedes Teilintervall $< \alpha, \beta >$ von $< a, b >$:

(17) $$\lim_{n=\infty}\int_\alpha^\beta\left(\int_{a'}^{b'}\overline{\varphi}(\xi, x, n)\, dx\right) d\xi = 0.$$

Setzen wir also:

$$\Phi(\xi, n) = \int_{a'}^{b'}\overline{\varphi}(\xi, x, n)\, dx,$$

so sehen wir aus (16) und (17), daß $\Phi(\xi, n)$ folgenden zwei Bedingungen genügt:

1. Es ist in $< a, b >$ abgesehen von einer Nullmenge:

$$|\Phi(\xi, n)| < M \quad \text{für alle } n.$$

2. Es ist für jedes Teilintervall $< \alpha, \beta >$ von $< a, b >$:

$$\lim_{n=\infty}\int_\alpha^\beta \Phi(\xi, n)\, d\xi = 0.$$

Nach einem Satze von Lebesgue[1] ist aber dann für jede in $< a, b >$ integrierbare Funktion $F(\xi)$:

$$\lim_{n=\infty}\int_a^b F(\xi)\,\Phi(\xi, n)\, d\xi = 0.$$

Nach (14) und (15) haben wir also auch:

(18) $$\lim_{n=\infty}\int_{a'}^{b'}\left|\int_{\mathfrak{B}_x} f(\xi)\,\varphi(\xi, x, n)\, d\xi\right| dx = 0.$$

Die Formeln (10), (11), (12), (18) aber ergeben zusammen: es ist für alle hinlänglich großen n:

$$\int_{a'}^{b'}|I_n(f, x) - f(x)|\, dx < \varepsilon\,(M(b' - a') + 1),$$

und da ε beliebig war, ist das die Behauptung (7) von Satz II.

Man kann sich vielfach eine Untersuchung, ob Bedingung 3. von Satz II erfüllt ist, ersparen durch die Bemerkung:

II'. Ist bekannt, daß, abgesehen von einer Nullmenge, für jedes in $< a, b >$ integrierbare f in jedem Punkte x von $< a', b' >$, in dem:

(19) $$\lim_{h=0}\frac{1}{h}\int_0^h |f(x+t) - f(x)|\, dt = 0$$

ist, die Beziehung:

(20) $$\lim_{n=\infty} I_n(f, x) = f(x)$$

gilt, so ist Bedingung 3. von Satz II erfüllt.

In der Tat, sei $\mathfrak{A}$ eine den Punkt x enthaltende Menge, die im Punkte x die Dichte 1 hat und sei $f(\xi) = 1$ in den Punkten von $\mathfrak{A}$, sonst $= 0$, so ist (19) erfüllt, es gilt somit wegen (20):

$$\lim_{n=\infty}\int_{\mathfrak{A}}\varphi(\xi, x, n)\, d\xi = 1,$$

somit ist Bedingung 3. von Satz II erfüllt.

---

[1] Annales de Toulouse, Serie 3, Bd. 1, p. 52 (Satz I meiner 1. Mitteilung.)

Von den zahlreichen Anwendungen von Satz II seien einige hervorgehoben.

Sei $r_n < 1$ und $\lim_{n=\infty} r_n = 1$. Wir setzen:

$$\varphi(\xi, x, n) = \frac{1}{2\pi} \frac{1-r_n^2}{1-2r_n \cos(\xi-x)+r_n^2}. \tag{21}$$

Da die Beziehung:

$$\lim_{r=1-0} \frac{1}{2\pi} \int_{-\pi}^{\pi} f(\xi) \frac{1-r^2}{1-2r\cos(\xi-x)+r^2} d\xi = f(x)$$

für jede in $<-\pi, \pi>$ integrierbare Funktion in jedem Punkte x von $(-\pi, \pi)$ gilt, in dem f Ableitung seines unbestimmten Integrales ist,[1] entnehmen wir aus II' sofort, daß für den Kern (21) alle Bedingungen von Satz II erfüllt sind für $-\pi \leqq \xi \leqq \pi$; $-\pi \leqq x \leqq \pi$.

Setzen wir noch, der Übersichtlichkeit halber:[2]

$$P(r, x) = \frac{1}{2\pi} \int_{-\pi}^{\pi} f(\xi) \frac{1-r^2}{1-2r\cos(\xi-x)+r^2} d\xi,$$

so haben wir also für die Annäherung des Poisson'schen Integrales an eine beliebige in $<-\pi, \pi>$ integrierbare Funktion $f(x)$:

$$\lim_{r=1-0} \int_{-\pi}^{\pi} |f(x) - P(r, x)|\, dx = 0.$$

Diese Eigenschaft des Poisson'schen Integrales wurde bewiesen von W. Groß.[3]

Setzen wir:

$$\varphi(\xi, x, n) = \frac{1}{2n\pi} \left(\frac{\sin n \dfrac{\xi-x}{2}}{\sin \dfrac{\xi-x}{2}} \right)^2,$$

so wird:

$$F_n(x) = \int_{-\pi}^{\pi} f(\xi)\, \varphi(\xi, x, n)\, d\xi$$

bekanntlich das Fejér'sche arithmetische Mittel aus den n ersten Gliedern der Fourier'schen Reihe von $f(x)$. Da die Fejér'schen Mittel für jedes in $<-\pi, \pi>$ integrierbare f überall in $(-\pi, \pi)$ konvergieren, wo:[4]

$$\lim_{h=0} \frac{1}{h} \int_0^h |f(x+t) - f(x)|\, dt = 0,$$

sind die Bedingungen von Satz II erfüllt, und wir erhalten folgenden Satz über die Konvergenz der Fejér'schen Mittel gegen $f(x)$: Es ist für jede in $<-\pi, \pi>$ integrierbare Funktion f:

$$\lim_{n=\infty} \int_{-\pi}^{\pi} |f(x) - F_n(x)|\, dx = 0.$$

---

[1] Zuerst bewiesen von P. Fatou; siehe meine I. Mitteilung, p. 69 [653].

[2] Es ist dies das sog. Poisson'sche Integral. Bekanntlich ist, wenn mit a_ν, b_ν die Koefizienten der Fourier'schen Reihe von $f(x)$ bezeichnet werden:

$$P(r, x) = \frac{a_0}{2} + \sum_{\nu=1}^{\infty} r^\nu (a_\nu \cos \nu x + b_\nu \sin \nu x) \quad (\text{für } 0 \leqq r < 1).$$

[3] Wien. Ber., Abt. II *a*, Bd. 124, p. 1024.

[4] H. Lebesgue Ann. de Toulouse, Serie 3, Bd. 1, p. 88.

Dasselbe gilt für die von de la Vallée Poussin angegebene Summierung der Fourier'schen Reihe: setzen wir:

$$\varphi(\xi, x, n) = \frac{1}{\pi} \cdot 2^{2n-1} \frac{(n!)^2}{(2n)!} \left(\cos \frac{\xi - x}{2}\right)^{2n},$$

so wird,[1] wenn mit a_ν, b_ν die Koeffizienten der Fourier'schen Reihe von $f(x)$ bezeichnet werden:

$$V_n(x) = \int_{-\pi}^{+\pi} f(\xi)\, \varphi(\xi, x, n)\, d\xi = \frac{a_0}{2} + \sum_{\nu=1}^{n} \frac{n(n-1)\ldots(n-\nu+1)}{(n+1)(n+2)\ldots(n+\nu)} (a_\nu \cos \nu x + b_\nu \sin \nu x),$$

und es ist für jedes in $< -\pi, \pi >$ integrierbare f:

$$f(x) = \lim_{n=\infty} V_n(x)$$

in jedem Punkte x von $(-\pi, \pi)$, in dem $f(x)$ Ableitung seines unbestimmten Integrales ist.[2] Es ist daher Satz II anwendbar, und ergibt für die de la Vallée Poussin'sche Summierung der Fourier'schen Reihe, bei beliebigem in $< -\pi, \pi >$ integrierbarem $f(x)$:

$$\lim_{n=\infty} \int_{-\pi}^{\pi} |f(x) - V_n(x)|\, dx = 0.$$

Setzen wir endlich, nach Landau und de la Vallée Poussin:

$$\varphi(\xi, x, n) = \sqrt{\frac{n}{\pi}} \{1 - (\xi - x)^2\}^n,$$

so wird:

$$L_n(x) = \int_0^1 f(\xi)\, \varphi(\xi, x, n)\, d\xi$$

ein Polynom in x, und es ist für jedes in $< 0, 1 >$ integrierbare f:

$$f(x) = \lim_{n=\infty} L_n(x)$$

in jedem Punkte x von $(0, 1)$, in dem $f(x)$ Ableitung seines unbestimmten Integrales ist.[3] Daher ist Satz II anwendbar und ergibt für die Annäherung der Landau'schen Polynome an die beliebige in $< -1, 1 >$ integrierbare Funktion $f(x)$:

$$\lim_{n=\infty} \int_0^1 |f(x) - L_n(x)|\, dx = 0.$$

§ 2.

Wir wollen nun die bisher bewiesenen Sätze auf unendliche Intervalle übertragen:

Ia. Sei $\varphi(\xi, x)$ eine in der ganzen ξx-Ebene meßbare Funktion. Als Funktion von x sei sie für alle ξ, abgesehen von einer Nullmenge, integrierbar in $(-\infty, +\infty)$, und es gebe ein M, so daß, wieder abgesehen von einer Nullmenge:

$$(1) \qquad \int_{-\infty}^{+\infty} |\varphi(\xi, x)|\, dx < M \quad \text{für alle } \xi.$$

---

[1] Ch. J. de la Vallée-Poussin, Bull. Bruxelles, Classe des sciences, 1908, p. 232.

[2] Zuerst bewiesen von H. Lebesgue, vgl. meine 1. Mitteilung, p. 47 [631].

[3] Zuerst bewiesen von Fr. Riesz; vgl. meine 1. Mitteilung, p. 47 [631].

Ist dann $f(\xi)$ integrierbar in $(-\infty, +\infty)$, so existiert das Integral:

(2) $$J(f, x) = \int_{-\infty}^{+\infty} f(\xi)\, \varphi(\xi, x)\, d\xi$$

für alle x, abgesehen von einer Nullmenge, und ist eine in $(-\infty, +\infty)$ integrierbare Funktion.

Es genügt wieder, den Beweis unter der Annahme:

(3) $$\varphi(\xi, x) \geqq 0, \quad f(\xi) \geqq 0$$

zu führen. Wir zeigen zunächst, daß für jedes endliche Intervall $\langle a, b \rangle$ das Integral:

$$J^*(f, x) = \int_a^b f(\xi)\, \varphi(\xi, x)\, d\xi$$

für alle x, abgesehen von einer Nullmenge, existiert und eine in $(-\infty, +\infty)$ integrierbare Funktion von x darstellt. Wird $\Phi_\nu(\xi, x)$ eingeführt, wie beim Beweise von Satz I, so wird es genügen, die Existenz des Doppelintegrales:

(4) $$\int_a^b \int_{-\infty}^{+\infty} \Phi_\nu(\xi, x)\, d\xi\, dx$$

nachzuweisen, da im übrigen der Beweis derselbe bleibt, wie für Satz I.

Für jedes endliche $A > 0$ existiert sicherlich das Doppelintegral:

$$\int_a^b \int_{-A}^{A} \Phi_\nu(\xi, x)\, d\xi\, dx.$$

Wegen (3) ist $\Phi_\nu(\xi, x) \geqq 0$, so daß das Integral:

$$\int_{-A}^{A} \Phi_\nu(\xi, x)\, dx$$

mit A monoton wächst. Es ist somit:

$$\int_a^b \left(\lim_{A=+\infty} \int_{-A}^{A} \Phi_\nu(\xi, x)\, dx \right) d\xi = \lim_{A=+\infty} \int_a^b \left(\int_{-A}^{A} \Phi_\nu(\xi, x)\, dx \right) d\xi,$$

wo das links stehende Integral sicher existiert, da wegen (1):

$$\int_{-A}^{A} \Phi_\nu(\xi, x)\, dx \leqq f(\xi) \int_{-\infty}^{+\infty} \varphi(\xi, x)\, dx \leqq M f(\xi)$$

ist. Da nun aber:

$$\lim_{A=+\infty} \int_a^b \left(\int_{-A}^{A} \Phi_\nu(\xi, x)\, dx \right) d\xi = \lim_{A=+\infty} \int_a^b \int_{-A}^{A} \Phi_\nu(\xi, x)\, d\xi\, dx = \int_a^b \int_{-\infty}^{+\infty} \Phi_\nu(\xi, x)\, d\xi\, dx$$

ist, so ist die Existenz des Doppelintegrales (4) bewiesen.

Um nun Satz I a selbst zu beweisen, wird es wieder genügen, die Existenz des Doppelintegrales:

(5) $$\int_{-\infty}^{+\infty} \int_{-\infty}^{+\infty} \Phi_\nu(\xi, x)\, d\xi\, dx$$

nachzuweisen. Nach dem eben Bewiesenen existiert für jedes endliche $A > 0$ das Integral:

$$\int_{-A}^{A} \int_{-\infty}^{+\infty} \Phi_\nu(\xi, x)\, d\xi\, dx = \int_{-A}^{A} \left(\int_{-\infty}^{+\infty} \Phi_\nu(\xi, x)\, dx \right) d\xi \leqq \int_{-\infty}^{+\infty} \left(f(\xi) \int_{-\infty}^{+\infty} \varphi(\xi, x)\, dx \right) d\xi,$$

und wegen (1) hat die rechte Seite dieser Ungleichung einen endlichen Wert. Daraus folgt die Existenz eines endlichen Grenzwertes:

$$\lim_{A=+\infty} \int_{-A}^{A} \int_{-\infty}^{+\infty} \Phi_\nu (\xi, x)\, d\xi\, dx,$$

da das unter dem Linuszeichen stehende Integral mit A monoton wächst. Dieser Grenzwert ist aber nichts anderes als das Doppelintegral (5), dessen Existenz hiemit bewiesen ist.

Wie Satz I, kann auch Satz I a ergänzt werden durch die Bemerkung:

I′ a. Unter den Voraussetzungen von Satz I a existiert das Doppelintegral:

$$\int_{-\infty}^{+\infty} \int_{-\infty}^{+\infty} f(\xi)\, \varphi (\xi, x)\, d\xi\, dx$$

und der Wert von

$$\int_{-\infty}^{+\infty} J(f, x)\, dx$$

ist gleich diesem Doppelintegrale.

Auch die zu I″ analoge Bemerkung bleibt richtig.

Wir lassen wieder $\varphi (\xi, x)$ abhängen von einer natürlichen Zahl n und setzen:

$$J_n (f, x) = \int_{-\infty}^{+\infty} f(\xi)\, \varphi (\xi, x, n)\, d\xi.$$

Dann können wir den zu II analogen Satz beweisen:

II a. Sei $\varphi (\xi, x, n)$ für jede natürliche Zahl n meßbar in der ganzen ξx-Ebene. Abgesehen von einer Nullmenge sei $\varphi (\xi, x, n)$ für jedes x nach ξ integrierbar in $(-\infty, +\infty)$; ebenso sei, abgesehen von einer Nullmenge, $\varphi (\xi, x, n)$ für jedes ξ integrierbar nach x in $(-\infty, +\infty)$. Ferner genüge $\varphi (\xi, x, n)$ folgenden Bedingungen:

1. Es gibt ein M, so daß für alle x (abgesehen von einer Nullmenge) und alle n:

$$\int_{-\infty}^{+\infty} |\varphi (\xi, x, n)|\, d\xi < M.$$

2. Es gibt ein M, so daß für alle ξ (abgesehen von einer Nullmenge) und alle n:

$$\int_{-\infty}^{+\infty} |\varphi (\xi, x, n)|\, dx < M.$$

3. Für jede geschränkte Punktmenge $\mathfrak{A}_x$ der Veränderlichen ξ, die im Punkte $\xi = x$ die Dichte 1 hat, gilt:

$$\lim_{n=\infty} \int_{\mathfrak{A}_x} \varphi (\xi, x, n)\, d\xi = 1.$$

4. Für jedes $h > 0$ und jedes ξ ist:

$$\lim_{n=\infty} \int_{-\infty}^{\xi-h} |\varphi (\xi, x, n)|\, dx = 0; \quad \lim_{n=\infty} \int_{\xi+h}^{+\infty} |\varphi (\xi, x, n)|\, dx = 0.$$

Dann gilt für jede in $(-\infty, +\infty)$ integrierbare Funktion $f(x)$ die Beziehung:

$$\lim_{n=\infty} \int_{-\infty}^{+\infty} |f(x) - J_n (f, x)|\, dx = 0. \tag{6}$$

Um das zu beweisen, können wir ganz ähnlich vorgehen, wie beim Beweise von Satz II. Sei $g(x)$ eine durchwegs positive, in $(-\infty, +\infty)$ integrierbare Funktion, und sei $\varepsilon > 0$ beliebig gegeben.

Wir wählen ein beliebiges $h > 0$, bezeichnen mit $\mathfrak{A}_x$ die Menge aller Punkte ξ von $< x-h, x+h >$, in denen:

$$|f(\xi)-f(x)| \leqq \varepsilon \cdot g(x), \tag{7}$$

und bezeichnen mit $\mathfrak{B}_x$ das Komplement von $\mathfrak{A}_x$. Nun schreiben wir:

$$J_n(f,x)-f(x)=\int_{\mathfrak{A}_x}(f(\xi)-f(x))\,\varphi(\xi,x,n)\,d\xi+f(x)\left\{\int_{\mathfrak{A}_x}\varphi(\xi,x,n)\,d\xi-1\right\}+ \tag{8}$$
$$+\int_{\mathfrak{B}_x}f(\xi)\,\varphi(\xi,x,n)\,d\xi.$$

Aus (7) folgt:

$$\left|\int_{\mathfrak{A}_x}(f(\xi)-f(x))\,\varphi(\xi,x,n)\,d\xi\right| \leqq \varepsilon\cdot g(x)\int_{-\infty}^{+\infty}|\varphi(\xi,x,n)|\,d\xi,$$

und mithin, wegen Bedingung 1.

$$\int_{-\infty}^{+\infty}\left|\int_{\mathfrak{A}_x}(f(\xi)-f(x))\,\varphi(\xi,x,n)\,d\xi\right| \leqq \varepsilon\cdot M\int_{-\infty}^{+\infty}g(x)\,dx. \tag{9}$$

Ganz so, wie Gleichung (12) in § 1 nachgewiesen wurde, wird hier gezeigt:

$$\lim_{n=\infty}\int_{-\infty}^{+\infty}\left|f(x)\left\{\int_{\mathfrak{A}_x}\varphi(\xi,x,n)\,d\xi-1\right\}\right|dx=0. \tag{10}$$

Wir setzen auch hier: $\bar{\varphi}(\xi,x,n)=|\varphi(\xi,x,n)|$ in den Punkten von $\mathfrak{B}_x$, sonst $=0$. Die durch (14) und (15) von § 1 ausgedrückte Schlußweise ergibt (bei Berufung auf Satz I'*a*):

$$\int_{-\infty}^{+\infty}\left|\int_{\mathfrak{B}_x}f(\xi)\,\varphi(\xi,x,n)\,d\xi\right|dx \leqq \int_{-\infty}^{+\infty}\left\{|f(\xi)|\int_{-\infty}^{+\infty}\bar{\varphi}(\xi,x,n)\,dx\right\}d\xi. \tag{11}$$

Wir setzen nun zur Abkürzung:

$$\Phi(\xi,n)=\int_{-\infty}^{+\infty}\bar{\varphi}(\xi,x,n)\,dx,$$

und zeigen, daß $\Phi(\xi,n)$ folgenden zwei Bedingungen genügt:

α) Es gibt ein M, sodaß, abgesehen von einer Nullmenge, für alle ξ und alle n:

$$|\Phi(\xi,n)| < M.$$

β) In jedem endlichen Intervalle $< \alpha, \beta >$ ist:

$$\lim_{n=\infty}\int_{\alpha}^{\beta}\Phi(\xi,n)\,d\xi=0. \tag{12}$$

Bedingung α) ist eine unmittelbare Folge von Bedingung 2. von Satz II *a*. Was Bedingung β) anlangt, so wählen wir A so groß, daß $< \alpha, \beta >$ in $(-A, A)$ liegt und setzen:

$$\Phi_1(\xi,n)=\int_{-A}^{A}\bar{\varphi}(\xi,x,n)\,dx$$
$$\Phi_2(\xi,n)=\int_{-\infty}^{-A}\bar{\varphi}(\xi,x,n)\,dx+\int_{A}^{+\infty}\bar{\varphi}(\xi,x,n)\,dx,$$

so daß wir haben:

$$\Phi(\xi,n)=\Phi_1(\xi,n)+\Phi_2(\xi,n).$$

Nun ist:

$$\int_{\alpha}^{\beta} \Phi_1(\xi, n)\, d\xi = \int_{-A}^{A}\left(\int_{\alpha}^{\beta} \overline{\varphi}(\xi, x, n)\, d\xi\right) dx.$$

Aus Bedingung 3. folgt unmittelbar für jedes x:

$$\lim_{n=\infty} \int_{\alpha}^{\beta} \overline{\varphi}(\xi, x, n)\, d\xi = 0;$$

aus Bedingung 1. folgt:

$$\left|\int_{\alpha}^{\beta} \overline{\varphi}(\xi, x, n)\, d\xi\right| < M \text{ für alle } n.$$

Es ist daher:

(13)
$$\lim_{n=\infty} \int_{\alpha}^{\beta} \Phi_1(\xi, n)\, d\xi = 0.$$

Ferner ist nach Bedingung 4. für alle ξ von $<\alpha, \beta>$:

$$\lim_{n=\infty} \int_{-\infty}^{-A} \overline{\varphi}(\xi, x, n)\, dx = 0; \quad \lim_{n=\infty} \int_{A}^{+\infty} \overline{\varphi}(\xi, x, n)\, dx = 0,$$

und wegen Bedingung 2. für alle ξ von $<\alpha, \beta>$ und alle n:

$$\left|\int_{-\infty}^{-A} \overline{\varphi}(\xi, x, n)\, dx\right| < M; \quad \left|\int_{A}^{+\infty} \overline{\varphi}(\xi, x, n)\, dx\right| < M,$$

und somit:

(14)
$$\lim_{n=\infty} \int_{\alpha}^{\beta} \Phi_2(\xi, n)\, d\xi = 0.$$

Die Beziehungen (13) und (14) zeigen nun, daß (12) erfüllt ist, das heißt, daß $\Phi(\xi, n)$ auch der Bedingung β) genügt.

Da nun $\Phi(\xi, n)$ den Bedingungen α) und β) genügt, so ist[1] für jedes in $(-\infty, +\infty)$ integrierbare F:

$$\lim_{n=\infty} \int_{-\infty}^{+\infty} F(\xi)\, \Phi(\xi, n)\, d\xi = 0.$$

Es ist also insbesondere auch:

$$\lim_{n=\infty} \int_{-\infty}^{+\infty} |f(\xi)| \left(\int_{-\infty}^{+\infty} \overline{\varphi}(\xi, x, n)\, dx\right) d\xi = 0,$$

und mithin nach (11):

(15)
$$\lim_{n=\infty} \int_{-\infty}^{+\infty} \left|\int_{\mathfrak{B}_x} f(\xi)\, \varphi(\xi, x, n)\, d\xi\right| dx = 0.$$

Die Beziehungen (8), (9), (10) und (15) ergeben aber (6), so daß II*a* bewiesen ist.

Es sei zunächst folgende Anwendung von Satz II*a* erwähnt: wir setzen nach Weierstraß:

$$\varphi(\xi, x, k) = \frac{k}{\sqrt{\pi}} e^{-k^2(\xi - x)^2}.$$

Dann wird für jedes $k > 0$, wenn $f(\xi)$ in $(-\infty, +\infty)$ integrierbar ist, der Ausdruck:

(16)
$$W(x, k) = \int_{-\infty}^{+\infty} f(\xi)\, \varphi(\xi, x, k)\, d\xi$$

[1] Nach Satz I *a* meiner 1. Mitteilung.

eine ganze transzendente Funktion [1] in x, und es gilt an jeder Stelle x, wo f Ableitung seines unbestimmten Integrales ist:[2]

$$f(x) = \lim_{k=+\infty} W(x, k).$$

Daraus entnimmt man, daß Bedingung 3 von Satz II*a* erfüllt ist, und da die anderen Bedingungen dieses Satzes offenkundig erfüllt sind, haben wir: das Weierstrass'sche Integral (16) liefert zu jeder in $(-\infty, +\infty)$ integrierbaren Funktion $f(x)$ eine ganze Funktion $W(x, k)$, für die die Beziehung gilt:

$$\lim_{k=+\infty} \int_{-\infty}^{+\infty} |f(x) - W(x, k)|\, dx = 0. \tag{17}$$

Man könnte glauben, daß Bedingung 4 für die Giltigkeit von Satz II*a* überflüssig ist. Es sei also noch an einem Beispiele gezeigt, daß eine Bedingung dieser Art jedenfalls erforderlich ist.

Sei $\varphi_n(u) = \frac{n}{2}$ in $< -\frac{1}{n}, \frac{1}{n} >$, $\varphi_n(u) = 1$ in $< n, n+1 >$, sonst $\varphi_n(u) = 0$. Wir setzen:

$$J_n(f, x) = \int_{-\infty}^{+\infty} f(\xi)\, \varphi_n(\xi - x)\, d\xi,$$

und haben offenbar, wenn $f(x)$ integrierbar ist in $(-\infty, +\infty)$, in jedem Punkte x, in dem f Ableitung seines unbestimmten Integrales ist:

$$f(x) = \lim_{n=\infty} J_n(f, x),$$

woraus man entnimmt, daß Bedingung 3 von Satz II*a* erfüllt ist. Daß Bedingung 1 und 2 erfüllt sind, ist offenkundig, ebenso daß Bedingung 4 nicht erfüllt ist.

Wir haben:

$$J_n(f, x) - f(x) = \frac{n}{2} \int_{x-\frac{1}{n}}^{x+\frac{1}{n}} f(\xi)\, d\xi - f(x) + \int_{x+n}^{x+n+1} f(\xi)\, d\xi.$$

Aus Satz II*a* folgt ohneweiteres, daß:

$$\lim_{n=\infty} \int_{-\infty}^{+\infty} \left| f(x) - \frac{n}{2} \int_{x-\frac{1}{n}}^{x+\frac{1}{n}} f(\xi)\, d\xi \right| dx = 0$$

ist. Hingegen haben wir:

$$\int_{-\infty}^{+\infty} \left(\int_{x+n}^{x+n+1} f(\xi)\, d\xi \right) dx = \int_{-\infty}^{+\infty} \left(\int_{x}^{x+1} f(\xi)\, d\xi \right) dx.$$

Diese Größe ist von n unabhängig, und im allgemeinen gewiß $\neq 0$, so daß (6) nicht bestehen kann.

Zum Schlusse seien noch von Satz II*a* zwei Anwendungen auf die Theorie der Fourier'schen Integrale gemacht.

Wie ich in meiner 1. Mitteilung gezeigt habe,[3] gilt für jede in $(-\infty, +\infty)$ integrierbare Funktion $f(x)$ die Beziehung:

$$f(x) = \lim_{\rho=+0} \frac{1}{\pi\rho} \int_{-\infty}^{+\infty} f(\xi) \frac{1}{1 + \left(\frac{\xi - x}{\rho}\right)^2} d\xi$$

[1] Siehe zum Beispiel É. Borel, Leçons sur les fonctions de variables réelles, p. 53.

[2] Siehe meine 1. Mitteilung, p. 50 [334].

[3]) p. 67 [651], 71 [655].

in jedem Punkte x, in dem $f(x)$ Ableitung seines unbestimmten Integrales ist. Daraus folgt, daß, wenn ρ_n irgend eine Folge positiver Zahlen mit $\lim\limits_{n=\infty} \rho_n = 0$ bezeichnet, der Kern

$$\varphi(\xi, x, n) = \frac{1}{\pi \cdot \rho_n} \frac{1}{1+\left(\frac{\xi - x}{\rho_n}\right)^2},$$

der offenbar den Bedingungen 1, 2 und 4 von Satz II*a* genügt, auch der Bedingung 3 dieses Satzes genügt. Setzen wir also:

$$\Pi(x, \rho) = \frac{1}{\pi \cdot \rho} \int_{-\infty}^{+\infty} f(\xi) \frac{1}{1+\left(\frac{\xi - x}{\rho}\right)^2} d\xi,$$

so lehrt Satz II*a* das Bestehen der Beziehung:

$$\lim_{\rho=+0} \int_{-\infty}^{+\infty} |f(x) - \Pi(x, \rho)|\, dx = 0 \tag{18}$$

für jede in $(-\infty, +\infty)$ integrierbare Funktion $f(x)$. Nun hat man aber,[1] wenn gesetzt wird:

$$A(\lambda) = \frac{1}{\pi} \int_{-\infty}^{+\infty} f(\xi) \cos \lambda\xi\, d\xi; \quad B(\lambda) = \frac{1}{\pi} \int_{-\infty}^{+\infty} f(\xi) \sin \lambda\xi\, d\xi, \tag{19}$$

die Beziehung:

$$\Pi(x, \rho) = \int_{-\infty}^{+\infty} e^{-\rho\lambda} (A(\lambda) \cos \lambda x + B(\lambda) \sin \lambda x)\, d\lambda,$$

und es kann die Formel:

$$f(x) = \lim_{\rho=+0} \Pi(x, \rho)$$

als eine Summationsformel für das Fourier'sche Integral betrachtet werden, die, da sie der Poisson'schen Summierung der Fourier'schen Reihe analog ist, auch hier als die Poisson'sche Summierung bezeichnet werden möge.

Wir haben also, in Analogie mit einem in § 1 für die Poisson'sche Summierung der Fourier'schen Reihe bewiesenen Satze: Für die Poisson'schen Summierungsausdrücke $\Pi(x, \rho)$ des Fourier'schen Integrales einer beliebigen in $(-\infty, +\infty)$ integrierbaren Funktion $f(x)$ gilt die Beziehung (18).

Um ein zweites Resultat ähnlicher Natur zu erhalten, knüpfen wir an die bekannte Formel[2] an:

$$\int_0^{+\infty} e^{-v^2} \cos 2\alpha v\, dv = \frac{\sqrt{\pi}}{2} e^{-\alpha^2}.$$

Setzen wir hierin:

$$\alpha = k(\xi - x); \quad 2kv = \lambda,$$

so erhalten wir:

$$\frac{1}{\pi} \int_0^{+\infty} e^{-\left(\frac{\lambda}{2k}\right)^2} (\cos \lambda\xi \cos \lambda x + \sin \lambda\xi \sin \lambda x)\, d\lambda = \frac{k}{\sqrt{\pi}} e^{-k^2 (\xi - x)^2}.$$

Führt man dies ins Weierstrass'sche Integral (16) ein, so erhält man, wenn man wieder von der Bezeichnungsweise (19) Gebrauch macht: für jede in $(-\infty, +\infty)$ integrierbare Funktion $f(x)$ gilt in jedem Punkte, in dem sie Ableitung ihres unbestimmten Integrales ist, die Formel:

$$f(x) = \lim_{\rho=+0} \int_{-\infty}^{+\infty} e^{-\rho\lambda^2} (A(\lambda) \cos \lambda x + B(\lambda) \sin \lambda x)\, d\lambda.$$

---

[1] Vgl. 1. Mitteilung, p. 71 [655].

[2] Siehe zum Beispiel Ch. J. de la Vallée-Poussin, Cours d'analyse II (2. éd.), p. 82.

Auch diese Formel kann als eine Summationsformel für das Fourier'sche Integral betrachtet werden. Da sie auch durch einen Grenzübergang aus der de la Vallée-Poussin'schen Summierung der Fourier'schen Reihe gewonnen werden kann, so möge sie als die de la Vallée-Poussin'sche Summierung des Fourier'schen Integrales bezeichnet werden.

Setzen wir noch:

$$V(x, \rho) = \int_{-\infty}^{+\infty} e^{-\rho\lambda^2} (A(\lambda) \cos \lambda x + B(\lambda) \sin \lambda x)\, d\lambda,$$

so lehrt die für das Weierstrass'sche Integral bewiesene Beziehung (17): für die de la Vallée-Poussin'schen Summierungsausdrücke $V(x, \rho)$ des Fourier'schen Integrales einer beliebigen in $(-\infty, +\infty)$ integrierbaren Funktion $f(x)$ gilt die Beziehung:

$$\lim_{\rho=+0} \int_{-\infty}^{+\infty} |(f(x) - V(x, \rho)|\, dx = 0.$$

§ 3.

Wir müssen nun zwei Hilfsätze beweisen, die wir im Folgenden benötigen werden. [1]

III. Ist $\varphi_n(\xi)$ $(n = 1, 2, \ldots)$ eine Folge in $< a, b >$ integrierbarer Funktionen, für die:

(1) $$\overline{\lim_{n=\infty}} \int_a^b |\varphi_n(\xi)| d\xi = +\infty$$

ist, so gibt es eine in $< a, b >$ stetige Funktion $g(\xi)$, für die:

(2) $$\overline{\lim_{n=\infty}} \int_a^b g(\xi)\, \varphi_n(\xi)\, d\xi = +\infty.$$

Sei $h_n(\xi)$ definiert durch:

$$h_n(\xi) = 1, \quad \text{wo } \varphi_n(\xi) \geqq 0$$
$$h_n(\xi) = -1, \quad \text{wo } \varphi_n(\xi) < 0.$$

Dann ist, wegen (1), auch:

$$\overline{\lim_{n=\infty}} \int_a^b h_n(\xi)\, \varphi_n(\xi)\, d\xi = +\infty.$$

Es hat keinerlei Schwierigkeit, [2] aus den, im allgemeinen unstetigen, $h_n(\xi)$ stetige Funktionen $g_n(\xi)$ herzuleiten, die der Ungleichung genügen:

(3) $$g_n(\xi)| \leqq 1,$$

und für die gleichfalls:

(4) $$\overline{\lim_{n=\infty}} \int_a^b g_n(\xi)\, \varphi_n(\xi)\, d\xi = +\infty$$

ist. — Wäre nun für eine dieser stetigen Funktionen $g_i(\xi)$ die Folge der Integrale:

(5) $$\int_a^b g_i(\xi)\, \varphi_n(\xi)\, d\xi \quad (n = 1, 2 \ldots)$$

---

[1] Der Beweis dieser Hilfssätze beruht auf einem von A. Haar und H. Lebesgue zu ähnlichem Zwecke verwendeten Gedanken. Vgl. § 2 meiner 1. Mitteilung.

[2] Vgl. meine 1. Mitteilung, p. 9 [593].

nicht geschränkt, so wäre damit unsere Behauptung bewiesen; wir hätten nur $g(\xi) = g_i(\xi)$ oder $= -g_i(\xi)$ zu wählen. Wir werden also annehmen, es sei für jedes einzelne i die Folge der Integrale (5) geschränkt:

$$\left| \int_a^b g_i(\xi)\, \varphi_n(\xi)\, d\xi \right| < M_i \quad (n = 1, 2 \ldots). \tag{6}$$

Wir setzen noch zur Abkürzung:

$$\int_a^b |\varphi_n(\xi)|\, d\xi = L_n \quad \text{und} \quad I_n(u) = \int_a^b u(\xi)\, \varphi_n(\xi)\, d\xi.$$

Wir können nun aus der Folge der Indizes $n = 1, 2, \ldots$, eine Teilfolge $n_1, n_2, \ldots, n_i, \ldots$ herausgreifen nach folgender Vorschrift:

Es sei $n_1 = 1$. Ist n_i gefunden, so werde n_{i+1} ($> n_i$) in nachstehender Weise bestimmt:

1. Es sei:

$$L_{n_{i+1}} > L_{n_i}. \tag{7}$$

2. Ist:

$$I_m\left(g_{n_1} + \frac{1}{2\,L_{n_1}}\, g_{n_2} + \ldots + \frac{1}{2^{i-1} L_{n_{i-1}}}\, g_{n_i}\right) < N_i \quad \text{für alle } m, \tag{8}$$

so werde n_{i+1} so groß gewählt, daß:

$$I_{n_{i+1}}(g_{n_{i+1}}) > (N_i + i + 1)\, 2^i \cdot L_{n_i}. \tag{9}$$

Forderung 1 kann erfüllt werden, wegen Voraussetzung (1). Was Forderung 2 anlangt, so ist zunächst eine Ungleichung der Form (8) tatsächlich erfüllt für alle m wegen (6), und es kann n_{i+1} so groß gewählt werden, daß (9) gilt, wegen (4).

Setzen wir nun:

$$g(\xi) = g_{n_1}(\xi) + \frac{1}{2\,L_{n_1}}\, g_{n_2}(\xi) + \ldots + \frac{1}{2^{i-1} L_{n_{i-1}}}\, g_{n_i}(\xi) + \ldots,$$

so ist, wegen (3) und (7), diese Reihe gleichmäßig konvergent, und es ist daher $g(\xi)$ stetig in $< a, b >$. Ferner ist:

$$I_m(g) = I_m(g_{n_1}) + \frac{1}{2\,L_{n_1}}\, I_m(g_{n_2}) + \ldots + \frac{1}{2^{i-1} L_{n_{i-1}}}\, I_m(g_{n_i}) + \ldots$$

Wir haben also:

$$I_{n_i}(g) \geqq \frac{1}{2^{i-1} L_{n_{i-1}}}\, I_{n_i}(g_{n_i}) - \left| I_{n_i}\left(g_{n_1} + \frac{1}{2\,L_{n_1}}\, g_{n_2} + \ldots + \frac{1}{2^{i-2} L_{n_{i-2}}}\, g_{n_{i-1}}\right)\right| - \left| \sum_{k=i+1}^{\infty} \frac{1}{2^{k-1} L_{n_{k-1}}}\, I_{n_i}(g_{n_k}) \right|. \tag{10}$$

Hierin ist, wegen (9):

$$\frac{1}{2^{i-1} L_{n_{i-1}}}\, I_{n_i}(g_{n_i}) > N_{i-1} + i; \tag{11}$$

sodann wegen (8):

$$I_{n_i}\left(g_{n_1} + \frac{1}{2\,L_{n_1}}\, g_{n_2} + \ldots + \frac{1}{2^{i-2} L_{n_{i-2}}}\, g_{n_{i-1}}\right) < N_{i-1}; \tag{12}$$

endlich folgt aus (3):

$$|I_{n_i}(g_{n_k})| \leqq L_{n_i},$$

und daher weiter, unter Berücksichtigung von (7):

(13) $$\left|\sum_{k=i+1}^{\infty} \frac{1}{2^{k-1} L_{n_{k-1}}} I_{n_i}(g_{n_k})\right| \leqq 1.$$

Aus (10) zusammen mit (11), (12), (13) aber ergibt sich:

(14) $$I_{n_i}(g) > i-1.$$

Damit ist unsere Behauptung (2) erwiesen. In ganz derselben Weise zeigt man:

III *a.* »Ist $\varphi_n(\xi)$ $(n=1, 2, \ldots)$ eine Folge in $(-\infty, +\infty)$ integrierbarer Funktionen, für die:

$$\overline{\lim_{n=\infty}} \int_{-\infty}^{+\infty} |\varphi_n(\xi)|\, d\xi = +\infty$$

ist, so gibt es eine in $(-\infty, +\infty)$ geschränkte, für jedes ξ stetige Funktion $g(\xi)$, für die:

$$\overline{\lim_{n=\infty}} \int_{-\infty}^{+\infty} g(\xi)\, \varphi_n(\xi)\, d\xi = +\infty«.$$

Wir kommen zum Beweise des zweiten Hilfsatzes:

IV. Ist $\varphi_n(\xi)$ $(n=1, 2, \ldots)$ eine Folge in $\langle a, b \rangle$ integrierbarer Funktionen, und gibt es in $\langle a, b \rangle$ eine Folge von Teilintervallen $\langle \alpha_i, \beta_i \rangle$ und eine Folge von Indizes n_i mit $\lim_{i=\infty} n_i = \infty$, so daß:

(15) $$\lim_{i=\infty} \int_{\alpha_i}^{\beta_i} \varphi_{n_i}(\xi)\, d\xi = +\infty,$$

so gibt es eine in $\langle a, b \rangle$ absolut stetige Funktion $g(\xi)$, für die:

$$\overline{\lim_{n=\infty}} \int_a^b g(\xi)\, \varphi_n(\xi)\, d\xi = +\infty.$$

Wir beschränken nicht die Allgemeinheit, wenn wir die Voraussetzung (15) ersetzen durch:

(15*a*) $$\lim_{n=\infty} \int_{\alpha_n}^{\beta_n} \varphi_n(\xi)\, d\xi = +\infty,$$

da das darauf hinausläuft, statt der vorgelegten Folge der $\varphi_n(\xi)$ eine Teilfolge zu betrachten.

Sei $h_n(\xi)$ definiert durch:

$$h_n(\xi) = 1 \quad \text{in } \langle \alpha_n, \beta_n \rangle$$
$$h_n(\xi) = 0 \quad \text{außerhalb } \langle \alpha_n, \beta_n \rangle$$

Dann ist wegen (15 *a*) auch:

$$\lim_{n=\infty} \int_a^b h_n(\xi)\, \varphi_n(\xi)\, d\xi = +\infty.$$

Wir wählen nun ein $\sigma_n > 0$ so klein, daß:

$$\sigma_n < \frac{\beta_n - \alpha_n}{2},$$

und so klein, daß:

$$\int_{\alpha_n}^{\alpha_n+\sigma_n} |\varphi_n(\xi)|\, d\xi + \int_{\beta_n-\sigma_n}^{\beta_n} |\varphi_n(\xi)|\, d\xi < 1.$$

Sodann definieren wir $g_n(\xi)$ durch die Festsetzung: in $\langle \alpha_n + \sigma_n, \beta_n - \sigma_n \rangle$ und außerhalb $\langle \alpha_n, \beta_n \rangle$ stimmt $g_n(\xi)$ mit $h_n(\xi)$ überein, in $\langle \alpha_n, \alpha_n + \sigma_n \rangle$ aber, sowie in $\langle \beta_n - \sigma_n, \beta_n \rangle$ mit derjenigen linearen Funktion, die in den Endpunkten des betreffenden Intervalles $= h_n(\xi)$ ist. Dann ist offenbar:

$$\left|\int_a^b g_n(\xi)\,\varphi_n(\xi)\,d\xi - \int_a^b h_n(\xi)\,\varphi_n(\xi)\,d\xi\right| \leqq 1,$$

und wir haben in den $g_n(\xi)$ eine Folge absolut stetiger Funktionen vor uns mit folgenden Eigenschaften: Es ist

$$0 \leqq g_n(\xi) \leqq 1;$$

die Totalvariation von $g_n(\xi)$ in $< a, b >$ hat den Wert 2, und es ist:

$$\lim_{n=\infty} \int_a^b g_n(\xi)\,\varphi_n(\xi)\,d\xi = +\infty. \tag{16}$$

Wäre nun für eine dieser Funktionen $g_i(\xi)$ die Folge der Integrale:

$$\int_a^b g_i(\xi)\,\varphi_n(\xi)\,d\xi \quad (n = 1, 2, \ldots)$$

nicht geschränkt, so wäre damit unsere Behauptung erwiesen, indem wir $g(\xi) = g_i(\xi)$ oder $= -g_i(\xi)$ setzen. Wir werden also annehmen, es gelte für jedes i eine Ungleichung der Form (6).

Die beim Beweise von Satz III eingeführten Abkürzungen L_n und $I_n(u)$ behalten wir bei. Wie dort bestimmen wir die Teilfolge der Indizes: $n_1, n_2, \ldots, n_i, \ldots$ Daß (7) erfüllt werden kann, folgt nun aus Voraussetzung (15), daß (9) erfüllt werden kann, aus (16).

Wie dort definieren wir die Funktion $g(\xi)$ und wollen uns überzeugen, daß sie absolut stetig ist in $< a, b >$. Das heißt wir haben zu zeigen: ist $\varepsilon > 0$ beliebig gegeben, so gehört dazu ein $\eta > 0$, so daß für jede Menge sich nicht überdeckender Teilintervalle $\delta_1, \delta_2, \ldots, \delta_k, \ldots$ von $< a, b >$, deren Gesamtinhalt:

$$\delta_1 + \delta_2 + \ldots + \delta_k + \ldots < \eta \tag{17}$$

ist, die Ungleichung besteht:[1]

$$\sum_{k=1}^{\infty} V(g, \delta_k) < \varepsilon. \tag{18}$$

Da jedes $g_n(\xi)$ in $< a, b >$ die Totalvariation 2 hat, kann zunächst i_0 so groß gewählt werden, daß:

$$g^*(\xi) = \sum_{i=i_0+1}^{\infty} \frac{1}{2^{i-1} L_{n_{i-1}}}\, g_{n_i}(\xi)$$

in $< a, b >$ eine Totalvariation $< \frac{\varepsilon}{2}$ hat. Dann ist erst recht, für jede Menge sich nicht überdeckender Teilintervalle δ_k von $< a, b >$:

$$\sum_{k=1}^{\infty} V(g^*, \delta_k) < \frac{\varepsilon}{2}. \tag{19}$$

Da jede der endlich vielen Funktionen $g_{n_1}(\xi), g_{n_2}(\xi), \ldots, g_{n_{i_0}}(\xi)$ absolut stetig ist, kann $\eta > 0$ so klein gewählt werden, daß aus (17) folgt:

$$\sum_{k=1}^{\infty} V(g_{n_i}, \delta_k) < \frac{\varepsilon}{2^{i_0+1}} \quad (i = 1, 2 \ldots, i_0). \tag{20}$$

Aus (19) und (20) aber folgt (18), womit die absolute Stetigkeit von $g(\xi)$ nachgewiesen ist.

---

[1] Es bedeutet $V(g, \delta_k)$ die Totalvariation von g im Intervalle δ_k.

Endlich beweist man wie oben das Bestehen von Ungleichung (14), womit der Beweis von Satz IV beendet ist.

Ganz ebenso beweist man:

IV *a.* »Ist $\varphi_n(\xi)$ $(n = 1, 2, \ldots)$ eine Folge in $(-\infty, +\infty)$ integrierbarer Funktionen, und gibt es eine Folge von Intervallen $< \alpha_i, \beta_i >$ und von Indizes n_i mit $\lim_{i=\infty} n_i = \infty$, so daß (15) gilt, so gibt es eine in $(-\infty, +\infty)$ geschränkte, in jedem endlichen Intervalle absolut stetige Funktion $g(\xi)$, für die:

$$\overline{\lim_{n=\infty}} \int_{-\infty}^{+\infty} g(\xi)\, \varphi_n(\xi)\, d\xi = +\infty.\text{«}$$

§ 4.

Sei $\omega_1(x), \omega_2(x), \ldots, \omega_\nu(x), \ldots$ ein normiertes Orthogonalsystem des Intervalles $< a, b >$, das heißt es sei:

$$\int_a^b \omega_\mu(x) \cdot \omega_\nu(x)\, dx = 0 \quad (\mu \neq \nu)$$

$$\int_a^b (\omega_\nu(x))^2\, dx = 1.$$

Das Orthogonalsystem heißt vollständig, wenn es keine Funktion $\omega(x)$ gibt, deren Quadrat in $< a, b >$ integrierbar ist, und für die:

$$\int_a^b \omega(x) \cdot \omega_\nu(x)\, dx = 0 \quad (\nu = 1, 2 \ldots); \quad \int_a^b (\omega(x))^2\, dx \neq 0$$

wäre.

Für jede Funktion $f(x)$, deren Quadrat in $< a, b >$ integrierbar ist, existieren die »Fourier'schen Konstanten« in bezug auf unser Orthogonalsystem:[1]

(1) $$f_\nu = \int_a^b f(\xi)\, \omega_\nu(\xi)\, d\xi \quad (\nu = 1, 2 \ldots),$$

und bekanntlich ist das Orthogonalsystem ein vollständiges dann und nur dann, wenn für jede Funktion $f(x)$, deren Quadrat in $< a, b >$ integrierbar ist, die Gleichung besteht:

$$\sum_{\nu=1}^{\infty} f_\nu^2 = \int_a^b (f(\xi))^2\, d\xi.$$

Eine unmittelbare Folge daraus ist, daß für jedes Paar in $< a, b >$ quadratisch integrierbarer Funktionen $f(x)$ und $g(x)$, deren Fourier'sche Konstanten in bezug auf unser Orthogonalsystem f_ν und g_ν seien, die Gleichung besteht:

(2) $$\sum_{\nu=1}^{\infty} f_\nu \cdot g_\nu = \int_a^b f(\xi) \cdot g(\xi)\, d\xi.$$

Wir wollen diese Formel als die Parseval'sche Formel bezeichnen,[2] weil sie die Verallgemeinerung des bekannten Parseval'schen Theoremes aus der Theorie der Fourier'schen Reihe[3] auf beliebige normierte vollständige Orthogonalsysteme darstellt.

---

[1] In der Tat ist in der Definition des normierten Orthogonalsystemes die Tatsache enthalten, daß jede seiner Funktion $\omega_\nu(x)$ in $< a, b >$ von integrierbarem Quadrate ist.

[2] D. Hilbert bezeichnet sie als »Vollständigkeitsrelation.«

[3] Siehe zum Beispiel Ch. J. de la Vallée Poussin, Cours d'analyse, tome II (2 éd), p. 165.

Es existieren die Fourier'schen Konstanten (1) nicht nur für jede in $< a, b >$ samt ihrem Quadrate integrierbare Funktion, sondern für jede in $< a, b >$ integrierbare Funktion dann und nur dann, wenn jede einzelne Funktion $\omega_\nu(x)$ unseres Orthogonalsystemes in $< a, b >$ geschränkt ist, abgesehen höchstens von einer Nullmenge.

In der Tat, ist ein $\omega_\nu(x)$ nicht geschränkt, und kann es auch nicht durch Abänderung seiner Werte in einer Nullmenge in eine geschränkte Funktion verwandelt werden, so gibt es sicher eine in $< a, b >$ integrierbare Funktion $f(x)$, für die das Integral:

$$\int_a^b f(x)\, \omega_\nu(x)\, dx$$

nicht existiert.[1]

Wir nehmen also an, es sei jedes einzelne $\omega_\nu(x)$ in $< a, b >$ geschränkt (abgesehen höchstens von einer Nullmenge). Ist dann $f(x)$ in $< a, b >$ integrierbar und $g(x)$ in $< a, b >$ geschränkt, so existiert das Integral:

$$\int_a^b f(\xi)\, g(\xi)\, d\xi,$$

und wir können die Frage aufwerfen, ob nun die Parseval'sche Formel (2) für jedes Funktionenpaar $f(x)$, $g(x)$ gilt, von denen in $< a, b >$ die eine integrierbar, die andere geschränkt ist.

Sei zunächst $g(x)$ eine gegebene in $< a, b >$ geschränkte Funktion. Dann gilt der Satz:

V. Damit bei gegebenem, in $< a, b >$ geschränkten $g(x)$ die Parseval'sche Formel (2) für jedes in $< a, b >$ integrierbare f gelte, ist notwendig und hinreichend, daß es ein M gibt, so daß für alle x von $< a, b >$, abgesehen von einer Nullmenge, und alle n:

$$\left| \sum_{\nu=1}^{n} g_\nu\, \omega_\nu(x) \right| < M. \tag{3}$$

In der Tat, die Frage nach der Giltigkeit von (2) ist gleichbedeutend mit der Frage nach der Giltigkeit von:

$$\lim_{n=\infty} \int_a^b f(\xi) \left\{ g(\xi) - \sum_{\nu=1}^{n} g_\nu\, \omega_\nu(\xi) \right\} d\xi = 0. \tag{4}$$

Wir setzen:

$$g(\xi) - \sum_{\nu=1}^{n} g_\nu\, \omega_\nu(\xi) = \varphi_n(\xi).$$

Nach einem Satze von Lebesgue[2] gilt nun (4) für jedes in $< a, b >$ integrierbare f dann und nur dann, wenn $\varphi_n(\xi)$ folgenden Bedingungen genügt:

1. Es gibt ein M, so daß, abgesehen von einer Nullmenge, für alle x von $< a, b >$ und alle n:

$$|\varphi_n(\xi)| < M.$$

2. Für jedes Teilintervall $< \alpha, \beta >$ von $< a, b >$ ist:

$$\lim_{n=\infty} \int_\alpha^\beta \varphi_n(\xi)\, d\xi = 0.$$

Hievon ist, da nach Voraussetzung $g(\xi)$, abgesehen von einer Nullmenge, geschränkt ist, Bedingung 1. gleichbedeutend mit (3), und Bedingung 2. ist sicher erfüllt. In der Tat setzen wir: $f(\xi) = 1$ in

[1] H. Lebesgue, Ann. de Toul. Serie 3, Bd. 1, p. 38.

[2] Satz I meiner 1. Mitteilung.

$< \alpha, \beta >$ und $= 0$ außerhalb $< \alpha, \beta >$, so können wir, da die Funktionen $f(\xi)$ und $g(\xi)$ nun beide in $< a, b >$ samt ihrem Quadrate integrierbar sind, Formel (2) anwenden und erhalten:

$$\int_\alpha^\beta g(\xi)\, d\xi = \sum_{\nu=1}^{\infty} g_\nu \int_\alpha^\beta \omega_\nu(\xi)\, d\xi.$$

Damit ist Satz V bewiesen. Wir sind nun auch in der Lage, die aufgeworfene Frage zu beantworten:

VI. Damit die Parseval'sche Formel (2) für jedes Funktionenpaar $f(x), g(x)$ gelte, von denen in $< a, b >$ die eine geschränkt, die andere integrierbar ist, ist notwendig und hinreichend, daß es ein M gibt, so daß für alle x von $< a, b >$ (abgesehen von einer Nullmenge) und alle n:

(5) $$\int_a^b \left| \sum_{\nu=1}^{n} \omega_\nu(\xi)\, \omega_\nu(x) \right| d\xi < M.$$

Die Bedingung ist hinreichend: in der Tat, sei $g(x)$ in $< a, b >$ geschränkt:

$$|g(x)| < G \quad \text{in} < a, b >.$$

Dann ist (abgesehen von einer Nullmenge):

$$\left| \sum_{\nu=1}^{n} g_\nu\, \omega_\nu(x) \right| = \left| \int_a^b \sum_{\nu=1}^{n} g(\xi)\, \omega_\nu(\xi)\, \omega_\nu(x)\, d\xi \right| \leqq G \int_a^b \left| \sum_{\nu=1}^{n} \omega_\nu(\xi)\, \omega_\nu(x) \right| d\xi < G \cdot M.$$

Es ist also für jedes einzelne geschränkte $g(x)$ Bedingung (3) von Satz V erfüllt.

Die Bedingung ist notwendig: angenommen, sie sei nicht erfüllt; es gäbe dann zu jeder natürlichen Zahl i eine in $< a, b >$ gelegene Menge M_i mit von Null verschiedenem Inhalte, so daß zu jedem Punkte x von M_i ein Index n_x gehört, für den:

(6) $$\int_a^b \left| \sum_{\nu=1}^{n_x} \omega_\nu(\xi)\, \omega_\nu(x) \right| d\xi > i.$$

Abgesehen von einer Nullmenge ist überall in $< a, b >$ $\omega_\nu(x)$ Ableitung seines unbestimmten Integrales. Es sind daher auch weiter überall in $< a, b >$, abgesehen von einer Nullmenge, sämtliche $\omega_\nu(x)$ Ableitung ihrer unbestimmten Integrale, und es gibt daher auch in jeder Menge M_i einen Punkt x_i, in dem jedes $\omega_\nu(x)$ Ableitung seines unbestimmten Integrales ist.

Wir entnehmen also aus (6): es gibt eine Punktfolge x_i in $< a, b >$, sowie eine Indizesfolge n_i, so daß:

$$\overline{\lim_{i=+\infty}} \int_a^b \left| \sum_{\nu=1}^{n_i} \omega_\nu(\xi)\, \omega_\nu(x_i) \right| d\xi = +\infty,$$

und so, daß in jedem Punkte x_i jedes $\omega_\nu(x)$ Ableitung seines unbestimmten Integrals ist.

Indem wir nun in Satz III setzen:

$$\varphi_i(\xi) = \sum_{\nu=1}^{n_i} \omega_\nu(\xi)\, \omega_\nu(x_i),$$

sehen wir: es gibt eine in $< a, b >$ geschränkte (und überdies stetige) Funktion $g(x)$, für die:

$$\overline{\lim_{i=\infty}} \int_a^b \sum_{\nu=1}^{n_i} g(\xi)\, \omega_\nu(\xi)\, \omega_\nu(x_i)\, d\xi = +\infty$$

ist. Da aber:

$$\int_a^b g(\xi)\, \omega_\nu(\xi)\, d\xi = g_\nu$$

ist, haben wir weiter:

$$\overline{\lim_{i=\infty}} \sum_{\nu=1}^{n_i} g_\nu\, \omega_\nu(x_i) = +\infty. \tag{7}$$

Zufolge der Wahl der Punkte x_i ist nun in jedem Punkte x_i die Funktion:

$$\sum_{\nu=1}^{n} g_\nu\, \omega_\nu(x) \tag{8}$$

Ableitung ihres unbestimmten Integrales, so daß (7) besagt: die rechtsseitige obere Ableitung des unbestimmten Integrales von (8) ist nicht nach oben geschränkt für alle n und alle x von $< a, b >$. Nun unterscheidet sich der Ausdruck (8) von dieser rechtsseitigen oberen Ableitung nur in einer Nullmenge; und da eine geschränkte rechtsseitige obere Ableitung ihre obere Grenze in einem Intervalle nicht ändert, wenn Nullmengen vernachlässigt werden,[1] so sehen wir, daß der Ausdruck (8) auch nicht durch bloße Wertänderung in einer Nullmenge in einen für alle x von $< a, b >$ und alle n geschränkten Ausdruck verwandelt werden kann; es genügt also die Funktion $g(x)$ der Bedingung (3) von Satz V nicht. Damit ist Bedingung (5) von Satz VI als notwendig erwiesen.

Als Beispiel eines die Bedingung von Satz VI erfüllenden Orthogonalsystemes diene das bekannte Haar'sche Orthogonalsystem.[2] Ein Beispiel eines die Bedingung von Satz VI nicht erfüllenden Orthogonalsystemes liefert das trigonometrische Orthogonalsystem; denn es ist bekanntlich:

$$\frac{1}{2\pi} + \frac{1}{\pi} \sum_{\nu=1}^{n} (\cos \nu\xi \cos \nu x + \sin \nu\xi \sin \nu x) = \frac{1}{2\pi} \frac{\sin \frac{2n+1}{2}(\xi - x)}{\sin \frac{\xi - x}{2}} \tag{9}$$

und:

$$\lim_{n=\infty} \frac{1}{2\pi} \int_{-\pi}^{\pi} \left| \frac{\sin \frac{2n+1}{2}(\xi - x)}{\sin \frac{\xi - x}{2}} \right| d\xi = +\infty.$$

Es gibt also in $< -\pi, \pi >$ eine integrierbare Funktion f und eine stetige Funktion g, für die die Parseval'sche Formel aus der Theorie der Fourier'schen Reihen nicht gilt.

VII. Damit die Parseval'sche Formel (2) für jedes Funktionenpaar $f(x)$, $g(x)$ gelte, von denen in $< a, b >$ die eine geschränkt und von geschränkter Variation, die andere integrierbar ist, ist notwendig und hinreichend, daß es ein M gibt, so daß für alle Teilintervalle $< \alpha, \beta >$ von $< a, b >$, alle x von $< a, b >$, abgesehen von einer Nullmenge, und alle n:

$$\left| \int_\alpha^\beta \sum_{\nu=1}^{n} \omega_\nu(\xi)\, \omega_\nu(x)\, d\xi \right| < M. \tag{10}$$

Die Bedingung ist hinreichend. Denn ist

$$|g(x)| < G \quad \text{in} < a, b >,$$

[1] Siehe zum Beispiel H. Lebesgue Leçons sur l'intégration, p. 80.

[2] Math. Ann. *69*, p. 361 ff.

und ist V die Totalvariation von g in $< a, b >$, so liefert der zweite Mittelwertsatz der Integralrechnung [1] (abgesehen von einer Nullmenge):

$$\left|\sum_{\nu=1}^{n} g_\nu \omega_\nu(x)\right| = \left|\int_a^b g(\xi) \sum_{\nu=1}^{n} \omega_\nu(\xi)\, \omega_\nu(x)\, d\xi\right| \leqq (G+V)\cdot M.$$

Es ist also für jedes einzelne $g(x)$ geschränkter Variation Bedingung (3) von Satz V erfüllt.

Die Bedingung ist notwendig. Angenommen, sie sei nicht erfüllt. Dann gibt es zu jeder natürlichen Zahl i eine Teilmenge M_i von $< a, b >$ mit von 0 verschiedenem Inhalte, zu deren jedem Punkte x ein Teilintervall $< \alpha_x, \beta_x >$ von $< a, b >$ und ein Index n_x gehört, so daß:

$$\left|\int_{\alpha_x}^{\beta_x} \sum_{\nu=1}^{n_x} \omega_\nu(\xi)\, \omega_\nu(x)\, d\xi\right| > i.$$

Wie beim Beweise von Satz VI sehen wir: es gibt in $< a, b >$ eine Folge von Teilintervallen $< \alpha_i, \beta_i >$ und eine Punktfolge x_i, sowie eine Indizesfolge n_i, so daß:

$$\lim_{i=\infty} \left|\int_{\alpha_i}^{\beta_i} \sum_{\nu=1}^{n_i} \omega_\nu(\xi)\, \omega_\nu(x_i)\, d\xi\right| = +\infty,$$

und so, daß in jedem Punkte x_i jedes $\omega_\nu(x)$ Ableitung seines unbestimmten Integrales ist.

Somit gibt es also nach Satz IV (oder IVa) eine Funktion $g(x)$, die in $< a, b >$ geschränkt und von geschränkter Variation (ja sogar absolut stetig) ist und für die:

$$\overline{\lim_{i=+\infty}} \int_a^b g(\xi) \sum_{\nu=1}^{n_i} \omega_\nu(\xi)\, \omega_\nu(x_i)\, d\xi = \overline{\lim_{i=+\infty}} \sum_{\nu=1}^{n_i} g_\nu\, \omega_\nu(x_i) = +\infty,$$

woraus man, wie beim Beweise von Satz VI weiter schließt, daß $g(x)$ der Bedingung (3) von Satz V nicht genügt. Damit ist gezeigt, daß Bedingung (10) von Satz VII notwendig ist.

Satz VII lehrt für das trigonometrische Orthogonalsystem, daß die Parseval'sche Formel in der Theorie der Fourier'schen Reihen gilt für jedes Paar von Funktionen f und g, von denen in $< -\pi, \pi >$ die eine integrierbar, die andere von geschränkter Variation ist;[2] es folgt dies nach Satz VII unmittelbar daraus, daß für den Kern (9) bekanntlich eine Ungleichung gilt:

$$\left|\frac{1}{2\pi}\int_\alpha^\beta \frac{\sin\frac{2n+1}{2}(\xi-x)}{\sin\frac{\xi-x}{2}}\, d\xi\right| < M$$

für alle n, alle x von $< -\pi, \pi >$ und alle Teilintervalle $< \alpha, \beta >$ von $< -\pi, \pi >$.

Wählt man für $g(x)$ speziell die Funktion, die $= 1$ ist in einem Teilintervalle $< \alpha, \beta >$ von $< -\pi, \pi >$, sonst $= 0$, so liefert die Parseval'sche Formel das bekannte Theorem [3] von der Integration der Fourier'schen Reihe einer beliebigen in $< -\pi, \pi >$ integrierbaren Funktion f.

Satz V liefert übrigens unmittelbar, indem man unter $g(x)$ die Funktion versteht, die $= 1$ ist in $< \alpha, \beta >$, sonst $= 0$:

Damit für jede in $< a, b >$ integrierbare Funktion f und jedes Teilintervall $< \alpha, \beta >$ von $< a, b >$ die Formel gelte:

$$\int_\alpha^\beta f(x)\, dx = \sum_{\nu=1}^{\infty} f_\nu \int_\alpha^\beta \omega_\nu(x)\, dx, \tag{11}$$

---

[1] Siehe zum Beispiel H. Lebesgue Ann. de Toul. Serie 3, Bd. 1, p. 37.

[2] Zuerst bemerkt von W. H. Young, Proc. Royal Soc. A, 85, p. 412.

[3] Siehe zum Beispiel de la Vallée Poussin, Cours d'analyse, tome II (2. éd), p. 134.

ist notwendig und hinreichend, daß zu jedem Teilintervalle $<\alpha, \beta>$ von $<a, b>$ ein M gehört, so daß für alle n, und alle x von $<a, b>$ (abgesehen von einer Nullmenge):

$$\left| \sum_{\nu=1}^{n} \int_{\alpha}^{\beta} \omega_\nu(\xi)\, d\xi \cdot \omega_\nu(x) \right| < M. \tag{12}$$

Diese Bedingung ist sicher erfüllt, wenn die Bedingung von Satz VII erfüllt ist.

Hingegen reicht Bedingung (12) noch keineswegs aus, um das Bestehen folgender Beziehung sicherzustellen, die wesentlich mehr besagt als (11):

$$\lim_{n=\infty} \int_a^b \left| f(x) - \sum_{\nu=1}^{n} f_\nu\, \omega_\nu(x) \right| dx = 0. \tag{13}$$

Das Bestehen dieser Beziehung für jedes in $<a, b>$ integrierbare f hat nämlich, wie aus einem Satze von Lebesgue[1] folgt, das Bestehen der Parseval'schen Formel für jedes integrierbare f und geschränkte g zur Folge. Es kann also, wenn es sich zum Beispiel um Fourier'sche Reihen handelt, (13) nicht für jedes integrierbare f gelten.

§ 5.

Nachdem wir gesehen haben, daß die Parseval'sche Formel keineswegs für jedes vollständige normierte Orthogonalsystem $\omega_\nu(x)$ und jedes Paar von Funktionen $f(x)$, $g(x)$ gilt, von denen die eine geschränkt, die andere integrierbar ist, wollen wir die Frage behandeln, ob sich nicht Summationsverfahren angeben lassen, durch die aus der Reihe:

$$\sum_{\nu=1}^{\infty} f_\nu g_\nu$$

eine Folge von Ausdrücken hergeleitet wird, die stets gegen:

$$\int_a^b f(x)\, g(x)\, dx$$

konvergiert, wenn von den beiden Funktionen f und g in $<a, b>$ die eine geschränkt, die andere integrierbar ist. Zur Lösung dieser Aufgabe werden wir geführt durch folgenden Satz:

VIII. Es genüge der Kern $\varphi(\xi, x, n)$ allen Voraussetzungen von Satz II. Ist dann $f(x)$ in $<a, b>$ integrierbar und $g(x)$ in $<a, b>$ geschränkt, so ist:

$$\int_a^b f(x)\, g(x)\, dx = \lim_{n=\infty} \int_a^b I_n(f, x)\, g(x)\, dx. \tag{1}$$

In der Tat, nach Satz II ist:

$$\lim_{n=\infty} \int_a^b |f(x) - I_n(f, x)|\, dx = 0,$$

daher, wenn $g(x)$ in $<a, b>$ geschränkt ist, auch:

$$\lim_{n=\infty} \int_a^b |g(x)| \cdot |f(x) - I_n(f, x)|\, dx = 0,$$

und somit auch:

$$\lim_{n=\infty} \int_a^b g(x)\, (f(x) - I_n(f, x))\, dx = 0, \tag{2}$$

wodurch (1) bewiesen ist.

[1] Satz III meiner ersten Mitteilung.

Dieser Satz wurde unter wesentlich engeren Voraussetzungen über $\varphi(\xi, x, n)$ bereits von H. Lebesgue bewiesen.[1]

Eine unmittelbare Folge aus Satz VIII ist folgende Bemerkung:

Sei $\omega_\nu(x)$ $(\nu = 1, 2, \ldots)$ ein vollständiges normiertes Orthogonalsystem des Intervalles $< a, b >$, und sei jedes einzelne $\omega_\nu(x)$ in $< a, b >$ geschränkt. Es existieren dann in bezug auf dieses Orthogonalsystem sowohl die Fourier'schen Konstanten f_ν von $f(x)$ als auch die Fourier'schen Konstanten $F_\nu(n)$ von $I_n(f, x)$ und zwar ist:

$$f_\nu = \lim_{n=\infty} F_\nu(n). \tag{3}$$

In der Tat, wir haben, um (3) zu erhalten, in (1) nur $g(x)$ durch $\omega_\nu(x)$ zu ersetzen. Dies legt den Gedanken nahe, daß man das gewünschte Summationsverfahren einfach erhält, indem man in der Reihe:

$$\sum_{\nu=1}^{\infty} f_\nu g_\nu$$

immer f_ν durch $F_\nu(n)$ ersetzt und den Grenzwert $\lim\limits_{n=\infty}$ bildet. Wir werden sehen, daß dies in der Tat, unter weiteren Einschränkungen über $\varphi(\xi, x, n)$ richtig ist, und zwar ohne jede Einschränkung bezüglich des Orthogonalsystems der $\omega_\nu(x)$.

IX. Es genüge $\varphi(\xi, x, n)$ für $a \leqq \xi \leqq b$, $a \leqq x \leqq b$ außer den Voraussetzungen 1., 2., 3., von Satz II noch folgender Voraussetzung:

4. Zu jedem n gibt es ein M_n, so daß im ganzen Quadrate: $a \leqq \xi \leqq b$, $a \leqq x \leqq b$ die Ungleichung gilt:

$$|\varphi(\xi, x, n)| < M_n.$$

Ist dann $\omega_\nu(x)$ ein vollständiges normiertes Orthogonalsystem in $< a, b >$, und wird gesetzt:

$$F_\nu(n) = \int_a^b \left(\int_a^b f(\xi)\, \varphi(\xi, x, n)\, d\xi \right) \cdot \omega_\nu(x)\, dx,$$

so gilt für jedes in $< a, b >$ integrierbare f und jedes in $< a, b >$ geschränkte g die Formel:

$$\int_a^b f(x)\, g(x)\, dx = \lim_{n=\infty} \sum_{\nu=1}^{\infty} F_\nu(n) \cdot g_\nu. \tag{4}$$

Unter Berufung auf Satz VIII genügt es nachzuweisen, daß:

$$\int_a^b I_n(f, x)\, g(x)\, dx = \sum_{\nu=1}^{\infty} F_\nu(n) \cdot g_\nu. \tag{5}$$

ist. Wegen unserer Voraussetzung 4. ist nun aber:

$$|I_n(f, x)| \leqq \int_a^b |f(\xi)\, \varphi(\xi, x, n)|\, d\xi \leqq M_n \int_a^b |f(\xi)|\, d\xi,$$

das heißt für jedes einzelne n ist $I_n(f, x)$ in $< a, b >$ geschränkt. Da nun die $F_\nu(n)$ nichts anderes sind, als die Fourier'schen Konstanten von $I_n(f, x)$ und für jedes Paar geschränkter (und somit samt ihrem Quadrate integrierbarer) Funktionen das Parseval'sche Theorem gilt, ist (5), und damit Satz IX bewiesen.

[1] Ann. de Toul. Ser. 3, Bd. 1, p. 105.

Um aus Satz IX speziell Resultate über die Koeffizienten der Fourier'schen Reihe zu erhalten, haben wir für die $\omega_\nu(x)$ das trigonometrische Orthogonalsystem für das Intervall $<-\pi, \pi>$ zu nehmen:

$$\frac{1}{\sqrt{2\pi}}, \quad \frac{1}{\sqrt{\pi}}\cos \nu x, \quad \frac{1}{\sqrt{\pi}}\sin \nu x \quad (\nu = 1, 2, \ldots).$$

Wir setzen:

$$a_\nu = \frac{1}{\pi}\int_{-\pi}^{\pi} f(\xi)\cos \nu\xi\, d\xi; \quad b_\nu = \frac{1}{\pi}\int_{-\pi}^{\pi} f(\xi)\sin \nu\xi\, d\xi,$$

$$\alpha_\nu = \frac{1}{\pi}\int_{-\pi}^{\pi} g(\xi)\cos \nu\xi\, d\xi; \quad \beta_\nu = \frac{1}{\pi}\int_{-\pi}^{\pi} g(\xi)\sin \nu\xi\, d\xi.$$

Wir nehmen weiter an, der Kern $\varphi(\xi, x, n)$ habe die Gestalt:

$$\varphi(\xi, x, n) = \varphi(\xi - x, n),$$

wo $\varphi(u, n)$ für jedes n in $< -2\pi, 2\pi >$ definiert sei und den Relationen:

$$\varphi(-u) = \varphi(u); \quad \varphi(u + 2\pi) = \varphi(u)$$

genüge. Setzen wir noch:

$$\int_{-\pi}^{\pi} \varphi(u, n)\cos \nu u\, du = \varphi_\nu(n),$$

so wird:

$$\frac{1}{\sqrt{\pi}}\int_{-\pi}^{\pi}\left(\int_{-\pi}^{\pi} f(\xi)\,\varphi(\xi - x, n)\, d\xi\right)\cos \nu x\, dx =$$

$$\frac{1}{\sqrt{\pi}}\int_{-\pi}^{\pi} f(\xi)\left(\int_{-\pi}^{\pi}\varphi(u, n)\cos \nu(\xi - u)\, du\right) d\xi = \sqrt{\pi}\cdot\varphi_\nu(n)\, a_\nu$$

$$\frac{1}{\sqrt{\pi}}\int_{-\pi}^{\pi}\left(\int_{-\pi}^{\pi} f(\xi)\,\varphi(\xi - x, n)\, d\xi\right)\sin \nu x\, dx = \sqrt{\pi}\,\varphi_\nu(n)\cdot b_\nu,$$

und (4) geht über in:

$$\text{(4a)} \qquad \frac{1}{\pi}\int_a^b f(x)\, g(x)\, dx = \lim_{n=\infty}\left\{\varphi_0(n)\frac{a_0\alpha_0}{2} + \sum_{\nu=1}^{\infty}\varphi_\nu(n)(a_\nu\alpha_\nu + b_\nu\beta_\nu)\right\}.$$

Hier wird also die Parseval'sche Reihe summiert, indem ihre Glieder mit den Konvergenz erzeugenden Faktoren $\varphi_\nu(n)$ multipliziert werden.

Wählen wir dann für $\varphi(u, n)$ den Poisson'schen Kern:

$$\varphi(u, n) = \frac{1}{2\pi}\,\frac{1 - r_n^2}{1 - 2r_n\cos u + r_n^2} \quad (r_n < 1, \lim_{n=\infty} r_n = 1),$$

so wird:

$$\varphi_\nu(n) = r_n^\nu,$$

und Formel (4a) ergibt:

$$\text{(6)} \qquad \frac{1}{\pi}\int_{-\pi}^{\pi} f(x)\, g(x)\, dx = \frac{1}{2}a_0\alpha_0 + \lim_{r=1-0}\sum_{\nu=1}^{\infty} r^\nu(a_\nu\alpha_\nu + b_\nu\beta_\nu)$$

für jedes Paar von Funktionen f, g, von denen in $< -\pi, \pi >$ die eine integrierbar, die andere geschränkt ist. Formel (6) wurde bewiesen von W. Groß.[1]

[1] Wien. Ber. Abt. IIa, Bd. 124, p. 1025. Sie ergibt sich übrigens, ebenso wie die Formeln (7) und (8) auch als Spezialfall einer Überlegung von W. H. Young, Proc. Royal Soc. A., *85*, p. 401 ff.

Wählt man für $\varphi(u, n)$ den Fejér'schen Kern:

$$\varphi(u, n) = \frac{1}{2n\pi}\left\{\frac{\sin\frac{nu}{2}}{\sin\frac{u}{2}}\right\}^2,$$

so geht (4*a*) über in:

$$\frac{1}{\pi}\int_{-\pi}^{\pi} f(x)g(x)\,dx = \frac{1}{2}a_0\alpha_0 + \lim_{n=\infty}\sum_{\nu=1}^{n}\left(1-\frac{\nu}{n}\right)(a_\nu\alpha_\nu + b_\nu\beta_\nu). \tag{7}$$

Wählt man für $\varphi(u, n)$ den Kern von de la Vallée Poussin:

$$\varphi(u\ n) = \frac{1}{\pi}2^{2n-1}\frac{(n!)^2}{(2n)!}\left(\cos\frac{u}{2}\right)^{2n},$$

so geht (4*a*) über in:

$$\frac{1}{\pi}\int_{-\pi}^{\pi} f(x)\,g(x)\,dx = \frac{1}{2}a_0\alpha_0 + \lim_{n=\infty}\sum_{\nu=1}^{n}\frac{(n!)^2}{(n-\nu)!\,(n+\nu)!}(a_\nu\alpha_\nu + b_\nu\beta_\nu), \tag{8}$$

und die Formeln (7), (8) gelten für jedes Paar von Funktionen, von denen in $< -\pi, +\pi >$ die eine integrierbar, die andere geschränkt ist.

§ 6.

Der im letzten Paragraphen zur Summation der Parseval'schen Reihe verwendete Gedanke kann auch zur Summation der Reihenentwicklung:

$$\sum_{\nu=1}^{\infty} f_\nu\,\omega_\nu(x)$$

einer gegebenen Funktion $f(x)$ nach den Funktionen $\omega_\nu(x)$ eines vollständigen normierten Orthogonalsystemes verwendet werden. Man erhält so augenblicklich folgendes Summationstheorem:

X. Ist $\varphi(\xi, x, n)$ für jedes n und jedes x von (a, b) nach ξ samt seinem Quadrate integrierbar in $< a, b >$, und wird gesetzt:

$$\Phi_\nu(x, n) = \int_a^b \varphi(\xi, x, n)\,\omega_\nu(\xi)\,d\xi, \tag{1}$$

so gilt, wenn $f(x)$ in $< a, b >$ samt seinem Quadrate integrierbar ist, die Formel:

$$f(x) = \lim_{n=\infty}\sum_{\nu=1}^{\infty} f_\nu\,\Phi_\nu(x, n) \tag{2}$$

in jedem Punkte x von (a, b), in dem:

$$f(x) = \lim_{n=\infty}\int_a^b f(\xi)\,\varphi(\xi, x, n)\,d\xi; \tag{3}$$

sie gilt gleichmäßig in jedem Teilintervalle $< \alpha, \beta >$ von (a, b), in dem (3) gleichmäßig gilt.

Bedingungen für $\varphi(\xi, x, n)$ unter denen (3) in jedem Punkte von (a, b) gilt, in dem die Funktion f stetig, oder erste Ableitung ihres unbestimmten Integrales, oder m-te Ableitung ihres m-fach iterierten unbestimmten Integrales ist, sowie Bedingungen, unter denen (3) gleichmäßig in jedem Teilintervalle $< \alpha, \beta >$ von (a, b) gilt, in dessen sämtlichen Punkten $f(x)$ stetig ist, findet man in der wiederholt zitierten, für diese Theorie grundlegenden Abhandlung von H. Lebesgue,[1] sowie in meiner 1. Mitteilung über diesen Gegenstand.

Es handelt sich noch darum, uns von der Voraussetzung frei zu machen, daß auch das Quadrat von f in $< a, b >$ integrierbar sei. Damit die Fourier'sche Konstanten f_ν für jedes integrierbare f existieren, müssen wir dabei wieder voraussetzen, daß jedes einzelne $\omega_\nu(\xi)$ geschränkt ist in $< a, b >$.

Satz V liefert uns dann folgendes Summationstheorem:

XI. Sei $\varphi(\xi, x, n)$ für jedes einzelne n und jedes einzelne x von (a, b) eine in $< a, b >$ geschränkte Funktion von ξ. Ferner gebe es zu jedem n und jedem x von (a, b) ein M, so daß für die durch (1) definierten Funktionen $\Phi_\nu(x, n)$ die Ungleichung:

$$\left|\sum_{\nu=1}^{\mu} \Phi_\nu(x, n)\, \omega_\nu(\xi)\right| < M$$

gilt für alle μ und alle ξ von $< a, b >$.

Dann gilt für jedes in $< a, b >$ integrierbare f die Formel (2) in jedem Punkte x von (a, b) in dem (3) gilt; sie gilt gleichmäßig in jedem Teilintervalle $< \alpha, \beta >$ von (a, b), in dem (3) gleichmäßig gilt.

Besonders einfach gestaltet sich, wie M. Schechter bemerkte,[2] die Summationsformel (2) für den Fall des trigonometrischen Orthogonalsystemes im Intervalle $< -\pi, \pi >$, wenn $\varphi(\xi, x, n)$ die Form hat:

$$\varphi(\xi, x, n) = \varphi(\xi - x, n),$$

wo $\varphi(u, n)$ eine in $< -2\pi, 2\pi >$ definierte, gerade Funktion der Periode 2π ist.

Sind dann a_ν, b_ν die Koeffizienten der Fourier'schen Reihe von f, und wird wieder (wie in § 5) gesetzt:

$$\varphi_\nu(n) = \int_{-\pi}^{\pi} \varphi(u, n) \cos \nu u \, du,$$

so nimmt (2) die Form an:

$$f(x) = \lim_{n=\infty} \left\{\frac{1}{2}\varphi_0(n)\, a_0 + \sum_{\nu=1}^{\infty} \varphi_\nu(n)(a_\nu \cos \nu x + b_\nu \sin \nu x)\right\}. \tag{4}$$

Alle gebräuchlichen Summationsformeln der trigonometrischen Reihen sind Spezialfälle dieser Formel.

In meiner 1. Mitteilung habe ich Bedingungen für $\varphi(\xi, x, n)$ entwickelt, unter denen überall, wo die m-te Ableitung $f^{(m)}(x)$ von $f(x)$ existiert, die Relation gilt:

$$f^{(m)}(x) = \lim_{n=\infty} \int_a^b f(\xi)\, \varphi(\xi, x, n)\, d\xi, \tag{5}$$

sowie Bedingungen, unter denen diese Relation gleichmäßig in jedem Teilintervalle $< \alpha, \beta >$ von (a, b) gilt, in dem $f(x)$ m-mal stetig differenzierbar ist. Und wir sehen:

XII. Gilt (5) in jedem Punkte x von (a, b), in dem $f^{(m)}(x)$ existiert, und zwar gleichmäßig in jedem Teilintervalle $< \alpha, \beta >$ von (a, b), in dem $f(x)$ m-mal stetig differenzierbar ist, so gilt, vorausgesetzt, daß $\varphi(\xi, x, n)$ den Bedingungen von Satz XI (beziehungsweise Satz X) genügt, dasselbe von der Formel:

$$f^{(m)}(x) = \lim_{n=\infty} \sum_{\nu=1}^{\infty} f_\nu\, \Phi_\nu(x, n) \tag{6}$$

[1] Sur les intégrales singulières, Ann. de Toul. Serie 3, Bd. 1, p. 25 ff.

[2] Monatsh. f. Math. Bd. 22, p. 224.

für jede in $< a, b >$ integrierbare (beziehungsweise samt ihrem Quadrate integrierbare) Funktion $f(x)$.

Handelt es sich um das trigonometrische Orthogonalsystem und hat $\varphi(\xi, x, n)$ die Form $\varphi(\xi - x, n)$, wo $\varphi(u, n)$ gerade und von der Periode 2π ist (dies kommt nur bei geradem m in Betracht), so reduziert sich (6) wieder auf:

$$f^{(m)}(x) = \lim_{n=\infty} \left\{ \frac{1}{2} \varphi_0(n) a_0 + \sum_{\nu=1}^{\infty} \varphi_\nu(n)(a_\nu \cos \nu x + b_\nu \sin \nu x) \right\}. \tag{7}$$

Hat hingegen (was bei ungeradem m in Betracht kommt), $\varphi(\xi, x, n)$ die Form $\varphi(\xi - x, n)$, wo $\varphi(u, n)$ eine ungerade Funktion der Periode 2π ist, so reduziert sich (6) auf:

$$f^{(m)}(x) = \lim_{n=\infty} \left\{ \sum_{\nu=1}^{\infty} \bar{\varphi}_\nu(n)(-a_\nu \sin \nu x + b_\nu \cos \nu x) \right\}, \tag{7a}$$

worin gesetzt ist:

$$\bar{\varphi}_\nu(n) = \int_{-\pi}^{\pi} \varphi(u, n) \sin \nu u \, du.$$

Wohl der einfachste Kern eines singulären Integrales ist der folgende, wo h eine beliebige positive Zahl bedeutet:

$$\varphi(\xi, x, h) = \begin{cases} \dfrac{1}{2h} & \text{in } (x-h, x+h) \\ 0 & \text{außerhalb } (x-h, x+h). \end{cases} \tag{8}$$

Setzt man:

$$F(x) = \int_a^x f(t)\, dt,$$

so wird:

$$\int_a^b f(\xi)\, \varphi(\xi, x, h)\, d\xi = \frac{F(x+h) - F(x-h)}{2h},$$

und somit gilt die Beziehung:

$$f(x) = \lim_{h=0} \int_a^b f(\xi)\, \varphi(\xi, x, h)\, d\xi \tag{9}$$

in jedem Punkte x von (a, b), wo:

$$f(x) = \lim_{h=0} \frac{F(x+h) - F(x-h)}{2h} \tag{10}$$

ist, insbesondere also dort, wo f Ableitung seines unbestimmten Integrales ist. Nach Satz XVIII meiner 1. Mitteilung gilt (9) gleichmäßig in jedem Teilintervalle $< \alpha, \beta >$ von (a, b), in dessen sämtlichen Punkten f stetig ist. Aus X und XI haben wir also:

XIII. Für jedes normierte Orthogonalsystem von $< a, b >$ gilt die Formel:

$$f(x) = \lim_{h=0} \frac{1}{2h} \sum_{\nu=1}^{\infty} f_\nu \int_{x-h}^{x+h} \omega_\nu(\xi)\, d\xi \tag{11}$$

für jede in $< a, b >$ samt ihrem Quadrate integrierbare Funktion f in jedem Punkte von (a, b), in dem (10) gilt; sie gilt gleichmäßig in jedem Teilintervalle $< \alpha, \beta >$ von (a, b), in dessen sämtlichen Punkten f stetig ist. — Dies gilt für alle in $<a, b>$ integrierbaren Funktionen, falls es zu jedem Teilintervall $<\alpha, \beta>$ von (a, b) ein M gibt, so daß;

$$\left|\sum_{\nu=1}^{\mu} \int_a^\beta \omega_\nu(\xi)\, d\xi \cdot \omega_\nu(x)\right| < M \tag{12}$$

für alle x von $<a, b>$ und alle μ.

Der einfachste Kern, der $f'(x)$ durch $f(x)$ darstellt, ist der folgende:

$$\varphi(\xi, x, h) = \frac{1}{4h^2} \text{ in } (x, x+2h), \quad = -\frac{1}{4h^2} \text{ in } (x-2h, x), \quad \text{sonst} = 0. \tag{13}$$

Wir haben hier:

$$\int_a^b f(\xi)\, \varphi(\xi, x, h)\, d\xi = \frac{F(x+2h) - 2F(x) + F(x-2h)}{4h^2},$$

und es gilt die Beziehung:

$$f'(x) = \lim_{h=0} \int_a^b f(\xi)\, \varphi(\xi, x, h)\, d\xi$$

in jedem Punkte x von (a, b), in dem $f'(x)$ existiert. Sie gilt nach Satz XI meiner 1. Mitteilung gleichmäßig in jedem Teilintervalle $<\alpha, \beta>$ von (a, b), in dem $f(x)$ stetig differenzierbar ist. Aus Satz XII haben wir also:

XIV. Für jedes normierte Orthogonalsystem von $<a, b>$ gilt die Formel:

$$f'(x) = \lim_{h=0} \frac{1}{4h^2} \sum_{\nu=1}^{\infty} f_\nu \left\{\int_x^{x+2h} \omega_\nu(\xi)\, d\xi - \int_{x-2h}^{x} \omega_\nu(\xi)\, d\xi\right\} \tag{14}$$

für jede in $<a, b>$ samt ihrem Quadrate integrierbare Funktion f, in jedem Punkte von (a, b), in dem $f'(x)$ existiert; sie gilt gleichmäßig in jedem Teilintervalle $<\alpha, \beta>$ von (a, b), in dem f stetig differenzierbar ist. — Ist Bedingung (12) erfüllt, so gilt dies auch für jede in $<a, b>$ integrierbare Funktion f.

Um zur Darstellung der höheren Ableitungen $f^{(m)}(x)$ zu gelangen, definieren wir $\varphi(\xi, x, h)$ für $h > 0$ durch die Vorschrift:

$$\varphi(\xi, x, h) = (-1)^k \frac{1}{(2h)^{m+1}} \binom{m}{k} \text{ in } (x+(m-2k-1)h,\ x+(m-2k+1)h)\ (k = 0, 1, \ldots, m), \tag{15}$$

außerhalb aller dieser Intervalle sei $\varphi(\xi, x, h) = 0$.

Wie man sieht, wird dann, wenn h so klein ist, daß das Intervall $<x-(m+1)h, x+(m+1)h>$ in $<a, b>$ liegt:

$$\begin{aligned}\int_a^b f(\xi)\, \varphi(\xi, x, h)\, d\xi &= \frac{1}{(2h)^{m+1}} \sum_{k=0}^{m} (-1)^k \binom{m}{k} \{F(x+(m-2k+1)h) - F(x+(m-2k-1)h)\} \\ &= \frac{1}{(2h)^{m+1}} \sum_{k=0}^{m+1} (-1)^k \binom{m+1}{k} F(x+(m-2k+1)h).\end{aligned} \tag{16}$$

Wählen wir $f(x) = x^i$, so wird:

$$F(x) = \frac{(x-a)^{i+1}}{i+1},$$

und da bekanntlich:

$$\text{(17)}\qquad \frac{1}{(2h)^{m+1}}\sum_{k=0}^{m+1}(-1)^k\binom{m+1}{k}F(x+(m-2k+1)h)=\begin{cases}0 \text{ für } F(x)=(x-a)^i\ (i=0,1,\ldots,m)\\ (m+1)! \text{ für } F(x)=(x-a)^{m+1}\end{cases}$$

ist, so sehen wir, daß der Kern (15) alle Voraussetzungen von Satz VII und Satz XI meiner 1. Mitteilung erfüllt. Das liefert den Satz:

XV. Für jedes normierte Orthogonalsystem von $<a, b>$ gilt die Formel:

$$\text{(18)}\qquad f^{(m)}(x)=\lim_{h=0}\frac{1}{(2h)^{m+1}}\sum_{\nu=1}^{\infty}f_\nu\cdot\left(\sum_{k=0}^{m}(-1)^k\binom{m}{k}\int_{x+(m-2k-1)h}^{x+(m-2k+1)h}\omega_\nu(\xi)\,d\xi\right)$$

für jede in $<a, b>$ samt ihrem Quadrate integrierbare Funktion f, in jedem Punkte von (a, b), in dem $f^{(m)}(x)$ existiert; sie gilt gleichmäßig in jedem Teilintervalle $<\alpha, \beta>$ von (a, b) in dem f m-mal stetig differenzierbar ist. – Ist Bedingung (12) erfüllt, so gilt dies für jede in $<a, b>$ integrierbare Funktion f.

Wir wollen nun die Formeln (11), (14), (18), speziell auf die trigonometrischen Reihen anwenden, für die Bedingung (12) bekanntlich erfüllt ist. Da der Kern (8) die für die Giltigkeit von (4) erforderliche Gestalt hat, kann (4) angewendet werden.[1] Dabei ist:

$$\varphi_\nu(u)=\frac{1}{2h}\int_{-h}^{h}\cos\nu u\,du=\frac{\sin\nu h}{\nu h},$$

so daß (11) hier lautet: In jedem Punkte x von $(-\pi, \pi)$ in dem f zu seinem unbestimmten Integrale F in der Beziehung steht:

$$f(x)=\lim_{h=0}\frac{F(x+h)-F(x-h)}{2h},$$

gilt für jedes in $<-\pi, \pi>$ integrierbare f die Formel:

$$\text{(19)}\qquad f(x)=\lim_{h=0}\left\{\frac{a_0}{2}+\sum_{\nu=1}^{\infty}\frac{\sin\nu h}{\nu h}(a_\nu\cos\nu x+b_\nu\sin\nu x)\right\};$$

sie gilt gleichmäßig in jedem Teilintervalle $<\alpha, \beta>$ von $(-\pi, \pi)$, in dessen sämtlichen Punkten f stetig ist.

Betrachten wir nun den Kern (13), so kann (7 a) für $m=1$ angewendet werden. Dabei ist:

$$\bar\varphi_\nu(u)=\frac{1}{4h^2}\int_0^{2h}\sin\nu u\,du-\frac{1}{4h^2}\int_{-2h}^{0}\sin\nu u\,du=\nu\cdot\left(\frac{\sin\nu h}{\nu h}\right)^2,$$

und wir erhalten: Für jedes in $<-\pi, \pi>$ integrierbare f gilt in jedem Punkte von $(-\pi, \pi)$, in dem die Ableitung f' existiert die Formel:

$$\text{(20)}\qquad f'(x)=\lim_{h=0}\sum_{\nu=1}^{\infty}\left(\frac{\sin\nu h}{\nu h}\right)^2(-\nu\cdot a_\nu\sin\nu x+\nu\cdot b_\nu\cos\nu x);$$

sie gilt gleichmäßig in jedem Teilintervalle $<\alpha, \beta>$ von $(-\pi, \pi)$, in dem f stetig differenzierbar ist.

Betrachten wir den Kern (15) für ungerades m, so können wir wieder (7 a) anwenden, und zwar erhalten wir nach (16):

$$\bar\varphi_\nu(u)=-\frac{1}{(2h)^{m+1}}\frac{1}{\nu}\sum_{k=0}^{m+1}(-1)^k\binom{m+1}{k}\cos(m-2k+1)\nu\cdot h.$$

[1] Vgl. M. Schechter a. a. O., p. 232.

Das ist der reelle Teil von:

$$-\frac{1}{(2h)^{m+1}}\frac{1}{\nu}\sum_{k=0}^{m+1}(-1)^k\binom{m+1}{k}e^{(m+1-2k)\nu h i}=-\frac{1}{(2h)^{m+1}}\frac{1}{\nu}(e^{\nu h i}-e^{-\nu h i})^{m+1}=$$

$$=-i^{m+1}\left(\frac{\sin\nu h}{\nu h}\right)^{m+1}\nu^m,$$

und da m ungerade ist, so ist i^{m+1} reell, und es ist somit:

$$\bar{\varphi}_\nu(n)=-i^{m+1}\left(\frac{\sin\nu h}{\nu h}\right)^{m+1}\cdot\nu^m.$$

Betrachten wir den Kern (15) für **gerades** m, so können wir (7) anwenden, und zwar erhalten wir nach (16):

$$\varphi_\nu(n)=\frac{1}{(2h)^{m+1}}\frac{1}{\nu}\sum_{k=0}^{m+1}(-1)^k\binom{m+1}{k}\sin(m-2k+1)\nu h.$$

Das ist der imaginäre Teil von:

$$\frac{1}{(2h)^{m+1}}\frac{1}{\nu}\sum_{k=0}^{m+1}(-1)^k\binom{m+1}{k}e^{(m+1-2k)\nu h i}=i^{m+1}\left(\frac{\sin\nu h}{\nu h}\right)^{m+1}\nu^m,$$

und da m gerade ist, so ist i^{m+1} rein imaginär, und es ist somit:

$$\varphi_\nu(n)=i^m\left(\frac{\sin\nu h}{\nu h}\right)^{m+1}\cdot\nu^m.$$

Wir haben somit das Resultat: Für jedes in $<-\pi,\pi>$ integrierbare f gilt in jedem Punkte von $(-\pi,\pi)$, in dem die m-te Ableitung $f^{(m)}$ existiert bei ungeradem m:

$$f^{(m)}(x)=-i^{m+1}\lim_{h=0}\sum_{\nu=1}^{\infty}\left(\frac{\sin\nu h}{\nu h}\right)^{m+1}(-\nu^m a_\nu\sin\nu x+\nu^m b_\nu\cos\nu x),\tag{21}$$

bei geradem m:

$$f^{(m)}(x)=i^m\lim_{h=0}\sum_{\nu=1}^{\infty}\left(\frac{\sin\nu h}{\nu h}\right)^{m+1}(\nu^m a_\nu\cos\nu x+\nu^m b_\nu\sin\nu x).\tag{21 a}$$

Diese Formeln gelten gleichmäßig in jedem Teilintervalle $<\alpha,\beta>$ von $(-\pi,\pi)$, in dem f m-mal stetig differenzierbar ist.

Wie man sieht, erhält man die rechte Seite dieser Formeln, indem man die Fourier'sche Reihe von f gliedweise m-mal differenziert, jedes Glied multipliziert mit $\left(\frac{\sin\nu h}{\nu h}\right)^{m+1}$ und den $\lim\limits_{h=0}$ bildet.

Bemerken wir zunächst, daß, wie die Herleitung der Formeln (21) und (21 *a*) zeigt, auf ihrer linken Seite $f^{(m)}(x)$ ersetzt werden kann durch den etwas allgemeineren Ausdruck:

$$\lim_{h=0}\frac{1}{(2h)^{m+1}}\sum_{k=0}^{m+1}(-1)^k\binom{m+1}{k}F(x+(m-2k+1)h),$$

in dem $F(x)$ das unbestimmte Integral von $f(x)$ bedeutet.

Setzen wir bei ungeradem m:

$$F_m^*(x)=-i^{m-1}\sum_{\nu=1}^{\infty}\frac{1}{\nu^m}(-a_\nu\sin\nu x+b_\nu\cos\nu x),$$

bei geradem m:

$$F_m^*(x) = i^m \sum_{\nu=1}^{\infty} \frac{1}{\nu^m} (a_\nu \cos \nu x + b_\nu \sin \nu x),$$

so ist $F_m^*(x)$ ein m-fach iteriertes unbestimmtes Integral von $f(x) - \frac{a_0}{2}$. Wenden wir auf $F_{m-1}^*(x)$ die entsprechende der beiden Formeln (21), (21 a) an, so erhalten wir den Satz:

Für jedes in $< -\pi, \pi >$ integrierbare f gilt in jedem Punkte von $(-\pi, \pi)$, in dem f mit seinem m-fach iterierten unbestimmten Integrale $F_m(x)$ in der Beziehung steht:

$$f(x) = \lim_{h=0} \frac{1}{(2h)^m} \sum_{k=0}^{m} (-1)^k \binom{m}{k} F_m(x + (m-2k)h), \tag{22}$$

insbesondere also überall dort, wo $f(x)$ m-te Ableitung von $F_m(x)$ ist, die Formel:[1]

$$f(x) = \frac{a_0}{2} + \lim_{h=0} \sum_{\nu=1}^{\infty} \left(\frac{\sin \nu h}{\nu h}\right)^m (a_\nu \cos \nu x + b_\nu \sin \nu x). \tag{23}$$

Diese Formel gilt gleichmäßig in jedem Teilintervalle $< \alpha, \beta >$ von $(-\pi, \pi)$, in dessen sämtlichen Punkten f stetig ist.

Ebenso wie die Formeln (21), (21 a) lediglich einen Spezialfall von (18) darstellen, ebenso ist (23) lediglich ein spezieller Fall eines allgemeinen Summationstheorems für Reihen nach Orthogonalfunktionen, das noch kurz erwähnt sei.

Wir bezeichnen den Kern (15) mit $\varphi^{(m)}(\xi, x, h)$:

$$\varphi^{(m)}(\xi, x, h) = (-1)^k \frac{1}{(2h)^{m+1}} \binom{m}{k} \quad \text{in } (x + (m-2k-1)h,\ x + (m-2k+1)h) \quad (k = 0, 1, \ldots, m) \tag{24}$$

$$\varphi^{(m)}(\xi, x, h) = 0 \quad \text{außerhalb dieser Intervalle,}$$

und führen die iterierten unbestimmten Integrale von $\varphi^{(m)}(\xi, x, h)$ ein durch:

$$\Phi_0^{(m)}(\xi, x, h) = \varphi^{(m)}(\xi, x, h); \quad \Phi_{i+1}^{(m)}(\xi, x, h) = \int_{x-(m+1)h}^{\xi} \Phi_i^{(m)}(\xi, x, h)\, d\xi. \tag{25}$$

Zunächst sehen wir, daß wir außerhalb $(x - (m+1)h,\ x + (m+1)h)$ haben:

$$\Phi_i^{(m)}(\xi, x, h) = 0 \quad (i = 0, 1, \ldots, m). \tag{26}$$

In der Tat, dies ist richtig für $i = 0$. Angenommen, es sei richtig für $i \leqq i_0$, so wird es auch für $i_0 + 1$ gelten, wenn:

$$\int_{x-(m+1)h}^{x+(m+1)h} \Phi_{i_0}^{(m)}(\xi, x, h)\, d\xi = 0 \tag{27}$$

ist. Durch mehrmalige partielle Integration finden wir aber:

$$\int_{x-(m+1)h}^{x+(m+1)h} \Phi_{i_0}^{(m)}(\xi, x, h)\, d\xi = \frac{(-1)^{i_0}}{i_0!} \int_{x-(m+1)h}^{x+(m+1)h} \varphi^{(m)}(\xi, x, h) \cdot \xi^{i_0}\, d\xi,$$

und wie wir in (17) gesehen haben, ist dies tatsächlich $= 0$ für $i_0 < m$, wodurch (27) bestätigt ist. Und damit ist (26) durch vollständige Induktion bewiesen.

---

[1] Für $m = 2$ ist dies die bekannte Riemann'sche Summationsmethode, vgl. M. Schechter a. a. O., p. 233.

Sei nun x ein Punkt von (a, b), und $h > 0$ so klein gewählt, daß $< x-(m+1)\,h,\ x+(m+1)\,h >$ in (a, b) liegt. Dann haben wir, unter Berücksichtigung von (26), durch partielle Integration (wobei mit $F_i(x)$ die iterierten unbestimmten Integrale[1] von $f(x)$ bezeichnet sind):

$$\int_a^b F_m(\xi)\,\varphi^{(m)}(\xi, x, h)\,d\xi = (-1)^m \int_a^b f(\xi)\,\Phi_m^{(m)}(\xi, x, h)\,d\xi.$$

Da nun, wie wir wissen:

$$f(x) = \lim_{h=0} \int_a^b F_m(\xi)\,\varphi^{(m)}(\xi, x, h)\,d\xi$$

ist in jedem Punkte von (a, b), in dem $f(x)$ die m-te Ableitung von $F_m(x)$ ist, oder allgemeiner, wo:

$$f(x) = \lim_{h=0} \frac{1}{(2\,h)^{m+1}} \sum_{k=0}^{m+1} (-1)^k \binom{m+1}{k} F_{m+1}(x+(m-2\,k+1)\,h) \tag{28}$$

ist, so ist in jedem solchen Punkte auch

$$f(x) = \lim_{h=0}\,(-1)^m \int_a^b f(\xi)\,\Phi_m^{(m)}(\xi, x, h)\,d\xi.$$

Wenden wir auf dieses Integral die Parseval'sche Formel an, so erhalten wir schließlich:

XVIII. Sei der Kern $\Phi_m^{(m)}(\xi, x, h)$ nach (25) aus dem Kerne (24) hergeleitet und es werde gesetzt:

$$(-1)^m \int_a^b \Phi_m^{(m)}(\xi, x, h)\,\omega_\nu(\xi)\,d\xi = \Omega_\nu^{(m)}(x, h).$$

Für jedes normierte Orthogonalsystem von $< a, b >$ gilt dann die Formel:

$$f(x) = \lim_{h=0} \sum_{\nu=1}^{\infty} f_\nu\,\Omega_\nu^{(m)}(x, h) \tag{29}$$

für jede in $< a, b >$ samt ihrem Quadrate integrierbare Funktion f, in jedem Punkte von (a, b), in dem (28) gilt, insbesondere also in jedem Punkte von (a, b), in dem $f\,(m+1)$te Ableitung seines $(m+1)$-fach iterierten unbestimmten Integrales ist; sie gilt gleichmäßig in jedem Teilintervalle $< \alpha, \beta >$ von (a, b), in dessen sämtlichen Punkten f stetig ist.

Dies gilt für alle in $< a, b >$ integrierbaren Funktionen f, falls es ein M gibt, so daß

$$\left| \sum_{\nu=1}^{\mu} \int_\alpha^\beta \omega_\nu(\xi)\,d\xi \cdot \omega_\nu(x) \right| < M \tag{30}$$

für alle Teilintervalle $< \alpha, \beta >$ von (a, b), alle x von $< a, b >$ und alle μ.

Eines Beweises bedarf nach dem Gesagten nur mehr der letzte Teil der Behauptung, und auch dieser Teil wird bewiesen sein, wenn wir zeigen, daß, wenn Voraussetzung (30) gilt, auf das Integral:

$$\int_a^b f(\xi)\,\Phi_m^{(m)}(\xi, x, h)\,d\xi$$

[1] Das heißt, es ist $F_0(x) = f(x)$; $F_{i+1}(x) = \int F_i(x)\,dx$

für jedes in $<a, b>$ integrierbare f das Parseval'sche Theorem angewendet werden kann. Dies aber wird, nach Satz V, der Fall sein, wenn es zu jedem $h > 0$ und jedem $\bar{x}$ von (a, b) ein N gibt, so daß:

$$\left|\sum_{\nu=1}^{\mu} \int_a^b \Phi_m^{(m)}(\xi, \bar{x}, h)\, \omega_\nu(\xi)\, d\xi \cdot \omega_\nu(x)\right| < N \tag{31}$$

für alle μ und alle x von $<a, b>$. Nun ist jedes $\Phi_m^{(m)}(\xi, x, h)$ in $<a, b>$ geschränkt und von geschränkter Variation. Bezeichnet $\Phi(\bar{x}, h)$ eine obere Schranke für $|\Phi_m^{(m)}(\xi, \bar{x}, h)|$ für alle ξ von $<a, b>$ und $V(\bar{x}, h)$ die Totalvariation von $\Phi_m^{(m)}(\xi, \bar{x}, h)$ in $<a, b>$, so liefert der zweite Mittelwertsatz der Integralrechnung, bei Berufung auf (30):

$$\left|\int_a^b \Phi_m^{(m)}(\xi, \bar{x}, h) \sum_{\nu=1}^{\mu} \omega_\nu(\xi)\, \omega_\nu(x)\, d\xi\right| \leqq (\Phi(\bar{x}, h) + V(\bar{x}, h)) \cdot M,$$

wodurch (31) bewiesen ist. Damit ist der Beweis von Satz XVIII beendet.

Die Berechnung der Ausdrücke $\Omega_\nu^{(m)}(x, h)$ in (29) ergibt für $m = 1$ und $m = 2$:

$$\Omega_\nu^{(1)}(x, h) = \frac{1}{(2h)^2} \int_{x-2h}^{x} (\xi - x + 2h)\, \omega_\nu(\xi)\, d\xi - \frac{1}{(2h)^2} \int_x^{x+2h} (\xi - x - 2h)\, \omega_\nu(\xi)\, d\xi.$$

$$\Omega_\nu^{(2)}(x, h) = \frac{1}{2} \frac{1}{(2h)^3} \int_{x-3h}^{x-h} (\xi - x + 3h)^2\, \omega_\nu(\xi)\, d\xi - \frac{1}{(2h)^3} \int_{x-h}^{x+h} \{(\xi - x)^2 - 3h^2\}\, \omega_\nu(\xi)\, d\xi +$$
$$+ \frac{1}{2} \frac{1}{(2h)^3} \int_{x+h}^{x+3h} (\xi - x - 3h)^2\, \omega_\nu(\xi)\, d\xi.$$

Endlich sei noch erwähnt, daß wir, unter Berufung auf Satz II und auf Formel (4a) von § 5, die durch (23) gegebenen Summationsformeln der trigonometrischen Reihen ergänzen können durch die Theoreme:

Setzt man:

$$R^{(m)}(x, h) = \frac{a_0}{2} + \sum_{\nu=1}^{\infty} \left(\frac{\sin \nu h}{\nu h}\right)^m (a_\nu \cos \nu x + b_\nu \sin \nu x),$$

so gilt die Beziehung:

$$\lim_{h=0} \int_{-\pi}^{\pi} |f(x) - R^{(m)}(x, h)|\, dx = 0.$$

Für jedes Paar von Funktionen f und g, von denen in $<-\pi, \pi>$ die eine integrierbar, die andere geschränkt ist, gilt:

$$\frac{1}{\pi} \int_{-\pi}^{\pi} f(x)\, g(x)\, dx = \frac{a_0\, \alpha_0}{2} + \lim_{h=0} \sum_{\nu=1}^{\infty} \left(\frac{\sin \nu h}{\nu h}\right)^m (a_\nu \alpha_\nu + b_\nu \beta_\nu).$$

Über Fejérs Summierung der Fourierschen Reihe.

Von HANS HAHN in Bonn.

Die Theorie der Fourierschen Reihen weist eine Anzahl von Summationsverfahren auf, die an allen Stetigkeitsstellen der darzustellenden Funktion $f(x)$ sicher gegen $f(x)$ konvergieren; die wichtigsten sind wohl die von Poisson, von Riemann, von de la Vallée Poussin und von Fejér. Für die drei erstgenannten ist bekannt[1]), daß sie allgemeiner für jeden Wert von x gegen $f(x)$ konvergieren, für den:

$$(1) \qquad \lim_{t=0} \frac{1}{t} \int_0^t \{f(x+2u) + f(x-2u) - 2f(x)\}\, du = 0$$

ist. Im Gegensatze hierzu wurde dies für das Fejérsche Verfahren nur für solche Werte von x bewiesen[2]), für welche:

$$(2) \qquad \lim_{t=0} \frac{1}{t} \int_0^t |f(x+2u) + f(x-2u) - 2f(x)|\, du = 0$$

1) Vgl. z. B. H. Lebesgue, Ann. de Toul. (3) 1, S. 88, 100.

2) H. Lebesgue a. a. O. S. 90.

ist. Die Frage aber, ob es sich hier wirklich um eine Verschiedenheit in der Tragweite der Methoden oder nur um einen Mangel der bisherigen Beweise handelt, scheint noch nicht beantwortet zu sein. Ich möchte deshalb diese Frage zur Entscheidung bringen, indem ich an einem Beispiel[1]) zeige, daß die Bedingung (1) in der Tat für die Konvergenz des Fejérschen Verfahrens *nicht* hinreicht.

Wir setzen zur Abkürzung:

(3) $$p_k = 1.\ 3.\ 5.\ \ldots\ (2k+1) \qquad (k=0,1,2,\ldots)$$

und bezeichnen mit J_k das Intervall: $< \frac{\pi}{p_k}, \frac{\pi}{p_{k-1}} >$. Mit c_k $(k=1,2,\ldots)$ bezeichnen wir eine Folge nicht negativer Zahlen mit $\lim\limits_{k=\infty} c_k = 0$, über die wir später noch näher verfügen werden.

Sodann definieren wir in $< 0, \frac{\pi}{2} >$ eine Funktion $\varphi(t)$ durch:

(4) $$\varphi(t) = c_k \sin p_k t \cdot \frac{\sin t}{t} \quad \text{in } J_k \qquad (k=1,2,\ldots)$$

und dehnen ihre Definition auf das Intervall $< -\frac{\pi}{2}, \frac{\pi}{2} >$ aus durch die Vorschriften:

(5) $$\varphi(0) = 0; \quad \varphi(-t) = \varphi(t).$$

Offenbar ist $\varphi(t)$ überall in $< -\frac{\pi}{2}, \frac{\pi}{2} >$ *stetig*, insbesondere auch für $t = 0$:

(6) $$\lim_{t=0} \varphi(t) = 0.$$

Wir zeigen nun weiter, daß die Funktion:

(7) $$F(t) = t \cdot \varphi\left(\frac{t}{2}\right)$$

in $< -\pi, \pi >$ *von geschränkter Variation* ist.

Bekanntlich gilt für die Variation $V_\alpha^\beta(f)$ im Intervalle $< \alpha, \beta >$ eines Produktes

$$f(t) = g(t) \cdot h(t),$$

wenn unter G und H die oberen Schranken von $|g(t)|$ und $|h(t)|$ in $< \alpha, \beta >$ verstanden werden, die Ungleichung:

$$V_\alpha^\beta(f) \leqq H \cdot V_\alpha^\beta(g) + G\, V_\alpha^\beta(h).$$

Die Variation von $\varphi(t) \cdot \frac{t}{\sin t}$ im Intervalle J_k, in dem ja:

$$\varphi(t) \cdot \frac{t}{\sin t} = c_k \sin p_k t$$

1) Dieses Beispiel weist eine gewisse Verwandtschaft auf mit einem von H. Lebesgue (Leç. s. l. series trigonométriques, s. 85) gegebenen Beispiele einer stetigen Funktion mit divergenter Fourierscher Reihe.

ist, beträgt $4k \cdot c_k$. Wir erhalten also für die Variation von:

$$F(2t) \cdot \frac{t}{\sin t} = 2t \cdot \varphi(t) \cdot \frac{t}{\sin t}$$

im Intervalle $J_k (k > 1)$:

$$V_k \leqq \frac{2\pi}{p_{k-1}} c_k \cdot 4k + \frac{2\pi}{p_{k-1}} \cdot c_k \cdot \frac{\pi}{2} < \frac{4\pi}{p_{k-1}} \cdot c_k (2k+1).$$

Daher erhält man für die Variation von $F(2t) \cdot \frac{t}{\sin t}$ in $< 0, \frac{\pi}{3} >$:

$$V_0^{\frac{\pi}{3}} = \sum_{k=2}^{\infty}{}' V_k \leqq 4\pi \sum_{k=2}^{\infty}{}' \frac{(2k+1)c_k}{p_{k-1}}.$$

Da die rechts stehende Reihe konvergiert, ist in $< 0, \frac{\pi}{3} >$ gewiß $F(2t)\frac{t}{\sin t}$, und somit auch $F(2t)$ von geschränkter Variation. Es ist also weiter $F(t)$ in $< 0, \frac{2\pi}{3} >$ von geschränkter Variation; und da $F(t)$ offenbar auch in $< \frac{2\pi}{3}, \pi >$ von geschränkter Variation ist, und da ferner nach (5) und (7):

$$F(-t) = -F(t)$$

ist, so ist tatsächlich $F(t)$ in $< -\pi, \pi >$ *von geschränkter Variation.*

Da aber $F(t)$ offenkundig in jedem, den Nullpunkt nicht enthaltenden, abgeschlossenen Teilintervalle von $< -\pi, \pi >$ *absolut stetig*, im Nullpunkte aber *stetig* ist, so folgt aus der Tatsache, daß $F(t)$ in $< -\pi, \pi >$ von geschränkter Variation ist, bekanntlich weiter, daß $F(t)$ *absolut stetig* ist. Daher besteht zwischen $F(t)$ und seiner Ableitung, die mit $f(t)$ bezeichnet werde, die Beziehung:

(8) $$F(t) = \int_0^t f(u)\,du.$$

Es sei noch erwähnt, daß:

$$f(0) = 0$$

ist; denn bei Berücksichtigung von (7) und (6) haben wir:

(9) $$f(0) = \lim_{t=0} \frac{1}{t} F(t) = \lim_{t=0} \varphi\left(\frac{t}{2}\right) = 0.$$

Ferner genügt $f(x)$ *im Punkte* $x = 0$ *der Beziehung* (1), die sich hier reduziert auf:

$$\lim_{t=0} \frac{2}{t} \int_0^t f(2u)\,du = \lim_{t=0} \frac{F(2t)}{t} = 0.$$

Wir werden nun zeigen, daß *trotzdem (bei geeigneter Wahl der Konstanten* c_k*) die Fejérsche Summierung der Fourierschen Reihe von* $f(x)$ *im Punkte* $x = 0$ *divergiert.*

Sei also:

$$f(x) \sim \frac{a_0}{2} + \sum_{\nu=1}^{\infty} (a_\nu \cos \nu x + b_\nu \sin \nu x)$$

die Fouriersche Reihe von $f(x)$. Wir betrachten nach Fejér die Ausdrücke:

$$S_n(x) = \frac{a_0}{2} + \sum_{\nu=1}^{n-1} \left(1 - \frac{\nu}{n}\right) (a_\nu \cos \nu x + b_\nu \sin \nu x),$$

und behaupten, *daß in* (4) *die Konstanten* c_k *so gewählt werden können, daß die Folge:*

$$S_n(0) \qquad (n = 1, 2, \ldots)$$

divergiert.

Bekanntlich ist:

$$S_n(0) = \frac{1}{2n\pi} \int_{-\pi}^{\pi} f(t) \frac{\sin^2 \frac{nt}{2}}{\sin^2 \frac{t}{2}} dt,$$

und durch partielle Integration findet man bei Berufung auf (8):

$$S_n(0) = \frac{1}{2n\pi} F(t) \frac{\sin^2 \frac{nt}{2}}{\sin^2 \frac{t}{2}} \Bigg]_{-\pi}^{\pi} - \frac{1}{2n\pi} \int_{-\pi}^{\pi} F(t) \frac{d}{dt} \frac{\sin^2 \frac{nt}{2}}{\sin^2 \frac{t}{2}} dt.$$

Es ist also:

$$\lim_{n=\infty} \left\{ S_n(0) + \frac{1}{2n\pi} \int_{-\pi}^{\pi} F(t) \frac{d}{dt} \frac{\sin^2 \frac{nt}{2}}{\sin^2 \frac{t}{2}} dt \right\} = 0,$$

und es genügt daher, die Folge:

$$(10) \qquad S_n^* = \frac{1}{2n\pi} \int_{-\pi}^{\pi} F(t) \frac{d}{dt} \frac{\sin^2 \frac{nt}{2}}{\sin^2 \frac{t}{2}} dt \qquad (n = 1, 2, \ldots)$$

zu betrachten. Nun ist aber, bei Ausführung der Differentiation:

$$S_n^* = \frac{1}{4\pi} \int_{-\pi}^{\pi} F(t) \frac{\sin nt}{\sin^2 \frac{t}{2}} dt - \frac{1}{2n\pi} \int_{-\pi}^{\pi} F(t) \frac{\sin^2 \frac{nt}{2} \cdot \cos \frac{t}{2}}{\sin^3 \frac{t}{2}} dt.$$

Der Subtrahend rechts ist nichts anderes als das Fejérsche Mittel aus den n ersten Gliedern der Fourierschen Reihe von $F(x) \cdot \operatorname{ctg} \frac{x}{2}$ im Punkte $x = 0$. Da aber nach (9):

$$\lim_{x=0} F(x) \cdot \operatorname{ctg} \frac{x}{2} = 0$$

ist, so ist nach dem Fejérschen Satze:

$$\lim_{n=\infty}\frac{1}{2n\pi}\int_{-\pi}^{\pi} F(t)\,\frac{\sin^2\frac{nt}{2}\cos\frac{t}{2}}{\sin^3\frac{t}{2}}\,dt = 0.$$

Es genügt also, statt der Folge (10) die Folge:

$$(11)\qquad \frac{1}{4\pi}\int_{-\pi}^{\pi} F(t)\,\frac{\sin nt}{\sin^2\frac{t}{2}}\,dt$$

zu betrachten. An Stelle dieser Folge kann ebensogut die Folge:

$$(12)\qquad \frac{1}{2\pi}\int_{-\pi}^{\pi} \varphi\left(\frac{t}{2}\right)\frac{\sin nt}{\sin\frac{t}{2}}\,dt$$

betrachtet werden, denn bei Berücksichtigung von (7) erhält man für die Differenz der Integrale (11) und (12), nach einem bekannten Satze[1]):

$$\lim_{n=\infty}\frac{1}{4\pi}\int_{-\pi}^{\pi} F(t)\,\frac{2\sin\frac{t}{2}-t}{t\sin^2\frac{t}{2}}\,\sin nt\,dt = 0,$$

da im Integranden der Faktor von $\sin nt$ integrierbar ist.

Die Folge der Integrale (12) kann endlich noch ersetzt werden durch die Folge:

$$(13)\qquad \frac{1}{2\pi}\int_{-\pi}^{\pi} \varphi\left(\frac{t}{2}\right)\frac{\sin\frac{2n+1}{2}t}{\sin\frac{t}{2}}\,dt;$$

in der Tat gilt für die Differenz der Integrale (12) und (13), nach dem eben zitierten Satz:

$$\lim_{n=\infty}\frac{1}{\pi}\int_{-\pi}^{\pi} \varphi\left(\frac{t}{2}\right)\frac{\sin\frac{t}{4}}{\sin\frac{t}{2}}\cos\left(n+\tfrac{1}{4}\right)t\,dt = 0,$$

da im Integranden der Faktor von $\cos(n+\frac{1}{4})t$ integrierbar ist.

Die Folge (13) kann endlich noch geschrieben werden:

$$(14)\qquad \frac{2}{\pi}\int_{0}^{\frac{\pi}{2}} \varphi(t)\,\frac{\sin(2n+1)t}{\sin t}\,dt.$$

---

1) Siehe z. B. de la Vallée-Poussin, Cours d'analyse, II (2. éd.), S. 140.

Wir haben also nur mehr zu zeigen, daß in (4) die Konstanten c_k als nicht negative Zahlen mit $\lim_{k=\infty} c_k = 0$ so gewählt werden können, daß die Folge (14) divergiert.

Diese Wahl geschieht in folgender Weise: Wir denken uns den Wert von c_1 beliebig gewählt. Nach (4) ist dadurch $\varphi(t)$ in J_1 und somit in $< \frac{\pi}{3}, \frac{\pi}{2} >$ gegeben. Nach einem schon wiederholt verwendeten Satze ist nun:

$$\lim_{n=\infty} \frac{2}{\pi} \int_{\frac{\pi}{3}}^{\frac{\pi}{2}} \frac{\varphi(t)}{\sin t} \cdot \sin(2n+1)t\,dt = 0.$$

Es gibt also einen Index $k_1 > 1$, so daß, wenn p_k wieder die Bedeutung (3) hat:

$$\left| \frac{2}{\pi} \int_{\frac{\pi}{3}}^{\frac{\pi}{2}} \varphi(t) \frac{\sin p_k t}{\sin t} dt \right| < 1 \quad \text{für} \quad k \geqq k_1.$$

Wir setzen nun:

$$c_k = 0 \quad \text{für} \quad 1 < k < k_1; \quad c_{k_1} = \frac{1}{\sqrt{\lg(2k_1+1)}}.$$

Dadurch ist $\varphi(t)$ nach (4) gegeben in $< \frac{\pi}{p_{k_1}}, \frac{\pi}{2} >\cdot$ Da wieder:

$$\lim_{n=\infty} \frac{2}{\pi} \int_{\frac{\pi}{p_{k_1}}}^{\frac{\pi}{2}} \frac{\varphi(t)}{\sin t} \sin(2n+1)t\,dt = 0$$

ist, gibt es einen Index $k_2 > k_1$, so daß:

$$\left| \frac{2}{\pi} \int_{\frac{\pi}{p_{k_1}}}^{\frac{\pi}{2}} \varphi(t) \frac{\sin p_k t}{\sin t} dt \right| < 1 \text{ für } k \geqq k_2.$$

Wir setzen wieder:

$$c_k = 0 \quad \text{für} \quad k_1 < k < k_2; \quad c_{k_2} = \frac{1}{\sqrt{\lg(2k_2+1)}}.$$

Indem man so weiter schließt, erhält man eine Folge wachsender Indizes k_i, und indem man immer setzt:

$$(15) \qquad c_k = 0 \quad \text{für} \quad k_i < k < k_{i+1}; \quad c_{ki} = \frac{1}{\sqrt{\lg(2k_i+1)}},$$

wird $\varphi(t)$ überall in $< 0, \frac{\pi}{2} >$ so definiert, daß:

$$(16) \qquad \left| \frac{2}{\pi} \int_{\frac{\pi}{p_{k_{i-1}}}}^{\frac{\pi}{2}} \varphi(t) \frac{\sin p_k t}{\sin t} dt \right| < 1 \quad \text{für } k \geqq k_i.$$

Wir greifen nun aus (14) die Teilfolge heraus:

$$\frac{2}{\pi} \int_0^{\frac{\pi}{2}} \varphi(t) \frac{\sin p_{k_i} t}{\sin t} dt,$$

und wenden auf dieses Integral die Zerlegung an:

$$(17) \qquad \int_0^{\frac{\pi}{2}} = \int_0^{\frac{\pi}{p_{k_i}}} + \int_{\frac{\pi}{p_{k_i}}}^{\frac{\pi}{p_{k_{i-1}}}} + \int_{\frac{\pi}{p_{k_{i-1}}}}^{\frac{\pi}{2}}.$$

Für den ersten Summanden erhalten wir:

$$(18) \qquad \lim_{i=\infty} \int_0^{\frac{\pi}{p_{k_i}}} \varphi(t) \frac{\sin p_{k_i} t}{\sin t} dt = 0;$$

In der Tat ist in $< 0, \frac{\pi}{p_{ki}} >$:

$$|\varphi(t)| < \frac{1}{\sqrt{\lg(2k_i + 1)}}; \quad 0 \leqq \frac{\sin p_{k_i} t}{\sin t} \leqq \tfrac{3}{2} p_{ki},$$

so daß durch Anwendung des ersten Mittelwertsatzes der Integralrechnung sofort (18) folgt.

Was den dritten Summanden in (17) anlangt, so ist wegen (16):

$$(19) \qquad \left| \frac{2}{\pi} \int_{\frac{\pi}{p_{k_{i-1}}}}^{\frac{\pi}{2}} \varphi(t) \frac{\sin p_{ki} t}{\sin t} dt \right| < 1 \qquad (i = 1, 2, \ldots).$$

Was endlich den zweiten Summanden in (17) anlangt, so haben wir zunächst, da nach (4) und (15) $\varphi(t) = 0$ ist in den Intervallen $J_k\,(k_{i-1} < k < k_i)$, wenn wir für p_k seinen Wert nach (3) einführen:

$$\int_{\frac{\pi}{p_{k_i}}}^{\frac{\pi}{p_{k_{i-1}}}} \varphi(t) \frac{\sin p_{ki} t}{\sin t} dt = \int_{\frac{\pi}{1 \cdot 3 \cdot \ldots (2k_i + 1)}}^{\frac{\pi}{1 \cdot 3 \ldots (2k_i - 1)}} \varphi(t) \frac{\sin p_{ki} t}{\sin t} dt,$$

und weiter, indem wir die Formel: $\sin^2 x = \frac{1}{2}(1 - \cos 2x)$ anwenden, wieder unter Berücksichtigung von (4) und (15):

$$\int\limits_{\frac{\pi}{p_{k_i}}}^{\frac{\pi}{p_{k_{i-1}}}} \varphi(t) \frac{\sin p_{k_i} t}{\sin t}\, dt = \tfrac{1}{2} \sqrt{\lg(2k_i + 1)} - \frac{1}{2\sqrt{\lg(2k_i + 1)}} \int\limits_{\frac{\pi}{1\cdot 3\cdot \ldots (2k_i+1)}}^{\frac{\pi}{1\cdot 3\cdot \ldots (2k_i-1)}} \frac{\cos 2p_{k_i} t}{t}\, dt;$$

und da bekanntlich das Integral im Subtrahenden bei unendlich wachsendem i endlich bleibt[1]), hat der Subtrahend bei unendlich wachsendem i den Grenzwert 0, und der mittlere Summand in (17) wächst mit i ins Unendliche. Diese Tatsache zusammen mit den Beziehungen (18) und (19) ergibt aber:

$$\lim_{i=\infty} \frac{2}{\pi} \int\limits_0^{\frac{\pi}{2}} \varphi(t) \frac{\sin p_{k_i} t}{\sin t}\, dt = +\infty.$$

Es ist also die Divergenz der Folge (14) und damit auch die Divergenz der Fejérschen Mittel für unsere Funktion $f(x)$ an der Stelle $x = 0$ nachgewiesen.

Wie angekündigt, ist also an einem Beispiele gezeigt, *daß Bedingung* (1) *für die Konvergenz der Fejérschen Summierung der Fourierschen Reihe nicht hinreicht.*

Einige Anwendungen der Theorie der singulären Integrale

Von

Hans Hahn in Bonn

(Vorgelegt in der Sitzung am 4. Juli 1918)

In einer in den Denkschriften der Akademie erschienenen Abhandlung[1] habe ich einige Anwendungen der Theorie der singulären Integrale auf die Lehre von den orthogonalen Funktionensystemen gemacht. Es sei mir gestattet, auf diesen Gegenstand noch einmal zurückzukommen.

Nachdem ich in der eben genannten Abhandlung eine notwendige und hinreichende Bedingung aufgestellt habe, der ein vollständiges normiertes Orthogonalsystem $\omega_\nu(x)$ $(\nu = 1, 2, \ldots)$ des Intervalles $<a, b>$ genügen muß, damit die Reihenentwicklung jeder (im Sinne von Lebesgue) integrierbaren Funktion f nach den Funktionen $\omega_\nu(x)$ in jedem Teilintervalle von $<a, b>$ gliedweise integrierbar sei, stelle ich nun eine notwendige und hinreichende Bedingung dafür auf, daß diese Reihenentwicklung auf jeder meßbaren Punktmenge von $<a, b>$ gliedweise integrierbar sei oder — wie wir dies in Anlehnung an eine von G. Vitali gebildete Bezeichnung ausdrücken wollen — daß diese Reihe in $<a, b>$ vollständig integrierbar gegen f konvergiere. Die Bedingungen für gliedweise Integrierbarkeit auf jeder meßbaren Menge einerseits, in jedem Teilintervalle andrerseits verhalten sich zueinander ungefähr so wie die Bedingungen dafür, daß die genannte Reihenentwicklung für jede geschränkte Funktion

[1] »Über die Darstellung gegebener Funktionen durch singuläre Integrale«, II. Mitteilung. Denkschr. d. Akad. d. Wiss. Wien, mathem.-naturw. Klasse, Bd. 93 (1916), p. 676 ff.

ohne Unstetigkeiten zweiter Art einerseits, für jede Funktion endlicher Variation andrerseits an allen Stetigkeitsstellen gegen f konvergiere.

Die Frage nach der vollständigen Integrierbarkeit der Reihenentwicklung:

$$f(x) \sim \sum_{\nu=1}^{\infty} f_\nu \, \omega_\nu (x)$$

steht in engster Beziehung zur Frage nach der Gültigkeit von:

$$\lim_{\nu=\infty} \int_a^b \left| f(x) - \sum_{\lambda=1}^{\nu} f_\lambda \, \omega_\lambda (x) \right| dx = 0.$$

Wir zeigen, daß im Falle der sogenannten asymptotischen Konvergenz diese beiden Fragen geradezu identisch sind, so daß wir im Falle einer (asymptotisch) gegen $g(x)$ konvergierenden Funktionenfolge $g_\nu (x)$ die von G. Vitali aufgestellte notwendige und hinreichende Bedingung für vollständige Integrierbarkeit durch die wohl wesentlich einfachere ersetzen können:

$$\lim_{\nu=\infty} \int_a^b |g(x) - g_\nu(x)| \, dx = 0.$$

Ich benütze endlich die Gelegenheit, um ein in meiner oben zitierten Abhandlung enthaltenes inkorrektes Resultat in richtiger Weise zu formulieren.

§ 1.

Wir erinnern an die beiden von H. Lebesgue bewiesenen Sätze:[1]

I. Damit für alle Funktionen $f(\xi)$, die im Intervalle $< a, b >$ geschränkt sind, dort keine Unstetigkeiten zweiter Art besitzen und im Punkte x von (a, b) stetig sind, die Beziehung bestehe:

$$f(x) = \lim_{\nu=\infty} \int_a^b f(\xi) \, \varphi_\nu(\xi) \, d\xi, \tag{1}$$

[1] Ann. Toulouse, Serie 3, Bd. 1, p. 70.

ist notwendig und hinreichend, daß die Folge integrierbarer Funktionen $\varphi_\nu(\xi)$ $(\nu=1, 2, \ldots)$ den beiden Bedingungen genüge:

1. Es gibt eine Zahl A, so daß:

$$\int_a^b |\varphi_\nu(\xi)|\, d\xi < A \text{ für alle } \nu.$$

2. Für jedes den Punkt x enthaltende Teilintervall (α, β) von $<a, b>$ ist:

$$\lim_{\nu=\infty} \int_\alpha^\beta \varphi_\nu(\xi)\, d\xi = 1. \tag{2}$$

II. Damit für alle Funktionen $f(\xi)$, die im Intervalle $<a, b>$ von endlicher Variation und im Punkte x von (a, b) stetig sind, die Beziehung (1) bestehe, ist notwendig und hinreichend, daß die Folge integrierbarer Funktionen $\varphi_\nu(\xi)$ $(\nu=1, 2, \ldots)$ den beiden Bedingungen genüge:

1. Es gibt eine Zahl A, so daß für alle Teilintervalle $<\alpha, \beta>$ von $<a, b>$ und alle ν:

$$\left| \int_\alpha^\beta \varphi_\nu(\xi)\, d\xi \right| < A.$$

2. Für jedes den Punkt x enthaltende Teilintervall (α, β) von $<a, b>$ gilt (2).

Sei nun:

$$\omega_\nu(\xi) \qquad (\nu = 1, 2, \ldots) \tag{3}$$

ein normiertes Orthogonalsystem des Intervalles $<a, b>$ und seien:

$$f_\nu = \int_a^b f(\xi)\, \omega_\nu(\xi)\, d\xi \tag{4}$$

die Fourier'schen Konstanten der Funktion f in Bezug auf das normierte Orthogonalsystem (3). Sei x ein Punkt von (a, b). Wir setzen:

$$\varphi_\nu(x, \xi) = \sum_{\lambda=1}^{\nu} \omega_\lambda(x)\, \omega_\lambda(\xi).$$

Dann ist die Frage nach der Gültigkeit der Entwicklung:

$$f(x) = \sum_{\nu=1}^{\infty} f_\nu \, \omega_\nu(x) \tag{5}$$

gleichbedeutend mit der Frage nach der Gültigkeit der Beziehung:

$$f(x) = \lim_{\nu=\infty} \int_a^b f(\xi)\, \varphi_\nu(x,\xi)\, d\xi.$$

Die angeführten Sätze I und II ergeben daher unmittelbar:

III. Damit für alle Funktionen $f(\xi)$, die im Intervalle $<a, b>$ geschränkt sind, dort keine Unstetigkeiten zweiter Art besitzen und im Punkte x von (a, b) stetig sind, die Entwicklung (5) nach den Funktionen des normierten Orthogonalsystems (3) gelte, ist notwendig und hinreichend, daß dieses Orthogonalsystem den beiden Bedingungen genüge:

1. Es gibt eine Zahl A, so daß:

$$\int_a^b \left| \sum_{\lambda=1}^{\nu} \omega_\lambda(x)\, \omega_\lambda(\xi) \right| d\xi < A \text{ für alle } \nu. \tag{6}$$

2. Es ist für alle y aus $<a, b>$:

$$\sum_{\nu=1}^{\infty} \int_a^y \omega_\nu(\xi)\, d\xi \,.\, \omega_\nu(x) = \begin{cases} 1 \text{ wenn } x \text{ in } (a, y) \\ 0 \text{ wenn } x \text{ in } (y, b). \end{cases} \tag{7}$$

IV. Damit für alle Funktionen $f(\xi)$, die im Intervalle $<a, b>$ von endlicher Variation und im Punkte x von (a, b) stetig sind, die Entwicklung (5) nach den Funktionen des normierten Orthogonalsystems (3) gelte, ist notwendig und hinreichend, daß dieses Orthogonalsystem den beiden Bedingungen genüge:

1. Es gibt eine Zahl A, so daß für alle y aus $<a, b>$ und alle ν:

$$\left| \sum_{\lambda=1}^{\nu} \int_a^y \omega_\lambda(\xi)\, d\xi \,.\, \omega_\lambda(x) \right| < A. \tag{8}$$

2. Es gilt für alle y aus $<a. b>$ die Beziehung (7).

Wiewohl die Sätze III und IV bloße Korollare der Lebesgue'schen Sätze I und II sind, scheinen sie wenig bekannt zu sein; daß die Bedingung 1) von Satz III notwendig ist, wurde von A. Haar ausgesprochen.[1] Ein spezieller Fall von Satz IV (enthaltend hinreichende, aber nicht notwendige Bedingungen) wurde von E. Helly bewiesen.[2] Ein den Bedingungen von Satz III genügendes Orthogonalsystem wurde zuerst angegeben von A. Haar;[3] Beispiele für Orthogonalsysteme, die den Bedingungen von Satz IV genügen, liefern das trigonometrische Orthogonalsystem und die (bei nicht singulären Differentialgleichungen zweiter Ordnung auftretenden) Sturm-Liouville'schen Orthogonalsysteme.

§ 2.

Wir setzen von nun an das normierte Orthogonalsystem $\omega_1(\xi), \omega_2(\xi), \ldots, \omega_\nu(\xi), \ldots$ als vollständig voraus.

Es existieren dann und nur dann für jede in $<a, b>$ integrierbare Funktion $f(\xi)$ die sämtlichen Fourier'schen Konstanten (4) in bezug auf das normierte Orthogonalsystem (3), wenn jede einzelne Funktion $\omega_\nu(\xi)$ geschränkt ist in $<a, b>$. Wir nehmen diese Bedingung als erfüllt an.

Mit f_ν bezeichnen wir auch weiterhin die durch (4) definierten Fourier'schen Konstanten von $f(\xi)$.

In meiner eingangs zitierten Arbeit habe ich gelegentlich die Frage nach der Gültigkeit der Formel:

$$\lim_{\nu=\infty} \int_a^b \left| f(x) - \sum_{\lambda=1}^{\nu} f_\lambda \, \omega_\lambda(x) \right| dx = 0 \qquad (9)$$

berührt.[4] Wir wollen nun diesbezüglich folgenden Satz beweisen:

---

[1] Math. Ann. 69 (1910), p. 335.

[2] Sitzungsber. d. Akad. d. Wiss. in Wien, Abt. IIa, Bd. 121 (1912), p. 1547.

[3] A. a. O., p. 361 ff.

[4] A. a. O., p. 681.

V. Damit für jede in $<a, b>$ integrierbare Funktion f die Formel (9) gelte, ist notwendig und hinreichend die Existenz einer Zahl A, so daß für alle ν und alle x von $<a, b>$ (abgesehen von einer Nullmenge) Ungleichung (6) erfüllt sei.

Die Bedingung ist notwendig. Nehmen wir in der Tat an, es sei Ungleichung (6) nicht im angegebenen Sinne erfüllt; es gibt dann zu jeder natürlichen Zahl i eine in $<a, b>$ gelegene Menge $\mathfrak{A}_i$ mit von 0 verschiedenem Inhalte, so daß zu jedem Punkte x von $\mathfrak{A}_i$ ein Index ν_x gehört, für den:

$$\int_a^b \left| \sum_{\lambda=1}^{\nu_x} \omega_\lambda(\xi)\, \omega_\lambda(x) \right| d\xi > i.$$

Dann aber gibt es, wie ich gezeigt habe,[1] eine stetige Funktion $g(\xi)$ von folgender Eigenschaft: Bedeuten

$$g_\nu = \int_a^b g(\xi)\, \omega_\nu(\xi)\, d\xi$$

die Fourier'schen Konstanten von g in bezug auf unser Orthogonalsystem, so gibt es zu jedem noch so großen B einen Index ν, für den die Ungleichung:

$$\left| \sum_{\lambda=1}^{\nu} g_\lambda\, \omega_\lambda(\xi) \right| > B$$

in einer Punktmenge positiven Inhaltes erfüllt ist.

Da $g(\xi)$ als stetige Funktion geschränkt ist, gilt dann auch: zu jedem noch so großen B gibt es einen Index ν, für den die Ungleichung:

$$\left| g(\xi) - \sum_{\lambda=1}^{\nu} g_\lambda\, \omega_\lambda(\xi) \right| > B$$

in einer Punktmenge positiven Inhaltes erfüllt ist.

[1] A. a. O., p. 678.

Daraus aber folgt nach einem von H. Lebesgue bewiesenen Satze:[1] Es gibt eine in $< a, b >$ integrierbare Funktion, für die die Beziehung:

$$\lim_{\nu=\infty} \int_a^b f(\xi) \left(g(\xi) - \sum_{\lambda=1}^{\nu} g_\lambda \, \omega_\lambda(\xi) \right) d\xi = 0$$

nicht gilt. Wegen

$$\int_a^b f(\xi) \left(g(\xi) - \sum_{\lambda=1}^{\nu} g_\lambda \, \omega_\lambda(\xi) \right) d\xi = \int_a^b g(\xi) \left(f(\xi) - \sum_{\lambda=1}^{\nu} f_\lambda \, \omega_\lambda(\xi) \right) d\xi$$

folgt daraus (da g geschränkt ist) sofort, daß (9) nicht gelten kann, womit die Behauptung bewiesen ist.

Die Bedingung ist hinreichend. Nehmen wir in der Tat an, sie sei erfüllt. Wir beweisen (9) zunächst für solche Funktionen $f(\xi)$, die samt ihrem Quadrate integrierbar sind.[2] Bekanntlich konvergieren dann die Teilsummen von $\sum_{\lambda=1}^{\infty} f_\lambda \, \omega_\lambda(x)$ im quadratischen Mittel gegen $f(x)$, d. h. es ist:

$$\lim_{\nu=\infty} \int_a^b \left(f(x) - \sum_{\lambda=1}^{\nu} f_\lambda \, \omega_\lambda(x) \right)^2 dx = 0. \tag{10}$$

---

[1] Ann. Toulouse, Serie 3, Bd. 1, p. 52. Es handelt sich um den dort mit I bezeichneten Satz. Durch eine leichte Verschärfung des dort geführten Beweises kann man übrigens sogar die Existenz einer integrierbaren Funktion f nachweisen, für die:

$$\overline{\lim_{\nu=\infty}} \int_a^b f(\xi) \left(g(\xi) - \sum_{\lambda=1}^{\nu} g_\lambda \, \omega_\lambda(\xi) \right) d\xi = +\infty.$$

und somit auch:

$$\overline{\lim_{\nu=\infty}} \int_a^b \left| f(\xi) - \sum_{\lambda=1}^{\nu} f_\lambda \, \omega_\lambda(\xi) \right| d\xi = +\infty.$$

[2] In dem Falle gilt (9) für jedes vollständige normierte Orthogonalsystem.

Aus (10) aber kann (9) leicht gefolgert werden. Setzen wir nämlich abkürzend:

$$f(x)-\sum_{\lambda=1}^{\nu} f_\lambda\, \omega_\lambda(x)=r_\nu(x),$$

so besagt (10):

$$\lim_{\nu=\infty}\int_a^b \left(r_\nu(x)\right)^2 dx=0. \qquad (11)$$

Sei $\varepsilon>0$ beliebig gegeben. Wir zerlegen das Intervall $<a, b>$ in zwei komplementäre Teilmengen $\mathfrak{A}_\nu^*$ und $\mathfrak{A}_\nu^{**}$, so daß:

$$|r_\nu|<\varepsilon \text{ auf } \mathfrak{A}_\nu^*; \quad |r_\nu| \geqq \varepsilon \text{ auf } \mathfrak{A}_\nu^{**}.$$

Dann ist:

$$\int_{(\mathfrak{A}_\nu^*)}|r_\nu(x)|\, dx \leqq \varepsilon(b-a) \text{ für alle } \nu, \qquad (12)$$

und da auf $\mathfrak{A}_\nu^{**}$:

$$|r_\nu(x)| \leqq \frac{1}{\varepsilon}\left(r_\nu(x)\right)^2,$$

so ist auch:

$$\int_{(\mathfrak{A}_\nu^{**})}|r_\nu(x)|\, dx \leqq \frac{1}{\varepsilon}\int_{(\mathfrak{A}_\nu^{**})}\left(r_\nu(x)\right)^2 dx.$$

Aus (11) folgt also unmittelbar:

$$\lim_{\nu=\infty}\int_{(\mathfrak{A}_\nu^{**})}|r_\nu(x)|\, dx=0. \qquad (13)$$

Und da:

$$\int_a^b=\int_{(\mathfrak{A}_\nu^*)}+\int_{(\mathfrak{A}_\nu^{**})}$$

ist, so folgt aus (12) und (13):

$$\lim_{\nu=\infty}\int_a^b|r_\nu(x)|\, dx=0.$$

Damit aber ist in der Tat (9) für jede samt ihrem Quadrat integrierbare, insbesondere also für jede geschränkte Funktion nachgewiesen.

Sei nun $f(x)$ eine beliebige in $<a, b>$ integrierbare Funktion. Mit $f_n(x)$ bezeichnen wir diejenige geschränkte Funktion, die aus f entsteht, indem man alle Funktionswerte, für die $|f| > n$ ist, durch 0 ersetzt.

Bekanntlich ist dann:

$$\lim_{n=\infty} \int_a^b |f(x) - f_n(x)|\, dx = 0. \tag{14}$$

Ferner ist, wenn mit $f_{n,\nu}$ die Fourier'schen Konstanten von f_n:

$$f_{n,\nu} = \int_a^b f_n(\xi)\, \omega_\nu(\xi)\, d\xi$$

bezeichnet werden, nach dem eben Bewiesenen:

$$\lim_{\nu=\infty} \int_a^b \left| f_n(x) - \sum_{\lambda=1}^{\nu} f_{n,\lambda}\, \omega_\lambda(x) \right| dx = 0. \tag{15}$$

Weiter ist:

$$\int_a^b \left| \sum_{\lambda=1}^{\nu} f_{n,\lambda}\, \omega_\lambda(x) - \sum_{\lambda=1}^{\nu} f_\lambda\, \omega_\lambda(x) \right| dx =$$

$$\int_a^b \left| \sum_{\lambda=1}^{\nu} \int_a^b \big(f_n(\xi) - f(\xi)\big)\, \omega_\lambda(\xi)\, d\xi \,.\, \omega_\lambda(x) \right| dx \leqq$$

$$\leqq \int_a^b \left(\int_a^b |f_n(\xi) - f(\xi)| \,.\, \left| \sum_{\lambda=1}^{\nu} \omega_\lambda(\xi)\, \omega_\lambda(x) \right| d\xi \right) dx.$$

In diesem letzten Integrale kann nach einem bekannten Satze[1] die Integrationsfolge vertauscht werden:

$$\int_a^b \left(\int_a^b |f_n(\xi) - f(\xi)| \,.\, \left| \sum_{\lambda=1}^{\nu} \omega_\lambda(\xi)\, \omega_\lambda(x) \right| d\xi \right) dx =$$

$$= \int_a^b \left\{ |f_n(\xi) - f(\xi)| \int_a^b \left| \sum_{\lambda=1}^{\nu} \omega_\lambda(\xi)\, \omega_\lambda(x) \right| dx \right\} d\xi.$$

[1] E. W. Hobson, Lond. Proc. Serie 2, Bd. 8 (1910), p. 31.

Und da nach Annahme für alle ν und alle x von $\langle a, b \rangle$ (ausgenommen eine Nullmenge) Ungleichung (6) besteht, so haben wir endlich für alle ν:

$$\int_a^b \left| \sum_{\lambda=1}^{\nu} f_{n,\lambda}\, \omega_\lambda(x) - \sum_{\lambda=1}^{\nu} f_\lambda\, \omega_\lambda(x) \right| d\,x \leqq$$

$$\leqq A \,.\, \int_a^b |f(x) - f_n(x)|\, d\,x. \tag{16}$$

Sei nun $\varepsilon > 0$ beliebig gegeben. Wir wählen, was wegen (14) sicher möglich ist, ein n so, daß

$$\int_a^b |f(x) - f_n(x)|\, d\,x < \varepsilon. \tag{17}$$

Es ist dann zufolge (15) für dieses n und fast alle ν:

$$\int_a^b \left| f_n(x) - \sum_{\lambda=1}^{\nu} f_{n,\lambda}\, \omega_\lambda(x) \right| d\,x < \varepsilon. \tag{18}$$

Endlich ist, wegen (16) und (17), für dieses n und alle ν:

$$\int_a^b \left| \sum_{\lambda=1}^{\nu} f_{n,\lambda}\, \omega_\lambda(x) - \sum_{\lambda=1}^{\nu} f_\lambda\, \omega_\lambda(x) \right| d\,x \leqq A \,.\, \varepsilon. \tag{19}$$

Aus (17), (18), (19) aber folgern wir für fast alle ν:

$$\int_a^b \left| f(x) - \sum_{\lambda=1}^{\nu} f_\lambda\, \omega_\lambda(x) \right| d\,x < \varepsilon\,(A+2).$$

Da hierin $\varepsilon > 0$ beliebig war, ist damit (9) nachgewiesen und der Beweis von Satz V beendet.

§ 3.

Wir sagen, die Folge in $\langle a, b \rangle$ meßbarer Funktionen $g_\nu(x)$ $(\nu = 1, 2, \ldots)$ konvergiere asymptotisch gegen die in $\langle a, b \rangle$ meßbare Funktion $g(x)$, wenn es zu jedem Paare positiver Zahlen ε und α ein ν_0 gibt, so daß für jedes

$\nu \geqq \nu_0$ die Menge aller Punkte von $< a, b >$, in denen $|g-g_\nu| \geqq \alpha$ ist, einen Inhalt $< \varepsilon$ hat. Konvergiert die Folge meßbarer Funktionen $g_\nu(x)$ überall in $< a, b >$ (abgesehen von einer Nullmenge) gegen $g(x)$, so konvergiert sie bekanntlich auch asymptotisch gegen $g(x)$.

Wir sagen, die Folge in $< a, b >$ integrierbarer Funktionen $g_\nu(x)$ $(\nu = 1, 2, \ldots)$ konvergiere vollständig integrierbar gegen die integrierbare Funktion $g(x)$, wenn für jede meßbare Teilmenge $\mathfrak{A}$ von $< a, b >$:

$$\int_{(\mathfrak{A})} g(x)\,dx = \lim_{\nu = \infty} \int_{(\mathfrak{A})} g_\nu(x)\,dx.$$

Wir sagen, die unbestimmten Integrale der Funktionen $g_\nu(x)$ seien gleichgradig absolut-stetig in $< a, b >$, wenn es zu jedem $\varepsilon > 0$ ein $\rho > 0$ von folgender Eigenschaft gibt: für jede meßbare Punktmenge $\mathfrak{A}$ aus $< a, b >$, deren Inhalt $< \rho$ ist, und für alle ν gilt die Ungleichung:[1]

$$\int_{(\mathfrak{A})} |g_\nu|\,dx < \varepsilon. \tag{20}$$

Von G. Vitali wurde der Satz bewiesen:[2]

VI. Damit die in $< a, b >$ asymptotisch gegen die integrierbare Funktion $g(x)$ konvergierende Folge integrierbarer Funktionen $g_\nu(x)$ vollständig integrierbar gegen $g(x)$ konvergiere, ist notwendig und hinreichend, daß die unbestimmten Integrale der $g_\nu(x)$ gleichgradig absolut-stetig seien in $< a, b >$.

Wir wollen nun, um Anschluß an die Untersuchungen von § 2 zu gewinnen, diese von Vitali gefundene Bedingung

---

[1] Diese Definition ändert ihren Sinn gar nicht, wenn (20) ersetzt wird durch:

$$\left| \int_{(\mathfrak{A})} {}_\nu \; dx \right| <$$

[2] Rend. Palermo, Bd. 23 (1907), p. 137. Genau gesprochen hat Vitali diesen Satz für Funktionenfolgen bewiesen, die (im gewöhnlichen Sinne) gegen $g(x)$ konvergieren. Doch hat die Übertragung auf asymptotisch konvergierende Funktionenfolgen keine Schwierigkeit.

für vollständig integrierbare Konvergenz durch eine andere ersetzen.

VII. Damit die in $< a, b >$ asymptotisch gegen die integrierbare Funktion $g(x)$ konvergierende Folge integrierbarer Funktionen $g_\nu(x)$ vollständig integrierbar gegen $g(x)$ konvergiere, ist notwendig und hinreichend das Bestehen der Beziehung:

$$\lim_{\nu=\infty} \int_a^b |g(x) - g_\nu(x)|\, dx = 0. \tag{21}$$

Die Bedingung ist notwendig. Angenommen in der Tat, es konvergiere die Folge der Funktionen $g_\nu(x)$ asymptotisch und vollständig integrierbar gegen $g(x)$. Nach Satz VI sind dann die unbestimmten Integrale der $g_\nu(x)$ gleichgradig absolut-stetig. Dasselbe gilt dann offenbar auch für die unbestimmten Integrale der Funktionen $g(x) - g_\nu(x)$. Ist $\varepsilon > 0$ beliebig gegeben, so gibt es also ein $\rho > 0$, so daß für alle Punktmengen $\mathfrak{A}$ aus $< a, b >$, deren Inhalt $< \rho$ ist, und für alle ν:

$$\int_{(\mathfrak{A})} |g - g_\nu|\, dx < \varepsilon. \tag{22}$$

Wir zerlegen nun das Intervall $< a, b >$ in zwei komplementäre Teilmengen $\mathfrak{A}_\nu^*$ und $\mathfrak{A}_\nu^{**}$, so daß:

$$|g(x) - g_\nu(x)| < \varepsilon \quad \text{auf } \mathfrak{A}_\nu^*; \qquad |g(x) - g_\nu(x)| \geqq \varepsilon \quad \text{auf } \mathfrak{A}_\nu^{**}$$

Dann ist, für alle ν:

$$\int_{(\mathfrak{A}_\nu^*)} |g(x) - g_\nu(x)|\, dx \leqq \varepsilon (b - a). \tag{23}$$

Ferner hat, wegen der asymptotischen Konvergenz der g_ν gegen g, für fast alle ν die Menge $\mathfrak{A}_\nu^{**}$ einen Inhalt $< \rho$, so daß wegen (22) für fast alle ν:

$$\int_{(\mathfrak{A}_\nu^{**})} |g(x) - g_\nu(x)|\, dx < \varepsilon. \tag{24}$$

Da aber:

$$\int_a^b = \int_{(\mathfrak{A}_\nu^*)} + \int_{(\mathfrak{A}_\nu^{**})},$$

so folgt aus (23) und (24) für fast alle ν:

$$\left|\int_a^b |g(x)-g_\nu(x)|\, dx < \varepsilon(b-a+1).\right.$$

Da hierin $\varepsilon > 0$ beliebig war, ist damit (21) bewiesen.

Die Bedingung ist hinreichend. In der Tat folgt aus (21) für jeden meßbaren Teil $\mathfrak{A}$ von $< a, b >$:

$$\lim_{\nu=\infty} \int_{(\mathfrak{A})} \big(g(x)-g_\nu(x)\big)\, dx = 0,$$

womit die vollständig integrierbare Konvergenz der g_ν gegen g bewiesen ist.

Nachdem so der Beweis von Satz VII beendet ist, sei noch ausdrücklich darauf hingewiesen, daß Beziehung (21) wohl bei asymptotisch konvergenten Folgen, nicht aber allgemein, für vollständig integrierbare Konvergenz notwendig ist. Wir überzeugen uns davon an einem Beispiel.

Wir zerlegen $< 0, 1)$ in die ν Teilintervalle $< \frac{i-1}{\nu}, \frac{i}{\nu})$ $(i = 1, 2, \ldots, \nu)$ und setzen:

$$g_\nu = 1 \text{ in} < \frac{2\,i}{\nu}, \frac{2\,i+1}{\nu}); \; g_\nu = -1 \text{ in} < \frac{2\,i-1}{\nu}, \frac{2\,i}{\nu})$$

Die so definierte Funktionenfolge $g_\nu(x)$ konvergiert in $< 0, 1 >$ vollständig integrierbar gegen $g(x) \equiv 0$. In der Tat, zunächst ist offenbar für jedes Teilintervall $< \alpha, \beta >$ von $< 0, 1 >$:

$$\lim_{\nu=\infty} \int_\alpha^\beta g_\nu(x)\, dx = 0. \tag{25}$$

Sei sodann $\mathfrak{A}$ irgend eine meßbare Punktmenge aus $< 0, 1 >$ Dann gibt es zu jedem $\varepsilon > 0$ endlich viele, zu je zweien fremde Teilintervalle $< \alpha_\lambda, \beta_\lambda > (\lambda = 1, 2, \ldots, l)$ von $< 0, 1 >$, so daß sowohl die Menge $\mathfrak{A}'$ aller nicht zu $\mathfrak{A}$ gehörigen Punkte dieser Teilintervalle als auch die Menge $\mathfrak{A}''$ der außerhalb dieser Teilintervalle liegenden Punkte von $\mathfrak{A}$ einen Inhalt $< \varepsilon$ hat. Aus der Definition der g_ν folgt daher für alle ν:

$$\left|\int_{(\mathfrak{A}')} g_\nu\, dx\right| < \varepsilon; \left|\int_{(\mathfrak{A}'')} g_\nu\, dx\right| < \varepsilon, \tag{26}$$

und wegen (25) ist:

$$\lim_{\nu=\infty} \sum_{\lambda=1}^{l} \int_{\alpha_\lambda}^{\beta_\lambda} g_\nu(x)\, dx = 0. \tag{27}$$

Da aber:

$$\int_{(\mathfrak{A})} = \sum_{\lambda=1}^{l} \int_{\alpha_\lambda}^{\beta_\lambda} - \int_{(\mathfrak{A}')} + \int_{(\mathfrak{A}'')}$$

ist, so folgt aus (26) und (27):

$$\lim_{\nu=\infty} \int_{(\mathfrak{A})} g_\nu(x)\, dx = 0,$$

so daß in der Tat die g_ν vollständig integrierbar gegen $g(x) = 0$ konvergieren. Beziehung (21) aber ist nicht erfüllt, denn für jedes ν ist:

$$\int_0^1 |g_\nu(x)|\, dx = 1.$$

§ 4.

Nach diesem Exkurs über vollständig integrierbare Konvergenz kehren wir zurück zur Entwicklung einer integrierbaren Funktion f nach den Funktionen eines vollständigen normierten Orthogonalsystems ω_ν, in dem jede einzelne Funktion ω_ν geschränkt ist. Wir wollen den Satz beweisen:

VIII. Damit für jede in $<a,b>$ integrierbare Funktion f die Folge der Teilsummen ihrer Entwicklung $\sum_{\nu=1}^{\infty} f_\nu\, \omega_\nu(x)$ nach den Funktionen des normierten Orthogonalsystems der $\omega_\nu(x)$ vollständig integrierbar in $<a,b>$ gegen $f(x)$ konvergiere, ist notwendig und hinreichend die Existenz einer Zahl A, so daß für alle ν und alle x von $<a,b>$ (abgesehen von einer Nullmenge) Ungleichung (6) erfüllt sei.

Die Bedingung ist **notwendig**. Nehmen wir in der Tat an, es konvergiere die Folge $\sum_{\lambda=1}^{\nu} f_\lambda\, \omega_\lambda(x)$ vollständig integrierbar in $< a, b >$ gegen $f(x)$, ohne daß Ungleichung (6) im angegebenen Sinne erfüllt wäre. Wir haben zu zeigen, daß dies auf einen Widerspruch führt. Zunächst gibt es dann, wie in § 2 bemerkt (p. 1769, Anmerkung [1]) eine in $< a, b >$ integrierbare Funktion f, so daß:

$$\overline{\lim_{\nu=\infty}} \int_a^b \left| f(\xi) - \sum_{\lambda=1}^{\nu} f_\lambda\, \omega_\lambda(\xi) \right| d\xi = +\infty. \tag{28}$$

Nach einer bekannten Eigenschaft der bestimmten Integrale gibt es ein $\sigma_\nu > 0$, so daß für jede meßbare Punktmenge $\mathfrak{A}$ aus $< a, b >$, deren Inhalt $< \sigma_\nu$ ist, die Ungleichung besteht:

$$\int_{(\mathfrak{A})} \left| f(\xi) - \sum_{\lambda=1}^{\nu} f_\lambda\, \omega_\lambda(\xi) \right| d\xi < 1. \tag{29}$$

Wir können ohne weiteres annehmen, es sei:

$$\sigma_{\nu+1} \leqq \frac{\sigma_\nu}{2}. \tag{30}$$

Aus (28) folgt nun sofort: es gibt eine wachsende Folge von Indizes $\nu_1 < \nu_2 < \ldots < \nu_i < \ldots$, und eine Folge von Teilintervallen aus $< a, b >$:

$$< \alpha_1, \beta_1 >, \; < \alpha_2, \beta_2 >, \; \ldots, \; < \alpha_i, \beta_i >, \; \ldots,$$

so daß:[1]

$$\beta_i - \alpha_i < \frac{\sigma_{\nu_{i-1}}}{2} \tag{31}$$

und:

$$\int_{\alpha_i}^{\beta_i} \left| f(\xi) - \sum_{\lambda=1}^{\nu_i} f_\lambda\, \omega_\lambda(\xi) \right| d\xi > 2i.$$

---

[1] Für $i = 1$ fällt diese Bedingung weg.

Es gibt daher in $<\alpha_i, \beta_i>$ eine meßbare Punktmenge $\mathfrak{A}_i$, so daß:

$$\left|\int_{(\mathfrak{A}_i)}\left(f(\xi)-\sum_{\lambda=1}^{\nu_i} f_\lambda\,\omega_\lambda(\xi)\right)d\,\xi\right|>i. \tag{32}$$

Wegen (31) hat $\mathfrak{A}_i$ einen Inhalt $<\frac{1}{2}\,\sigma_{\nu_{i-1}}$. Wir bezeichnen nun mit $\mathfrak{B}_i$ die Vereinigung der Mengen $\mathfrak{A}_{i+1}, \mathfrak{A}_{i+2}, \ldots$. Wegen (30) hat dann $\mathfrak{B}_i$ einen Inhalt $<\sigma_{\nu_i}$. Bezeichnen wir weiter mit $\mathfrak{A}_i^*$ die Menge aller nicht zu $\mathfrak{B}_i$ gehörigen Punkte von $\mathfrak{A}_i$, so sind je zwei Mengen $\mathfrak{A}_i^*$ fremd. Da die Menge $\mathfrak{A}_i - \mathfrak{A}_i^*$ einen Inhalt $<\sigma_{\nu_i}$ hat, ist zufolge (29):

$$\left|\int_{(\mathfrak{A}_i-\mathfrak{A}_i^*)}\left(f(\xi)-\sum_{\lambda=1}^{\nu_i} f_\lambda\,\omega_\lambda(\xi)\right)d\,\xi\right|<1,$$

und daher wegen (32):

$$\left|\int_{(\mathfrak{A}_i^*)}\left(f(\xi)-\sum_{\lambda=1}^{\nu_i} f_\lambda\,\omega_\lambda(\xi)\right)d\,\xi\right|>i-1. \tag{33}$$

Sei nun weiter $\mathfrak{B}_i$ die Vereinigung der Mengen $\mathfrak{A}_1^*, \mathfrak{A}_2^*, \ldots, \mathfrak{A}_i^*$. Da nach Annahme die Funktionenfolge $\sum_{\lambda=1}^{\nu} f_\lambda\,\omega_\lambda(\xi)$ vollständig integrierbar gegen $f(\xi)$ konvergiert, so ist:

$$\lim_{\nu=\infty}\int_{(\mathfrak{B}_i)}\left(f(\xi)-\sum_{\lambda=1}^{\nu} f_\lambda\,\omega_\lambda(\xi)\right)d\,\xi=0.$$

Es gibt daher einen Index $\nu_i' > \nu_i$, so daß für alle $\nu \geqq \nu_i'$:

$$\left|\int_{(\mathfrak{B}_i)}\left(f(\xi)-\sum_{\lambda=1}^{\nu} f_\lambda\,\omega_\lambda(\xi)\right)d\,\xi\right|<1. \tag{34}$$

Indem wir nötigenfalls von der Indizesfolge ν_i zu einer Teilfolge übergehen, können wir immer annehmen, es sei $\nu_i' = \nu_{i+1}$, so daß dann (34) für $\nu \geqq \nu_{i+1}$ gilt.

Sei nun endlich $\mathfrak{B}$ die Vereinigung aller Mengen $\mathfrak{A}_i^*$ $(i = 1, 2, \ldots)$ und $\mathfrak{B}_i^*$ die Vereinigung der Mengen $\mathfrak{A}_{i+1}^*, \mathfrak{A}_{i+2}^*, \ldots$. Da $\mathfrak{B}_i^*$ Teil von $\mathfrak{B}_i$ ist, so hat, ebenso wie $\mathfrak{B}_i$, auch $\mathfrak{B}_i^*$ einen Inhalt von $< \sigma_{\nu_i}$. Es ist also wegen (29):

$$\left| \int_{(\mathfrak{B}_i^*)} \left(f(\xi) - \sum_{\lambda=1}^{\nu_i} f_\lambda \, \omega_\lambda(\xi) \right) d\,\xi \right| < 1. \tag{35}$$

Da (34) für alle $\nu \geqq \nu_{i+1}$, insbesonders also für $\nu = \nu_{i+1}$ gilt, so ist:

$$\left| \int_{(\mathfrak{B}_{i-1})} \left(f(\xi) - \sum_{\lambda=1}^{\nu_i} f_\lambda \, \omega_\lambda(\xi) \right) d\,\xi \right| < 1. \tag{36}$$

Und da:

$$\int_{(\mathfrak{B})} = \int_{(\mathfrak{B}_{i-1})} + \int_{(\mathfrak{A}_i^*)} + \int_{(\mathfrak{B}_i^*)}$$

ist, so ergibt (33) zusammen mit (35) und (36):

$$\left| \int_{(\mathfrak{B})} \left(f(\xi) - \sum_{\lambda=1}^{\nu_i} f_\lambda \, \omega_\lambda(\xi) \right) d\,\xi \right| > i - 3.$$

Dies aber widerspricht der gemachten Voraussetzung, daß die Funktionenfolge $\sum_{\lambda=1}^{\nu} f_\lambda \, \omega_\lambda(\xi)$ vollständig integrierbar gegen $f(\xi)$ konvergiert, derzufolge:

$$\lim_{\nu = \infty} \int_{(\mathfrak{B})} \left(f(\xi) - \sum_{\lambda=1}^{\nu} f_\lambda \, \omega_\lambda(\xi) \right) d\,\xi = 0$$

sein müßte. Die Bedingung von Satz VIII ist also als notwendig erwiesen.[1]

[1] Es ist sogar darüber hinaus bewiesen: Ist die Bedingung von Satz VIII nicht erfüllt, so gibt es in $\langle a, b \rangle$ eine integrierbare Funktion $f(\xi)$ und eine meßbare Punktmenge $\mathfrak{A}$, so daß:

$$\overline{\lim_{\nu = \infty}} \int_{(\mathfrak{A})} \left(f(\xi) - \sum_{\lambda=1}^{\nu} f_\lambda \, \omega_\lambda(\xi) \right) d\,\xi = +\infty.$$

Die Bedingung ist hinreichend. In der Tat, ist sie erfüllt, so ist nach Satz V für jede in $< a, b >$ integrierbare Funktion f:

$$\lim_{\nu=\infty} \int_a^b \left| f(\xi) - \sum_{\lambda=1}^{\nu} f_\lambda \, \omega_\lambda(\xi) \right| d\xi = 0,$$

woraus sofort für jede meßbare Punktmenge $\mathfrak{A}$ von $< a, b >$ folgt:

$$\lim_{\nu=\infty} \int_{(\mathfrak{A})} \left(f(\xi) - \sum_{\lambda=1}^{\nu} f_\lambda \, \omega_\lambda(\xi) \right) d\xi = 0. \tag{37}$$

Das aber ist die Behauptung.[1]

Nachdem wir uns in § 3 überzeugt haben, daß bei asymptotisch konvergenten Folgen vollständig integrierbare Konvergenz und Beziehung (21) völlig gleichbedeutend sind, können wir nun auf Grund der Sätze V und VIII feststellen, daß auch für die Teilsummen der Entwicklung einer integrierbaren Funktion nach den Funktionen eines vollständigen normierten Orthogonalsystems[2] diese beiden Tatsachen in

[1] Übrigens folgt (37) auch unmittelbar aus der von mir (a. a. O., p. 678, Satz VI) bewiesenen Tatsache, daß, wenn die Bedingung von Satz VIII erfüllt ist, für jedes Paar von Funktionen $f(\xi)$ und $g(\xi)$, von denen die erste integrierbar, die zweite geschränkt ist in $< a, b >$, die Parseval'sche Formel gilt:

$$\int_a^b f(\xi)\, g(\xi)\, d\xi = \sum_{\lambda=1}^{\infty} f_\lambda \, g_\lambda.$$

Man hat in der Tat nur für g die Funktion zu wählen, die $= 1$ ist auf $\mathfrak{A}$, sonst $= 0$, wobei man für g_λ erhält:

$$g_\lambda = \int_{(\mathfrak{A})} \omega_\lambda(\xi)\, d\xi.$$

[2] Diese Entwicklungen stehen den asymptotisch konvergenten Reihen sehr nahe: die Entwicklung einer samt ihrem Quadrate integrierbaren Funktion ist stets im quadratischen Mittel und daher auch asymptotisch konvergent.

naher Beziehung zueinander stehen: sowohl dafür, daß vollständig integrierbare Konvergenz, als auch dafür, daß Beziehung (9) für jedes integrierbare f vorhanden sei, erwies sich als notwendig und hinreichend das Bestehen von Ungleichung (6) für alle ν und alle x von $< a, b >$, abgesehen von einer Nullmenge. Eben diese Bedingung ist aber, wie ich früher gezeigt habe,[1] auch notwendig und hinreichend dafür, daß für jedes Paar von Funktionen, deren eine integrierbar, deren andere geschränkt ist, die Parseval'sche Formel gelte.

Es ist nicht ohne Interesse, die im vorstehenden gewonnenen Resultate der von mir früher[2] aufgestellten Bedingung für gliedweise Integrierbarkeit in beliebigen Teilintervallen gegenüberzustellen. Wir sehen so:

Damit für jedes Teilintervall $< \alpha, \beta >$ von $< a, b >$ die Formel gelte:

$$\int_\alpha^\beta f(x)\,dx = \sum_{\nu=1}^{\infty} f_\nu \int_\alpha^\beta \omega_\nu(x)\,dx \tag{38}$$

ist notwendig und hinreichend, daß es zu jedem y von $< a, b >$ eine Zahl A gebe, so daß für alle ν und alle x von $< a, b >$, abgesehen von einer Nullmenge, Ungleichung (8) gelte

Damit für jede meßbare Teilmenge $\mathfrak{A}$ von $< a, b >$ die Formel gelte:

$$\int_{(\mathfrak{A})} f(x)\,dx = \sum_{\nu=1}^{\infty} f_\nu \int_{(\mathfrak{A})} \omega_\nu(x)\,dx \tag{39}$$

ist notwendig und hinreichend, daß es eine Zahl A gebe, so daß für alle ν und alle x von $< a, b >$, abgesehen von einer Nullmenge, Ungleichung (6) gelte.

Ein ähnliches Doppeltheorem erhält man, wenn man (38) ersetzt durch:

---

1 A. a. O., p. 678.

2 A. a. O., p. 680.

$$\lim_{\nu=\infty} \int_{\alpha}^{\beta} \left(f(x) - \sum_{\lambda=1}^{\nu} f_\lambda \, \omega_\lambda(x) \right) dx = 0$$

und (39) durch:

$$\lim_{\nu=\infty} \int_{a}^{b} \left| f(x) - \sum_{\lambda=1}^{\nu} f_\lambda \, \omega_\lambda(x) \right| dx = 0.$$

Endlich sei noch bemerkt, daß durch Satz VIII die Existenz von Fourier'schen Reihen integrierbarer Funktionen:

$$f(x) \sim \frac{a_0}{2} + \sum_{\nu=1}^{\infty} (a_\nu \cos \nu x + b_\nu \sin \nu x)$$

nachgewiesen ist, die nicht vollständig integrierbar gegen $f(x)$ konvergieren.

§ 5.

Die in § 2 behandelte Frage nach der Gültigkeit der Beziehung (9) ist ein Spezialfall der allgemeinen Frage: sei

$$I_\nu(f, x) = \int_a^b f(\xi) \, \varphi_\nu(\xi, x) \, d\xi$$

ein singuläres Integral; unter welchen Umständen gilt:

$$\lim_{\nu=\infty} \int_a^b |f(x) - I_\nu(f, x)| \, dx = 0?$$

Es sei mir gestattet, an dieser Stelle ein offenkundiges Versehen richtigzustellen, das sich in meinen dieser Frage gewidmeten Untersuchungen [1] findet. Es gelten die Sätze: [2]

IX. Sei $< a', b' >$ ein Teilintervall von $< a, b >$ und sei $\varphi_\nu(\xi, x)$ meßbar im Rechtecke $a \leqq \xi \leqq b$, $a' \leqq x \leqq b'$. Abgesehen von einer Nullmenge sei $\varphi_\nu(\xi, x)$ für jedes x von

[1] A. a. O., p. 658 ff.

[2] Diese beiden Sätze haben an Stelle von Satz II meiner obenzitierten Abhandlung (p. 660) zu treten.

$< a', b' >$ nach ξ integrierbar in $< a, b >$. Ferner genüge $\varphi_\nu(\xi, x)$ folgenden Bedingungen:

1. Es gibt ein A, so daß für alle ν und für alle x von $< a', b' >$, abgesehen von einer Nullmenge:

$$\int_a^b |\varphi_\nu(\xi, x)| \, d\xi < A. \tag{40}$$

2. Es gibt ein A, so daß für alle ν und alle ξ von $< a, b >$, abgesehen von einer Nullmenge:

$$\int_{a'}^{b'} |\varphi_\nu(\xi, x)| \, dx < A. \tag{41}$$

3. Abgesehen von einer Nullmenge sei für jedes x von $< a', b' >$ und jede im Intervalle $< a, b >$ der Veränderlichen ξ gelegene Punktmenge A_x, die im Punkte x die Dichte 1 hat:

$$\lim_{\nu = \infty} \int_{(\mathfrak{A}_x)} \varphi_\nu(\xi, x) \, d\xi = 1. \tag{42}$$

Dann gilt für jedes Teilintervall $< a'', b'' >$ von $< a', b' >$:

$$\lim_{\nu = \infty} \int_{a''}^{b''} \bigl(f(x) - I_\nu(f, x)\bigr) \, dx = 0.$$

X. Wird hierin zu (42) noch die Bedingung hinzugefügt:

$$\lim_{\nu = \infty} \int_{(\mathfrak{A}_x)} |\varphi_\nu(\xi, x)| \, d\xi = 1, \tag{43}$$

so gilt auch

$$\lim_{\nu = \infty} \int_{a'}^{b'} |f(x) - I_\nu(f, x)| \, dx = 0.$$

Der Beweis dieser beiden Sätze bleibt im wesentlichen derselbe, wie ich ihn a. a. O. geführt habe. Ich gebe daher hier nur die erforderlichen Abänderungen an. Es folgt nicht, wie irrtümlicherweise behauptet wurde,[1] aus (42) die dort mit (8) bezeichnete Gleichung:

---

[1] Der Irrtum liegt darin, daß die beim Beweise vorkommenden Mengen $\mathfrak{B}_x^{(1)}$ und $\mathfrak{B}_x^{(2)}$ nicht von dem (dort mit n bezeichneten) Index unabhängig sind.

$$\lim_{\nu=\infty} \int_{(\mathfrak{B}_x)} |\varphi_\nu(\xi, x)|\, d\xi = 0, \tag{44}$$

sondern lediglich:

$$\lim_{\nu=\infty} \int_{(\mathfrak{B}_x)} \varphi_\nu(\xi, x)\, d\xi = 0. \tag{45}$$

Um (44) behaupten zu können, bedarf es der schärferen Voraussetzung (43). Indem man im übrigen den Beweis gänzlich ungeändert läßt, erhält man Satz X.

Der Beweis von Satz IX bleibt bis einschließlich Gleichung (12) (a. a. O., p. 662) ungeändert.

Von da an schließt man so weiter:

Es handelt sich darum, an Stelle der dortigen Gleichung (18) die folgende Gleichung zu beweisen: Ist $< a'', b'' >$ ein beliebiges Teilintervall von $< a', b' >$, so ist:

$$\lim_{\nu=\infty} \int_{a''}^{b'} \int_{(\mathfrak{B}_x)} \left(f(\xi)\, \varphi_\nu(\xi, x)\, d\xi \right) dx = 0. \tag{46}$$

Wir setzen für jedes x aus $< a', b' >$:

$$\bar{\varphi}_\nu(\xi, x) = \varphi_\nu(\xi, x) \quad \text{auf } \mathfrak{B}_x, \quad \text{sonst: } \bar{\varphi}_\nu(\xi, x) = 0.$$

Wir gehen aus von:

$$\int_{a''}^{b''} \left(\int_a^b f(\xi)\, \bar{\varphi}_\nu(\xi, x)\, d\xi \right) dx = \int_a^b \left(f(\xi) \int_{a''}^{b''} \bar{\varphi}_\nu(\xi, x)\, dx \right) d\xi.$$

Setzen wir:

$$\int_{a''}^{b''} \bar{\varphi}_\nu(\xi, x)\, dx = \Phi_\nu(\xi),$$

so wird also (46) gleichbedeutend mit:

$$\lim_{\nu=\infty} \int_a^b f(\xi)\, \Phi_\nu(\xi)\, d\xi = 0. \tag{47}$$

Nun ist wegen (41) (abgesehen von einer Nullmenge):

$$|\Phi_\nu(\xi)| < A. \tag{48}$$

Ferner ist, wenn $< \alpha, \beta >$ irgend ein Teilintervall von $< a, b >$ bedeutet:

$$\lim_{\nu=\infty} \int_{\alpha}^{\beta} \Phi_\nu(\xi)\, d\xi = 0. \tag{49}$$

In der Tat, es ist:

$$\int_{\alpha}^{\beta} \Phi_\nu(\xi)\, d\xi = \int_{\alpha}^{\beta} \left(\int_{a''}^{b''} \overline{\varphi}_\nu(\xi, x)\, dx \right) d\xi =$$

$$= \int_{a''}^{b''} \left(\int_{\alpha}^{\beta} \overline{\varphi}_\nu(\xi, x)\, d\xi \right) dx.$$

Hierin aber ist, wegen (45) und (40) (abgesehen von Nullmengen):

$$\lim_{\nu=\infty} \int_{\alpha}^{\beta} \overline{\varphi}_\nu(\xi, x)\, d\xi = 0\,; \quad \left| \int_{\alpha}^{\beta} \overline{\varphi}_\nu(\xi, x)\, d\xi \right| < A,$$

woraus bekanntlich folgt:

$$\lim_{\nu=\infty} \int_{a''}^{b''} \left(\int_{\alpha}^{\beta} \overline{\varphi}_\nu(\xi, x)\, d\xi \right) dx = 0,$$

womit (49) bewiesen ist. Aus (48) und (49) aber folgt bekanntlich für jedes integrierbare f das Bestehen von (47) und damit von (46).

Sobald aber (46) bewiesen ist, erledigt sich der Beweis von Satz IX wie bisher.

Freitag, den 26. September 1924, 4 Uhr nachmittags.

Vorsitz: Pringsheim.

1. H. Hahn, Wien: Über Fouriersche Reihen und Integrale.

Setzt man:

$$\Phi(\mu) = \int_{-\infty}^{+\infty} f(x) \frac{\sin \mu x}{x} dx$$

$$\Psi(\mu) = \int_{-1}^{1} f(x) \frac{1 - \cos \mu x}{x} dx - \int_{1}^{+\infty} f(x) \frac{\cos \mu x}{x} dx - \int_{-\infty}^{-1} f(x) \frac{\cos \mu x}{x} dx,$$

so erhält man die folgende, unter sehr allgemeinen Voraussetzungen gültige Darstellung von $f(x)$:

$$(*) \qquad f(x) = \frac{1}{\pi} \int_{0}^{+\infty} \cos \mu x d\, \Phi(\mu) + \sin \mu x d\, \Psi(\mu).$$

Ist $f(x)$ eine für alle reellen x definierte Funktion, für die die Fouriersche Integraldarstellung gilt, so reduziert sich die rechte Seite von (*) auf das Fouriersche Integral; ist $f(x)$ eine in die Fouriersche Reihe entwickelbare periodische Funktion, so reduziert sich die rechte Seite von (*) auf die Fouriersche Reihe von $f(x)$. Nennen wir die rechte Seite von (*) das Fouriersche Stieltjes-Integral von $f(x)$, so erscheint die Fouriersche Reihe einer periodischen Funktion als ihr Fouriersches Stieltjes-Integral.

2. A. Hammerstein, Berlin: Über die asymptotische Darstellung der Eigenfunktionen linearer Integralgleichungen.

Das asymptotische Verhalten der zu positiven Eigenwerten λ_n gehörigen Eigenfunktionen einer linearen Integralgleichung zweiter Art bei wachsendem λ_n wird zu Stetigkeitseigenschaften des Kerns in Beziehung gebracht. So können beispielsweise die Eigenfunktionen eines Kerns $K(x,y)$, der neben unwesentlichen Voraussetzungen die Eigenschaft hat, daß seine Ableitung $\frac{\partial K(x,y)}{\partial x}$ für $x = y$ einen positiven Sprung aufweist, durch trigonometrische asymptotisch dargestellt werden, während die eines Kerns $K(x, y; \xi, \eta)$, der für $x = \xi$, $y = \eta$ wie $\frac{-1}{2\pi} \log r$, wo $r = \sqrt{(x-\xi)^2 + (y-\eta)^2}$ ist, unendlich wird, mit gewissen Lösungen der Differentialgleichung $\Delta w + \lambda_n w = 0$ asymptotisch übereinstimmen.

Erscheint in den Mathematischen Annalen.

Gedruckt mit Unterstützung aus dem Jerome und Margaret Stonborough-Fonds

Über die Methode der arithmetischen Mittel in der Theorie der verallgemeinerten Fourier'schen Integrale

Von

Hans Hahn in Wien

k. M. d. Akad. d. Wiss.

(Vorgelegt in der Sitzung am 3. Dezember 1925)

In einem Vortrag in der Wiener Mathematischen Gesellschaft im Februar 1924 sowie in einem Vortrag auf der Innsbrucker Naturforscherversammlung 1924 habe ich mich mit einer Verallgemeinerung des Fourier'schen Integraltheoremes beschäftigt, und die betreffenden Untersuchungen erscheinen demnächst an andrer Stelle in ausführlicher Darstellung. Diese Verallgemeinerung lautet folgendermaßen: Es gilt unter geeigneten Annahmen über die Funktion $f(x)$ die Formel:

$$f(x_0) = \frac{1}{\pi}\int_0^{+\infty}(\cos\mu x_0\, d\Phi_1(\mu) + \sin\mu x_0\, d\Psi_1(\mu)), \tag{0}$$

wo:

$$\Phi_1(\mu) = \int_{-\infty}^{+\infty} f(x)\frac{\sin\mu x}{x}\,dx;\quad \Psi_1(\mu) = \int_{-1}^{1} f(x)\frac{1-\cos\mu x}{x}\,dx -$$

$$-\int_1^{+\infty} f(x)\frac{\cos\mu x}{x}\,dx - \int_{-\infty}^{-1} f(x)\frac{\cos\mu x}{x}\,dx.$$

Insbesondere verlangt diese Formel zu ihrer Gültigkeit, daß $f(x)$ an der Stelle x_0 in die Fourier'sche Reihe entwickelbar sei.

Im folgenden soll nun eine noch weitertragende Formel entwickelt werden, die zu ihrer Gültigkeit nichts weiter verlangt, als daß $f(x)$ im Unendlichen beschränkt und an der Stelle x_0 stetig sei. Sie lautet:

$$f(x_0) = \lim_{\mu\to+\infty}\frac{1}{\mu\pi}\int_0^{\mu}\left(\int_0^{\lambda}\cos\mu x_0\frac{d^2\Phi_2(\tau)}{d\tau} + \sin\mu x_0\frac{d^2\Psi_2(\tau)}{d\tau}\right)d\lambda, \tag{00}$$

wo:

$$\Phi_2(\mu) = 2\int_{-\infty}^{+\infty} f(x)\sin^2\frac{\frac{\mu}{2}x}{x^2}\,dx;\ \Psi_2(\mu) = \int_{-1}^{1} f(x)\frac{\mu x - \sin\mu x}{x^2}\,dx -$$

$$-\int_{1}^{+\infty} f(x)\frac{\sin\mu x}{x^2}\,dx - \int_{-\infty}^{-1} f(x)\frac{\sin\mu x}{x^2}\,dx.$$

Der darin auftretende Integralbegriff wird in § 1 erläutert.

Während Formel (0) in der Theorie der harmonischen Analyse beschränkter Funktionen dieselbe Stellung einnimmt, wie die Fourier-sche Reihe in der harmonischen Analyse periodischer Funktionen, so nimmt Formel (00) in der harmonischen Analyse beschränkter Funktionen dieselbe Stellung ein wie Fejér's Summierung durch arithmetische Mittel in der harmonischen Analyse periodischer Funktionen. In der Tat reduziert sich Formel (00), wenn $f(x)$ periodisch ist, auf Fejér's Summationsformel, so wie sich (0), wenn $f(x)$ periodisch ist, auf die Fourier'sche Reihe reduziert.

Ich hoffe, später noch eingehender auf diese Fragen zurückkommen zu können, da ich mich für diesmal begnügen muß, die nächstliegenden Folgerungen aus (00) zu ziehen.[0]

§ 1.

Seien die Funktionen $f(x)$ und $g(x)$ definiert und stetig im Intervalle $[a, b]$. Wir nehmen mit diesem Intervalle eine Zerlegung Z vor durch Einschalten der Punkte:

$$a = x_0 < x_1 < x_2 < \ldots < x_{n-1} < x_n = b$$

und bilden den Ausdruck:

$$S(Z) = \sum_{i=1}^{n-1} f(x_i)\left(\frac{g(x_{i+1})-g(x_i)}{x_{i+1}-x_i} - \frac{g(x_i)-g(x_{i-1})}{x_i-x_{i-1}}\right).$$

Wir lassen nun Z eine ausgezeichnete Zerlegungsfolge[1] $\{Z_\nu\}$ durchlaufen und zeigen:

Hat $f(x)$ in $[a, b]$ eine Ableitung $f'(x)$, die stetig und von endlicher Variation ist, existiert ferner im Punkte a die rechtsseitige Ableitung $g'_+(a)$, im Punkte b die linksseitige Ableitung $g'_-(b)$, so existiert ein endlicher Grenzwert $\lim_{\nu\to\infty} S(Z_\nu)$, der von der Wahl der ausgezeichneten Zerlegungsfolge $\{Z_\nu\}$ unabhängig ist.

In der Tat, es ist:

$$S(Z) = f(x_{n-1})\frac{g(x_n)-g(x_{n-1})}{x_n-x_{n-1}} - f(x_1)\frac{g(x_1)-g(x_0)}{x_1-x_0} -$$

[0] (Zusatz bei der Korrektur.) Es sei auf eine mittlerweile erschienene höchst bedeutungsvolle Arbeit von N. Wiener verwiesen: Math. Zeitschr. 24, p. 575.

[1] Im Sinne der Terminologie von G. Kowalewski, Grundz. d. Diff.- und Integr.-Rechnung, p. 171.

$$-\sum_{i=1}^{n-2} \frac{f(x_{i+1})-f(x_i)}{x_{i+1}-x_i}(g(x_{i+1})-g(x_i)) =$$

$$= f(x_{n-1})\frac{g(x_n)-g(x_{n-1})}{x_n-x_{n-1}} - f(x_1)\frac{g(x_1)-g(x_0)}{x_1-x_0} +$$

$$+ f'(\xi_1)(g(x_1)-g(x_0)) + f'(\xi_n)(g(x_n)-g(x_{n-1})) -$$

$$-\sum_{i=1}^{n} f'(\xi_i)(g(x_i)-g(x_{i-1})),$$

wo ξ_i einen Punkt des Intervalles (x_{i-1}, x_i) bedeutet.

Durchläuft Z die ausgezeichnete Zerlegungsfolge $\{Z_\nu\}$, so gilt $x_1 \to a$, $x_{n-1} \to b$, $\xi_1 \to a$, $\xi_n \to b$, und somit wegen der gemachten Voraussetzungen:

$$f(x_{n-1})\frac{g(x_n)-g(x_{n-1})}{x_n-x_{n-1}} \to f(b)\,g'_-(b);\; f(x_1)\frac{g(x_1)-g(x_0)}{x_1-x_0} \to f(a)\,g'_+(a);$$

$$f'(\xi_1)(g(x_1)-g(x_0)) \to 0;\; f'(\xi_n)(g(x_n)-g(x_{n-1})) \to 0,$$

und gleichzeitig gilt:[2]

$$\sum_{i=1}^{n} f'(\xi_i)(g(x_i)-g(x_{i-1})) \to \int_a^b f'(x)\,dg(x).$$

Damit ist die Behauptung bewiesen.

Bezeichnen wir noch den Grenzwert von $S(Z_\nu)$ mit $\int_a^b f(x)\,\frac{d^2 g(x)}{dx}$, so haben wir gefunden:

$$\int_a^b f(x)\,\frac{d^2 g(x)}{dx} = f(b)\,g'_-(b) - f(a)\,g'_+(a) - \int_a^b f'(x)\,dg(x). \quad (1)$$

Aus dieser Formel folgt sofort, wenn c gemäß $a < c < b$ so gewählt ist, daß $g'_+(c)$ und $g'_-(c)$ existieren:

$$\int_a^b f(x)\,\frac{d^2 g(x)}{dx} = \int_a^c f(x)\,\frac{d^2 g(x)}{dx} + \int_c^b f(x)\,\frac{d^2 g(x)}{dx} +$$

$$+ f(c)\,(g'_+(c) - g'_-(c)), \quad (2)$$

---

[2] Vgl. hierzu H. Hahn, Monatsh. f. Math. u. Phys., *32*, p. 69 ff.

also insbesondere wenn $f(c)=0$, oder wenn in c eine Ableitung $g'(c)$ existiert:

$$\int^b f(x)\frac{d^2g(x)}{dx}=\int_a^c f(x)\frac{d^2g(x)}{dx}+\int_c^b f(x)\frac{d^2g(x)}{dx}. \tag{3}$$

Besitzt $g(x)$ in $[a,b]$ eine erste Ableitung $g'(x)$ und eine zweite Ableitung $g''(x)$, und gilt für jedes Teilintervall $[\alpha,\beta]$ von $[a,b]$:

$$g'(\beta)-g'(\alpha)=\int_\alpha^\beta g''(x)\,dx,$$

so ist:

$$\int_a^b f(x)\frac{d^2g(x)}{dx}=\int_a^b f(x)g''(x)\,dx. \tag{4}$$

In der Tat, in (1) ist dann:

$$\int_a^b f'(x)\,dg(x)=\int_a^b f'(x)g'(x)\,dx=f(b)g'(b)-f(a)g'(a)-\int_a^b f(x)g''(x)\,dx,$$

und durch Einsetzen in (1) ergibt sich die Behauptung (4).

Besitzt $g(x)$ in $[a,b]$ eine stetige Ableitung $g'(x)$, so ist:

$$\int_a^b f(x)\frac{d^2g(x)}{dx}=\int^b f(x)\,dg'(x). \tag{5}$$

In der Tat, es ist:

$$\int_a^b f(x)\frac{d^2g(x)}{dx}=\lim\sum_{i=1}^{n-1}f(x_i)\left(\frac{g(x_{i+1})-g(x_i)}{x_{i+1}-x_i}-\frac{g(x_i)-g(x_{i-1})}{x_i-x_{i-1}}\right)=$$

$$=\lim\sum_{i=1}^{n-1}f(x_i)\,(g'(\xi_{i+1})-g'(\xi_i)),$$

wo ξ_i einen Punkt des Intervalles (x_{i-1},x_i) bedeutet. Hierin ist weiter:

$$\sum_{i=1}^{n-1}f(x_i)\,(g'(\xi_{i+1})-g'(\xi_i))=$$

$$=f(x_n)g'(\xi_n)-f(x_0)g'(\xi_1)-\sum_{i=1}^{n}g'(\xi_i)\,(f(x_i)-f(x_{i-1})),$$

also:

$$\lim \sum_{i=1}^{n-1} f(x_i)\,(g'(\xi_{i+1})-g'(\xi_i)) = f(b)g'(b)-f(a)g'(a) - \int_a^b g'(x)\,df(x) =$$

$$= \int_a^b f(x)\,dg'(x),$$

womit die Behauptung bewiesen ist.

Wir beweisen noch den Satz: Genügen $f(x)$ und $g_\nu(x)$ den eingangs gemachten Voraussetzungen, konvergiert $g_\nu(x)$ in $[a,b]$ gleichmäßig gegen $g(x)$, und ist:

$$g'_+(a) = \lim_{\nu\to\infty} g'_{\nu+}(a),\quad g'_-(b) = \lim_{\nu\to\infty} g'_{\nu-}(b), \tag{6}$$

so ist:

$$\int_a^b f(x)\frac{d^2 g(x)}{dx} = \lim_{\nu\to\infty} \int_a^b f(x)\frac{d^2 g_\nu(x)}{dx}. \tag{7}$$

In der Tat, nach (1) ist:

$$\int_a^b f(x)\frac{d^2 g_\nu(x)}{dx} = f(b)g'_{\nu-}(b)-f(a)\,g'_{\nu+}(a) - \int_a^b f'(x)\,dg_\nu(x).$$

Wegen (6) und wegen:[3]

$$\lim_{\nu\to\infty} \int_a^b f'(x)\,dg_\nu(x) = \int_a^b f'(x)\,dg(x)$$

geht hierin die rechte Seite für $\nu \to \infty$ in die rechte Seite von (1) über, womit (7) bewiesen ist.

§ 2.

Bekanntlich gilt der Satz: Existieren die Integrale

$\int_{-\infty}^{+\infty} |g(x)|\,dx$ und $\int_{-\infty}^{+\infty} (g(x))^2\,dx$ und setzt man:

$$A(\mu) = \frac{1}{\pi}\int_{-\infty}^{+\infty} g(x)\cos\mu x\,dx;\quad B(\mu) = \frac{1}{\pi}\int_{-\infty}^{+\infty} g(x)\sin\mu x\,dx,$$

so ist

$$\int_{-\infty}^{+\infty} (g(x))^2\,dx = \pi\int_0^{+\infty} \{(A(\mu))^2+(B(\mu))^2\}\,d\mu. \tag{8}$$

---

[3] L. c. [2], p. 77 ff.

Sei nun $\{g_\nu(x)\}$ eine Folge von Funktionen, für die die Integrale $\int_{-\infty}^{+\infty} |g_\nu(x)|\,dx$ und $\int_{-\infty}^{+\infty} (g_\nu(x))^2\,dx$ existieren; außerdem sei die Folge $\{g_\nu(x)\}$ in $(-\infty, +\infty)$ im Mittel konvergent, d. h. zu jedem $\varepsilon > 0$ gehört ein N, so daß:

$$\int_{-\infty}^{+\infty} (g_{\nu'}(x) - g_{\nu''}(x))^2\,dx < \varepsilon \text{ für } \nu' \geqq N,\ \nu'' \geqq N. \tag{9}$$

Wir setzen:

$$A_\nu(\mu) = \frac{1}{\pi}\int_{-\infty}^{+\infty} g_\nu(x)\cos\mu x\,dx;\ B_\nu(\mu) = \frac{1}{\pi}\int_{-\infty}^{+\infty} g_\nu(x)\sin\mu x\,dx$$

und erkennen sofort: die beiden Folgen $\{A_\nu(\mu)\}$ und $\{B_\nu(\mu)\}$ sind in $[0, +\infty)$ im Mittel konvergent.

In der Tat, zufolge (8) ist:

$$\pi\int_0^{+\infty}\{(A_{\nu'}(\mu) - A_{\nu''}(\mu))^2 + (B_{\nu'}(\mu) - B_{\nu''}(\mu))^2\}\,d\mu =$$
$$= \int_{-\infty}^{+\infty}(g_{\nu'}(x) - g_{\nu''}(x))^2\,dx,$$

also zufolge (9):

$$\int_0^{+\infty}(A_{\nu'}(\mu) - A_{\nu''}(\mu))^2\,d\mu < \varepsilon,\ \int_0^{+\infty}(B_{\nu'}(\mu) - B_{\nu''}(\mu))^2\,d\mu < \varepsilon$$
$$\text{für } \nu' \geqq N,\ \nu'' \geqq N,$$

womit die Behauptung bewiesen ist.

Sei nun $f(x)$ eine in jedem endlichen Intervalle integrierbare Funktion, die im Unendlichen beschränkt bleibt; d. h. es gebe ein p und ein q, so daß

$$|f(x)| \leqq p \text{ für } |x| \geqq q. \tag{10}$$

Wir setzen für $\mu \geqq 0$:

$$\Phi_2(\mu) = \int_{-\infty}^{+\infty} f(x)\frac{1 - \cos\mu x}{x^2}\,dx = 2\int_{-\infty}^{+\infty} f(x)\frac{\sin^2\frac{\mu}{2}x}{x^2}\,dx$$

$$\Psi_2(\mu) = \int_{-1}^{1} f(x) \frac{\mu x - \sin \mu x}{x^2} dx - \int_{1}^{+\infty} f(x) \frac{\sin \mu x}{x^2} dx - \int_{-\infty}^{-1} f(x) \frac{\sin \mu x}{x^2} dx. \tag{11}$$

Hierdurch sind $\Phi_2(\mu)$ und $\Psi_2(\mu)$ für alle $\mu \geqq 0$ definiert, und zwar als stetige Funktionen von μ. Um dies einzusehen, setzen wir:

$$\begin{aligned} \Phi_{2,n}(\mu) &= \int_{-n}^{n} f(x) \frac{1-\cos \mu x}{x^2} dx \\ \Psi_{2,n}(\mu) &= \int_{-1}^{1} f(x) \frac{\mu x - \sin \mu x}{x^2} dx - \int_{1}^{n} f(x) \frac{\sin \mu x}{x^2} dx - \int_{-n}^{-1} f(x) \frac{\sin \mu x}{x^2} dx. \end{aligned} \tag{12}$$

Offenbar ist jedes $\Phi_{2,n}(\mu)$ und jedes $\Psi_{2,n}(\mu)$ eine stetige Funktion von μ. Für hinlänglich große n gilt wegen (10):

$$|\Phi_{2,n}(\mu) - \Phi_2(\mu)| \leqq 2p \int_{n}^{+\infty} \frac{dx}{x^2} + 2p \int_{-\infty}^{-n} \frac{dx}{x^2} < \varepsilon,$$

so daß die $\Phi_{2,n}(\mu)$ gleichmäßig für alle $\mu \geqq 0$ gegen $\Phi_2(\mu)$ konvergieren; ebenso konvergieren die $\Psi_{2,n}(\mu)$ gleichmäßig gegen $\Psi_2(\mu)$. Damit aber ist die Stetigkeit von $\Phi_2(\mu)$ und $\Psi_2(\mu)$ nachgewiesen.

Wir zeigen darüber hinaus, daß $\Phi_2(\mu)$ und $\Psi_2(\mu)$ für alle $\mu \geqq 0$, abgesehen von einer Nullmenge, eine Ableitung besitzen. Wir führen den Beweis etwa für $\Phi_2(\mu)$, da er für $\Psi_2(\mu)$ ganz analog verläuft.

Aus (12) entnehmen wir:

$$\Phi'_{2,n}(\mu) = \int_{-n}^{n} f(x) \frac{\sin \mu x}{x} dx.$$

Wir zeigen zunächst, daß die Folge der $\Phi'_{2,n}(\mu)$ in $[0, +\infty)$ im Mittel konvergiert. Es genügt, dies für die Folge

$\int_q^n f(x) \frac{\sin \mu x}{x} dx$ und für die Folge $\int_{-n}^{-q} f(x) \frac{\sin \mu x}{x} dx$ zu zeigen, wo q die in (10) auftretende Größe bedeutet; wir zeigen es etwa für die erste.

Wir setzen:

$$g_n(x) = \frac{f(x)}{x} \text{ für } q \leqq x \leqq n; \; g_n(x) = 0 \text{ für } x < q \text{ und } x > n.$$

Wegen (10) ist dann die Folge der $g_n(x)$ in $(-\infty, +\infty)$ im Mittel konvergent, und wie zu Beginn dieses Paragraphen gezeigt, ist somit auch die Folge

$$\int_{-\infty}^{+\infty} g_n(x) \sin \mu x \, dx = \int_q^n f(x) \frac{\sin \mu x}{x} dx$$

in $[0, +\infty)$ im Mittel konvergent, wie behauptet.

Da sowohl die Folge der $\Phi'_{2,n}(\mu)$, als auch die Folge der $\Psi'_{2,n}(\mu)$ in $[0, +\infty)$ im Mittel konvergiert, so konvergiert jede von ihnen (in jedem endlichen Intervalle aus $[0, +\infty)$) im Mittel gegen eine bestimmte Funktion $\Phi^*(\mu)$, beziehungsweise $\Psi^*(\mu)$, und es ist bekanntlich für jedes $\mu > 0$:

$$\lim_{n \to \infty} \int_0^\mu \Phi'_{2,n}(\mu) d\mu = \int_0^\mu \Phi^*(\mu) d\mu; \quad \lim_{n \to \infty} \int_0^\mu \Psi'_{2,n}(\mu) d\mu = \int_0^\mu \Psi^*(\mu) d\mu,$$

oder da:

$$\lim_{n \to \infty} \int_0^\mu \Phi'_{2,n}(\mu) d\mu = \lim_{n \to \infty} \Phi_{2,n}(\mu) = \Phi_2(\mu),$$

und ebenso:

$$\lim_{n \to \infty} \int_0^\mu \Psi'_{2,n}(\mu) d\mu = \Psi_2(\mu),$$

so haben wir:

$$\Phi_2(\mu) = \int_0^\mu \Phi^*(\mu) d\mu; \quad \Psi_2(\mu) = \int_0^\mu \Psi^*(\mu) d\mu. \tag{13}$$

Also gilt für alle $\mu \geqq 0$, abgesehen von einer Nullmenge:

$$\Phi'_2(\mu) = \Phi^*(\mu); \quad \Psi'_2(\mu) = \Psi^*(\mu), \tag{14}$$

und die Behauptung ist bewiesen.

Da die Folge der $\Phi'_{2,n}(\mu)$ in jedem endlichen Intervalle aus $[0,+\infty)$ im Mittel gegen $\Phi^*(\mu)$ konvergiert und die Folge $\Psi'_{2,n}(\mu)$ gegen $\Psi^*(\mu)$, so gibt es in ihnen Teilfolgen $\Phi'_{2,n_i}(\mu)$, beziehungsweise $\Psi'_{2,n_i}(\mu)$, die in $[0,+\infty)$ überall, abgesehen von einer Nullmenge, gegen $\Phi^*(\mu)$, beziehungsweise $\Psi^*(\mu)$ konvergieren. Es gilt also dann auch für alle $\mu \geqq 0$, abgesehen von einer Nullmenge:

$$\lim_{i\to\infty} \Phi'_{2,n_i}(\mu) = \Phi'_2(\mu); \quad \lim_{i\to\infty} \Psi'_{2,n_i}(\mu) = \Psi'_2(\mu). \tag{15}$$

§ 3.

Sei wieder $f(x)$ eine in jedem endlichen Intervalle integierbare, im Unendlichen beschränkte Funktion. Außerdem sei $f(x)$ an der Stelle x_0 stetig. Dann gilt:

$$f(x_0) = \lim_{\mu\to+\infty} \frac{2}{\mu\pi}\int_{-\infty}^{+\infty} f(x)\frac{\sin^2\frac{\mu}{2}(x-x_0)}{(x-x_0)^2}\,dx. \tag{16}$$

In der Tat, wegen:

$$\frac{2}{\mu\pi}\int_{-\infty}^{+\infty}\frac{\sin^2\frac{\mu}{2}(x-x_0)}{(x-x_0)^2}\,dx = 1 \tag{17}$$

genügt es zu beweisen:

$$\lim_{\mu\to+\infty}\frac{2}{\mu\pi}\int_{-\infty}^{+\infty}(f(x)-f(x_0))\frac{\sin^2\frac{\mu}{2}(x-x_0)}{(x-x_0)^2}\,dx = 0. \tag{18}$$

Wir wählen $h > 0$ so klein, daß:

$$|f(x)-f(x_0)| < \varepsilon \text{ für } |x-x_0| \leqq h.$$

Dann ist wegen (17):

$$\left|\frac{2}{\mu\pi}\int_{x_0-h}^{x_0+h}(f(x)-f(x_0))\frac{\sin^2\frac{\mu}{2}(x-x_0)}{(x-x_0)^2}\,dx\right| \leqq \varepsilon \text{ für alle } \mu. \tag{19}$$

Sodann gibt es, weil $f(x)$ im Unendlichen beschränkt bleibt, ein M, so daß für $\mu \geqq M$:

$$\left|\frac{2}{\mu\pi}\int_{x_0+h}^{+\infty}(f(x)-f(x_0))\frac{\sin^2\frac{\mu}{2}(x-x_0)}{(x-x_0)^2}\,dx\right| < \varepsilon; \tag{20}$$

$$\left| \frac{2}{\mu\pi} \int_{-\infty}^{x_0-h} (f(x)-f(x_0)) \frac{\sin^2 \frac{\mu}{2}(x-x_0)}{(x-x_0)^2} dx \right| < \varepsilon. \tag{20}$$

Aus (19) und (20) folgt:

$$\left| \frac{2}{\mu\pi} \int_{-\infty}^{+\infty} (f(x)-f(x_0)) \frac{\sin^2 \frac{\mu}{2}(x-x_0)}{(x-x_0)^2} dx \right| < 3\varepsilon \text{ für } \mu \geqq M.$$

Es ist also (18) und somit auch (16) bewiesen.

Nun ist:

$$2\int_{-n}^{n} f(x) \frac{\sin^2 \frac{\mu}{2}(x-x_0)}{(x-x_0)^2} dx = \int_{-n}^{n} \left(f(x) \int_0^{\mu} \frac{\sin \lambda (x-x_0)}{x-x_0} d\lambda \right) dx =$$

$$= \int_{-n}^{n} \left\{ f(x) \int_0^{\mu} \left(\int_0^{\lambda} \cos \tau (x-x_0)\, d\tau \right) d\lambda \right\} dx =$$

$$= \int_0^{\mu} \left\{ \int_0^{\lambda} \left(\cos \tau x_0 \int_{-n}^{n} f(x) \cos \tau x\, dx + \sin \tau x_0 \int_{-n}^{n} f(x) \sin \tau x\, dx \right) d\tau \right\} d\lambda.$$

Machen wir wieder von der Bezeichnungsweise (12) Gebrauch, so ist hierin:

$$\int_{-n}^{n} f(x) \cos \tau x\, dx = \Phi''_{2,n}(\tau); \quad \int_{-n}^{n} f(x) \sin \tau x\, dx = \Psi''_{2,n}(\tau).$$

Unter Berufung auf (4) können wir auch schreiben:

$$2\int_{-n}^{n} f(x) \frac{\sin^2 \frac{\mu}{2}(x-x_0)}{(x-x_0)^2} dx = \int_0^{\mu} \left\{ \int_0^{\lambda} \left(\cos \tau x_0 \frac{d^2 \Phi_{2,n}(\tau)}{d\tau} + \sin \tau x_0 \frac{d^2 \Psi_{2,n}(\tau)}{d\tau} \right) \right\} d\lambda. \tag{21}$$

Wir setzen nun für $\tau < 0$:

$$\Phi_{2,n}(\tau) = 0, \quad \Psi_{2,n}(\tau) = 0. \tag{22}$$

Da für $\tau \geqq 0$

$$\Phi'_{2,n}(\tau) = \int_{-n}^{-n} f(x) \frac{\sin \tau x}{x} dx,$$

also:

$$\Phi'_{2,n}(0) = 0,$$

ist zufolge (3) in (21) für jedes $h > 0$:

$$\int_0^\lambda \left(\cos \tau x_0 \frac{d^2 \Phi_{2,n}(\tau)}{d\tau} + \sin \tau x_0 \frac{d^2 \Psi_{2,n}(\tau)}{d\tau} \right) =$$

$$= \int_{-h}^{\lambda} \left(\cos \tau x_0 \frac{d^2 \Phi_{2,n}(\tau)}{d\tau} + \sin \tau x_0 \frac{d^2 \Psi_{2,n}(\tau)}{d\tau} \right),$$

so daß wir (21) ersetzen können durch.

$$2 \int_{-n}^{n} f(x) \frac{\sin^2 \frac{\mu}{2}(x-x_0)}{(x-x_0)^2} dx = \int_0^\mu \left\{ \int_{-h}^{\lambda} \left(\cos \tau x_0 \frac{d^2 \Phi_{2,n}(\tau)}{d\tau} + \right.\right. \tag{23}$$

$$\left.\left. + \sin \tau x_0 \frac{d^2 \Psi_{2,n}(\tau)}{d\tau} \right) \right\} d\lambda.$$

Hierin ist nun der Grenzübergang $n \to \infty$ zu vollziehen. Wir setzen

$$\Phi_2(\tau) = 0, \ \Psi_2(\tau) = 0 \text{ für } \tau < 0. \tag{24}$$

Es konvergiert, wie wir in § 2 sahen, in $[0, \lambda]$ die Folge $\Phi_{2,n}(\tau)$ gleichmäßig gegen $\Phi_2(\tau)$, und $\Psi_{2,n}(\tau)$ gegen $\Psi_2(\tau)$ und wegen (22) und (24) gilt dies auch in $[-h, \lambda]$. Ferner ist für alle n:

$$\Phi'_{2,n}(-h) = 0, \ \Psi'_{2,n}(-h) = 0, \ \Phi'_2(-h) = 0, \ \Psi'_2(-h) = 0;$$

ferner gibt es nach (15) eine Folge n_i, so daß für alle $\lambda \geqq 0$, abgesehen von einer Nullmenge:

$$\lim_{i \to \infty} \Phi'_{2,n_i}(\lambda) = \Phi'_2(\lambda); \ \lim_{i \to \infty} \Psi'_{2,n_i}(\lambda) = \Psi'_2(\lambda).$$

Zufolge (7) gilt also für alle $\lambda \geqq 0$, abgesehen von einer Nullmenge:

$$\lim_{i \to \infty} \int_{-h}^{\lambda} \left(\cos \tau x_0 \frac{d^2 \Phi_{2,n_i}(\tau)}{d\tau} + \sin \tau x_0 \frac{d^2 \Psi_{2,n_i}(\tau)}{d\tau} \right) =$$

$$= \int_{-h}^{\lambda} \left(\cos \tau x_0 \frac{d^2 \Phi_2(\tau)}{d\tau} + \sin \tau x_0 \frac{d^2 \Psi_2(\tau)}{d\tau} \right).$$

Wir zeigen weiter, daß auch:

$$\lim_{i\to\infty}\int_0^\mu\left\{\int_{-h}^\lambda \cos\tau x_0 \frac{d^2\Phi_{2,n_i}(\tau)}{d\tau}+\sin\tau x_0 \frac{d^2\Psi_{2,n_i}(\tau)}{d\tau}\right\}d\lambda=$$
$$=\int_0^\mu\left\{\int_{-h}^\lambda \cos\tau x_0 \frac{d^2\Phi_2(\tau)}{d\tau}+\sin\tau x_0 \frac{d^2\Psi_2(\tau)}{d\tau}\right\}d\lambda. \tag{25}$$

In der Tat, unter Benützung von (1) finden wir (abgesehen von einer Nullmenge):

$$\int_{-h}^\lambda\left(\cos\tau x_0 \frac{d^2\Phi_{2,n_i}(\tau)}{d\tau}+\sin\tau x_0 \frac{d^2\Psi_{2,n_i}(\tau)}{d\tau}\right)-$$
$$-\int_{-h}^\lambda\left(\cos\tau x_0 \frac{d^2\Phi_2(\tau)}{d\tau}+\sin\tau x_0 \frac{d^2\Psi_2(\tau)}{d\tau}\right)= \tag{26}$$
$$=\cos\lambda x_0\,(\Phi'_{2,n_i}(\lambda)-\Phi'_2(\lambda))+\sin\lambda x_0\,(\Psi'_{2,n_i}(\lambda)-\Psi'_2(\lambda))+$$
$$+x_0\int_{-h}^\lambda \sin\tau x_0\, d\,(\Phi_{2,n_i}(\tau)-\Phi_2(\tau))-$$
$$-x_0\int_{-h}^\lambda \cos\tau x_0\, d\,(\Psi_{2,n_i}(\tau)-\Psi_2(\tau)).$$

Da wegen (13) und (14):

$$\Phi_2(\mu)=\int_0^\mu \Phi'_2(\mu)\,d\mu$$

ist, so ist:

$$\int_0^\mu \cos\lambda x_0\,(\Phi'_{2,n_i}(\lambda)-\Phi'_2(\lambda))\,d\lambda=\cos\lambda x_0\,(\Phi_{2,n_i}(\lambda)-\Phi_2(\lambda))\Big]_0^\mu+$$
$$+x_0\int_0^\mu \sin\lambda x_0\,(\Phi_{2,n_i}(\lambda)-\Phi_2(\lambda))\,d\lambda,$$

und mithin, da die $\Phi_{2,n}(\lambda)$ gleichmäßig gegen $\Phi_2(\lambda)$ konvergieren:

$$\lim_{i\to\infty}\int_0^\mu \cos\lambda x_0\,(\Phi'_{2,n_i}(\lambda)-\Phi'_2(\lambda))\,d\lambda=0. \tag{27}$$

Analog findet man:

$$\lim_{i \to \infty} \int_0^{\mu} \sin \lambda x_0 \left(\Psi'_{2, n_i}(\lambda) - \Psi'_2(\lambda) \right) d\lambda = 0. \tag{28}$$

Bezeichnet M_n die obere Schranke von $|\Phi_{2,n}(\tau) - \Phi_2(\tau)|$ in $[-h, \lambda]$ und $V(\lambda)$ die Variation von $\sin \tau x_0$ in $[-h, \lambda]$, so gilt die Abschätzungsformel:[4]

$$\left| \int_{-h}^{\lambda} \sin \tau x_0 \, d\left(\Phi_{2,n}(\tau) - \Phi_2(\tau)\right) \right| \leqq (V(\lambda) + 1)\, M_n.$$

Wegen $\lim\limits_{n \to \infty} M_n = 0$ gilt also:

$$\lim_{n \to \infty} \int_0^{\mu} \left\{ \int_{-h}^{\lambda} \sin \tau x_0 \, d\left(\Phi_{2,n}(\tau) - \Phi_2(\tau)\right) \right\} d\lambda = 0, \tag{29}$$

und ebenso findet man:

$$\lim_{n \to \infty} \int_0^{\mu} \left\{ \int_{-h}^{\lambda} \cos \tau x_0 \, d\left(\Psi_{2,n}(\tau) - \Psi_2(\tau)\right) \right\} d\lambda = 0. \tag{30}$$

Aus (26) zusammen mit (27), (28), (29), (30) folgt nun:

$$\lim_{i \to \infty} \int_0^{\mu} \left\{ \int_{-h}^{\lambda} \left(\cos \tau x_0 \frac{d^2 \Phi_{2, n_i}(\tau)}{d\tau} + \sin \tau x_0 \frac{d^2 \Psi_{2, n_i}(\tau)}{d\tau} \right) - \right.$$

$$\left. - \int_{-h}^{\lambda} \left(\cos \tau x_0 \frac{d^2 \Phi_2(\tau)}{d\tau} + \sin \tau x_0 \frac{d^2 \Psi_2(\tau)}{d\tau} \right) \right\} d\lambda = 0;$$

damit ist (25) bewiesen.

Ersetzen wir in (23) n durch n_i und vollziehen den Grenzübergang $i \to \infty$, so erhalten wir:

$$2 \int_{-\infty}^{+\infty} f(x) \frac{\sin^2 \frac{\mu}{2} (x - x_0)}{(x - x_0)^2} \, dx = \int_0^{\mu} \left\{ \int_{-h}^{\lambda} \cos \tau x_0 \frac{d^2 \Phi_2(\tau)}{d\tau} + \right. \tag{31}$$

$$\left. + \sin \tau x_0 \frac{d^2 \Psi_2(\tau)}{d\tau} \right\} d\lambda.$$

---

[4] L. c. [2]), p. 74.

Hierin kann h noch eine beliebige Zahl > 0 bedeuten, insbesondere beliebig klein gewählt werden.

Wir schreiben deshalb statt (31):

$$2\int_{-\infty}^{+\infty} f(x)\,\frac{\sin^2\frac{\mu}{2}(x-x_0)}{(x-x_0)^2}\,dx = \int_0^{\mu}\left\{\int_{-0}^{\lambda}\cos\tau x_0\,\frac{d^2\Phi_2(\tau)}{d\tau}+\right.$$

$$\left.+\sin\tau x_0\,\frac{d^2\Psi_2(\tau)}{d\tau}\right\}d\lambda.$$

Formel (16) ergibt nun:

$$f(x_0) = \frac{1}{\pi}\lim_{\mu\to+\infty}\frac{1}{\mu}\int_0^{\mu}\left\{\int_{-0}^{\lambda}\cos\tau x_0\,\frac{a^2\Phi_2(\tau)}{d\tau}+\right.$$

$$\left.+\sin\tau x_0\,\frac{d^2\Psi_2(\tau)}{d\tau}\right\}d\lambda. \qquad (32)$$

Diese Formel gilt für jede Funktion $f(x)$, die in allen endlichen Intervallen integrierbar, im Unendlichen beschränkt und an der Stelle x_0 stetig ist.

§ 4.

Wir setzen nun für $\mu \geqq 0$ (vorausgesetzt, daß diese Ausdrücke einen Sinn haben):

$$\Phi_1(\mu) = \int_{-\infty}^{+\infty} f(x)\,\frac{\sin\mu x}{x}\,dx$$

$$\Psi_1(\mu) = \int_{-1}^{1} f(x)\,\frac{1-\cos\mu x}{x}\,dx - \int_1^{+\infty} f(x)\,\frac{\cos\mu x}{x}\,dx -$$

$$-\int_{-\infty}^{-1} f(x)\,\frac{\cos\mu x}{x}\,dx.$$

An anderer Stelle habe ich gezeigt, daß für eine in jedem endlichen Intervalle integrierbare Funktion $f(x)$, die im Unendlichen geeignete Bedingungen erfüllt und an der Stelle x_0 in die Fouriersche Reihe entwickelbar ist, die Formel gilt:

$$f(x_0) = \frac{1}{\pi}\int_0^{+\infty}(\cos\lambda x_0\,d\Phi_1(\lambda)+\sin\lambda x_0\,d\Psi_1(\lambda)).$$

Wir sind nun in der Lage, diesem Resultate ein anderes zur Seite zu stellen, das lediglich voraussetzt, daß $f(x)$ an der Stelle x_0 stetig sei.

Wir zeigen zunächst: Wenn $\Phi_1(\mu)$ und $\Psi_1(\mu)$ für $\mu \geqq 0$ existieren und stetig sind, so gilt für $\mu \geqq 0$:[5]

$$\Phi_1(\mu) = \Phi_2'(\mu), \quad \Psi_1(\mu) = \Psi_2'(\mu). \tag{33}$$

In der Tat, es ist, wenn wir von der Bezeichnungsweise (12) Gebrauch machen:

$$\Phi_1(\mu) = \lim_{n \to \infty} \int_{-n}^{n} f(x) \frac{\sin \mu x}{x} dx = \lim_{n \to \infty} \Phi_{2,n}'(\mu).$$

Also gilt, wegen (15), abgesehen von einer Nullmenge:

$$\Phi_1(\mu) = \Phi_2'(\mu). \tag{34}$$

Aus (13) und (14) folgt daher:

$$\Phi_2(\mu) = \int_0^{\mu} \Phi_1(\mu) d\mu,$$

und da $\Phi_1(\mu)$ stetig ist, gilt (34) überall. Damit ist die erste Gleichung (33) bewiesen, und analog beweist man die zweite.

Beachtet man die Festsetzung (24), so ergibt Formel (2), da (wie eben bewiesen, vgl. Anmerkung [5]) $\Phi_{2+}'(0)$ und $\Psi_{2+}'(0)$ existieren, und zufolge (33) $\Phi_{2+}'(0) = 0$ ist:

$$\int_{-0}^{\lambda} \left(\cos \tau x_0 \frac{d^2 \Phi_2(\tau)}{d\tau} + \sin \tau x_0 \frac{d^2 \Psi_2(\tau)}{d\tau} \right) =$$

$$= \int_{0}^{\lambda} \left(\cos \tau x_0 \frac{d^2 \Phi_2(\tau)}{d\tau} + \sin \tau x_0 \frac{d^2 \Psi_2(\tau)}{d\tau} \right);$$

und Formel (5) ergibt weiter, da $\Phi_1(\mu)$, $\Psi_1(\mu)$ für $\mu \geqq 0$ als stetig vorausgesetzt sind:

$$\int_0^{\lambda} \left(\cos \tau x_0 \frac{d^2 \Phi_2(\tau)}{d\tau} + \sin \tau x_0 \frac{d^2 \Psi_2(\tau)}{d\tau} \right) =$$

$$= \int_0^{\lambda} (\cos \tau x_0 \, d\Phi_1(\tau) + \sin \tau x_0 \, d\Psi_1(\tau)).$$

Und wir haben somit das Resultat:

---

[5] Für $\mu = 0$ ist unter $\Phi_2'(0)$ und $\Psi_2'(0)$ die rechtsseitige Ableitung $\Phi_{2+}'(0)$, beziehungsweise $\Psi_{2+}'(0)$ zu verstehen.

Ist die in jedem endlichen Intervalle integrierbare, im Unendlichen beschränkte Funktion $f(x)$ stetig an der Stelle x_0, existieren für $\mu \geqq 0$ die Ausdrücke $\Phi_1(\mu)$, $\Psi_1(\mu)$ und sind sie stetige Funktionen von μ, so gilt die Formel:

$$f(x_0) = \frac{1}{\pi} \lim_{\mu \to +\infty} \frac{1}{\mu} \int_0^{\mu} \left\{ \int_0^{\lambda} (\cos \tau x_0 \, d\Phi_1(\tau) + \sin \tau x_0 \, d\Psi_1(\tau)) \right\} d\lambda. \quad (35)$$

Wir setzen nun für $\mu \geqq 0$ (falls diese Ausdrücke existieren):

$$\varphi(\mu) = \int_{-\infty}^{+\infty} f(x) \cos \mu x \, dx; \quad \psi(\mu) = \int_{-\infty}^{+\infty} f(x) \sin \mu x \, dx.$$

Nehmen wir an, diese Ausdrücke existieren für alle $\mu \geqq 0$, abgesehen von einer Nullmenge, und es sei für jedes Intervall $0 \leqq \alpha \leqq \mu \leqq \beta$:

$$\Phi_1(\beta) - \Phi_1(\alpha) = \int_{\alpha}^{\beta} \varphi(\mu) \, d\mu; \quad \Psi_1(\beta) - \Psi_2(\beta) = \int_{\alpha}^{\beta} \psi(\mu) \, d\mu,$$

so kann (35) auch in der Form geschrieben werden:

$$f(x_0) = \frac{1}{\pi} \lim_{\mu \to \infty} \frac{1}{\mu} \int_0^{\mu} \left\{ \int_0^{\lambda} (\cos \tau x_0 \, \varphi(\tau) + \sin \tau x_0 \, \psi(\tau)) \, d\tau \right\} d\lambda.$$

§ 5.

Sei nun $f(x)$ eine integrierbare und beschränkte periodische Funktion; der Einfachheit halber nehmen wir an, sie habe die Periode 2π. Ihre Fourier'sche Reihe sei gegeben durch:

$$f(x) \sim \frac{a_0}{2} + \sum_{\nu=1}^{\infty} (a_\nu \cos \nu x + b_\nu \sin \nu x). \quad (36)$$

Wir wollen zeigen, daß die zugehörigen Funktionen $\Phi_2(\mu)$ und $\Psi_2(\mu)$ aus dieser Fourier'schen Reihe durch gliedweise Integrationen berechnet werden können. Es wird genügen, dies für $\Phi_2(\mu)$ nachzuweisen. Offenbar können wir dabei ohneweiteres $a_0 = 0$ annehmen, so daß:

$$f(x) \sim \sum_{\nu=1}^{\infty} (a_\nu \cos \nu x + b_\nu \sin \nu x).$$

Wir haben nachzuweisen:

$$\lim_{n \to \infty} \int_{-\infty}^{+\infty} \left(f(x) - \sum_{\nu=1}^{n} (a_\nu \cos \nu x + b_\nu \sin \nu x) \right) \frac{\sin^2 \frac{\mu}{2} x}{x^2} \, dx = 0.$$

Dazu genügt es, zu zeigen: zu jedem $\varepsilon > 0$ gibt es ein N, so daß für alle $n \geqq N$ und alle endlichen Intervalle $[p, q]$:

$$\left| \int_p^q \left(f(x) - \sum_{\nu=1}^{n} (a_\nu \cos \nu x + b_\nu \sin \nu x) \right) \frac{\sin^2 \frac{\mu}{2} x}{x^2} \, dx \right| < \varepsilon.$$

Nun gilt, wenn $g(x)$ integrierbar, $h(x)$ von endlicher Variation in $[p, q]$ ist, die Abschätzung:

$$\left| \int_p^q g(x) \, h(x) \, dx \right| \leqq G(H' + H''),$$

wenn H' die Variation von $h(x)$, H'' die obere Schranke von $|h(x)|$ und G die obere Schranke von $\left| \int_p^x g(x) dx \right|$ in $[p, q]$ bedeutet. Bezeichnen wir also mit V die (sicherlich endliche) Variation von $\frac{1}{x^2} \sin^2 \frac{\mu}{2} x$ in $(-\infty, +\infty)$, mit M_n die obere Schranke von

$$\left| \int_0^x \left(f(x) - \sum_{\nu=1}^{n} (a_\nu \cos \nu x + b_\nu \sin \nu x) \, dx \right| \right.$$

in $(-\infty, +\infty)$, so erhalten wir:

$$\left| \int_p^q \left(f(x) - \sum_{\nu=1}^{n} (a_\nu \cos \nu x + b_\nu \sin \nu x) \right) dx \right| \leqq 2 \, M_n \left(V + \frac{\mu^2}{4} \right),$$

und da hierin bekanntlich $\lim\limits_{n \to \infty} M_n = 0$ ist, so ist die Behauptung bewiesen.

Nun ist, wie unmittelbar zu sehen:

$$2 \int_{-\infty}^{+\infty} \cos \nu x \frac{\sin^2 \frac{\mu}{2} x}{x^2} \, dx = \begin{cases} 0 & \text{für } 0 \leqq \mu \leqq \nu \\ \pi(\mu - \nu) & \text{für } \quad \mu \geqq \nu \end{cases}$$

$$-\int_{-\infty}^{+\infty} \sin \nu x \frac{\sin \mu x}{x^2} \, dx = \begin{cases} -\pi \mu & \text{für } 0 \leqq \mu \leqq \nu \\ -\pi \nu & \text{für } \quad \mu \geqq \nu \end{cases}$$

Berechnet man zu der durch (36) gegebenen Funktion $f(x)$ die zugehörigen Funktionen $\Phi_2(\mu)$ und $\Psi_2(\mu)$ durch gliedweise Integration, so ergibt sich leicht:

$$\left.\begin{aligned}\Phi_2(\mu) &= \Phi_2(n) + \pi\left(\frac{a_0}{2} + \sum_{\nu=1}^{n} a_\nu\right)(\mu - n)\\ \Psi_2(\mu) &= \Psi_2(n) + \pi\left(C + \sum_{\nu=1}^{n} b_\nu\right)(\mu - n)\end{aligned}\right\} \begin{array}{c}\text{für } n \leqq \mu \leqq n+1\\ (n = 0, 1, 2, \ldots),\end{array}$$

wo mit C eine Konstante bezeichnet ist, auf deren Wert es weiter nicht ankommt.

Auf Grund von (2) wird nun in Formel (32) für $n-1 < \lambda \leqq n$

$$\int_{-0}^{\lambda}\left(\cos\tau x_0 \frac{d^2\Phi_2(\tau)}{d\tau} + \sin\tau x_0 \frac{d^2\Psi_2(\tau)}{d\tau}\right) = \int_{-h}^{0}\left(\cos\tau x_0 \frac{d^2\Phi_2(\tau)}{d\tau} +\right.$$

$$\left. + \sin\tau x_0 \frac{d^2\Psi_2(\tau)}{d\tau}\right) + \sum_{\nu=1}^{n-1}\int_{\nu-1}^{\nu}\left(\cos\tau x_0 \frac{d^2\Phi_2(\tau)}{d\tau} +\right.$$

$$\left. + \sin\tau x_0 \frac{d^2\Psi_2(\tau)}{d\tau}\right) + \int_{n-1}^{\lambda}\left(\cos\tau x_0 \frac{d^2\Phi_2(\tau)}{d\tau} + \sin\tau x_0 \frac{d^2\Psi_2(\tau)}{d\tau}\right) +$$

$$+ \pi\left\{\frac{a_0}{2} + \sum_{\nu=1}^{n-1}(a_\nu \cos\nu x_0 + b_\nu \sin\nu x_0)\right\},$$

und da in den einzelnen Intervallen $[-h, 0]$, $[0, 1]$, …, $[n-2, n-1]$, $[n-1, \lambda]$ die Funktionen $\Phi_2(\tau)$ und $\Psi_2(\tau)$ linear sind, fallen die betreffenden Integrale fort. Wir haben also für $n-1 < \lambda \leqq n$:

$$\int_{-0}^{\lambda}\left(\cos\tau x_0 \frac{d^2\Phi_2(\tau)}{d\tau} + \sin\tau x_0 \frac{d^2\Psi_2(\tau)}{d\tau}\right) = \qquad (37)$$

$$= \pi\left\{\frac{a_0}{2} + \sum_{\nu=1}^{n-1}(a_\nu \cos\nu x_0 + b_\nu \sin\nu x_0)\right\}.$$

Lassen wir in Formel (32) μ speziell die Folge $1, 2, \ldots, n, \ldots$ durchlaufen, so liefert sie:

$$f(x_0) = \lim_{n\to\infty} \frac{1}{n}\left\{\frac{n a_0}{2} + \sum_{\nu=1}^{n-1}(n-\nu)(a_\nu \cos\nu x_0 + b_\nu \sin\nu x_0)\right\}. \quad (37)$$

Wir sehen also: Ist $f(x)$ eine periodische Funktion, so reduziert sich Formel (32) auf Fejér's Summationsformel für die Fourier'sche Reihe.

Dies ist nur ein spezieller Fall des folgenden allgemeinen Resultates (das übrigens leicht noch weiter verallgemeinert werden könnte): Sei $q_1 < q_2 < \ldots < q_n < \ldots$ eine ins Unendliche wachsende Folge positiver Zahlen und sei $f(x)$ eine in jedem endlichen Intervalle integrierbare, im Unendlichen beschränkte Funktion, deren zugehörige Funktionen $\Phi_2(\mu)$ und $\Psi_2(\mu)$ in den Intervallen $[0, q_1]$, $[q_1, q_2]$, $\ldots$, $[q_{n-1}, q_n]$, $\ldots$ linear sind. Setzt man:

$$\begin{gathered}\Phi'_{2+}(0) = \pi a_0, \quad \Phi'_{2+}(q_n) - \Phi'_{2-}(q_n) = \pi a_n, \\ \Psi'_{2+}(q_n) - \Psi'_{2-}(q_n) = \pi b_n,\end{gathered} \tag{38}$$

so gilt an jeder Stelle x_0, an der $f(x)$ stetig ist, die Darstellung:

$$f(x_0) = \lim_{n\to\infty} \frac{1}{q_n} \left\{ q_n a_0 + \sum_{\nu=1}^{n-1} (q_n - q_\nu)(a_\nu \cos q_\nu x_0 + b_\nu \sin q_\nu x_0) \right\}. \tag{39}$$

Der Beweis ergibt sich ganz ebenso wie der von (37) aus Formel (32).

Insbesondere gilt Formel (39), wenn $f(x)$ eine fastperiodische Funktion[6] ist, deren Fourierexponenten q_ν sich nirgends im endlichen häufen. Wir denken die q_ν wachsend geordnet:

$$0 < q_1 < q_2 < \ldots < q_\nu < \ldots$$

Wie H. Bohr gezeigt hat, kann eine Folge endlicher trigonometrischer Summen:

$$f_n(x) = a_0^{(n)} + \sum_{\nu=1}^{n} (a_\nu^{(n)} \cos q_\nu x + b_\nu^{(n)} \sin q_\nu x)$$

gefunden werden, die in $(-\infty, +\infty)$ gleichmäßig gegen $f(x)$ konvergieren. Ist:

$$f(x) \sim a_0 + \sum_{\nu=1}^{\infty} (a_\nu \cos q_\nu x + b_\nu \sin q_\nu x)$$

die verallgemeinerte Fourierreihe von $f(x)$, so gelten die Grenzbeziehungen:

$$\lim_{n\to\infty} a_\nu^{(n)} = a_\nu, \quad \lim_{n\to\infty} b_\nu^{(n)} = b_\nu.$$

Bezeichnen $\Phi_2^{(n)}(\mu)$, $\Psi_2^{(n)}(\mu)$ die zu $f_n(x)$ gehörigen Funktionen Φ_2, Ψ_2, so konvergieren die $\Phi_2^{(n)}(\mu)$ und $\Psi_2^{(n)}(\mu)$ in jedem endlichen

[6] H. Bohr, Acta Math., *45*, p. 29 ff.; *46*, p. 101 ff.

Intervalle gleichmäßig gegen $\Phi_2(\mu)$ und $\Psi_2(\mu)$. In jedem Intervalle $[0, q_1]$, $[q_1, q_2], \ldots$, $[q_\nu, q_{\nu+1}], \ldots$ sind aber alle $\Phi_2^{(n)}(\mu)$ und $\Psi_2^{(n)}(\mu)$ linear, dasselbe gilt daher für $\Phi_2(\mu)$ und $\Psi_2(\mu)$. Und zwar ist für $n > \nu$ in $[q_\nu, q_{\nu+1}]$:

$$\Phi_2^{(n)}(\mu) = \Phi_2^{(n)}(q_\nu) + \pi \sum_{i=0}^{\nu} a_i^{(n)} . (\mu - q_\nu);$$

$$\Psi_2^{(n)}(\mu) = \Psi_2^{(n)}(q_\nu) + \pi \left(C_n + \sum_{i=1}^{\nu} b_i^{(n)} \right) (\mu - q_\nu),$$

also

$$\Phi_2(\mu) = \Phi_2(q^\nu) + \pi \sum_{i=0}^{\nu} a_i(\mu - q_\nu);$$

$$\Psi_2(\mu) = \Psi_2(q_\nu) + \pi \left(C + \sum_{i=1}^{\nu} b_i \right) (\mu - q_\nu).$$

Es gelten somit die Formeln (38), womit (39) bewiesen ist.

§ 6.

Wir wollen zum Schlusse noch folgenden Satz beweisen:

Ist $f(x)$ in jedem endlichen Intervalle integrierbar und im Unendlichen beschränkt, und existiert für jede Folge von Intervallen $[p_\nu, q_\nu]$, deren Länge ins Unendliche wächst, der Grenzwert:

$$A(\mu) = \lim_{\nu \to \infty} \frac{1}{q_\nu - p_\nu} \int_{p_\nu}^{q_\nu} f(x) \cos \mu x \, dx, \tag{40}$$

so ist:

$$A(\mu) = \frac{1}{2\pi} \lim_{h \to 0} \frac{\Phi_2(\mu + h) - 2\,\Phi_2(\mu) + \Phi_2(\mu - h)}{h}. \tag{41}$$

In der Tat, es ist:

$$\Phi_2(\mu + h) - \Phi_2(\mu) = \int_{-\infty}^{+\infty} f(x) \frac{\cos \mu x - \cos (\mu + h) x}{x^2} dx =$$

$$= 2 \int_{-\infty}^{+\infty} f(x) \frac{\sin (\mu + \frac{h}{2}) x \sin \frac{h}{2} x}{x^2} dx =$$

$$= 2\int_{-\infty}^{+\infty} f(x)\cos\mu x \frac{\sin^2 \frac{h}{2}x}{x^2}\,dx + \int_{-\infty}^{+\infty} f(x)\sin\mu x \frac{\sin hx}{x^2}\,dx.$$

Mithin ist:

$$\frac{\Phi_2(\mu+h)-2\,\Phi_2(\mu)+\Phi_2(\mu-h)}{h} = \frac{4}{h}\int_{-\infty}^{+\infty} f(x)\cos\mu x \frac{\sin^2\frac{h}{2}x}{x^2}\,dx =$$

$$= 4\int_{-\infty}^{+\infty} f\left(\frac{x}{h}\right)\cos\frac{\mu x}{h}\,\frac{\sin^2\frac{x}{2}}{x^2}\,dx.$$

Wir haben also zu zeigen:

$$\lim_{h\to 0} \frac{2}{\pi}\int_{-\infty}^{+\infty} f\left(\frac{x}{h}\right)\cos\frac{\mu x}{h}\,\frac{\sin^2\frac{x}{2}}{x^2}\,dx = A(\mu),$$

oder was dasselbe heißt:

$$\lim_{h\to\infty}\int_{-\infty}^{+\infty}\left(f\left(\frac{x}{h}\right)\cos\frac{\mu x}{h} - A(\mu)\right)\frac{\sin^2\frac{x}{2}}{x^2}\,dx = 0.$$

Ist $\varepsilon > 0$ beliebig gegeben, so gibt es, da $f(x)$ im Unendlichen beschränkt ist, ein a, so daß für alle $|h| \leqq 1$:

$$\left|\int_{a}^{+\infty}\left(f\left(\frac{x}{h}\right)\cos\frac{\mu x}{h} - A(\mu)\right)\frac{\sin^2\frac{x}{2}}{x^2}\,dx\right| < \varepsilon,$$

$$\left|\int_{-\infty}^{-a}\left(f\left(\frac{x}{h}\right)\cos\frac{\mu x}{h} - A(\mu)\right)\frac{\sin^2\frac{x}{2}}{x^2}\,dx\right| < \varepsilon.$$

Es genügt also, zu zeigen, daß

$$\lim_{h\to 0}\int_{-a}^{a}\left(f\left(\frac{x}{h}\right)\cos\frac{\mu x}{h} - A(\mu)\right)\frac{\sin^2\frac{x}{2}}{x^2}\,dx = 0. \tag{42}$$

Nun ist für jedes Intervall $[\alpha, \beta]$ wegen (40):

$$\lim_{\to 0} \int_{\alpha}^{\beta} f\left(\frac{x}{h}\right) \cos \frac{\mu x}{h}\, dx = \lim_{h \to 0} h \int_{\frac{\alpha}{h}}^{\frac{\beta}{h}} f(x) \cos \mu x\, dx = (\beta - \alpha)\, A(\mu),$$

also:

$$\lim_{h \to 0} \int_{\alpha}^{\beta} \left(f\left(\frac{x}{h}\right) \cos \frac{\mu x}{h} - A(\mu) \right) dx = 0. \tag{43}$$

Ferner können wir leicht beweisen: Es gibt ein M und ein η, so daß für alle x von $[-a, a]$ und alle $|h| \leqq \eta$:

$$\left| \int^{x} \left(f\left(\frac{x}{h}\right) \cos \frac{\mu x}{h} - A(\mu) \right) dx \right| \leqq M. \tag{44}$$

In der Tat, andernfalls gäbe es eine Folge h_ν mit $\lim\limits_{\nu \to \infty} h_\nu = 0$ und eine Folge $\{x_\nu\}$ in $[-a, a]$, so daß:

$$\left| \int_{0}^{x_\nu} \left(f\left(\frac{x}{h_\nu}\right) \cos \frac{\mu x}{h_\nu} - A(\mu) \right) dx \right| > \nu,$$

oder was dasselbe heißt:

$$\left| h_\nu \int_{0}^{\frac{x_\nu}{h_\nu}} f(x) \cos \mu x\, dx - A(\mu) . x_\nu \right| > \nu \tag{45}$$

und somit:

$$\left| \frac{h_\nu}{x_\nu} \int_{0}^{\frac{x_\nu}{h_\nu}} f(x) \cos \mu x\, dx - A(\mu) \right| > \frac{\nu}{|x_\nu|};$$

da aber aus (45) folgt, daß $\left| \frac{x_\nu}{h_\nu} \right|$ über alle Grenzen wächst, steht dies in Widerspruch zu (40).

Aus (43) und (44) aber folgt bekanntlich (42), womit die Behauptung (41) bewiesen ist.

In ganz analoger Weise zeigt man:

Ist $f(x)$ in jedem endlichen Intervalle integrierbar und im Unendlichen beschränkt, und existiert für jede Folge von Intervallen $[p_\nu, q_\nu]$, deren Länge ins Unendliche wächst, der Grenzwert:

$$B(\mu) = \lim_{\nu \to \infty} \frac{1}{q_\nu - p_\nu} \int_{p_\nu}^{q_\nu} f(x) \sin \mu x\, dx,$$

so ist:

$$B(\mu) = \frac{1}{2\pi} \lim_{h \to 0} \frac{\Psi_2(\mu + h) - 2\,\Psi_2(\mu) + \Psi_2(\mu - h)}{h}.$$

ÜBER EINE VERALLGEMEINERUNG DER FOURIERSCHEN INTEGRALFORMEL.

VON

HANS HAHN

in WIEN.

Im Folgenden soll das Fouriersche Integraltheorem in einer Gestalt bewiesen werden, die infolge Heranziehung von Stieltjesintegralen den Geltungsbereich dieses Theorems beträchtlich erweitert. Es fallen z. B. in ihren Geltungsbereich alle Funktionen, die an der betrachteten Stelle x_0 in die Fourierreihe entwickelbar sind, und im Unendlichen eine der folgenden Bedingungen erfüllen: 1) es ist $\left|\frac{f(x)}{x}\right|$ im Unendlichen integrierbar; 2) $f(x)$ ist das Produkt einer periodischen Funktion und einer Funktion, die im Unendlichen beschränkt und monoton ist; 3) $[f(x)]^2$ ist im Unendlichen integrierbar.[1] Zum Falle 2) gehören insbesondere alle an der Stelle x_0 in die Fourierreihe entwickelbaren periodischen Funktionen, und für diese reduziert sich unsere Formel einfach auf die Fourierreihe, die so als Stieltjessches Fourierintegral erscheint. In den Fällen, in denen $f(x)$ durch die klassische Fouriersche Integralformel darstellbar ist, reduziert sich unsere Formel auf diese, und gestattet so zugleich, nicht nur die wichtigsten Bedingungen für die Giltigkeit der klassischen Fourierschen Integralformel[2] einfach und durchsichtig zu begründen, sondern auch neue, sehr allgemeine Bedingungen für die Giltigkeit dieser Formel aufzustellen. An andrer Stelle hoffe

---

[1] Dieser Teil unsrer Untersuchungen berührt sich enge mit Untersuchungen von M. PLANCHEREL, Rend. Pal. 30 (1910), S. 289, Math. Ann. 74 (1913), S. 573; Math. Ann. 76 (1915), S. 315.

[2] Etwa die von A. PRINGSHEIM angegebenen: Math. Ann. 68 (1910), S. 367; Math. Ann. 71 (1911), S. 289.

ich, bald zeigen zu können, wie sich die hier vorgebrachten Untersuchungen auf alle Funktionen ausdehnen lassen, die im Unendlichen beschränkt sind.

Die vorliegende Arbeit ist (mit Ausnahme des später hinzugefügten § 19) die ausführliche Wiedergabe eines auf der Innsbrucker Versammlung (September 1924) gehaltenen Vortrages, von dem eine kurze Inhaltsangabe bereits in den Jahresberichten der Deutschen Mathematikervereinigung erschienen ist. Zwei kurzen Noten[1] entnehme ich, dass sich Herr N. WIENER mit demselben Probleme befasst hat. Ob seine Methoden mit den meinen in irgend einer Beziehung stehen, entzieht sich meiner Kenntnis.

§ 1.

Sei $f(x)$ eine in jedem endlichen Intervalle integrierbare[2] Funktion, die obendrein folgende Eigenschaften besitze:

A) An der Stelle x_0 sei $f(x)$ in die Fouriersche Reihe entwickelbar (d. h. für jede mit $f(x)$ in einer Umgebung von x_0 übereinstimmende Funktion der Periode 2π konvergiere die Fouriersche Reihe an der Stelle x_0 und stelle den Funktionswert $f(x_0)$ dar).

B) Es sei $\frac{f(x)}{x}$ im Unendlichen absolut integrierbar (d. h. für $a > 0$ fallen die beiden Grenzwerte $\lim_{q \to +\infty} \int_a^q \left|\frac{f(x)}{x}\right| dx$ und $\lim_{q \to +\infty} \int_{-q}^{-a} \left|\frac{f(x)}{x}\right| dx$ endlich aus). Unter diesen Voraussetzungen gilt die Formel:

$$f(x_0) = \frac{1}{\pi} \lim_{\lambda \to +\infty} \int_{-\infty}^{+\infty} f(x) \frac{\sin \lambda (x - x_0)}{x - x_0} dx. \tag{1}$$

In der Tat, zunächst ist wegen Eigenschaft B), da ausserhalb des Inter-

---

[1] Am. Bull. 31 (1925), S. 106, 221. (Zusatz bei der Korrektur: die sehr bedeutungsvollen Untersuchungen des Herrn WIENER sind mittlerweile ausführlich erschienen: Math. Zeitschr. 24, S. 575).

[2] Integrierbarkeit ist im Folgenden stets im Sinne von LEBESGUE zu verstehen; die auftretenden Integrale sind, wo nicht ausdrücklich das Gegenteil gesagt ist, Lebesguesche Integrale. Dabei bedeutet in üblicher Weise $\int_{-\infty}^{+\infty}$ so viel wie $\lim_{p \to -\infty,\, q \to +\infty} \int_p^q$. Von allen auftretenden Funktionen wird vorausgesetzt, dass sie in jedem endlicher Intervalle integrierbar sind.

valles $(x_0-1,\ x_0+1)$ der Ausdruck $\frac{x}{x-x_0}\sin\lambda(x-x_0)$ beschränkt ist, das in (1) auftretende Integral sicher vorhanden; ist ferner $\varepsilon>0$ beliebig gegeben, so kann $b>0$ so gewählt werden, dass für alle λ:

$$(2)\qquad \left|\frac{1}{\pi}\int_{x_0+b}^{+\infty} f(x)\frac{\sin\lambda(x-x_0)}{x-x_0}\,dx\right|<\varepsilon;\quad \left|\frac{1}{\pi}\int_{-\infty}^{x_0-b} f(x)\frac{\sin\lambda(x-x_0)}{x-x_0}\,dx\right|<\varepsilon.$$

Wegen Eigenschaft A) aber ist:

$$\frac{1}{\pi}\lim_{\lambda\to+\infty}\int_{x_0-b}^{x_0+b} f(x)\frac{\sin\lambda(x-x_0)}{x-x_0}\,dx=f(x_0),$$

d. h. es gibt ein λ_0, so dass:

$$(3)\qquad \left|\frac{1}{\pi}\int_{x_0-b}^{x_0+b} f(x)\frac{\sin\lambda(x-x_0)}{x-x_0}\,dx-f(x_0)\right|<\varepsilon \text{ für } \lambda>\lambda_0.$$

Aus (2) und (3) folgt:

$$\left|\frac{1}{\pi}\int_{-\infty}^{+\infty} f(x)\frac{\sin\lambda(x-x_0)}{x-x_0}\,dx-f(x_0)\right|<3\,\varepsilon \text{ für } \lambda>\lambda_0,$$

womit (1) bewiesen ist.

Bezeichnet n eine natürliche Zahl, so ist:

$$(4)\qquad \int_{-\infty}^{+\infty} f(x)\frac{\sin\lambda(x-x_0)}{x-x_0}\,dx=\int_{-\infty}^{+\infty}\left(f(x)\int_0^{\lambda}\cos\mu(x-x_0)\,d\mu\right)dx$$

$$=\lim_{n\to\infty}\int_{-n}^{n}\left(f(x)\int_0^{\lambda}\cos\mu(x-x_0)\,d\mu\right)dx.$$

Hierin nun ist:

$$(5)\qquad \int_{-n}^{n}\left(f(x)\int_{0}^{\lambda}\cos\mu(x-x_0)\,d\mu\right)dx=\int_{0}^{\lambda}\left(\cos\mu x_0\int_{-n}^{n}f(x)\cos\mu x\,dx\right)d\mu+$$

$$+\int_{0}^{\lambda}\left(\sin\mu x_0\int_{-n}^{n}f(x)\sin\mu x\,dx\right)d\mu.$$

Setzen wir:

$$\Phi_n(\mu)=\int_{-n}^{n}f(x)\frac{\sin\mu x}{x}\,dx;\quad \Psi_n(\mu)=\int_{-n}^{n}f(x)\frac{1-\cos\mu x}{x}\,dx,$$

so sind $\Phi_n(\mu)$ und $\Psi_n(\mu)$ stetige, differenzierbare Funktionen, und es ist:

$$\frac{d\,\Phi_n(\mu)}{d\mu}=\int_{-n}^{n}f(x)\cos\mu x\,dx;\qquad \frac{d\,\Psi_n(\mu)}{d\mu}=\int_{-n}^{n}f(x)\sin\mu x\,dx,$$

so dass (5) auch geschrieben werden kann[1]:

$$(6)\qquad \int_{-n}^{n}\left(f(x)\int_{0}^{\lambda}\cos\mu(x-x_0)\,d\mu\right)dx=\int_{0}^{\lambda}\cos\mu x_0\,d\,\Phi_n(\mu)+\int_{0}^{\lambda}\sin\mu x_0\,d\,\Psi_n(\mu).$$

Hierin nun ist der Grenzübergang $n\to\infty$ zu vollziehen.

Aus Eigenschaft B) folgt die Existenz der Integrale:

$$(7)\qquad \Phi(\mu)=\int_{-\infty}^{+\infty}f(x)\frac{\sin\mu x}{x}\,dx;\qquad \Psi(\mu)=\int_{-\infty}^{+\infty}f(x)\frac{1-\cos\mu x}{x}\,dx,$$

und es ist:

$$(8)\qquad \lim_{n\to\infty}\Phi_n(\mu)=\Phi(\mu);\qquad \lim_{n\to\infty}\Psi_n(\mu)=\Psi(\mu).$$

Die Konvergenz hierin ist gleichmässig für alle μ des Intervalles $[0, \lambda]$; denn ist $\varepsilon>0$ beliebig gegeben, so kann man wegen Eigenschaft B) die Zahl b so gross wählen, dass:

---

[1] Über den hiebei zur Verwendung kommenden Integralbegriff vgl. Monatsh. f. Math. u. Phys. 32 (1922), S. 69 ff.

$$\int_b^{+\infty}\left|\frac{f(x)}{x}\right|dx<\varepsilon;\qquad \int_{-\infty}^{-b}\left|\frac{f(x)}{x}\right|dx<\varepsilon;$$

dann aber ist für alle $n>b$ und alle μ:

$$|\Phi_n(\mu)-\Phi(\mu)|=\left|\int_{-\infty}^{-n}f(x)\frac{\sin\mu x}{x}dx+\int_n^{+\infty}f(x)\frac{\sin\mu x}{x}dx\right|<2\varepsilon,$$

$$|\Psi_n(\mu)-\Psi(\mu)|=\left|\int_{-\infty}^{-n}f(x)\frac{1-\cos\mu x}{x}dx+\int_n^{+\infty}f(x)\frac{1-\cos\mu x}{x}dx\right|<4\varepsilon.$$

Aus der Stetigkeit von $\Phi_n(\mu)$ und $\Psi_n(\mu)$ *folgt also die Stetigkeit von* $\Phi(\mu)$ *und* $\Psi(\mu)$, und wir sehen ferner, dass die Gesamtheit der $\Phi_n(\mu)$ und $\Psi_n(\mu)$ im Intervalle $[0, \lambda]$ zwischen endlichen Schranken bleibt:

(9) $\qquad |\Phi_n(\mu)|\leqq M, \qquad |\Psi_n(\mu)|\leqq M$ für $0\leqq\mu\leqq\lambda$ und alle n.

Aus (8) und (9) aber folgt[1]:

$$\lim_{n\to\infty}\int_0^\lambda\cos\mu x_0\,d\Phi_n(\mu)=\int_0^\lambda\cos\mu x_0\,d\Phi(\mu),$$

$$\lim_{n\to\infty}\int_0^\lambda\sin\mu x_0\,d\Psi_n(\mu)=\int_0^\lambda\sin\mu x_0\,d\Psi(\mu).$$

Aus (1), (4) und (6) folgt somit:

$$f(x_0)=\frac{1}{\pi}\lim_{\lambda\to+\infty}\left(\int_0^\lambda\cos\mu x_0\,d\Phi(\mu)+\int_0^\lambda\sin\mu x_0\,d\Psi(\mu)\right),$$

wofür wir auch schreiben:

(10) $$f(x_0)=\frac{1}{\pi}\int_0^{+\infty}(\cos\mu x_0\,d\Phi(\mu)+\sin\mu x_0\,d\Psi(\mu)).$$

[1] l. c. S. 304, S. 77—80.

Wir haben also den Satz[1]:

Satz I. *Hat $f(x)$ die Eigenschaften* A) *und* B), *so gilt* (10), *worin* $\Phi(\mu)$ *und* $\Psi(\mu)$ *die Bedeutung* (7) *haben und stetige Funktionen von* μ *sind.*

§ 2.

Formel (10) steht in naher Beziehung zur *Fourierschen Integralformel:*

$$f(x_0) = \frac{1}{\pi}\int_0^{+\infty} (\varphi(\mu)\cos \mu x_0 + \psi(\mu) \sin \mu x_0)\, d\mu, \tag{11}$$

worin gesetzt ist:

$$\varphi(\mu) = \int_{-\infty}^{+\infty} f(x)\cos \mu x\, dx; \qquad \psi(\mu) = \int_{-\infty}^{+\infty} f(x) \sin \mu x\, dx. \tag{12}$$

Doch ist die Tragweite der Formel (10) eine bei weitem grössere. Wir wollen zunächst an einem Beispiele zeigen, dass die für die Giltigkeit von (10) hinreichenden Bedingungen A) und B) für die Giltigkeit der Fourierschen Integralformel (11) keineswegs ausreichen.[2]

Die Funktion $f(x)$ sei gerade, sie sei konstant in jedem Intervalle $(\nu, \nu+1)$:

$$f(x) = c_\nu \text{ in } (\nu, \nu+1) \qquad (\nu = 0, 1, 2, \ldots),$$

und es sei:

$$f(0) = c_0, \qquad f(\nu) = \frac{1}{2}(c_{\nu-1} + c_\nu) \qquad (\nu = 1, 2, \ldots).$$

Dann hat $f(x)$ die Eigenschaft A) in jedem Punkte x_0. Wegen:

$$\int_1^{+\infty} \left|\frac{f(x)}{x}\right| dx = \sum_{\nu=1}^{\infty} |c_\nu| \int_\nu^{\nu+1} \frac{dx}{x} = \sum_{\nu=1}^{\infty} |c_\nu| \lg \frac{\nu+1}{\nu}$$

---

[1] Dieser Satz findet sich, wie mich Herr Hilb freundlichst aufmerksam machte, bereits bei H. Weyl, Jahresber. Math. Ver. 20 (1911), S. 134.

[2] Das folgende Beispiel verdanke ich Herrn W. Wirtinger. Ein andres Beispiel findet man bei A. Pringsheim, Math. Ann. 71 (1911), S. 296.

wird $f(x)$ die Eigenschaft B) dann und nur dann haben, wenn die Reihe $\sum_{\nu=1}^{\infty} |c_\nu| \lg \frac{\nu + 1}{\nu}$ konvergiert, oder, was dasselbe heisst, wenn die Reihe

$$\sum_{\nu=1}^{\infty} |c_\nu| \cdot \frac{1}{\nu} \tag{13}$$

konvergiert.

Falls die durch (12) gegebene Funktion $\varphi(\mu)$ existiert, so muss sein:

$$\varphi(\mu) = 2 \int_0^{+\infty} f(x) \cos \mu x \, dx = 2 \sum_{\nu=0}^{\infty} c_\nu \int_\nu^{\nu+1} \cos \mu x \, dx = 4 \frac{\sin \frac{\mu}{2}}{\mu} \sum_{\nu=0}^{\infty} c_\nu \cos \left(\nu + \frac{1}{2}\right) \mu, \tag{14}$$

und diese letzte Reihe muss konvergent sein. Setzen wir also $c_\nu = 1$, wenn ν das Quadrat einer natürlichen Zahl ist: $\nu = k^2$ $(k = 1, 2, \ldots)$, sonst aber $c_\nu = 0$, so wird:

$$\sum_{\nu=1}^{\infty} |c_\nu| \cdot \frac{1}{\nu} = \sum_{k=1}^{\infty} \frac{1}{k^2},$$

so dass die Reihe (13) konvergiert; *die Funktion $f(x)$ hat also dann die Eigenschaften* A) *und* B). Die in (14) auftretende Reihe aber wird: $\sum_{k=1}^{\infty} \cos \left(k^2 + \frac{1}{2}\right) \mu$ und ist als trigonometrische Reihe, deren Koeffizienten nicht gegen 0 konvergieren, für alle μ abgesehen von einer Nullmenge divergent.[1] *Für die Funktion $f(x)$ gilt also das Fouriersche Integraltheorem* (11) *nicht*, weil die darin auftretende Funktion $\varphi(\mu)$ (abgesehen von einer Nullmenge) nicht existiert. Unsre Formel (10) hingegen ist zur Darstellung von $f(x_0)$ anwendbar, und zwar ist in ihr $\Psi(\mu) = 0$ und:

$$\Phi(\mu) = 2 \sum_{k=1}^{\infty} \int_{k^2}^{k^2+1} \frac{\sin \mu x}{x} \, dx,$$

welche Reihe offenbar konvergiert. Wie man sieht, entsteht sie aus der in (14) auftretenden (divergenten) Reihe $2 \sum_{k=1}^{\infty} \int_{k^2}^{k^2+1} \cos \mu x \, dx$ durch gliedweise Integration nach μ von 0 bis μ.

[1] Vgl. z. B. H. Lebesgue, Leçons sur les séries trigonométriques, S. 110.

§ 3.

Obwohl, wie wir eben sahen, die Tragweite in Formel (10) eine grössere ist, als die der Fourierschen Integralformel (11), so kann doch in manchen Fällen aus der Giltigkeit von (10) auf die Giltigkeit von (11) geschlossen werden, wodurch man zu sehr allgemeinen Kriterien für die Giltigkeit der Fourierschen Integralformel gelangt. Wir zeigen zunächst:

Satz II. *Gilt für eine Funktion* $f(x)$ *Formel* (10), *existieren ferner für alle* $\mu \geqq 0$ *(abgesehen von einer Nullmenge) die durch* (12) *definierten Funktionen* $\varphi(\mu)$ *und* $\psi(\mu)$, *und bestehen zwischen diesen und den durch* (7) *definierten Funktionen* $\Phi(\mu)$ *und* $\Psi(\mu)$ *die Beziehungen:*

$$(15) \qquad \Phi(\mu) = \int_0^\mu \varphi(\mu)\, d\mu; \qquad \Psi(\mu) = \int_0^\mu \psi(\mu)\, d\mu \qquad (\mu \geqq 0)$$

so gilt für $f(x)$ *auch die Fouriersche Integralformel* (11).

In der Tat, auf grund von (15) kann geschrieben werden:

$$(16) \qquad \int_0^\lambda \cos \mu x_0\, d\Phi(\mu) = \int_0^\lambda \cos \mu x_0\, \varphi(\mu)\, d\mu;$$

$$\int_0^\lambda \sin \mu x_0\, d\Psi(\mu) = \int_0^\lambda \sin \mu x_0\, \psi(\mu)\, d\mu,$$

so dass (10) unmittelbar in (11) übergeht.

Erinnern wie uns an Satz I, so sehen wir, *dass die Eigenschaften* A) *und* B), *zusammen mit den Beziehungen* (15) *für die Giltigkeit der Fourierschen Integralformel* (11) *hinreichen.*

Ein Beispiel zu diesem sehr allgemeinen Kriterium liefert die Funktion $\frac{\sin x}{x}$[1], die offenbar die Eigenschaften A) und B) hat. Hier ist $\Psi(\mu) = 0$ und:

---

[1] Wie A. Pringsheim (Math. Ann. 68 (1910), S. 368) mitteilt, wurde er von H. Weber aufmerksam gemacht, dass für die Funktion $\frac{\sin x}{x}$ die Fouriersche Integralformel gilt, während sie keiner der üblichen Bedingungen für die Giltigkeit dieser Formel genügt.

$$(17)\qquad \varphi(\mu)=2\int_0^{+\infty}\frac{\sin x\cos\mu x}{x}\,dx=$$

$$\int_0^{+\infty}\frac{\sin(1+\mu)x}{x}\,dx+\int_0^{+\infty}\frac{\sin(1-\mu)x}{x}\,dx=\begin{cases}\pi & (0\leqq\mu<1)\\ \frac{\pi}{2} & (\mu=1)\\ 0 & (\mu>1)\end{cases}$$

Die zweite Formel (15) ist in trivialer Weise erfüllt ($\Psi(\mu)=0$, $\psi(\mu)=0$). Um die erste Formel (15) zu beweisen, beachten wir, dass:

$$2\int_b^{+\infty}\frac{\sin x\cos\mu x}{x}\,dx=\int_{(1+\mu)b}^{+\infty}\frac{\sin x}{x}\,dx\pm\int_{|1-\mu|b}^{+\infty}\frac{\sin x}{x}\,dx.$$

Das Integral $\int_b^{+\infty}\frac{\sin x\cos\mu x}{x}\,dx$ bleibt also für alle b und alle μ zwischen endlichen Grenzen. Daher ist:

$$\int_0^{\mu}\varphi(\mu)\,d\mu=2\int_0^{\mu}\left(\lim_{b\to+\infty}\int_0^{b}\frac{\sin x\cos\mu x}{x}\,dx\right)d\mu$$

$$=2\lim_{b\to+\infty}\int_0^{\mu}\left(\int_0^{b}\frac{\sin x\cos\mu x}{x}\,dx\right)d\mu=2\lim_{b\to+\infty}\int_0^{b}\left(\int_0^{\mu}\frac{\sin x\cos\mu x}{x}\,d\mu\right)dx$$

$$=2\int_0^{+\infty}\frac{\sin x}{x}\cdot\frac{\sin\mu x}{x}\,dx=\Phi(\mu),$$

womit auch die erste Formel (15) bewiesen ist. — Aus (17) entnehmen wir, dass hier:

$$\Phi(\mu)=\begin{cases}\pi\mu & (0\leqq\mu\leqq 1)\\ \pi & (\mu\geqq 1);\end{cases}$$

die Fouriersche Integralformel lautet hier einfach:

$$\frac{\sin x_0}{x_0}=\int_0^1\cos\mu x_0\,d\mu.$$

§ 4.

Satz II kann noch verallgemeinert werden:

Satz III. *Gilt für eine Funktion $f(x)$ Formel* (10), *sind die durch* (7) *definierten Funktionen $\Phi(\mu)$ und $\Psi(\mu)$ stetig für $\mu \geqq 0$, existieren für alle $\mu \geqq 0$ (abgesehen von einer Nullmenge) die durch* (12) *definierten Funktionen $\varphi(\mu)$ und $\psi(\mu)$, bestehen für jedes Intervall* $(0 \leqq)\ \alpha \leqq \mu \leqq \beta$, *das keinen Punkt einer gewissen reduziblen*[1] *Punktmenge Q enthält, die Beziehungen:*

$$(18) \qquad \Phi(\beta) - \Phi(\alpha) = \int_\alpha^\beta \varphi(\mu)\, d\mu; \qquad \Psi(\beta) - \Psi(\alpha) = \int_\alpha^\beta \psi(\mu)\, d\mu,$$

so gilt für $f(x)$ auch die Fouriersche Integralformel (11).

Wir haben zu zeigen, dass auch unter den Voraussetzungen von Satz III die Formeln (16) gelten. Dabei wird es genügen, anzunehmen, dass ins Intervall $[0, \lambda]$ nur ein einziges Punkt q der Menge Q fällt, da der Übergang zum allgemeinen Falle nach ganz bekannten Schlüssen vor sich geht; und zwar wollen wir etwa annehmen, der Punkt q falle ins Innere von $[0, \lambda]$. Dann ist, wegen der Stetigkeit von $\Phi(\mu)$:

$$\int_0^\lambda \cos \mu x_0\, d\Phi(\mu) = \lim_{h \to +0} \int_0^{q-h} \cos \mu x_0\, d\Phi(\mu) + \lim_{h \to +0} \int_{q+h}^\lambda \cos \mu x_0\, d\Phi(\mu).$$

In jedem der beiden Intervalle $[0, q-h]$ und $[q+h, \lambda]$ ist nun aber nach Voraussetzung $\Phi(\mu)$ unbestimmtes Integral der integrierbaren Funktion $\varphi(\mu)$, so dass:

$$\int_0^{q-h} \cos \mu x_0\, d\Phi(\mu) = \int_0^{q-h} \cos \mu x_0\, \varphi(\mu)\, d\mu; \qquad \int_{q+h}^\lambda \cos \mu x_0\, d\Phi(\mu) = \int_{q+h}^\lambda \cos \mu x_0\, \varphi(\mu)\, d\mu.$$

Es ist also:

[1] d. h. einer Punktmenge Q, die eine leere Ableitung $Q^{(\nu)}$ (von endlicher oder transfiniter Ordnung ν) hat. Das Wort »reduzibel» ist also hier in dem Sinne gebraucht, in dem es von G. Cantor ursprünglich gebraucht wurde.

$$\int_0^\lambda \cos \mu x_0 \, d\Phi(\mu) = \lim_{h \to +0} \int_0^{q-h} \cos \mu x_0 \, \varphi(\mu) \, d\mu + \lim_{h \to +0} \int_{q+h}^\lambda \cos \mu x_0 \, \varphi(\mu) \, d\mu .$$

Es existiert also der rechts stehende Grenzwert, d. h. das (eventuell) *uneigentliche* Integral $\int_0^\lambda \cos \mu x_0 \, \varphi(\mu) \, d\mu$, und es ist:

$$\int_0^\lambda \cos \mu x_0 \, d\Phi(\mu) = \int_0^\lambda \cos \mu x_0 \, \varphi(\mu) \, d\mu ,$$

womit die erste Formel (16) bewiesen ist; und ganz ebenso beweist man die zweite; nur ist festzuhalten, dass nun die auf den rechten Seiten der Formeln (16) auftretenden Integrale *uneigentliche* Integrale (mit der singulären Stelle q) sein können.

Ebenso wird im allgemeinen Falle das in (11) auftretende Integral ein *uneigentliches* Lebesguesches Integral sein können, dessen singuläre Punkte die Punkte von Q und deren Häufungspunkte sind.

Aus Satz III entnehmen wir, *dass die Fouriersche Integralformel für jede Funktion $f(x)$ gilt, die die Eigenschaften* A) *und* B) *hat, für die (für $\mu \geqq 0$) die Funktionen $\varphi(\mu)$ und $\psi(\mu)$ abgesehen von einer Nullmenge existieren, und in jedem Intervalle $[\alpha, \beta]$, das keinen Punkt einer gewissen reduziblen Menge Q enthält, die Beziehungen* (18) *gelten.*

Als Beispiel betrachten wir die Funktion:

$$f(x) = g(x) \cos qx \qquad (q > 0),$$

wo $g(x)$ die Eigenschaft B) habe und im Unendlichen monoton sei (d. h. es gebe ein a, sodass $g(x)$ für $x > a$ und $x < -a$ monoton ist). Offenbar ist dann:

$$(19) \qquad \lim_{x \to +\infty} g(x) = 0; \qquad \lim_{x \to -\infty} g(x) = 0.$$

Die Funktion $f(x)$ hat dann ebenfalls die Eigenschaft B). Wir nehmen an, sie habe auch die Eigenschaft A). Nach Satz I gilt dann für $f(x)$ Formel (10) und $\Phi(\mu)$ und $\Psi(\mu)$ sind (für $\mu \geqq 0$) stetig.

Wir untersuchen die zugehörige Funktion $\varphi(\mu)$ (für $\mu \geqq 0$). Wählen wir $b > a$ und $p > b$, so ist nach dem zweiten Mittelwertsatz:

$$\int_b^p f(x)\cos\mu x\,dx = \int_b^p g(x)\cos qx\cos\mu x\,dx =$$

$$= \frac{1}{2}\int_b^p g(x)(\cos(\mu+q)x + \cos(\mu-q)x)\,dx =$$

$$= \frac{1}{2}g(b)\int_b^{p^*}(\cos(\mu+q)x + \cos(\mu-q)x)\,dx,$$

und somit für $\mu \neq q$:

$$\left|\int_b^p f(x)\cos\mu x\,dx\right| \leqq |g(b)|\left(\frac{1}{\mu+q} + \frac{1}{|\mu-q|}\right).$$

Ganz ebenso findet man:

$$\left|\int_{-p}^{-b} f(x)\cos\mu x\,dx\right| \leqq |g(-b)|\left(\frac{1}{\mu+q} + \frac{1}{|\mu-q|}\right).$$

Wegen (19) folgt daraus, dass für $\mu \geqq 0$ und $\neq q$ das Integral:

$$\varphi(\mu) = \int_{-\infty}^{+\infty} f(x)\cos\mu x\,dx$$

existiert, und dass die Konvergenz von:

$$\varphi_n(\mu) = \int_{-n}^{n} f(x)\cos\mu x\,dx \tag{20}$$

gegen $\varphi(\mu)$ in jedem Intervalle $(0 \leqq)\ \alpha \leqq \mu \leqq \beta$, das den Punkt q nicht enthält, gleichmässig erfolgt. Darum folgt weiter für jedes solche Intervall:

$$\int_\alpha^\beta \varphi(\mu)\,d\mu = \lim_{n\to\infty}\int_\alpha^\beta \varphi_n(\mu)\,d\mu = \lim_{n\to\infty}\int_{-n}^{n}\left(f(x)\int_\alpha^\beta \cos\mu x\,d\mu\right)dx$$

$$= \int_{-\infty}^{+\infty} f(x)\frac{\sin\beta x - \sin\alpha x}{x}\,dx = \Phi(\beta) - \Phi(\alpha).$$

Ganz ebenso finden wir, dass $\psi(\mu)$ für $\mu \geqq 0$ und $\neq q$ existiert und dass für jedes Intervall $(0 \leqq) \alpha \leqq \mu \leqq \beta$, das den Punkt q nicht enthält, die zweite Formel (18) gilt.

Für die Funktion $f(x)=g(x) \cos qx$ sind also die sämtlichen Voraussetzungen von Satz III erfüllt; wir sehen also, dass für sie die Fouriersche Integralformel (11) gilt, wobei aber das Integral auf der rechten Seite von (11) ein uneigentliches mit der singulären Stelle q sein kann. — Ganz dasselbe gilt, wie analoge Rechnungen zeigen, für $f(x)=g(x) \sin qx$.

§ 5.

Das eben behandelte Beispiel ist ein Spezialfall des folgenden Satzes, der gleichfalls in Satz III enthalten ist:

Satz IV. *Sei $f(x)$ eine Funktion der Eigenschaft* A), *die die Gestalt hat:*

$$f(x)=g(x)\,h(x),$$

wo $g(x)$ eine im Unendlichen monotone Funktion der Eigenschaft B) *und $h(x)$ eine periodische Funktion bedeutet. Dann gilt für $f(x)$ die Fouriersche Integralformel* (11).

Sei $g(x)$ etwa monoton für $x > a$ und $x < -a$. Ohne Beschränkung der Allgemeinheit können wir annehmen, $h(x)$ habe die Periode 2π.

Zunächst überzeugen wir uns leicht, dass auch $f(x)$ die Eigenschaft B) hat. In der Tat ist für $b > a$:

$$(21) \qquad \int_b^{+\infty} \left|\frac{f(x)}{x}\right| dx \leqq \sum_{k=0}^{\infty} \left|\frac{g(b+2k\pi)}{b+2k\pi}\right| \cdot \int_b^{b+2\pi} |h(x)|\,dx.$$

Da aber:

$$\sum_{k=1}^{\infty} \left|\frac{g(b+2k\pi)}{b+2k\pi}\right| \leqq \frac{1}{2\pi} \int_b^{+\infty} \left|\frac{g(x)}{x}\right| dx$$

ist, so folgt aus Eigenschaft B) von $g(x)$ die Konvergenz der in (21) auftretenden Reihe, und damit die Eigenschaft B) für $f(x)$.

Nach Satz I gilt also für $f(x)$ Formel (10) und $\Phi(\mu)$ und $\Psi(\mu)$ sind (für $\mu \geqq 0$) stetig.

Sodann zeigen wir: Zu jedem $\delta > 0$ gibt es ein M, so dass für alle Intervalle $[\xi, \eta]$ und alle $\mu \geqq 0$, die keinem der Intervalle $|\mu - k| < \delta$ $(k = 0, 1, 2, \ldots)$ angehören:

$$(22) \qquad \left|\int_{\xi}^{\eta} h(x) \cos \mu x \, dx\right| < M, \qquad \left|\int_{\xi}^{\eta} h(x) \sin \mu x \, dx\right| < M.$$

Sei etwa $\eta > \xi$ und sei die ganze Zahl n so gewählt, dass $\xi + 2n\pi \leqq \eta < \xi + + 2(n+1)\pi$. Dann ist:

$$(23) \qquad \int_{\xi}^{\eta} h(x) e^{i\mu x} dx = \sum_{k=0}^{n-1} \int_{\xi}^{\xi+2\pi} h(x) e^{i\mu(x+2k\pi)} dx + \int_{\xi+2n\pi}^{\eta} h(x) e^{i\mu x} dx$$

$$= \int_{\xi}^{\xi+2\pi} h(x) e^{i\mu x} dx \cdot \frac{e^{2\pi i\mu n} - 1}{e^{2\pi i\mu} - 1} + \int_{\xi+2n\pi}^{\eta} h(x) e^{i\mu x} dx.$$

Da hierin:

$$\left|\int_{\xi}^{\xi+2\pi} h(x) e^{i\mu x} dx\right| \leqq \int_{\xi}^{\xi+2\pi} |h(x)| \, dx, \qquad \left|\int_{\xi+2n\pi}^{\eta} h(x) e^{i\mu x} dx\right| \leqq \int_{\xi}^{\xi+2\pi} |h(x)| \, dx,$$

$$|e^{2\pi i\mu n} - 1| \leqq 2,$$

ist (22) bewiesen.[1]

Wir gehen nun über zur Untersuchung von $\varphi(\mu)$. Sei $\mu > 0$, $b > a$, $p > b$. Nach dem zweiten Mittelwertsatze ist:

$$\int_{b}^{p} f(x) \cos \mu x \, dx = \int_{b}^{p} g(x) h(x) \cos \mu x \, dx = g(b) \int_{b}^{p^*} h(x) \cos \mu x \, dx.$$

Da auch hier die Beziehungen (19) gelten, entnehmen wir aus (22): Ist $\varepsilon > 0$ beliebig gegeben, und enthält das Intervall $(0 <) \alpha \leqq \mu \leqq \beta$ keine natürliche Zahl, so gibt es ein b_0, so dass für alle $b \geqq b_0$, alle $p > b$ und alle μ von $[\alpha, \beta]$:

$$(24) \qquad \left|\int_{b}^{p} f(x) \cos \mu x \, dx\right| < \varepsilon; \qquad \left|\int_{-p}^{-b} f(x) \cos \mu x \, dx\right| < \varepsilon.$$

[1] Dieser einfache Beweis wurde mir von Herrn W. Wirtinger mitgeteilt.

Daraus folgt sofort die Existenz von $\varphi(\mu)$ für $\mu \neq 0, 1, 2, \ldots$ und die gleichmässige Konvergenz der durch (20) definierten Funktion $\varphi_n(\mu)$ gegen $\varphi(\mu)$ in jedem Intervalle $(0 <)\ \alpha \leqq \mu \leqq \beta$, das keine natürliche Zahl enthält. Daraus folgt wie in § 4, dass in jedem solchen Intervalle die erste Gleichung (18) gilt, und ebenso zeigt man dies für die zweite.

Für die Funktion $f(x) = g(x)\,h(x)$ sind also sämtliche Voraussetzungen von Satz III erfüllt (wobei Q die Menge der Zahlen $0, 1, 2, \ldots$ ist), und damit ist Satz IV bewiesen. Das in der Fourierschen Integralformel (11) für $f(x)$ auftretende Integral wird hier im allgemeinen ein uneigentliches sein mit den singulären Stellen $0, 1, 2, \ldots$.

Ein verwandter Satz ist der folgende[1]:

Satz IV a. *Satz IV bleibt richtig, wenn darin unter $h(x)$ ein Funktion der Gestalt:*

$$h(x) = \sum_{\nu=1}^{\infty} (a_\nu \cos q_\nu x + b_\nu \sin q_\nu x)$$

verstanden wird, vorausgesetzt dass die Reihe $\sum_{\nu=1}^{\infty}(|a_\nu| + |b_\nu|)$ *konvergiert und die nicht-negativen Zahlen q_ν eine reduzible Menge Q bilden.*

Da hier $h(x)$ beschränkt ist, folgt aus der Eigenschaft B) von $g(x)$ sofort auch die Eigenschaft B) für $f(x)$. Weiter ist (wenn $\mu \geqq 0$ und von allen q_ν verschieden ist):

$$\int_\xi^\eta h(x) \cos \mu x\, dx = \sum_{\nu=1}^{\infty} \left(a_\nu \int_\xi^\eta \cos q_\nu x \cos \mu x\, dx + b_\nu \int_\xi^\eta \sin q_\nu x \cos \mu x\, dx \right)$$

$$= \frac{1}{2} \sum_{\nu=1}^{\infty} \frac{a_\nu}{q_\nu + \mu} (\sin (q_\nu + \mu)\eta - \sin (q_\nu + \mu)\xi) + \frac{1}{2} \sum_{\nu=1}^{\infty} \frac{a_\nu}{q_\nu - \mu} (\sin (q_\nu - \mu)\eta - \sin (q_\nu - \mu)\xi)$$

$$- \frac{1}{2} \sum_{\nu=1}^{\infty} \frac{b_\nu}{q_\nu + \mu} (\cos(q_\nu + \mu)\eta - \cos(q_\nu + \mu)\xi) - \frac{1}{2} \sum_{\nu=1}^{\infty} \frac{b_\nu}{q_\nu - \mu} (\cos (q_\nu - \mu)\eta - \cos (q_\nu - \mu)\xi),$$

und somit:

---

[1] Dieser Satz findet sich im Wesentlichen schon bei A. Pringsheim, Math. Ann. 68 (1910), S. 367–408; 71 (1911), S. 289–298.

$$\left|\int_{\xi}^{\eta} h(x)\cos\mu x\,dx\right| \leqq \sum_{\nu=1}^{\infty}(|a_\nu|+|b_\nu|)\left(\frac{1}{q_\nu+\mu}+\frac{1}{|q_\nu-\mu|}\right),$$

und dieselbe Ungleichung erhält man für $\left|\int_{\xi}^{\eta} h(x)\sin\mu x\,dx\right|$. Wir entnehmen daraus, dass hier die Ungleichungen (22) gelten für alle Intervalle $[\xi, \eta]$ und alle $\mu \geqq 0$, die keinem der Intervalle $|\mu - q_\nu| < \delta$ $(\nu = 1, 2, \ldots)$ angehören. Wie beim Beweise von Satz IV folgt daraus weiter, dass $\varphi(\mu)$ und $\psi(\mu)$ für jedes μ existieren, das weder Punkt noch Häufungspunkt von Q ist, und dass die Beziehungen (18) in jedem Intervalle $[\alpha, \beta]$ gelten, das weder einen Punkt noch einen Häufungspunkt von Q enthält. Damit sind aber wieder alle Voraussetzungen von Satz III als erfüllt nachgewiesen, und somit ist auch Satz IV a bewiesen.

§ 6.

Ersetzen wir nun Eigenschaft B) durch die folgende:

C) Es sei:

$$\lim_{x\to+\infty}\frac{f(x)}{x}=0; \qquad \lim_{x\to-\infty}\frac{f(x)}{x}=0,$$

und es gebe ein a, so dass $\frac{f(x)}{x}$ monoton ist für $x > a$ und $x < -a$.

Wir erkennen leicht, dass (1) auch für jede Funktion $f(x)$ der Eigenschaften A) und C) gilt. In der Tat ist dann (wenn $a > |x_0|$ gewählt wurde) auch $\frac{f(x)}{x-x_0}$ monoton für $x > a$ und $x < -a$, und es ist auch:

$$\lim_{x\to+\infty}\frac{f(x)}{x-x_0}=0; \qquad \lim_{x\to-\infty}\frac{f(x)}{x-x_0}=0.$$

Nach dem zweiten Mittelwertsatze ist für $q > a$, $p > q$:

$$\left|\int_q^p f(x)\frac{\sin\lambda(x-x_0)}{x-x_0}\,dx\right| = \left|\frac{f(q)}{q-x_0}\int_q^{p^*}\sin\lambda(x-x_0)\,dx\right| \leqq \frac{2}{\lambda}\left|\frac{f(q)}{q-x_0}\right|;$$

ist also $\varepsilon > 0$ beliebig gegeben und bedeutet λ_0 irgend eine positive Zahl, so gibt es ein b, so dass für $p > q \geqq b$ und alle $\lambda \geqq \lambda_0$:

$$\left| \int_q^p f(x) \frac{\sin \lambda (x-x_0)}{x-x_0} \, dx \right| < \varepsilon, \qquad \left| \int_{-p}^{-q} f(x) \frac{\sin \lambda (x-x_0)}{x-x_0} \, dx \right| < \varepsilon.$$

Es existiert also das Integral $\int_{-\infty}^{+\infty} f(x) \frac{\sin \lambda (x-x_0)}{x-x_0} dx$, und es ist für $\lambda \geqq \lambda_0$:

$$\left| \int_b^{+\infty} f(x) \frac{\sin \lambda (x-x_0)}{x-x_0} \, dx \right| \leqq \varepsilon; \qquad \left| \int_{-\infty}^{-b} f(x) \frac{\sin \lambda (x-x_0)}{x-x_0} \, dx \right| \leqq \varepsilon,$$

und indem man von hieraus weiter schliesst wie in § 1, beweist man die Giltigkeit von (1).

Man könnte daher versucht sein zu glauben, dass Formel (10) auch für alle Funktionen $f(x)$ der Eigenschaften A) und C) gilt. Das ist aber nicht der Fall. Sei etwa $f(x) = |x|^{\frac{1}{2}}$. Dann hat $f(x)$ die Eigenschaften A) und C). Nach (7) ergibt sich $\Phi(0) = 0$ und:

$$\Phi(\mu) = 2 \int_0^{+\infty} \frac{\sin \mu x}{x^{\frac{1}{2}}} \, dx = 2 \mu^{-\frac{1}{2}} \int_0^{+\infty} \frac{\sin x}{x^{\frac{1}{2}}} \, dx \qquad \text{für } \mu > 0.$$

Das Integral $\int_0^{+\infty} \cos \mu x_0 \, d\Phi(\mu)$ hat also hier keinen Sinn. Wohl aber werden wir im Folgenden zeigen, dass (10) für jede Funktion $f(x)$ der Eigenschaft A) gilt, die im Unendlichen *beschränkt und monoton* ist. Wir schicken folgende Bemerkungen voran.

Wir setzen im Folgenden stets:

$$(24) \qquad \Phi(\mu) = \lim_{n \to \infty} \int_{-n}^{n} f(x) \frac{\sin \mu x}{x} \, dx$$

$$(25)\qquad \Psi(\mu)=\int_{-1}^{1} f(x)\frac{1-\cos\mu x}{x}\,dx-\lim_{n\to\infty}\left(\int_{1}^{n} f(x)\frac{\cos\mu x}{x}\,dx+\int_{-n}^{-1} f(x)\frac{\cos\mu x}{x}\,dx\right),$$

vorausgesetzt, dass diese Grenzwerte existieren. Für jede beliebige ungerade Funktion $f(x)$ ist dann $\Phi(\mu)=0$, für jede gerade Funktion $f(x)$ ist $\Psi(\mu)=0$. Abgesehen von diesen Spezialfällen werden wir es aber nur mit dem Falle zu haben, dass die Integrale:

$$\int_{0}^{+\infty} f(x)\frac{\sin\mu x}{x}\,dx,\quad \int_{-\infty}^{0} f(x)\frac{\sin\mu x}{x}\,dx,\quad \int_{1}^{+\infty} f(x)\frac{\cos\mu x}{x}\,dx,\quad \int_{-\infty}^{-1} f(x)\frac{\cos\mu x}{x}\,dx$$

existieren. Die Grenzwerte (24) und (25) sind dann sicher vorhanden.

Die durch (25) definierte Funktion $\Psi(\mu)$ unterscheidet sich von der bisher verwendeten, durch (7) definierten, falls sie alle beide existieren, nur durch eine additive Konstante, auf die es nicht ankommt, da die Funktion $\Psi(\mu)$ nur in der Verbindung $\int \sin\mu x_0\, d\Psi(\mu)$ auftritt.

§ 7.

Wir werden nun zeigen (§§ 7—16), dass Formel (10) für jede Funktion der Eigenschaft A) gilt, die im Unendlichen beschränkt und monoton ist.

Sei zunächst $f(x)$ eine Funktion der Eigenschaft A), die im Unendlichen *monoton gegen* 0 *konvergiert:*

$$(26)\qquad \lim_{x\to+\infty} f(x)=0;\qquad \lim_{x\to-\infty} f(x)=0.$$

Wir untersuchen zuerst die zugehörige Funktion $\Phi(\mu)$. Ist $f(x)$ monoton für $x>a$ und $x<-a$, so haben wir nach dem zweiten Mittelwertsatz für $p>q>a$, wenn M eine geeignete Konstante bedeutet:

$$\left|\int_{q}^{p} f(x)\frac{\sin\mu x}{x}\,dx\right|=\left|f(q)\int_{q}^{p^*}\frac{\sin\mu x}{x}\,dx\right|=\left|f(q)\int_{\mu q}^{\mu p^*}\frac{\sin x}{x}\,dx\right|\leqq |f(q)|\,.\,M.$$

Wegen (26) gibt es also zu jedem $\varepsilon>0$ ein b, so dass für $b\leqq q<p$ und alle μ:

$$\left|\int_q^p f(x)\frac{\sin\mu x}{x}\,dx\right| < \varepsilon.$$

Daraus folgt: Für alle μ existiert das Integral $\int_0^{+\infty} f(x)\frac{\sin\mu x}{x}\,dx$, und das Integral $\int_0^n f(x)\frac{\sin\mu x}{x}\,dx$ konvergiert mit $n\to\infty$ gleichmässig für alle μ gegen $\int_0^{+\infty} f(x)\frac{\sin\mu x}{x}\,dx$. Dies letztere Integral ist also eine stetige Funktion von μ. Da genau dasselbe von $\int_{-\infty}^0 f(x)\frac{\sin\mu x}{x}\,dx$ gilt, so sehen wir: $\Phi(\mu)$ *existiert und ist stetig für alle* $\mu \geqq 0$.

Nun untersuchen wir die Funktion $\Psi(\mu)$. Wie vorhin erhalten wir:

$$\left|\int_q^p f(x)\frac{\cos\mu x}{x}\,dx\right| = \left|f(q)\int_{\mu q}^{\mu p^*}\frac{\cos x}{x}\,dx\right|,$$

woraus wir entnehmen: zu jedem $\varepsilon > 0$ und $\delta > 0$ gibt es ein b, so dass für $b \leqq q < p$ und $\mu \geqq \delta$:

$$\left|\int_q^p f(x)\frac{\cos\mu x}{x}\,dx\right| < \varepsilon.$$

Daraus folgt: für alle $\mu > 0$ existiert $\int_1^{+\infty} f(x)\frac{\cos\mu x}{x}\,dx$, und für alle $\mu \geqq \delta\ (> 0)$ konvergiert $\int_1^n f(x)\frac{\cos\mu x}{x}\,dx$ gleichmässig gegen $\int_1^{+\infty} f(x)\frac{\cos\mu x}{x}\,dx$. Analoges gilt für $\int_{-\infty}^{-1} f(x)\frac{\cos\mu x}{x}\,dx$, so dass wir sehen: $\Psi(\mu)$ *existiert und ist stetig für alle* $\mu > 0$.

Zugleich haben wir bewiesen: setzen wir:

(27) $$\Phi_n(\mu) = \int_{-n}^{n} f(x) \frac{\sin \mu x}{x} dx;$$

$$\Psi_n(\mu) = \int_{-1}^{1} f(x) \frac{1-\cos \mu x}{x} dx - \int_{1}^{n} f(x) \frac{\cos \mu x}{x} dx - \int_{-n}^{-1} f(x) \frac{\cos \mu x}{x} dx,$$

so konvergiert $\Phi_n(\mu)$ gleichmässig für alle $\mu \geqq 0$ gegen $\Phi(\mu)$ und $\Psi_n(\mu)$ gleichmässig für alle $\mu \geqq \delta\ (>0)$ gegen $\Psi(\mu)$.

Wie wir in § 6 gesehen haben, gilt auch hier Formel (1). Da ferner:

(28) $$\int_{-\infty}^{+\infty} f(x) \frac{\sin \lambda(x-x_0)}{x-x_0} dx = \int_{-\infty}^{+\infty} f(x_0+x) \frac{\sin \lambda x}{x} dx,$$

ist dieses Integral nichts anderes als die zur Funktion $f(x_0+x)$ gehörige Funktion $\Phi(\lambda)$, und daher, nach dem eben Bewiesenen stetig für alle $\lambda \geqq 0$.

Ganz wie in § 1 finden wir nun für $0 < h < \lambda$:

$$\int_{-\infty}^{+\infty} f(x) \frac{\sin \lambda(x-x_0)}{x-x_0} dx - \int_{-\infty}^{+\infty} f(x) \frac{\sin h(x-x_0)}{x-x_0} dx =$$

$$\lim_{n\to\infty} \int_{-n}^{n} \left(f(x) \int_{h}^{\lambda} \cos \mu(x-x_0)\, d\mu \right) dx = \lim_{n\to\infty} \int_{h}^{\lambda} (\cos \mu x_0\, d\Phi_n(\mu) + \sin \mu x_0\, d\Psi_n(\mu))$$

$$= \int_{h}^{\lambda} (\cos \mu x_0\, d\Phi(\mu) + \sin \mu x_0\, d\Psi(\mu)).$$

Wegen der Stetigkeit des Integrales (28) konvergiert für $h \to 0$ auf der linken Seite der Subtrahend gegen 0. Wir haben daher:

$$\int_{-\infty}^{+\infty} f(x) \frac{\sin \lambda(x-x_0)}{x-x_0} dx = \lim_{h\to+0} \int_{h}^{\lambda} (\cos \mu x_0\, d\Phi(\mu) + \sin \mu x_0\, d\Psi(\mu)),$$

und wenn wir schreiben:

$$(29)\qquad \lim_{h\to+0}\int_h^\lambda (\cos\mu x_0\, d\Phi(\mu)+\sin\mu x_0\, d\Psi(\mu))=\int_0^\lambda (\cos\mu x_0\, d\Phi(\mu)+\sin\mu x_0\, d\Psi(\mu))$$

so können wir den Satz aussprechen:

Satz V. *Für jede Funktion $f(x)$ der Eigenschaft* A), *die im Unendlichen monoton gegen* o *konvergiert, gilt Formel* (10); *darin ist die durch* (24) *definierte Funktion $\Phi(\mu)$ stetig für $\mu\geqq 0$, die durch* (25) *definierte Funktion $\Psi(\mu)$ stetig für $\mu>0$ und das in* (10) *auftretende Integral ist im Sinne von* (29) *zu verstehen.*

Um auch hier den Übergang zur Fourierschen Integralformel (11) zu vollziehen, betrachten wir die durch (12) definierte Funktion $\varphi(\mu)$. Wir haben hier für $a<q<p$:

$$\left|\int_q^p f(x)\cos\mu x\,dx\right|=\left|f(q)\int_q^{p^*}\cos\mu x\,dx\right|\leqq\frac{2}{\mu}|f(q)|,$$

woraus wir in gewohnter Weise schliessen, dass $\varphi(\mu)$ für $\mu>0$ existiert, und dass die durch (20) definierte Funktion $\varphi_n(\mu)$ für $\mu\geqq\delta\,(>0)$ gleichmässig gegen $\varphi(\mu)$ konvergiert; es ist also $\varphi(\mu)$ für $\mu>0$ *stetig*, und dasselbe gilt für $\psi(\mu)$. Ganz wie in § 4 können wir auch schliessen, dass in jedem Intervalle $0<\alpha\leqq\mu\leqq\beta$ die Relationen (18) bestehen. Daraus folgt für $0<h<\lambda$:

$$\int_h^\lambda (\cos\mu x_0\, d\Phi(\mu)+\sin\mu x_0\, d\Psi(\mu))=\int_h^\lambda (\cos\mu x_0\,\varphi(\mu)+\sin\mu x_0\,\psi(\mu))\,d\mu.$$

Vollzieht man noch den Grenzübergang $h\to+0$, so wird aus (10) die Fouriersche Integralformel (11), in der nun aber das auftretende Integral ein *uneigentliches* (mit der singulären Stelle $\mu=0$) sein kann. Wir haben also[1]:

Satz V a. *Für jede Funktion $f(x)$ der Eigenschaft* A), *die im Unendlichen monoton gegen* o *konvergiert, gilt die Fouriersche Integralformel* (11).

---

[1] Dieser Satz stammt von A. PRINGSHEIM, l. c., not. 1 S. 315.

41—2661. *Acta mathematica.* 49. Imprimé le 29 juillet 1926.

§ 8.

Sei nun

$$f(x) = 1.$$

Dann wird $\Psi(\mu)=0$ und:

(30) $$\Phi(\mu) = \int_{-\infty}^{+\infty} \frac{\sin \mu x}{x}\, dx = \begin{cases} 0 & \text{für } \mu = 0 \\ \pi & \text{für } \mu > 0. \end{cases}$$

Es ist also für jedes $\lambda > 0$:

$$\int_0^{\lambda} (\cos \mu x_0\, d\Phi(\mu) + \sin \mu x_0\, d\Psi(\mu)) = \pi,$$

woraus man ersieht: *für die Funktion* $f(x)=1$ *gilt Formel* (10); es ist in ihr $\Phi(\mu)$ die durch (30) gegebene unstetige Funktion, und $\Psi(\mu)=0$. Die Fouriersche Integralformel (11) verliert natürlich hier jeden Sinn.

Sei sodann $f(x)$ die Funktion:

(31) $$f_0(x) = \begin{cases} 1 & \text{für } x > 0 \\ 0 & \text{für } x = 0 \\ -1 & \text{für } x < 0. \end{cases}$$

Hier wird $\Phi(\mu)=0$, und (für $\mu > 0$):

$$\Psi(\mu) = 2\int_0^1 \frac{1-\cos \mu x}{x}\, dx - 2\int_1^{+\infty} \frac{\cos \mu x}{x}\, dx$$

$$= 2\int_0^1 \frac{1-\cos \mu x}{x}\, dx - 2\int_{\mu}^{+\infty} \frac{\cos x}{x}\, dx.$$

Es ist also:

$$\Psi'(\mu) = 2\int_0^1 \sin \mu x\, dx + 2\,\frac{\cos \mu}{\mu} = \frac{2}{\mu},$$

und somit (für $\mu > 0$):

$$\Psi(\mu) = \lg \mu^2 + C.$$

Es ist also für $0 < h < \lambda$:

$$\int_h^\lambda \sin \mu x_0 \, d\Psi(\mu) = 2 \int_h^\lambda \frac{\sin \mu x_0}{\mu} \, d\mu .$$

Vollziehen wir hierin die Grenzübergänge $\lambda \to +\infty$, $h \to +0$, so wird daraus:

$$\int_0^{+\infty} \sin \mu x_0 \, d\Psi(\mu) = 2 \int_0^{+\infty} \frac{\sin \mu x_0}{\mu} \, d\mu = \begin{cases} \pi & \text{für } x_0 > 0 \\ 0 & \text{für } x_0 = 0 \\ -\pi & \text{für } x_0 < 0. \end{cases}$$

Daraus ersehen wir: *auch für die Funktion* (31) *gilt Formel* (10); das in ihr auftretende Integral $\int_0^{+\infty}$ ist dabei im Sinne des Grenzwertes $\lim\limits_{h \to +0} \int_h^{+\infty}$ zu verstehen.

Zugleich sehen wir, dass das bekannte Integral $\dfrac{2}{\pi} \displaystyle\int_0^{+\infty} \frac{\sin \mu x_0}{\mu} \, d\mu$ in gewissem Sinne als (verallgemeinerte) Fouriersche Integraldarstellung der Funktion (31) aufgefasst werden kann.

Nun können wir ganz allgemein den Satz aussprechen:

Satz VI. *Für jede im Unendlichen beschränkte und monotone Funktion* $f(x)$ *der Eigenschaft* A) *gilt Formel* (10); *darin bedeutet* $\Phi(\mu)$ *eine für* $\mu \geqq 0$ *definierte und für* $\mu > 0$ *stetige Funktion, die den Grenzwert* $\Phi(+0)$ *besitzt, während* $\Psi(\mu)$ *für* $\mu > 0$ *definiert und stetig ist; das in* (10) *auftretende Integral ist im folgenden Sinne zu verstehen:*

$$\int_0^1 \cos \mu x_0 \, d\Phi(\mu) + \lim_{h \to +0} \int_h^1 \sin \mu x_0 \, d\Psi(\mu) + \int_1^{+\infty} (\cos \mu x_0 \, d\Phi(\mu) + \sin \mu x_0 \, d\Psi(\mu)).$$

In der Tat, da $f(x)$ im Unendlichen monoton und beschränkt ist, existieren endliche Grenzwerte:

$$\lim_{x \to +\infty} f(x) = c_1, \qquad \lim_{x \to -\infty} f(x) = c_2.$$

Wir bezeichnen wieder mit $f_0(x)$ die Funktion (31) und setzen:

$$f(x) = \frac{c_1 + c_2}{2} + \frac{c_1 - c_2}{2} f_0(x) + f_1(x). \tag{32}$$

Dann ist $f_1(x)$ im Unendlichen monoton mit:

$$\lim_{x \to +\infty} f_1(x) = 0, \qquad \lim_{x \to -\infty} f_1(x) = 0.$$

Nach Satz V und den zu Beginn dieses Paragrafen gefundenen Resultaten ist jeder der drei Summanden auf der rechten Seite von (32) der Darstellung (10 fähig; daraus erhält man sofort Satz VI. Offenbar ist, wenn man (30) beachtet:

$$\Phi(+0) = \frac{c_1 + c_2}{2} \pi.$$

§ 9.

Sei nun:

$$f(x) = \cos qx \qquad (q > 0).$$

Dann ist $\Psi(\mu) = 0$ und:

$$(33) \qquad \Phi(\mu) = \int_{-\infty}^{+\infty} \cos qx \frac{\sin \mu x}{x} dx$$

$$= \frac{1}{2} \int_{-\infty}^{+\infty} \frac{\sin (\mu + q)x}{x} dx + \frac{1}{2} \int_{-\infty}^{+\infty} \frac{\sin (\mu - q)x}{x} dx = \begin{cases} 0 & \text{für } 0 \leqq \mu < q \\ \frac{\pi}{2} & \text{für } \mu = q \\ \pi & \text{für } \mu > q. \end{cases}$$

Es ist also für jedes $\lambda > q$:

$$\int_0^\lambda (\cos \mu x_0 \, d\Phi(\mu) + \sin \mu x_0 \, d\Psi(\mu)) = \pi \cos qx_0,$$

d. h. *für die Funktion* $\cos qx \, (q > 0)$ *gilt Formel* (10); es ist in ihr $\Phi(\mu)$ die durch (33) gegebene unstetige Funktion, und $\Psi(\mu) = 0$.

Sei sodann:

$$f(x) = \sin qx \qquad (q > 0).$$

Dann ist $\Phi(\mu) = 0$ und:

$$\Psi(\mu) = \int_{-1}^{1} \sin qx \frac{1 - \cos \mu x}{x} dx - \int_{1}^{+\infty} \sin qx \frac{\cos \mu x}{x} dx - \int_{-\infty}^{-1} \sin qx \frac{\cos \mu x}{x} dx$$

$$= \int_{-1}^{1} \frac{\sin qx}{x} dx - \int_{-\infty}^{+\infty} \frac{\sin qx \cos \mu x}{x} dx$$

$$= \int_{-1}^{1} \frac{\sin qx}{x} dx - \frac{1}{2} \int_{-\infty}^{+\infty} \frac{\sin (q+\mu)x}{x} dx - \frac{1}{2} \int_{-\infty}^{+\infty} \frac{\sin (q-\mu)x}{x} dx,$$

mithin:

$$\Psi(\mu) = \int_{-1}^{1} \frac{\sin qx}{x} dx - \begin{cases} \pi & \text{für } 0 \leqq \mu < q \\ \frac{\pi}{2} & \text{für } \mu = q \\ 0 & \text{für } \mu > q. \end{cases} \tag{34}$$

Es ist also für jedes $\lambda > q$:

$$\int_{0}^{\lambda} (\cos \mu x_0 \, d\Phi(\mu) + \sin \mu x_0 \, d\Psi(\mu)) = \pi \sin q x_0,$$

d. h. *für die Funktion* $\sin qx$ $(q > 0)$ *gilt Formel* (10); es ist in ihr $\Phi(\mu) = 0$ und $\Psi(\mu)$ die unstetige Funktion (34).

§ 10.

Wir bezeichnen wieder mit $f_0(x)$ die Funktion (31), setzen

$$f(x) = f_0(x) \cos qx \qquad (q > 0),$$

und wollen zeigen, dass auch für diese Funktion Formel (10) gilt. Zunächst beweisen wir die Giltigkeit der Formel (1). Für $|x_0| < \xi < \eta$ ist:

$$\int_{x_0+\xi}^{x_0+\eta} f(x) \frac{\sin \lambda(x-x_0)}{x-x_0} dx = \int_{\xi}^{\eta} f(x_0+x) \frac{\sin \lambda x}{x} dx = \int_{\xi}^{\eta} \cos q(x_0+x) \frac{\sin \lambda x}{x} dx$$

$$= \frac{1}{2} \cos q x_0 \left(\int_{\xi}^{\eta} \frac{\sin (\lambda+q)x}{x} dx + \int_{\xi}^{\eta} \frac{\sin (\lambda-q)x}{x} dx \right) +$$

$$+ \frac{1}{2} \sin q x_0 \left(\int_{\xi}^{\eta} \frac{\cos (\lambda+q)x}{x} dx - \int_{\xi}^{\eta} \frac{\cos (\lambda-q)x}{x} dx \right).$$

Wir entnehmen daraus: zu jedem $\varepsilon > 0$ und $\delta > 0$ gibt es ein b, so dass für $b \leqq \xi < \eta$ und alle der Bedingung $|\lambda - q| \geqq \delta$ genügenden nicht-negativen λ die Ungleichung gilt:

$$\left| \int_{x_0+\xi}^{x_0+\eta} f(x) \frac{\sin \lambda(x-x_0)}{x-x_0} dx \right| < \varepsilon .$$

Da Analoges für $\int_{x_0-\eta}^{x_0-\xi} f(x) \frac{\sin \lambda(x-x_0)}{x-x_0} dx$ gilt, sehen wir, dass für alle nicht-negativen $\lambda \neq q$ das Integral (1) existiert, und dass für $\lambda \geqq q + \delta$ die Ungleichung gilt:

$$\left| \int_{x_0+b}^{+\infty} f(x) \frac{\sin \lambda(x-x_0)}{x-x_0} dx \right| \leqq \varepsilon .$$

Indem man von hier aus weiter schliesst, wie in § 1, beweist man die Giltigkeit von (1).

Das Verhalten des Integrales (1) an der Stelle $\lambda = q$ wird charakterisiert durch die Beziehung:

$$(35) \qquad \lim_{h \to 0} \left(\int_{-\infty}^{+\infty} f(x) \frac{\sin (q+h)(x-x_0)}{x-x_0} dx - \int_{-\infty}^{+\infty} f(x) \frac{\sin (q-h)(x-x_0)}{x-x_0} dx \right) = 0 .$$

In der Tat, jedenfalls ist (für jedes $b > 0$):

$$(36) \qquad \lim_{h \to 0} \left(\int_{x_0-b}^{x_0+b} f(x) \frac{\sin (q+h)(x-x_0)}{x-x_0} dx - \int_{x_0-b}^{x_0+b} f(x) \frac{\sin (q-h)(x-x_0)}{x-x_0} dx \right) = 0 .$$

Ferner ist für $b > |x_0|$:

$$\int_{x_0+b}^{+\infty} + \int_{-\infty}^{x_0-b} f(x) \frac{\sin (q+h)(x-x_0) - \sin (q-h)(x-x_0)}{x-x_0} dx$$

$$= \int_{b}^{+\infty} (\cos q(x_0+x) - \cos q(x_0-x)) \frac{\sin (q+h)x - \sin (q-h)x}{x} dx$$

$$= -4 \sin q x_0 \int_b^{+\infty} \frac{\sin qx \cos qx \sin hx}{x} dx$$

$$= \sin q x_0 \left(\int_b^{+\infty} \frac{\cos (2q+h)x}{x} dx - \int_b^{+\infty} \frac{\cos (2q-h)x}{x} dx \right).$$

Daraus entnehmen wir: zu jedem $\varepsilon > 0$ gibt es ein b, so dass für $0 < h \leqq q$:

$$\left| \int_{x_0+b}^{+\infty} + \int_{-\infty}^{x_0-b} f(x) \frac{\sin (q+h)(x-x_0) - \sin (q-h)(x-x_0)}{x-x_0} dx \right| < \varepsilon.$$

Diese Ungleichung zusammen mit (36) aber ergibt die Behauptung (35).

Nun wenden wir uns zur Untersuchung der zu unsrer Funktion $f(x)$ gehörigen Funktionen $\Phi(\mu)$ und $\Psi(\mu)$. Da $f(x)$ ungerade, ist $\Phi(\mu) = 0$. Für $\mu \geqq 0$ und $\neq q$ ist:

$$\Psi(\mu) = 2 \int_0^1 \cos qx \frac{1 - \cos \mu x}{x} dx - 2 \int_1^{+\infty} \cos qx \frac{\cos \mu x}{x} dx$$

$$= 2 \int_0^1 \cos qx \frac{1 - \cos \mu x}{x} dx - \int_1^{+\infty} \left(\frac{\cos (\mu+q)x}{x} + \frac{\cos (\mu-q)x}{x} \right) dx$$

$$= 2 \int_0^1 \cos qx \frac{1 - \cos \mu x}{x} dx - \int_{\mu+q}^{+\infty} \frac{\cos x}{x} dx - \int_{|\mu-q|}^{+\infty} \frac{\cos x}{x} dx,$$

woraus sich ergibt:

$$(37) \qquad \Psi'(\mu) = 2 \int_0^1 \cos qx \sin \mu x \, dx + \frac{\cos (\mu+q)}{\mu+q} + \frac{\cos (\mu-q)}{\mu-q} = \frac{2\mu}{\mu^2 - q^2},$$

und somit:

$$(38) \qquad \Psi(\mu) = \lg |\mu^2 - q^2| + C.$$

Für das Verhalten von $\Psi(\mu)$ an der Stelle q folgt hieraus:

$$\lim_{h \to 0} (\Psi(q+h) - \Psi(q-h)) = 0.$$

Betrachten wir die durch (27) definierte Funktion $\Psi_n(\mu)$, so ist:

$$\Psi_n(\mu)-\Psi(\mu)=2\int_n^{+\infty}\cos qx\frac{\cos\mu x}{x}\,dx=\int_{n(\mu+q)}^{+\infty}\frac{\cos x}{x}\,dx+\int_{n|\mu-q|}^{+\infty}\frac{\cos x}{x}\,dx,$$

woraus sich ergibt, dass $\Psi_n(\mu)$ gleichmässig gegen $\Psi(\mu)$ konvergiert in jedem Intervalle $(0\leqq)\ \alpha\leqq\mu\leqq\beta$, das den Punkt q nicht enthält.

Nun ist nach Schlüssen, die schon wiederholt durchgeführt wurden, für jedes Intervall $(0\leqq)\ \alpha\leqq\mu\leqq\beta$:

$$\int_{-\infty}^{+\infty}f(x)\frac{\sin\beta(x-x_0)}{x-x_0}\,dx-\int_{-\infty}^{+\infty}f(x)\frac{\sin\alpha(x-x_0)}{x-x_0}\,dx=\lim_{n\to\infty}\int_\alpha^\beta\sin\mu x_0\,d\Psi_n(\mu),$$

mithin wenn $[\alpha,\beta]$ den Punkt q nicht enthält:

$$\int_{-\infty}^{+\infty}f(x)\frac{\sin\beta(x-x_0)}{x-x_0}\,dx-\int_{-\infty}^{+\infty}f(x)\frac{\sin\alpha(x-x_0)}{x-x_0}\,dx=\int_\alpha^\beta\sin\mu x_0\,d\Psi(\mu).$$

Für $\lambda>q$ und $0<h<q$ erhalten wir, indem wir diese Formel auf die Intervalle $[0,\ q-h]$ und $[q+h,\ \lambda]$ anwenden:

$$\int_{-\infty}^{+\infty}f(x)\frac{\sin\lambda(x-x_0)}{x-x_0}\,dx-\left(\int_{-\infty}^{+\infty}f(x)\frac{\sin(q+h)(x-x_0)}{x-x_0}\,dx-\int_{-\infty}^{+\infty}f(x)\frac{\sin(q-h)(x-x_0)}{x-x_0}\,dx\right)$$
$$=\int_0^{q-h}\sin\mu x_0\,d\Psi(\mu)+\int_{q+h}^{\lambda}\sin\mu x_0\,d\Psi(\mu).$$

Vollziehen wir hierin den Grenzübergang $h\to+0$ und setzen:

$$\lim_{h\to+0}\left(\int_0^{q-h}\sin\mu x_0\,d\Psi(\mu)+\int_{q+h}^{\lambda}\sin\mu x_0\,d\Psi(\mu)\right)=\int_0^{\lambda}\sin\mu x_0\,d\Psi(\mu),\tag{39}$$

so erhalten wir bei Beachtung von (35):

$$\int_{-\infty}^{+\infty}f(x)\frac{\sin\lambda(x-x_0)}{x-x_0}\,dx=\int_0^{\lambda}\sin\mu x_0\,d\Psi(\mu).$$

Und da, wie bewiesen, Formel (1) gilt, folgt daraus durch den Grenzübergang $\lambda \to +\infty$ der Satz:

Bedeutet $f_0(x)$ *die Funktion* (31), *so gilt für* $f_0(x) \cos qx$ $(q>0)$ *die Formel* (10), *in deren Integral gemäss* (39) *an der Stelle* $\lambda=q$ *der Cauchysche Hauptwert zu nehmen ist*; es ist in ihr $\Phi(\mu)=0$ und $\Psi(\mu)$ die Funktion (38).

Wegen (37) kann diese Formel auch so geschrieben werden:

$$(40) \qquad \int_0^{+\infty} \sin \mu x \frac{\mu}{\mu^2-q^2} d\mu = \begin{cases} \frac{\pi}{2} \cos qx & \text{für } x>0 \\ 0 & \text{für } x=0, \\ -\frac{\pi}{2} \cos qx & \text{für } x<0, \end{cases}$$

wobei links wieder an der Stelle q der Cauchysche Hauptwert zu nehmen ist. Formel (40), die natürlich auch durch direkte Rechnung gewonnen werden kann, ist in gewissem Sinne die Fouriersche Integraldarstellung der auf ihrer rechten Seite stehenden Funktion.

§ 11.

In ganz derselben Weise kann der Fall:

$$f(x)=f_0(x) \sin qx \qquad (q>0)$$

behandelt werden, wo wieder $f_0(x)$ die Funktion (31) bedeutet. Ganz wie in § 10 beweist man die Giltigkeit der Formeln (1) und (35).

Da diesmal $f(x)$ gerade, ist $\Psi(\mu)=0$. Für $\mu \geqq 0$ und $\neq q$ ist:

$$\Phi(\mu)=2\int_0^{+\infty} \sin qx \frac{\sin \mu x}{x} dx = 2\int_0^1 \frac{\sin qx \sin \mu x}{x} dx + \int_{|\mu-q|}^{+\infty} \frac{\cos x}{x} dx - \int_{\mu+q}^{+\infty} \frac{\cos x}{x} dx,$$

woraus sich ergibt:

$$(41) \quad \Phi'(\mu)=2\int_0^1 \sin qx \cos \mu x \, dx + \frac{\cos(\mu+q)}{\mu+q} - \frac{\cos(\mu-q)}{\mu-q} = \frac{1}{q+\mu} + \frac{1}{q-\mu} = \frac{2q}{q^2-\mu^2},$$

und somit:

$$(42) \qquad \Phi(\mu) = \lg\left|\frac{q+\mu}{q-\mu}\right| + C.$$

Für das Verhalten von $\Phi(\mu)$ an der Stelle q gilt auch hier:

$$\lim_{h\to 0}(\Phi(q+h)-\Phi(q-h))=0.$$

Indem man weiter schliesst, wie in § 10, findet man:

Bedeutet $f_0(x)$ *die Funktion* (31), *so gilt für* $f(x)=f_0(x)\sin qx (q>0)$ *die Formel* (10), *in deren Integral an der Stelle* $\lambda=q$ *der Cauchysche Hauptwert zu nehmen ist;* es ist in ihr $\Phi(\mu)$ die Funktion (42) und $\Psi(\mu)=0$.

Wegen (41) kann diese Formel auch geschrieben werden:

$$\int_0^{+\infty}\cos\mu x\frac{q}{q^2-\mu^2}d\mu=\begin{cases}\frac{\pi}{2}\sin qx \text{ für } x\geqq 0\\ -\frac{\pi}{2}\sin qx \text{ für } x\leqq 0\end{cases}\qquad (q>0)$$

wobei links wieder an der Stelle q der Cauchysche Hauptwert zu nehmen ist.

§ 12.

Nun nehmen wir $f(x)$ in der Form an:

$$f_1(x)=g(x)\cos qx \quad \text{oder} \quad f_2(x)=g(x)\sin qx \qquad (q>0),$$

wo $g(x)$ *im Unendlichen monoton gegen* 0 *konvergiere:*

$$\lim_{x\to+\infty}g(x)=0;\ \lim_{x\to-\infty}g(x)=0.$$

Sei $g(x)$ monoton für $x>a$ und $x<-a$. Wir untersuchen die zu $f_1(x)$ gehörige Funktion $\Phi(\mu)$:

$$\Phi^{(1)}(\mu)=\int_{-\infty}^{+\infty}g(x)\cos qx\frac{\sin\mu x}{x}dx.$$

Da für $a<\xi<\eta$:

$$\int_\xi^\eta g(x)\cos qx\frac{\sin\mu x}{x}dx=\frac{1}{2}g(\xi)\left(\int_\xi^{\xi^*}\frac{\sin(\mu+q)x}{x}dx+\int_\xi^{\xi^*}\frac{\sin(\mu-q)x}{x}dx\right),$$

gibt es zu jedem $\varepsilon>0$ ein b, so dass für $b\leqq\xi<\eta$ und alle $\mu\geqq 0$:

$$\left|\int_{\xi}^{\eta} g(x)\cos qx \frac{\sin \mu x}{x}\,dx\right| < \varepsilon; \quad \left|\int_{-\eta}^{-\xi} g(x)\cos qx \frac{\sin \mu x}{x}\,dx\right| < \varepsilon.$$

Daraus folgt: für alle μ existiert $\Phi^{(1)}(\mu)$, und die durch (27) definierte Funktion $\Phi_n(\mu)$ konvergiert für alle μ gleichmässig gegen $\Phi^{(1)}(\mu)$. *Es ist also* $\Phi^{(1)}(\mu)$ *eine für alle* $\mu \geqq 0$ *definierte und stetige Funktion.*

Nun untersuchen wir die zu $f_1(x)$ gehörige Funktion $\Psi(\mu)$:

$$\Psi^{(1)}(\mu) = \int_{-1}^{1} g(x)\cos qx \frac{1-\cos \mu x}{x}\,dx - \int_{1}^{+\infty} g(x)\cos qx \frac{\cos \mu x}{x}\,dx$$
$$- \int_{-\infty}^{-1} g(x)\cos qx \frac{\cos \mu x}{x}\,dx.$$

Hier haben wir für $a<\xi<\eta$ und $\mu \geqq 0$ und $\neq q$:

$$\int_{\xi}^{\eta} g(x)\cos qx \frac{\cos \mu x}{x}\,dx = \frac{1}{2}g(\xi)\left(\int_{\xi}^{\xi^*} \frac{\cos(\mu+q)x}{x}\,dx + \int_{\xi}^{\xi^*} \frac{\cos(\mu-q)x}{x}\,dx\right)$$
$$= \frac{1}{2}g(\xi)\left(\int_{(\mu+q)\xi}^{(\mu+q)\xi^*} \frac{\cos x}{x}\,dx + \int_{|\mu-q|\xi}^{|\mu-q|\xi^*} \frac{\cos x}{x}\,dx\right);$$

und somit gibt es zu jedem $\varepsilon>0$ und $\delta>0$ ein b, so dass für $b \leqq \xi < \eta$ und $\mu \geqq 0$, $|\mu-q| \geqq \delta$:

$$\left|\int_{\xi}^{\eta} g(x)\cos qx \frac{\cos \mu x}{x}\,dx\right| < \varepsilon, \quad \left|\int_{-\eta}^{-\xi} g(x)\cos qx \frac{\cos \mu x}{x}\,dx\right| < \varepsilon.$$

Daraus folgt: für alle $\mu \geqq 0$ und $\neq q$ existiert $\Psi^{(1)}(\mu)$, und die durch (27) definierte Funktion $\Psi_n(\mu)$ konvergiert in jedem Intervalle $(0 \leqq)\,\alpha \leqq \mu \leqq \beta$, das den Punkt q nicht enthält, gleichmässig gegen $\Psi^{(1)}(\mu)$. *Es ist also* $\Psi^{(1)}(\mu)$ *eine für alle* $\mu \geqq 0$ *und* $\neq q$ *definierte und stetige Funktion.*

Für das Verhalten von $\Psi^{(1)}(\mu)$ an der Stelle q finden wir:

$$\lim_{h\to 0}\left(\Psi^{(1)}(q+h) - \Psi^{(1)}(q-h)\right) = 0. \tag{43}$$

In der Tat, es ist:

$$\Psi^{(1)}(q+h)-\Psi^{(1)}(q-h)=-\int_{-\infty}^{+\infty} g(x) \cos qx \frac{\cos(q+h)x-\cos(q-h)x}{x} dx$$

$$=-\frac{1}{2}\int_{-\infty}^{+\infty} g(x) \frac{\cos(2q+h)x-\cos(2q-h)x}{x} dx.$$

Bezeichnen wir die zu $g(x)$ gehörige Funktion Ψ mit $\Psi^{(0)}(\mu)$, so ist also:

$$(44) \qquad \Psi^{(1)}(q+h)-\Psi^{(1)}(q-h)=\frac{1}{2}(\Psi^{(0)}(2q+h)-\Psi^{(0)}(2q-h)).$$

Da $g(x)$ im Unendlichen monoton gegen 0 konvergiert, ist aber zufolge § 7 $\Psi^{(0)}(\mu)$ stetig für $\mu>0$. Durch (44) ist also (43) bewiesen.

Völlig analoge Rechnungen ergeben für die zu $f_2(x)=g(x) \sin qx$ gehörigen Funktionen $\Phi(\mu)$ und $\Psi(\mu)$: *es ist* $\Phi^{(2)}(\mu)$ *definiert und stetig für* $\mu \geqq 0$ *und* $\neq q$, *während* $\Psi^{(2)}(\mu)$ *für alle* $\mu \geqq 0$ *definiert und stetig ist; und es ist:*

$$(45) \qquad \lim_{h\to 0} (\Phi^{(2)}(q+h)-\Phi^{(2)}(q-h))=0.$$

Nun wollen wir zeigen, dass auch für die jetzt betrachteten Funktionen $f_1(x)$ und $f_2(x)$, (falls sie die Eigenschaft A) haben) Formel (1) gilt. Wir setzen zu dem Zwecke:

$$I^{(1)}(\lambda)=\int_{-\infty}^{+\infty} f_1(x)\frac{\sin\lambda(x-x_0)}{x-x_0}dx; \quad I^{(2)}(\lambda)=\int_{-\infty}^{+\infty} f_2(x)\frac{\sin\lambda(x-x_0)}{x-x_0}dx.$$

Wir führen die Untersuchung für $f_1(x)$ durch; für $f_2(x)$ geht sie analog vor sich. Es ist:

$$I^{(1)}(\lambda)=\int_{-\infty}^{+\infty} g(x_0+x)\cos q(x_0+x)\frac{\sin\lambda x}{x}dx$$

$$=\cos qx_0\int_{-\infty}^{+\infty} g(x_0+x)\cos qx\frac{\sin\lambda x}{x}dx-\sin qx_0\int_{-\infty}^{+\infty} g(x_0+x)\sin qx\frac{\sin\lambda x}{x}dx.$$

Da ebenso wie $g(x)$ auch $g(x_0+x)$ im Unendlichen monoton gegen o konvergiert, folgt aus dem oben über $\Phi^{(1)}$ und $\Phi^{(2)}$ bewiesenen, dass das Integral

$$\int_{-\infty}^{+\infty} g(x_0+x) \cos qx \frac{\sin \lambda x}{x} dx \qquad \text{für alle } \lambda \geqq 0,$$

das Integral

$$\int_{-\infty}^{+\infty} g(x_0+x) \sin qx \frac{\sin \lambda x}{x} dx \qquad \text{für alle } \lambda \geqq 0 \text{ und } \neq q$$

existiert und stetig ist. Es existiert also das Integral $I^{(1)}(\lambda)$ und ist stetig für alle $\lambda \geqq 0$ und $\neq q$. Hinsichtlich seines Verhaltens für $\lambda = q$ entnehmen wir aus (45):

$$\lim_{h \to 0} (I^{(1)}(q+h) - I^{(1)}(q-h)) = 0.$$

Endlich zeigen wir mit Hilfe wiederholt durchgeführter Schlüsse: zu jedem $\varepsilon > 0$ und $\delta > 0$ gehört ein b, so dass für alle der Ungleichung $\lambda \geqq q + \delta$ genügenden λ:

$$\left| \int_{b}^{+\infty} f_1(x) \frac{\sin \lambda(x-x_0)}{x-x_0} dx \right| < \varepsilon; \qquad \left| \int_{-\infty}^{-b} f_1(x) \frac{\sin \lambda(x-x_0)}{x-x_0} dx \right| < \varepsilon;$$

von hier aus schliesst man wie in § 1: *genügt* $f_1(x)$ *(oder* $f_2(x)$*) der Bedingung* A), *so gilt für* $f_1(x)$ *(bzw.* $f_2(x)$*) die Formel* (1).

Ganz wie in § 10 erhalten wir nun den Satz: *Bedeutet* $g(x)$ *eine im Unendlichen monoton gegen* o *konvergierende Funktion, so gilt für jede der beiden Funktionen* $g(x) \cos qx$ *und* $g(x) \sin qx (q > 0)$, *falls sie die Eigenschaft* A) *hat, Formel* (10), *in der an der Stelle* $\lambda = q$ *der Cauchysche Hauptwert zu nehmen ist.*[1]

Wie schon wiederholt, können wir hier weiter schliessen, dass unter den genannten Voraussetzungen auch die Fouriersche Integralformel (11) gilt, in der wieder an der Stelle $\lambda = q$ der Cauchysche Hauptwert zu nehmen ist.[2]

[1] Genauer gesprochen: da im Falle $g(x) \cos qx$ die Funktion $\Phi(\mu)$, im Falle $g(x) \sin qx$ die Funktion $\Psi(\mu)$ für $\mu = q$ stetig bleibt, muss im ersten Falle nur vom zweiten, im zweiten Falle nur vom ersten Summanden des Integranden $\cos \mu x_0 \, d\Phi(\mu) + \sin \mu x_0 \, d\Psi(\mu)$ der Cauchysche Hauptwert gebildet werden.

[2] A. Pringsheim, Math. Ann. 68 (1910), S. 398, 399.

§ 13.

Wir fassen nun die Resultate der §§ 9—12 zusammen. Sei:

$$f(x) = g(x) \cos qx \qquad (q > 0),$$

wo $g(x)$ im Unendlichen *monoton* und *beschränkt* sei. Wir setzen wie in (32):

$$g(x) = \frac{c_1 + c_2}{2} + \frac{c_1 - c_2}{2} f_0(x) + g_1(x),$$

wo $f_0(x)$ die Funktion (31) bedeutet, und:

$$c_1 = \lim_{x \to +\infty} g(x), \qquad c_2 = \lim_{x \to -\infty} g(x);$$

dann konvergiert $g_1(x)$ im Unendlichen monoton gegen 0.

Wir bezeichnen nun die zu $\cos qx, f_0(x) \cos qx, g_1(x) \cos qx$ gehörigen Funktionen Φ und Ψ der Reihe nach mit $\Phi^*(\mu)$, $\Psi^*(\mu)$; $\Phi^{**}(\mu)$, $\Psi^{**}(\mu)$; $\Phi^{***}(\mu)$, $\Psi^{***}(\mu)$. Da $\Psi^*(\mu) = 0$ und $\Phi^{**}(\mu) = 0$ ist, erhalten wir für die zu $f(x)$ gehörigen Funktionen Φ und Ψ:

$$\Phi(\mu) = \frac{c_1 + c_2}{2} \Phi^*(\mu) + \Phi^{***}(\mu); \qquad \Psi(\mu) = \frac{c_1 - c_2}{2} \Psi^{**}(\mu) + \Psi^{***}(\mu),$$

und die Resultate der §§ 9—12 ergeben: $\Phi^*(\mu)$ ist die Funktion (33); $\Phi^{***}(\mu)$ ist stetig für $\mu \geqq 0$; $\Psi^{**}(\mu)$ und $\Psi^{***}(\mu)$ sind stetig für $\mu \geqq 0$ und $\neq q$, und es ist:

$$\lim_{h \to 0} (\Psi^{**}(q+h) - \Psi^{**}(q-h)) = 0; \qquad \lim_{h \to 0} (\Psi^{***}(q+h) - \Psi^{***}(q-h)) = 0;$$

und wir erhalten den Satz:

Satz VII. *Für jede Funktion $f(x)$ der Eigenschaft* A), *die die Gestalt hat $f(x) = g(x) \cos qx$ $(q > 0)$, wo $g(x)$ im Unendlichen beschränkt und monoton ist, gilt Formel* (10):

$$f(x_0) = \frac{1}{\pi} \left(\int_0^{+\infty} \cos \mu x_0 \, d\Phi(\mu) + \sin \mu x_0 \, d\Psi(\mu) \right),$$

wo im zweiten Summanden an der Stelle q der Cauchysche Hauptwert zu nehmen ist.

Genau so finden wir:

Satz VII a. *Die Aussage von Satz VII gilt auch für* $f(x)=g(x)\sin qx$ $(q>0)$, *nur dass nun im ersten Summanden an der Stelle* q *der Cauchysche Hauptwert zu nehmen ist.*

§ 14.

Wir betrachten nun allgemein den Fall, dass $f(x)$ die Gestalt hat:

$$f(x)=g(x)\,h(x),$$

wo $g(x)$ im Unendlichen *monoton gegen* 0 *konvergiert*, und $h(x)$ eine *periodische Funktion* bedeutet.

Ohne Beschränkung der Allgemeinheit können wir annehmen, $h(x)$ habe die Periode 2π. Die Funktion $g(x)$ sei für $x>a$ monoton.

Wir untersuchen zunächst die zu $f(x)$ gehörige Funktion $\Phi(\mu)$. Sei:

$$h(x)\sim\frac{a_0}{2}+\sum_{k=1}^{\infty}(a_k\cos kx+b_k\sin kx) \tag{46}$$

die Fouriersche Reihe von $h(x)$. Sei N irgend eine natürliche Zal. Wir setzen:

$$h^*(x)=\frac{a_0}{2}+\sum_{k=1}^{N}(a_k\cos kx+b_k\sin kx)$$

und nehmen die Zerlegung vor:

$$h(x)=h^*(x)+h^{**}(x).$$

Entsprechend zerlegt sich dann $\Phi(\mu)$ in:

$$\Phi(\mu)=\Phi^*(\mu)+\Phi^{**}(\mu).$$

Über $\Phi^*(\mu)$ sind wir durch § 7 und § 12 völlig orientiert: es existiert $\Phi^*(\mu)$ und ist stetig für alle $\mu\geqq 0$ und $\neq 1, 2, \ldots, N$, und es ist:

$$\lim_{h\to 0}(\Phi^*(k+h)-\Phi^*(k-h))=0 \qquad (k=1, 2, \ldots, N).$$

Wir haben noch $\Phi^{**}(\mu)$ zu untersuchen.

Es ist nach dem zweiten Mittelwertsatze für $a<\xi<\eta$:

$$\int_{\xi}^{\eta} g(x)\,h^{**}(x)\frac{\sin\mu x}{x}\,dx=\frac{g(\xi)}{\xi}\int_{\xi}^{\xi^*}h^{**}(x)\sin\mu x\,dx. \tag{47}$$

Nach (23) aber ist, wenn n eine geeignete ganze Zahl bedeutet:

$$\int_{\xi}^{\xi^*} h^{**}(x)\, e^{i\mu x}\, dx = \int_{\xi}^{\xi+2\pi} h^{**}(x)\, e^{i\mu x}\, dx \cdot \frac{e^{2\pi i\mu n} - 1}{e^{2\pi i\mu} - 1} + \int_{\xi+2n\pi}^{\xi^*} h^{**}(x)\, e^{i\mu x}\, dx,$$

und hierin ist:

$$|e^{2\pi i\mu n} - 1| \leqq 2; \qquad \left| \int_{\xi+2n\pi}^{\xi^*} h^{**}(x)\, e^{i\mu x}\, dx \right| \leqq \int_{\xi}^{\xi+2\pi} |h^{**}(x)|\, dx.$$

Ferner ist offenbar:

$$\int_{\xi}^{\xi+2\pi} h^{**}(x)\, e^{ikx}\, dx = 0 \qquad (k = 0, 1, 2, \ldots, N).$$

Es ist also

$$\int_{\xi}^{\xi+2\pi} h^{**}(x)\, e^{i\mu x}\, dx \cdot \frac{1}{e^{2\pi i\mu} - 1}$$

eine analytische Funktion von μ, die für $\mu = 0, 1, 2, \ldots, N$ regulär ist. Aus der Periodizität von $e^{2\pi i\mu}$ und der Ungleichung

$$\left| \int_{\xi}^{\xi+2\pi} h^{**}(x)\, e^{i\mu x}\, dx \right| \leqq \int_{\xi}^{\xi+2\pi} |h^{**}(x)|\, dx$$

folgt nun sofort: Zu jedem $\delta > 0$ gibt es ein M, sodass für alle Intervalle $[\xi, \xi^*]$ und alle $\mu \geqq 0$, die keinem der Intervalle $|\mu - k| < \delta$ $(k = N+1, N+2, \ldots)$ angehören, die Ungleichungen gelten:

$$(48) \qquad \left| \int_{\xi}^{\xi^*} h^{**}(x) \cos \mu x\, dx \right| < M; \qquad \left| \int_{\xi}^{\xi^*} h^{**}(x) \sin \mu x\, dx \right| < M.$$

Aus (47) ergibt sich nun: zu jedem $\varepsilon > 0$ und $\delta > 0$ gibt es ein b, sodass für $b \leqq \xi < \eta$ und für alle $\mu \geqq 0$, die keinem der Intervalle $|\mu - k| < \delta$ $(k = N+1, N+2 \ldots)$ angehören, die Ungleichungen gelten:

$$\left| \int_{\xi}^{\eta} g(x)\, h^{**}(x) \frac{\sin \mu x}{x}\, dx \right| < \varepsilon, \qquad \left| \int_{-\eta}^{-\xi} g(x)\, h^{**}(x) \frac{\sin \mu x}{x}\, dx \right| < \varepsilon.$$

Daraus schliessen wir in bekannter Weise: die Funktion $\Phi^{**}(\mu)$ existiert und ist stetig für alle $\mu \geqq 0$, die $\neq N+1, N+2, \ldots$ sind.

Zusammen mit dem vorhin über $\Phi^*(\mu)$ festgestellten ergibt das, da in den obigen Überlegungen die natürliche Zahl N ganz beliebig war: *Die Funktion $\Phi(\mu)$ existiert und ist stetig für alle $\mu \geqq 0$, die $\neq 1, 2, \ldots$ sind, und es ist:*

$$\lim_{h\to 0}(\Phi(k+h)-\Phi(k-h))=0 \qquad (k=1, 2, \ldots).$$

Ganz ebenso beweist man: *Die Funktion $\Psi(\mu)$ existiert und ist stetig für alle $\mu>0$, die $\neq 1, 2, \ldots$ sind, und es ist:*

$$\lim_{h\to 0}(\Psi(k+h)-\Psi(k-h))=0 \qquad (k=1, 2, \ldots).$$

Nun beweisen wir wieder die Giltigkeit der Formel (1). Das Integral:

$$I(\lambda)=\int_{-\infty}^{+\infty} f(x)\frac{\sin\lambda(x-x_0)}{x-x_0}\,dx=\int_{-\infty}^{+\infty} f(x_0+x)\frac{\sin\lambda x}{x}\,dx$$

ist die zu $f(x_0+x)$ gehörige Funktion Φ. Nach dem eben Bewiesenen existiert es also und ist stetig für $\lambda \geqq 0$ und $\neq 1, 2 \ldots$, und es ist:

$$\lim_{h\to 0}(I(k+h)-I(k-h))=0 \qquad (k=1, 2, \ldots). \tag{49}$$

Ferner gibt es zu jedem $\varepsilon>0$ und $\delta>0$ ein b, sodass für alle λ, die keinem der Intervalle $|\lambda-k|<\delta$ $(k=1, 2, \ldots)$ angehören:

$$\left|\int_{b}^{+\infty} f(x)\frac{\sin\lambda(x-x_0)}{x-x_0}\,dx\right|<\varepsilon, \qquad \left|\int_{-\infty}^{-b} f(x)\frac{\sin\lambda(x-x_0)}{x-x_0}\,dx\right|<\varepsilon.$$

Also gilt hier, falls $f(x)$ die Eigenschaft A) hat, Formel (1) im folgenden Sinne: *Bedeutet δ irgend eine positive Zahl und ist $\{\lambda_\nu\}$ eine Folge positiver Zahlen mit $\lim_{\nu\to\infty}\lambda_\nu=+\infty$, die keinem der Intervalle $|\lambda-k|<\delta$ $(k=1, 2, \ldots)$ angehören, so ist:*

$$f(x_0)=\frac{1}{\pi}\lim_{\nu\to\infty}\int_{-\infty}^{+\infty} f(x)\frac{\sin\lambda_\nu(x-x_0)}{x-x_0}\,dx. \tag{50}$$

43—2661. *Acta mathematica.* 49. Imprimé le 30 juillet 1926.

Wieder gilt für jedes Intervall $0 < \alpha \leqq \mu \leqq \beta$, das keinen der Punkte 1, 2, ... enthält:

$$I(\beta) - I(\alpha) = \int_{\alpha}^{\beta} (\cos \mu x_0 \, d\Phi(\mu) + \sin \mu x_0 \, d\Psi(\mu)).$$

Sei nun $\lambda > 0$ und $\neq 1, 2, \ldots$. Ist die ganze Zahl n so gewählt, dass $n < \lambda < n+1$, und bedeuten $h_0, h_1, \ldots, h_n$ positiven Zahlen $< \frac{1}{2}$, so gilt also:

$$I(\lambda) - \sum_{k=1}^{n} (I(k+h_k) - I(k-h_k)) - I(h_0) = \sum_{k=0}^{n-1} \int_{k+h_k}^{k+1-h_{k+1}} (\cos \mu x_0 \, d\Phi(\mu) + \sin \mu x_0 \, d\Psi(\mu))$$
$$+ \int_{n+h_n}^{\lambda} (\cos \mu x_0 \, d\Phi(\mu) + \sin \mu x_0 \, d\Psi(\mu)).$$

Vollziehen wir hierin den Grenzübergang $h_k \to +0$ ($k = 0, 1, \ldots, n$), so erhalten wir wegen (49) und der Stetigkeit von $I(\lambda)$ für $\lambda = 0$:

$$I(\lambda) = \int_{0}^{\lambda} \cos \mu x_0 \, d\Phi(\mu) + \sin \mu x_0 \, d\Psi(\mu),$$

worin $\int_{0}^{\lambda}$ soviel bedeutet, wie $\lim\limits_{h \to +0} \int_{h}^{\lambda}$, und an den Stellen $\lambda = k$ ($k = 1, 2, \ldots, n$) der Cauchysche Hauptwert zu nehmen ist.

Beachten wir (50), so können wir also den Satz aussprechen:

Satz VIII. *Für jede Funktion $f(x)$ der Eigenschaft* A), *die die Gestalt hat $f(x) = g(x)\,h(x)$, wo $g(x)$ im Unendlichen monoton gegen* 0 *konvergiert, und $h(x)$ eine Funktion der Periode p bedeutet, gilt Formel* (10). *Es ist in ihr $\int_{0}^{\lambda}$ im Sinne von $\lim\limits_{h \to 0} \int_{h}^{\lambda}$ zu verstehen, für $\lambda = \frac{2k\pi}{p}$ ($k = 1, 2, \ldots$) der Cauchysche Hauptwert zu nehmen,* *und der Grenzwert* $\int\limits_0^{+\infty} = \lim\limits_{\lambda \to +\infty} \int\limits_0^{\lambda}$ *in der Weise zu bilden, dass* λ *ausserhalb der Intervalle* $\left|\lambda - \frac{2k\pi}{p}\right| < \delta$ $(k=1, 2, \ldots)$ *bleibt, wo* $\delta > 0$ *beliebig klein sein kann.*

Auch hier kann der Übergang zur Fourierschen Integralformel (11) vollzogen werden. Es ist für $\alpha < \xi < \eta$:

$$\int\limits_\xi^\eta f(x) \cos \mu x \, dx = g(\xi) \int\limits_\xi^{\xi^*} h(x) \cos \mu x \, dx.$$

Aus (22) folgt also: zu jedem $\varepsilon > 0$ und $\delta > 0$ gibt es ein b, sodass für $b \leqq \xi < \eta$ und alle $\mu \geqq 0$, die keinem der Intervalle $|\mu - k| < \delta$ $(k=0, 1, 2, \ldots)$ angehören:

$$\left| \int\limits_\xi^\eta f(x) \cos \mu x \, dx \right| < \varepsilon, \qquad \left| \int\limits_{-\eta}^{-\xi} f(x) \cos \mu x \, dx \right| < \varepsilon.$$

Daraus folgt, dass die durch (12) definierte Funktion $\varphi(\mu)$ für alle $\mu > 0$ und $\neq 1, 2, \ldots$ existiert, und dass die durch (20) definierte Funktion $\varphi_n(\mu)$ in jedem Intervalle $0 < \alpha \leqq \mu \leqq \beta$, das keinen der Punkte $1, 2, \ldots$ enthält, gleichmässig gegen $\varphi(\mu)$ konvergiert. Ganz analoges gilt für $\psi(\mu)$. In bekannter Weise folgt daraus, dass in jedem solchen Intervalle $[\alpha, \beta]$ die Formeln (18) gelten, woraus wir weiter schliessen:

Satz VIII a. *Für jede Funktion* $f(x)$ *der Eigenschaft* A), *die die Gestalt hat* $f(x) = g(x)\, h(x)$, *wo* $g(x)$ *im Unendlichen monoton gegen* 0 *konvergiert, und* $h(x)$ *eine Funktion der Periode* p *bedeutet, gilt die Fouriersche Integralformel* (11); *das in ihr auftretende Integral ist so zu verstehen wie in Satz VIII.*

§ 15.

Nunmehr sei $f(x)$ selbst eine *periodische Funktion*. Ohne Beschränkung der Allgemeinheit können wir annehmen, sie habe die Periode 2π. Um die Bezeichnungen von § 14 verwenden zu können, schreiben wir $f(x) = h(x)$, und zerlegen wie in § 14 $h(x)$ in $h^*(x) + h^{**}(x)$, und bezeichnen die zugehörigen Funktionen Φ mit $\Phi^*(\mu)$, $\Phi^{**}(\mu)$.

Was $\Phi^*(\mu)$ anlangt, entnehmen wir unmittelbar aus §§ 8, 9:

$$(51)\begin{cases}\Phi^*(0)=0,\ \Phi^*(\mu)=\pi\frac{a_0}{2}\text{ in }(0,1),\ \Phi^*(1)=\pi\left(\frac{a_0}{2}+\frac{a_1}{2}\right),\ \Phi^*(\mu)=\pi\left(\frac{a_0}{2}+a_1\right)\text{ in }(1,2),\\ \Phi^*(2)=\pi\left(\frac{a_0}{2}+a_1+\frac{a_2}{2}\right),\ \Phi^*(\mu)=\pi\left(\frac{a_0}{2}+a_1+a_2\right)\text{ in }(2,3),\dots,\\ \Phi^*(N)=\pi\left(\frac{a_0}{2}+a_1+\dots+a_{N-1}+\frac{a_N}{2}\right),\ \Phi^*(\mu)=\pi\left(\frac{a_0}{2}+a_1+\dots+a_N\right)\text{ für }\mu>N.\end{cases}$$

Es ist also $\Phi^*(\mu)$ konstant in $(0, 1), (1, 2), \dots (N-1, N)$ und für $\mu>N$, und es ist:

$$\Phi^*(+0)-\Phi^*(0)=\pi\frac{a_0}{2},\qquad \Phi^*(k+0)-\Phi^*(k)=\pi\frac{a_k}{2},\qquad \Phi^*(k)-\Phi^*(k-0)=\pi\frac{a_k}{2}$$
$$(k=1, 2, \dots, N).$$

Es handelt sich noch um die Untersuchung von $\Phi^{**}(\mu)$. Für jedes Intervall $(0<)\xi\leqq\mu\leqq\eta$ ist:

$$\int_\xi^\eta h^{**}(x)\frac{\sin\mu x}{x}dx=\frac{1}{\xi}\int_\xi^{\xi^*} h^{**}(x)\sin\mu x\,dx.$$

Aus den Ungleichungen (48) folgt also: zu jedem $\varepsilon>0$ gibt es ein b, sodass für $b\leqq\xi<\eta$ und $0\leqq\mu\leqq N+\frac{1}{2}$:

$$\left|\int_\xi^\eta h^{**}(x)\frac{\sin\mu x}{x}dx\right|<\varepsilon;\qquad \left|\int_{-\eta}^{-\xi} h^{**}(x)\frac{\sin\mu x}{x}dx\right|<\varepsilon.$$

Es existiert also $\Phi^{**}(\mu)$ für $0\leqq\mu\leqq N+\frac{1}{2}$.

Wir gehen nun an die Berechnung von $\Phi^{**}(\mu)$, und wollen zeigen, dass sie aus:

$$h^{**}(x)\sim\sum_{k=N+1}^{\infty}(a_k\cos kx+b_k\sin kx)$$

durch gliedweise Integration erfolgen kann.

Bekanntlich ist durch

$$(52) \qquad H(x) = \sum_{k=N+1}^{\infty} \frac{1}{k} (a_k \sin kx - b_k \cos kx)$$

ein unbestimmtes Integral von $h^{**}(x)$, durch:

$$(53) \qquad \overline{H}(x) = -\sum_{k=N+1}^{\infty} \frac{1}{k^2} (a_k \cos kx + b_k \sin kx)$$

ein unbestimmtes Integral von $H(x)$ gegeben, und die Reihen (52), (53) konvergieren gleichmässig (als Fouriersche Reihen totalstetiger Funktionen).

Durch zweimalige partielle Integration findet man:

$$\int_{\xi}^{\eta} h^{**}(x) \frac{\sin \mu x}{x} dx = H(x) \frac{\sin \mu x}{x}\Bigg]_{\xi}^{\eta} + \overline{H}(x) \left(\frac{\sin \mu x}{x^2} - \mu \frac{\cos \mu x}{x}\right)\Bigg]_{\xi}^{\eta}$$

$$+ \int_{\xi}^{\eta} \overline{H}(x) \left(2 \frac{\sin \mu x}{x^3} - 2\mu \frac{\cos \mu x}{x^2} - \mu^2 \frac{\sin \mu x}{x}\right) dx$$

$$= \sum_{k=N+1}^{\infty} \frac{1}{k} (a_k \sin kx - b_k \cos kx) \frac{\sin \mu x}{x}\Bigg]_{\xi}^{\eta}$$

$$- \sum_{k=N+1}^{\infty} \frac{1}{k^2} (a_k \cos kx + b_k \sin kx) \left(\frac{\sin \mu x}{x^2} - \mu \frac{\cos \mu x}{x}\right)\Bigg]_{\xi}^{\eta}$$

$$- \sum_{k=N+1}^{\infty} \frac{1}{k^2} \int_{\xi}^{\eta} (a_k \cos kx + b_k \sin kx) \left(2 \frac{\sin \mu x}{x^3} - 2\mu \frac{\cos \mu x}{x^2} - \mu^2 \frac{\sin \mu x}{x}\right) dx.$$

Aufgefasst als Funktionen von ξ oder von η sind die auf der rechten Seite auftretenden Reihen gleichmässig konvergent: die beiden ersten wegen der gleichmässigen Konvergenz der Reihen (52) und (53), die dritte, weil die darin auftretenden Integrale $\left(\text{für } 0 \leqq \mu \leqq N + \frac{1}{2}\right)$ zwischen endlichen Schranken bleiben. Es kann also der Grenzübergang $\xi \to +\infty$ gliedweise vollzogen werden:

$$\int_{-\infty}^{+\infty} h^{**}(x) \frac{\sin \mu x}{x} dx =$$

$$- \sum_{k=N+1}^{\infty} \frac{1}{k^2} \int_{-\infty}^{+\infty} (a_k \cos kx + b_k \sin kx) \left(2 \frac{\sin \mu x}{x^3} - 2\mu \frac{\cos \mu x}{x^2} - \mu^2 \frac{\sin \mu x}{x}\right) dx.$$

Wie zweimalige partielle Integration lehrt, ist aber $\left(\text{für } k>N,\ 0\leqq\mu\leqq N+\frac{1}{2}\right)$:

$$\int_{-\infty}^{+\infty}(a_k\cos kx+b_k\sin kx)\frac{\sin\mu x}{x}\,dx=$$

$$-\frac{1}{k^2}\int_{-\infty}^{+\infty}(a_k\cos kx+b_k\sin kx)\left(2\frac{\sin\mu x}{x^3}-2\mu\frac{\cos\mu x}{x^2}-\mu^2\frac{\sin\mu x}{x}\right)dx,$$

sodass wir schliesslich haben:

$$\int_{-\infty}^{+\infty}h^{**}(x)\frac{\sin\mu x}{x}\,dx=\sum_{k=N+1}^{\infty}\int_{-\infty}^{+\infty}(a_k\cos kx+b_k\sin kx)\frac{\sin\mu x}{x}\,dx,$$

womit die Behauptung, dass $\Phi^{**}(\mu)$ durch gliedweise Integration berechnet werden kann, bewiesen ist.

Nach § 9 ist aber für $k=N+1,\ N+2,\ldots$ und $0\leqq\mu\leqq N+\frac{1}{2}$:

$$\int_{-\infty}^{+\infty}(a_k\cos kx+b_k\sin kx)\frac{\sin\mu x}{x}\,dx=0,$$

es ist also:

$$\Phi^{**}(\mu)=0 \text{ für } 0\leqq\mu\leqq N+\frac{1}{2}.$$

Für diese Werte von μ stimmen also die Werte von $\Phi(\mu)$ mit den durch (51) gegebenen Werten von $\Phi^{*}(\mu)$ überein. Und da dabei N ganz beliebig war, haben wir ganz allgemein:

$$\Phi(0)=0,\quad \Phi(\mu)=\pi\frac{a_0}{2} \text{ für } 0<\mu<1,\quad \Phi(\mu)=\pi\left(\frac{a_0}{2}+a_1+\cdots+a_k\right) \text{ für } k<\mu<k+1 \qquad (k=1,2,\ldots)$$

$$\Phi(k)=\pi\left(\frac{a_0}{2}+a_1+\cdots+a_{k-1}+\frac{a_k}{2}\right) \qquad (k=1,2,\ldots).$$

Ganz ebenso findet man (abgesehen von einer additiven Konstante, auf die es nicht ankommt):

$\Psi(\mu)=0$ für $0\leqq\mu<1$; $\Psi(\mu)=\pi(b_1+b_2+\cdots+b_k)$ für $k<\mu<k+1$ $(k=1, 2, \ldots)$

$$\Psi(1)=\pi\frac{b_1}{2},\ \Psi(k)=\pi\left(b_1+\cdots+b_{k-1}+\frac{b_k}{2}\right) \qquad (k=1, 2, \ldots).$$

Infolgedessen ist:

$$\int_0^\lambda (\cos\mu x_0\, d\Phi(\mu)+\sin\mu x_0\, d\Psi(\mu))=$$

$$=\pi\left(\frac{a_0}{2}+\sum_{k=1}^{n}(a_k\cos kx_0+b_k\sin kx_0)\right) \text{ für } n<\lambda<n+1,$$

$$\int_0^n (\cos\mu x_0\, d\Phi(\mu)+\sin\mu x_0\, d\Psi(\mu))=$$

$$=\pi\left(\frac{a_0}{2}+\sum_{k=1}^{n-1}(a_k\cos kx_0+b_k\sin kx_0)+\frac{1}{2}(a_n\cos nx_0+b_n\sin nx_0)\right).$$

Hat nun $f(x)$ die Eigenschaft A), so gilt:

$$f(x_0)=\frac{a_0}{2}+\sum_{k=1}^{\infty}(a_k\cos kx_0+b_k\sin kx_0),$$

und wir erhalten somit:

Satz IX. *Ist $f(x)$ eine periodische Funktion der Eigenschaft* A), *so gilt Formel* (10); *und zwar reduziert sich dann die rechte Seite von* (10) *auf die Fouriersche Reihenentwicklung von $f(x_0)$.*

Wir sehen also, dass unsere Formel (10) die sämtlichen konvergenten Fourierschen Reihen umfasst.

§ 16.

Nun behandeln wir noch den Fall:

$$f(x)=f_0(x)\,h(x),$$

wo $f_0(x)$ die Funktion (31) bedeutet, und $h(x)$ *periodisch* ist. Wieder können wir annehmen, $h(x)$ habe die Periode 2π. Wir stellen dieselben Überlegungen wie in § 14 an.

Aus §§ 8, 10 und 11 entnehmen wir: $\Phi^*(\mu)$ ist stetig für $\mu \geqq 0$ und $\neq 1, 2, \ldots, N$, und es ist:

$$\lim_{h \to 0} (\Phi^*(k+h) - \Phi^*(k-h)) = 0 \qquad (k = 1, 2, \ldots, N);$$

$\Psi^*(\mu)$ ist stetig für $\mu > 0$ und $\neq 1, 2, \ldots, N$, und es ist:

$$\lim_{h \to 0} (\Psi^*(k+h) - \Psi^*(k-h)) = 0 \qquad (k = 1, 2, \ldots, N).$$

Wie in § 14 sehen wir: $\Phi^{**}(\mu)$ und $\Psi^{**}(\mu)$ existieren und sind stetig für $0 \leqq \mu < N+1$. Daraus folgt (da N beliebig ist): $\Phi(\mu)$ existiert und ist stetig für $\mu \geqq 0$ und $\neq 1, 2, \ldots$; $\Psi(\mu)$ existiert und ist stetig für $\mu > 0$ und $\neq 1, 2, \ldots$, und es ist:

$$\lim_{h \to 0} (\Phi(k+h) - \Phi(k-h)) = 0, \quad \lim_{h \to 0} (\Psi(k+h) - \Psi(k-h)) = 0 \qquad (k = 1, 2, \ldots).$$

Da auch alles über das Integral $I(\lambda)$ in § 14 Bewiesene hier giltig bleibt, erhalten wir:

Satz X. *Die Aussage von Satz VIII gilt auch für jede Funktion der Eigenschaft* A), *die die Gestalt hat:* $f(x) = f_0(x)\, h(x)$, *wo* $f_0(x)$ *die Funktion* (31) *bedeutet, und* $h(x)$ *die Periode* p *hat.*

Die Überlegungen von § 8 gestatten es nun, die Sätze VIII, IX, X zusammenzufassen. Sei $f(x) = g(x)\, h(x)$, wo $g(x)$ im Unendlichen beschränkt und monoton, $h(x)$ periodisch ist. Wie in § 8 setzen wir:

$$g(x) = \frac{c_1 + c_2}{2} + \frac{c_1 - c_2}{2} f_0(x) + g_1(x),$$

wo:

$$c_1 = \lim_{x \to +\infty} g(x), \qquad c_2 = \lim_{x \to +\infty} g(x).$$

Dann konvergiert $g_1(x)$ im Unendlichen monoton gegen 0. Wir finden, wenn zunächst wieder $h(x)$ die Periode 2π und die Fourierreihe (46) hat: Die zu $f(x)$ gehörige Funktion $\Phi(\mu)$ existiert für $\mu \geqq 0$ und ist stetig für $\mu \neq 0, 1, 2, \ldots$; die Funktion $\Psi(\mu)$ existiert und ist stetig für $\mu > 0$ und $\neq 1, 2, \ldots$. Es ist

$$\Phi(+0) - \Phi(0) = \pi \cdot \frac{c_1 + c_2}{2} \cdot \frac{a_0}{2} \quad \lim_{h \to +0} (\Phi(k+h) - \Phi(k-h)) = \pi \cdot \frac{c_1 + c_2}{2} \cdot a_k \qquad (k = 1, 2, \ldots)$$

$$\lim_{h \to +0} (\Psi(k+h) - \Psi(k-h)) = \pi \cdot \frac{c_1 + c_2}{2} b_k,$$

und es gilt:

Satz XI. *Ist $f(x)$ eine Funktion der Eigenschaft* A), *die die Gestalt hat:* $f(x)=g(x)\,h(x)$, *wo* $g(x)$ *im Unendlichen beschränkt und monoton ist und* $h(x)$ *die Periode* p *hat, so gilt Formel* (10). *Dabei ist in ihr* $\int_0^{+\infty} = \lim_{\nu\to\infty}\int_0^{\lambda_\nu}$ *zu setzen, wo* λ_ν *eine beliebige Folge positiver Zahlen mit* $\lambda_\nu \to +\infty$ *bedeutet, deren Glieder keinem der Intervalle* $\left|\lambda - \frac{2k\pi}{p}\right| < \delta$ $(k=1, 2, \ldots)$ *angehören* (δ *eine beliebige positive Zahl*), *und unter* $\int_0^\lambda$, *wenn* λ *zwischen* $\frac{2n\pi}{p}$ *und* $\frac{2(n+1)\pi}{p}$ *liegt, der Grenzwert zu verstehen ist:*

$$\int_0^\lambda = \lim_{h_0\to+0,\ldots,h_n\to+0}\left(\sum_{k=1}^{n}\int_{\frac{2(k-1)\pi}{p}+h_{k-1}}^{\frac{2k\pi}{p}-h_k} + \int_{\frac{2n\pi}{p}+h_n}^{\lambda}\right)$$

$$+\lim_{h\to+0}(\Phi(h)-\Phi(0)) + \sum_{k=1}^{n}\left\{\cos\frac{2k\pi}{p}x_0\lim_{h\to+0}\left(\Phi\left(\frac{2k\pi}{p}+h\right)-\Phi\left(\frac{2k\pi}{p}-h\right)\right)\right.$$

$$\left.+\sin\frac{2k\pi}{p}x_0\lim_{h\to+0}\left(\Psi\left(\frac{2k\pi}{p}+h\right)-\Psi\left(\frac{2k\pi}{p}-h\right)\right)\right\}.$$

§ 17.

Nun wollen wir annehmen, $f(x)$ habe die Gestalt:

$$f(x) = \sum_{k=1}^{\infty}(a_k\cos q_k x + b_k\sin q_k x), \tag{54}$$

wo alle $q_k > 0$, und die Reihe

$$\sum_{k=1}^{\infty}(|a_k| + |b_k|) \tag{55}$$

konvergent sei.

Wir zerlegen $f(x)$ in $f_1(x)+f_2(x)$, wo

$$f_1(x) = \sum_{k=1}^{\infty} a_k\cos q_k x; \qquad f_2(x) = \sum_{k=1}^{\infty} b_k\sin q_k x.$$

Entsprechend zerlegt sich $\Phi(\mu)$ in $\Phi_1(\mu)+\Phi_2(\mu)$; da aber $f_2(x)$ ungerade, ist $\Phi_2(\mu)=0$, mithin $\Phi(\mu)=\Phi_1(\mu)$. Da die Reihe für $f_1(x)$ gleichmässig konvergiert, ist für jedes endliche Intervall $[\xi, \eta]$:

$$\int_{\xi}^{\eta} f_1(x)\frac{\sin\mu x}{x}\,dx=\sum_{k=1}^{\infty} a_k\int_{\xi}^{\eta}\cos q_k x\,\frac{\sin\mu x}{x}\,dx,$$

wo die rechts stehende Reihe wieder gleichmässig für alle ξ und η konvergiert, so dass die Grenzübergänge $\xi\to-\infty$, $\eta\to+\infty$ gliedweise vollzogen werden können. So erhalten wir:

$$\Phi(\mu)=\int_{-\infty}^{+\infty} f_1(x)\frac{\sin\mu x}{x}\,dx=\sum_{k=1}^{\infty} a_k\int_{-\infty}^{+\infty}\cos q_k x\,\frac{\sin\mu x}{x}\,dx,$$

wo die rechts stehende Reihe wieder gleichmässig in μ konvergiert.

Aufgrund der Resultate von § 8 und 9 ersehen wir daraus: die Funktion $\Phi(\mu)$ ist (für $\mu\geqq 0$) konstant in jedem Intervalle, das keinen Punkt q_k $(k=1, 2\ldots)$ enthält; sie ist stetig in jedem von allen q_k verschiedenem Punkte, während sie in den Punkten q_k unstetig ist gemäss:

$$\Phi(q_k+0)-\Phi(q_k)=\frac{\pi}{2}a_k;\qquad \Phi(q_k)-\Phi(q_k-0)=\frac{\pi}{2}a_k\qquad (k=1, 2, \ldots).$$

Ebenso findet man: die Funktion $\Psi(\mu)$ ist (für $\mu\geqq 0$) konstant in jedem Intervalle, das keinen Punkt $q_k (k=1, 2, \ldots)$ enthält, sie ist stetig in jedem von allen q_k $(k=1, 2, \ldots)$ verschiedenem Punkte, während sie in den Punkten q_k unstetig ist gemäss:

$$\Psi(q_k+0)-\Psi(q_k)=\frac{\pi}{2}b_k;\qquad \Psi(q_k)-\Psi(q_k-0)=\frac{\pi}{2}b_k.$$

Daraus folgt unmittelbar:

Satz XII. *Ist $f(x)$ gegeben durch* (54), *wo die Reihe* (55) *konvergiert, so gilt Formel* (10); *und zwar reduziert sich die rechte Seite von* (10) *auf die Reihe* (54) *für $x=x_0$.*

§ 18.

Nunmehr nehmen wir an, $f(x)$ habe die Gestalt:

$$f(x)=g(x)\,h(x),$$

wo $g(x)$ im Unendlichen beschränkt und monoton sei, und $h(x)$ gegeben sei durch:

$$h(x)=\sum_{k=1}^{\infty}(a_k\cos q_k x+b_k\sin q_k x), \tag{56}$$

wo die q_k eine monoton ins Unendlichen wachsende Folge positiver Zahlen bedeuten, und wieder die Reihe (55) konvergiere.

Wir nehmen mit $g(x)$ wieder die Zerlegung (32) vor:

$$g(x)=\frac{c_1+c_2}{2}+\frac{c_1-c_2}{2}f_0(x)+g_1(x),$$

und können uns, da wir über den Beitrag des Summanden $\frac{c_1+c_2}{2}$ durch § 17 völlig informiert sind, auf die Betrachtung der beiden anderen Summanden beschränken. Es bedeute also im Folgenden $g(x)$, entweder die Funktion $f_0(x)$, oder eine Funktion, die im Unendlichen monoton gegen 0 konvergiert. In beiden Fällen konvergiert $\frac{g(x)}{x}$ im Unendlichen monoton gegen 0.

Ist $g(x)$ monoton für $x>a$, so gilt für $a<\xi<\eta$:

$$\int_\xi^\eta f(x)\frac{\sin\mu x}{x}\,dx=\frac{g(\xi)}{\xi}\int_\xi^{\xi^*}h(x)\sin\mu x\,dx$$

$$=\frac{g(\xi)}{\xi}\sum_{k=1}^{\infty}\left(a_k\int_\xi^{\xi^*}\cos q_k x\sin\mu x\,dx+b_k\int_\xi^{\xi^*}\sin q_k x\sin\mu x\,dx\right).$$

Nun gehört zu jedem $\delta>0$ ein M, so dass, wenn $\mu\geqq 0$ allen Ungleichungen genügt $|\mu-q_k|\geqq\delta$ $(k=1, 2, \ldots)$, für alle ξ und ξ^* die Ungleichungen bestehen:

$$\left|\int_\xi^{\xi^*}\cos q_k x\sin\mu x\,dx\right|<M,\qquad\left|\int_\xi^{\xi^*}\sin q_k x\sin\mu x\,dx\right|<M,$$

und mithin für alle ξ und η:

$$\left|\int_{\xi}^{\eta} f(x)\frac{\sin\mu x}{x}dx\right| < \left|\frac{g(\xi)}{\xi}\right|\cdot M\cdot\sum_{k=1}^{\infty}(|a_k|+|b_k|).$$

Es gibt also zu jedem $\varepsilon>0$ und $\delta>0$ ein b, sodass für $b\leqq\xi<\eta$ und alle $\mu\geqq 0$, die den sämtlichen Ungleichungen $|\mu-q_k|\geqq\delta\,(k=1, 2, \ldots)$ genügen, die Ungleichung gilt:

$$\left|\int_{\xi}^{\eta} f(x)\frac{\sin\mu x}{x}dx\right| < \varepsilon.$$

Wie immer entnehmen wir hieraus: Für alle $\mu\geqq 0$ und $\neq q_k$ existiert $\Phi(\mu)$ und ist stetig. In jedem Intervalle $(0\leqq)\,\alpha\leqq\mu\leqq\beta$, das keinen der Punkte q_k enthält, konvergiert die durch (27) definierte Funktion $\Phi_n(\mu)$ gleichmässig gegen $\Phi(\mu)$. Und die analogen Aussagen gelten für $\Psi(\mu)$.

Um das Verhalten von $\Phi(\mu)$ und $\Psi(\mu)$ an der Stelle q_k festzustellen, setzen wir:

$$h^*(x)=h(x)-(a_k\cos q_k x+b_k\sin q_k x).$$

Die zu $f^*(x)=g(x)\,h^*(x)$ gehörige Funktion Φ bleibt dann nach dem eben Bewiesenen für $\mu=q_k$ stetig; es verhält sich also die zu $f(x)$ gehörige Funktion $\Phi(\mu)$ für $\mu=q_k$ so wie $\int_{-\infty}^{+\infty} g(x)\,(a_k\cos q_k x+b_k\sin q_k x)\frac{\sin\mu x}{x}dx$, woraus nach § 10, 11, 12 sofort folgt:

$$\lim_{h\to 0}(\Phi(q_k+h)-\Phi(q_k-h))=0,$$

und ebenso beweist man:

$$\lim_{h\to 0}(\Psi(q_k+h)-\Psi(q_k-h))=0.$$

In gewohnter Weise folgern wir aus diesen Resultaten, dass das Integral:

$$I(\lambda)=\int_{-\infty}^{+\infty} f(x)\frac{\sin\lambda(x-x_0)}{x-x_0}dx$$

für alle $\lambda\geqq 0$ und $\neq q_k$ $(k=1, 2, \ldots)$ existiert und stetig ist, dass:

$$\lim_{h \to 0} (I(q_k+h)-I(q_k-h))=0$$

ist, dass falls $f(x)$ die Eigenschaft A) hat, für jede Folge positiver Zahlen λ_ν mit $\lambda_\nu \to +\infty$, die keinem der Intervalle $|\lambda-q_k| < \delta$ $(k=1, 2, \ldots)$ angehören:

$$f(x_0) = \frac{1}{\pi} \lim_{\nu \to \infty} I(\lambda_\nu)$$

ist, und dass für jedes Intervall $(0 \leqq)\, \alpha \leqq \mu \leqq \beta$, das keinen der Punkte q_k enthält:

$$I(\beta)-I(\alpha) = \int_\alpha^\beta (\cos \mu x_0 \, d\Phi(\mu) + \sin \mu x_0 \, d\Psi(\mu))$$

ist. Daraus ergibt sich:

Satz XIII. *Ist $f(x)=g(x).h(x)$, wo $g(x)$ entweder die durch (31) definierte Funktion $f_0(x)$ bedeutet, oder im Unendlichen monoton gegen 0 konvergiert, und $h(x)$ die Bedeutung (56) mit konvergenter Reihe (55) hat, so gilt, falls $f(x)$ die Eigenschaft A) hat, und $\delta>0$ so klein gewählt werden kann, dass es eine Folge positiver Zahlen λ_ν mit $\lambda_\nu \to +\infty$ gibt, die keinem der Intervalle $|\lambda-q_k| < \delta$ $(k=1, 2, \ldots)$ angehören, Formel (10). Es ist in ihr $\int_0^{+\infty} = \lim_{\nu\to\infty} \int_0^{\lambda_\nu}$ zu setzen und für $\lambda=q_k$ $(k=1, 2, \ldots)$ der Cauchysche Hauptwert zu nehmen.*

Durch Kombination von Satz XII und XIII erhalten wir:

Satz XIV. *Ist $f(x)=g(x)h(x)$, wo $g(x)$ im Unendlichen beschränkt und monoton ist, und $h(x)$ die Bedeutung (56) mit konvergenter Reihe (55) hat, so gilt, falls $f(x)$ die Eigenschaft A) hat, und $\delta>0$ so klein gewählt werden kann, dass es eine Folge positiver Zahlen λ_ν mit $\lambda_\nu \to +\infty$ gibt, die keinem der Intervalle $|\lambda-q_k| < \delta$ $(k=1, 2, \ldots)$ angehören, Formel (10). Es ist in ihr $\int_0^{+\infty} = \lim_{\nu\to\infty} \int_0^{\lambda_\nu}$ zu setzen und unter $\int_0^\lambda$, wenn $q_n<\lambda<q_{n+1}$ ist, der Grenzwert zu verstehen:*

$$\int_0^\lambda = \lim_{h_1\to+0,\ldots,h_n\to+0}\left(\int_0^{q_1-h_1} + \sum_{k=1}^{n-1}\int_{q_k+h_k}^{q_{k+1}-h_{k+1}} + \int_{q_n+h_n}^{\lambda}\right)$$

$$+\sum_{k=1}^{n}\{\cos q_k x_0 \lim_{h\to+0}(\Phi(q_k+h)-\Phi(q_k-h))+\sin q_k x_0 \lim_{h\to+0}(\Psi(q_k+h)-\Psi(q_k-h))\}.$$

§ 19.

Zum Schlusse betrachten wir noch Funktionen $f(x)$, denen folgende Eigenschaft zukommt:

D) Es sei $f(x)$ im Unendlichen quadratisch integrierbar (d. h. es gebe ein a, so dass die beiden Grenzwerte $\lim_{q\to+\infty}\int_a^q (f(x))^2 dx$ und $\lim_{q\to+\infty}\int_{-q}^{-a}(f(x))^2 dx$ endlich ausfallen).

Wir erkennen sofort: *hat* $f(x)$ *die Eigenschaften* A) *und* D), *so gilt Formel* (1). In der Tat, die Schwarzsche Ungleichung ergibt für jedes Intervall $[\xi, \eta]$:

$$\left|\int_\xi^\eta f(x)\frac{\sin\lambda(x-x_0)}{x-x_0}dx\right| \leqq \left(\int_\xi^\eta (f(x))^2 dx\right)^{\frac{1}{2}}\cdot\left(\int_\xi^\eta\left(\frac{\sin\lambda(x-x_0)}{x-x_0}\right)^2 dx\right)^{\frac{1}{2}}$$

$$< M\cdot\left(\int_\xi^\eta (f(x))^2 dx\right)^{\frac{1}{2}},$$

wo M eine geeignete Konstante bedeutet. Daraus folgt: zu jedem $\varepsilon>0$ gibt es ein b, so dass für $b\leqq\xi<\eta$:

$$\left|\int_\xi^\eta f(x)\frac{\sin\lambda(x-x_0)}{x-x_0}dx\right|<\varepsilon,\qquad \left|\int_{-\eta}^{-\xi} f(x)\frac{\sin\lambda(x-x_0)}{x-x_0}dx\right|<\varepsilon,$$

woraus in bekannter Weise die Behauptung folgt.

Wir definieren wieder $\Phi(\mu)$ und $\Psi(\mu)$ durch (7). Genau wie eben ergibt die Schwarzsche Ungleichung, dass $\Phi(\mu)$ und $\Psi(\mu)$ für alle $\mu\geqq 0$ existieren, und dass auch hier die Beziehungen (8) gleichmässig erfüllt sind. Daraus folgt, dass

$\Phi(\mu)$ und $\Psi(\mu)$ für alle $\mu \geqq 0$ stetig sind, und ganz wie in § 1 gelangt man zum Satze:

Satz XV. *Für jede Funktion der Eigenschaften* A) *und* D) *gilt Formel* (10).

Sei $\mu_1', \mu_2', \ldots, \mu_n', \ldots$ eine monotone Folge negativer Zahlen mit $\mu_n' \to -\infty$, und $\mu_1'', \mu_2'', \ldots, \mu_n'', \ldots$ eine monotone Folge positiver Zahlen mit $\mu_n'' \to +\infty$. Wir setzen:

$$\Phi^{(n)}(\mu) = \int_{\mu_n'}^{\mu_n''} f(x) \frac{\sin \mu x}{x} dx; \qquad \Psi^{(n)}(\mu) = \int_{\mu_n'}^{\mu_n''} f(x) \frac{1 - \cos \mu x}{x} dx,$$

$$(57) \qquad \varphi^{(n)}(\mu) = \int_{\mu_n'}^{\mu_n''} f(x) \cos \mu x \, dx; \qquad \psi^{(n)}(\mu) = \int_{\mu_n'}^{\mu_n''} f(x) \sin \mu x \, dx;$$

dann ist·

$$(58) \qquad \Phi^{(n)}(\mu) = \int_0^{\mu} \varphi^{(n)}(\mu) \, d\mu, \qquad \Psi^{(n)}(\mu) = \int_0^{\mu} \psi^{(n)}(\mu) \, d\mu.$$

$$(59) \qquad \lim_{n \to \infty} \Phi^{(n)}(\mu) = \Phi(\mu), \qquad \lim_{n \to \infty} \Psi^{(n)}(\mu) = \Psi(\mu).$$

Wir zeigen: *die Folge der* $\varphi^{(n)}(\mu)$, *sowie die Folge der* $\psi^{(n)}(\mu)$ *ist im Intervalle* $(0, +\infty)$ *im Mittel konvergent.*

In der Tat, nach dem Parsevalschen Theorem für Fouriersche Integrale ist:

$$\frac{1}{\pi} \int_0^{+\infty} \{(\varphi^{(n_1)}(\mu) - \varphi^{(n_2)}(\mu))^2 + (\psi^{(n_1)}(\mu) - \psi^{(n_2)}(\mu))^2\} \, d\mu = \int_{\mu_{n_2}'}^{\mu_{n_1}'} (f(x))^2 \, dx + \int_{\mu_{n_1}''}^{\mu_{n_2}''} (f(x))^2 \, dx.$$

Aus Eigenschaft D) von $f(x)$ ergibt sich hieraus unmittelbar: zu jedem $\varepsilon > 0$ gibt es ein N, so dass für $n_1 \geqq N$, $n_2 \geqq N$:

$$\int_0^{+\infty} (\varphi^{(n_1)}(\mu) - \varphi^{(n_2)}(\mu))^2 \, d\mu < \varepsilon, \qquad \int_0^{+\infty} (\psi^{(n_1)}(\mu) - \psi^{(n_2)}(\mu))^2 \, d\mu < \varepsilon,$$

womit die Behauptung bewiesen ist.

Es konvergieren also die $\varphi^{(n)}(\mu)$ im Mittel gegen eine Funktion $\varphi^*(\mu)$, die $\psi^{(n)}(\mu)$ gegen eine Funktion $\psi^*(\mu)$. Bekanntlich folgt daraus für alle $\mu \geqq 0$:

$$\lim_{n\to\infty} \int_0^\mu \varphi^{(n)}(\mu)\,d\mu = \int_0^\mu \varphi^*(\mu)\,d\mu; \qquad \lim_{n\to\infty} \int_0^\mu \psi^{(n)}(\mu)\,d\mu = \int_0^\mu \psi^*(\mu)\,d\mu.$$

Wegen (58) und (59) kann das geschrieben werden:

$$\Phi(\mu) = \int_0^\mu \varphi^*(\mu)\,d\mu, \qquad \Psi(\mu) = \int_0^\mu \psi^*(\mu)\,d\mu,$$

und durch Einsetzen in (10) erhalten wir:

Satz XVI. *Für jede Funktion $f(x)$ der Eigenschaften* A) *und* D) *gilt:*

$$f(x_0) = \frac{1}{\pi} \int_0^{+\infty} (\varphi^*(\mu) \cos \mu x_0 + \psi^*(\mu) \sin \mu x_0)\,d\mu;$$

dabei bedeuten $\varphi^*(\mu)$ und $\psi^*(\mu)$ die Funktionen, gegen die die durch (57) *definierten Funktionen $\varphi^{(n)}(\mu)$ bzw. $\psi^{(n)}(\mu)$ im Intervalle* $(0, +\infty)$ *im Mittel konvergieren, wenn* $\mu_n' \to -\infty$, $\mu_n'' \to +\infty$.

Der Unterschied gegenüber der Fourierschen Integralformel (11) ist der, dass wir nicht behaupten können, dass $\varphi^*(\mu)$ und $\psi^*(\mu)$ durch die Integrale (12) gegeben sind, da diese letzteren nicht zu existieren brauchen. Doch ergeben bekannte Sätze:

Satz XVII. *Für jede Funktion $f(x)$ der Eigenschaften A) und* D) *gilt die Fouriersche Integralformel* (11), *falls die durch* (12) *definierten Funktionen $\varphi(\mu)$ und $\psi(\mu)$ abgesehen von einer Nullmenge existieren.*

In der Tat, da die $\varphi^{(n)}(\mu)$ im Mittel gegen $\varphi^*(\mu)$ konvergieren, kann aus ihnen eine Teilfolge $\varphi^{(n_i)}(\mu)$ herausgegriffen werden, die für alle $\mu \geqq 0$, abgesehen von einer Nullmenge, gegen $\varphi^*(\mu)$ konvergiert:

$$\varphi^*(\mu) = \lim_{i\to\infty} \int_{\mu_{n_i}'}^{\mu_{n_i}''} f(x) \cos \mu x\,dx.$$

Wenn aber $\varphi(\mu)$ existiert, stimmt dieser Grenzwert mit $\varphi(\mu)$ überein; und dasselbe gilt für $\psi^*(\mu)$ und $\psi(\mu)$. Also folgt Satz XVII aus Satz XVI.

Nach einem Satze von M. PLANCHEREL[1] existieren $\varphi(\mu)$ und $\psi(\mu)$, abgesehen von einer Nullmenge, sicher dann, wenn das Integral $\int_0^{+\infty} (f(x))^2 (\log x)^2 \, dx$ existiert.

[1] Math. Ann. 76 (1915), S. 315—326.

45 — 2661. *Acta mathematica.* 49. Imprimé le 11 août 1926.

Comments on Hans Hahn's work in complex analysis

Ludger Kaup

Konstanz

Hahn's article is an example for the enormous impact of Poincaré's work on the theory of several complex variables. Let us use the terminology and the notations of nowadays. In [Po], Poincaré wrote that his method of solution for the so-called "Poincaré problem" could be used to prove the following result: *If f is an (not necessarily univalent, but with discrete values for every fixed argument) entire holomorphic function of one variable, then its graph $\mathbb{C}^2$ can be given by one global holomorphic equation.*

The Poincaré problem is this: Given a meromorphic function $m \in \mathcal{M}(X)$, can it be written as $m = f/g$, where $f, g \in \mathcal{O}(X)$ are relatively prime holomorphic functions? If, for every $x \in X$, even the germs f_x and g_x are required to be relatively prime, then one calls it the "strong Poincaré problem". For simplicity we restrict formulations to the case that X is a complex manifold (for a more general version of this and the following theory see for instance [KpKp, §§ 53–54 A]).

The solution of that problem is intimately related to the so-called multiplicative Cousin problem: Assume that $\mathfrak{U} = (U_j)_{j\in J}$ is an open cover of X and that, for each $j \in J$, there is a meromorphic function $m_j \in \mathcal{M}(U_j)$ such that m_i / m_j is an invertible holomorphic function on $U_i \cap U_j$. For such a multiplicative Cousin-distribution D, does there exist a global meromorphic function $m \in \mathcal{M}(X)$ such that m/m_j is holomorphic and without zeros on U_j for every j?

In his thesis [Cou], Cousin claimed in 1895 that the answer is in the affirmative if X is a product of one-dimensional domains in $\mathbb{C}$. But in 1917 Gronwall [Gr] published counterexamples, using multiplicative automorphic functions. He showed that all factors of X but one have to be simply connected. It then was a challenge for many years to find a satisfactory solution to the multiplicative Cousin problem. Only in 1953, using the machinery of cohomology of coherent analytic sheaves, a definite and in that language surprisingly simple answer was achieved in [Se]:

If a multiplicative Cousin distribution D is solvable by a global meromorphic function, then the associated Chern class $c(D) \in H^2(X,\mathbb{Z})$ *vanishes. If* $H^1(X,\mathcal{O}) = 0$*, then this necessary condition is sufficient.*

The condition $H^1(X,\mathcal{O}) = 0$ is satisfied for the so-called Stein manifolds and thus in particular for products of one-dimensional domains. On the other hand, $H^2(X,\mathbb{Z})$ vanishes for such a product iff at most one factor is not simply connected.

As a consequence, on a Stein manifold X with $H^2(X,\mathbb{Z}) = 0$ the strong Poincaré problem is always solvable.

Hahn followed the idea of Poincaré's mentioned above. As an easy consequence of the Weierstrass Preparation Theorem he derived as a necessary condition that the multivalued function f has to be given locally by a finite branched covering. So let us use here the language of weakly holomorphic correspondences [St_2, section 1], covering a somewhat more general situation:

Let Z and W be one-dimensional domains. A correspondence, denoted by $f : Z \overset{c}{\to} W$, is called *weakly holomorphic* if its graph $\Gamma(f)$ is an analytic subset of $Z \times W$. Assume moreover that f has only one-dimensional irreducible components. Then Hahn essentially derives from Cousin's results [Cou] that $\Gamma(f)$ is a hypersurface in $Z \times W$, i. e., the solution set of one global holomorphic equation. But that was hard luck; in 1905 he could not yet know Gronwall's contribution. A counterexample to Hahn's general result is given by the multivalued function $z^i : \mathbb{C}^* \overset{c}{\to} \mathbb{C}^*$, where $i^2 = -1$, see [St, § 4] or [FoRa, Beispiel 2]. In fact, the Chern class $c(\Gamma(z^i))$ is a generator of $H^2(\mathbb{C}^* \times \mathbb{C}^*, \mathbb{Z}) \cong \mathbb{Z}$ and thus non-zero. For more general Z and W Hahn's result remains correct if one of the factors is simply connected; hence, it holds in particular for $Z = \mathbb{C} = W$.

As an application, Hahn proves that meromorphic functions on $\mathbb{C}^2$ admit a decomposition into a product of powers of at most countably many holomorphic prime factors. That is not at all obvious, since $\mathcal{O}(\mathbb{C}^2)$ is not factorial. The modern methods, mentioned above, imply more generally for Stein manifolds X with $H^2(X,\mathbb{Z}) = 0$ that every meromorphic function $m \in \mathcal{M}(X)$ has such a factorization, at least if X is simply connected ([LaLü, Satz 3.30]). It is unique up to order and units.

References

[AnGr] Andreotti A, Grauert H (1962) Théorèmes de finitude pour la cohomologie des espaces complexes. Bull Soc Math France 90: 193–259

[Cou] Cousin P (1895) Sur les fonctions de *n* variables complexes. Acta Math 19

[FoRa] Forster O, Ramspott KJ (1964) Über die Darstellung analytischer Mengen. Bayer Akad Wiss Math Nat Kl: 89–99

[Gr] Gronwall TH (1917) On Expressibility of Uniform Functions of Several Complex Variables as Quotient of two Functions of Entire Character. Am Math Soc Trans 18

[KpKp] Kaup L, Kaup B (1983) Holomorphic Functions of Several Variables. An Indroduction to the Fundamental Theory. De Gruyter, Berlin

[LaLü] Langmann K, Lütkebohmert W (1974) Cousinverteilungen und Fortsetzungssätze. Springer Lecture Notes 367. Springer, Berlin Heidelberg New York

[Po] Poincaré H (1883) Sur les fonctions de deux variables indépendentes. Acta math 2: 97–113

[Se] Serre JP (1953) Quelques problèmes globaux relatifs aux variétés de Stein. Coll Plus Var, Bruxelles: 57–68

[St] Stein K (1941) Topologische Bedingungen für die Existenz analytischer Funktionen zu vorgegebenen Nullstellenflächen. Math Ann 117: 727–757

[St_2] Stein K (1968) Fortsetzung holomorpher Korrespondenzen. Invent math 6: 78–90

Hahn's work in complex analysis
Hahns Arbeit zur Funktionentheorie

Über Funktionen zweier komplexer Veränderlicher.

Von **Hans Hahn** in Wien.

In Band 2 der Acta mathematica spricht Poincaré am Schlusse seiner bekannten Abhandlung „Sur les fonctions de deux variables indépendentes" die Ansicht aus, die Methode, welche er daselbst zum Beweise des Satzes benützt, daß jede Funktion zweier voneinander unabhängiger komplexer Veränderlicher, welche im Endlichen keine wesentlich singulären Stellen hat, als Quotient zweier ganzer Funktionen darstellbar ist, könne noch zum Beweise anderer Theoreme aus der Theorie der Funktionen zweier komplexer Veränderlicher verwendet werden, und er führt insbesondere den Satz an: „Bedeutet Y irgend eine, nicht eindeutige Funktion von X, welche im Endlichen keine wesentlich singuläre Stelle hat und nicht für einen und denselben Wert von X unendlich viele unendlich benachbarte Werte annehmen kann, so kann sie als Lösung einer Gleichung $G(X, Y) = o$ betrachtet werden, wo G eine ganze Funktion bedeutet." Ein Beweis dieses Satzes wurde meines Wissens weder auf dem von Poincaré vorgeschlagenen Wege noch sonstwie erbracht;[1] ich hielt es daher nicht für überflüssig, einen Beweis mitzuteilen, der nicht nur den genannten Poincaréschen Satz, sondern noch eine Reihe ähnlicher Sätze aus einem Resultate allgemeinerer Natur abzuleiten gestattet. Allerdings erleidet dabei der Satz gewisse einschränkende Bedingungen, die aber, wie ich im folgenden zeige, nicht nur hinreichend, sondern auch notwendig sind. Endlich verwende ich noch die auf diese Weise gewonnenen Resultate zum Beweise eines Satzes, der als das Analogon, zu Weierstraß' Satz über die Zerlegung einer ganzen Funktion einer Veränderlichen in Primfaktoren für das Gebiet zweier Veränderlicher betrachtet werden kann und der einen von Biermann[2]) im Anschlusse an Appell[3]) bewiesenen Satz als speziellen Fall enthält.

[1]) Doch vergleiche man eine während der Drucklegung dieses Aufsatzes erschienene Note von P. Boutroux (Comptes rendus 1904), sowie H. Baker, Proc. of London Mathem. Soc. ser II. vol. 1. 1903.

[2]) O. Biermann: Wiener Ber., Bd. 89.

[3]) Appell: Acta mathem., Bd. 2.

§ 1.

Wir werden uns im folgenden mit Funktionen zweier voneinander unabhängiger komplexer Veränderlicher x und y zu beschäftigen haben. Die Werte einer jeden dieser beiden Veränderlichen mögen in der bekannten Weise auf je einer Ebene dargestellt werden. Das Wertesystem x_0, y_0 wollen wir auch als den Punkt (x_0, y_0) bezeichnen und die beiden Werte x_0 und y_0 als die Koordinaten dieses Punktes. Als ein Gebiet der beiden Veränderlichen x und y wollen wir die Gesamtheit aller jener Punkte bezeichnen, deren x-Koordinate einem bestimmten Gebiete der x-Ebene und deren y-Koordinate einem bestimmten Gebiete der y-Ebene angehört, wobei das Gebiet in der x-Ebene von der Lage des Punktes y im Gebiete seiner Ebene gänzlich unabhängig sein soll und ebenso das Gebiet in der y-Ebene unabhängig sei von der Lage des Punktes x. Hat dabei das Gebiet der x-Ebene die Bezeichnung R, das der y-Ebene die Bezeichnung S, so werden wir das Gebiet des Punktes (x, y) mit (R, S) bezeichnen. Wir wollen weiter von einem Punkte (x, y) dann, und nur dann, sagen, er liege im Innern von (R, S), wenn sowohl x im Innern von R als auch y im Innern von S liegt; ist auch nur einer der beiden Punkte x und y ein Randpunkt seines Bereiches, so werde auch der Punkt (x, y) als Randpunkt von (R, S) bezeichnet. Wo schlechtweg von Funktionen der beiden Veränderlichen x und y die Rede ist, sind immer Funktionen zu verstehen, die im betrachteten Gebiete eindeutig und analytisch sind.

Ich werde die folgenden Entwicklungen auf einen von P. Cousin im 19. Bande der Acta mathematica bewiesenen Satz stützen.[1]) Derselbe lautet:

Seien zwei beliebige, ganz im Endlichen gelegene Bereiche T_x und T_y gegeben,[2]) der eine in der x-Ebene, der andere in der y-Ebene, ferner zwei beliebige zugehörige Bereiche T'_x und T'_y. Jedem Punkte (x_0, y_0) von (T_x, T_y) mögen entsprechen:

1. ein Kreis $\Gamma_{x_0 y_0}$ mit dem Mittelpunkte x_0 und ein Kreis $\gamma_{x_0 y_0}$ mit dem Mittelpunkte y_0, klein genug, um ganz innerhalb T_x beziehungsweise T_y zu liegen;

2. eine innerhalb $(\Gamma_{x_0 y_0}, \gamma_{x_0 y_0})$ reguläre Funktion $v_{x_0 y_0}(x, y)$ von der Eigenschaft, daß, wenn (x'_0, y'_0) einen anderen Punkt des Gebietes $(\Gamma_{x_0 y_0}, \gamma_{x_0 y_0})$ bedeutet, der Quotient $\frac{v_{x_0 y_0}(x, y)}{v_{x'_0 y'_0}(x, y)}$ im Punkte (x'_0, y'_0) regulär und von Null verschieden ist:

---

[1]) P. Cousin. Acta math. 19. Sur les fonctions de n variables complexes. S. 25.

[2]) Die Bezeichnung ist die von Osgood in Enzykl. der math. Wiss., II. B. 1, festgehaltene. (Siehe daselbst Seite 10.) Die Bereiche T_x und T_y enthalten ihre Randpunkte nicht, während die ganz im Innern von T_x beziehungsweise T_y gelegenen Bereiche T'_x und T'_y ihre Randpunkte enthalten

dann existiert eine in jedem Punkte von (T'_x, T'_y) reguläre Funktion $V(x, y)$ von der Eigenschaft, daß in jedem Punkte (x_0, y_0) von (T'_x, T'_y) der Quotient $\frac{V(x, y)}{v_{x_0 y_0}(x, y)}$ regulär und von Null verschieden ist.

Wir werden diesen Satz benützen, um das nachstehende Theorem zu beweisen:

Sei T_x ein ganz im Endlichen gelegenes Gebiet der x-Ebene und T_y ein ebensolches Gebiet der y-Ebene. Mit $y = f(x)$ werde eine in T_x definierte (im allgemeinen unendlich vieldeutige) analytische Funktion bezeichnet. Damit es zu jedem in (T_x, T_y) gelegenen Bereiche (T'_x, T'_y) eine in demselben eindeutige und analytische Funktion $G(x, y)$ von x und y gebe, von der Art, daß die in (T'_x, T'_y) gelegenen Nullstellen von $G(x, y)$ genau übereinstimmen mit den durch die Gleichung $y = f(x)$ gelieferten Wertepaaren, sofern sie in (T'_x, T'_y) liegen, — mit anderen Worten, damit $y = f(x)$ in (T'_x, T'_y) als Lösung einer Gleichung $G(x, y) = o$ betrachtet werden könne, — ist notwendig und hinreichend, daß die Funktion $y = f(x)$ der folgenden Bedingung genüge: Zu jedem Bereiche (T'_x, T'_y) und zu jedem Punkte x_0 von T'_x läßt sich eine Konstante ρ so definieren, daß alle durch die Gleichung $y = f(x)$ gelieferten Wertepaare die in (T'_x, T'_y) liegen und deren x-Koordinaten der Ungleichung $|x - x_0| < \rho$ genügen, durch eine endliche Anzahl von Entwicklungen der Form:

$$y = \sum_{\nu=0}^{\infty} a_\nu (x - x_0)^{\frac{\nu}{p}}$$

geliefert werden, wo unter p positive ganze Zahlen zu verstehen sind.

Wir zeigen zunächst, daß diese Bedingung eine notwendige ist. Es sei eine Folge von Wertepaaren (x, y) gegeben, die der Gleichung $y = f(x)$ genügen und gegen einen Punkt (x_0, y_0) von (T_x, T_y) konvergieren. Sei:

$$(x_1, y_1), (x_2, y_2), \ldots, (x_\nu, y_\nu), \ldots$$

diese Folge. Soll dann $y = f(x)$ einer Gleichung $G(x, y) = 0$ genügen, so muß:

$$G(x_1, y_1) = G(x_2, y_2) = \cdots = G(x_\nu, y_\nu) = \cdots = 0$$

sein, und hieraus folgt, da $G(x, y)$ an der Stelle (x_0, y_0) regulär und somit stetig sein soll, auch:

$$G(x_0, y_0) = 0.$$

Dann gilt aber nach einem fundamentalen Satze von Weierstraß für eine Umgebung $|x - x_0| < r$, $|y - y_0| < r'$ der Stelle (x_0, y_0) die Identität:

$$G(x, y) = \left(y^m + \varphi_1(x)\, y^{m-1} + \cdots + \varphi_m(x)\right) e^{H(x, y)}$$

wo $\varphi_1(x), \cdots, \varphi_m(x)$ an der Stelle x_0, und $H(x, y)$ an der Stelle (x_0, y_0) regulär sind. Hieraus folgt aber, daß alle durch die Gleichung $y = f(x)$ gelieferten Wertepaare (x, y), die gegen (x_0, y_0) konvergieren, auch der Gleichung:

$$y^m + \varphi_1(x)\, y^{m-1} + \cdots + \varphi_m(x) = 0$$

genügen müssen, und somit in der Tat durch eine endliche Anzahl Entwicklungen von der Form:

$$y = \sum_{\nu=0}^{\infty} a_\nu (x - x_0)^{\frac{\nu}{r}}$$

geliefert werden müssen. Unsere Bedingung wird sich also als notwendig erwiesen haben, sobald gezeigt ist, daß es in jedem T_y' nur eine endliche Anzahl von Werten y_0 geben kann von der Art, daß wenn x_0 einen Punkt von T_x und $x_1, x_2, \cdot \cdot, x_\nu, \cdots$ eine gegen x_0 konvergierende Folge bedeuten, sich unter den Funktionswerten:

$$f(x_1),\ f(x_2), \cdots,\ f(x_\nu), \cdots$$

eine gegen y_0 konvergierende Folge:

$$y_1,\ y_2, \cdots,\ y_\nu, \cdots$$

auswählen läßt.

Um dies zu beweisen, setzen wir zunächst:

$$G(x, y) = (x - x_0)^m\, G_1(x, y),$$

wo $(x - x_0)^m$ die höchste in $G(x, y)$ aufgehende Potenz von $(x - x_0)$ sei, so daß $G_1(x, y)$ für $x = x_0$ nicht mehr identisch verschwinde. Gäbe es nun in T_y' unendlich viele Werte y_0 von der oben angegebenen Art, so müßten sie mindestens eine in T_y' gelegene Häufungsstelle y_0^0 haben. Der Punkt y_0^0 wäre aber dann für die nicht identisch verschwindende, für $y = y_0^0$ reguläre analytische Funktion $G(x_0, y)$ eine Häufungsstelle von Nullstellen, was unmöglich ist.

Es können also in T_y' nur eine endliche Anzahl Werte y_0 von der angegebenen Beschaffenheit liegen, und unser Beweis ist somit geführt.

Wir gehen nunmehr an den Beweis, daß die angeführte Bedingung auch hinreichend ist; und zwar soll der Beweis mit Hilfe des eingangs zitierten Satzes von Cousin geführt werden.

Es sei also ein beliebiger zu (T_x, T_y) gehöriger Bereich (T'_x, T'_y) gegeben.

Wir teilen die sämtlichen Punkte von T_x in vier Klassen. Die erste möge alle Punkte x_0 von der Art enthalten, daß von sämtlichen Zweigen von $f(x)$ für $x = x_0$ keiner einen in T'_y gelegenen Wert annimmt. In die zweite nehmen wir alle diejenigen Punkte x_0 auf, für welche die zugehörigen in T'_y liegenden Funktionswerte $f(x_0)$ sämtlich verschieden sind und kein einziger derselben auf den Rand von T'_y zu liegen kommt. Die dritte enthalte jene Punkte, für welche mehrere der Zweige von $f(x)$ denselben in T'_y gelegenen Wert y_0 annehmen, während keiner der Funktionswerte auf den Rand von T'_y falle; und endlich die vierte Klasse enthalte diejenigen Werte x_0 für welche mindestens einer der zugehörigen Funktionswerte auf den Rand von T'_y falle.

Wir ordnen nun gemäß dem Satze von Cousin jedem Punkte (x_0, y_0) von (T'_x, T'_y) ein Gebiet $(\Gamma_{x_0 y_0}, \gamma_{x_0 y_0})$ und eine Funktion $v_{x_0 y_0}(x, y)$ zu; und zwar zunächst denjenigen Punkten, deren x-Koordinate der Klasse (1) angehört. Sei also (x_0, y_0) ein solcher Punkt. Zunächst läßt sich aus der Bedingung, der die Funktion $f(x)$ unterliegt, leicht folgern, daß dann eine Umgebung der Stelle x_0 existiert von der Art, daß auch für keinen Punkt dieser Umgebung ein Zweig von $f(x)$ einen in T'_y gelegenen Wert annimmt. Wir ordnen nun dem Punkte (x_0, y_0) zu:

1. Einen Kreis $\Gamma_{x_0 y_0}$ in der x-Ebene vom Radius ρ, wo ρ so klein zu wählen ist, daß dieser Kreis ganz in die eben genannte Umgebung der Stelle x_0 fällt; und einen Kreis $\gamma_{x_0 y_0}$ in der y-Ebene vom Radius ρ', der so klein zu wählen ist, daß der Kreis ganz in T'_y liegt.

2. Eine Funktion $v_{x_0 y_0}(x, y) = e^{H(x, y)}$, wo $H(x, y)$ irgend eine innerhalb dieser Kreise reguläre Funktion bedeutet.

Man sieht sofort, daß die Bedingung des Cousinschen Satzes, der die Funktionen $v_{x_0 y_0}(x, y)$ unterliegen, hier erfüllt ist. Denn jeder in $(\Gamma_{x_0 y_0}, \gamma_{x_0 y_0})$ liegende Punkt (x'_0, y'_0) ist wieder von der Klasse (1), die ihm zugeordnete Funktion $v_{x'_0 y'_0}(x, y)$ hat also die Form $e^{H'(x, y)}$ und somit ist der Quotient:

$$\frac{v_{x_0 y_0}(x, y)}{v_{x'_0 y'_0}(x, y)} = e^{H(x, y) - H'(x, y)}$$

im Punkte (x'_0, y'_0) regulär und von Null verschieden.

Wir gehen nun über zur Betrachtung der Punkte, deren x-Koordinate der zweiten Klasse angehört. Es werde mit:

$$y^{(1)} = f_1(x),\ y^{(2)} = f_2(x),\ \cdots,\ y^{(n)} = f_n(x)$$

die Gesamtheit aller jener Zweige bezeichnet, die für $x = x_0$ einen in T'_y gelegenen Wert $y_0^{(\nu)}$ annehmen. Zufolge der über $f(x)$ gemachten Voraussetzung läßt sich dann jede der Funktionen $f_\nu(x)$ $(\nu = 1, 2, \cdots, n)$ nach positiven ganzzahligen Potenzen von $x - x_0$ entwickeln.

Wir bestimmen nun irgendwie n Teilgebiete $T_y^{(\nu)}$ von T_y gemäß den folgenden Vorschriften:

1. Die Randpunkte von $T_y^{(\nu)}$ gehören nicht zu $T_y^{(\nu)}$.

2. Jeder Punkt von T_y gehört mindestens einem der Gebiete $T_y^{(\nu)}$ an (also gemäß der ersten Bedingung als innerer Punkt).[1])

3. $T_y^{(\nu)}$ enthält den Punkt $y_0^{(\nu)}$ (in seinem Innern), während von den übrigen $(n-1)$ Punkten $y_0^{(1)}, \cdots, y_0^{(n)}$ kein einziger weder im Innern noch am Rande von $T_y^{(\nu)}$ liegen soll.

Jedem Punkte (x_0, y_0) von (T'_x, T'_y), dessen x-Koordinate der zweiten Klasse angehört, ordnen wir nun zu:

1. Einen Kreis $\Gamma_{x_0 y_0}$ in der x-Ebene, dessen Radius ρ so klein zu wählen ist, daß für $|x - x_0| < \rho$:

α) außer den n Zweigen $f_\nu(x)$ $(\nu = 1, 2, \cdots, n)$ kein weiterer Zweig von $f(x)$ einen in T'_y gelegenen Wert annimmt (was unter unserer Voraussetzung immer möglich ist),

β) keiner der n Zweige $f_\nu(x)$ $(\nu = 1, 2, \cdots, n)$ einen Wert y annimmt, der sei es am Rande von T'_y, sei es am Rande oder im Innern eines der Gebiete $T_y^{(\nu')}$ liegt, wenn $\nu' \neq \nu$ ist;[2])

ferner einen Kreis $\gamma_{x_0 y_0}$ in der y-Ebene vom Radius ρ', der so zu wählen ist, daß, wenn y_0 in $T_y^{(\nu)}$ liegt, auch der ganze Kreis $\gamma_{x_0 y_0}$ in $T_y^{(\nu)}$ liegt.

2. Die Funktion:

$$v_{x_0 y_0}(x, y) = y - f_\nu(x),$$

die im Gebiete $(\Gamma_{x_0 y_0}, \gamma_{x_0 y_0})$ regulär ist; dabei ist, wenn y_0 mehreren Gebieten $T_y^{(\nu)}$ angehört, unter den zugehörigen Funktionen $f_\nu(x)$ eine beliebige auszuwählen. Es ist nun wieder zu zeigen, daß diese Zuordnung der Bedingung des Cousinschen Satzes entspricht. Es gehört offenbar jeder Punkt (x'_0, y'_0), der in $(\Gamma_{x_0 y_0}, \gamma_{x_0 y_0})$ liegt, ebenfalls der zweiten Klasse an. Für $x = x'_0$ nehmen nun genau dieselben n Zweige von $f(x)$ in T'_y gelegene Werte an, wie für $x = x_0$; nehmen wir also auch für $x = x'_0$ nach

---

[1]) Diese Forderung hat zur Folge, daß sich die $T_y^{(\nu)}$ teilweise überdecken werden.

[2]) Daraus folgt unmittelbar, daß für $|x - x_0| < \rho$ die Potenzreihen: $f_\nu(x) = \sum_{i=0}^{\infty} a_i^{(\nu)} (x - x_0)^i$ $(\nu = 1, 2, \ldots, n)$ alle konvergieren.

den obigen Regeln eine Teilung von T_y' in Teilintervalle $\overline{T}_y^{(\nu)}$ vor, so lassen sich die Indices der $\overline{T}_y^{(\nu)}$ so wählen, daß gerade der oben mit $f_\nu(x)$ bezeichnete Zweig für $x = x_0'$ einen in $\overline{T}_y^{(\nu)}$ gelegenen Wert annimmt. Wir haben nun zwei Fälle zu unterscheiden; entweder liegt y_0 in $T_y^{(\nu)}$ und y_0' in $\overline{T}_y^{(\nu)}$; dann ist:

$$v_{x_0 y_0}(x, y) = v_{x_0' y_0'}(x, y) = y - f_\nu(x),$$

somit:

$$\frac{v_{x_0 y_0}(x, y)}{v_{x_0' y_0'}(x, y)} = 1;$$

oder es liegt y_0 in $T_y^{(\nu)}$ und y_0' in $\overline{T}_y^{(\nu')}$, ohne daß $\nu' = \nu$ wäre. Dann kann für $y = y_0'$ der Ausdruck $y - f_{\nu'}(x_0')$ nicht verschwinden, da y_0' in $\gamma_{x_0 y_0}$ und somit in $T_y^{(\nu)}$ liegt. Ebensowenig kann aber der Ausdruck $y_0' - f_\nu(x_0')$ verschwinden, da y_0' in $\overline{T}_y^{(\nu')}$ liegt. Es ist also in der Tat in allen Fällen für jeden Punkt (x_0', y_0') von $(\Gamma_{x_0 y_0}, \gamma_{x_0 y_0})$ der Quotient $\dfrac{v_{x_0 y_0}(x, y)}{v_{x_0' y_0'}(x, y)}$ regulär und von Null verschieden.

Wir wenden uns nunmehr den Punkten (x_0, y_0) zu, deren x-Koordinaten der dritten Klasse angehören. Ist etwa $y_0^{(\nu)}$ ein Wert aus dem Innern von T_y', der für $x = x_0$ von mehreren Zweigen von $f(x)$ angenommen wird,[1]) so gilt für jeden dieser Zweige eine Entwicklung:

$$y - y_0^{(\nu)} = \sum_{i=1}^{\infty} a_i (x - x_0)^{\frac{i}{r}}.$$

Werden also mit $y^{(1)}, y^{(2)}, \cdots, y^{(m)}$ die sämtlichen Zweige von $f(x)$ bezeichnet, die für $x = x_0$ den Wert $y_0^{(\nu)}$ annehmen, so ist:

$$\begin{aligned} \varphi_\nu(x, y) = (y - y^{(1)})(y - y^{(2)}) \cdots (y - y^{(m)}) = \\ y^m + \varphi_1(x)\, y^m = + \cdots + \varphi_m(x), \end{aligned}$$

wo $\varphi_1(x), \cdots, \varphi_m(x)$ Funktionen von x bedeuten, die für $x = x_0$ regulär (und eindeutig) sind.

Wir führen nun hier genau wie oben die Teilgebiete $T_y^{(\nu)}$ ein und ordnen dem Punkte (x_0, y_0) zu:

1. einen Kreis $\Gamma_{x_0 y_0}$ in der x-Ebene, dessen Radius ρ so klein zu wählen ist, daß für $|x - x_0| < \rho$:

α) außer den n Zweigen $f_1(x), f_2(x), \cdots, f_n(x)$ kein weiterer Zweig von $f(x)$ einen in T_y' gelegenen Wert annimmt,

[1]) Wegen der für $f(x)$ geltenden Einschränkung ist die Anzahl dieser Zweige immer eine endliche.

β) der Zweig $f_\nu(x)$ $(\nu = 1, 2, \cdots, n)$ nur solche Werte y annimmt, die weder im Innern noch am Rande eines anderen Teilgebietes liegen, als desjenigen, in dessen Innern der Wert $y_0^{(\nu)}$ liegt, den $f_\nu(x)$ für $x = x_0$ annimmt, oder aber am Rande von T_y' liegen;

γ) außer für $x = x_0$ keine zwei unserer n Zweige gleiche Werte annehmen;

ferner einen Kreis $\gamma_{x_0 y_0}$ in der y-Ebene vom Radius ρ', der so zu wählen ist, daß, wenn y_0 in $T_y^{(\nu)}$ liegt, auch der ganze Kreis $\gamma_{x_0 y_0}$ in $T_y^{(\nu)}$ liegt.

2. Wenn der Punkt y_0 in $T_y^{(\nu)}$ liegt und der Wert $y_0^{(\nu)}$ nur von einem Zweige angenommen wird, so ordnen wir dem Punkte (x_0, y_0) die Funktion zu:

$$v_{x_0 y_0}(x, y) = y - f_\nu(x);$$

wird hingegen der Wert $y_0^{(\nu)}$ von mehreren (etwa m) Zweigen angenommen, so werde dem Punkte (x_0, y_0) die oben definierte Funktion:

$$v_{x_0 y_0}(x, y) = \varphi_\nu(x, y) = y^m + \varphi_1(x)\, y^{m-1} + \cdots + \varphi_m(x)$$

zugeordnet.

Wir zeigen auch hier, daß die Zuordnung entsprechend dem Satze von Cousin vorgenommen wurde. Es bedeute also (x_0', y_0') einen von (x_0, y_0) verschiedenen Punkt des Gebietes $(\Gamma_{x_0 y_0}, \gamma_{x_0 y_0})$. Der Punkt x_0' gehört dann offenbar der zweiten Klasse an. Es liege nun der Punkt y_0 in einem zu dem mehrfachen Punkte $y_0^{(\nu)}$ gehörigen Bereiche $T_y^{(\nu)}$, und $y^{(i)}$ sei einer jener Zweige, die für $x = x_0$ den Wert $y_0^{(\nu)}$ annehmen. Der Wert dieses Zweiges für $x = x_0'$ werde mit $(y_0^{(i)})'$ bezeichnet und liege im Gebiete $\overline{T}_y^{(i)}$. Wir nehmen nun zunächst an, der Punkt y_0' liege auch in $\overline{T}_y^{(i)}$. Dann ist:

$$v_{x_0 y_0}(x, y) = \varphi_\nu(x, y); \quad v_{x_0' y_0'}(x, y) = y - f_i(x)$$

und mithin:

$$\frac{v_{x_0 y_0}(x, y)}{v_{x_0' y_0'}(x, y)} = \frac{\varphi_\nu(x, y)}{y - f_i(x)} = \prod_1^{m}{}' (y - f_k(x)),$$

wo der Akzent am Produktzeichen andeutet, daß der Faktor $(y - f_i(x))$ auszulassen ist.

Dieses Produkt ist aber offenbar im Punkte (x_0', y_0') regulär und von Null verschieden, was man wie bei Behandlung der zweiten Punktklasse erkennen kann. Liegt hingegen der Punkt y_0 nicht in dem zu dem mehrfachen Punkte $y_0^{(\nu)}$ gehörigen Bereiche

$T_y^{(\nu)}$, oder gehört der Bereich $\overline{T}_y^{(i)}$ nicht zu einem Zweige der für $x = x_0$ den Wert $y_0^{(\nu)}$ annimmt so unterscheidet sich die anzuwendende Argumentation überhaupt nicht von der für die zweite Klasse verwendeten.

Wir haben nun nur noch den Fall zu untersuchen, daß die Abszisse x_0 von (x_0, y_0) der vierten Klasse angehört. Wir bestimmen die Teilgebiete von T_y wie bisher. Auch die Kreise $\Gamma_{x_0 y_0}$, $\gamma_{x_0 y_0}$, sowie die Funktionen $v_{x_0 y_0}(x, y)$ sind zu definieren wie bisher. Nur wird bei Bestimmung des Radius ρ von $\Gamma_{x_0 y_0}$ für alle Zweige $f_\nu(x)$, die für $x = x_0$ einen am Rande von T_y' gelegenen Wert annehmen, die Bedingung wegfallen, daß sie keine am Rande von T_y' gelegenen Werte annehmen sollen. Jeder Punkt (x_0', y_0') aus dem Inneren von $(\Gamma_{x_0 y_0}, \gamma_{x_0 y_0})$ gehört dann, bei genügend kleiner Wahl des Radius von $\Gamma_{x_0 y_0}$ der ersten, zweiten oder vierten Klasse an. Daß $\dfrac{v_{x_0 y_0}(x, y)}{v_{x_0' y_0'}(x, y)}$ im Punkte (x_0', y_0') regulär und von Null verschieden ist, wird bewiesen wie bisher.

Wir haben jetzt jedem Punkte von (T_x', T_y') zugewiesen:

1. einen Kreis $\Gamma_{x_0 y_0}$ in der x-Ebene und einen Kreis $\gamma_{x_0 y_0}$ in der y-Ebene,

2. eine in $(\Gamma_{x_0 y_0}, \gamma_{x_0 y_0})$ definierte Funktion $v_{x_0 y_0}(x, y)$; wir haben uns weiter überzeugt, daß die Funktionen $v_{x_0 y_0}(x, y)$ der Voraussetzung des zu Beginn dieses Aufsatzes zitierten Theorems von Cousin entsprechen.

Die Anwendung dieses Theorems zeigt also, daß, wenn (T_x'', T_y'') einen ganz im Innern von (T_x', T_y') gelegenen Bereich von der Art dieses letzteren bezeichnet, es immer eine in (T_x'', T_y'') reguläre Funktion gibt, die, durch $v_{x_0 y_0}(x, y)$ dividiert, in $(\Gamma_{x_0 y_0}, \gamma_{x_0 y_0})$ regulär und von Null verschieden ist. Diese Funktion ist offenkundig diejenige, deren Existenz wir beweisen wollten. (T_x'', T_y'') bedeutet dabei einen beliebigen ganz in (T_x, T_y) gelegenen Bereich, denn offenbar läßt sich zu einem beliebig gegebenen (T_x'', T_y'') immer ein (T_x', T_y') finden, das einerseits (T_x'', T_y'') ganz in seinem Innern enthält, anderseits aber selbst in (T_x, T_y) enthalten ist. Unser Satz ist somit bewiesen.

§ 2.

Durch Spezialisierung des in § 1 erhaltenen Resultates erhält man eine Reihe bemerkenswerter Sätze, deren einer im Folgenden bewiesen werden möge.

Wir wollen annehmen, das Gebiet T_x sowohl wie das Gebiet T_y seien Kreise, deren Mittelpunkt der Koordinatenursprung der betreffenden Ebene sei, und die Bedingungen unseres Satzes seien erfüllt, wie groß wir auch die Radien dieser Kreise wählen. Unter Anwendung einer von Weierstraß herrührenden, bereits klassisch gewordenen Beweismethode, die auch auf Funktionen mehrerer Veränderlicher schon wiederholt angewendet wurde,[1]) gehen wir nun folgendermaßen vor. Es bezeichne K_x^1, K_x^2, $\cdots$, K_x^ν, $\cdots$ eine Folge von Kreisen in der x-Ebene, deren gemeinsamer Mittelpunkt der Koordinatenursprung dieser Ebene sei und deren Radien mit ν ununterbrochen bis ins Unendliche wachsen mögen. Dieselbe Bedeutung habe K_y^1, K_y^2, $\cdots$, K_y^ν, $\cdots$ für die y-Ebene. Wir betrachten die Gebiete (K_x^1, K_y^1), (K_x^2, K_y^2), $\cdots$, (K_x^ν, K_y^ν), $\cdots$. Auf jedes dieser Gebiete ist der Satz des vorigen Paragraphen anwendbar und liefert eine im Innern desselben reguläre Funktion $G^{(\nu)}(x, y)$. In (K_x^ν, K_y^ν) ist sowohl $G^{(\nu+1)}(x, y)$ als $G^{(\nu)}(x,y)$ regulär, und der Quotient $\dfrac{G^{(\nu+1)}(x, y)}{G^{(\nu)}(x, y)}$ ist daselbst regulär und von Null verschieden. Es läßt sich daher $lg\, G^{(\nu+1)}(x, y) - lg\, G^{(\nu)}(x, y)$ in eine in (K_x^ν, K_y^ν) konvergierende, nach Potenzen von x und y fortschreitende Potenzreihe entwickeln. Unter $\varepsilon_1, \varepsilon_2, \cdots, \varepsilon_\nu, \cdots$ verstehen wir eine Folge positiver, gegen Null konvergierender Zahlen von der Art, daß $\sum\limits_{\nu=1}^{\infty} \varepsilon_\nu$ einen endlichen Wert hat. Dann läßt sich immer eine Folge von Polynomen $P^{(\nu)}(x, y)$ so bestimmen, daß im ganzen Gebiete (K_x^ν, K_y^ν) die Ungleichung besteht.

$$| lg\, G^{(\nu+1)}(x, y) - lg\, G^{(\nu)}(x,y) - P^{(\nu)}(x,y) | < \varepsilon_\nu.$$

Bestimmt man nun weiter eine Folge von Polynomen $Q^{(\nu)}(x, y)$ aus den Bedingungen:

$$P^{(\nu)}(x, y) = Q^{(\nu+1)}(x, y) - Q^{(\nu)}(x, y),$$

so existiert $\lim\limits_{\nu=\infty} \dfrac{G^{(\nu)}(x, y)}{e^{Q^{(\nu)}(x, y)}}$ für jedes endliche Wertepaar (x, y) und stellt eine ganze Funktion $G(x, y)$ dar, von der Art, daß überall in (K_x^ν, K_y^ν) der Quotient $\dfrac{G(x, y)}{G^{(\nu)}(x, y)}$ regulär und von Null verschieden ist.

Wir können daher den folgenden Satz aussprechen (Satz von Poincaré)[2]):

Damit es eine ganze Funktion $G(x, y)$ von x und y gebe, deren Nullstellen zusammenfallen mit den durch

[1]) Poincaré. Acta mathem, Bd. 2; Appell, Acta mathem., Bd. 2; Biermann, Wiener Ber., Bd. 89; Cousin, Acta mathem, Bd. 19.

[2]) Poincaré. Acta mathem. Bd. 2. S. 113.

die Gleichung $y=f(x)$ gelieferten Wertepaaren, ist notwendig und hinreichend, daß die Funktion $f(x)$ der Bedingung genüge: Zu jeder beliebigen positiven Zahl A sowie zu jedem Punkte x_0 läßt sich eine Konstante ρ so definieren, daß alle der Gleichung $y=f(x)$ genügenden Wertepaare (x, y), deren x-Koordinate der Ungleichung $|x-x_0|<\rho$ genügt, und deren y-Koordinate dem absoluten Betrage nach unter A liegt, durch eine endliche Anzahl von Entwicklungen der Form:

$$y=\sum_{\nu=0}^{\infty} a_\nu (x-x_0)^{\frac{\nu}{r}}$$

geliefert werden.

Man kann diese eine Bedingung auch durch die drei folgenden ersetzen:

1. Für keinen endlichen Wert von x haben die Funktionswerte y eine im Endlichen gelegene Häufungsstelle (was z. B. bei $y=x^\varrho$ der Fall wäre, wenn ρ eine irrationale oder komplexe Zahl bedeutet), noch können für einen endlichen Wert von x unendlich viele Zweige von y einen und denselben endlichen Wert annehmen (was etwa bei der Funktion $y=\frac{1}{lg\,x}$ für $x=0$ eintritt).

2. Haben für $x=x_0$ einer oder mehrere Zweige von y singuläre Stellen, ohne daß für dieselben Entwicklungen von der Form:

$$y=\sum_{\nu=0}^{\infty} a_\nu (x-x_0)^{\frac{\nu}{r}}$$

gelten, so muß für alle diese Zweige $\lim_{x=x_0} y=\infty$ sein, und zwar so, daß sich zu jeder noch so großen Zahl A eine zweite ρ so finden läßt, daß für $|x-x_0|<\rho$ für alle diese Zweige die Ungleichung besteht:

$$|y|>A.$$

(Dadurch sind alle Funktionen ausgeschlossen, welche für einen endlichen Wert von x eine Stelle der Unbestimmtheit, z. B. eine isolierte wesentlich singuläre Stelle haben; auch läßt sich aus diesen Bedingungen folgern, daß die Funktion $y=f(x)$ keine Linien von singulären Stellen haben kann.)

3. Bedeuten R und R' positive Konstante, von denen R' die kleinere sei, so muß sich zu jedem Punkte x_0 eine Konstante η definieren lassen, so daß alle Zweige von $f(x)$, deren Absolutwert für $x=x_0$ größer als R ist, für $|x-x_0|<\eta$ größer als R' bleiben.

Die dritte dieser Bedingungen kommt in dem in der Einleitung zitierten Satze von Poincaré nicht vor. Es ist denkbar, daß sie eine Folge der beiden ersten Bedingungen ist, doch ist es mir nicht gelungen, dies zu zeigen. Daß alle drei Bedingungen auch notwendig sind, geht aus den Überlegungen des ersten Paragraphen hervor.

Daß dieses Resultat verschiedener Verallgemeinerungen fähig ist, liegt auf der Hand. So kann man annehmen, daß die Funktionswerte $y = f(x)$ sich außer im Unendlichen, noch in einer Anzahl im Endlichen gelegener Punkte $y_1, \ldots, y_n$ häufen können und daß alle Zweige von $f(x)$, die sich an der Stelle x_0 nicht verhalten wie eine algebraische Funktion, daselbst unendlich werden oder einen der Werte $y_1, \cdots, y_n$ zur Grenze haben. Bei Hinzunahme einer der dritten Bedingung unseres obigen Satzes analogen Bedingung kann man dann beweisen, daß unsere Funktion als Wurzel einer Gleichung $F(x, y) = 0$ aufgefaßt werden kann, wo $F(x, y)$, als Funktion von y betrachtet, außer im Unendlichen nur in den Punkten $y_1, \cdots, y_n$ singuläre Stellen haben kann, als Funktion von x betrachtet aber eine ganze Funktion ist.

Der Beweis dieses Satzes sowie einiger ähnlicher Sätze läßt sich unschwer führen bei Benützung der Sätze, die in der schon mehrfach erwähnten ausgezeichneten Arbeit von Cousin enthalten sind. Ich brauche hier nicht näher darauf einzugehen, zumal wir im folgenden von diesen Sätzen keinen Gebrauch machen.

§ 3.

Die bisher gewonnenen Resultate ermöglichen es uns nun, einen Satz zu beweisen, der als Übertragung des bekannten Satzes von Weierstraß über die Zerlegung einer ganzen Funktion einer Veränderlichen in Primfaktoren auf die Theorie der ganzen Funktionen zweier komplexer Veränderlicher betrachtet werden kann.

Zu diesem Zwecke muß ich zunächst auf eine spezielle Klasse von ganzen Funktionen hinweisen. Sei $G(x, y)$ eine beliebige ganze Funktion von x und y, die für (x_0, y_0) verschwinde. Bekanntlich gilt dann für eine Umgebung von (x_0, y_0) die Identität:

$$G(x, y) = \left(y^m + \varphi_1(x)\, y^{m-1} + \cdots + \varphi_m(x)\right) e^{H(x, y)}$$

woraus sich ergibt, daß alle übrigen in der Umgebung von (x_0, y_0) gelegenen Nullstellen von $G(x, y)$ durch eine oder mehrere Entwicklungen von der Form:

$$y - y_0 = \sum_{\nu=1}^{\infty} a_\nu (x - x_0)^{\frac{\nu}{r}}$$

geliefert werden. Wir greifen eine dieser Entwicklungen heraus und betrachten sie als Element einer analytischen Funktion $y = f(x)$ von x. Es ist klar, daß alle durch diese Funktion gelieferten Wertepaare (x, y) wieder Nullstellen von $G(x, y)$ darstellen. Es sind nun zweierlei Fälle zu unterscheiden: es ist möglich, daß die durch die Gleichung $y = f(x)$ gelieferten Wertepaare die Nullstellen von $G(x, y)$ erschöpfen; es ist aber auch möglich, daß dies nicht der Fall ist. Schon bei den Polynomen von x und y treten ja bekanntlich beide Fälle auf.

Die Funktionen der ersten Art — bei denen, wie wir uns auch ausdrücken können, die Nullstellen auf einer einzigen analytischen Kurve liegen — wollen wir nun noch weiter einschränken. Wir wollen nämlich diejenigen unter ihnen besonders hervorheben, bei welchen im Produkte

$$y^m + \varphi_1(x)\, y^{m-1} + \cdots + \varphi_m(x) = (y - y^{(1)}) \cdots (y - y^{(m)})$$

keine zwei identisch gleichen Faktoren vorkommen, und dieselben als Primfunktionen von zwei Veränderlichen bezeichnen.[1])

Es gilt zunächst der Satz, daß jede ganze Funktion von zwei Veränderlichen, deren Nullstellen auf einer einzigen analytischen Kurve liegen, abgesehen von einem Exponentialfaktor $e^{G_1(x,y)}$ [wo $G_1(x, y)$ wieder eine ganze Funktion bedeutet], als Potenz einer Primfunktion aufgefaßt werden kann, wie sich sofort aus den Überlegungen von § 2 ableiten läßt. Wir können demgemäß, genau wie für Funktionen einer Veränderlichen, von einem einfachen oder mehrfachen Verschwinden ganzer Funktionen zweier Veränderlicher in den Punkten einer analytischen Kurve sprechen.

Der Satz, den ich beweisen will, ist der folgende:

Jede beliebige ganze Funktion von zwei Veränderlichen läßt sich darstellen als Produkt einer endlichen oder unendlichen Anzahl von Primfunktionen.

Dieser Satz steht zu dem eingangs erwähnten Satze von Weierstraß über die Zerlegung einer ganzen Funktion einer komplexen Veränderlichen in Primfaktoren in demselben Verhältnisse, wie der Satz, daß sich jedes Polynom in x und y als Produkt von unzerlegbaren Polynomen darstellen läßt, zu dem Satze, daß jedes Polynom in x ein Produkt von Linearfaktoren ist. Denn in der Tat sind die unzerlegbaren Polynome in x und y dadurch charakterisiert, daß die durch Nullsetzen eines solchen Polynoms definierte Funktion y eine monogene Funktion von x ist.

Ich schicke zunächst die Bemerkung voraus, daß, wenn (T_x, T_y) irgend einen ganz im Endlichen gelegenen Bereich bedeutet, es unter den durch die Gleichung:

$$G(x, y) = 0$$

definierten monogenen Funktionen $y = f(x)$ nur eine endliche Anzahl geben kann, die in (T_x, T_y) gelegene Wertepaare liefern; denn angenommen, es gäbe unendlich viele solcher Funktionen:

$$y = f_1(x); \quad y = f_2(x), \cdots, \quad y = f_n(x), \cdots,$$

[1]) Man erkennt, daß die ganze Funktion $G(x, y)$, die wir in § 2 hergestellt haben, eine solche Primfunktion ist.

die die in (T_x, T_y) gelegenen Wertepaare $(x_1, y_1), (x_2, y_2) \cdots (x_n, y_n) \cdot \cdot$ annehmen. Diese Wertepaare müssen mindestens einen Punkt von (T_x, T_y) (oder einen Randpunkt dieses Bereiches) zur Häufungsstelle haben; er werde bezeichnet mit (x_0, y_0). Aus der Stetigkeit von $G(x, y)$ in der Umgebung von (x_0, y_0) folgt dann:

$$G(x_0, y_0) = 0.$$

Aus dem nun schon mehrfach angewendeten Satze von Weierstraß folgt weiter, daß alle in der Umgebung von (x_0, y_0) gelegenen Nullstellen von $G(x, y)$ durch eine Gleichung:

$$y^m + \varphi_1(x) y^{m-1} + \cdots + \varphi_m(x) = 0$$

geliefert werden, entgegen unserer Annahme, daß unendlich viele monogene Funktionen Nullstellen liefern, die gegen (x_0, y_0) konvergieren.

Die hiemit bewiesene Eigenschaft der ganzen Funktionen $G(x, y)$ kann als Analogon betrachtet werden zur Tatsache, daß eine ganze Funktion $G(x)$ in jedem endlichen Bereiche nur eine endliche Anzahl von Nullstellen hat. Eine unmittelbare Folge davon ist, daß die durch eine Gleichung:

$$G(x, y) = 0$$

definierten monogenen Funktionen $y = f(x)$ nur eine abzählbare Folge bilden können.

Die durch die Gleichung $G(x, y) = 0$ definierten monogenen Funktionen ordnen wir nun in der folgenden Weise an. Es bezeichne:

$$K_x^1, K_x^2, \cdots, K_x^\nu, \cdots$$

eine Folge von Kreisen in der x-Ebene, deren gemeinsamer Mittelpunkt der Nullpunkt dieser Ebene sei und von denen immer der folgende den vorhergehenden einschließt. Die Radien dieser Kreise mögen ins Unendliche wachsen. Dieselbe Bedeutung habe:

$$K_y^1, K_y^2, \cdots, K_y^\nu, \cdots$$

für die y-Ebene.

Es kann unter den durch $G(x, y) = 0$ definierten monogenen Funktionen nur eine endliche Anzahl geben, die in (K_x^1, K_y^1) gelegene Wertepaare liefern; etwa:

$$y = f_1(x),\ y = f_2(x), \cdots, y = f_m(x).$$

Wir bilden nun nach den Vorschriften von § 2 Primfunktionen, deren Nullstellen mit den durch diese Funktionen gelieferten Wertepaaren übereinstimmen und bezeichnen sie mit:

$$G_1(x, y),\ G_2(x, y), \cdots, G_m(x, y).$$

Der Quotient:

$$\frac{G(x, y)}{G_1(x, y)}$$

ist dann selbst wieder eine ganze Funktion, die keine anderen Nullstellen haben kann als $G(x, y)$. Es sind zwei Fälle zu unterscheiden. Entweder verschwindet dieser Quotient für alle Nullstellen von $G_1(x, y)$, oder er kann nicht längs eines noch so kleinen Stückes der analytischen Kurve:

$$G_1(x, y) = 0$$

verschwinden. Im ersten Falle ist dann auch noch:

$$\frac{G(x, y)}{\left(G_1(x, y)\right)^2}$$

eine ganze Funktion. Indem man in der angegebenen Weise weiterschließt, kommt man schließlich zu einer ganzen Funktion

$$\frac{G(x, y)}{\left(G_1(x, y)\right)^{\lambda_1}}$$

die für kein noch so kleines Stück der Kurve $G_1(x, y) = 0$ mehr verschwindet. Alle in (K_x^1, K_y^1) gelegenen Nullstellen dieser ganzen Funktion müssen auf einer der Kurven:

$$G_2(x, y) = 0, \ldots, G_m(x, y) = 0$$

liegen. Indem wir alle diese Kurven so behandeln wie $G_1(x, y) = 0$, erhalten wir eine ganze Funktion:

$$\frac{G(x, y)}{\left(G_1(x, y)\right)^{\lambda_1}\left(G_2(x, y)\right)^{\lambda_2}\cdots\left(G_m(x, y)\right)^{\lambda_m}}$$

die in (K_x^1, K_y^1) überhaupt keine Nullstelle hat.

Wir betrachten nun die gleichfalls nur in endlicher Anzahl vorhandenen monogenen Funktionen:

$$y = f_{m+1}(x), \cdots \quad y = f_{m'}(x),$$

welche nur solche Nullstellen von $G(x, y)$ liefern, die wohl in (K_x^2, K_y^2), nicht aber in (K_x^1, K_y^1) liegen. Wir werden dadurch auf eine ganze Funktion:

$$\frac{G(x, y)}{\left(G_1(x, y)\right)^{\lambda_1}\left(G_2(x, y)\right)^{\lambda_2}\cdots\left(G_{m'}(x, y)\right)^{\lambda_{m'}}}$$

geführt, die in (K_x^2, K_y^2) keine Nullstelle mehr hat. Diese Schlußweise kann beliebig fortgesetzt werden. Man beachte nun, daß die Funktionen $G_1(x, y)$, $G_2(x, y)$, $\cdots$ so gewählt werden können, daß das Produkt:

$$\Gamma(x, y) = \prod_1^\infty \left(G_\nu(x, y)\right)^{\lambda_\nu}$$

eine ganze Funktion darstellt, die nur für solche Wertepaare (x, y) verschwindet, für die mindestens einer ihrer Primfaktoren verschwindet. Der Beweis kann genau so geführt werden, wie er in § 2 geführt wurde, und bietet keine Schwierigkeit.

Der Ausdruck:

$$\frac{G(x, y)}{\Gamma(x, y)}$$

ist somit eine ganze Funktion, die im Endlichen nirgends verschwindet und daher in der Form darstellbar:

$$\frac{G(x, y)}{\Gamma(x, y)} = e^{H(x, y)}$$

wo $H(x, y)$ ebenfalls eine ganze Funktion bedeutet. Wir haben also die definitive Formel:

$$G(x, y) = e^{H(x, y)} \prod_1^\infty \left(G_\nu(x, y)\right)^{\lambda_\nu}$$

das heißt, wir haben die willkürliche ganze Funktion $G(x, y)$ dargestellt als Produkt von Primfunktionen $G_\nu(x, y)$ worunter wir ganze Funktionen verstanden haben von der Art, daß durch die Gleichung: $G_\nu(x, y) = 0$ nur eine einzige monogene Funktion y von x definiert wird.

Man sieht, wie auch dieses Resultat gewisser Verallgemeinerungen fähig ist. Man kann aus demselben durch Berufung auf einen Satz von Poincaré das folgende ableiten: Jede Funktion zweier Veränderlicher, die im Endlichen keine wesentlich singulären Stellen hat, ist in der Form darstellbar:

$$Q(x, y) = e^{H(x, y)} \frac{\prod_1^\infty \left(F_\nu(x, y)\right)^{\lambda_\nu}}{\prod_1^\infty \left(G_\nu(x, y)\right)^{\lambda_\nu}},$$

wo sowohl $F_\nu(x, y)$ als $G_\nu(x, y)$ Primfunktionen bedeuten.

Durch diese Darstellung sind also sowohl die analytischen Kurven, längs deren die meromorphe Funktion $Q(x, y)$ verschwindet, als diejenigen, auf welchen ihre Pole liegen, in Evidenz gesetzt.

Comments on Hans Hahn's philosophical writings

Christian Thiel
Erlangen

When the collection *Alte Probleme – Neue Lösungen in den exakten Wissenschaften,*[1] containing Hahn's paper „Gibt es Unendliches?" as well as a few commemorative words on its recently deceased author by Karl Menger, appeared in 1934, Richard von Mises wrote a review for *Die Naturwissenschaften*[2] and remembered Hahn by mentioning what he thought were his two most characteristic features: Hahn „belonged to those men, rare today as they have been rare at any time, who do not know of any other goal in their life than the *emendatio intellectus,* the improvement of knowledge", and he was „the master of lucid style". Only a man of this disposition and abilities could have become „the real founder of the Vienna Circle" and „its permanent centre", as Philipp Frank described Hahn in his obituary in volume 4 of the journal *Erkenntnis.* Indeed, over and above this personal rôle and influence, Hahn will be remembered for some of the clearest and finest expositions of the analytical approach to philosophical problems that was typical for the Vienna Circle.

While the necessity of providing for a „philosophical section" in Hahn's *Collected Works* will therefore be evident, the reader will perhaps welcome a few words on the choice of philosophical and foundational texts for the present edition, compared with that made by McGuinness for his editions of Hahn's philosophical papers.[3] Neither our collection nor

[1] *Alte Probleme – Neue Lösungen in den exakten Wissenschaften. Fünf Wiener Vorträge. Zweiter Zyklus.* Franz Deuticke: Leipzig/Wien 1934.

[2] von Mises, Richard: Review of *Alte Probleme – Neue Lösungen* [. . .] (cf. the preceding footnote). *Die Naturwissenschaften* 23, no. 29 (19. Juli 1935), 517–518.

[3] Hahn, Hans: *Empiricism, Logic, and Mathematics. Philosophical Papers.* Edited by Brian McGuinness, with an introduction by Karl Menger. Reidel: Dordrecht etc. 1980 *(Vienna*

those of McGuinness include the manifesto *Wissenschaftliche Weltauffassung. Der Wiener Kreis,*[4] usually quoted as co-authored by Hans Hahn, Otto Neurath, and Rudolf Carnap. Actually, the pamphlet was published anonymously, the names of Hahn, Neurath, and Carnap appearing only as signatures at the end of the „Geleitwort". Karl Menger in his introduction to the McGuinness editions assures us that Hahn only „received the final draft", and that he „signed even though he would have written the pamphlet somewhat differently and was not in complete agreement with all details" (Hahn *1980,* xiv and xviii). So it was a well-founded decision to include this text in Otto Neurath's *Gesammelte philosophische and methodologische Schriften,*[5] as well as an English translation in the collection of papers edited by Marie Neurath and Robert S. Cohen under the title *Empiricism and Sociology.*[6]

The editors of the present collection have purposely omitted Hahn's „Reflexionen über Max Plancks *Positivismus und reale Außenwelt*" which was posthumously discovered in Hahn's *Nachlaß* but turned out to be a fragment which, while containing an interesting discussion of sensory perception and illusions, does not advance to the anti-empiricist contentions in Planck's famous lecture of 1930 which had stimulated Hahn's reflections. Interested readers will find the German text on pp. 59–65 in Hahn *1980,* and an English translation on pp. 43–50 in Hahn *1988.*

„Logik, Mathematik und Naturerkennen", an English version of which had been missing in Hahn *1980* since it had already been assigned for publication in the English edition of the *Einheitswissenschaft* collection,[7] was included in Hahn *1988* and is also included here.

So the only item the addition of which to the present selection might have been considered is Hahn's „Anmerkungen" to Alois Höfler's edition

Circle Collection, vol. 13); idem, *Empirismus, Logik, Mathematik.* Mit einer Einleitung von Karl Menger ed. Brian McGuinness. Suhrkamp: Frankfurt a. M. 1988 (*stw* 645); henceforth quoted, respectively, as „Hahn *1980*" and „Hahn *1988*".

[4] *Wissenschaftliche Weltauffassung. Der Wiener Kreis.* Herausgegeben vom Verein Ernst Mach. Artur Wolf: Wien 1929 *(Veröffentlichungen des Vereines Ernst Mach.)*

[5] Neurath, Otto: *Gesammelte philosophische und methodologische Schriften,* ed. Rudolf Haller/Heiner Rutte, I–II. Hölder/Pichler/Tempsky: Wien 1981.

[6] Neurath, Otto: *Empiricism and Sociology. With a Selection of Autobiographical Sketches,* ed. Marie Neurath/Robert S. Cohen. Reidel: Boston 1973 (*Vienna Circle Collection,* vol. 1).

[7] Hahn, Hans: „Logic, Mathematics, and Knowledge of Nature", in: *Unified Science,* ed. Brian McGuinness (Reidel: Dordrecht etc. 1987, *Vienna Circle Collection,* vol. 19), 24–45 and (endnotes) 278–282.

of Bernard Bolzano's *Paradoxien des Unendlichen.*[8] There is no doubt that many of these notes are illuminating and of interest for the philosophy of mathematics, but it is equally obvious that they can unfold their explanatory potential only jointly with Bolzano's text the reproduction of which, together with Hahn's notes, was out of the question for manifest reasons.

All of Hahn's philosophical papers and public lectures are concerned with or touch upon the nature of mathematical knowledge, the reach of human intuition and intellect, the meaning of mathematics for culture in general, and the motivation and details of the position taken by logical empiricism on these questions. Thanks to their directness and perspicuity of style, the texts are not in need of any commentary on their purpose or contents. Comments have, therefore, been restricted to the origin and context of the topic, to the location (and sometimes the background) of quotations, and to bibliographical references; in some cases where the lasting or growing importance of a problem discussed by Hahn seemed particularly striking, I have added a few comments on their present status.

The collection begins with three items which are of interest mainly for mathematics, its methodology and philosophy. Only two thirds of the first item, an unusually long review of Alfred Pringsheim's *Lectures on the Theory of Numbers and Functions* in the *Göttingische gelehrte Anzeigen* of 1919,[9] was included in Hahn *1980* and Hahn *1988* since the last part was considered to be of purely mathematical interest. Hahn concentrates on problems connected with the introduction of the concept of natural number and with the use of the so-called principle of permanence for the extension of domains of numbers. In view of the famous lecture „The Crisis in Intuition" of 1933 (cf. *infra*), particular attention should be paid to Hahn's discussion of intuition which he splits into geometrical and arithmetical intuition, and to his preference of „logicization" („Logisierung") to „arithmetization". But even though arithmetic, analysis and geometry may all be logicized, all progress in the more subtle parts of analysis has first and foremost been achieved with the help of intuition, the use of which is regarded by Hahn as an indispensable means of mathematical research. It is for good reasons that McGuinness has placed these considerations which

[8] Bolzano, Bernard: *Paradoxien des Unendlichen,* ed. Alois Höfler, mit Anmerkungen versehen von Hans Hahn. Felix Meiner: Leipzig 1920 (*Philosophische Bibliothek,* vol. 99; repr. Hamburg 1955). Hahn's annotations are on pp. 133–153.

[9] Hahn, Hans: Review of Alfred Pringsheim, *Vorlesungen über Zahlen- und Funktionenlehre.* Erster Band, erste Abteilung [. . .] Zweite Abteilung, *Göttingische gelehrte Anzeigen* 181 (1919), Nr. 9 and Nr. 10 (September and October), 321–347.

conclude the part of the review selected for inclusion in Hahn *1980* and Hahn *1988*, immediately before the reprint of „Die Krise der Anschauung".

The next two items are part of a quarrel provoked by Hahn's (at one point rather harsh) criticism on details of Otto Stolz's and Josef Anton Gmeiner's *Theoretische Arithmetik* (Leipzig ^{2}1911/1915) in the Pringsheim review. Gmeiner responded with „Arithmetische Bemerkungen; insbesondere über die Peanoschen Axiome. (Entgegnung auf eine Bemerkung des Herrn H. Hahn.)" in the *Jahresbericht der Deutschen Mathematiker-Vereinigung* 30 (1921), 82–91, and Hahn renewed his criticism in (our second item) „Arithmetische Bemerkungen. (Entgegnung auf Bemerkungen des Herrn J. A. Gmeiner)", ibid. 30 (1921), 170–175. Gmeiner was not satisfied and published another reply „Zu den ‚Arithmetischen Bemerkungen' des Herrn Hahn", loc. cit. 175–178, to which Hahn wrote „Schlußbemerkungen hierzu von H. Hahn" on pp. 178–179 (our third item). Philosophically, the gain is scanty.

„Empirismus, Mathematik, Logik"[10] is a very short exposition of the empiricist point of view concerning the epistemological status of logic and mathematics, stating that the answer to the question „Is a consistent empiricism possible?" will depend on the outcome of the dispute about the foundations of mathematics. This issue will be taken up again (and commented on) in „Logik, Mathematik and Naturerkennen".

Few comments are required by „Mengentheoretische Geometrie",[11] i. e. „Set-theoretical Geometry", a beautifully perspicuous survey of new results and logically satisfactory clarifications of notions achieved by applying set-theoretical concepts and methods to geometry; in the more abstract setting of today, all this is topology. The discussion of concepts like continuity, local connectedness, dimension etc. leads to considerable overlapping with Hahn's „Die Krise der Anschauung", to be discussed later, but the present article gives more details (as well as bibliographical references) due to the fact that, unlike the lecture of 1933, it is directed to readers with a substantial background in mathematics.

„Die Bedeutung der wissenschaftlichen Weltauffassung, insbesondere für Mathematik und Physik" („The Significance of the Scientific World

[10] Hahn, Hans: „Empirismus, Mathematik, Logik", *Forschungen und Fortschritte* 5 Nr. 35/36 (10. und 20. Dezember 1929), 409–410, also in Hahn *1988*, 39–42; English translation „Empiricism, Logic, and Mathematics" in Hahn *1980*, 39–42.

[11] Hahn, Hans: „Mengentheoretische Geometrie", *Die Naturwissenschaften* 17 (1929), 916–919. This text was not included in Hahn *1980* and Hahn *1988*.

View, especially for Mathematics and Physics") was published in volume 1 of *Erkennntis.*[12] Hahn is canvassing for a scientific world-view, which he sees as a confession of faith in the methods of the exact sciences, in careful logical reasoning, in the patient observation of phenomena – all this in contrast to philosophy in its customary understanding as a doctrine about the world, claiming a standing equal to or even higher than that of the scientific disciplines. „Everything that can be said meaningfully at all, is a proposition of some special science", and serves, in this position, as a standard for testing other propositions and possibly unveiling them as mere pseudo-propositions. Experience is the only source of our knowledge – not experience and thinking, as traditional philosophy assumed; all thought is nothing but tautological transformation, although this is by no means unimportant since otherwise not even the fundamental equations of the electromagnetic field would be amenable to experimental verification.

Hahn's statement that scientific philosophy aims at establishing an axiom-system suitable for the logicization of all physics by embedding it into a theory of relations, exhibits the most striking contrast to some present-day approaches which would rather try to find a firm foundation for the sciences in unquestionable parts of our life-world (thereby keeping its distance from traditional foundationalism as well as from scientistic naïvetés).

Hahn's opposition to Plato on page 99 refers to the allegedly Platonic dictum „God eternally geometrizes" which, however, does not appear in the extant writings of Plato's.[13] Wittgenstein's tenet of the tautological character of logic (page 100) can be found in his *Tractatus Logico-Philosophicus.*[14]

„La Divina Commedia" is the now customary designation of Dante Alighieri's epos „La Comedia" which was composed c. 1307/1321 and first printed in 1472.

[12] Hahn, Hans: „Die Bedeutung der wissenschaftlichen Weltauffassung, insbesondere für Mathematik und Physik", *Erkenntnis* 1 (1930–1931), 96–105; reprinted in Hahn *1988*, 38–47; English translation „The Significance of the Scientific World View, especially for Mathematics and Physics" in Hahn *1980*, 20–30.

[13] It is ascribed to Plato in Plutarch's *Symposiaka* (VIII 2, 718 b–c), cf. *Plutarch's Moralia in Sixteen Volumes*, IX, ed. E. L. Milnar et al. (Cambridge, Mass./London 1969), 118–119. C. F. Gauß is reported to have used the parallel version „God arithmetizes".

[14] Wittgenstein, Ludwig: *Tractatus Logico-Philosophicus* (Routledge & Kegan Paul: London 1922, quoted from the 1958 impression), 6.1: „The propositions of logic are tautologies"; 6.11 (1): „The propositions of logic therefore say nothing"; 4.0312 (2–3): „My fundamental thought is that the ‚logical constants' do not represent. That the *logic* of the facts cannot be represented".

The „axiom of reducibility", mentioned on page 101, was introduced by A. N. Whithehead and B. Russell in their *Principia Mathematica* (ch. II, § VI) in order to avoid some consequences of the theory of types for classical mathematics (e. g. the illegitimacy of the usual definition of the least upper bound of a non-empty bounded set of real numbers, because of its violation of the vicious-circle principle). It postulates that every function is coextensive with a predicative one, i. e. with a function the highest order of an argument of which is just one less than that of the function. Whitehead and Russell admitted that this axiom „has a purely pragmatic justification" and is „not the kind of axiom with which we can rest content" (*PM* ²1925, xiv). Wittgenstein reflects critically on the axiom of reducibility in his *Tractatus.*[15] There is an excellent discussion of the axiom in chapter XI of W. V. Quine's *Set Theory and Its Logic* (The Belknap Press of Harvard University Press, Cambridge, Mass. 1963). On Wittgenstein's conception of identity see his *Tractatus,* 4.241–4.243 and 6.23–6.2323.

Hahn's unconcealed scepticism towards Hilbert's metamathematics or proof theory seems unnecessary today since the syntactic approach has received strong philosophical and methodological support from operative logic („protologic") and operative mathematics.

The reference on page 102 is to Russell's *Unser Wissen von der Außenwelt* (Felix Meiner: Leipzig 1926), Vierte Vorlesung: „Die Welt der Naturwissenschaft und die Sinnenwelt", 131–169.[16] The quotations on page 104 are from Hedwig Conrad-Martius' „Realontologie. I. Buch", *Jahrbuch für Philosophie und phänomenologische Forschung* 6 (1923), 159–333.[17]

Überflüssige Wesenheiten (Occams Rasiermesser), i. e. *Superfluous Entities (Occam's razor)*[18], appeared as a pamphlet of twenty-four pages in 1930. Aiming at philosophical enlightenment, Hahn, in an amusing way, inverts the usual stereotype of the reflective thinker vs. the thoughtless worldling to the equally ideological dichotomy between „world-affirming" and „world-denying" philosophies. While the former are optimistic, em-

[15] Op. cit., 6.1232 (2)–6.1233: „Propositions like Russell's ‚axiom of reducibility' are not logical propositions, and this explains our feeling that, if true, they can only be true by a happy chance. We can imagine a world in which the axiom of reducibility is not valid. But it is clear that logic has nothing to do with the question whether our world is really of this kind or not."

[16] This corresponds to pp. 106–134 in the original English edition *Our Knowledge of the External World as a Field for Scientific Method in Philosophy* (Allen & Unwin, London, and The Open Court, Chicago/London, 1914).

[17] The passages quoted are from pp. 321–322, or pp. 163–164 of the separate edition.

[18] Reprinted in Hahn *1988*, 21–37; English translation in Hahn *1980*, 1–19.

brace an empiristic or sensualistic epistemology, feel at home in the sensible world and will not look out for entities behind it, the latter are essentially pessimistic, offer an idealistic epistemology and assume „another" world populated with entities different from those we are familiar with. Among them are „superfluous entities" like Platonic ideas, metaphysical substances, impossible objects, points of space, moments of time, numbers and logical classes. All of them, in Hahn's opinion, should be removed by „Occam's razor", the principle that tells us not to assume more entities than absolutely necessary.

Hahn's short characterization of the doctrine of ideas on page 5 follows the usual scheme modelled upon Aristoteles' discussion in his *Metaphysics* (I 6, 987a ff., I 9, 990a ff., XIII 4, 1078b ff.). There is no single comprehensive exposition of the doctrine in Plato (but cf. his *Parmenides* 132 d).

The formula „Entia non sunt multiplicanda praeter necessitatem" (page 5), the customary formulation of Occam's razor, does not occur in Occam's writings, although we find related ones like „Pluralitas non est ponenda sine necessitate", and „frustra fit per plura, quod potest fieri per pauciora" (cf. Oswald Schwemmer, „Ockham", *Enzyklopädie Philosophie und Wissenschaftstheorie* 2, 1984, 1057–1062). For a thorough investigation into the origins of the „razor" formulation cf. Wolfgang Hübener, „Occam's razor not mysterious", *Archiv für Begriffsgeschichte* 27 (1983), 73–92.

I have not been able to locate, in the time given, Seneca's complaint about translating the Greek τὸ ὄν into Latin (page 7).

In contrast to the „entia"- quotation just mentioned, the formula „Sufficiunt singularia, et ita tales res universales omnino frustra ponuntur" on page 8 *does* occur in Occam: cf. *Scriptum in librum primum sententarium – Ordinatio,* ed. St. Brown (St. Bonaventure, N. Y. 1970 = *Opp. theol.* II), 143.

„Hie Waiblingen!" on page 9 is an historical allusion calling on for taking sides: the party-cry „Hie Welf, hie Waiblingen!" is said to have been shouted at the battle of Weinsberg in 1140, where Conrad III. fought Count Welf (Waiblingen was a Hohenstaufen castle near Stuttgart). The aim of this imaginative backdating, as well as the purpose of Hahn's allusion, is to conjure up the later battle between secular and ecclesiastic powers (in Italy: between the Ghibellini and the Guelfi) beginning in the middle of the 13th century.

An interesting feature, in view of the Vienna Circle's involvement in the „linguistic turn", is Hahn's warning against an overestimation of language that may lead to a mistaken inference from linguistic structures to the „structure of the world", and has already misguided renowned thinkers into developing a doctrine of „impossible objects". As mistaken as their introduction is the retention of pseudo-entities like space, time, number, substance, and – in the ultimate analysis – even classes, all of which call for an application of Occam's razor.

The „Diskussion zur Grundlegung der Mathematik" („Discussion about the Foundations of Mathematics") was printed in *Erkenntnis* 2 (1931), 135–141, as part of the discussion following the papers on the same subject read by Carnap, Heyting, von Neumann and Waismann[19] on September 7, 1930, at the „2. Tagung für Erkenntnislehre der exakten Wissenschaften" in Königsberg. For Hahn, the empiricist, it is the foremost task of a foundation of mathematics to explain how the applicability of logic and mathematics to reality is compatible with the empiricist position. Heyting's and von Neumann's papers have made it clear to him that intuitionism and formalism do not meet this requirement – Brouwer's investigations as well as Hilbert's are highly significant within mathematics, but they do not provide a foundation of mathematics. Hahn proceeds to a sketch of his own position which concurs with Carnap's logistic (i. e. logicist) standpoint. He is quite aware of the problem posed by some axioms of mathematics which, like the axiom of choice, appear to be non-tautological and as having a „real content"; at least this seems to Hahn to have been Russell's position,[20] and he thinks that Ramsey's attempt[21] to ascribe a tautological character to the

[19] Cf. Carnap, Rudolf: „Die logizistische Grundlegung der Mathematik", *Erkenntnis* 2 (1931), 91–105: Heyting, Arend: „Die intuitionistische Grundlegung der Mathematik", ibid., 106–115; von Neumann, Johann: „Die formalistische Grundlegung der Mathematik", ibid., 116–121. A prefatory notice to the conference report states that Waismann, whose paper on Wittgenstein's position had been announced in the programme, did not have it ready for printing and that it would be published in a later issue. It seems that this plan was not realized either; Waismann's paper „Über den Begriff der Identität" in *Erkenntnis* 5 (1935), 56–64, which refers to some of Wittgenstein's views on mathematical knowledge, is not marked as the belated conference paper and has only a footnote stating Waismann's indebtedness to Wittgenstein.

[20] This claim of Hahn's is not easily confirmed, unless he has in mind Russell's illustration of the multiplicative axiom (which is a consequence of „Zermelo's axiom", i. e. of the axiom of choice) by the problem of picking left boots and right boots from a set of pairs of boots, in „On some difficulties in the theory of transfinite numbers and order types", *Proceedings of the London Mathematical Society* (2) 4 (1907), 29–53, especially 47 f.

[21] Cf. Ramsey, Frank Plumpton: „The Foundations of Mathematics", *Proceedings of the London Mathematical Society* (2) 25 Pt. 5 (1925), 338–384, also in F. P. Ramsey, *Foundations. Essays in Philosophy, Logic, Mathematics and Economics,* ed. D. H. Mellor (Routledge & Kegan Paul: London/Henley 1978), 152–212. See especially 175 and 208 f.; „the Multiplicative Axiom seems to me the most evident tautology", 208).

axiom of choice has failed. Hahn closes with some remarks on Wittgenstein's criticism of Russell's position, as reported to the participants of the conference by Waismann (cf. footnote 19).

Logik, Mathematik und Naturerkennen appeared as the second monograph of the series *Einheitswissenschaft* in 1933; it comprises two public lectures given in 1932, and was translated into French in 1935, and into English in 1987.[22] As in his *Erkenntnis* paper of 1930/31, Hahn argues extensively for the neopositivist position, in his usual clear manner that does not call for a commentary, so that a few comments on details will suffice.

The sensualist motto on page 7, „Nihil est in intellectu, quod non prius fuerit in sensu" appears in this wording in Aquinas's *Quaestio disputata de veritate.*[23] It is mainly known through Leibniz's slightly modified and augmented version „Nihil est in intellectu quod non fuerit in sensu, excipe: nisi ipse intellectus" in his *Nouveaux Essais sur l'Entendement Humain* (Bk. II, ch. I, § 2),[24] which could explain the frequent ascription of the phrase to Locke who, in the corresponding passage of his *Essay Concerning Human Understanding* (to which Leibniz's *Nouveaux Essais* refer throughout), maintains the underlying doctrine without, however, using or quoting Aquinas's Latin version or a translation thereof.

The discussion of colour predicates on pages 12 ff. has remained a topic of modern analytic philosophy, in the constructivistic branch of which propositions like „no object is red as well as blue at the same time" are considered as based on predicator rules, and called „materially analytic".

The example in lines 6 to 13 on page 14 is taken from the game of taroc or tarot which was (and still is) quite customary in Austria as well as in Southern Germany. The „Skieß" (also „Sküs") is the „fool", a card like the joker that can neither take nor be taken, so that an attempt to take it with the moon would simply violate the rules that constitute the game.

A discussion of the „tautological" character of true arithmetical statements on pp. 17 ff. could profit from the more recent distinction of „properly" analytic propositions and formally synthetic ones, the latter involving

[22] Cf. *Logique, Mathématiques et Connaissance de la Réalité.* Hermann & Cie.: Paris 1935 (with an introduction by Marcel Boll; *Actualités scientifiques et industrielles,* 266). For the bibliographical data of the English translation of 1987 see footnote 7.

[23] S. Thomae Aquinatis *Opera Omnia,* ed. Robertus Busa SI, t. 3: *Quaestiones disputatae, Quaestiones quodlibetales, Opuscula* (Frommann – Holzboog: Stuttgart – Bad Cannstatt 1980), qu 2, ar 2, ag 19.

[24] Cf. Gottfried Wilhelm Leibniz, *Sämtliche Schriften und Briefe, Sechste Reihe: Philosophische Schriften. Sechster Band: Nouveaux Essais* (Akademie-Verlag: Berlin 1962), 111.

at least one („synthetic") construction rule in their proofs (which may happen indirectly by reference to such rules in proofs by induction, e. g. in a proof of Leibniz's example „3 = 2 + 1"). It should also be noted that on pages 17 and 18, the phrase „dasselbe meinen" must be understood as „having the same reference", and not as „having the same sense". For God's „mathematizing" see footnote 13 above.

Information on the astronomical examples on page 20 can be found in Morton Grosser's *The Discovery of Neptune*[25] and in Fred Hoyle's *Astronomy and Cosmology.*[26]

Hahn's emphasis on rules as „directions for use" of unconstitutable terms will remind contemporary readers of Weyl-Lorenzen-style „modern abstraction" which makes use of transformation rules for almost the same purpose.

The term „paper currency" was presumably taken over from Weyl's article on the „new foundational crisis" in mathematics.[27]

Hahn's dismissal of prediction as the *characterizing* feature of science in contrast to history (which, incidentally, I have failed to find in Poincaré) is, of course, due to this strong defence of the idea of a unified science. Even though critical readers may wish (unlike Hahn on page 26) to distinguish between the *criterion* and the *meaning* of an historical statement, Hahn's proposal to interpret a historical statement as a prediction of its corroboration by all future findings of documents, new evidence etc., seems to be all his own and is indeed a highly original contribution that would merit a closer methodological discussion.

There are, finally, a few corrigenda to the endnotes. In note 2, the publication date of Kant's *Prolegomena* is wrong, since the first edition was published only in 1783; one might conjecture that „1749" is a misprint for „1794", a year in which the *Prolegomena* were indeed reprinted in Frankfurt and Leipzig. In note 11, Russell's co-author in the *Principia* should be, of course, not „A. Wittgenstein", but Alfred North Whitehead.

[25] Grosser, Morton: *The Discovery of Neptune,* Harvard University Press: Cambridge, Mass. 1962; translated into German as *Entdeckung des Planeten Neptun,* Suhrkamp: Frankfurt a. M. 1970.

[26] Hoyle, Fred: *Astronomy and Cosmology. A Modern Course.* W. H. Freeman: San Francisco 1975.

[27] Weyl, Hermann: „Über die neue Grundlagenkrise der Mathematik", *Mathematische Zeitschrift* 10 (1961), 39–79, reprinted separately by Wissenschaftliche Buchgesellschaft: Darmstadt 1965, also in Weyl's *Gesammelte Abhandlungen,* ed. K. Chandrasekharan (Springer: Berlin/Heidelberg/New York 1968), II, 143–179. The „Papiergeld" on pp. 54–55, and pp. 156–157, respectively.

In note 14, the article „Die“ should be prefixed to the title of Mach's *Mechanik,* and in note 17, the title of Dewey's book should read „Studies in Logical Theory“.

Jointly with the pamphlet on Occam's razor, the two last items of our selection belong to Hahn's most famous philosophical writings. Their style, maintaining an exemplary clearness on a high level of exposition, has made them real „classics of presentation“ worth being read even more than fifty years after their first publication. The popular lecture „Die Krise der Anschauung“[28] („The Crisis in Intuition“), a didactical masterpiece, outlines the progress of a „physicalization of physics“ by purging it from metaphysical elements since Kant's time (and Kant's teachings). With an eye to the relation between geometry and intuition, some fundamental concepts and results of analysis, topology, and the theory of curves are explained. Hahn strongly backs up the depreciation of intuition which is connected with this development, and he emphasizes the successful application of many-dimensional and non-Euclidean constructions in physics where (to quote Hahn's concluding sentence) „it is not true, as Kant urged, that intuition is a pure *a priori* means of knowledge, but rather that it is force of habit rooted in psychological inertia“ (62, i. e. 101 in Hahn *1980*). This extreme position, which met with mild criticism already in von Mises's review (cf. footnote 2), has recently been discussed (and rejected) by Volkert[29] who ascribes its impetus to Hahn's favouring logicism, a position that banishes intuition from the foundations of mathematics altogether.

The famous passage from Kant with which Hahn starts his argument is from the *Kritik der reinen Vernunft* (A 50 = B 74; for „jene“ read „diese“); its interpretation by Hahn, and the adequacy of his outline of Kant's doctrine in general, are open to some doubt. The reference to Karl Menger on page 44 is to his lecture „Die neue Logik“ in the same volume *Krise und Neuaufbau* [. . .], 93–122. The square-filling curve on page 51 *f.* was published by Giuseppe Peano („a master of the counter-example“ according to

[28] Hahn, Hans: „Die Krise der Anschauung“. In: *Krise und Neuaufbau in den exakten Wissenschaften. Fünf Wiener Vorträge.* [Erster Zyklus] (Franz Deuticke: Leipzig/Wien 1933), 41–64; reprinted in Hahn *1988,* 86–114. English translation in James R. Newman (ed., comm.), *The World of Mathematics* [. . .], vol. III (Simon and Schuster: New York 1956), 1956–1976, reprinted in Hahn *1980,* 73–102.

[29] Volkert, Klaus Thomas: *Die Krise der Anschauung. Eine Studie zu formalen und heuristischen Verfahren in der Mathematik seit 1850.* Vandenhoeck & Ruprecht: Göttingen 1986 (*Studien zur Wissenschafts-, Sozial- und Bildungsgeschichte der Mathematik,* Band 3), 4.3.1: Hahns Vortrag „Die Krise der Anschauung“, 251–260.

his biographer Hubert C. Kennedy) in 1890.[30] The concept of „Local connectedness" („Zusammenhang im Kleinen"), mentioned on page 53, was introduced independently by Hahn and by Stefan Mazurkiewicz. Brouwer's construction explained on pages 54 f. is from his „Zur Analysis Situs",[31] Wada's Construction was reported in K. Yoneyama's „Theory of continuous set of points" in 1917;[32] I have so far not been able to obtain any further information on Wada.

Sierpiński's curve all of whose points are branch points (a construction remindful of that of „Sierpiński's carpet")[33] is the last of the detailed examples on which, together with many-dimensional, non-Euclidean and non-Archimedean geometries, Hahn draws in order to demonstrate the incapability of intuition of acting as a reliable guide within mathematics: if we had relied on it in our examples, we „would have been led into heavy error" (56). While this concurs with Poincaré's reaction to Peano's curve,[34] the original excitement has settled by now, and contemporary attitude tends to take the „critical cases" simply as evidence that intuition is in need of being made more precise by resort to formal methods – even if one joins Hahn's judgement that „the space of geometry is not a form of pure intuition, but a logical construction" (60).[35] Hahn, who in 1933 stated that „We have as yet no indication that the inclusion of non-Archimedean geometry might prove useful" (61), would have rejoiced at modern applications of

[30] Peano, Giuseppe: „Sur une courbe qui remplit toute une aire plane", *Mathematische Annalen* 36 (1890), 157–160. Cf. Hubert C. Kennedy, *Peano. Life and Works of Giuseppe Peano* (Reidel: Dordrecht etc. 1980), 31–32, 173.

[31] Brouwer, Luitzen Egbertus Jan: „Zur Analysis Situs", *Mathematische Annalen* 68 (1910), 422–434, repr. in his *Collected Works. 2. Geometry, Analysis, Topology and Mechanics*, ed. Hans Freudenthal (North-Holland: Amsterdam/Oxford, American Elsevier: New York, 1976), 352–366.

[32] Yoneyama, K.: „Theory of continuous set [sic!] of points" [Part I] *Tôhoku Mathematical Journal* 12 (1917), 43–158; the pages 60–62 with Wada's construction have been reprinted on pp. 367–369 in volume 2 of Brouwer's *Collected Works* (see the preceding footnote).

[33] Sierpiński, Wacław: „Sur une courbe dont tout point est un point de ramification", *Comptes Rendus hebdomadaires des séances de l'Académie des Sciences [Paris]* 160 (1915), 302–305. For the carpet, see e. g. N. Ya. Vilenkin, *Stories about Sets* (Academic Press: New York/London 1968), 109.

[34] See e. g. Vilenkin (*op. cit.* in footnote 32), 123, and A. G. Konforowitsch, *Logischen Katastrophen auf der Spur* (Fachbuchverlag: Leipzig/Köln [2]1992), 189.

[35] An alternative approach to geometry which tries to keep a „reasonable core" of the traditional conceptions was not developed until the sixties. For an introduction to the idea, cf. C. Thiel, *Philosophie und Mathematik* (Wissenschaftliche Buchgesellschaft: Darmstadt 1995), chapter 13: „Geometrie als Theorie der Formen".

non-Archimedean systems in Nonstandard Analysis and, e. g., to the theory of Brownian motion. Perhaps he would even have taken this as confirming his prediction that, since intuition is not „a pure *a priori* means of knowledge", but rather a „force of habit rooted in psychological inertia" (62), mankind would once regard the „deviant" geometries as intuitively acceptable.

On page 61, lines 16–17, „Anschauungsmöglichkeiten" should be „Anschauungsunmöglichkeiten" (a typographical error already corrected tacitly in the translation in Hahn *1980,* 100).

On Urysohn, see A. Paplauskas in the *Dictionary of Scientific Biography* XIII, 548 *f.,* and Hahn in „Mengentheoretische Geometrie" *(v. supra),* 917–918.

The lecture „Gibt es Unendliches?"[36] of 1934, after a short outline of the history of the problem, discusses in plain language and in a clear line of thought the so-called paradoxes of the infinite, the foundational positions on the question of mathematical existence, the aims of Hilbert's Programme (essentially, the consistency of mathematics), and the idea of answering the question whether physical space is infinite by incorporating physical events into a *mathematical* space. Hahn is convinced that even a future „finite physics", should it become necessary, would not lessen the value and beauty of the logic and mathematics of the infinite. Once more, there is no need for a commentary beyond references.

The quotations on pages 93–94 and their context can be found in the following places: René Descartes, *Œuvres de Descartes,* ed. Charles Adam/Paul Tannery, VIII-1 *Principia Philosophiae,* Nouvelle Présentation (J. Vrin: Paris 1982), 14 *f.;* God. Guil. Leibnitii *Opera Philosophica quae exstant Latina Gallica Germanica Omnia,* ed. J. Erdmann, Pars Prior (G. Eichler: Berlin 1840), 17–118. The passage is taken from Leibniz's reply to a letter by Foucher; its first sentence appears as motto on the title page of Bolzano's *Paradoxien des Unendlichen* (cf. footnote 8, and *infra*). The second quotation from Leibniz is from his *Nouveaux Essais* (*Philosophische Schriften* VI-6, Akademie-Verlag: Berlin 1962), 159. The words quoted are from the mouth of Philalethes who speaks for John Locke, who, in the corresponding passage of § 8 of *An Essay Concerning Human Understanding* (1690; ed. Peter H. Nidditch, Clarendon Press: Oxford 1975)

[36] „Gibt es Unendliches?", in: *Alte Probleme – Neue Lösungen in den exakten Wissenschaften. Fünf Wiener Vorträge. Zweiter Zyklus* (Franz Deuticke: Leipzig/Wien 1934), 93–116; reprinted in Hahn *1988,* 115–140. English translation: „Does the Infinite exist?" in Hahn *1980,* 103–131.

writes: „The infinity of Numbers, to the end of whose addition every one perceives there is no approach, easily appears to any one that reflects on it: But how clear soever this *Idea* of the Infinity of Number be, there is nothing yet more evident, than the absurdity of the actual *Idea* of an Infinite Number." Carl Friedrich Gauß's words are from a letter to Schumacher,[37] Bolzano's from his *Paradoxien des Unendlichen.*[38]

On pages 97 ff., Hahn speaks unhesitatingly of the cardinal numbers of the integers as well as of the reals, even though he is careful to say that what Cantor's procedure establishes is the fact that the totality of real numbers *is not denumerable.* Modern foundational critique has pointed out (beginning with Thoralf Skolem in the twenties) that the concepts of denumerability, cardinality, finiteness and infiniteness are relative to the formal language employed, i. e. they are not absolute concepts, so that it is at least doubtful whether the reals have a unique cardinality. For an introduction to this problem field, see chapter 8 of Thiel *1995* quoted in footnote 35.

In the discussion of the axiom of choice on page 100, the reference is to Karl Menger's paper „Ist die Quadratur des Kreises lösbar?" in the same volume, pp. 1–28. Menger, who in turn refers to Hahn's later paper, insists on accepting the axiom of choice. For the Banach-Tarski paradox and its context, see Stan Wagon's *The Banach-Tarski Paradox* (Cambridge University Press: Cambridge etc. 1985, corr. 1986), for the continuum problem P. Schroeder-Heister's article „Kontinuumhypothese", *Enzyklopädie Philosophie und Wissenschaftstheorie* 2 (1984), 460–461, and Hahn on page 105 of the present lecture.

In the remaining parts of the lecture, Hahn takes sides with „classical" logic and mathematics, and against intuitionistic or constructivistic revisionism – beginning with the presentation of the different views on the concept of mathematical existence. The introductory „quot capita – tot sententiae" („as many positions as people") is a hybrid of Terence's „Quot homines, tot sententiae" (*Phormio* II, 4, 14) and Horace's „Quot capitum vivunt, totidem studiorum milia" (*Saturnalia* II, 1, 27 *f.*). Today, many aspects of the ensuing discussion would appear in a different light, and the classification of positions would seem incomplete since modern constructivist approaches can no longer be identified with „the intuitionistic or

[37] C. F. Gauß to H. C. Schumacher, July 12, 1831, cited after C. F. Gauß, *Werke. Achter Band* (B. G. Teubner [on Commission]: Leipzig 1900), 216.

[38] Bolzano, Bernard: *Paradoxien des Unendlichen,* ed. Alois Höfler, mit Anmerkungen versehen von Hans Hahn (see footnote 8 above), 1. Hahn refers to this edition in footnote 2 of this text.

Kantian position" (103) and would perhaps no longer be simply dismissed as „resting on a rather shaky foundation" (104). Hahn's attitude towards the axiom of choice shows his adoption of Carnap's Principle of Tolerance:[39] mathematics with the axiom of choice and mathematics without it (a „non-Zermelian" mathematics) are equally justified – a conclusion arrived at without knowledge of the independence of the axiom of choice, i. e. the relative consistency of the adjunction of the axiom of choice as well as that of its negation (proved by Paul J. Cohen only in 1963). The decision that has to be taken is on our use of the word „set" (a statement that would at once be accepted by an intuitionist, who would however strongly disagree with Hahn's claim that this decision is merely a convention).

As to Hahn's conviction that it is not possible to arrive at *absolute* consistency proofs, and that this situation is only a special case of the impossibility of absolute certainty in any area of human knowledge, it should perhaps be kept in mind that a purely syntactical consistency proof of (Zermelo – Fraenkel, Neumann – Bernays – Gödel, or any other) set theory *would* be a proof of that kind, but that our present mastery of metamathematical techniques is evidently too poor to hold out hopes for finding such a proof. On page 107, the „reform of logic" addressed by Menger the year before[40] is of course the introduction of the theory of types. The „German translation of *Principia Mathematica*" to which we are referred in endnote 10 on page 116 covers only a small part of that work, viz. the (important) „Introduction" together with the preface and the introduction to the second edition.

The claim (on page 110) that Gauß made experiments for finding out whether physical space is Euclidean or not, has been called in question: whereas there is no doubt that Gauß measured the sum of the angles in a „physical" triangle, it is pretty clear that he did not measure them for the purpose of testing the Euclidicity of physical space.[41]

„Gibt es Unendliches?" is the last lecture that Hahn could give to a lar-

[39] Cf. Rudolf Carnap: *Logische Syntax der Sprache* (Julius Springer: Wien 1934; *Schriften zur wissenschaftlichen Weltauffassung,* 8), 44 f. („Wir wollen nicht Verbote aufstellen, sondern Festsetzungen treffen", i. e., „We do not want to set up prohibitions, but to arrive at conventions"), and 45 („In der Logik gibt es keine Moral", i. e., „In logic, there are no morals").

[40] Menger, Karl: „Die neue Logik", in: *Krise und Neuaufbau in den exakten Wissenschaften. Fünf Wiener Vorträge* (Franz Deuticke: Wien 1933), 93–122.

[41] Cf. Arthur I. Miller, „The Myth of Gauss' Experiment on the Euclidean Nature of Physical Space", *Isis* 63 (1972), 345–348, and the comments by George Goe and B. L. van der Waerden, with Miller's reply, ibid. 65 (1974), 83–85, 85, and 86 f., respectively.

ger public before his early death on July 24, 1934. It shows, for the last time, his unique gift of crystal-clear formulation, of coherent and aesthetically engaging exposition, and of bringing home to his audience the significance of the subject. Unlike mathematics, where Hahn's name found its way into eponyms like the „Hahn-Banach theorem", the pantheon of the history of philosophy has not held a place for Hans Hahn, although his importance as founder, centre and teacher of the Vienna Circle has been acknowledged. Hahn's engagement for the intellectual liberation of a broad public is, however, worth being remembered in the world of philosophy as vividly as his convincing endeavours to use his superior mental powers for the benefit of a mankind in distress. There is much we can learn from his teachings, and from his life.

Hahn's philosophical writings
Hahns philosophische Schriften

Nr. 9 u. 10 **September u. Oktober 1919**

Vorlesungen über Zahlen und Funktionenlehre von **Alfred Pringsheim** (B. G. Teubners Sammlung von Lehrbüchern auf dem Gebiete der mathematischen Wissenschaften mit Einschluß ihrer Anwendungen, Band XL). Leipzig und Berlin, B. G. Teubner 1916. Erster Band, erste Abteilung XII, 292 S. Geh. 12 M., geb. 13.40 M. Zweite Abteilung VIII, 222 S. Geh. 10.80 M., geb. 12.40 M.

Von A. Pringsheims seit geraumer Zeit angekündigten Vorlesungen über Zahlen- und Funktionenlehre liegen nun zwei Abteilungen des ersten Bandes vor; die erste trägt den Untertitel: Reelle Zahlen und Zahlenfolgen; die zweite: Unendliche Reihen mit reellen Gliedern; eine dritte Abteilung wird die Einführung der komplexen Zahlen, die dadurch notwendig werdende Vervollständigung der Reihenlehre, die Theorie der unendlichen Produkte und Kettenbrüche bringen. Den Inhalt des zweiten Bandes soll bilden ›eine Einführung in die Theorie der eindeutigen analytischen Funktionen einer komplexen Veränderlichen und der einfachsten mehrdeutigen Umkehrungsfunktionen auf Grund der Weierstraßschen Methoden und deren weiterer Ausbildung, namentlich in bezug auf die Theorie der ganzen transzendenten Funktionen und der analytischen Fortsetzung‹.

Wenn vielleicht manche Autoren ihren Werken den Titel ›Vorlesungen‹ geben, um damit entschuldigend anzudeuten, daß sie etwas nicht ganz abgerundetes, nicht in sich abgeschlossenes und vollendetes bringen, so trifft dies auf das vorliegende Werk ganz und gar nicht zu. Es waren schon die Vorlesungen, die Pringsheim an der Münchner Universität wiederholt über unsren Gegenstand hielt, völlig druckfertige, mit vollster Beherrschung von Stoff und Form sorgfältigst durchgearbeitete Kunstwerke; und das ›durch Zusammenfassung und teilweise weitere Ausführung‹ dieser Vorlesungen entstandene vorliegende Buch hat denn einen Grad der Vollendung erreicht, wie er bei mathematischen Werken — leider, aber begreiflicherweise —

heute selten ist. Wenn wir uns trotzdem bei Besprechung dieses Buches nicht auf bloßes Referieren beschränken, sondern auch veranlaßt sehen, in einigen Punkten Kritik zu üben, so sei von vorneherein betont, daß es sich dabei — abgesehen von gelegentlichem Hinweise auf das eine oder andre kleine Versehen, wie sie bei Erstauflagen größerer Werke unvermeidlich sind — fast ausnahmslos um jene die Grundlagen betreffenden allgemeinsten Fragen handelt, über die sich die Mathematiker noch bei weitem nicht einig sind, und um Fragen methodischer Natur, die bis zu einem gewissen Grade Fragen persönlichen Geschmackes sind, und als solche wohl zu allen Zeiten von verschiedenen Mathematikern werden verschieden beantwortet werden.

Um gleich auf eine die Grundlagen betreffende Frage zu sprechen zu kommen, sei an eine Stelle des Vorwortes angeknüpft; dort heißt es: ›Daß trotz des elementaren Charakters der Darstellung durchweg möglichste Strenge der Beweisführung angestrebt wird, bedarf wohl kaum der Erwähnung, da dies nach meinem Dafürhalten als selbstverständliche Forderung jeder mathematischen Darstellung gelten sollte‹. Es erscheint auffällig, daß hier nur ›möglichste‹ Strenge, nicht absolute Strenge verlangt wird; wer Pringsheims Leistungen kennt, weiß von vorneherein, daß dies nicht ein Hintertürchen sein soll, um auf grund irgendwelcher pädagogischer Erwägungen da und dort, wo Strenge unbequem und schwierig wird, der Unexaktheit Einlaß zu gewähren. Und doch mag jenes Wörtchen ›möglichst‹ nicht zufällig gewählt sein; vielleicht soll es andeuten, daß absolute Strenge des Aufbaues der Zahlenlehre eine recht schwer erfüllbare Forderung ist, der im gegebenen Rahmen gar nicht hätte Genüge getan werden können. Denn was ist absolute Strenge? Wann darf ein Beweis als völlig streng gelten? Unbedenklich werden viele darauf antworten: wenn er nur rein logische Hilfsmittel benutzt. Doch dann kommt die weitere Frage: Welches sind rein logische Hilfsmittel? Einen Hinweis auf die üblichen Lehrbücher der Logik könnte man da nicht als befriedigende Antwort gelten lassen, denn das ist wohl unter allen Mathematikern, die sich mit den Grundlagen ihrer Wissenschaft beschäftigen, unbestritten, daß die Sätze der üblichen Logik zum Aufbau der Mathematik nicht ausreichen, ganz abgesehen davon, daß die gewöhnlichen Darstellungen der Logik durchaus die Präzision vermissen lassen, die erforderlich wäre, wenn man die mathematische Beweisführung auf sie gründen wollte. Also: sachlich nicht ausreichend, formal nicht hinlänglich präzise, so muß der Mathematiker, dem der folgerichtige Aufbau seiner Wissenschaft am Herzen liegt, die üblichen Darstellungen der Logik charakterisieren. Bekanntlich hat diese Lage der Dinge dazu geführt, daß es von mathematischer Seite unternommen wurde, der Logik jene Gestalt zu geben, wie sie der Mathematiker für seine Zwecke bedarf; es sei hier nur auf das System der symbolischen Logik hingewiesen, das Peano und seine Schüler ausgearbeitet haben. Kann nun die oben gestellte Frage nach den rein logischen Hilfsmitteln etwa durch einen Hinweis auf Peano's Formulario beantwortet werden? Wir fürchten, daß auch dies nicht der Fall ist. Wir dürfen wohl heute daran glauben, daß die Aufnahme von G. Cantor's Mengenlehre in das System der Mathematik eine endgiltige und inappellable Tatsache darstellt. Bekanntlich treten aber in dieser Disziplin gewisse Widersprüche auf, die offenbar von irgend einem Mangel in den logischen Grundlagen herrühren. Diese Widersprüche treten aber nicht nur dann auf, wenn man sich der üblichen Logik bedient, auch Peanos symbolische Logik vermag es nicht, diese Widersprüche zu vermeiden; insolange aber sie das nicht vermag, kann auch sie nicht als ausreichende logische Grundlage der Mathematik gelten.

Nun ist wohl neuerdings ein groß angelegter Versuch [1]) gemacht worden, die symbolische Logik so zu gestalten, daß alle Widersprüche vermieden werden, und daß sie uns die vollständige Tafel aller jener Grundbegriffe und Grundsätze darstelle, deren die Mathematik zu ihrem Aufbau bedarf. Ob dieser Versuch gelungen ist, darüber kann ich zur Zeit kein Urteil abgeben; eines aber scheint sicher; dieses logische System ist so schwierig und umfangreich geworden, daß es aus praktischen Erwägungen gänzlich ausgeschlossen ist, es zum Ausgangspunkt für eine mathematische Darstellung elementareren Charakters zu wählen, und die oben aufgeworfene Frage nach den Kriterien absoluter Strenge durch einen Hinweis auf dieses System zu beantworten. Haben sich die Mathematiker seit den Tagen Euklids damit abgefunden, daß es zu ihrer Wissenschaft keinen Königsweg gibt, so darf doch andrerseits der Zugang zu ihr nicht über die langwierigsten und halsbrecherischesten Hochgebirgspfade führen, sodaß die Mehrzahl aller Beschreiter unterwegs scheitern müßte, die wenigen aber, denen es gelänge, deren Schwierigkeiten zu überwinden, zu Tode erschöpft ankämen, und zwar nicht am Ziele, sondern dort, wo nun die eigentliche Mathematik erst beginnen soll.

Es scheint also tatsächlich, daß die Forderung nach absoluter Strenge, wie die Dinge heute liegen, nicht gut gestellt werden kann;

1) Principia mathematica von A. N. Whitehead und B. Russel, Cambridge.

21*

wir müssen uns mit »möglichster Strenge« begnügen, wobei es denn freilich dem einzelnen überlassen bleibt, was er als »möglichst« betrachten will. Das hat, sobald man einmal über die Grundlegung hinaus ist, wenig zu bedeuten, macht sich aber, wie wir gleich sehen werden, eben bei der Grundlegung recht unangenehm fühlbar. Der Verfasser selbst ist sich dessen offenbar bewußt, daß was er in diesen Fragen für ausreichend streng hält, anderen nicht so erscheinen könnte; spricht er doch im Vorworte selbst von Ueberlegungen, auf grund deren für ihn die Existenz der natürlichen Zahlen außer Zweifel stehe, »selbst auf die Gefahr hin, daß unerbittliche Logiker, Axiomatiker oder Mengentheoretiker hiergegen Widerspruch erheben sollten«.

Wenn wir also notgedrungen darauf verzichten müssen, in der logischen Grundlegung usque ad initium, bis zum bittern Anfang zurückzugehen — wo soll nun der Aufbau der Mathematik einsetzen? Die meisten Darstellungen, und so auch die vorliegende, wählen — und auch mir scheint dies das naturgemäße — als Ausgangspunkt den Begriff der natürlichen Zahl. Damit aber sind wir wieder bei einer Frage angelangt, die seit langem mathematisch interessierte Philosophen und philosophisch interessierte Mathematiker in Atem hält: kann die Lehre von den natürlichen Zahlen (und damit weiter Arithmetik und Analysis) aus rein logischen Grundbegriffen und Grundsätzen aufgebaut werden, oder bedarf es dazu spezifisch mathematischer Grundbegriffe und Grundsätze? Natürlich hat diese Frage einen präzisen Sinn nur dann, wenn eine Tafel der rein logischen Grundbegriffe und Grundsätze aufgestellt ist; verzichtet man darauf, so kann auch unsre Frage nicht mehr recht präzise beantwortet werden. Und so kommt es, daß wir nicht mit voller Sicherheit sagen können, welchen Standpunkt nun das vorliegende Werk zu dieser Frage einnimmt.

Herr Pringsheim hat jederzeit mit Erfolg und Temperament den Standpunkt vertreten, daß Berufung auf die Anschauung kein zulässiges Mittel mathematischer Beweisführung sei; er selbst hat in schönen und wichtigen Arbeiten zur Säuberung der Analysis von alogischen, anschaulichen Pseudobeweisen beigetragen. Dabei handelt es sich allerdings um geometrische Anschauung; aber was für die Geometrie recht ist, muß für die Arithmetik billig sein; auch in der reinen Arithmetik ist kein außerlogisches Beweismittel zulässig. Kann die Arithmetik außerlogischer Elemente nicht entraten, so hat sie die Verpflichtung, diese, als arithmetische Grundbegriffe und Grundsätze, an die Spitze zu stellen; als Quelle für die Gewißheit dieser Grundsätze mag sie sich — wenn sie sich überhaupt verpflichtet fühlt, nach dieser Quelle zu fragen — auf reine Anschauung, oder auf sonst eine Erkenntnisquelle berufen, ihren weiteren Aufbau aber hat sie sodann allein mit Hilfe der an die Spitze gestellten arithmetischen und der rein logischen Grundbegriffe und Grundsätze zu bewerkstelligen. Diesen Weg haben auch verschiedene moderne Darstellungen eingeschlagen, vielfach im Anschlusse an ein von Peano aufgestelltes System von Grundbegriffen und Grundsätzen der Arithmetik[1]).

Im vorliegenden Werke finden wir nirgends einen außerlogischen Grundbegriff oder Grundsatz formuliert und als solchen kenntlich gemacht. Wir gehen also wohl mit der Ansicht nicht irre, daß der Verfasser die Arithmetik rein logisch aufzubauen beansprucht; und wir werden hierin noch bestärkt durch einen Passus der Vorrede, wo er betont, er lasse eine kanonische Form für die unbegrenzt fortsetzbare, geordnete Folge der natürlichen Zahlen gewissermaßen vor den Augen der Leser entstehen, sodaß die Existenz dieser letzteren für ihn außer Zweifel stehe. Hieran schließt sich der oben erwähnte Passus über etwaigen Widerspruch unerbittlicher Logiker. Wir müssen also zunächst die Einführung der natürlichen Zahlen näher ins Auge fassen.

Es wird ausgegangen von einer einfach geordneten Menge[2]), die folgenden Forderungen genügt: 1) sie selbst und jede durch Weglassung von Anfangsgliedern entstehende Teilmenge hat ein erstes Element[3]). 2) Zu jedem Elemente, ausgenommen das erste, gibt es ein unmittelbar vorhergehendes. 3) Es gibt in ihr kein letztes Element. Es ist gewiß, daß sobald die Existenz einer solchen Menge feststeht, die Lehre von den natürlichen Zahlen entwickelt werden kann. Es handelt sich also vor allem darum, die Existenz einer solchen Menge nachzuweisen, was am besten durch effektive Angabe einer solchen Menge geschähe. Das ist denn auch das Ziel, daß der Verf. zunächst verfolgt. Er sagt, man könnte eine solche Menge »auf primitivste Art aus einem einzigen Fundamentalzeichen, etwa |, durch sukzessive Wiederholung herstellen, also:

1) Arithmetices principia nova e methodo exposita. Torino 1889. Man findet dieses System auch in Anhang II von Genocchi-Peano Differentialrechnung und Grundzüge der Integralrechnung.

2) Im Buche heißt es: »eine endlose Folge«. Ich möchte glauben, daß unter dem Wort »Folge« von Dingen ziemlich allgemein eine Belegung der geordneten Menge der natürlichen Zahlen mit diesen Dingen verstanden wird, so daß ich an dieser Stelle, wo der Begriff der natürlichen Zahl erst entwickelt werden soll, das Wort »Folge« lieber vermeide.

3) Das heißt, in der in der Mengenlehre üblichen Terminologie: die Menge ist wohlgeordnet.

(*) |, ||, |||, ||||, |||||, ...,

doch wäre sie wegen ihrer außerordentlichen Unübersichtlichkeit gänzlich unbrauchbar«. An Stelle dieses Vorganges wird deshalb ein andrer gewählt, der zugleich die dekadische Schreibweise mit Hilfe der arabischen Ziffern liefert. Da es uns hier aber nur auf das Prinzipielle ankommt, und wir unseren Einwand, der beide Verfahren in gleicher Weise trifft, durchsichtiger an das erstgenannte anknüpfen können, kehren wir zu diesem zurück. Wir können nicht verhehlen, daß uns hier eine petitio principii vorzuliegen scheint, die sich in den in (*) auftretenden Punkten .... verbirgt. Denn wollte man das, was diese Punkte kurz andeuten sollen, explizit aussprechen, was könnte es anders heißen als: die Operation des Hinzufügens eines Elementes | ist nach dem Typus ω durchzuführen, d. h. diesen Punkten kann ein präziser Sinn nur beigelegt werden, wenn der Ordnungstypus der Menge der natürlichen Zahlen bereits als bekannt angenommen wird[1]). Oder anders ausgedrückt: durch sukzessive Wiederholung des Fundamentalzeichens | kann doch auch eine nach dem Typus ω (oder nach dem Typus irgend einer transfiniten Ordinalzahl) geordnete Menge solcher Fundamentalzeichen hergestellt werden. Das ist aber hier nicht gemeint: die Punkte sollen andeuten, daß die Widerholung des Zeichens | nur *endlich oft* vorgenommen werden darf; damit aber ist die petitio principii offenbar[2]). Der Versuch,

1) Ganz denselben Einwand haben wir gegen die Darstellung der Lehre von den natürlichen Zahlen im kürzlich erschienenen Lehrbuch der Algebra von A. Loewy zu erheben. Dort werden (S. 2) die Peano'schen Axiome zugrunde gelegt; dabei wird aber Peanos fünftes Axiom; »enthält die Klasse s die Zahl 1, und neben jeder in ihr vorkommenden Zahl x auch die (unmittelbar folgende) Zahl x^+, so enthält sie die Klasse $\mathfrak{N}$ aller natürlichen Zahlen« durch das Axiom ersetzt: »Jedes Element von $\mathfrak{N}$ ist in dem Systeme 1, 1^+, 1^{++}, 1^{+++}, ... enthalten«. Durch diese Abänderung verliert meines Erachtens das Peano'sche System seinen ganzen Sinn. — Auch in der neuesten Auflage der Theoretischen Arithmetik von O. Stolz und J. A. Gmeiner haben die Peano'schen Axiome einige Abänderungen erdulden müssen, durch die diesem so wohldurchdachten und klar formulierten Systeme ein Todesstoß versetzt wird; es ist in der Gestalt, die Gmeiner diesen Axiomen gibt, (S. 15) die Rede von »den einer Zahl a in der Zahlenlinie nachfolgenden Zahlen«, ein Begriff der erläutert wird durch a^+, $(a^+)^+$ usf. Hier haben wir wieder das ominöse »usf.«, das an dieser Stelle, wo der Begriff der natürlichen Zahl, oder — wenn man will — der Typus ω, nicht zur Verfügung steht, keinen Sinn hat und eine petitio principii enthält. Aus den richtig wiedergegebenen Axiomen Peano's kann aber eine Anordnung der natürlichen Zahlen *deduziert* werden, und dadurch dem Begriffe: »die der Zahl a nachfolgenden Zahlen« ein präziser Sinn beigelegt werden.

2) Man wende hiegegen nicht ein, daß eine unendlich-oftmalige Wiederholung unausführbar sei. Auch eine $10^{10^{10}}$-malige Wiederholung ist unausführbar.

eine Menge mit den oben postulierten Eigenschaften zu konstruieren, scheint mir also nicht gelungen. Doch sei gleich betont, daß ich dies nicht für ein großes Unglück halte. Wir denken uns den Schaden gut gemacht, indem wir die Existenz einer Menge unterscheidbarer Dinge, der die gewünschten Eigenschaften zukommen, als unbewiesenen Grundsatz hinnehmen.

Wie nun gelangt man von der Existenz einer solchen Menge zum Begriff der natürlichen Zahl? Hier muß ich abermals auf eine Meinungsverschiedenheit zwischen mir und dem Verfasser hinweisen, die diesmal allerdings nicht so sehr logischer als allgemein philosophischer Natur ist. Pr. erklärt nun schlechtweg die Elemente unsrer Menge als natürliche Zahlen [1]). Ich bin mir nicht völlig darüber klar, was damit gemeint ist. Gewiß ist Herr Pr. nicht der Ansicht, daß die jetzt von mir mit Tinte auf Papier gemalte 1 eine natürliche Zahl sei; so konkret ist die Sache offenbar nicht gemeint, sondern irgendwie anders, es dürfte nicht leicht fallen, den zugrunde liegenden Gedanken präzise zu formulieren. Meiner Ansicht nach kann 1 nur als konventionelles Zeichen für die Zahl eins aufgefaßt werden, das sich von Zeichen, wie a, b, x, y nur dadurch unterscheidet, daß alle Menschen unsres Kulturkreises darin einig sind, daß ein solches Zeichen (ohne gegenteilige Abmachung) stets die Zahl eins bedeuten soll — so wie man stets die irrationale Zahl, die das Verhältnis von Kreisumfang zum Durchmesser angibt, mit π bezeichnet, ohne daß doch jemandem beifallen wird, zu sagen das Zeichen π sei diese irrationale Zahl. Meiner Ansicht nach verhält sich die Sache folgendermaßen. Man kann beweisen, daß je zwei Mengen, die alle oben geforderten Eigenschaften besitzen, gleichen Ordnungstypus haben, d. h. umkehrbar eindeutig und ähnlich auf einander abgebildet werden können. Man kann weiter beweisen, daß es zwischen zwei solchen Mengen nur eine einzige ähnliche Abbildung geben kann. Da-

Wollte man aber behaupten, die unendlich oftmalige Wiederholung sei noch in anderem Sinne, gewissermaßen in höherem Grade unausführbar als die $10^{10^{10}}$-malige, so hätte man die Verpflichtung das zu begründen, d. h. den genauen Sinn der Termini »ausführbar« und »unausführbar« darzulegen, was zu äußerst schwierigen Fragen psychologischer, erkenntnistheoretischer, ja wohl sogar methaphysischer Natur führen würde, die man bei Begründung der Arithmetik wohl unbedingt wird vermeiden wollen.

1) Um genau zu referieren, möchte ich nicht unerwähnt lassen, daß Pr. nicht von einer Menge irgendwelcher Elemente spricht, sondern von diesen Elementen verlangt, sie sollen »Zeichen« sein. Nun kann doch wohl jedes Ding als Zeichen für jedes andre Ding verwendet werden, sodaß es anscheinend gleichgiltig ist, ob wir von Mengen beliebiger Elemente, oder Mengen von Zeichen sprechen. Für die Erörterungen des Textes ist dies sicherlich gleichgiltig.

durch ist jedem Elemente einer solchen Menge in jeder andern solchen Menge in eindeutiger Weise ein bestimmtes Element zugeordnet. Jedes Element einer solchen Menge steht also — rein ordinal betrachtet — zu seiner Menge in derselben Relation, wie jedes der ihm zugeordneten Elemente zu der seinen. Diese Relation nun ist die natürliche (Ordinal-)Zahl des treffenden Elementes[1]). Das Element selbst kann dann als Zeichen für diese Ordinalzahl verwendet werden.

Soviel über den Begriff der natürlichen Zahl als Ordinalzahl. Aber die Arithmetik kommt mit diesem Begriffe allein nicht aus, sie benötigt auch den Begriff der Anzahl, der Kardinalzahl. Schon das allgemeine assoziative Gesetz kann ohne diesen Begriff gar nicht formuliert werden. Dieser Sachverhalt — der anscheinend keineswegs allen Autoren, die über die Grundlegung der Arithmetik geschrieben haben, zum Bewußtsein gekommen ist — wird vom Verf. mit dankenswerter Klarheit formuliert; es muß also nun aus dem bereits vorliegenden Begriff der natürlichen Ordinalzahl der der natürlichen Kardinalzahl hergeleitet werden. Die Grundlage hierfür bietet bekanntlich der Satz: In der Menge $\mathfrak{N}$ der natürlichen Zahlen ist kein Abschnitt einem anderen Abschnitte oder der Menge $\mathfrak{N}$ selbst äquivalent. Dieser Satz ist es denn auch im Wesentlichen, dem die Ueberlegungen von § 3 gelten. Daß aber diese Ueberlegungen einen logischen Beweis des fraglichen Satzes liefern, vermag ich nicht anzuerkennen. Der Gedankengang ist der folgende: sei 1, 2, 3, ..., n ein Abschnitt $\mathfrak{A}$ von $\mathfrak{N}$, und seien a, b, c, ... die irgendwie umgeordneten Elemente dieses Abschnittes. Man denke sie sich der Reihe nach den natürlichen Zahlen 1, 2, 3, ... zugeordnet. Nun bringe man durch eine Transposition zuerst 1 wieder auf den ursprünglichen Platz, dann 2, und »so fortfahrend gelangt man schließlich dazu, auch jedem der übrigen Elemente 3, ..., n den mit der entsprechenden Nummer versehenen Platz zuzuteilen«. Da bei diesem Verfahren niemals ein von einem der Elemente a, b, c ... besetzter Platz leer, ebensowenig ein anfänglich leerer Platz besetzt wird, im Schlußresultate aber gerade die Plätze 1, 2, ..., n besetzt erscheinen, so muß dies auch anfänglich der Fall gewesen sein; auch in der Umordnung a, b, c, ... waren also die Zahlen 1, 2, ..., n gerade dem Abschnitte $\mathfrak{A}$ von $\mathfrak{N}$ zugeordnet. Dieser Beweis scheint auf den ersten Blick einwandfrei,

1) Hätten wir an unsre Mengen nur die erste der drei Forderungen gestellt, so würden wir so auch Cantor's transfinite Ordinalzahlen erhalten. — Eine solche rein ordinale Definition der rationalen (oder der reellen) Zahlen kann es nicht geben. Denn zwei Mengen, deren Ordnungstypus der der natürlich geordneten Rationalzahlen ist, können nicht nur auf eine sondern auf unendlich viele Weisen ähnlich aufeinander abgebildet werden.

und doch läßt sich an einem Beispiele zeigen, daß er nicht bindend sein kann. Nehmen wir statt des Abschnittes $\mathfrak{A}$ von $\mathfrak{N}$ die Menge $\mathfrak{N}$ selbst, d. h. die nach dem Typus ω geordnete Menge der natürlichen Zahlen:

$$(*) \qquad 1, 2, 3, \ldots\ldots$$

Durch folgende Umordnung denken wir sie uns auf den Typus $\omega . 2$ gebracht:

$$(**) \qquad 1, 3, 5, \ldots\ldots; \; 2, 4, 6, \ldots\ldots$$

Wie oben können wir nun argumentieren: man bringe durch Transposition zuerst das Element 2 auf seinen ursprünglichen Platz, dann das Element 3 usf. Man erhält so der Reihe nach:

$$1, 2, 5, \ldots\ldots; \qquad 3, 4, 6, \ldots\ldots$$
$$1, 2, 3, 7, \ldots\ldots; \qquad 5, 4, 6, \ldots\ldots$$
$$1, 2, 3, 4, 9, \ldots\ldots; \; 5, 7, 6, \ldots\ldots$$
$$\cdot \quad \cdot \quad \cdot \quad \cdot \quad \cdot \quad \cdot \quad \cdot \quad \cdot \quad \cdot \quad \cdot \quad \cdot \quad \cdot \quad \cdot$$

Auch hier wird im Laufe der Operationen niemals ein in (**) besetzter Platz leer, nie ein leerer Platz besetzt; auch hier kommt man »schließlich« (d. h. hier nach einer abzählbaren, nach dem Typus ω geordneten Menge von Transpositionen) zur ursprünglichen Anordnung (*) zurück, obwohl doch gewiß die Elemente in (**) nicht der Reihe nach den Elementen (*) zugeordnet sind. Offenbar handelt es sich dabei um die Bedeutung des von mir oben durch Anführungszeichen hervorgehobenen Wörtchens »schließlich«. Der Beweis ist bindend nur dann, wenn dieses »schließlich« heißt: »durch endlich viele Transpositionen«, damit aber ist auch klar geworden, daß es sich wieder um eine petitio principii handelt, denn der Begriff »endlich viele« soll ja erst gewonnen werden.

Keineswegs soll durch diese Kritik der vom Verf. eingeschlagene Weg, der vom Begriffe der natürlichen Ordinalzahl zu dem der natürlichen Kardinalzahl führen soll, für ungangbar erklärt werden: dieser Weg ist gangbar, und sogar unschwer gangbar. Aber auch der umgekehrte Weg kann eingeschlagen werden, und dürfte, wie auch Pr. hervorhebt, mit der natürlichen Entwicklung des Zahlbegriffes besser übereinstimmen. Und nicht nur das: ganz im Gegensatze zu Pr. möchte ich diesen umgekehrten Weg auch für den logisch näherliegenden und befriedigenderen halten. Für den Mengentheoretiker ist der Kardinalzahlbegriff (die Mächtigkeit), als das gegenüber beliebiger eineindeutiger Abbildung invariante Merkmal einer Menge, der einfachere Begriff als die Ordinalzahl (der Ordnungstypus), die das nur gegenüber eineindeutiger und ähnlicher

Abbildung invariante Merkmal einer einfach geordneten Menge ist[1]). Wenn dem gegenüber Herr Pr. »es für ein wenig aussichtsreiches Unternehmen« hält, »auf der Grundlage des Anzahlbegriffes zu einer befriedigenden Ausgestaltung der Lehre von den reellen Zahlen zu gelangen« und noch hinzufügt, »neuere, zum Teil äußerst verkünstelte Versuche dieser Art« hätten seine Ansicht nur in vollstem Maße bestätigt, so möchte ich mir doch gestatten, mit knappen Worten anzudeuten, wie mir eine solche Theorie durchaus naturgemäß durchführbar erscheint[2]).

Man definiere zunächst (rein logisch) die Einheitsmengen durch die Eigenschaft: ist a Element von M, und b Element von M, so ist a mit b identisch. Die Kardinalzahl 1 werde nun definiert als die Mächtigkeit der Einheitsmengen. Man definiere sodann, wenn $\mathfrak{a}$ die Mächtigkeit irgend einer Menge $\mathfrak{A}$ bedeutet, $\mathfrak{a}+1$ als die Mächtigkeit der Vereinigungsmenge von $\mathfrak{A}$ mit einer zu $\mathfrak{A}$ fremden Einheitsmenge; und man definiere schließlich die endlichen (oder natürlichen) Kardinalzahlen als diejenigen Mächtigkeiten, die in jeder Menge vorkommen, die die Zahl 1 enthält, und neben jeder in ihr enthaltenen Mächtigkeit $\mathfrak{a}$ auch die Mächtigkeit $\mathfrak{a}+1$ enthält. Man hat damit die vollständige Induktion zur Verfügung, und beweist mit ihrer Hilfe leicht: eine Menge, deren Mächtigkeit die endliche Kardinalzahl n ist, kann nur nach einem einzigen Ordnungstypus einfach geordnet werden. Diese so den endlichen Kardinalzahlen eindeutig zugeordneten Ordnungstypen sind die endlichen (natürlichen) Ordinalzahlen. Ich kann nicht finden, daß dieser Gedankengang irgend etwas gekünsteltes an sich habe. Er scheint mir im Gegenteil geradezu die reinlich logische Einkleidung der anschaulichen, aber

1) Die symbolische Logik unterscheidet nicht zwischen dem für alle Dinge einer Klasse charakteristischen Merkmale und dieser Klasse selbst. Und da sie die Worte »Menge« und »Klasse« synonym verwendet, so definierte sie einfach: Mächtigkeit der Klasse $\mathfrak{A}$ ist die Klasse aller zu $\mathfrak{A}$ äquivalenten Klassen. Diese Definition wurde verlassen, weil sich der Begriff der Klasse aller zu $\mathfrak{A}$ äquivalenten Klassen als widerspruchsvoll erwies. Ich glaube, daß sich diese Definition durch eine geringe Modifikation halten läßt. Man gehe aus von einem als gegeben angenommenen Bereich $\mathfrak{B}$ von Dingen. Zusammenfassungen dieser Dinge bezeichne man als Mengen, und statuiere den logischen Grundsatz: keine Menge ist ein Ding von $\mathfrak{B}$. Nun erweitere man den Bereich $\mathfrak{B}$ zu $\mathfrak{B}'$ durch Hinzufügung der Mengen als uneigentlicher Dinge. Man kann nun Mengen aus Dingen von $\mathfrak{B}'$ bilden, nennen wir sie »Mengen zweiter Stufe« und zum Unterschiede die bisherigen Mengen »Mengen erster Stufe«. Die Definition der Mächtigkeit hat dann zu lauten: Mächtigkeit der Menge erster Stufe $\mathfrak{A}$ ist die Menge zweiter Stufe aller zu $\mathfrak{A}$ äquivalenten Mengen erster Stufe. Dabei treten, so viel ich sehe, Widersprüche nicht mehr auf. Vgl. M. Pasch, Grundlagen der Analysis S. 94.

2) Vgl. B. Russell, The principles of mathematics S. 128.

unpräzisen Vorstellungen, die wohl jedermann naiver Weise mit den Begriffen der endlichen Kardinalzahl und Ordinalzahl verknüpft.

Wir haben uns nun eingehend mit den Gedankengängen auseinandergesetzt, durch die Pr. die natürlichen Zahlen einführt, weil wir hier ziemlich weitgehende Verschiedenheiten zwischen den Anschauungen des Verf. und den eigenen feststellen mußten. Nehmen wir aber einmal das feste Fundament der natürlichen Zahl für den weiteren Aufbau als gewonnen an, so können wir zu den nun folgenden Entwicklungen fast durchweg rückhaltlose Zustimmung äußern. Die wenigen Punkte, in denen wir uns dem Verf. nicht völlig anschließen können, betreffen einzelne Fragen methodischer Natur, von denen hier nur eine einzige schärfer ins Auge gefaßt werde: das sogenannte Prinzip der Permanenz. Hören wir, was der Verf. darüber im Vorworte (S. VII) sagt. Er spricht von der Einführung der Brüche, der Null und der negativen Zahlen und fährt sodann fort: ›Für die Feststellung, wie diese neuen Zahlen zu ordnen bzw. in die Folge der schon vorhandenen einzuordnen sind, und wie mit ihnen gerechnet werden muß, dient das gewöhnlich nach dem Vorgange von Haukel (nicht besonders glücklich) als Prinzip der ›Permanenz‹ formaler Gesetze bezeichnete Uebertragungsprinzip und zwar in einer nach meinem Dafürhalten merklich verbesserten Form, welche ihm den Charakter einer gewissen logischen Notwendigkeit verleiht. Es werden nämlich allemal neue Zahlzeichen in solchem Umfange eingeführt, daß eine Teilmenge derselben lediglich Zeichen für bereits vorhandene Zahlen vorstellt. Für diese letzteren bestehen also schon ganz bestimmte, auf die Feststellung ihrer Sukzession und die Definition der Rechnungsoperationen bezügliche Regeln, die sich ohne weiteres in die neuen Bezeichnungen umschreiben lassen. Soll dann in der Handhabung des gesamten neu geschaffenen Zeichenvorrats nicht eine vollständige Verwirrung eintreten, so bleibt kaum etwas andres übrig, als jene für einen Teil derselben bereits zu Recht bestehenden Regeln definitionsweise auf die Gesamtheit auszudehnen und diesen Schritt durch den Nachweis zu legitimieren, daß die so getroffenen Festsetzungen den an sie zu stellenden Anforderungen widerspruchslos genügen.‹

Wir müssen, um zu diesen Ausführungen Stellung nehmen zu können, einige Worte über das Prinzip der Permanenz vorausschicken. Kann dieses Prinzip irgendwie präzise formuliert werden? H. Schubert hat in der Encyklopädie der mathematischen Wissenschaften (I, 1, S. 11) einen solchen Versuch gemacht, doch hat Peano in überzeugender Weise dargetan, daß dieser Versuch mißlungen ist[1]:

1) Rev. de math. 8 (1903).

nach Schubert verlangt das Prinzip der Permanenz, die Erweiterungen des Zahlengebietes seien so vorzunehmen, »daß für die Zahlen im erweiterten Sinne dieselben Sätze gelten, wie für die Zahlen im noch nicht erweiterten Sinne«. Diese Forderung aber ist unerfüllbar. Würden für die Zahlen im erweiterten Sinne wirklich dieselben Gesetze weiterbestehen, so wären sie von den Zahlen im noch nicht erweiterten Sinne logisch nicht unterscheidbar: es läge keine Erweiterung des Zahlgebietes vor. Daß also alle Sätze des ursprünglichen Zahlengebietes im erweiterten Gebiete fortbestehen, kann nicht verlangt werden; es darf nur verlangt werden, daß etwa die wichtigsten Sätze weiterbestehen. Was aber die wichtigsten Sätze sind, das liegt im Ermessen des einzelnen. Das Prinzip der Permanenz hat damit aufgehört logischer Natur zu sein, es ist bestenfalls ein methodologischer Ratschlag geworden, der ein Moment der Willkür in sich enhält. Dieses Moment der Willkür wurde nun vielfach stark in den Vordergrund gerückt; man betonte geflissentlich, es seien willkürliche Festsetzungen, wenn bei Ausdehnung der Multiplikation auf negative Zahlen $(-1).(-1) = 1$ gesetzt werde, wenn man bei Ausdehnung des Potenzbegriffes $e^0 = 1$ setze. Erfahrungsgemäß lehnt sich dagegen der Intellekt der Lernenden auf; sie haben das Gefühl, daß in den »Beweisen«, die sie auf der Schule hatten, daß $(-1).(-1) = 1$ ist, daß $e^0 = 1$ ist, doch etwas dran sei. Und dies gibt einen Fingerzeig, daß mit Betonung der Willkür noch nicht alles geleistet sei, daß neben dem willkürlichen Momente auch ein gesetzmäßiges Moment mit in Frage kommt, das nun auch seinerseits reinlich herausgearbeitet werden muß. In diesem Sinne ist denn auch die Pringsheim'sche Darstellung gehalten, und mit dieser Tendenz möchte ich mich rückhaltlos einverstanden erklären; nicht ganz aber mit der Art ihrer Durchführung. Es scheint mir, als hätte Herr Pr. selbst das Gefühl, es sei mit seiner Darstellung nicht das letzte Wort gesprochen: er beansprucht für sie, wie wir oben zitierten, nur den Charakter »einer gewissen logischen Notwendigkeit«, und daß hier dieses böse Wörtchen »gewiß« notwendig wird, das liegt meines Erachtens daran, daß das willkürliche und das gesetzmäßige Moment, die bei den üblichen Erweiterungen des Zahlgebietes zusammenwirken, noch nicht so scharf getrennt sind, als es wohl möglich wäre. Wir wollen dies etwa an der Erweiterung des Gebietes der natürlichen Zahlen zu dem der positiven rationalen Zahlen erläutern. Der Gedankengang bei Pr. ist der folgende: Ein »Bruch« $\frac{b}{a}$ soll, immer wenn b ein Vielfaches von a ist, $b = na$, nur als neues Zeichen für die natürliche Zahl n angesehen werden. Daraus ergibt sich der

Satz: zwei solche ›uneigentliche‹ Brüche $\frac{b}{a}$ und $\frac{b'}{a'}$ sind gleich dann und nur dann, wenn:

$$(0) \qquad ba' = ab'.$$

Ebenso erhält man für solche uneigentliche Brüche Additions- und Multiplikationsregel als beweisbare Lehrsätze. Nun heißt es (S. 41): ›Während nun die uneigentlichen Brüche lediglich als andere Zeichen für die natürlichen Zahlen auftraten, so sind die eigentlichen Brüche[1]) vollkommen neue Zeichen, die wir dadurch zu neuen Zahlzeichen machen wollen, daß wir ihre Sukzession innerhalb der Reihe der natürlichen Zahlen, ..., sodann die Grundoperationen der Addition und Multiplikation auf sie auszudehnen suchen — und zwar das alles auf die Weise, daß mit den bisherigen Festsetzungen und Rechnungsregeln keinerlei Widerspruch entsteht. Ist dieses Ziel überhaupt erreichbar, so ist die Möglichkeit eines Erfolges nur gegeben, wenn wir diejenigen Regeln, die im vorigen Paragraphen für die uneigentlichen Brüche als direkte Folgerungen der für natürliche Zahlen geltenden sich ergaben, nunmehr als entsprechende Definitionen für die Beziehungen der eigentlichen Brüche unter sich und in Verbindung mit uneigentlichen Brüchen einführen‹. Damit wäre nun in der Tat die ›gewisse logische Notwendigkeit‹ erreicht; wir hätten nun (wenn wir nicht auf die Erweiterung des Zahlgebietes überhaupt verzichten wollen) keine andre Wahl, als die Gleichheit zweier beliebiger Brüche durch (0) zu definieren, jede Willkür wäre ausgeschaltet.

So aber, möchte ich glauben, verhält sich die Sache nicht. Die Bedingungen für die Gleichheit zweier uneigentlicher Brüche kann mit demselben Rechte wie durch (0), auch ausgedrückt werden durch:

$$(00) \qquad (ba' - ab')^2 + (r_a(b) - r_{a'}(b'))^2 = 0,$$

wo $r_a(b)$ (und analog $r_{a'}(b')$) den absolut kleinsten Rest von b modulo a bedeutet. Wäre es wahr, daß die Erweiterung des Zahlgebietes nur so vorgenommen werden kann, das (0) weiter gilt, so würde das gleiche, mit demselben Rechte, für (00) gelten. Das aber würde zu einer gänzlich verschiedenen Gleichheitsdefinition für die uneigentlichen Brüche führen. Es ist also offenbar nicht so, daß wir ohne weiteres schließen dürfen, durch (0) sei die einzig mögliche Gleichheitsdefinition auch für uneigentliche Brüche gegeben. Und wenn wir (0) gegenüber der durch (00) gegebenen, und an und für sich ebenso möglichen Gleichheitsdefinition bevorzugen, so liegt dafür anscheinend

1) Das sind diejenigen, in denen der Zähler nicht Vielfaches des Nenners ist.

kein logischer Zwang vor Die Willkür scheint wieder Alleinherrscherin zu sein[1]).

Es sei uns nun noch gestattet kurz eine Darstellung anzudeuten, die unsres Erachtens wirklich auseinanderhält, was Willkür und was logische Notwendigkeit ist, und jedem dieser beiden Momente sein volles Recht zuteil werden läßt[2]). Präzisieren wir zunächst die Aufgabe: es soll das System der natürlichen Zahlen durch Hinzufügung neuer »Zahlen« so erweitert werden, daß im erweiterten Systeme die Multiplikation stets eindeutig umkehrbar wird. Unter »Multiplikation« soll dabei eine stets ausführbare assoziative und kommutative Verknüpfung verstanden werden, die sich auf die bereits definierte Multiplikation der natürlichen Zahlen reduziert, wenn beide Faktoren gleich natürlichen Zahlen werden. Daß wir gerade das Fortbestehen dieser Eigenschaften der Multiplikation verlangen (und nicht z. B. auch das Fortbestehen der Ungleichung $a.b \geqq b$) ist willkürlich; von nun an aber bleibt für eine Willkür kein Raum. Da nach Forderung im erweiterten System die Division stets ausführbar ist, muß darin der Quotient zweier natürlichen Zahlen $b:a$ vorhanden sein; wir bezeichnen ihn mit $\frac{b}{a}$, es ist also:

$$(1) \qquad \frac{b}{a} \cdot a = b.$$

Wann sind nun zwei Zeichen $\frac{b}{a}$, $\frac{b'}{a'}$ als gleich zu definieren? Ist

$$(2) \qquad \frac{b}{a} = \frac{b'}{a'},$$

so ist zufolge des Begriffes der eindeutigen Verknüpfung auch:

$$\frac{b}{a} \cdot (a \,.\, a') = \frac{b'}{a'} \cdot (a \,.\, a'),$$

also wegen des assoziativen und kommutativen Charakters der Multiplikation auch:

$$\left(\frac{b}{a} \cdot a\right) \cdot a' = \left(\frac{b'}{a'} \cdot a'\right) . a$$

1) Selbstverständlich ist der Begriff »Willkür« in dieser ganzen Erörterung in rein logischem Sinne, also lediglich als Gegensatz zum Begriffe »logische Notwendigkeit« zu verstehen. Es kann sehr wohl sein, daß von mehreren Fällen, die logisch möglich sind, zwischen denen eine Auswahl also logisch willkürlich ist, aus Gründen der Durchführbarkeit oder Anwendbarkeit nur ein einziger Fall praktisch möglich ist.

2) In ganz derselben Weise ließe sich die allgemeine Theorie der Erweiterung eines Größensystems umgestalten, die von O. Stolz und J. A. Gmeiner in »Theoretische Arithmetik«, Dritter Abschnitt, 7, (S. 67) vorgetragen wird.

und somit wegen (1) auch:

$$(3) \qquad b \, . \, a' = a \, . \, b'.$$

Ist umgekehrt (3) erfüllt, so folgt wegen

$$\frac{b}{a} \cdot (a \, . \, a') = b \, . \, a' \quad \text{und} \quad \frac{b'}{a'} \cdot (a \, . \, a') = b' \, . \, a$$

aus der geforderten Eindeutigkeit der Division auch (2). Wir haben also tatsächlich keine andre Wahl, als die Gleichheit zweier Brüche (2) durch (3) zu definieren.

Um nun zu erkennen, wie die Multiplikation zweier Brüche auszuführen ist, gehen wir aus von der aus den geforderten Eigenschaften der Multiplikation und aus (1) folgenden Beziehung:

$$\left(\frac{b}{a} \cdot \frac{b'}{a'}\right) \cdot (a \cdot a') = \left(\frac{b}{a} \cdot a\right) \cdot \left(\frac{b'}{a'} \cdot a'\right) = b \, . \, b'.$$

Die geforderte eindeutige Ausführbarkeit der Division zeigt also, daß wir keine andre Wahl haben, als zu definieren:

$$\frac{b}{a} \cdot \frac{b'}{a'} = \frac{b \, . \, b'}{a \, . \, a'}.$$

Und nun können wir weiter beweisen, daß es eine und nur eine zur Multiplikation distributive Verknüpfung gibt, die wenn beide verknüpften Brüche gleich natürlichen Zahlen werden, sich auf die Addition der natürlichen Zahlen reduziert. Verlangen wir also auch von der Addition der Brüche, sie solle distributiv sein (daß wir dies verlangen, ist wieder willkürlich), so müssen wir sie notwendig definieren durch:

$$\frac{b}{a} + \frac{b'}{a'} = \frac{a' b + a b'}{a a'}.$$

Durch diese kurze Entwicklung hoffe ich genügend deutlich gemacht zu haben, was mir an Pringsheim's Darstellung der sukzessiven Erweiterungen des Zahlengebietes nicht ganz befriedigend erscheint, und nach welcher Richtung hin mir eine Vervollkommnung möglich scheint.

Noch zwei Fragen prinzipieller Natur möchte ich berühren, bevor ich mich einem kurzen Referate über den Inhalt des zu besprechenden Buches zuwende. Herr Pringsheim ist heute der hervorragendste Vertreter einer arithmetischen Schule, einer Schule die in der Abweisung jedes geometrischen Elementes in der Analysis so weit geht, daß sie sogar auf die Hilfe der so suggestiven geometrischen Terminologie zur Erleichterung des Verständnisses ihrer Sätze und Beweise verzichtet. Niemand wird sich daher wundern, in Pr.'s ›Vorlesungen‹ auch nicht eine Spur von Geometrie zu finden. Da diese

›Vorlesungen‹ nun wohl auf lange hinaus als der klassische Repräsentant dieser arithmetischen Richtung reinster Observanz werden zu gelten haben, so mag es wohl am Platze sein, in einer Besprechung dieses Werkes auch zur Frage der völligen Verbannung alles Geometrischen aus der Analysis Stellung zu nehmen.

Es ist allbekannt, wie im vergangenen Jahrhunderte der große Prozeß der Arithmetisierug der Analysis einsetzte und großenteils durchgeführt wurde; man hatte lange Zeit in allzu naiver, allzu unkritischer Weise vermeintliche geometrische Evidenzen als Beweismittel benutzt, bis die infolgedessen auftretenden Unklarheiten und Fehler zu einer Umkehr zwangen. Die geometrische Anschauung wurde nun als Beweismittel für unzulässig erklärt, es wurde vollständige ›Arithmetisierung« verlangt. Dieses Schlagwort der Arithmetisierung scheint mir nun nicht glücklich, oder zumindest scheint es mir nicht das auszudrücken, was man heute verlangt. So wie es geometrische Evidenzen alogischer Natur gibt, die man dann als anschaulich bezeichnet, so gibt es zweifellos auch arithmetische Evidenzen alogischer Natur (es sei mir gestattet, auch diese als ›anschaulich‹ zu bezeichnen): so erkennen wir z. B. die kommutative Eigenschaft der Addition natürlicher Zahlen auch ohne logische Analyse mit voller Evidenz. Ohne weitere Begründung nun die geometrische Anschauung als Beweismittel ausschließen, die arithmetische Anschauung aber zulassen, scheint mir ein dogmatischer Willkürakt. Also nicht Arithmetisierung ist das Ideal, sondern Logisierung, und in dieser Form gilt die Forderung in gleicher Weise für Arithmetik, Analysis und Geometrie. Die Logisierung ist erreicht, wenn die ganze Disziplin aus einer Reihe von Grundbegriffen und Grundsätzen mit rein logischen Hilfsmitteln entwickelt wird, sie ist erst recht erreicht, wenn die Disziplin zu ihrer Entwicklung keiner anderen als der rein logischen Grundbegriffe und Grundsätze bedarf.

Mit welchem Rechte könnte nun, von einem rein logischen Standpunkte aus, die Analysis eine in diesem Sinne logisierte Geometrie ablehnen? Doch wohl nur dann, wenn diese Geometrie zu ihrem Aufbaue irgendwelcher Grundbegriffe oder Grundsätze bedürfte, deren die Analysis nicht bedarf. Ist dies nun der Fall? Offenbar nein! Dies ist evident, wenn man die Geometrie so begründet, wie dies E. Study in seinem Buche ›Die realistische Weltansicht und die Lehre vom Raume‹ tut[1]), wo etwa der Punkt der Ebene als geordnetes Paar reeller Zahlen definiert wird; es ist aber nicht minder richtig bei sogenanntem axiomatischen Aufbau der Geometrie, wo die

1) Kap. V, S. 81 ff.

geometrischen Begriffe implizit definiert werden durch das System der »Axiome«[1]), und nachträglich der Beweis für die Existenz und Einzigartigkeit dieser Begriffe durch Berufung auf die Analysis erbracht wird[2]). Rein logisch genommen scheint mir also die Verbannung aller Geometrie aus der Analysis nicht gerechtfertigt.

In der Tat scheint es sich dabei auch mehr um eine psychologisch-didaktische Frage zu handeln: man fürchtet, daß die Verwendung geometrischer Begriffe eine zu starke Verlockung zur (vielleicht unbewußten) Verwendung anschaulicher, alogischer Beweismittel mit sich bringen könnte. Daß eine solche Gefahr besteht, muß unbedingt zugegeben werden. Ob dieser Vorteil durch die, meinem Dafürhalten nach sehr bedeutenden Nachteile aufgewogen wird, die die asketische Fernhaltung von allem Geometrischen mit sich bringt, wird immer Ansichtssache bleiben. Ich halte diese Nachteile für so bedeutend, weil ich überzeugt bin, daß fast alle Fortschritte in den subtileren Teilen der Analysis, wie etwa der Lehre von den reellen Funktionen, zunächst, subjektiv, auf anschaulichem Wege gewonnen werden, daß also der Gebrauch der Anschauung ein unentbehrliches Forschungsmittel ist. Freilich nicht ein so roher und unkritischer Gebrauch, wie der, durch den die Analysis seinerzeit in die Irre geleitet wurde, sondern ein durch die Erkenntnisse eines Jahrhunderts geläuterter und verfeinerter Gebrauch der Anschauung. Und dieses Forschungsmittel weiter zu stärken und zu verfeinern, nicht es durch Beiseitestellung immer mehr verkümmern zu lassen, scheint mir — unbeschadet der unter allen Umständen unerbittlich zu fordernden logischen Strenge aller Beweisführung — eine wichtige Aufgabe des mathematischen Unterrichtes, in mündlichen Vorlesungen wie in Lehrbüchern.

Dies wäre die erste prinzipielle Frage, zu der ich noch Stellung nehmen wollte. Die zweite ist die, betreffend freiwillige Beschränkung auf »elementare Methoden«. »Und als leitenden Grundgedanken«,

1) Diese »Axiome« sind also keineswegs »Grundsätze« in dem oben erwähnten Sinne, ebensowenig, wie etwa beim Studium der allgemeinsten assoziativen und kommutativen Verknüpfung in einem Größensystem (wie es z. B. bei Stolz-Gmeiner, Theoretische Arithmetik 2. Aufl. S. 50 ff. betrieben wird) die Forderungen, die betrachtete Verknüpfung solle assoziativ und kommutativ sein, »Grundsätze« sind.

2) Für die Zwecke der Analysis wird wohl der zuerst genannten Auffassungsweise der Vorzug vor der axiomatischen zu geben sein, da diese, wenn auch nicht logisch, so doch methodologisch der Analysis gegenüber fremdartig ist, was bei jener nicht der Fall ist: wer sich weigern wollte, für den Begriff des geordneten Paares reeller Zahlen den Namen »Punkt« zuzulassen, müßte sich konsequenterweise auch weigern, für eine konvergente Folge rationaler Zahlen den eigenen Namen: »Reelle Zahl« zuzulassen.

so schreibt der Verf. im Vorworte, ›der mir in gleicher Weise bei Abfassung des arithmetischen, wie des funktionentheoretischen Teiles vorschwebte, möchte ich die Durchführung der Absicht bezeichnen, die elementaren Methoden nach Möglichkeit auszunützen bzw. weit genug auszubilden, um den Leser mit den modernen Verschärfungen und Vertiefungen der Begriffe und Fragestellungen vertraut machen zu können und ihn so nahe, wie es die Einfachheit der aufgewendeten Hilfsmittel irgend gestattet, an die Grenzen unsrer heutigen Erkenntnis heranzuführen‹. Ich habe mich schon oft vergeblich nach einer Begriffsbestimmung der ›elementaren« Methoden gefragt; ich hatte gehofft, sie in diesem Werke zu finden; doch wurde meine Hoffnung, wenigstens in den bisherigen Lieferungen, getäuscht. Es würde sich bei einer solchen Begriffsbestimmung natürlich nicht um eine bloße Aufzählung derjenigen Methoden handeln, die als elementar gelten sollen, sondern vielmehr um Angabe der Prinzipien, um derentwillen die eine Methode als elementare betrachtet wird, und somit verwendet werden darf, die andre aber, weil sie nicht elementar sei, von der Verwendung ausgeschlossen bleiben soll. Was z. B. ist der charakteristische Unterschied zwischen den Grenzprozessen, die im Begriffe der Potenzreihe und in dem des Integrales stecken, und der es bewirkt, daß jener das Fundament einer ›elementaren‹ Funktionentheorie abgeben kann, dieser aber nicht? Insolange aber ein solcher charakteristischer Unterschied zwischen elementaren und nicht elementaren Methoden nicht aufgestellt ist, werden sich wohl viele des Gefühles nicht erwehren können, daß in der freiwilligen Beschränkung auf elementare Methoden eine gewisse Willkür, man könnte fast sagen, eine Art von Selbstverstümmelung steckt, der zuliebe die Einfachheit auch nicht eines Beweises geopfert werden sollte. Freilich kann wieder andrerseits nicht geleugnet werden, daß durch Beschränkung in der Auswahl der Methoden eine gewisse ästhetische Wirkung erzielt, eine gewisse harmonische Einheitlichkeit erreicht wird, die gerade dem vorliegendem Werke in hohem Maße zueigen ist. Und so möchten wir zu diesem Punkte nur den Wunsch äußern, es mögen die an den Terminus ›elementare Methoden‹ sich knüpfenden prinzipiellen Fragen bald völlig geklärt werden, wozu denn wohl niemand kompetenter wäre beizutragen als der hervorragendste und erfolgreichste Anhänger dieser elementaren Methoden: Herr Pringsheim selbst.

Nach dieser vielleicht allzulangen Erörterung über prinzipielle Fragen, sei nun kurz über den Inhalt des Buches referiert. Abschnitt I ›Reelle Zahlen und Zahlenfolgen‹. Kapitel I ›Die rationalen Zahlen‹. Es wird ausgegangen von der oben ausführlich besprochenen

Einführung der natürlichen Zahlen als Ordinalzahlen (§ 1). Nach dem Vorgange von Graßmann wird ihre Addition rekursorisch definiert durch die beiden Festsetzungen: 1) $a+1$ ist die auf a folgende Zahl; 2) $a+(n+1) = (a+n)+1$, aus denen alle Eigenschaften der Addition hergeleitet werden (§ 2). Um aber insbesondere die assoziative und kommutative Eigenschaft von Summen aus beliebig vielen Summanden beweisen zu können, muß zunächst der Begriff der Anzahl entwickelt werden (§ 3), wovon schon eingehend die Rede war. Es werden sodann die Eigenschaften der Multiplikation der natürlichen Zahlen aus den beiden definierenden Festsetzungen:

$$1)\quad a.1 = a; \qquad 2)\quad a.(n+1) = a.n+a$$

hergeleitet (§ 4). Es folgt eine kurze Besprechung von Subtraktion und Division im Gebiete der natürlichen Zahlen (§ 5), woran sich naturgemäß die Herleitung der Teilbarkeitseigenschaften natürlicher Zahlen schließt, soweit sie für das Folgende benötigt werden (§ 6). Nun kommt die schon oben ausführlich besprochene Einführung der positiven rationalen Zahlen (§§ 7—9), sodann die ganz analog durchgeführte Einführung der Null und der negativen Zahlen (§§ 10—12). Die Definition der Potenz mit ganzzahligen Exponenten (§ 13) und ein Beweis des binomischen Lehrsatzes für positive ganzzahlige Exponenten (§ 14) schließen dieses erste Kapitel.

Kapitel II, ›Begrenzte und unbegrenzte Systembrüche. — Rational-konvergente Zahlenfolgen‹ soll die Erweiterung des Systemes der rationalen Zahlen zu dem der reellen entsprechend vorbereiten. Es geht aus von der systematischen Darstellung der natürlichen Zahlen:

$$g = a_m b^m + a_{m-1} b^{m-1} + \cdots + a_1 b + a_0 \qquad (0 \leqq a_i < b)$$

durch eine Grundzahl $b > 1$ (§ 15), und bespricht sodann ausführlich (§ 16, 17, 18) die Entwicklung der rationalen Zahlen in begrenzte oder periodische unbegrenzte ›Systembrüche‹ (wie sehr glücklich statt des schleppenden Ausdruckes: ›systematische Brüche‹ gesagt wird). Die Beziehungen, in denen der einer rationalen Zahl zugeordnete unbegrenzte Systembruch zu dieser Zahl steht, geben den Anlaß, allgemeiner gewisse unbegrenzte Zahlenfolgen als neue Zeichen, als ›Ersatz‹ für rationale Zahlen anzusehen; es sind dies diejenigen unbegrenzten Folgen rationaler Zahlen, die einen rationalen Grenzwert besitzen, und für die hier, wieder in sehr glücklicher Weise, der neue Terminus ›rational konvergente Zahlenfolgen‹ geprägt wird. Für diese rational konvergenten Zahlenfolgen können nun, da sie lediglich als neue Zeichen für die bereits vorhandenen rationalen Zahlen aufgefaßt werden, die Kriterien für die Gleichheit, für die Relationen

22*

$>$ und $<$, und Regeln für die Ausführung der rationalen Verknüpfungen *deduziert* werden (§ 19, 20). Am Radizierungsproblem wird schließlich das Bedürfnis nach einer neuerlichen Erweiterung des Zahlgebietes dargetan und durch den sich dabei von selbst einstellenden *nicht periodischen* unbegrenzten Systembruch gleichzeitig ein Fingerzeig gegeben, in welcher Weise die Erweiterung des Zahlengebietes vorzunehmen sein wird.

Kapitel III ›Konvergente Zahlenfolgen, reelle Zahlen und Grenzwerte reeller Zahlen‹ führt nun diese Erweiterung durch, und zwar mit völlig analoger Benützung des Permanenzprinzipes, wie bei den bereits vorgenommenen Erweiterungen des Zahlengebietes. Es werden nun an Stelle der bereits betrachteten rational konvergenten Zahlenfolgen beliebige konvergente Zahlenfolgen eingeführt (die rational konvergenten, sowie die durch einen beliebigen unbegrenzten Systembruch gelieferten Zahlenfolgen erweisen sich sofort als Spezialfälle solcher konvergenter Zahlenfolgen), und jede konvergente Zahlenfolge wird als neues Zahlzeichen betrachtet. Die Relationen $=$, $>$, $<$ sowie die rationalen Verknüpfungen werden für diese Zahlzeichen eingeführt, indem die in Kap. II bei rational konvergenten Zahlenfolgen *deduzierten* Regeln nunmehr bei beliebigen Zahlenfolgen als *Definitionen* gelten (§ 21—24). Nachdem so die Irrationalzahlen als konvergente, aber nicht rational-konvergente Folgen rationaler Zahlen eingeführt sind und so das Gesammtgebiet der reellen Zahlen gewonnen ist, wird nun der Beweis geführt, daß es sozusagen ›unvergleichlich‹ mehr irrationale als rationale Zahlen gibt: daß die Menge der rationalen Zahlen abzählbar, die der irrationalen nicht abzählbar ist; auch die Abzählbarkeit der Menge aller algebraischen, und damit die Existenz transzendenter Zahlen wird bewiesen (§ 25). Sodann (§ 26) werden wie bisher Folgen rationaler Zahlen, nun allgemein Folgen reeller Zahlen betrachtet; es wird der Begriff des Grenzwertes eingeführt (wobei auch der durchaus notwendige Nachweis nicht vergessen wird, daß jede eine reelle Zahl definierende Folge rationaler Zahlen diese Zahl zum Grenzwert hat); auch die uneigentlichen Grenzwerte $+\infty$, $-\infty$ treten hier auf, freilich ohne daß sich der Verf. entschließen könnte, auch sie als Zahlen aufzufassen, gewonnen aus einer neuerlichen Erweiterung des Zahlengebietes. Obwohl bei dieser Erweiterung die Rechenregeln nicht in dem Umfange erhalten bleiben, wie bei den früheren, so lohnt sie sich doch durchaus, da viele Sätze dadurch von Ausnahmefällen befreit werden, und einen viel einfacheren Wortlaut erhalten. — Unter den Folgen reeller Zahlen werden nun insbesondere die *konvergenten* näher untersucht (§ 27), und das Kapitel schließt mit dem ›Fundamentalsatz der Analysis‹, daß jede konvergente Folge einen Grenzwert besitzt, und einigen allgemeinen Grenzwertbeziehungen (§ 28).

Es sei uns an dieser Stelle gestattet, unser Referat für einen Augenblick zu unterbrechen und eine kurze kritische Zwischenbemerkung einzufügen. Die eben besprochene Lehre von den reellen Zahlen ist durchaus klar aufgebaut, konsequent durchgeführt und in sich abgeschlossen; in dieser Hinsicht haben wir zu einer Kritik auch nicht den entferntesten Anlaß. Doch hat der Verf. im Vorwort angekündigt, er suche »bei Behandlung der einzelnen Gegenstände eine gewisse über den nächstliegenden Bedarf des Anfängers wesentlich hinausgehende Vollständigkeit zu erreichen«. Und von diesem Standpunkte aus darf vielleicht kritisch angemerkt werden, daß der Begriff des Dedekind'schen Schnittes keinerlei Erwähnung gefunden hat; ich glaube, daß der Studierende die beiden klassischen Einführungen der Irrationalzahl, durch die konvergente Folge und durch den Schnitt, kennen lernen soll, denn für die Anwendungen eignet sich bald die eine, bald die andre besser. Daß ich die Dedekind'sche Methode sogar der Cantor-Méray'schen vorziehe, ist selbstverständlich subjektiv; der Grund dafür ist der, daß der Begriff des Schnittes rein ordinaler, der Begriff der konvergenten Folge aber metrischer Natur ist, und dies hat zur Folge, daß zu einer Erweiterung des Zahlengebietes, die der Willkür so wenig Raum gibt, wie die oben skizzierte Einführung der positiven rationalen Zahlen, sich, soviel ich sehe, der Schnitt besser eignet, als die konvergente Folge. Die Erweiterung des Zahlengebietes wird vorgenommen auf Grund der Forderung, jeder Schnitt im Gebiete der rationalen Zahlen solle durch eine Zahl hervorgerufen werden. Jeder Schnitt definiert also eine Zahl, und definiert gleichzeitig ihre Anordnungsbeziehungen zu allen übrigen Zahlen. Fordert man für Addition und Multiplikation der neuen Zahlen nun lediglich Fortbestehen der diese Operationen im Gebiete der rationalen Zahlen beherrschenden Anordnungsbeziehungen (Ungleichungen), so ist dadurch Addition und Multiplikation (und damit auch Subtraktion und Division) der reellen Zahlen völlig und ohne weitere Willkür definiert.

Gehen wir weiter in unserem Referate. Kap. IV »Potenzen mit beliebigem Exponenten und Logarithmen« beginnt mit der Lösung des Radizierungsproblemes und der Einführung der Potenz mit rationalem Exponenten (§ 29). Es folgt eine sehr schöne Herleitung eines Systemes von Abschätzungsformeln für diese Potenzen, das man in solcher Vollständigkeit nirgends anders finden wird (§ 30), und das sogleich bei Definition der Potenz mit beliebigem reellen Exponenten seine Dienste leistet (§ 31). Es folgen noch die Lehre von den Logarithmen (§ 32), von der Zahl e (einschließlich Irrationalitätsbeweis) (§ 33) und schließlich der Zusammenhang zwischen natürlichem Logarithmus und harmonischer Reihe (§ 34). Wenn ich auch am Schlusse dieses die üblichen Darstellungen weit überragenden Kapitels mir eine kritische Bemerkung erlauben darf, so sei darauf hingewiesen, daß ähnlich wie die Erweiterungen der Zahlgebiete, so auch die Erweiterung des Potenzbegriffes meinem Dafürhalten nach durch Scheidung des Willkürlichen vom Notwendigen gewonnen hätte; bekanntlich wird dies für die Erweiterung auf rationale Exponenten erreicht durch die Forderung nach Fortbestehen der Eigenschaft $a^{x+y} = a^x . a^y$, für die Erweiterung auf irrationale Exponenten aber durch Forderung nach Fortbestehen der Monotonieeigenschaft.

Das nun folgende Kapitel V ›Erweiterungen des Grenzwertbegriffes. Null und Unendlichkeitstypen‹ bringt zunächst die Einführung der Begriffe obere und untere Grenze (§ 35), oberer und unterer Limes, Häufungszahl (§ 36) (wobei unseres Erachtens dieser letztere, für die Anwendungen der Mengenlehre auf die Funktionenlehre so fundamentale Begriff etwas zu kurz wegkommt). Es folgt das Studium von Grenzwerten der Form $\lim\limits_{\nu \to \infty} \frac{a_\nu}{b_\nu}$, wo $\lim\limits_{\nu \to \infty} a_\nu = \lim\limits_{\nu \to \infty} b_\nu = \pm \infty$ oder $= 0$, mit ausführlicher Besprechung der Cauchy'schen Sätze über den Zusammenhang zwischen Folgen $\frac{a_\nu}{M_\nu}$ und $\frac{a_\nu - a_{\nu-1}}{M_\nu - M_{\nu-1}}$ (wo die M_ν monoton ins Unendliche wachsende oder gegen 0 abnehmende Zahlen bedeuten). Hier werden auch die zum Vergleiche des asymptotischen Verhaltens der Glieder zweier Folgen dienenden Begriffe ›infinitär gleich, ähnlich, größer, kleiner‹ eingeführt (§ 37), die sogleich zur Konstruktion der exponentiellen und logarithmischen Unendlichkeitsskalen verwendet werden (§ 38).

Kap. VI ›Zweifach unendliche Zahlenfolgen (Doppelfolgen)‹ weist zunächst die Abzählbarkeit der Elemente einer Doppelfolge und allgemeiner einer p-fach unendlichen Zahlenfolge nach (§ 39), führt sodann den Begriff des eigentlichen und uneigentlichen Grenzwertes (Doppellimes) einer Doppelfolge ein (§ 40) und behandelt ausführlich den Begriff des oberen und unteren Doppellimes (§ 41). Und nun wird (§ 42, 43) auf die Beziehungen zwischen Doppellimes und iterierten Limites eingegangen. Die Ausdrücke [1]):

1) Hier macht sich die schon angemerkte Tatsache unangenehm fühlbar, das $+\infty$ und $-\infty$ nicht als ›Zahlen‹ eingeführt wurden. Denn in (A) kann $\underline{\lim}_{\mu \to \infty} a_\mu^{(\nu)}$ und ebenso $\overline{\lim}_{\mu \to \infty} a_\mu^{(\nu)}$ unendlich oft den Wert $+\infty$ oder $-\infty$ haben,

(A) $$\lim_{\nu\to\infty} (\lim_{\mu\to\infty} a_\mu^{(\nu)}) \quad \text{und} \quad \overline{\lim_{\nu\to\infty}} (\overline{\lim_{\mu\to\infty}} a_\mu^{(\nu)})$$

werden als iterierter unterer und oberer Zeilenlimes bezeichnet. Sind sie einander gleich, so heißt ihr gemeinsamer Wert der iterierte Zeilenlimes. Um über die fundamentale Tatsache hinauszukommen, daß die beiden iterierten Limites (A) stets zwischen unterem und oberem Doppellimes liegen, wird ein Begriff eingeführt, der in gewissem Sinne eine Verallgemeinerung des Begriffes der gleichmäßigen Konvergenz aller Zeilen darstellt. Da die von Pr. gegebene Definition dieses Begriffes aber zu umständlich ist, als daß sie hier wiedergegeben werden könnte, so sei es mir gestattet, sie durch die folgende völlig äquivalente aber kürzere zu ersetzen: Die Doppelfolge $a_\mu^{(\nu)}$ mit endlichem iterierten oberen Zeilenlimes habe folgende Eigenschaft: zu jedem $\varepsilon > 0$ gibt es ein n und ein m, so daß:

$$a_\mu^{(\nu)} < \overline{\lim_{\nu\to\infty}} (\overline{\lim_{\mu\to\infty}} a_\mu^{(\nu)}) + \varepsilon \quad \text{für } \mu \geqq m,\ \nu \geqq n.$$

Sie wird dann von Pr. als gleichmäßig beschränkt nach oben bezeichnet. Analog ist die Definition von gleichmäßig beschränkt nach unten. Eine Doppelfolge, die beide Eigenschaften hat, wird als gleichmäßig beschränkt bezeichnet[1]). Dieser Terminologie kann ich mich nicht anschließen. Der Ausdruck ›gleichmäßig beschränkt‹ hat bereits eine feste Bedeutung, die mit der hier eingeführten nicht übereinstimmt, und derzufolge unendlich viele Zahlenfolgen gleichmäßig beschränkt heißen, wenn es zwei endliche Zahlen gibt, zwischen denen sämtliche Zahlen sämtlicher Folgen liegen. Sollte sich nun auch die von Pr. angewendete Terminologie einbürgern, so könnte das die Quelle zahlloser Mißverständnisse werden. Da ich mich aber nicht berufen fühle, an dieser Stelle eine andre Terminologie vorzuschlagen, so sei der Satz, in dem die Untersuchungen dieses Kapitels gipfeln, in der Terminologie des Verf. wieder gegeben: ›Für die Endlichkeit und das Zusammenfallen des iterierten unteren bzw. oberen Zeilenlimes mit dem unteren bzw. oberen Doppellimes ist notwendig und hinreichend, daß die Zeilen[2])

sodaß also oberer und unterer Limes von Folgen zu bilden ist, in denen unendlich oft $+\infty$ oder $-\infty$ auftritt. Von solchen Folgen war aber nie die Rede, und es ist eine gewisse logische Härte, wenn nun plötzlich als selbstverständlich betrachtet wird, was unter den Hauptlimites solcher Folgen zu verstehen sei.

1) Auch der Begriff »gleichmäßig unbeschränkt« wird eingeführt; und zwar ist die Definition die Pr. für diesen Begriff gibt, das Gegenstück zu der von mir vorgeschlagenen Definition von »gleichmäßig beschränkt«, nicht zu der von Pr. gegebenen.

2) Hier steht im Texte noch der Zusatz »mit eventuellem Ausschlusse einer endlichen Anzahl«. Dieser Zusatz ist pleonastisch, weil schon im Begriffe »gleichmäßig beschränkt« enthalten.

nach unten bzw. nach oben gleichmäßig beschränkt sind‹. Leider enthält der Beweis dieses schönen Theoremes ein Versehen (das einzige ernstliche Versehen, auf das ich in dem ganzen Buche gestoßen bin). Auf S. 289 heißt es beim Beweise, daß die Bedingung notwendig ist: ›so hat man bei passender Wahl von $n \geqq n'$:

$$\underline{a} - \varepsilon \leqq \underline{a}^{(\nu)} \leqq \underline{a} \quad \text{bzw.} \quad \bar{a} \leqq \bar{a}^{(\nu)} \leqq \bar{a} + \varepsilon \quad \text{für} \quad \nu \geqq n\text{‹.}$$

Hierin hängt nun n von ε ab, und wird ins Unendliche wachsen, wenn ε gegen 0 geht. Wenn also nun, wie gezeigt wird, für $\nu \geqq n$ und $\mu \geqq m$ die Ungleichungen bestehen:

$$\text{(B)} \qquad \underline{a}^{(\nu)} - \varepsilon \leqq a_\mu^{(\nu)} \quad \text{bzw.} \quad a_\mu^{(\nu)} \leqq \bar{a}^{(\nu)} + \varepsilon,$$

so ist damit die Pringsheim'sche Definition der gleichmäßigen Beschränktheit nicht als erfüllt nachgewiesen, da sie das Bestehen von (B) für alle $\nu \geqq n$ und $\mu \geqq m$ verlangt, wobei zwar m, nicht aber n von ε abhängen darf. Verwendet man die von mir vorgeschlagene Definition, so ist der Beweis der Notwendigkeit augenblicklich erbracht.

Abschnitt II ›Unendliche Reihen mit reellen Gliedern‹. Kap. I. ›Allgemeine Grundlagen‹. Nach Einführung des Begriffes der Konvergenz und Divergenz einer unendlichen Reihe, werden eingehend die verschiedenen Formulierungen für diese Begriffe erörtert (§ 44) und sodann einige Anwendungen der in § 37 bewiesenen Cauchy'schen Grenzwertsätze auf unendliche Reihen gemacht, insbesondere auf das Verhalten der arithmetischen Mittel aus den Teilsummen (§ 45).

Kap. II ›Reihen mit lauter positiven Gliedern‹. Man kennt die großen Verdienste, die sich Pr. um dieses Gebiet erworben hat. Hier ist es ihm nun gelungen, seine schöne Theorie der Konvergenz und Divergenzkriterien mit unübertrefflicher Klarheit und Präzision, in meisterhaft abgerundeter Form darzustellen. Nach Feststellung der elementaren Eigenschaften von Reihen aus positiven Gliedern (§ 46), wird das Prinzip der Reihenvergleichung eingeführt und der Unterschied zwischen den Divergenz- und Konvergenzkriterien erster und zweiter Art besprochen (§ 47). Es werden sodann typische Formen für die Glieder divergenter und konvergenter Reihen angegeben und Skalen immer schwächer divergierender und konvergierender Reihen hergestellt (§ 48, 49). Die gewonnenen Resultate dienen zur Aufstellung von Kriterien erster Art: die Skala von A. de Morgan und Ossian Bonnet, die entsprechende disjunktive Skala von Bertrand, endlich ein von Pr. selbst herrührendes Kriterium, das durch seine Allgemeinheit ein Seitenstück zum Kummer'schen Kriterium zweiter Art bildet (§ 50). Es folgen einige Anwendungen und asymptotische Abschätzungen von Teilsummen einiger konvergenter

Reihen (§ 51), woran sich Untersuchungen über die Tragweite der Kriterien erster Art schließen: Nachweis, daß durch bloße Umordnung der Glieder einer Reihe die Wirksamkeit eines solchen Kriteriums zerstört werden kann[1]); Konstruktion von Reihen, die so schwach konvergieren und divergieren, daß sie auf kein Kriterium der logarithmischen Skala reagieren (§ 52). Nun werden insbesondere Reihen mit monoton abnehmenden Gliedern betrachtet. Nachdem gezeigt ist: »daß es keinesfalls gleichzeitig irgend eine allgemein gültige Schranke für die Divergenz und eine andre für die Konvergenz in dem Sinne geben kann, daß die Glieder aller divergenten Reihen (mit monotonen Gliedern) von irgendeinem bestimmten Index ab durchweg oberhalb der einen (unteren) Schranke, die aller konvergenten Reihen unterhalb der andern (oberen) Schranke liegen müßten«, wird gezeigt, »daß eine solche Schranke für die Konvergenz allein existiert«: für jede konvergente Reihe mit monoton abnehmenden Gliedern ist $\lim_{\nu\to\infty} \nu a_\nu = 0$, während, wenn die M_ν auch noch so langsam mit ν ins Unendliche wachsen, es stets konvergente Reihen mit monoton abnehmenden Gliedern gibt, für die $\lim_{\nu\to\infty} \nu M_\nu a_\nu = \infty$ (§ 53). Nun werden die Kriterien zweiter Art aufgestellt, die zu den in § 50 aufgestellten Kriterien erster Art gehören (§ 54), und es werden aus ihnen die Gauß'schen Kriterien sammt Verallgemeinerungen hergeleitet (§ 55). Den Schluß bilden Untersuchungen über die Tragweite der Kriterien zweiter Art, insbesondere der Nachweis, daß die Kriterien zweiter Art weniger wirksam sind, als die entsprechenden Kriterien erster Art (§ 56).

Kap. III »Reihen mit positiven und negativen Gliedern«. Es werden zunächst die Begriffspaare: absolute und nicht absolute, unbedingte und bedingte Konvergenz eingeführt und ihre Identität dargetan, sodann die grundlegenden Eigenschaften absolut-konvergenter Reihen entwickelt (§ 57, 58). Nun wird eine Reihe, »von der nur soviel feststeht, daß sie überhaupt konvergiert«, als effektiv konvergent bezeichnet[2]), und es werden Kriterien »für effektive, d. h. eventuell nur bedingte Konvergenz« aufgestellt: der Satz von den alternierenden Reihen, seine auf der Abel'schen Umformung beruhenden Verallgemeinerungen und einiges andre aus diesem Ideenkreise (§ 59).

1) Der Satz auf S. 353 gewinnt an Klarheit, wenn daraus die überflüssige Bedingung $\lim_{\nu\to\infty} P_\nu = \infty$ weggelassen wird.

2) Da sich also der Begriff der effektiven Konvergenz von dem der Konvergenz überhaupt nicht unterscheidet, sehe ich nicht ein, wozu hier dieser Terminus eingeführt wird.

Es folgen sehr eingehende Untersuchungen über die Wertveränderungen bedingt konvergenter Reihen bei Umordnung ihrer Glieder, expemplifiziert insbesondere auf die Fälle

$$a_\nu = \pm \frac{1}{\nu}, \quad = \pm \frac{1}{\nu \lg \nu}, \quad = \pm \frac{1}{\nu^{1-\varrho}} \text{ (§ 60)},$$

und ein Paragraph über Methoden schlecht konvergierende Reihen in besser konvergierende zu verwandeln (§ 61).

Kap. IV ›Unendliche Doppelreihen mit reellen Gliedern‹ bildet das Gegenstück zu Kap. VI des ersten Abschnittes. Den Begriffen ›Doppellimes‹, und ›iterierter Limes‹ entsprechen hier die Begriffe ›Doppelreihe‹, ›iterierte Reihe‹, die Beziehungen zwischen jenen ergeben Beziehungen zwischen diesen (§ 62). Nachdem so die Summation einer Doppelreihe nach Zeilen oder Kolonnen untersucht ist, wird in die Summation nach Diagonalen eingetreten (§ 63). Das Hauptresultat wird so ausgesprochen: ›Besitzt die konvergente Doppelreihe $\sum_0^\infty{}_{\mu,\nu}\, a_\mu^{(\nu)} = S$ die Eigenschaft, daß jede einzelne Zeile und Kolonne konvergiert oder innerhalb endlicher Grenzen oszilliert, so kann die Reihe[1]) $\sum_0^\infty{}_\nu\, w_\nu$ nur konvergieren oder oszillieren, und zwar ist im Falle der Konvergenz stets auch: $\sum_0^\infty{}_\nu\, w_\nu = S$‹. In Wirklichkeit wird aber mehr bewiesen; die über die einzelnen Zeilen und Kolonnen gemachte Voraussetzung dient nur dazu, nachzuweisen, daß die Doppelfolge der Teilsummen $S_\mu^{(\nu)}$ unsrer Doppelreihe beschränkt ist, und der geführte Beweis zeigt, daß unter dieser Voraussetzung[2]) die Reihe der w_ν, nach der Methode der arithmetischen Mittel summiert, stets S ergibt. Nun wird die Identität von absoluter und unbedingter Konvergenz auch für Doppelreihen dargetan (§ 64) und eine für alle (wenn auch nur bedingt konvergenten) Doppelreihen mit konvergenten Zeilen und Kolonnen giltige Methode zur Umwandlung in eine einfache Reihe vorgetragen und auf die Lambertsche Reihe angewendet (§ 65). Als weitere Anwendung der Theorie der Doppelreihen wird die Multiplikation zweier einfacher Reihen behandelt, wobei außer dem klassischen Resultate, daß die

1) Dabei ist w_ν die Summe der Glieder der ν-ten Diagonale:

$$w_1 = a_0^{(\nu)} + a_1^{(\nu-1)} + \cdots + a_\nu^{(0)}.$$

2) Daß die Voraussetzung, die Doppelfolge der $S_\mu^{(\nu)}$ sei beschränkt, für die Giltigkeit des Resultates durchaus wesentlich ist, wird an einem Beispiele gezeigt.

bekannte Multiplikationsformel immer gilt, wenn wenigstens eine der beiden Reihen absolut konvergiert, insbesondere für alternierende Reihen mit abnehmenden Gliedern sehr weitgehende teils notwendige, teils hinreichende Bedingungen für die Giltigkeit der Multiplikationsformel aufgestellt werden (§ 66). Der letzte Paragraph endlich behandelt Konvergenz- und Divergenzkriterien für Doppelreihen mit nicht negativen Gliedern (§ 67).

Unser Referat ist zu Ende. Wir möchten es nicht schließen, ohne unsrer Bewunderung für eine Darstellung Ausdruck zu geben, die es verstand, Exaktheit, Einfachheit und Schönheit zu vereinen, und wenn der Verf., allzu bescheiden, die Hoffnung ausspricht, ›daß diese Vorlesungen auch fortgeschrittenen Lesern merklichen Nutzen gewähren und, zum mindesten in bezug auf die Art der Darstellung, auch der Beachtung der eigentlichen Fachgenossen nicht unwert sein dürften‹, so sei die Ueberzeugung ausgesprochen, daß für jeden wahren Mathematiker das Studium des Pringsheim'schen Buches eine Quelle reinsten Genusses sein wird.

Bonn Hans Hahn

Arithmetische Bemerkungen.

(Entgegnung auf Bemerkungen des Herrn J. A. Gmeiner.)

Von HANS HAHN in Wien.

Es sei mir gestattet, auf die im letzten Hefte dieser Zeitschrift erschienene Note des Herrn J. A. Gmeiner: „Arithmetische Bemerkungen; insbesondere über die Peanoschen Axiome (Entgegnung auf eine Bemerkung des Herrn H. Hahn)" kurz zu erwidern.

1. In meiner Besprechung von Pringsheims Vorlesungen über Zahlen- und Funktionenlehre habe ich, gelegentlich der Einführung der Reihe der natürlichen Zahlen, bemerkt[1]), daß die Verwendung eines „usf." oder, was auf dasselbe hinauskommt, die Verwendung von Punkten ... die natürliche Zahlenreihe bereits voraussetzt, also unzulässig ist, solange es sich erst um die *Einführung* dieser Zahlenreihe handelt. Was das anlangt, scheint mir Herr Gmeiner zuzustimmen. Ich habe nun in einer Fußnote bemerkt, daß auch in der zweiten Auflage der *Theoretischen Arithmetik* von Stolz und Gmeiner sich ein solches „usf." findet[2]); der vierte der Grundsätze, auf die dort (S. 14) die Lehre von den natürlichen Zahlen aufgebaut wird, lautet nämlich:

4. „Jede Zahl ist größer als jede ihr in der Zahlenreihe vorausgehende und kleiner als jede ihr nachfolgende Zahl

d. i. $$a^+ > a\,(a < a^+), \quad (a^+)^+ > a\,(a < (a^+)^+) \quad \text{usf.}$$

Demnach kehrt in der Zahlenreihe niemals dieselbe Zahl wieder, sondern es sind je zwei Zahlen derselben ungleich."

Die Worte „vorausgehend" und „nachfolgend" sind hier nur durch Berufung auf die Symbole a^+, $(a^+)^+$ und das beanstandete „usf." erklärt.[3]) Herr Gmeiner entgegnet nun, daß man diesen Worten auch auf andere Weise einen Sinn verleihen kann, der das zirkelhafte „usf." nicht verwendet. Das habe ich nie bestritten und auch nie bezweifelt, sondern lediglich beanstandet, daß es *nicht geschehen* ist. Doch auch

1) Götting. gel. Anz. 1919, S. 326.

2) Ich möchte noch eigens feststellen, daß mir bei meiner gegen diese einzelne Stelle gerichteten Kritik eine Herabsetzung dieses ausgezeichneten und höchst zuverlässigen Werkes gänzlich ferne lag.

3) Auch die Verwendung des nicht erklärten Wortes „Zahlenreihe" an dieser Stelle scheint mir zu beanstanden.

mit der Art und Weise, wie es Herr Gmeiner in seiner neuen Note macht, kann ich mich nicht ganz einverstanden erklären.

Herr Gmeiner führt (auf S. 83) eine Menge $R(a)$ ein, die drei Forderungen genügen soll, die er mit a) b) c) bezeichnet.

Die an diese Menge $R(a)$ geknüpften Erörterungen scheinen mir nun nicht einwandfrei. Es wird von ihr ohne weitere Begründung gesagt, sie sei wohlgeordnet vom Ordnungstypus ω, also eine unendliche Reihe oder Folge von Zahlen, die man in der Gestalt: $a, a^+, (a^+)^+$ usf., oder $a, a^+, (a^+)^+, \ldots$ anschreiben kann, lauter Aussagen, die — vor Kenntnis der natürlichen Zahlenreihe — ohne petitio principii nicht gemacht werden können. Dann aber entfällt das Mittel, auf das sich Herr Gmeiner stützt, um, falls in der genannten Folge ein und dieselbe Zahl wiederholt auftritt, die entsprechenden Glieder voneinander zu unterscheiden. Und da durch das Axiom 2), auf das allein Herr Gmeiner sich stützen will, nicht ausgeschlossen ist, daß $p^+ = p$, so kann $R(a)$ auch eine endliche Menge sein, ja sogar nur das eine Element a enthalten. Sobald man aber die (meiner Ansicht nach unzulässige) Schreibweise $a, a^+, (a^+)^+$ usw. (in der ein und dasselbe Element an mehreren Stellen auftreten könnte) nicht zur Verfügung hat, kann man nicht fortfahren: „Von der Zahl a sage ich, daß sie jeder anderen Zahl des Systemes $R(a)$ vorausgehe, während diese Zahlen selbst der Zahl a nachfolgen.“ Denn definieren wir $1^+ = 1$, so besteht $R(1)$ nur aus der Zahl 1, zur Zahl 1 gäbe es daher keine vorausgehende oder nachfolgende, und das aus der einzigen Zahl 1 (mit der Regel $1^+ = 1$) bestehende System würde allen Forderungen 1) bis 5) der Theoretischen Arithmetik genügen, was natürlich nicht gemeint ist.

Wohl aber könnte man so vorgehen: an Stelle des Axiomes 4) von Gmeiner stellen wir das Axiom auf:

4) *In $R(a)$ gibt es kein p, so daß $a = p^+$.*

In der Tat können die Axiome 3) und 4) von Peano durch dieses Axiom ersetzt werden, da sie aus ihm folgen, wie noch kurz gezeigt sei:

Nennen wir eine Zahl p, die zu a in der Beziehung $a = p^+$ steht, einen Vorgänger von a, so sehen wir zunächst: In $R(a)$ kann es außer a keine Zahl geben, die nicht in $R(a)$ einen Vorgänger hätte. Angenommen nämlich, b wäre eine solche, so lasse man b aus $R(a)$ weg; die übrigbleibende Menge enthält immer noch a und neben p auch p^+, im Widerspruche zur Definition von $R(a)$.

Es ist $R(a^+)$ Teil von $R(a)$: in der Tat, man lasse aus $R(a^+)$ alle Elemente weg, die nicht in $R(a)$ vorkommen: die übrigbleibende Menge enthält immer noch a^+ und neben p auch p^+, muß also mit $R(a^+)$ identisch sein.

$R(a^+)$ enthält a nicht; denn da nach unserem Axiome 4) $a^+ \neq a$ ist, müßte, wie schon bewiesen, a in $R(a^+)$ einen Vorgänger p haben, und da $R(a^+)$ Teil von $R(a)$, käme p auch in $R(a)$ vor, was unserem Grundsatze 4) widerspricht.[1])

Die Menge $R(a)$ entsteht aus $R(a^+)$ durch Hinzufügen des Elementes a. In der Tat, die so entstehende Menge M ist jedenfalls Teil von $R(a)$, da $R(a)$ sowohl a als $R(a^+)$ enthält. Umgekehrt enthält M das Element a und neben p auch p^+, sie enthält also $R(a)$.

Bezeichnen wir nun mit $D(a)$ die Menge aller nicht in $R(a^+)$ vorkommenden Elemente, so entsteht nun offenbar $D(a^+)$ aus $D(a)$ durch Hinzufügen von a^+.

Ist p ein von a verschiedenes Element von $D(a)$, so kommt auch p^+ in $D(a)$ vor. Dies beweisen wir durch vollständige Induktion. Die Behauptung ist trivial für $a = 1$. Angenommen, sie sei richtig für a; wir haben zu zeigen, daß sie dann auch für a^+ gilt. Sei also p ein von a^+ verschiedenes Element von $D(a^+)$, d. h. ein Element von $D(a)$; ist p das Element a, so ist p^+ das Element a^+, kommt daher in $D(a^+)$ vor; ist hingegen p von a verschieden, so kommt p^+ nach Annahme in $D(a)$, mithin auch in $D(a^+)$ vor, und die Behauptung ist bewiesen.

Nach diesen Vorbereitungen beweisen wir die Axiome 3) und 4) von Peano. Axiom 3) lautet: *Sind a^+ und b^+ dieselbe Zahl, so auch a und b.* Angenommen, es wären a^+ und b^+ dieselbe, a und b verschiedene Zahlen. Da dann die Mengen $R(a^+)$ und $R(b^+)$ übereinstimmen, und b nicht in $R(b^+)$ stehen kann, muß es in $D(a)$ stehen. Da aber b von a verschieden ist, steht dann, wie bewiesen, auch b^+ in $D(a)$, das aber ist unmöglich, da b^+ dasselbe wie a^+ ist, und a^+ nicht zu $D(a)$ gehört.

Das Axiom 4) von Peano besagt: *Es gibt keine Zahl b, so daß* 1 *die Zahl b^+ wäre.* In der Tat, es wäre sonst $R(b^+)$ die Menge $R(1)$, d. h. nach Axiom 5) die Menge aller Zahlen, und es müßte b in $R(b^+)$ vorkommen, entgegen einer Tatsache, die wir oben bewiesen haben.

Es können also tatsächlich die Axiome 3) und 4) von Peano durch unser oben formuliertes Axiom 4) ersetzt werden.

2. Auf S. 90, 91 seiner Note will Herr Gmeiner durch ein Beispiel folgende Behauptung widerlegen, die ich in der zitierten Besprechung von Pringsheims Vorlesungen aufgestellt habe[2]): Soll das System der natürlichen Zahlen durch Hinzufügung neuer Zahlen so er-

1) Auf Grund des eben Bewiesenen könnte man definieren: die in der Menge $R(a^+)$ vorkommenden Zahlen sind die der Zahl a *nachfolgenden*.

2) A. a. O. S. 334.

weitert werden, daß im erweiterten Systeme die Multiplikation stets eindeutig umkehrbar wird (unter Multiplikation eine stets ausführbare, assoziative und kommutative Verknüpfung verstanden, die sich auf die Multiplikation der natürlichen Zahlen reduziert, wenn beide Faktoren gleich natürlichen Zahlen werden), so bleibt für eine Willkür kein Raum. Gemeint ist damit folgendes: Wird das System der natürlichen Zahlen durch Hinzufügung nur so viel neuer Zahlen erweitert, als erforderlich sind, um die Multiplikation umkehrbar zu machen[1]), so sind die Regeln der Vergleichung, die Regeln für Multiplikation und Division im erweiterten Systeme durch die genannten Forderungen eindeutig bestimmt.

Ich wiederhole zunächst den Beweis meiner Behauptung: Da nach Forderung im erweiterten System die Division stets ausführbar ist, muß darin der Quotient zweier natürlicher Zahlen $b:a$ vorhanden sein. Ich habe ihn mit $\frac{b}{a}$ bezeichnet. Herr Gmeiner beanstandet dies, ich will ihn also[2]), da er durch a und b eindeutig bestimmt, d. h. eine Funktion von a und b ist, mit $f(a, b)$ bezeichnen. Es ist also:

(1) $$f(a, b) \cdot a = b.$$

Angenommen nun es sei:

(2) $$f(a, b) = f(a', b'),$$

dann ist zufolge des Begriffes der eindeutigen Verknüpfung auch:

$$f(a, b) \cdot (a \cdot a') = f(a', b') \cdot (a \cdot a'),$$

und wegen des assoziativen und kommutativen Charakters der Multiplikation auch:

$$(f(a, b) \cdot a) \cdot a' = (f(a', b') \cdot a') \cdot a,$$

mithin wegen (1) auch:

(3) $$b \cdot a' = b' \cdot a.$$

Ist umgekehrt (3) erfüllt, so folgt wegen:

$$f(a, b)(a \cdot a') = b \cdot a' \quad \text{und} \quad f(a', b') \cdot (a \cdot a') = b' \cdot a$$

aus der geforderten Eindeutigkeit der Division auch (2). Es gilt also (2) dann und nur dann, wenn (3) gilt, und die Vergleichungsregel ist

1) Die Multiplikation wird natürlich auch umkehrbar, wenn man durch Hinzufügung aller positiven reellen Zahlen, oder darüber hinaus durch Hinzufügung geeigneter nichtarchimedischer Größen erweitert. Auf solche Erweiterungen bezieht sich die Behauptung selbstverständlich nicht. Auf diesen Punkt beziehen sich aber auch die Einwendungen des Herrn Gmeiner nicht.

2) Obwohl ich nicht einzusehen vermag, wieso durch eine konsequent festgehaltene, eindeutige Bezeichnungsweise ein Irrtum entstehen kann.

gewonnen. Ihr Sinn ist der: Die durch den Quotienten $b:a$ und durch den Quotienten $b':a'$ gelieferten Größen des erweiterten Systemes sind dann und nur dann gleich, wenn zwischen den natürlichen Zahlen a, b, a', b' die Relation (3) besteht. *Dies ist also die einzig mögliche Vergleichungsregel für das erweiterte System.*

In ähnlicher Weise beweist man, wie a. a. O. näher ausgeführt, daß:

$$f(a, b) \cdot f(a', b') = f(aa', bb')$$

sein muß, d. h. das Produkt der Quotienten $b:a$ und $b':a'$ muß gleich dem Quotienten $bb':aa'$ sein. *Dies ist also die einzig mögliche Multiplikationsregel für das erweiterte System.*

Sehen wir uns nun das vermeintliche Gegenbeispiel des Herrn Gmeiner an. Er läßt das erweiterte System bestehen aus den Paaren (a, b) natürlicher Zahlen mit der Gleichheitsdefinition:

$$(a, b) = (a', b') \quad \text{wenn} \quad a \cdot b'^2 = a' \cdot b^2$$

und der Multiplikationsregel:

$$(a, b) \cdot (a', b') = (a \cdot a', b \cdot b').$$

Der Quotient zweier natürlicher Zahlen $b:a$ ist dann gegeben durch das Paar (ab, a). Es ist also in meinen obigen Formeln zu setzen:

$$f(a, b) = (ab, a).$$

Nehmen wir nun zwei gleiche Quotienten $a:b$ und $a':b'$, also:

$$(4) \qquad (ab, a) = (a'b', a')$$

und wenden wir die Gleichheitsdefinition des Herrn Gmeiner an. Sie ergibt:

$$ab \cdot a'^2 = a^2 \cdot a'b', \quad \text{d. h.} \quad ba' = ab',$$

in Übereinstimmung mit (3). Sei umgekehrt (3) erfüllt, so gilt (4). Durch das vermeintliche Gegenbeispiel ist also meine Behauptung nicht widerlegt, sondern bestätigt.

Was aus dem Beispiele des Herrn Gmeiner folgt, ist nur die selbstverständliche Tatsache, daß, wenn ich zur Bezeichnung der Größen des erweiterten Systemes Zahlenpaare benütze und nicht dafür Sorge trage, daß der Quotient $b:a$ durch das Zahlenpaar (b, a) bezeichnet wird, die Vergleichungsregel für diese Zahlenpaare im allgemeinen nicht lauten wird:

$$(\alpha, \beta) = (\alpha', \beta'), \quad \text{wenn} \quad \alpha\beta' = \alpha'\beta,$$

und die Multiplikationsregel nicht:

$$(\alpha, \beta) \cdot (\alpha', \beta') = (\alpha \cdot \alpha', \beta \cdot \beta').$$

Ich kann ja selbstverständlich zur Bezeichnung des Quotienten $b:a$ das Zahlenpaar $(\varphi(a,b), \psi(a,b))$ verwenden, wo φ und ψ beliebige Funktionen von a und b bedeuten können, mit der einzigen Einschränkung, daß durch die Funktionswerte

$$\alpha = \varphi(a,b), \quad \beta = \psi(a,b)$$

die Argumentwerte a, b eindeutig festgelegt sind:

$$a = \Phi(\alpha,\beta), \quad b = \Psi(\alpha,\beta).$$

Dann aber kann auf Grund meiner Behauptungen die Vergleichungsregel und die Multiplikationsregel für diese Zahlenpaare sofort hingeschrieben werden; sie müssen lauten:

$$(\alpha,\beta) = (\alpha',\beta'), \quad \text{wenn} \quad \Phi(\alpha,\beta)\cdot\Psi(\alpha',\beta') = \Phi(\alpha',\beta')\cdot\Psi(\alpha,\beta),$$
$$(\alpha,\beta)\cdot(\alpha',\beta') = (\varphi(\Phi(\alpha,\beta)\cdot\Phi(\alpha',\beta'),\ \Psi(\alpha,\beta)\cdot\Psi(\alpha',\beta')),$$
$$\psi(\Phi(\alpha,\beta)\cdot\Phi(\alpha',\beta'),\ \Psi(\alpha,\beta)\cdot\Psi(\alpha',\beta'))),$$

was man wieder am Beispiele des Herrn Gmeiner sofort bestätigt findet.

Meine Behauptungen werden also durch das Beispiel des Herrn Gmeiner in keiner Weise berührt, sondern bleiben in vollem Umfange aufrecht. Sie lauten: *im erweiterten System sind Gleichheit und Multiplikation eindeutig festgelegt.* Daß sie sich bei verschiedener Wahl der Bezeichnungsweise für die Größen des erweiterten Systemes verschieden ausdrücken, ist dagegen kein Einwand, ebensowenig, wie etwa in der Geometrie die Behauptung, eine Gerade sei durch gewisse Forderungen eindeutig festgelegt, dadurch widerlegt wird, daß ihre Gleichung bei verschiedener Wahl des Koordinatensystemes verschieden ausfällt.

Schlußbemerkungen hierzu von H. Hahn.

Ich darf vielleicht hierzu noch folgendes bemerken:

Zu 1. Durch die neuerlichen Ausführungen des Herrn Gmeiner sind meine Bedenken gegen seine ursprünglichen Ausführungen nicht zerstreut.

Zu 2. Hier möchte ich nur kurz auf die Bemerkung des Herrn Gmeiner eingehen, man könne verschiedener Meinung darüber sein, ob es logisch zulässig ist, von dem Quotienten $a : b$ *vor* der Vergleichung und *vor* der Definition der Multiplikation der neuen Größen zu sprechen. Ich kann nicht sehen, daß dies zu einem Bedenken Anlaß gibt. Meine Behauptung ist doch hypothetischer Natur und lautet:

„*Wenn* es möglich ist, das System der natürlichen Zahlen so zu erweitern, daß im erweiterten Systeme die Multiplikation stets ausführbar, assoziativ, kommutativ und eindeutig umkehrbar ist, und sich auf die Multiplikation der natürlichen Zahlen reduziert, wenn beide Faktoren natürliche Zahlen sind, so ist für die Größen des erweiterten Systemes (soweit sie Quotienten natürlicher Zahlen sind) die Vergleichung und die Multiplikation eindeutig festgelegt.“ *Ob* eine solche Erweiterung möglich ist, ist damit nicht gesagt; *daß* sie möglich ist, wird hinterher in gewohnter Weise durch die Methode der Zahlenpaare gezeigt. All dies habe ich in einer im Winter 1920 in Bonn gehaltenen Vorlesung bis in alle Details vorgetragen.

Empirismus, Mathematik, Logik

Von Prof. Dr. Hans Hahn, Universität Wien

Die Grundthese des Empirismus ist diese: die einzige Quelle, die uns ein Wissen über die Welt, ein Wissen über Tatsachen, ein Wissen, dem Inhalt zukommt, liefern kann, ist die Erfahrung; alles derartige Wissen entstammt dem unmittelbar Erlebten. Die Stellung der Mathematik hat seit jeher diesem Standpunkt große Schwierigkeit bereitet; denn die Erfahrung kann uns kein allgemeines Wissen verschaffen, die Mathematik aber scheint allgemeines Wissen zu sein; jedes der Erfahrung entstammende Wissen bleibt mit einem Koeffizienten der Unsicherheit behaftet, an der Mathematik bemerken wir keinerlei Unsicherheit. Ich weiß aus der Erfahrung, daß ein sich selbst überlassener Stein zur Erde fällt; aber da ich dies nur aus der Erfahrung weiß, so kann es sein, daß bei meinem nächsten Versuche es sich anders verhält, daß ein sich selbst überlassener Stein frei in der Luft schweben bleibt; und wir haben nicht die geringste Schwierigkeit, uns dies anschaulich vorzustellen. Wüßten wir nur aus der Erfahrung, daß zwei mal zwei vier ist, so müßte es sich mit diesem Satze ebenso verhalten — aber das ist nicht der Fall; es ist völlig unvorstellbar, daß morgen zwei mal zwei fünf sei; wir haben die Gewißheit, daß dieser Satz jederzeit und überall galt und gilt und gelten wird. Und darum ist jeder Versuch, den Satz „zwei mal zwei ist vier" auf die Erfahrung zu gründen, von vornherein aussichtslos, und auch die Argumentation, dieser Satz beruhe auf besonders einfachen, besonders oft gemachten Erfahrungen und habe deshalb einen so hohen Grad von Gewißheit gewonnen, vermag dabei nicht zu helfen.

Es scheint also zunächst tatsächlich, als müßte am Bestehen der Mathematik der reine Empirismus scheitern, als hätten wir in der Mathematik ein Wissen über die Welt, das nicht aus der Erfahrung stammt, ein Wissen a priori; und bekanntlich ist dies tatsächlich die Meinung mancher rein rationalistisch gerichteter Philosophen, insbesondere aber der an Kant orientierten Philosophie. Und die hier für den Empirismus vorliegende Schwierigkeit ist so auffällig, so einschneidend, daß jeder, der konsequenten Empirismus vertreten will, zu dieser Schwierigkeit Stellung nehmen muß. Ein solcher

Versuch sei hier — in knappen Andeutungen — unternommen.

Bemerken wir zunächst, daß das über die Sätze der Mathematik Gesagte auch auf die Sätze der Logik zutrifft; der Satz „wenn alle a b sind, und alle b c sind, so sind auch alle a c“ ist allgemein gültig wie der Satz „zwei mal zwei ist vier“; auch bei ihm ist es unvorstellbar, daß er etwa morgen nicht gelten könnte, auch er kann daher nicht aus der Erfahrung stammen. Fragen wir also zunächst: wie läßt sich der Empirismus mit dem Bestehen der reinen Logik vereinbaren?

Wollte man die Logik — wie dies geschehen ist — auffassen als die Lehre von den allgemeinsten Eigenschaften der Gegenstände, als die Lehre von den Gegenständen überhaupt, so stünde hier der Empirismus tatsächlich von einer unüberwindlichen Schwierigkeit. In Wirklichkeit aber sagt die Logik überhaupt nichts über Gegenstände aus; Logik ist nicht etwas, das sich in der Welt vorfindet, Logik entsteht vielmehr erst dadurch, daß — vermöge einer Symbolik — über die Welt gesprochen wird, und zwar vermöge einer Symbolik, deren Zeichen keineswegs, wie man zunächst vermuten möchte, umkehrbar eindeutig und isomorph dem zu Bezeichnenden zugeordnet sind (die Einführung einer Symbolik durch eine umkehrbar eindeutige und isomorphe Abbildung hätte sehr geringes Interesse). Nur ein Beispiel: in der Logik wird neben jeder Aussage p auch deren Negat non-p betrachtet; in der Welt aber ist von den beiden mit p und non-p gemeinten Sachverhalten immer nur einer vorhanden. Unsere Symbolik nun gestattet es, den einen in der Welt vorhandenen Sachverhalt doppelt auszudrücken: durch Bejahung von p, durch Verneinung von non-p. Wäre es so, daß in der Welt sowohl der Sachverhalt p, als auch der Sachverhalt non-p vorhanden wäre, und hätte der Satz des Widerspruches zum Inhalt, daß die beiden vorhandenen Sachverhalte p und non-p in irgendeiner (näher zu präzisierenden) Weise niemals vereinigt anzutreffen sind, so würde der Satz des Widerspruches etwas über die Welt aussagen und wir stünden vor dem für den Empirismus verhängnisvollen Dilemma: entweder stammt der Satz des Widerspruches aus der Erfahrung, dann kann er nicht sicher sein; oder er ist sicher, dann kann er nicht aus der Erfahrung stammen. Da aber den beiden Aussagen p und non-p nicht verschiedene Sachverhalte in der Welt entsprechen, sondern lediglich einer, der nur durch sie verschieden bezeichnet wird, so sagt der Satz des Widerspruchs über die Welt gar nichts aus, er handelt vielmehr von der Art, wie die verwendete Symbolik bezeichnen soll.

Und ebenso wie der Satz des Widerspruches sagen auch alle anderen Sätze der Logik über die Welt nichts aus. Die Logik entsteht dadurch, daß die Symbolik, die wir verwenden, um über die Welt zu sprechen, es gestattet, dasselbe auf verschiedene Arten zu sagen, und die sogenannten Sätze der Logik sind Anweisungen, wie man — innerhalb der verwendeten Symbolik — etwas Gesagtes auch anders sagen kann, einen auf eine Art bezeichneten Sachverhalt noch auf eine andere Art bezeichnen kann.

Es wäre nun verlockend, zu zeigen, wie bei dieser Auffassung der Logik das anscheinend so merkwürdige Problem vom Parallelismus im Ablauf unseres Denkens und des Weltgeschehens sich auflöst — jenem Parallelismus, demzufolge wir imstande wären, durch Denken etwas über die Welt auszumachen, wie es scheinbar in so großartiger Weise in der theoretischen Physik geschieht. Wir müssen es uns versagen und den wenigen noch verfügbaren Raum dazu verwenden, um zu dem Problem zurückzukehren, von dem wir ausgingen: wie ist reiner Empirismus mit dem Bestehen der Mathematik verträglich?

Wir haben schon angedeutet, wie reiner Empirismus mit dem Bestehen der Logik verträglich ist. Die Frage: „wie ist reiner Empirismus mit dem Bestehen der Mathematik verträglich?" ist also miterledigt, wenn es gelingt, zu zeigen, daß Mathematik ein Teil der Logik ist, daß also auch die Sätze der Mathematik nichts über die Welt aussagen, sondern lediglich Anweisungen sind, etwas Gesagtes anders zu sagen. Bekanntlich hat Kant aufs entschiedenste den rein logischen Charakter der Mathematik bestritten, und durch den Einfluß Kants wurden die Bestrebungen, Mathematik in Logik aufzulösen, in den Hintergrund gedrängt. Neuerdings aber sind diese Bestrebungen, unter Führung von B. Russell, wieder gewaltig erstarkt und scheinen in siegreichem Vordringen. Vielleicht kann den knappen Andeutungen, die ich hier in wenigen Zeilen zu geben versuchte, entnommen werden, warum gerade der Empirismus besonders interessiert ist an der rein logischen Grundlegung der Mathematik. Die Untersuchungen über die

Grundlagen der Mathematik — scheinbar das Werk einiger Spezialisten — sind in Wahrheit von gewaltigster Bedeutung für das gesamte System unserer Erkenntnis: am Ausgange des Streites über die Grundlagen der Mathematik hängt die Beantwortung der Frage: „Ist konsequenter Empirismus möglich?".

Mengentheoretische Geometrie.

Von HANS HAHN, Wien.

Die Geometrie hat im Laufe der Jahrhunderte ihren Forschungsbereich gewaltig erweitert. Im Altertum waren es noch recht wenige Gebilde, mit denen sich die Geometer beschäftigten: Gerade und Ebenen, die aus geradlinigen Strecken zusammengesetzten Polygone und die aus ebenen Polygonen zusammengesetzten Polyeder, dann einige wenige krumme Linien, vor allem der Kreis und die anderen Kegelschnitte, und einige wenige krumme Flächen, vor allem die Kugel und der Kreiskegel. Die Erfindung der analytischen Geometrie erweiterte diesen Bereich gewaltig: alle durch irgendwelche analytischen Gleichungen erfaßbaren geometrischen Gebilde wurden zu Objekten geometrischer Forschung, vor allem die sämtlichen algebraischen Kurven und Flächen. Die Anwendung der Methoden der Differential- und Integralrechnung auf die Geometrie brachte dann neuerdings eine Erweiterung von größter Tragweite; es entstand die sog. Differentialgeometrie, in der Längen-, Oberflächen- und Volumenbestimmungen krummer Gebilde, Berührungs- und Krümmungseigenschaften und vieles andere systematisch behandelt werden konnten. All dies ist jedem, der ein wenig Mathematik kennt, durchaus geläufig. Aber wenigen dürfte es bewußt sein, daß in den letzten Jahrzehnten, anschließend an die bahnbrechenden Forschungen von G. CANTOR (1845—1918) ein ganz neuer Zweig der Geometrie entstanden ist, dessen Objekte ungleich mannigfaltiger und subtiler sind, als die aller früheren geometrischen Disziplinen und der uns Analysen räumlicher Verhältnisse gegeben hat von einer Feinheit, wie sie früher nicht einmal geahnt werden konnten; es ist dies die *mengentheoretische Geometrie*. Aber nicht nur eine außerordentliche Vermehrung ihrer Forschungsobjekte hat die mengentheoretische Methode der Geometrie gebracht; dieser Methode war es vorbehalten, eine ganze Reihe grundlegender geometrischer Begriffe, die — mehr oder weniger unbestimmt — durch die Anschauung gegeben waren, die immerzu verwendet wurden, ohne daß irgendeine zureichende Definition für sie möglich gewesen wäre, zum erstenmal logisch scharf zu erfassen. Und von dieser Rolle der mengentheoretischen Geometrie soll hier die Rede sein.

Nachdem die mengentheoretische Geometrie zunächst im Anschluß an CANTOR beliebige Punktmengen einer Geraden, einer Ebene, eines euklidischen Raumes in den Kreis ihrer Betrachtung gezogen hatte, bemerkte M. FRÉCHET[1], daß die mengentheoretischen Methoden weitgehend von der Natur des zugrunde gelegten Raumes unabhängig sind. Seither ist es üblich geworden, mengentheoretische Geometrie in „abstrakten" Räumen großer Allgemeinheit zu treiben. Die wichtigste Klasse dieser Räume sind wohl die „topologischen" Räume.

Als topologischen *Raum* bezeichnet die mengentheoretische Geometrie irgendeine Menge E aus irgendwelchen Elementen; diese Elemente heißen *Punkte*, die Teilmengen von E heißen *Punktmengen*. Dabei wird nur angenommen, daß jedem Punkte x gewisse, einfachen Forderungen genügende Punktmengen zugeordnet seien, die *Umgebungen* des Punktes x genannt werden; am meisten hat sich da durchgesetzt die Festlegung des Umgebungsbegriffes durch ein System von 4 Forderungen, die HAUSDORFF formuliert hat[2] (HAUSDORFFS Umgebungsaxiome) und das so lautet: 1. Jedem Punkte x entspricht mindestens eine Umgebung U_x; jede Umgebung U_x enthält den Punkt x. 2. Sind U_x, V_x 2 Umgebungen desselben Punktes x, so gibt es eine Umgebung W_x, die Teilmenge von beiden ist. 3. Liegt der Punkt y in U_x, so gibt es eine Umgebung U_y, die Teilmenge von U_x ist. 4. Für 2 verschiedene Punkte x, y gibt es 2 Umgebungen U_x, U_y ohne gemeinsamen Punkt[3]. Auf Grund dieses Umgebungsbegriffes können nun die grundlegenden Begriffe der mengentheoretischen Geometrie definiert werden; wir geben als Beispiel die Definition des Begriffes *Häufungspunkt*, der — wenn er auch gelegentlich schon früher auftrat — seit G. CANTOR eine fundamentale Rolle spielt: Der Punkt x heißt Häufungspunkt der Punktmenge A, wenn jede Umgebung U_x von x unendlich viele Punkte von A enthält. Eine endliche (d. h. aus endlich vielen Punkten

[1] Rend. Pal. **22** (1906). Eine systematische Darstellung: M. FRÉCHET, Les espaces abstraits et leur théorie considérée comme introduction à l'analyse générale. Paris, Gauthier-Villars 1928.

[2] F. HAUSDORFF, Grundzüge der Mengenlehre. Leipzig 1914, S. 213.

[3] Beispiel: Ist E die Menge aller Punkte x einer Geraden, so kann man jedem Punkte x als Umgebungen zuordnen die Strecken, deren Mittelpunkt x ist (ohne ihre Endpunkte).

bestehende) Punktmenge hat demnach nie einen Häufungspunkt, eine unendliche Punktmenge kann Häufungspunkte haben, oder auch nicht (z. B. hat auf der Geraden die Menge G der Punkte mit ganzzahliger Abszisse keinen Häufungspunkt); die aus allen Punkten der Strecke $\overline{0\,1}$, mit Ausnahme der beiden Endpunkte 0 und 1 bestehende Menge J hat unendlich viele Häufungspunkte, nämlich sämtliche Punkte dieser Strecke, einschließlich ihrer Endpunkte; teils gehören diese Häufungspunkte der Menge J an (nämlich die inneren Punkte der Strecke), teils nicht (nämlich die beiden Endpunkte). Dies gibt Anlaß, unter allen möglichen Punktmengen eine Klasse herauszuheben, die durch besonders einfache Eigenschaften ausgezeichnet sind: man nennt eine Punktmenge A *in sich kompakt*[1], wenn jede unendliche Teilmenge von A einen zu A gehörigen Häufungspunkt hat; die oben genannten Mengen G und J sind nicht in sich kompakt, hingegen ist z. B. die aus allen Punkten der Strecke $\overline{0\,1}$, einschließlich der beiden Endpunkte bestehende Menge in sich kompakt.

Mit Hilfe des Umgebungsbegriffes gelingt leicht eine scharfe Definition des Begriffes *Begrenzung* einer Menge: sei A eine Punktmenge des Raumes E; ein Punkt x heißt Begrenzungspunkt von A, wenn jede Umgebung U_x des Punktes x sowohl Punkte von A enthält, als auch Punkte, die nicht zu A gehören; die Gesamtheit aller Begrenzungspunkte von A heißt nun die Begrenzung von A. Ein anderer der Anschauung entstammender Begriff, der nun scharf logisch gefaßt werden kann, ist der Begriff *zusammenhängend*[2]: nennen wir eine Menge spaltbar, wenn sie so in 2 Teile B und C zerlegt werden kann, daß kein Punkt von B Häufungspunkt von C, kein Punkt von C Häufungspunkt von B ist; eine Menge, die nicht in dieser Weise spaltbar ist, heißt zusammenhängend; eine Strecke (mit oder ohne Endpunkte) ist zusammenhängend, tilgt man aus ihr aber auch nur einen inneren Punkt, so ist die übrigbleibende Menge nicht mehr zusammenhängend[3]. Dieser Begriff des Zusammenhanges führt zu einer Fülle geometrischer Erkenntnisse, die teils von HAUSDORFF selbst, teils von W. SIERPIŃSKI und seinen Schülern (vor allem B. KNASTER und C. KURATOWSKI) entwickelt wurden. Ein ganz einfacher Satz, der einen durchaus anschaulichen Sachverhalt formuliert, lautet: Ist A eine ganz beliebige Punktmenge, B eine zusammenhängende Punktmenge, die sowohl einen zu A gehörigen, als auch einen nicht zu A gehörigen Punkt enthält, so enthält B mindestens einen Begrenzungspunkt von A. Auch der Begriff der *Trennung* läßt sich nun scharf fassen: seien A, B und C 3 Punktmengen des Raumes E; dann heißen A und B getrennt durch C, wenn jede zusammenhängende Punktmenge, die einen Punkt von A und einen Punkt von B enthält, auch einen Punkt von C enthält; der eben vorhin angeführte Sachverhalt kann nun auch so ausgesprochen werden: jede Menge A wird von ihrem Komplemente (d. h. der Menge der nicht zu A gehörigen Punkte) durch die Begrenzung von A getrennt. Noch eine Begriffsbestimmung sei angeführt: eine mehr als einen Punkt enthaltende Menge, die in sich kompakt und zusammenhängend ist, heißt ein *Kontinuum*; z. B. ist jede Strecke (einschließlich ihrer Endpunkte) ein Kontinuum. Eine Menge, die kein Kontinuum als Teil hat, heißt *diskontinuierlich* (Beispiele: jede nur aus endlich vielen Punkten bestehende Menge, ferner die schon oben angeführte Menge G, die aus allen Punkten einer Geraden mit ganzzahligen Abszissen besteht).

Nun einiges über den Begriff *Kurve!* Wohl hat jedermann eine einigermaßen anschauliche Vorstellung davon, was mit dem Worte „Kurve" gemeint ist, aber an einer präzisen Definition dieses Begriffes fehlte es bis vor kurzem gänzlich. Es waren wohl verschiedene Begriffsbestimmungen versucht worden, aber alle hatten sich als unzulänglich erwiesen; erinnern wir nur an eine dieser Begriffsbestimmungen, die besagt: „Kurve ist, was durch stetige Bewegung eines Punktes erzeugt wird" oder präziser: Eine Punktmenge heißt Kurve, wenn sie von einem sich stetig bewegenden Punkte in einem endlichen Zeitintervall durchlaufen werden kann. Diese Kurvendefinition erwies sich als unbrauchbar in dem Augenblicke, als G. PEANO die Entdeckung machte, daß auch eine Quadratfläche von einem sich stetig bewegenden Punkte in einem endlichen Zeitintervall durchlaufen werden kann; denn niemand wird eine Quadratfläche als Kurve bezeichnen wollen. Es erwuchs nun zwar die Aufgabe, festzustellen, welche Punktmengen es sind, die durch jene vermeintliche Kurvendefinition getroffen werden — eine Aufgabe, die gleichzeitig von S. MAZURKIEWICZ und mir gelöst wurde und die zur Aufdeckung einer bis dahin unbeachteten geometrischen Eigenschaft führte, des *Zusammenhanges im Kleinen*, der sich seither bei vielen Aufgaben gestaltlicher Analyse als bedeutungsvoll erwies — aber für eine brauchbare Definition des Begriffes „Kurve" war damit nichts gewonnen. Eine solche Definition wurde erst in den letzten Jahren von K. MENGER und P. URYSOHN aufgestellt; sie lautet: Ein Kontinuum K heißt Kurve, wenn es zu jedem Punkte x von K in jeder Umgebung U_x eine Umgebung V_x von x gibt, deren Begrenzung mit K eine diskontinuierliche Menge gemein hat. Diese Definition hat sich als äußerst fruchtbar erwiesen. Nachdem analytische Geometrie und Differentialgeometrie

[1] Diese Begriffsbildung stammt im wesentlichen von M. FRÉCHET.

[2] Die folgende Definition findet sich zuerst bei N. J. LENNES, Amer. J. of Math. 33, 303 (1911), dann bei HAUSDORFF, Grundz. d. Mengenlehre 1914, 244.

[3] In der Tat ist es im Grunde genommen die Forderung des Zusammenhanges, welche zur CANTOR-DEDEKINDschen *Definition* des (linearen) Kontinuums führt.

nur ganz spezielle Arten von Kurven hatten untersuchen können, konnten nun mit mengentheoretischen Methoden zum ersten Male Untersuchungen angestellt werden, die die allgemeinsten Kurven betrafen; es entstand in wenigen Jahren eine umfang- und inhaltreiche allgemeine Kurventheorie[1], an deren Ausbau sich neben MENGER und URYSOHN vorwiegend eine Reihe junger amerikanischer Forscher (W. L. AYRES, H. M. GEHMANN, R. L. WILDER, G. T. WHYBURN) erfolgreich beteiligten. Nur ein interessantes Resultat sei angeführt: es gelang MENGER im dreidimensionalen euklidischen Raum eine „Universalkurve" K zu konstruieren, die folgendes leistet: jede beliebige Kurve C eines beliebigen (nur ganz geringfügigen Einschränkungen unterliegenden) Raumes kann umkehrbar eindeutig und stetig auf einen geeigneten Teil von K abgebildet werden; oder kürzer gesprochen: die Universalkurve K enthält ein topologisches Urbild jeder beliebigen Kurve.

Einer der größten Erfolge der mengentheoretischen Geometrie war es, daß es ihr gelang, den Begriff der *Dimension* zu erfassen. Seit jeher bezeichnet man eine Gerade, einen Kreis, eine Ellipse usw. als eindimensional, eine Ebene, eine Kugelfläche, eine Kegelfläche usw. als zweidimensional, den Raum der Anschauung, einen Würfel, einen Kugelkörper usw. als dreidimensional, eine präzise Definition der Begriffe „eindimensional", „zweidimensional", „dreidimensional" gab es aber nicht. Vielleicht mochte mancher meinen, der Unterschied liege darin, daß eine zweidimensionale Menge viel mehr Punkte enthalte als eine eindimensionale, eine dreidimensionele mehr als eine zweidimensionale; diese Meinung wurde widerlegt, als CANTOR — gleich zu Beginn seiner Forschungen über Mengenlehre — die Entdeckung machte, daß Gerade, Ebene und Raum gleichviel Punkte enthalten: es gibt eine umkehrbar eindeutige Zuordnung der Punkte der Ebene (des Raumes) zu den Punkten einer Geraden. Der erste Ansatz zu einer wirklichen Definition des Dimensionsbegriffes findet sich, wie es scheint, bei B. BOLZANO[2], der in so vielen Punkten ein Vorläufer der kritischen Mathematik unserer Tage war; es heißt bei ihm: „. . . so sage ich, ein räumlich Ausgedehntes sei einfach ausgedehnt, oder eine Linie, wenn jeder Punkt für jede hinlänglich kleine Entfernung einen oder mehrere, keinesfalls aber so viele Nachbarn hat, daß deren Inbegriff für sich allein schon ein Ausgedehntes wäre; ich sage ferner, ein räumlich Ausgedehntes sei doppelt ausgedehnt, oder eine Fläche, wenn jeder Punkt für jede hinlänglich kleine Entfernung eine ganze Linie von Punkten zu seinen Nachbarn hat; ich sage endlich, ein räumlich Ausgedehntes sei dreifach ausgedehnt oder ein Körper, wenn jeder Punkt für jede hinlänglich kleine Entfernung eine ganze Fläche voll Punkten zu seinen Nachbarn hat." Wie man sieht, wird hier eine ***rekursive*** Definition des Dimensionsbegriffes versucht: es soll zuerst festgestellt werden, was „eindimensional" bedeutet, der Begriff „zweidimensional" soll sodann auf den Begriff „eindimensional" zurückgeführt werden usw. Den Gedanken einer rekursiven Dimensionsdefinition finden wir dann wieder bei H. POINCARÉ[1]: „Ein Kontinuum besitzt n Dimensionen, wenn man es in mehrere getrennte Teile zerlegen kann, dadurch, daß man einen oder mehrere Schnitte führt, die selbst Kontinua von $n - 1$ Dimensionen sind." Dieser POINCARÉsche Ansatz wurde von L. E. J. BROUWER[2] verbessert und präzisiert und dadurch zu einer allen Anforderungen logischer Strenge genügenden rekursiven Dimensionsdefinition ausgebaut, und BROUWER führte auch den Nachweis, daß das, was man in der analytischen Geometrie immer als n-dimensionalen Raum bezeichnet hatte, auch n-dimensional ist im Sinne der neuen Definition. Damit war zum ersten Male ein Resultat gewonnen, das zum Lehrgebäude der heutigen Dimensionstheorie gehört. Allerdings blieb dieses Resultat zunächst isoliert, bis in den Jahren 1921—1922 der seither auf tragische Weise ums Leben gekommene Russe P. URYSOHN[3] und K. MENGER[4] — voneinander und von BROUWER unabhängig — eine rekursive Dimensionsdefinition gaben, die ***weitgehend*** — für die wichtigsten Räume — ***aber nicht vollständig*** mit der BROUWERschen Definition äquivalent ist, und an die nun eine schnelle und bedeutsame Entwicklung der Dimensionstheorie anknüpft. Diese Definition will ich, in der Form anführen, die von K. MENGER stammt[5]. Die Menge A heißt ***nulldimensional,*** wenn es zu jedem Punkt x von A in jeder Umgebung U_x eine Umgebung V_x des Punktes x gibt, deren Begrenzung mit A keinen Punkt gemein hat. Die Menge A heißt ***höchstens n-dimensional*** ($n > 0$), wenn es zu jedem Punkt x von A in jeder Umgebung U_x eine Umgebung V_x gibt, deren Begrenzung mit A eine höchstens $(n - 1)$-dimensionale Menge gemein hat. Ist A höchstens n-dimensional, aber nicht höchstens $(n - 1)$-

[1] K. MENGER, Math. Ann. **95**, Fund. math. **10**; P. URYSOHN, Amst. Verh. **1927**.

[2] B. BOLZANO, Paradoxien des Unendlichen (1851), (Phil. Bibliothek F. Meiner, Bd. 99, S. 80. Leipzig 1920).

[1] H. POINCARÉ, Dernières pensées. Paris 1913, S. 67.

[2] L. E. J. BROUWER, J. f. Math. **142**, 146 (1913); vgl. auch Proc. Akad. Amsterdam **27**, 636 (1924).

[3] Erste Veröffentlichung (ohne Beweise) C. r. Acad. Sci. Paris **175**, 440 (1922). Eine ausführliche Darstellung erschien erst nach dem Tode des Autors in Fund. Math. **7**, 8.

[4] Die ersten Veröffentlichungen MENGERS sind: Monatsh. f. Math. u. Physik **33**, 157 (1923); **34**, 138 (1924). Einige frühere, unveröffentlicht gebliebene, aus den Jahren 1921—1923 stammende Manuskripte und Hinterlegungen bei der Wiener Akademie kommen zum Abdruck in Monatsh. f. Math. u. Physik **36**.

[5] Während BROUWERS Definition eine Präzisierung des POINCARÉschen Ansatzes bedeutet, kann diese Definition aufgefaßt werden als Präzisierung von BOLZANOS Ansatz.

dimensional, so heißt *A n-dimensional*. Die oben angegebene Kurvendefinition kann nun auch so ausgesprochen werden: Kurve ist ein eindimensionales Kontinuum. Um den Ausbau der Dimensionstheorie haben sich neben URYSOHN und MENGER vor allem P. ALEXANDROFF, W. HUREWICZ, ST. MAZURKIEWICZ, L. TUMARKIN verdient gemacht. Von dem Reichtum an bedeutungsvollen Resultaten, über die die Dimensionstheorie heute schon verfügt, gibt ein von K. MENGER verfaßtes Buch „Dimensionstheorie"[1] Kunde. Der Raum verbietet es, hier auch nur die wichtigsten Resultate dieser Theorie vorzuführen und zu erläutern; es sei nur erwähnt, daß MENGER den Nachweis erbracht hat, daß jede n-dimensionale Menge umkehrbar eindeutig und stetig auf eine Teilmenge des euklidischen Raumes von $2n + 1$ Dimensionen abgebildet werden kann, daß also topologische Urbilder aller möglichen Mengen endlicher Dimensionszahl sich in den euklidischen Räumen endlicher Dimensionszahl finden; und es sei noch der merkwürdige und überraschende „Zerspaltungssatz" angeführt, der besagt, daß jede n-dimensionale Menge aus $n + 1$, aber nicht aus weniger als $n + 1$ nulldimensionalen Mengen zusammengesetzt werden kann, wofür ein einfaches Beispiel die Gerade liefert, die selbst eindimensional ist und sich zusammensetzt aus der Menge ihrer rationalen und der Menge ihrer irrationalen Punkte, die beide nulldimensional sind. Noch manche dimensionstheoretischen Probleme harren ihrer Lösung; man findet am Schlusse des MENGERschen Buches eine Reihe solcher Probleme zusammengestellt.

[1] K. MENGER, Dimensionstheorie. Leipzig und Berlin: B. G. Teubner 1928, IV und 320 S.

Hans Hahn
Wien

Die Bedeutung der wissenschaftlichen Weltauffassung, insbesondere für Mathematik und Physik

Der Name „wissenschaftliche Weltauffassung" soll ein Bekenntnis geben und eine Abgrenzung:

Ein Bekenntnis zu den Methoden der exakten Wissenschaft, insbesondere der Mathematik und Physik, ein Bekenntnis zu sorgfältigem logischem Schließen (im Gegensatze zu kühnem Gedankenfluge, zu mystischer Intuition, zu gefühlsmäßigem Bemächtigen), ein Bekenntnis zu geduldiger Beobachtung möglichst isolierter Vorgänge, mögen sie an sich noch so geringfügig und bedeutungslos erscheinen (im Gegensatze zu dichterisch-phantastischem Erfassenwollen möglichst bedeutungsvoller, möglichst weltumspannender Ganzheiten und Komplexe);

eine Abgrenzung gegen Philosophie im üblichen Sinne, als einer Lehre von der Welt, die beansprucht, gleichberechtigt neben den einzelnen Fachwissenschaften oder gar höher berechtigt über ihnen zu stehen. — Denn wir sind der Meinung: was sich überhaupt sinnvoll sagen läßt, ist Satz einer Fachwissenschaft, und Philosophie treiben heißt nur: Sätze der Fachwissenschaften kritisch danach prüfen, ob sie nicht Scheinsätze sind, ob sie wirklich die Klarheit und Bedeutung besitzen, die die Vertreter der betreffenden Wissenschaft ihnen zuschreiben; und heißt weiter: Sätze, die eine andersartige, höhere Bedeutung vortäuschen, als die Sätze der Fachwissenschaften, als Scheinsätze entlarven.

Wir bekennen uns als Fortsetzer der empiristischen Richtung in der Philosophie, stehen somit in entschiedenem Gegensatze zu allem Rationalismus, der sei es das Denken als alleinige Erkenntnisquelle ansieht, sei es als höher berechtigt gegenüber der Erfahrung. Wir stehen aber auch in Gegensatz zu den dualistischen Richtungen, die in Denken und Erfahrung zwei selbständige und gleichberechtigte Erkenntnisquellen sehen wollen. Vielmehr glauben wir, daß nur die Erfahrung, nur die Beobachtung uns Kenntnis vermittelt von den Tatsachen, die die Welt bilden, während alles Denken nichts ist als tautologisches Umformen. Dies wird noch näher auszuführen sein; doch sei schon jetzt gesagt, daß dieses tautologische Umformen für unsere Erkenntnis höchst bedeutungsvoll ist, daß die Bezeichnung „tautologisch" keine Herabsetzung, keine Beschimpfung bedeutet.

Mit dieser Auffassung des Denkens stehen wir im Gegensatz nicht nur zu Rationalismus und Dualismus, sondern auch zur Auffassung einiger Empiristen, die glaubten, Logik (und Mathematik) aus der Erfahrung herleiten zu können, indem sie lehrten, die Sätze der Logik und der Mathematik drückten Erfahrungstatsachen aus — nicht anders als etwa die Sätze der Physik — nur daß es sich dabei um besonders häufig gemachte Erfahrungen handle, wodurch diese Sätze einen besonders hohen Grad von Sicherheit erlangt hätten.

In der Tat, das Verständnis von Logik und Mathematik war immer das Hauptkreuz des Empirismus; denn jeder allgemeine Satz, der der Erfahrung entstammt, bleibt mit einem Momente von Unsicherheit behaftet — bei den Sätzen der Logik und der Mathematik finden wir nichts von solcher Unsicherheit. Wir können uns vorstellen, daß morgen die Sonne nicht aufgeht, aber wir können uns nicht vorstellen, daß $2 \times 2 = 5$ ist, oder daß $p \rightarrow q$ und p richtig, q aber falsch ist; und darüber hilft auch die eben erwähnte Argumentation nicht hinweg, daß es sich hier um besonders oft gemachte Erfahrungen drehe. Durch die (erst der jüngsten Zeit entstammende) Aufklärung der Stellung von Logik und Mathematik, die wir besprechen werden, wurde erst konsequenter Empirismus möglich.

Beobachtung und die tautologischen Umformungen des Denkens, das sind die einzigen Mittel der Erkenntnis, die wir anerkennen. Eine Erkenntnis a priori erkennen wir nicht an, schon deshalb nicht, weil wir sie nirgends benötigen: wir kennen kein einziges synthetisches Urteil a priori, wüßten auch nicht wie es zustande kommen könnte; und was die sogenannten analytischen Urteile der Logik (und Mathematik) anlangt, so sind sie Anweisungen zu tautologischen Umformungen.

Das einzig Gegebene ist für uns das individuell Wahrgenommene, das unmittelbar von mir Erlebte, und Sinn kommt nur dem zu, was in letzter Linie auf Gegebenes zurückführbar ist, aus Gegebenem konstituiert werden kann. So müssen die sogenannten „beharrenden" Gegenstände der Außenwelt, die das tägliche Leben kennt, (Tische, Berge, usw.), erst konstituiert werden, und dies gilt auch von den Körpern der Lebewesen, Tieren und Menschen. Auch die Psychen (die eigene und die fremden) müssen konstituiert werden — und die Erfahrung lehrt, daß sie an Körper gebunden sind; doch wäre die Konstituierung nicht an Körper gebundener Psychen prinzipiell nicht undenkbar, nur hat sie sich (entgegen den Lehren der Religionen und des Spiritismus) bisher nicht als notwendig, nicht einmal als zweckmäßig gezeigt. Auch die Gegenstände der Physik (Atome, Elektronen, Wellen aller Art) müssen durch Konstituierung auf das Gegebene zurückgeführt werden. Sowie für einen Term eine solche Konstituierung aus Gegebenem nicht als möglich erscheint, muß er als sinnlos betrachtet werden und mit ihm jeder Satz, in dem er vorkommt; und dies ist das Schicksal eines Großteiles dessen, was in Philosophie, Metaphysik und Theologie abgehandelt wird.

Die Notwendigkeit des Konstituierens zeigt schon, daß beim Prozesse unserer Erkenntnis zum Gegebenen noch eine Verarbeitung dieses Gegebenen hinzutritt. Dies gibt uns Gelegenheit, betreffs der Stellung der Logik eine Auffassung zu skizzieren, die in unserm Kreise eingehend diskutiert wird. Infolge dieser Auffassung ist Logik nicht etwas, das sich im Gegebenen, oder sagen wir: in der Welt findet; Logik ist nicht, wie geglaubt wurde, die Lehre von den allgemeinsten Eigenschaften der Gegenstände, die Lehre von den Gegenständen überhaupt; vielmehr entsteht Logik erst durch Verarbeitung des Gegebenen, erst dadurch, daß das erkennende Subjekt dem Gegebenen gegenüber tritt, sich ein Bild davon zu machen sucht, eine Symbolik einführt: Logik ist daran geknüpft, daß etwas über die Welt gesagt wird. Diese Symbolik ist nun aber nicht eine umkehrbar eindeutige, isomorphe Abbildung — eine solche Symbolik hätte nur sehr geringes Interesse, und bei Einführung einer solchen Symbolik würde auch keine Logik entstehen. Die Logik entsteht gerade dadurch, daß das Abzubildende und seine Bilder, die Symbole, verschiedene Struktur aufweisen. Ein ganz einfaches Beispiel hierfür liefert schon die Negation: in der Symbolik gibt es zu jeder Aussage p die Negation $\bar{p}$, in der Welt kommt von den beiden entsprechenden Sachverhalten nur einer vor; ist dies etwa der p entsprechende Sachverhalt so können wir ihn auf zwei Arten ausdrücken: durch Behaupten von p, durch Verneinen von $\bar{p}$.

Ein anderes Beispiel: In der Symbolik haben wir neben den beiden Symbolen p und q als durchaus selbständiges Symbol auch das logische Produkt $p \, . \, q$; in der Welt besteht nicht neben den Sachverhalten p und q noch ein selbständiger, durch ein reales „und" gekoppelter Sachverhalt $p \, . \, q$. (Der „molekulare" Sachverhalt $p \cdot q$ ist schon etwas ins Gegebene Hineingetragenes, Resultat einer Verarbeitung). Das Symbol $p \cdot q$ nun ermöglicht die gleichzeitige Behauptung von p und q; indem ich $p \cdot q$ sage, habe ich p mitgesagt und ebenso q. Aus dem Umstande nun, daß die verwendete Symbolik es gestattet, dasselbe verschieden zu sagen, und daß sie es gestattet, dadurch daß man etwas sagt, etwas anderes mitzusagen, entsteht die Logik. Sie ist eine Anweisung, wie man etwas Gesagtes anders sagen kann, oder wie man aus etwas Gesagtem ein anderes Mitgesagtes herausholen kann. Dies ist es, was wir als den tautologischen Charakter der Logik bezeichnen wollen.

Und der Anlaß dafür, eine Symbolik einzuführen, deren Struktur von der des Abzubildenden abweicht, eine Symbolik, die es gestattet, dasselbe verschieden zu sagen, liegt darin, daß wir nicht allwissend sind, daß wir die Sachverhalte, die die Welt bilden, nur sehr fragmentarisch kennen. Dies sieht man wieder deutlich am Beispiel der Negation: es bestünde keinerlei Anlaß, eine Negation einzuführen, wenn wir jeden einzelnen Sachverhalt der Welt kennen würden. Ein allwissendes Subjekt braucht keine Logik, und im Gegensatze zu Plato können wir sagen: Niemals treibt Gott Mathematik.

Also, nicht eine Lehre über das Verhalten der Welt ist die Logik — ein logischer Satz sagt vielmehr über die Welt gar nichts aus — sondern eine Anweisung zu gewissen Transformationen innerhalb der verwendeten Symbolik. Bei dieser Auffassung der Logik löst sich auch ganz von selbst das vielbehandelte Problem vom scheinbar rätselhaften Parallelismus im Ablauf unseres Denkens und des Weltgeschehens, der scheinbar prästabilierten Harmonie zwischen Denken und Weltgeschehen, der zufolge wir imstande wären, durch Denken etwas über die Welt auszumachen. Das ist nie und nirgends der Fall. Das Denken kann nur Sätze tautologisch umformen, und dadurch in eine Gestalt bringen, die für uns besser überblickbar ist, oder aus ihnen Aussagen hervorholen, die in ihnen mitbehauptet waren, und die sich zur Kontrolle besser eignen als die ursprünglichen Sätze.

Dies ist die Rolle des Denkens im Systeme unserer Erkenntnis, dies ist insbesondere auch die Rolle des Denkens in der Physik, und ein wenig Besinnung, ein kurzer Blick auf die Geschichte der Wissenschaft zeigt, daß diese Rolle des Denkens — trotz seines tautologischen Charakters — keine nebensächliche ist. Nur ein Beispiel: Die Grundgleichungen des elektromagnetischen Feldes sind unmittelbar durch Beobachtung kaum kontrollierbar; durch tautologisches Umformen aber erkennt man, daß in ihnen Aussagen mitbehauptet sind über unmittelbar kontrollierbare Verhältnisse, wie sie z. B. beim Michelsonversuch vorliegen, und nun konnte dieser Versuch zur Kontrolle der Grundgleichungen dienen.

Es war Wittgenstein, der den tautologischen Charakter der Logik erkannte, und der betonte, daß den sogenannten logischen Konstanten (wie „und", „oder", usw.) in der Welt nichts entspricht, während Russell früher den logischen Charakter einer Aussage darin zu sehen glaubte, daß sie sich allein mit Hilfe von Variablen und logischen Konstanten ausdrücken lasse; eine solche Aussage wäre nun zum Beispiel diese: „es gibt zwanzig Individuen in der Welt". Wir können aber eine solche Aussage nicht als eine logische betrachten, da sie tatsächlich etwas über die Welt aussagt und der Nachprüfung durch die Erfahrung prinzipiell zugänglich ist.

Nachdem ich nun einiges über die Stellung der Logik im Systeme unserer Erkenntnis gesagt habe, muß ich auch ein wenig über die Stellung der Mathematik sprechen. Da wir uns schon zur Auffassung bekannt haben, unsere einzigen Erkenntnismittel seien die Erfahrung und das tautologische Umformen der Logik, und da in die Mathematik von Erfahrung nichts eingeht, ist die Antwort von vornherein gegeben: die Mathematik muß ebenfalls tautologischen Charakter haben, d. h. die Mathematik ist ein Teil der Logik. Das ist bekanntlich der Standpunkt Russells. Indem wir uns aber zu diesem Standpunkt bekennen, wollen wir nicht sagen, daß wir die Principia Mathematica als unantastbares Evangelium betrachten; wir glauben vielmehr, daß — trotz der gewaltigen Verdienste von Russell und Whitehead — die Aufgabe, die Mathematik als reine Logik zu entwickeln, noch nicht restlos gelöst ist — schon deshalb, weil die Aufgabe einer befriedigenden Darstellung der Logik selbst noch nicht restlos gelöst ist. Doch sind wir himmelweit vom Standpunkt jener Kritiker entfernt, die das Russellsche System für abgetan erklären, weil es in ihm noch Schwierigkeiten gibt, und die diesem Werke nicht anders gegenüberstehen, als der Divina Commedia ein Mann gegenübersteht, der nicht italienisch kann und dem von dieser Dichtung nur eine knappe Inhaltsangabe zur Verfügung steht.

Abgesehen von einer etwas anderen Auffassung der Logik selbst, scheint uns das Werk Russells verbesserungsbedürftig hinsichtlich der verwendeten Typentheorie, die — wie es scheint — durch eine einfachere zu ersetzen ist, bei der auch das vielumstrittene und wohl endgültig als extralogisch erkannte Reduzibilitätsaxiom entbehrlich wird — ferner hinsichtlich des von Wittgenstein bekämpften, absoluten Identitätsbegriffes — wenn auch hier die Remedur (wie mir scheint) nicht so einschneidend wird sein müssen, als Wittgenstein dies annimmt.

Diese Stellungnahme für Russell mag überraschen inmitten eines Deutschland, das nur auf den Streit Intuitionismus — Formalismus — „hie Brouwer", „hie Hilbert" — hört. Doch scheint ein Bekenntnis zum Brouwerschen Intuitionismus für einen Bekenner der von uns zugrunde gelegten Prinzipien schwierig; allzu verwandt scheint dieser Ausgangspunkt der Kantschen reinen Anschauung und dem Kantschen a priori. Das hindert nicht, die intern mathematische Bedeutung der Brouwerschen Forschungen sehr hoch einzuschätzen als Feststellungen dessen, was sich in der Mathematik mit vorgegebenen Hilfsmitteln erreichen läßt, was nicht. Die Einordnung dieser Forschungen in das logizistische System der Mathematik ist eine ebenso wichtige als dringliche Aufgabe, die noch zur Gänze zu leisten ist.

Was Hilberts Formalismus anlangt, so müssen wir von unserem Standpunkte aus vor allem auf die ungeklärte Rolle der metamathematischen Überlegungen hinweisen. Woher stammen die metamathematischen Erkenntnisse? Aus der Erfahrung doch wohl nicht! Dagegen streiten eben die Schwierigkeiten, die wir bei den Versuchen, Logik und Mathematik auf Erfahrung zu gründen, schon besprachen. Sind es logische Umformungen? Wohl auch nicht! Denn sie sollen ja zur Begründung dieser Umformungen dienen. Was also sind sie? Doch soll durch diese skeptische Stellungnahme dem Ausgangspunkt gegenüber nichts gegen die Bedeutung der Hilbertschen Untersuchungen gesagt sein. Ich bin vielmehr der Überzeugung, daß in die Weiterbildung und Verbesserung des Russellschen Systems viel von Hilberts konkreten Resultaten eingehen wird.

Das bisher zur Mathematik Gesagte bezog sich auf Arithmetik und Analysis. Ganz anders — unbestrittener und zeitlich früher —

vollzog sich die Logisierung der Geometrie. Hier ging die Einordnung in die Logik vor sich durch die sogenannte Axiomatisierung. Jeder Lehrsatz der Geometrie erscheint so als (tautologische) Implikation $P \rightarrow Q$, deren Vorderglied P das logische Produkt der Axiome, deren Hinterglied Q der betreffende Lehrsatz ist. Die Axiome erscheinen dabei nicht mehr als von selbst einleuchtende aber unbeweisbare Wahrheiten, sondern als Annahmen, aus denen deduziert wird; die Grundbegriffe erscheinen nicht mehr als durch Definition nicht weiter zerlegbare, aber durch Anschauung unmittelbar erfaßbare Gegenstände, sondern lediglich als logische Variable. Da jedes einzelne Axiom eine Relation zwischen den die Grundbegriffe repräsentierenden Variablen ist, so erscheint Geometrie als spezielles Kapitel der Relationstheorie, als Untersuchung gewisser spezieller Relationssysteme.

Aber damit ist nicht alles gesagt, was sich über Geometrie sagen läßt, und das wurde von den Mathematikern in der ersten Freude über die Logisierung der Geometrie allzusehr übersehen. Anders als Logik und Arithmetik, deren Sätze nichts über die Welt aussagen, kann Geometrie den begründeten Anspruch erheben, auch eine Tatsachenwissenschaft zu sein. Sie wird es dadurch, daß für ihre Grundbegriffe, die bisher für uns logische Variable waren, eine Deutung in der Wirklichkeit gegeben wird, daß sie aus dem Gegebenen konstituiert werden. Mag diese Konstituierung der geometrischen Grundbegriffe auch nicht zur Mathematik gehören (die mit der Wirklichkeit nichts zu tun hat), so ist sie doch eine unabweisbare Aufgabe der Forschung. Weitgehende Ansätze dazu findet man im Anschluß an Whitehead, im IV. Kapitel von Russells Buch, „Unser Wissen von der Außenwelt". — Ob dann die so konstituierten Grundbegriffe den Axiomen genügen, ist eine Frage der Empirie, Geometrie in diesem Sinne also eine empirische Wissenschaft.

Durch die letzten Betrachtungen ist die Brücke zur Physik geschlagen. Manche Kapitel der Physik wurden bereits axiomatisiert, im selben Sinne wie die Geometrie, und dadurch zu Spezialkapiteln der Relationstheorie gemacht. Doch bleiben sie Kapitel der Physik, und damit einer empirischen, einer Tatsachenwissenschaft dadurch, daß die in ihnen auftretenden Grundbegriffe aus dem Gegebenen konstituiert werden.

Als Ziel mag hier vorschweben: Aufstellung eines Axiomensystemes, durch das die ganze Physik logisiert, in die Relationstheorie eingeordnet wird. Dabei wird es sich wohl zeigen, daß je umfassender die Axiomensysteme werden, je mehr sie gleichzeitig vom Gesamtgebiet der Physik erfassen, ihre Grundbegriffe immer wirklichkeitsferner werden, durch immer längere, immer kompliziertere Konstitutionsketten mit dem Gegebenen zusammenhängen. Wir können nur: dies konstatieren, als eine Eigentümlichkeit des Gegebenen; keine Brücke aber führt von da zur Behauptung, hinter der Sinnenwelt liege eine zweite, die „reale" Welt, die ein selbständiges Dasein führe, anders geartet als die Welt unserer Sinne, eine Welt, die wir niemals direkt wahrnehmen können, die sich in der Sinnenwelt nur in verzerrter, schwer enträtselbarer Weise manifestiere, und die wir nun aus diesen spärlichen und verworrenen Andeutungen zu rekonstruieren hätten. Eine solche, durchaus metaphysische Behauptung scheint uns inhaltleer, ein bloßer Scheinsatz; denn es gelingt nicht, die in ihr auftretenden Terme durch Konstituierung an das Gegebene zu knüpfen.

Und hiermit sind wir wieder bei der Grundthese wissenschaftlicher Weltauffassung angelangt: nur zwei Mittel der Erkenntnis gibt es: Erfahrung und logisches Denken; dieses aber ist nichts als tautologisches Umformen, und somit gänzlich außerstande, aus sich heraus etwas über ein Dasein auszumachen, aus der Welt des Gegebenen heraus zu einer anders gearteten Welt wahren Seins zu führen. Jederlei Metaphysik ist daher unmöglich, jeder metaphysische Einschlag ist als sinnlose Wortkombination aus der Wissenschaft zu entfernen.

Dem Eindringen metaphysischer Elemente in die Wissenschaft wurde durch zwei Momente Vorschub geleistet: durch Überschätzung des Denkens, die — den tautologischen Charakter des Denkens verkennend — annahm, das Denken könne aus sich heraus zu Neuem führen, und durch Überschätzung der Sprache. Die Wortsprache ist eine sehr unvollkommene Symbolik, deren Syntax schlecht mit der Syntax der Logik übereinstimmt; insbesondere die in den uns geläufigen Sprachen vorherrschende Subjekt-Prädikatstruktur und ihre Vorliebe für Substantiva hat in einer Philosophie, die aus der Struktur der Sprache auf die Struktur der Welt schloß, viel Unheil angerichtet, zur Einführung mannigfacher überflüssiger Scheinwesenheiten metaphysischen Charakters geführt, wie Substanz, Raum, Zeit, Zahl. Dieser Mangel der Wortsprachen ist einer der Gründe, die die Anhänger wissenschaftlicher Weltauffassung veranlassen, sich auch als Anhänger der sogenannten symbolischen Logik zu bekennen.

Ein weiterer Grund ist dieser: die Worte der Sprache führen, neben dem Hinweise auf das, was sie ihrem wörtlichen Sinne nach symbolisieren sollen, noch die verschiedenartigsten Begleitvorstellungen mit sich. Diese Begleitvorstellungen begünstigen nun ein Hin- und Herschwanken zwischen der wörtlichen Bedeutung und „übertragenen", „bildlichen" Bedeutungen. Die lyrische Dichtung beruht fast zur Gänze auf dieser Eigenschaft der Wortsprachen. So berechtigt nun die Lyrik als Mittel zum Ausdruck und zur Erzeugung von Gefühlen ist, so heterogen ist Lyrik dem Prozesse wissenschaftlicher Erkenntnis, dessen Wesen Eindeutigkeit und Klarheit ist. Und doch haben sich durch unvorsichtige und mißbräuchliche Verwendung der Sprache neben vielen metaphysischen auch viele lyrische Elemente in die Wissenschaft eingeschlichen — oft versteckt, verschämt und kaum aufspürbar, keineswegs immer so ersichtlich, wie in folgenden Sätzen, die ich einer in Husserls Jahrbuch der Phänomenologie erschienenen Abhandlung über „Realontologie" entnehme: „Stoffexstase setzt Licht. Wo ein Stoff aus der Immanenz zur Transzendenz hervorbricht, wird er dadurch und damit lichthaft. ... Wo Lichthaftigkeit auftritt (dagegen), handelt es sich (wie wir wissen) um ein wirkliches „Außer-sich-geraten". Nicht die Totalität in ihrer äußeren Fixiertheit, sondern die innere Gebundenheit des Stoffes selbst und als solchen wird durchbrochen... Das selbstverständliche „nach innen" schlichter Stoffgegebenheit verwandelt sich in den in sich über sich selbst hinausgehobenen Zustand des „nach außen".

Um nun alle „lyrischen" Elememte aus der Wissenschaft wegzuschaffen, bedarf es einer Symbolik, an deren Symbole sich nicht die verschiedenartigsten Assoziationen knüpfen, wie an die Worte der Sprache. Um wieviel in dieser Hinsicht die symbolische Logik der Wortsprache überlegen ist, leuchtet ein, wenn man sich ein in den Symbolen der Principian Mathematica geschriebenes lyrisches Gedicht vorzustellen sucht! Und so macht der Wunsch, die Wissenschaft sowohl von allen metaphysischen als auch von allen lyrischen Einschlägen zu befreien, die Anhänger wissenschaftlicher Weltauffassung auch zu Anhängern symbolischer Logik.

Mag es einem oberflächlichen Beobachter scheinen, als stünde die geschilderte wissenschaftliche Weltauffassung im Gegensatze zum Zeitgeist der Gegenwart, der nach Metaphysik tendiere, nach Zusammenhängen, die nur mystischer Intuition zugänglich, nur durch das Gefühl erfaßbar seien, der aufs Ganze gehe, nicht auf minutiöse Kleinarbeit — wir sind uns bewußt, daß das ein Oberflächenurteil ist. Der wahre Ausdruck unserer Zeit, der Zeit der Organisationen mit ihrem festen Gefüge, das nur fest ist durch Arbeit am Einzelnen, der Zeit der Rationalisierung der Industrie, die in einem wohlumgrenzten Systeme wirksam ist, indem sie das kleinste erfaßt, der Zeit der Sachlichkeit in Architektur und Gewerbe, der wahre Ausdruck dieser Zeit ist die wissenschaftliche Weltauffassung mit ihrer liebevollen, sorgfältigen, ins Einzelne gehenden Beobachtung des Gegebenen, mit ihren vorsichtigen, logischen Konstruktionen Schritt für Schritt, mit ihrer schlichten Sprache, die keine andere Aufgabe kennt, als: das klar zu sagen, was gesagt werden soll.

In dem wirren Vielerlei philosophischer Systeme kann man, wie mir scheint, zwei Haupttypen unterscheiden: Systeme *weltzugewandter* und Systeme *weltabgewandter* Philosophie. Die *weltzugewandte* Philosophie baut ganz auf die uns durch die Sinne kundgetane Welt, sie nimmt diese Welt, wie sie sich darbietet, in ihrer Unbeständigkeit, ihrer Regellosigkeit, ihrer Buntheit, und sucht sich in ihr zurechtzufinden, sich mit ihr abzufinden, sie zu genießen. Das einzig Wesenhafte ist ihr das durch die Sinne Kundgetane; sie verabscheut es, außerhalb dieser Sinnenwelt nach andersgearteten Wesenheiten zu fahnden. Die *weltabgewandte* Philosophie hingegen mißtraut den Sinnen, hält die Sinnenwelt für Lug und Trug, für bloßen *Schein*, und sucht nach wahren Wesenheiten, nach wahrhaftem *Sein*, hinter der durch die Sinne vorgetäuschten Welt des Scheins. Letzte Ursache für diese so verschiedene Art des Philosophierens dürften wohl gewisse psychologische Grundstimmungen sein: ein gewisser Optimismus auf der einen, ein gewisser Pessimismus auf der anderen Seite. Wer sich in der Sinnenwelt wohl fühlt, wer imstande ist, sich ihrer zu freuen, wird nicht nach Wesenheiten hinter ihr Ausschau halten; wer aber in ihr seine Befriedigung nicht findet, wem es versagt ist, diese Welt der Sinne zu genießen, der will Zuflucht finden in einer Welt anders gearteter Wesenheiten. Und so erweist sich die weltabgewandte Philosophie auch als ein immer wieder benütztes Mittel, um die Menge derer, die mit Recht nicht sehr zufrieden sind in *dieser* Welt, auf eine an-

3

dere Welt zu vertrösten; und diesem Umstande zumeist verdankt es vielleicht die weltabgewandte Philosophie, daß sie durch zwei Jahrtausende eine ziemlich unbestrittene Herrschaft führen konnte, die erst in unseren Tagen ernstlich ins Wanken gerät. Doch von dieser, ich möchte sagen politischen und ökonomischen Rolle der weltabgewandten Philosophie soll hier nicht die Rede sein. Wohl aber müssen wir uns — ganz kurz nur und ganz schematisch — die Gedankengänge ansehen, die weggeführt haben von der Sinnenwelt.

Da beruft man sich auf Fälle, wo uns die Sinne anscheinend täuschen. Ein ins Wasser gesteckter Stab z. B. scheint uns geknickt, aber in Wirklichkeit ist er doch gerade; die Fata Morgana zeigt uns herrliche Palmenhaine mitten in der Wüste, kommt man aber hin an den Ort der vermeintlichen Oase, so findet man nichts dort als Sand; man sieht einen Regenbogen in deutlichen Farben an einer ganz bestimmten Stelle, begibt man sich aber an diese Stelle, so regnet es dort, sonst nichts. Und nun heißt es: wer einmal lügt, dem glaubt man nicht! Da uns die Sinne *manchmal* betrügen, vielleicht betrügen sie uns *immer*. Schaut man durch rotes Glas, so sieht alles rot aus; hielte ein Dämon jedem von uns jederzeit eine rote Scheibe vor die Augen, würden wir nicht glauben, daß alles rot *ist*? Vielleicht ist wirklich ein solcher Dämon da, der uns fortwährend ein Blendwerk vormacht, weil es ihn freut, uns zum Narren zu halten? Vielleicht ist es auch kein Dämon, sondern ein allgütiger und allgerechter Gott, der uns bestraft, weil wir einmal etwas angestellt haben, das ihn ärgert?

Gedankengänge solcher Art mögen es gewesen sein, die zum Mißtrauen gegen die Sinnenwelt geführt haben. Wenn man aber den Sinnen mißtraut, wem sollte man denn trauen? Und da wird uns geantwortet: dem *Denken*. Das Denken, so heißt es, erfaßt nicht gleich den Sinnen

4

bloßen Schein, es erfaßt das wahre wesenhafte Sein; und drum müssen die wahren Wesenheiten so sein, wie die Begriffe unseres Denkens. Diese Begriffe aber sehen ganz anders aus, als die Gegenstände der Sinnenwelt. In der Sinnenwelt finden wir viele Pferde — aber es gibt nur *einen* Begriff „Pferd"; ein Pferd der Sinnenwelt wird geboren, ist erst jung, dann wird es alt, dann stirbt es — der Begriff „Pferd" aber wird nicht geboren, wird nicht älter und stirbt nicht; die einzelnen Pferde der Sinnenwelt bewegen sich, verändern sich, entstehen, vergehen — der Begriff „Pferd" aber ist unveränderlich und unbeweglich, ist nicht dem Werden und Vergehen unterworfen. Und so schloß nun *Plato*, die Welt unsrer Sinne, die bunte, vielgestaltige, wechselnde, in der alles wird und vergeht und nichts Bestand hat, sei nicht die Welt des wahren Seins; die Welt wahren Seins sei eine Welt der *Ideen*, von denen unsere Begriffe uns ein Abbild liefern; in der thront irgendwo und irgendwie die Idee „Pferd", unentstanden und unvergänglich, unbewegt und unveränderlich und einheitlich; den Pferden der Sinnenwelt aber kommt ein Sein nur insoferne zu, als sie an der Idee „Pferd" Teil haben — was das allerdings heißen soll, ist schwer zu sagen.

Eine noch abstrusere Lehre stellten die mit Plato ungefähr gleichzeitigen Philosophen der *eleatischen Schule* auf. Während Plato im Gegensatze zu den vielen Pferden der Sinnenwelt nur *eine* Idee „Pferd" als wahrhaft seiend annahm, so lehrte er doch, daß es in der Welt des wahren Seins viele verschiedene Ideen gebe: eine Idee des Menschen und des Löwen, und des Wahren, des Guten, des Schönen, und eine Idee der Zwei, und der Drei usw. Den Eleaten aber schien schon jede Vielheit dem wahren Sein zu widersprechen und sie lehrten, das wahrhaft Seiende sei Eines, unentstanden, unvergänglich, unbewegt, unveränderlich und undifferenziert.

5

So sucht die weltabgewandte Philosophie hinter der Flucht der Erscheinungen, der sie mißtraut, nach dem ruhenden Pol, dem sie vertrauen kann. Aber wo immer man außerhalb der Mathematik und Geographie das Wort Pol hört, ist größte Vorsicht am Platze. Worte wie Pol, polar, kosmisch und viele andere gehören in ein Verbrecheralbum der philosophischen Terminologie, und Sätze, in denen diese Worte vorkommen, klingen zwar sehr tief, aber in Wirklichkeit sind sie meist völlig bodenlos.

Damals nun, in den Tagen Platos, vierhundert Jahre vor Christi Geburt, trug die weltabgewandte Philosophie den Sieg davon über die weltzugewandte Philosophie des Demokrit und seit damals blieb sie — wenn auch etwas gemildert durch Aristoteles — trotz starker Gegenströmungen, wie der der Epikuräer, herrschend durchs ganze Altertum, als kirchlich-scholastische Philosophie durchs ganze Mittelalter, als Rationalismus in der neueren Zeit, und in den Systemen des deutschen Idealismus bis auf unsere Tage — und wie sollte es auch anders sein: die Deutschen sind ja bekanntlich das Volk der Denker und der Dichter. Allmählich aber dämmert doch der Tag, und die Befreiung kommt von dort, von wo auch die politische Befreiung in die Welt kam, von England: die Engländer sind ja bekanntlich das Volk der Krämer. Die leuchtendsten Namen am Wege dieser Befreiung sind: John Locke, David Hume, und als unser Zeitgenosse Bertrand Russell, der englische Adelssproß, der während des Krieges wegen antimilitaristischer Gesinnung im Gefängniß saß. Und es ist gewiß kein Zufall, daß es dasselbe Volk war, das der Welt die Demokratie und die Wiedergeburt der weltzugewandten Philosophie schenkte, und es ist kein Zufall, daß in dem Lande, in dem die Metaphysik hingerichtet wurde, auch ein Königshaupt fiel. Denn alle die hinterweltlichen Wesenheiten der Metaphysik: die Ideen Platos, und das Eine der Eleaten, die reine Form und der erste Beweger des

6

Aristoteles, und die Götter und Dämonen der Religionen, und die Könige und Fürsten auf Erden, sie alle bilden eine Schicksalsgemeinschaft — und wenn der Purpur fällt, muß auch der Herzog nach.

Doch die Waffen der weltzugewandten Philosophie sind nicht das Schwert und das Beil des Henkers — so blutdürstig ist sie nicht — und doch sind ihre Waffen scharf genug. Und von einer dieser Waffen will ich heute sprechen: von Occam's Rasiermesser. Wilhelm von Occam, der doctor invincibilis (der unbesiegbare Doktor) der scholastischen Philosophie, der venerabilis inceptor (der ehrwürdige Wiedererneuerer) der Nominalistenschule war — welch merkwürdiger Zufall — ein Engländer, wie John Locke, wie David Hume, wie Bertrand Russell. Er lebte 1280—1340, also im tiefsten Mittelalter. Und sein Rasiermesser, das die hinterweltlichen Wesenheiten wegfegt wie die Sense das Gras auf der Wiese, ist der Satz: „Entia non sunt multiplicanda praeter necessitatem" — zu deutsch: Man soll nicht mehr Wesenheiten annehmen, als unbedingt nötig ist. Das Wort „entia" (Seiendes) würde einem Gymnasiasten in der Schularbeit rot angestrichen; denn das klassische Latein, bekanntlich die logischeste aller Sprachen, hatte kein Participium praesentis von „esse" (sein), sie konnte „seiend" nicht sagen; und Seneca klagt, daß wenn man das ὄν (seiend) der griechischen Philosophen übersetzen wolle, man sagen muß: „quod est" (das, was ist); das Wort „ens" ist eine späte künstliche Bildung.

Kehren wir zurück zu Occams Leitsatz. Um den Sinn dieses Satzes aus der Zeit seiner Entstehung zu verstehen, müssen wir ein Wort über den sogenannten Universalienstreit des Mittelalters sagen. Die Universalien, das sind die Allgemeinbegriffe: der Begriff „Pferd" im Gegensatz zu den Einzelpferden, die es wirklich gibt, die sich in der Welt vorfinden. Die der Sinnenwelt abgewandte Philosophie behauptete nun, als Fortsetzerin der Ideenlehre Platos, daß diesen Allgemeinbegriffen Wesenheiten entsprechen, die in irgend einer Ideenwelt reale Existenz besitzen; ihre Vertreter hießen deshalb Realisten. Die der Sinnenwelt zugewandte Philosophie aber sagt: es ist genug mit den Pferden, die sich in der Sinnenwelt vorfinden; eine Wesenheit, die dem Begriffe Pferd entspricht, eine Idee „Pferd" noch daneben als existent anzunehmen, ist überflüssig. „Sufficiunt singularia, et ita tales res universales omnino frustra ponuntur" — zu deutsch: „Es genügen die Einzeldinge, und so werden diese Allgemeingegenstände gänzlich überflüssigerweise angenommen". Und weil sie überflüssig sind — und dies ist die Anwendung von Occams Rasiermesser — weg mit ihnen! — Nach dieser Auffassung der weltzugewandten Philosophie entsprechen also den Universalien keine Wesenheiten, diese Universalien sind vielmehr bloße Namen, „nomina", und deswegen heißen die Anhänger dieser Richtung Nominalisten. Also: die Realisten waren die Weltabgewandten, die Nominalisten waren die Weltzugewandten, und Occam war der Führer der Nominalisten.

In diesem Zusammenhange müssen wir auch einige Worte über das Leben Occams sagen. Er war Franziskaner — die Orden waren damals die Träger des geistigen Lebens — und Occam war nicht ungläubig, das kann man von einem Philosophen der damaligen Zeit nicht verlangen. Aber trotzdem war er Freigeist; und seine Freigeistigkeit äußerte sich darin, daß er sagte, das Denken könne niemals die Dogmen der Kirche, die Existenz Gottes ergeben. Wir werden sofort sehen, in welch enger Beziehung diese Lehre zum Gegensatz von weltabgewandter und weltzugewandter Philosophie steht. In einem französischen Dictionnaire de philosophie las ich: „Téméraire en philosophie, il fut insubordonné en religion", zu deutsch: „Verwegen in der Philosophie, war er unbotmäßig in der Religion."

Er wurde aus dem Orden der Franziskaner ausgestoßen, von den Universitäten vertrieben und vor das Gericht des Papstes nach Avignon geladen; er wurde sogar zur Strafe ewigen Gefängnisses verurteilt, die er allerdings nicht abgesessen hat. Die damaligen Päpste beanspruchten den Vorrang der geistlichen Macht des Papstes vor der weltlichen Macht der Kaiser und Könige. Occam schloß sich Ludwig dem Bayern, dem deutschen Kaiser, dem Gegner des Papstes an. Sie sehen: Hie Waiblingen! Und der Kampf der Waiblinger, die für die weltliche Macht des Kaisers waren, gegen die Welfen, die für die geistliche Macht des Papstes waren, dieser jahrhundertelange Kampf der Ghibellinen und der Guelfen, auch er ist nur ein Teil des ewigen Kampfes der Weltzugewandten und der Weltabgewandten, so wie in der Philosphie der Kampf der Nominalisten und Realisten.

Sprechen wir weiter vom Gegensatz zwischen weltzugewandter und weltabgewandter Philosophie! Wir haben gesehen: die erste vertraut den Sinnen und sieht in der Sinnenwelt die wahren Wesenheiten; die zweite mißtraut den Sinnen und verläßt sich auf das Denken, das ihrer Meinung nach den Zugang zur Welt der wahren Wesenheiten liefert. Wir sind der Ueberzeugung, daß das eine ungeheuerliche Ueberschätzung des Denkens ist. Denn alles Denken ist bloßes Umformen, nie kann das Denken zu etwas Neuem führen. Eine grundlegende Erkenntnis der jüngsten Zeit besagt: alles logische Denken ist tautologisch; es kann nur dazu verhelfen, schon Gesagtes anders zu sagen, nie kann es dazu verhelfen, etwas Neues zu sagen. ,,Alle Menschen sind sterblich, Alexander ist ein Mensch, also ist Alexander sterblich" — ungefähr wie dieses Beispiel sieht alles Denken aus. Wenn aber Alexander ein Mensch ist, und ich sage: „Alle Menschen sind sterblich," so habe ich doch schon mit ausgesprochen, daß Alexander sterblich ist, und der Schlußsatz: „Also ist Alexander sterblich," liefert somit nichts Neues. Das Denken ist also ganz außerstande, uns aus der Sinnenwelt in eine Welt anders gearteter Wesenheiten zu führen; es kann keine logische Gedankenkette geben, die in der Sinnenwelt beginnt, und in einer Welt andersgearteter Wesenheiten endigen würde. Und nun sehen wir wieder, wie sehr Occam Geist von unserem Geiste war. So wie wir heute lehren, daß kein Denken aus der Sinnenwelt zu hinterweltlichen Wesenheiten führen kann, so lehrte er damals, daß kein Denken zu Gott und den Dogmen der Kirche führen kann; die bleiben einer anderen Domäne überlassen: der des Glaubens. Er glaubte daran, heute glauben viele nicht daran; der Unterschied aber ist erkenntnistheoretisch genommen nicht groß, so wie es, erkenntnistheoretisch genommen, keinen Unterschied ausmacht, daß der eine Narkotika liebt, und der andere nicht.

Ist so der erste Grundfehler der weltabgewandten Philosophie die Ueberschätzung des Denkens, so ist ihr zweiter Grundfehler die Ueberschätzung der Sprache. Die Sätze unserer Sprache sind im Wesentlichen so gebaut: sie sagen von einem Subjekt, eventuell mehreren Subjekten, ein Prädikat aus. „Dieses Kreidestück ist weiß." „Dieses Kreidestück und dieses Kreidestück sind weiß." So weit wäre alles in Ordnung. Die Sprache sagt aber auch: „Dieses Kreidestück und dieses Kreidestück sind gleichfarbig." Die Sprache tut also so, als ob — wie früher jedem der beiden Kreidestücke die Eigenschaft „weiß" — so jetzt jedem von beiden die Eigenschaft „gleichfarbig" zugeschrieben würde. Das ist aber offenbar Unsinn: „Gleichfarbig" ist nicht eine Eigenschaft eines Individuums, wie etwa „weiß", sondern ist eine Beziehung, eine Relation zwischen zwei Individuen. Die Welt hat nicht die einfache Subjekt-Prädikat-Struktur, wie die Sprache sie vortäuscht. Im Gegenteil, es scheint, daß mit fortschreitender Erkenntnis die Subjekt-Prädikat-Struktur immer mehr zurückgedrängt wird zugunsten der Relationsstruktur. Einsteins Relativitätstheorie ist ein gewaltiger Schritt vorwärts auf diesem Wege. Auf die Spitze getrieben ist die Subjekt-Prädikatform in Sätzen wie: „Es regnet", „Es schneit", „Es donnert".

Jedenfalls ist es ganz verfehlt, aus der Struktur der Sprache auf die Struktur der Welt zu schließen. Man hört zwar oft vom Tiefsinn der Sprache reden, aber das ist größtenteils Phrase. Die Sprache ist in Urzeiten entstanden, in Zeiten, da unsere Vorfahren sich auf einem nichts weniger als tiefsinnigen Niveau befanden. Unsere Vettern, die Menschenaffen, sind gewiß nicht sehr tiefsinnig, und es ist absurd, anzunehmen, daß der Entwicklungsgang der Menschheit von einem affenähnlichen Zustand zum heutigen durch eine Epoche besonderen Tiefsinns geführt habe. Die Sprache ist demnach ein außerordentlich unvollkommenes Instrument, aus dem immer wieder die Primitivität der Urzeiten hervorgrinst, wie etwa, wenn ein höchst aufgeklärter Freigeist sich eines Unbehagen nicht erwehren kann, wenn er als Dreizehnter bei Tische sitzt oder auf dem Wege zu einer wichtigen Unternehmung einem alten Weibe begegnet.

So wie nun die weltabgewandte Philosophie die Macht des Denkens überschätzt, so überschätzt sie auch die Bedeutung der Sprachformen. Man kann von ihr wohl sagen: Wo ein Subjekt sich findet, stellt eine Wesenheit zur rechten Zeit sich ein. Und man könnte sich fast wundern, daß sich in der Geschichte der Philosophie nicht ein Kapitel findet, in dem die „Es" der Sätze „Es regnet", „Es donnert", „Es schneit", zu Wesenheiten hypostasiert sind und wo mit allem Scharfsinn und Tiefsinn untersucht wird, ob das regnende Es, das donnernde Es und das schneiende Es dieselbe Wesenheit seien, oder verschiedene Wesenheiten, und welche Beziehungen zwischen ihnen bestehen.

Als Beispiel für diese Ueberschätzung der Sprache

wollen wir uns ganz kurz die Lehre von den sogenannten unmöglichen Gegenständen ansetzen. „Ein hölzernes Eisen existiert nicht." Das ist, so sagte man, ein wahrer Satz; von seinem Subjekte „hölzernes Eisen" wird in giltiger Weise ein Prädikat ausgesagt, nämlich die Nichtexistenz. Da sich nun, so wurde weiter geschlossen, vom hölzernen Eisen in giltiger Weise etwas aussagen läßt, so muß es doch in irgend einer Weise irgend etwas sein. Trotz seiner Nichtexistenz muß es also doch irgend einen Gegenstand, einen unmöglichen Gegenstand hölzernes Eisen geben. So behauptet es die Lehre von den unmöglichen Gegenständen; und sie beansprucht sogar Bedeutung für die Mathematik, da die Mathematik in den indirekten Beweisen sinnvoll mit unmöglichen Gegenständen operiere. Diese Lehre ruht durchaus auf einer Irreführung durch die Sprachform. Wir haben aber heute eine Sprache, die viel weniger unvollkommen ist, als die verschiedenen Wortsprachen: das ist die Sprache der symbolischen Logik. Sie drückt den in Rede stehenden Sachverhalt in unsere Sprache rückübersetzt so aus: Die Aussage „X ist Eisen und X ist hölzern", ist für jeden Gegenstand X falsch. Nach dieser Korrektur der Sprache sehen wir, daß in der neuen Formulierung keinerlei unmöglicher Gegenstand mehr vorkommt. Gehe ich sämtliche wirklichen Gegenstände X der wirklichen Welt durch, so stelle ich fest: keiner von ihnen ist zugleich Eisen und hölzern. Ich brauche also keinen unmöglichen Gegenstand anzunehmen, um den Satz: „Ein hölzernes Eisen existiert nicht", zu verstehen; und nun setzt Occams Rasiermesser an: die unmöglichen Gegenstände sind überflüssige Wesenheiten, also weg mit ihnen! Wir brauchen keine Sphäre, in der unmögliche Gegenstände eine Art von Sein, eine Art von Bestand haben, ein schattenhaftes Dasein führen.

Als nächstes Beispiel für die Tätigkeit von Occams Rasiermesser betrachten wir die Lehre von Zeit und

R a u m. Sprechen wir zuerst von der Zeit! Die methaphysisch gerichteten Philosophen nehmen eine Wesenheit „Zeit" an, in der die von uns sinnlich erlebten Ereignisse hinfließen. Diese Wesenheit Zeit, so lehren sie, setzt sich zusammen aus einfachen Wesenheiten, den ausdehnungslosen Zeitmomenten oder Zeitpunkten. Sehen wir uns das vom Standpunkt einer weltzugewandten Philosophie an. Da haben wir die Vorgänge in der Sinnenwelt, die wir erleben, sonst nichts. Manche solche Vorgänge erleben wir als (ganz oder teilweise) gleichzeitig. Von manchen anderen erleben wir den einen als früher, den anderen als später. Und diese Gleichzeitigkeits- und Sukzessionserlebnisse sind ebenso wirkliche und ursprüngliche Erlebnisse, wie die Erlebnisse etwa unseres Gesichtssinnes oder unseres Tastsinnes. Man darf sich da nicht durch die Sprache irremachen lassen, durch die Worte „zur gleichen Zeit", als müßte ich erst Kenntnis von einer Zeit und von Zeitmomenten haben, um festzustellen, daß zwei Ereignisse gleichzeitig, im gleichen Zeitmoment vor sich gehen. Die Gleichzeitigkeit ist das Ursprüngliche, die erleben wir, wie wir Farben und Töne erleben; einen Zeitmoment hat noch kein Mensch erlebt und wird auch kein Mensch erleben.

Man darf sich auch nicht durch die weltabgewandte Philosophie Kants irremachen lassen, derzufolge Gleichzeitigkeit und zeitliche Aufeinanderfolge subjektive, vom Menschen hineingetragene Zutaten zu einer zeitlosen Welt der „Dinge an sich" wären. Die Gleichzeitigkeit, das Früher und das Später von Erlebnissen sind real, wie rot und grün, wie der Hunger und die Liebe; von einem „Ding an sich" aber hat kein Mensch je etwas erlebt und wird auch kein Mensch etwas erleben.

Denken wir uns nun eine Gruppe von Vorgängen, von Erlebnissen, die allesamt ganz oder teilweise gleichzeitig sind, das heißt sich zeitlich (ganz oder teilweise) überdecken. Während einer gewissen Spanne Zeit (um zunächst in der gewöhnlichen Ausdrucksweise zu sprechen), werden also diese Erlebnisse allesamt gemeinsam andauern, — einzelne werden schon früher begonnen haben, einzelne noch später andauern. Betrachten wir nun außerdem noch weitere Erlebnisse, die allesamt mit allen den eben betrachteten Erlebnissen (ganz oder teilweise) gleichzeitig sind: die Spanne Zeit, während der sowohl alle die vorhin betrachteten, als auch alle die nun außerdem noch betrachteten Erlebnisse andauern, wird kürzer geworden sein. Betrachten wir nun eine nicht mehr erweiterbare Klasse solcher teilweise gleichzeitig stattfindender Erlebnisse, d. h. eine Klasse von Erlebnissen, von denen jede beliebige Anzahl (teilweise) gleichzeitig stattfindet, die aber so umfassend ist, daß es kein weiteres Erlebnis gibt, das mit allen Erlebnissen dieser Klasse (teilweise) gleichzeitig wäre: die Zeit, während deren sie allesamt stattfinden, reduziert sich dann auf einen einzigen Zeitmoment. Bisher haben wir in der üblichen Ausdrucksweise gesprochen, nun aber sagen wir mit Russell: Ein Zeitmoment ist nichts anderes, als eine solche nicht mehr erweiterbare Klasse gleichzeitiger Erlebnisse.

Natürlich ist das Gesagte nur eine Andeutung. Auf zwei naheliegende Einwände müssen wir aber kurz eingehen. Ein erster Einwand sagt: Es könnte doch sein, daß die sämtlichen Erlebnisse einer solchen nicht mehr erweiterbaren Klasse gleichzeitiger Erlebnisse nicht nur einen einzigen Zeitmoment, sondern eine ganze Zeitspanne lang gemeinsam andauern. Dieser Einwand beruht ganz darauf, daß er, im Sinne der weltabgewandten Philosophie, von vorneherein die Zeit als selbständige Wesenheit annimmt. Wir aber antworten darauf: würden wirklich die sämtlichen Erlebnisse der genannten Erlebnisklasse während einer ganzen Zeitspanne gemeinsam andauern, so hieße das, daß in dieser ganzen Zeitspanne kein Erlebnis von mir beginnt und keines endet; ich würde also in dieser ganzen Zeitspanne keinerlei Veränderung erleben: und so hätte ich gar kein Mittel, um festzustellen, daß eine wirkliche Zeitspanne verflossen ist und nicht nur ein einziger Moment. Was aber in keiner wie immer gearteten Weise festgestellt werden kann, das hat auch keinen wie immer gearteten Sinn.

Ein zweiter Einwand sagt: Ein Zeitmoment ist doch offenkundig etwas ganz anderes, als eine Klasse von Erlebnissen! Aber auch dieser Einwand hat keinerlei Sinn. Wüßten wir, was ein Zeitmoment ist, könnten wir einen Zeitmoment erleben, so könnte man das Erlebnis „Zeitmoment" vergleichen mit der vorhin betrachteten Klasse von Erlebnissen und könnte vielleicht mit Recht sagen, es ist etwas anderes — so wie wir durch Erleben die Farbe Rot kennen, und wenn jemand sagen wollte: „Die Farbe Rot ist ein Haufen von Bananen", wir mit Recht erwidern könnten: „Das ist nicht wahr." Da wir aber einen Zeitmoment n i c h t erleben können, so verliert der gemachte Einwand jeden Sinn.

In dieser Russell'schen Definition des Zeitmomentes kommen nur Erlebnisse vor, sie bleibt durchaus weltzugewandt, sie beansprucht keinerlei hinterweltliche, außersinnliche Wesenheiten. Und es läßt sich zeigen, daß diese Russell'schen Zeitmomente alles das leisten, was die Physik von Zeitmomenten verlangt. Also ist es überflüssig, neben den Erlebnissen noch Zeitmomente als existierend anzunehmen, in denen diese Erlebnisse stattfinden, und eine Zeit als existierend anzunehmen, deren ausdehnungslose Teile diese Zeitmomente wären. Nichts anderes brauchen wir als unsere Erlebnisse mit ihren erlebbaren Relationen des gleichzeitig oder früher und später. Und weil es überflüssig ist, die eigene Existenz der Zeit und ihrer Zeitmomente anzunehmen, setzen wir wieder Occams Rasiermesser an und sagen: Weg mit ihnen, und weg mit allen philosophischen Scheinproblemen, die sich an die metaphysische Existenz einer Zeit knüpften.

Aehnliches wie über die Zeit können wir nun auch über den Raum sagen. Gewöhnlich nimmt man an, es gäbe eine Wesenheit „Raum“, zusammengesetzt aus unteilbaren, unausgedehnten Wesenheiten: den Punkten; dieser Raum sei das Reservoir, in dem sich die Vorgänge abspielen, von denen unsere Sinne uns Zeugnis geben; in diesem an sich existierenden Raume schwimmen die Körper der Sinnenwelt herum, wie die Fische in einem Aquarium. Als Kind sagte man mir, es sei sehr schwer zu begreifen, was so ein ausdehnungsloser Punkt sei, und es sei sehr schwer, sich einen solchen vorzustellen. Sehr schwer? Das ist eine außerordentlich milde Ausdrucksweise. Es ist nicht schwer, sondern es ist ganz und gar unmöglich. Ebenso wie kein Mensch je etwas wie einen Zeitmoment erlebt hat, ebenso hat noch keiner etwas wie einen Raumpunkt erlebt.

Man war bis vor kurzem der Meinung, die Geometrie handle vom Raum und seinen Punkten; und die Laien meinten wohl, wenn es schon ihnen selbst nicht gelänge zu begreifen, was ein Raumpunkt sei, so begriffen es wenigstens die Mathematiker ganz genau. Aber das war ein gewaltiger Irrtum; die Mathematiker begriffen es ebensowenig und waren ebensowenig wie der nächstbeste Laie imstande, zu sagen, was denn eigentlich ein Punkt sei. Und da es ihnen recht hoffnungslos schien, es jemals herauszubekommen, während doch — abgesehen von dem Schönheitsfehler, daß man nicht recht sagen konnte, was ein Punkt ist — die Wissenschaft von den Punkten, die Geometrie, tadellos funktionierte, schlugen die Mathematiker einen verblüffend einfachen Ausweg ein. Sie sagten: Die Frage, was ein Punkt sei, gehört nicht zur Mathematik, das geht uns nichts an, und man kann tadellos eine Wissenschaft von den Punkten treiben, ohne zu wissen, was ein Punkt ist. Das klingt zwar zunächst etwas sonderbar, ist aber ganz und gar nicht unsinnig, es brachte sogar einen recht beträchtlichen Fortschritt in unserer Erkenntnis von der Struktur der Wissenschaft: es entstand die sogenannte axiomatische Methode, mit ihrer Lehre von den impliziten Definitionen und den hypothetisch-deduktiven Systemen. Aber auf diese Dinge können wir hier nicht eingehen; für uns ist nur folgendes wichtig: Dadurch, daß die Mathematiker sagen: „ein Problem geht uns nichts an", ist dieses Problem weder gelöst, noch aus der Welt geschafft. Es bleibt also die leidige Frage bestehen: „Was ist ein Punkt?" Und da gehen wir ebenso vor wie bei der Untersuchung der Zeitmomente. Wir finden in der Sinnenwelt zwar keine Punkte vor, wohl aber Körper, und wir wissen, was es heißt, zwei Körper haben ein Stück gemein, sie überdecken sich ganz oder teilweise. Wir wollen wenigstens annehmen, wir wissen es, obwohl auch da noch schwierige Probleme lauern, aber Probleme etwas anderer Art, als das jetzt zu lösende. Nun betrachten wir wieder eine Klasse von Körpern, die allesamt ein Stück gemein haben. Je mehr solcher Körper wir betrachten, umso kleiner wird das allen gemeinsame Stück sein. Wir betrachten wieder eine nicht mehr erweiterbare Klasse solcher Körper und sagen mit Russell: Eine solche Klasse von Körpern *ist* ein Punkt. Natürlich sind wir jetzt himmelweit von der früheren Definition: „Ein Punkt ist ein ausdehnungsloser Raumteil" entfernt, bei der sich nichts denken ließ. Es ist jetzt gar keine Wesenheit „Raum" da, von der irgendwelche, insbesondere ausdehnungslose Teile zu betrachten wären, es sind nur da die Körper, die uns die Sinnenwelt liefert; und gewisse Klassen solcher Körper nennen wir Punkte. Also keinerlei von der Sinnenwelt wesensverschiedene Wesenheit muß angenommen werden; kein an sich existierender Raum, in dem die Körper herumschwimmen, keine an sich existierenden Punkte, aus denen dieser Raum sich zusammensetzt. Wir brauchen sie nicht, und also setzt wieder Occams Rasiermesser an: „Weg mit ihnen!" Nichts anderes benötigen wir, als die Körper der Sinnenwelt, um alles verstehen zu können, was die Geometrie braucht.

Und da wir schon bei der Mathematik sind, wollen wir auch von den Zahlen sprechen. Metaphysisch gerichtete Philosophen faßten die Zahlen als Wesenheiten auf, denen irgend eine mystische Existenz zukommt. Nach Plato gibt es in der Welt der Ideen eine Idee der Zwei, eine Idee der Drei usf., und manche Metaphysiker glaubten an geheime Wunderkräfte dieser Zahlenwesen. Da war vor allen die Zahl Sieben sehr beliebt; sie soll besonders heilig sein: Es gibt sieben Sakramente, allerdings — wenn ich nicht irre — auch sieben Hauptsünden, aber das ist vermutlich Polarität. Und tatsächlich hat die Wesenheit Sieben einmal schöpferisch gewirkt, sie hat die Spektralfarbe Indigo zustandegebracht; denn ohne Indigo hätten wir nur sechs Spektralfarben, und es wäre ein sehr bedauerlicher Mangel der Welt, wenn es nicht ebensoviel Spektralfarben als heilige Sakramente gäbe; und deshalb lernen in der Optik die Kinder von einer Farbe Indigo, die ihnen sonst wohl nirgends mehr unterkommt.

Mit den Zahlen steht es ähnlich wie mit Zeit und Raum. Wir haben gesehen: wir erleben gleichzeitige Ereignisse, aber nie und nimmer erleben wir die Zeit selbst oder einen Zeitmoment. Ebenso finden wir in unseren Erlebnissen Mengen, die aus drei oder aus fünf Gegenständen bestehen, nie und nimmer aber erleben wir die Zahl Drei oder die Zahl Fünf. Was meinen wir nun eigentlich, wenn wir sagen: eine Menge von Gegenständen besteht aus fünf Gegenständen? Wir meinen damit: wir können sie abzählen mit Hilfe der Finger einer Hand; es sind soviele Gegenstände, als eine Hand Finger hat, oder präziser: die Gegenstände dieser Menge können so den Fingern einer Hand zugeordnet werden, daß jedem Gegenstand genau ein Finger, jedem Finger genau ein Gegenstand entspricht. Die Mengen, die aus fünf Gegenständen bestehen, sind also gerade diejenigen, deren Gegenstände in der angegebenen Weise den Fingern einer Hand zugeordnet werden können, und dadurch unterscheiden sie sich von allen anderen Mengen, die mehr oder weniger als fünf Gegenstände enthälten. Und nun können wir sagen — und es ist das ganz deselbe Gedanke, der uns zur Definition der Zeitmomente und der Raumpunkte führte: die Zahl Fünf *ist* gar nichts anderes, als die Klasse aller Mengen, deren Gegenstände in der angegebenen Weise den Fingern einer Hand zugeordnet werden können. Man wende auch hier nicht ein: eine Zahl ist doch um Gotteswillen etwas anderes als eine Klasse von Mengen. Da wir Zahlen nicht erleben, wie wir Farben und Töne erleben, so sind die Zahlen an und für sich gar nichts; sie sind uns nicht gegeben, wie uns Farben und Töne gegeben sind; sie müssen also aus dem Gegebenen, d. h. aus der Sinnenwelt konstruiert werden, und zwar so konstruiert werden, daß sie alles leisten, was eben die Zahlen im täglichen Leben und in der Wissenschaft zu leisten haben. Wir haben, nach *Frege* und *Russell*, eine solche Konstruktion der Zahlen aus der gegebenen Sinnenwelt angedeutet. Sie leistet — was ich hier natürlich nicht ausführlich zeigen kann — tatsächlich alles, was im täglichen Leben und in der Wissenschaft die Zahlen zu leisten haben, und mehr brauchen wir nicht. Wir kommen vollständig mit den Gegenständen der Sinnenwelt aus, und haben es nie und nirgends nötig, neben oder hinter der Sinnenwelt noch Zahlen als eigene Wesenheiten einer eigenen Existenzart anzunehmen. Und weil wir Zahlen als eigene Wesenheiten nicht brauchen, weil wir ohne solche Wesenheiten ausgekommen, so — und dies ist wieder eine Anwendung von Occams Rasiermesser — wollen wir solche Wesenheiten auch nicht annehmen. Es gibt nur Gegenstände, die wir zählen, aber es gibt nicht daneben Zahlen als eigene mystische Wesenheiten.

Noch genug andere Beispiele gibt es, in denen uns Occams Rasiermesser von mehr oder weniger schattenhaften, mehr oder weniger aufdringlichen, überflüssigen Wesenheiten befreit. Von besonders aufdringlichen solchen Wesenheiten will ich noch ein paar Worte sagen. Ich sprach schon von der Subjekt-Prädikat-Struktur der Sprache und ihrer unheilvollen Wirkung auf eine weltabgewandte Philosophie, die aus der Struktur der Sprache auf die Struktur der Welt schloß. Wir sagen: „Dieses Stück Papier ist jetzt glatt, jetzt zusammengefaltet, jetzt zu einem Ballen zerdrückt," und wir finden in diesem Satze: ein beharrendes Subjekt (das Stück Papier) und wechselnde Prädikate (jetzt glatt, jetzt zusammengefaltet, jetzt zu einem Ballen zerdrückt). Daraus schloß die von der Sprache irregeleitete Philosophie, die Welt sehe so aus: entsprechend den „beharrenden" Subjekten der Sprache existieren in der Welt beharrende Substanzen, die — entsprechend den wechselnden Prädikaten — wechselnde Eigenschaften (Qualitäten) tragen. Wo aber erleben wir die beharrende Substanz? Nie und nirgends! Das Kreidestück als beharrende Substanz, als Träger seiner Qualitäten erlebe ich nicht. Was ich erlebe, ist: seine weiße Farbe, das Gefühl des Widerstandes, der Härte, wenn ich hingreife, seine Gestalt, die ich sowohl sehen als tasten kann. Aber die gesehene Gestalt wechselt, wenn ich das Kreidestück anderswohin lege oder wenn ich selbst anderswohin trete. Daß das, was ich früher sah, und das, was ich jetzt sehe, dasselbe ist, erlebe ich nicht. Kurz, die uns vertrauten Gegenstände der Umwelt, wir erleben sie nicht als beharrende substanzielle Wesenheiten; was ich erlebe, sind wechselnde Farben, Gestalten, Härten usf.

Sind wir nun vielleicht hier an einem Punkte angelangt, wo wir außer dem von uns Erlebten, den Farben, den Töne etc. noch andere Wesenheiten annehmen müssen, die wir nicht erleben können: die beharrenden Substanzen, welche die wechselnden Farben, Gestalten, Härten, die wir erleben, als Qualitäten an sich tragen? Wieder hilft uns auch hier derselbe Gedanke, der uns Raum, Zeit und Zahl begreifen ließ, ohne daß wir über das Erlebbare hinausgehen mußten, ohne daß wir eigene Wesenheiten neben oder hinter denen unserer Sinnenwelt annehmen mußten. Was wir „dies Stück Kreide" nennen, ist nichts anderes, als die Klasse der angeblich von ihm getragenen Farben, Gestalten, Härten usf. Die Durchführung dieses Gedankens, den wir der weltzugewandten englischen Philosophie verdanken, und dem mit besonderem Nachdruck Ernst Mach nachhing, der große Wiener Physiker und Philosoph, dessen Name unsere Gesellschaft trägt, lehrt uns, daß es nie und nirgends nötig ist, hinter oder unter den von uns erlebbaren Wesenheiten der Sinnenwelt noch andere unerlebbare Wesenheiten, die Substanzen anzunehmen; und nun fegt Occams Rasiermesser mit einem Strich auch noch die Substanzen weg: ihr gepriesenes Beharren, der Schärfe dieses Rasiermessers hält es nicht stand. Doch hierauf brauche ich nicht näher einzugehen, da diese Frage gewiß in einem späteren Vortrag unserer Gesellschaft eingehend behandelt werden wird.

Aber nun haben wir uns noch mit einem letzten, tiefen und schwierigen Probleme auseinanderzusetzen. Blikken wir zurück! Wir haben die Zeitmomente erkannt als gewisse Klassen von Erlebnissen, die Raumpunkte als gewisse Klassen von Körpern, die Zahlen als Klassen von Mengen (und Mengen sind selbst Klassen von Gegenständen); die Gegenstände der Umwelt haben wir erkannt als Klassen von Farben, Gestalten, Härten etc. Immer wieder werden wir also auf Klassen geführt. Es scheint also, daß wir doch nicht mit den Wesenheiten der Sinnenwelt allein auskommen können, daß wir vielmehr neben diesen Wesenheiten erster Stufe noch andere Wesenheiten höherer Stufe annehmen müssen, die nun dem Rasiermesser Occams standhalten, nämlich Klassen von Wesenheiten erster Stufe, dann Klassen solcher Klassen (die Zahlen waren solche Klassen von Klassen) und vielleicht noch weiter hinauf. Ich kann hier nur erwähnen, daß uns die weltzugewandte Philosophie auch hier den rechten Weg gewiesen hat: Russell hat gezeigt, wie jede sinnvolle Aussage, in der das Wort „Klasse" vorkommt, verwandelt werden kann in eine Aussage, in der nur mehr die Gegenstände selbst vorkommen, und nicht mehr Klassen von Gegenständen. Um hiefür nur ein ganz nahegelegenes Beispiel zu geben: Mit dem Satze: „Die Klasse der Löwen ist enthalten in der Klasse der Säugetiere" meinen wir doch gar nichts anderes als: „jeder Löwe ist ein Säugetier", d. h. etwas ausführlicher gesprochen: jeder Gegenstand, der die definierenden Eigenschaften des Begriffes „Löwe" hat, hat auch die definierenden Eigenschaften des Begriffes „Säugetier", und in dieser Formulierung kommen nur mehr die erlebbaren Gegenstände der Sinnenwelt vor, und gar nichts mehr von „Klassen". In letzter Analyse erweisen sich also auch die Klassen als überflüssige Wesenheiten und es bleibt nichts übrig, als die erlebbaren Wesenheiten unserer Sinnenwelt.

Wir haben nun Occams Rasiermesser an der Arbeit gesehen und wir dürfen sagen: es arbeitet gut — den das Rasieren fördernden Schaum hat die weltabgewandte, metaphysische Philosophie in ausgiebiger Weise geschlagen. Und zurück bleibt eine Welt, gesäubert von allen schattenhaften Halbwesenheiten, die irgendwo zwischen Sein und Nichtsein wohnen sollen, wie unmögliche Gegenstände, Universalien, leerer Raum, leere Zeit; gesäubert aber auch von allen präpotenten Wesenheiten, die ein stärkeres, beharrenderes Sein beanspruchen als die bunten wechselnden Wesenheiten unserer Sinnenwelt, wie: Ideen und Substanzen und wie sie sonst heißen mögen.

Aber, so hören wir sagen, diese weltzugewandte Philosophie, die sich bescheidet mit der Sinnenwelt und nicht nach Höherem und Tieferem fragt, ist — entsprechend ihrer Herkunft — eine richtige Krämerphilosophie, eine Philosophie der Banausen und Philister, eine Philosophie für phantasieunbegabte Nichtdichter; schön ist doch nur eine Welt, in der unter den flüchtigen Erscheinungen wuchtig beharrende Substanzen ruhen, über ihnen — aufs höchste vergeistigt und sublimiert — Ideen walten, wie sie Goethe schildert:

> Göttinnen thronen hehr in Einsamkeit,
> um sie kein Ort, noch weniger eine Zeit;
> von ihnen sprechen — — ist Verlegenheit!

welch letzteres wir bereitwilligst zugestehen.

Mag sein, daß es sehr schön war auf der Welt, als man überall Götter und Dämonen witterte; mag sein, daß es eine sehr poetische Zeit war, als man unentwegt alle möglichen höheren Wesen durch Gebete und Opfer bei halbwegs guter Laune erhalten, alle möglichen bösen Geister durch Zaubersprüche im Zaum halten mußte. Und auch heute noch mögen die Leute sehr glücklich sein, die überzeugt sind, durch dreimaliges Ausspucken ein Unglück bannen zu können, und ich vermag durchaus das Hochgefühl eines Mannes zu würdigen, der einen Spiegel zerbrochen hat, und sich nun in der Ueberzeugung wiegt, er werde sieben Jahre nicht heiraten. Aber wir meinen eben doch, all das sei Aberglaube, und wir müssen es doch nun einmal für richtig halten, an diese Dinge nicht zu glauben, mögen sie auch noch so hübsch und poetisch sein. Und so ist es auch kein Argument gegen Occams Rasiermesser, es sei doch jammerschade um all die weggefegten überflüssigen Wesenheiten, weil sie so schön und poetisch gewesen sind.

Aber — und nun kehren wir zum Ausgangspunkt zurück — der Geschmack ist verschieden. Der eine freut sich der Mannigfaltigkeit, des bunten, mutwilligen Wechsels, des schwer zu enträtselnden Durcheinander in der Sinnenwelt, und bleibt darum ihr zugewandt — und der andere, dem es versagt ist, sich dieser Sinnenwelt zu erfreuen, wendet sich ab von ihr, und sucht, sich Hinterwelten zu erfabeln. Wir aber, die Anhänger weltzugewandter Philosophie, sehen darin einen Abweg, einen Irrweg, eine Krankheit, an der die Menschheit durch Jahrtausende gelitten hat, und unser Wille ist es, die Menschheit von diesem Alpdruck zu befreien. Und dem Menschen, der aus dem mystischen Düster weltabgewandter Philosophie sich zu den einfachen, durchsichtig klaren Lehren der weltzugewandten Philosophie bekehrt, ihm sagen wir als Gruß:

> Du tauchst aus Tod, aus ungewissen Leiden,
> aus allem auf, was feig und halb und vag,
> und lernst mit freien Blicken unterscheiden
> die kranke Dämmerung vom reinen Tag!

Diskussion zur Grundlegung der Mathematik

am Sonntag, dem 7. Sept. 1930

HAHN: Die folgenden Ausführungen sind lediglich Diskussionsbemerkungen, also notwendigerweise nur skizzenhaft; ich muß also bitten, es zu entschuldigen, daß ich keineswegs mit der in diesen Fragen gebotenen Präzision sprechen werde.

Will man sich für einen der Standpunkte bei Grundlegung der Mathematik entscheiden, die hier ausführlich begründet wurden, so muß man sich vor allem fragen: Was ist von einer Grundlegung der Mathematik zu verlangen? Und um zu dieser Frage Stellung zu nehmen, muß ich einige Worte philosophischen Inhalts vorausschicken.

Der einzig mögliche Standpunkt der Welt gegenüber scheint mir der *empiristische* zu sein, den man ganz roh so charakterisieren kann: Irgendeine Erkenntnis, der Inhalt zukommt, die wirklich etwas über die Welt besagt, kann nur durch Beobachtung, durch Erfahrung zustande kommen; durch reines Denken kann eine Erkenntnis über die Wirklichkeit in keiner Weise gewonnen werden; und ein einmaliges Hinsehen kann keine Erkenntnis liefern, die über den betreffenden Einzelfall hinausreicht (welch letztere Bemerkung sich gegen alle Lehren von reiner Anschauung und von Wesensschau richtet). Ich stelle mich auf diesen empiristischen Standpunkt nicht auf Grund einer Auswahl unter verschiedenen möglichen Standpunkten, sondern weil er mir als der einzig mögliche erscheint, weil mir jede Realerkenntnis durch reines Denken, durch reine Anschauung, durch Wesensschau als etwas durchaus Mystisches erscheint.

Der Durchführung dieses empiristischen Standpunktes scheint nun eine sehr einfache Tatsache entgegenzustehen: die Tatsache nämlich, daß es eine Logik und eine Mathematik gibt, die uns doch anscheinend absolut sichere und allgemeine Erkenntnisse über die Welt liefern. So entsteht die Grundfrage: *Wie ist der empiristische Standpunkt mit der Anwendbarkeit von Logik und Mathematik auf Wirk-*

liches verträglich? Und im Sinne dieser Frage ist meiner Ansicht nach von einer Grundlegung der Mathematik vor allem zu verlangen, daß sie dartut, wieso die Anwendbarkeit der Mathematik auf Wirkliches mit dem empiristischen Standpunkt verträglich ist.

Die Vertreter des Intuitionismus und des Formalismus, die hier zu Worte kamen, haben ihre Standpunkte so deutlich dargelegt, daß man wohl mit Bestimmtheit sagen kann: weder Intuitionismus noch Formalismus erfüllen diese Forderung. Ich halte sowohl die Untersuchungen Brouwers als die Hilberts für höchst bedeutungsvoll innerhalb der Mathematik, aber ich halte sie nicht für Grundlegungen der Mathematik. Herr Heyting ging in seinem Referate aus von einer Urintuition der Zahlenreihe; diese Urintuition hat für mich, so wie reine Anschauung oder Wesensschau, etwas Mystisches, und eignet sich daher nicht als Ausgangspunkt für die Grundlegung der Mathematik. Und Herr v. Neumann hat mit aller Deutlichkeit gesagt, daß der Formalismus die gesamte finite Arithmetik voraussetzt, um von da aus die klassische Mathematik zu rechtfertigen; ein Standpunkt aber, der die finite Arithmetik voraussetzt, kann nicht als Grundlegung der Mathematik angesehen werden.

Der Darlegung meines eigenen Standpunktes sei eine kleine Erörterung vorausgeschickt. Sei irgendein Bereich von Gegenständen gegeben, zwischen denen irgendwelche Relationen bestehen; dieser Bereich werde abgebildet auf einen Bildbereich, so daß den Gegenständen und Relationen des ursprünglichen Bereiches Gegenstände und Relationen des Bildbereiches entsprechen; die Gegenstände und Relationen des Bildbereiches können wir dann als *Symbole* für die Gegenstände und Relationen des ursprünglichen Bereiches auffassen. Ist die vorgenommene Abbildung nicht ein-eindeutig, sondern ein-mehrdeutig, so werden einunddemselben Sachverhalte im ursprünglichen Bereiche verschiedene Symbolkomplexe im Bildbereiche entsprechen; es wird also Transformationen dieser Symbolik in sich geben, und es entsteht die Aufgabe, Regeln anzugeben für die Umformung eines Symbolkomplexes in einen anderen, der denselben Sachverhalt des ursprünglichen Bereiches abbildet. So nun steht meiner Meinung nach die Sprache der Wirklichkeit gegenüber: die Sprache ordnet den Sachverhalten der Welt Symbolkomplexe zu, und zwar nicht in ein-eindeutiger Weise (was wenig Zweck hätte), sondern in ein-mehrdeutiger Weise; und die Logik gibt die Regeln an, wie ein Symbolkomplex der Sprache umgeformt werden kann in einen anderen, der denselben Sachverhalt bezeichnet; das ist es, was als der „tautologische" Charakter der Logik bezeichnet wird; ein ganz einfaches Beispiel ist die doppelte Negation: der Satz p und der Satz *non-non-p* bezeichnen denselben Sachverhalt. Immer, wenn eine ein-mehrdeutige Abbildung vorliegt, gibt es in diesem Sinne eine „Logik" dieser Abbildung; was man gewöhnlich *Logik* nennt ist der Spezialfall, in dem es sich um die Zuordnung der Sprachsymbole zu den Sachverhalten der Welt handelt.

Die Logik sagt also über die Welt gar nichts aus, sondern sie bezieht sich nur auf die Art, wie ich über die Welt spreche, und es leuchtet wohl ein, daß bei dieser Auffassung das Bestehen der Logik ohne weiteres mit dem empiristischen Standpunkte verträglich ist, während die Auffassung der Logik als Lehre von den allgemeinsten Eigenschaften der Gegenstände mit dem empiristischen Standpunkte durchaus unverträglich ist. Nehmen wir als Beispiel den logischen Grundsatz $(x)\,\varphi(x) \cdot \supset \cdot \varphi(y)$, der besagt: was für alle gilt, gilt für jedes einzelne. Dieser Grundsatz besagt nichts über die Welt; es ist nicht eine Eigenschaft der Welt, daß, was für alle gilt, auch für jedes einzelne gilt; sondern die Sätze: „$\varphi(x)$ gilt für alle Individuen" und „$\varphi(y)$ gilt für jedes einzelne Individuum" sind nur verschiedene sprachliche Symbole für denselben Sachverhalt; der angeführte logische Grundsatz drückt also nur eine Ein-mehrdeutigkeit der als Sprache verwendeten Symbolik aus; er drückt aus, in welchem Sinne das Symbol „alle" verwendet wird.

Nun kommen wir auf die Grundlegung der Mathematik zurück. Der von Herrn Carnap dargelegte logistische Standpunkt behauptet, daß kein Unterschied zwischen Mathematik und Logik besteht. Ist dieser Standpunkt durchführbar, so ist mit obiger Aufklärung der Stellung der Logik im Systeme unserer Erkenntnis auch die Stellung der Mathematik aufgeklärt; ebenso wie das Bestehen der Logik, ist dann auch das Bestehen der Mathematik mit dem empiristischen Standpunkte verträglich. Und dies ist der Grund, warum ich unter den drei hier vorgebrachten Auffassungen über die Grundlegung der Mathematik für die logizistische Auffassung optiere.

Nun kann man tatsächlich einsehen, daß die Sätze der finiten Arithmetik, wie $3 + 5 = 5 + 3$, denselben tautologischen Charakter haben wie die Sätze der Logik; man hat nur auf die Definition der Symbole $3, 5, +$ und $=$ zurückzugehen. Die finite Arithmetik bereitet also dem logizistischen Standpunkte keine Schwierigkeit. Nicht so klar liegen die Dinge bezüglich der transzendenten Schluß-

10*

weisen der Mathematik, wie der Lehre von der vollständigen Induktion, der Mengenlehre und mancher Kapitel der Analysis. Hier scheinen Grundsätze eine Rolle zu spielen, die nicht tautologisch sind; so scheint z. B. das Auswahlaxiom einen realen Inhalt zu haben, wirklich etwas über die Welt auszusagen; das war zumindest der Standpunkt Russells, und Ramseys Versuch, auch dem Auswahlaxiom tautologischen Charakter zuzuschreiben, ist sicherlich nicht geglückt.

Russells absolutistisch-realistischer Standpunkt nimmt an, die Welt bestehe aus Individuen, Eigenschaften von Individuen, Eigenschaften solcher Eigenschaften usf.; und die logischen Axiome seien nun Aussagen über diese Welt. Daß diese Auffassung mit konsequentem Empirismus unvereinbar ist, habe ich schon gesagt, und ich halte die Polemik Wittgensteins und der Intuitionisten gegen diese Auffassung für durchaus gerechtfertigt; ebenso scheint mir die realistisch-metaphysische Auffassung Ramseys, gegen die auch Herr Carnap sich gewendet hat, unmöglich.

Wenn ich so Russells philosophische Interpretation seines Systems bekämpfe, so glaube ich doch, daß die formale Seite dieses Systems großenteils in Ordnung und zur Begründung der Mathematik weitgehend geeignet ist; es muß nur nach einer anderen philosophischen Interpretation gesucht werden. Bevor ich eine solche Interpretation anzudeuten versuche, möchte ich, des leichteren Verständnisses halber, auf etwas Ihnen Wohlbekanntes hinweisen: Denken Sie an irgendein System von Axiomen der euklidischen Geometrie, z. B. an das von Hilbert. Dieses Axiomsystem ist zur Beschreibung der Welt ausgezeichnet verwendbar; und doch glaubt niemand, daß in der Welt Gegenstände aufweisbar seien, die sich wie die Punkte, Geraden, Ebenen der euklidischen Geometrie verhalten; es handelt sich dabei eben nur um Idealisierungen, um Annahmen, die man zum Zwecke einer geeigneten Weltbeschreibung macht.

Nun nehme ich wie Russell an, es stünde uns zur Beschreibung der Welt (oder besser: eines Ausschnittes der Welt) ein System von Aussagefunktionen, von Aussagefunktionen über Aussagefunktionen usf. zur Verfügung (wobei ich aber im Gegensatze zu Russell nicht glaube, diese Aussagefunktionen seien etwas absolut Gegebenes, in der Welt Aufweisbares). Die Beschreibung der Welt wird nun verschieden ausfallen, je nach der Reichhaltigkeit dieses Systemes von Aussagefunktionen; wir machen also gewisse *An-*

nahmen über diese Reichhaltigkeit; z. B. werden wir fordern, daß, wenn $\varphi(x)$ und $\psi(x)$ in dem System vorkommen, dies auch von $\varphi(x) \vee \psi(x)$ und $\varphi(x) \wedge \psi(x)$ gilt; wir werden auch annehmen, daß neben $\varphi(x, y)$ auch $(y)\, \varphi(x, y)$ in dem Systeme vorkomme; wir können etwa auch annehmen, das System sei sogar so reichhaltig, daß auch eine Bildung wie $(\varphi)\, \varphi(x)$ nicht aus ihm herausführe; auch die Forderung, es solle das Unendlichkeitsaxiom oder das Auswahlaxiom gelten, sind in diesem Sinne Forderungen an die Reichhaltigkeit des Systems von Aussagefunktionen, mit Hilfe dessen ich die Welt beschreiben will. Die ganze Mathematik entsteht nun durch tautologische Umformung der an die Reichhaltigkeit unseres Systems von Aussagefunktionen gestellten Forderungen. Ob ein bestimmter Satz gilt oder nicht (z. B. der Satz von der Mächtigkeit der Potenzmenge, oder der Wohlordnungssatz), hängt von den an die Reichhaltigkeit des zugrunde gelegten Systems von Aussagefunktionen gestellten Forderungen ab, die man, wenn man will, *Axiome* nennen kann; die Frage nach einer *absoluten* Gültigkeit solcher Sätze ist gänzlich sinnlos.

Nun wird man vielleicht die Frage stellen wollen: *Gibt es* ein solches System von Aussagefunktionen, wie es hier gefordert wird? Im empirischen Sinne (oder im realistischen Sinne Russells) gibt es ein solches System gewiß nicht: es ist ausgeschlossen, ein solches System in der Welt aufzuweisen. Im konstruktiven Sinne der Intuitionisten gibt es ein solches System auch nicht. Aber daran liegt nichts; so wie die euklidische Geometrie sehr nützlich ist zur Beschreibung der Welt, obwohl ihre Punkte, Geraden, Ebenen nicht aufweisbar sind, ebenso ist die Annahme eines Systems von Aussagefunktionen, wie wir es besprachen, sehr nützlich zur Beschreibung der Welt, obwohl ein solches System weder empirisch noch konstruktiv aufweisbar ist. Die so aufgebaute Analysis hat nur hypothetischen Charakter: Nehme ich an, zur Beschreibung der Welt stehe mir ein System von Aussagefunktionen zur Verfügung, das gewissen Reichhaltigkeitsanforderungen genügt, so gelten in einer so gearteten Beschreibung der Welt die Sätze der Analysis. Tatsächlich geht auch die Beschreibung der Welt mit den Hilfsmitteln der Analysis über jede empirische Kontrollmöglichkeit weit hinaus. Natürlich muß aber von den an das postulierte System von Aussagefunktionen gestellten Reichhaltigkeitsforderungen *Widerspruchslosigkeit* vorausgesetzt werden, wodurch Anschluß an die Gedanken Hilberts gewonnen ist.

Was bedeutet nun in der so aufgefaßten Analysis eine Existentialbehauptung? Sicherlich behauptet sie keinerlei Konstruierbarkeit im intuitionistischen Sinne; ist sie aber deshalb so bedeutungsleer, wie die Intuitionisten meinen? Nehmen wir an, es sei mit transzendenten (also nicht konstruktiven) Hilfsmitteln irgendein Existentialsatz bewiesen worden, z. B., nur um konkreter zu sprechen, der Satz: „Es gibt eine stetige Funktion ohne Ableitung"; wird dann noch irgendwer versuchen, den Satz zu beweisen: „Jede stetige Funktion hat eine Ableitung?" Ich glaube, nein. Und damit hat dieser bloße Existentialsatz eine faktische Bedeutung; nicht die, daß irgendwie eine solche Funktion in der Welt empirisch aufweisbar sei; auch nicht die, daß sie „konstruierbar" sei; wohl aber die, ich möchte sagen „wissenschaftstechnische" Bedeutung einer Warnungstafel: Suche nicht den Satz: „Jede stetige Funktion hat eine Ableitung", zu beweisen, denn es wird dir nicht gelingen. Daß dies tatsächlich die Rolle der bloßen „Existentialsätze" ist, werden — denke ich — die meisten Fachgenossen zugeben, die sich aktiv an Forschungen, wie sie etwa in der Theorie der reellen Funktionen getrieben werden, beteiligen.

Zum Schluß noch einige Worte zu Wittgensteins Kritik an Russell, über die Herr Waismann hier referiert hat. Daß mir diese Kritik in sehr wesentlichen Punkten berechtigt scheint, habe ich schon gesagt. Doch glaube ich, daß der Unterschied hier nicht durchweg so groß ist, als es nach Waismanns Referat scheinen mag. Nach Russell sind die natürlichen Zahlen Klassen von Klassen; Wittgensteins Auffassung ist anscheinend eine ganz andere; beachtet man aber, daß nach Russell Klassensymbole unvollständige Symbole sind, die erst eliminiert werden müssen, wenn man die wirkliche Bedeutung eines Satzes erkennen will, und führt man diese Eliminierung nach den von Russell angegebenen Regeln durch, so sieht man, daß die beiden Auffassungen durchaus nicht so verschieden sind. — Sicherlich besteht der von Wittgenstein betonte Unterschied zwischen System und Gesamtheit, zwischen Operation und Funktion, und es ist richtig, daß im Russellschen System diese Unterscheidung nicht gemacht wird. Doch haben Operationen und Funktionen, Systeme und Gesamtheiten viel Gemeinsames und können deshalb sicherlich weitgehend gemeinsam, mit derselben Symbolik, behandelt werden. Um die in diesem Punkte geübte Kritik wirksam zu gestalten, müßte also aufgewiesen werden, daß Russell in dieser gemeinsamen Behandlung zu weit geht, sie auch noch in Fällen anwendet, wo sie wegen effektiver Unterschiede nicht mehr angewendet werden kann, und dadurch in Irrtum gerät.

Beobachtung und Denken

I.

Schon ein flüchtiger Blick auf die Sätze der Physik zeigt, daß sie offenbar von sehr verschiedener Natur sind. Da haben wir Sätze wie: „Wenn man eine gespannte Saite zupft, so hört man einen Ton"; „Läßt man einen Sonnenstrahl durch ein Glasprisma gehen, so sieht man auf einem dahinter stehenden Schirme ein farbiges, von dunklen Linien unterbrochenes Band", die jederzeit unmittelbar durch Beobachtung kontrolliert werden können; wir haben aber auch Sätze wie: „In der Sonne gibt es Wasserstoff"; „Der Begleiter des Sirius hat etwa die Dichte 60.000"; „Ein Wasserstoffatom besteht aus einem positiv geladenen Kern, der von einem negativ geladenen Elektron umkreist wird", die keineswegs durch unmittelbare Beobachtung kontrolliert werden können, die nur auf Grund theoretischer Erwägungen aufgestellt und auch nur mit Hilfe theoretischer Erwägungen kontrolliert werden. Und damit stehen wir vor der brennenden Frage: welches ist die gegenseitige Stellung von *Beobachtung* und *Theorie* in der Physik — und nicht nur in der Physik, sondern in der Wissenschaft überhaupt, denn es gibt nur *eine* Wissenschaft, und wo immer Wissenschaft getrieben wird, wird sie im letzten Grunde nach denselben Methoden getrieben; nur daß wir bei der Physik als der vorgeschrittensten, der saubersten, der wissenschaftlichsten Wissenschaft alles am klarsten sehen.[1]) Und bei der Physik sehen wir denn auch das Zusammenwirken von Beobachtung und Theorie am augenfälligsten vor uns, sogar behördlich anerkannt durch Systemisierung von eigenen Professuren für Experimentalphysik und für theoretische Physik.

Die übliche Auffassung ist nun wohl, ganz schematisch gesprochen, die: wir haben eben zwei Erkenntnisquellen, durch die wir „die Welt", „die Realität", in die wir „hineingestellt" sind, der wir „gegenübergestellt" sind, erfassen: die *Erfahrung,* die *Beobachtung* einerseits, das *Denken* andererseits; je nachdem man z. B. in der Physik gerade von der einen oder der anderen dieser beiden Erkenntnisquellen Gebrauch macht, treibt man Experimentalphysik oder theoretische Physik.

Seit altersher wogt nun in der Philosophie der Streit um diese beiden Erkenntnisquellen: Welche Teile unseres Wissens entstammen der Beobachtung, sind „a posteriori", und welche entstammen dem Denken, sind „a priori"? Ist eine dieser beiden Erkenntnisquellen der anderen überlegen, und wenn ja, welche?

Schon in den Anfängen der Philosophie sehen wir Zweifel an der Zuverlässigkeit der *Beobachtung* auftreten (ja vielleicht sind diese Zweifel sogar die Quelle aller Philosophie). Daß solche Zweifel entstehen konnten, ist recht naheliegend: man glaubte sich in manchen Fällen durch die Sinneswahrnehmung getäuscht. Bei Morgen- oder Abendbeleuchtung sehen wir den Schnee auf fernen Bergen rot, aber „in Wirklichkeit" ist er doch weiß! Einen in Wasser gesteckten Stab sehen wir gebrochen, aber „in Wirklichkeit" ist er doch gerade! Entfernt sich ein Mensch von mir, so sehe ich ihn immer kleiner und kleiner, aber „in Wirklichkeit" bleibt er doch gleich groß!

Obwohl nun alle die Phänomene, von denen ich eben sprach, längst in physikalische Theorien eingeordnet sind, und es keinem Menschen mehr einfällt, in ihnen eine Täuschung durch die Beobachtung zu erblicken, so wirken die Konsequenzen, die aus dieser primitiven, längst abgetanen Auffassung flossen, auch heute noch mächtig nach. Man sagte sich: wenn die Beobachtung manchmal täuscht, so täuscht sie vielleicht immer! Vielleicht ist alles, was uns die Sinne liefern, bloßer Schein, Lug und Trug! Jeder Mensch kennt das Phänomen des Traumes, und jeder Mensch weiß, wie schwer es gelegentlich ist, zu entscheiden, ob wir etwas „wirklich erlebt" oder „nur geträumt" haben; vielleicht ist alles, was wir beobachten nur ein Traum? Jedermann weiß, daß Halluzinationen vorkommen, und daß sie so lebhaft sein können, daß der Betroffene nicht davon abzubringen ist, das Halluzinierte für „wirklich" zu halten; vielleicht ist alles, was wir beobachten, nur Halluzination? Blicken wir durch geeignet geschliffene Linsen, so sehen wir alles verzerrt; wer weiß, ob wir nicht, ohne es zu wissen, immerzu die Welt gewissermaßen durch verzerrende Gläser anblicken, und deshalb alles verzerrt sehen, anders als es wirklich ist? Das ist eines der Grundmotive der Philosophie *Kant's*.[2])

Doch kehren wir zurück zu den alten Zeiten! Man glaubte sich, wie gesagt, in vielen Fällen durch die Beobachtung getäuscht; nie aber war mit dem Denken so etwas passiert: es gab *Sinnestäuschungen* in Hülle und Fülle, aber es gab keine *Denktäuschungen!* Und so mag, als das Vertrauen in die Beobachtung erschüttert war, die Meinung aufgekommen sein, im *Denken* hätten wir ein der Beobachtung unbedingt übergeordnetes, ja das einzig zuverlässige Erkenntnismittel: die Beobachtung liefert bloßen Schein, das Denken erfaßt das wahre Sein.

Diese, sagen wir kurz, „rationalistische" Lehre: das Denken sei die der Beobachtung überlegene, ja die einzig zuverlässige Erkenntnisquelle,[3]) blieb seit der Hochblüte der griechischen Philosophie bis in die Neuzeit vorherrschend. Ich kann nicht einmal andeuten, welche absonderlichen Früchte am Baume dieser Erkenntnis reiften; jedenfalls aber: diese Früchte erwiesen sich als außerordentlich wenig nahrhaft; und so kam langsam, von England ausgehend, gestützt auf die mächtigen Erfolge der neuzeitlichen Naturwissenschaft, die „empiristische" Gegenströmung hoch, welche lehrt, die Beobachtung sei die dem Denken überlegene, ja die einzige Erkenntnisquelle:[4]) „nihil est in intellectu, quod non prius fuerit in sensu", zu deutsch: „Nichts ist im Verstande, was nicht vorher in den Sinnen gewesen wäre".

Aber diese empiristische Auffassung sieht sich bald einer scheinbar unüberwindlichen Schwierigkeit gegenüber: wie soll sie von der Gültigkeit der logischen und mathematischen Sätze in der Wirklichkeit Rechenschaft geben? Die Beobachtung lehrt mich nur ein Einmaliges, sie greift nicht über das Beobachtete hinaus; es gibt kein Band, das von einer beobachteten Tatsache zu einer anderen führte, das künftige Beobachtungen zwänge, ebenso auszufallen, wie schon gemachte — die Gesetze der Logik und Mathematik aber beanspruchen *absolut allgemeine* Gültigkeit: daß die Türe meines Zimmers jetzt geschlossen ist, weiß ich durch Beobachtung, bei meiner nächsten Beobachtung wird sie vielleicht offen sein; daß erwärmte Körper sich ausdehnen, weiß ich durch Beobachtung, schon die nächste Beobachtung kann ergeben, daß ein erwärmter Körper sich nicht ausdehnt; daß aber zweimal zwei vier ist, gilt nicht nur in dem Falle, in dem ich es eben nachzähle, ich weiß bestimmt, daß es immer und überall gilt. Was ich durch Beobachtung weiß, könnte auch anders sein: die Türe meines Zimmers könnte jetzt auch offen sein, ich kann mir das auch ohneweiters vorstellen; ich kann mir auch ohneweiters vorstellen, daß ein Körper sich bei Erwärmung nicht ausdehnt; es könnte aber nicht zweimal zwei gelegentlich auch fünf sein, ich kann mir keinerlei Vorstellung bilden, wie es zugehen müßte, daß zweimal zwei gleich fünf wäre.

Also: weil die Sätze der Logik und Mathematik absolut allgemein gelten, weil sie apodiktisch sicher sind, weil es so sein muß, wie sie sagen, und nicht anders sein kann, können diese Sätze nicht aus der Erfahrung stammen. Bei der ungeheuren Rolle, die Logik und Mathematik im Systeme unserer Erkenntnis spielen, scheint damit der Empirismus endgültig widerlegt. Wohl haben trotz alledem ältere Empiristen den Versuch gemacht, Logik und Mathematik auf die Erfahrung zu gründen:[5]) sie lehrten, auch alles logische und mathematische Wissen stamme aus der Erfahrung, nur handle es sich dabei um so uralte Erfahrung, um so unzähligemale wiederholte Beobachtungen daß wir nun glaubten, es müsse so sein und könne nicht anders sein; nach dieser Auffassung wäre es also durchaus denkbar, daß, so wie eine Beobachtung ergeben könnte, daß ein erwärmter Körper sich nicht ausdehnt, gelegentlich auch zweimal zwei fünf sein könnte, nur daß uns dies bisher entgangen ist, weil es so ungeheuer selten vorkommt; und Abergläubische könnten vielleicht glauben, wenn es schon Glück bringt, ein vierblättriges Kleeblatt zu finden, was doch nicht einmal gar so selten vorkommt — wie glückverheißend müßte es erst sein, wenn man auf einen Fall stößt, in dem zweimal zwei gleich fünf ist! Man kann wohl sagen, daß bei näherem Zusehen diese Versuche, Logik und Mathematik aus der Erfahrung herzuleiten, von Grund auf unbefriedigend sind, und es dürfte heute kaum irgendwer diese Meinung ernstlich verfechten.

Waren so Rationalismus und Empirismus gleichermaßen gescheitert — der Rationalismus, weil seinen Früchten jeder Nährwert mangelte, der Empirismus, weil er sich mit Logik und Mathematik nicht zurecht finden konnte — so gewannen *dualistische* Auffassungen die Oberhand, die lehrten: Denken und Beobachtung sind zwei gleichberechtigte Erkenntnisquellen, die beide für uns zum Erfassen der Welt unentbehrlich sind und jede ihre eigene Rolle im Systeme unserer Erkenntnis spielen. Das *Denken* erfaßt die allgemeinsten Gesetze alles Seins, wie sie etwa in Logik und Mathematik niedergelegt sind; die *Beobachtung* füllt diesen Rahmen im einzelnen aus. Über die Grenzen, die den beiden Erkenntnisquellen gezogen sind, gehen die Meinungen auseinander.

So wird z. B. gestritten über die Frage, ob die Geometrie a priori oder a posteriori sei, ob sie auf reinem Denken oder auf Erfahrung beruhe. Und bei einzelnen besonders grundlegenden physikalischen Gesetzen finden wir denselben Streit, z. B. beim Trägheitsgesetze, den Gesetzen von der Erhaltung der Masse und der Energie, beim Gesetze von der Massenanziehung: alle diese Gesetze wurden schon von einzelnen Philosophen als a priori, als Denknotwendigkeiten reklamiert[6]) — aber immer erst, nachdem sie von der Physik als empirische Gesetze aufgestellt worden waren und sich gut bewährt hatten. Das mußte skeptisch stimmen, und tatsächlich ist unter Physikern wohl die Tendenz vorherrschend, den durch das Denken erfaßten Rahmen möglichst weit, möglichst allgemein anzunehmen, und für alles einigermaßen Konkretere die Erfahrung als Erkenntnisquelle anzuerkennen.

Die übliche Auffassung kann man dann etwa so schildern: Aus der Erfahrung entnehmen wir gewisse Tatsachen, die wir als „Naturgesetze“ formulieren; da wir aber durch das Denken die allgemeinsten in der Realität herrschenden gesetzmäßigen Zusammenhänge (logischer und mathematischer Natur) erfassen, so beherrschen wir auf Grund der der Beobachtung entnommenen Tatsachen die Natur in viel weiterem Umfange, als sie beobachtet wurde: wir wissen nämlich, daß sich auch alles realisiert finden muß, das aus Beobachtetem durch Anwendung von Logik und Mathematik gefolgert werden kann. Der Experimentalphysiker verschafft uns nach dieser Auffassung durch direkte Beobachtung Kenntnis von Naturgesetzen; der theoretische Physiker erweitert dann durch das Denken diese Kenntnisse in ungeheurem Maßstabe, so daß wir in der Lage sind, auch Aussagen zu machen über Vorgänge, die sich weit von uns in Raum und Zeit entfernt abspielen, und über Vorgänge, die sich durch ihre Größe oder ihre Kleinheit jeder direkten Beobachtung entziehen, die aber an das direkt Beobachtete gekettet sind durch die allgemeinsten, im Denken erfaßten Gesetze alles Seins, die Gesetze der Logik und der Mathematik. Eine mächtige Stütze scheint diese Auffassung zu finden durch die zahlreichen auf theoretischem Wege gemachten Entdeckungen, wie — um nur einige der berühmtesten zu nennen — die Errechnung des Planeten Neptun durch *Leverrier,* die Errechnung elektrischer Wellen durch *Maxwell,* die Errechnung der Ablenkung der Lichtstrahlen im Gravitationsfelde der Sonne durch *Einstein* und die Errechnung der Rotverschiebung im Spektrum der Sonne gleichfalls durch *Einstein.*

Trotzdem aber vertreten wir die Meinung, daß diese Auffassung gänzlich unhaltbar ist. Denn bei näherer Besinnung zeigt sich, daß die Rolle des Denkens eine ungleich bescheidenere ist, als sie ihm bei dieser Auffassung zugeschrieben wird. Die Auffassung, wir hätten im Denken ein Mittel zur Hand, mehr über die Welt zu wissen, als beobachtet wurde, etwas zu wissen, was immer und überall in der Welt unbedingte Geltung haben muß, ein Mittel, allgemeine Gesetze alles Seins zu erfassen, scheint uns durchaus mysteriös. Wie soll es zugehen, daß wir von irgend einer Beobachtung im vorhinein sagen könnten, wie sie ausfallen muß, bevor wir sie noch angestellt haben? Woher sollte unser Denken eine Exekutivgewalt nehmen, durch die es eine Beobachtung zwänge, so und nicht anders auszugehen? Warum sollte das, was für unser Denken zwingend ist, auch für den Ablauf der Welt zwingend sein? Es bliebe nur übrig, an eine wundersame prästabilierte Harmonie zwischen dem Ablauf unseres Denkens und dem Ablauf der Welt zu glauben, eine Vorstellung, die reichlich mystisch und in letzter Linie theologisierend ist.

Aus dieser Situation zeigt sich kein anderer Ausweg, als Rückkehr zu einem rein *empiristischen* Standpunkt, Rückkehr zur Auffassung, daß die einzige Quelle eines Wissens über Tatsachen die Beobachtung ist: Es gibt kein Wissen a priori über Tatsächliches, *es gibt kein „materiales" a priori.*[7]) Nur werden wir den Fehler früherer Empiristen vermeiden müssen, in den Sätzen der Logik und Mathematik lediglich Erfahrungstatsachen sehen zu wollen; wir müssen uns nach einer anderen Auffassung von Logik und Mathematik umsehen.

II.

Beginnen wir mit der Logik! Die alte Auffassung der Logik wäre etwa die: die Logik ist die Lehre von den allgemeinsten Eigenschaften der Gegenstände, die Lehre von den allen Gegenständen gemeinsamen Eigenschaften; so wie die Ornithologie die Lehre von den Vögeln, die Zoologie die Lehre von allen Tieren, die Biologie schon die Lehre von allen Lebewesen, so ist die Logik die Lehre von *allen* Gegenständen, die Lehre von den Gegenständen überhaupt. Wäre dem so, so bliebe es ganz unverständlich, woher die Logik ihre Sicherheit nimmt, denn wir kennen doch nicht alle Gegenstände, wir haben nicht alle Gegenstände beobachtet, können also nicht wissen, wie sich alle Gegenstände verhalten.

Unsere Auffassung hingegen besagt: die Logik handelt keineswegs von sämtlichen Gegenständen, sie handelt überhaupt nicht von irgendwelchen Gegenständen, sondern *sie handelt nur von der Art, wie wir über die Gegenstände sprechen;* die Logik entsteht erst durch die Sprache. Und gerade daraus, daß ein Satz der Logik überhaupt nichts über irgendwelche Gegenstände aussagt, fließt seine Sicherheit und Allgemeingültigkeit oder, besser gesagt, seine Unwiderleglichkeit.

Ein Beispiel möge uns das näher bringen. Ich spreche von einer wohlbekannten Pflanze: ich beschreibe sie, wie es in den botanischen Bestimmungsbüchern geschieht, durch Anzahl, Farbe und Form ihrer Blütenblätter, ihrer Kelchblätter, ihrer Staubgefäße, Gestalt ihrer Blätter, ihres Stengels, ihrer Wurzel etc., und treffe die Festsetzung: jede solche Pflanze wollen wir „Schneerose" nennen, wir wollen sie aber außerdem auch „Helleborus niger" nennen. Dann kann ich mit absoluter Sicherheit und Allgemeingültigkeit den Satz aussprechen: „Jede Schneerose ist ein Helleborus niger"; er trifft ganz bestimmt zu, immer und überall, er ist durch keinerlei Beobachtung widerlegbar; aber er sagt gar nichts über Tatsachen aus; ich erfahre nichts aus ihm über die Pflanze, von der die Rede ist, nicht in welcher Jahreszeit sie blüht, nicht wo ich sie finden kann, nicht ob sie häufig oder selten ist, er sagt mir gar nichts über die Pflanze, aber gerade deshalb, weil er gar nichts über Tatsächliches aussagt, kann er durch keine Beobachtung widerlegt werden, gerade daraus zieht er seine Sicherheit und Allgemeingültigkeit. Dieser Satz drückt lediglich eine Verabredung aus, wie wir über die fragliche Pflanze sprechen wollen.

Ähnlich nun steht es mit den Sätzen der Logik. Wir wollen uns das zunächst an den beiden berühmtesten Sätzen der Logik überlegen: dem Satze vom Widerspruch und dem Satze vom ausgeschlossenen Dritten. Sprechen wir etwa von Gegenständen, denen Farbe zukommt. Wir lernen, ich möchte sagen: durch Dressur, gewissen dieser Gegenstände die Bezeichnung „rot" beizulegen, und treffen die Vereinbarung, jedem andern dieser Gegenstände die Bezeichnung „nicht rot" beizulegen. Auf Grund dieser Vereinbarung können wir nun mit absoluter Sicherheit den Satz aussprechen: keinem Gegenstand wird sowohl die Bezeichnung „rot" als auch die Bezeichnung „nicht rot" beigelegt, was man gewöhnlich kurz so ausspricht: kein Gegenstand ist sowohl rot als nicht rot. Das ist der Satz vom Widerspruch. Und da wir die Vereinbarung getroffen haben, gewissen Gegenständen die Bezeichnung „rot" und *jedem* andern die Bezeichnung „nicht rot" beizulegen, können wir ebenso mit absoluter Sicherheit den Satz aussprechen: jedem Gegenstande wird entweder die Bezeichnung „rot" oder die Bezeichnung „nicht rot" beigelegt, gewöhnlich kurz so ausgesprochen: jeder Gegenstand ist entweder rot oder nicht rot. Das ist der Satz vom ausgeschlossenen Dritten. Diese beiden Sätze, der Satz vom Widerspruch und vom ausgeschlossenen Dritten, sagen gar nichts über irgendwelche Gegenstände aus; von keinem einzigen erfahre ich durch diese Sätze ob er rot ist, ob er nicht rot ist, welche Farbe er hat, noch auch sonst irgend etwas: diese Sätze legen lediglich etwas über die Art und Weise fest, wie wir den Gegenständen die Bezeichnungen „rot" und „nicht rot" beilegen wollen, d. h. sie setzen etwas darüber fest, *wie wir über die Gegenstände sprechen wollen.* Und gerade aus dem Umstande, daß sie gar nichts über Gegenstände aussagen, fließt ihre Allgemeingültigkeit und Sicherheit, fließt ihre Unwiderlediglichkeit.

So wie mit diesen beiden Sätzen, steht es nun auch mit den andern Sätzen der Logik; wir werden alsbald darauf zurückkommen. Vorher wollen wir noch eine andere Betrachtung einschalten.

Wir haben vorhin festgestellt, daß es kein materiales a priori, d. h. kein Wissen a priori über Tatsächliches geben kann; denn wir können von keiner

Beobachtung, bevor sie angestellt wurde, wissen, wie sie ausgehen muß. Wir haben uns überlegt, daß in den Sätzen vom Widerspruch und vom ausgeschlossenen Dritten kein materiales a priori vorkommt, denn sie sagen nichts über Tatsachen aus. Manche Leute aber, die vielleicht zugeben würden, daß es sich mit den Sätzen der Logik so verhalten mag, wie wir es sagen, beharren darauf, daß es doch anderswo ein materiales a priori gebe, z. B. im Satze: „kein Gegenstand ist sowohl rot als blau" (natürlich ist gemeint: zur selben Zeit und an derselben Stelle); hier handle es sich um ein wirkliches Wissen a priori über das Verhalten von Gegenständen; noch bevor man eine Beobachtung angestellt habe, könne man mit absoluter Sicherheit sagen, daß bei ihr sich nicht herausstellen kann, ein Gegenstand sei sowohl rot als blau; und es wird behauptet, daß man dieses Wissen a priori durch „Wesensschau" erhalte, durch Erfassen des Wesens der Farben.[8]) Will man unsere These, daß es keinerlei materiales a priori gibt, aufrecht erhalten, so muß man irgendwie zu einem Satze wie: „kein Gegenstand ist sowohl rot als blau" Stellung nehmen, und ich will das mit einigen andeutenden Worten versuchen, die freilich keineswegs dieses nicht leichte Problem erschöpfen können. Es ist gewiß richtig, daß noch bevor wir eine Beobachtung angestellt haben, wir mit völliger Sicherheit sagen können: sie wird nicht ergeben, daß ein Gegenstand sowohl rot als blau ist — so wie wir mit völliger Sicherheit sagen können, daß keine Beobachtung ergeben wird, ein Gegenstand sei sowohl rot als nicht rot, oder eine Schneerose sei kein Helleborus niger; aber ebenso wenig wie im zweiten und dritten Falle handelt es sich auch im ersten um ein materiales a priori; ebenso wie die Sätze: „Jede Schneerose ist ein Helleborus niger", „kein Gegenstand ist sowohl rot als nicht rot", sagt auch der Satz „kein Gegenstand ist sowohl rot als blau" gar nichts über das Verhalten von Gegenständen aus; auch er handelt nur davon, wie wir über die Gegenstände sprechen wollen, wie wir ihnen Bezeichnungen beilegen wollen. Wie wir früher sagten: gewisse Gegenstände nennen wir „rot", jeden andern Gegenstand nennen wir „nicht rot", woraus die Sätze vom Widerspruch und vom ausgeschlossenen Dritten flossen, so sagen wir jetzt: gewisse Gegenstände nennen wir „rot", gewisse *andere* Gegenstände nennen wir „blau", wieder gewisse *andere* Gegenstände nennen wir „grün" etc. Wenn wir aber in dieser Art die Farbbezeichnungen den Gegenständen zuweisen, so können wir mit Sicherheit von vornherein sagen: keinem Gegenstand wird dabei sowohl die Bezeichnung „rot" als auch die Bezeichnung „blau" zugewiesen, oder, kürzer ausgedrückt: kein Gegenstand ist sowohl rot als blau; und zwar können wir es deshalb mit Sicherheit sagen, weil wir die Zuweisung der Farbbezeichnungen an die Gegenstände eben so eingerichtet haben.[9])

Wie wir sehen, gibt es zwei ganz verschiedene Arten von Sätzen: solche, die wirklich etwas über Gegenstände aussagen, und solche, die nichts über Gegenstände aussagen, sondern nur Regeln festlegen, wie wir über die Gegenstände sprechen wollen. Frage ich: „Welche Farbe hat das neue Kleid von Fräulein Erna?" und erhalte ich die Antwort: „Das neue Kleid von Fräulein Erna ist nicht (in seiner Gänze) sowohl rot als blau", so wurde mir gar nichts über dieses Kleid mitgeteilt, ich weiß nahher genau so viel wie vorher; erhalte ich aber die Antwort „Das neue Kleid von Fräulein Erna ist rot", so wurde mir wirklich etwas über dieses Kleid mitgeteilt.

Wir wollen uns diesen Unterschied noch an einem Beispiel klarmachen. Ein Satz, der wirklich etwas über die Gegenstände aussagt, von denen er spricht, ist der folgende: „Wenn du dieses Stück Eisen auf 800° erwärmst, wird es rot, wenn du es auf 1300° erwärmst, wird es weiß werden". Worauf beruht der Unterschied dieses Satzes von den eben angeführten Sätzen, die nichts Tatsächliches aussagen? Die Zuweisung der Temperaturbezeichnungen an die Gegenstände geschieht *unabhängig* von der Zuweisung der Farbbezeichnungen, die Farbbezeichnungen „rot" und „nicht rot", oder „rot" und „blau" hingegen werden *in Abhängigkeit von einander* den Gegenständen zugewiesen: die Sätze „Das neue Kleid von Fräulein Erna ist entweder rot oder nicht rot", „Das neue Kleid von Fräulein Erna ist nicht sowohl rot als blau" drücken lediglich diese Abhängigkeit aus, sagen darum nichts über dieses Kleid aus, und sind deshalb absolut sicher und unwiderlegbar; der obige Satz über das Stück Eisen hingegen bringt von einander unabhängig gegebene Bezeichnungen in Beziehung, sagt deshalb wirklich etwas über dieses Stück Eisen aus und ist eben deshalb nicht sicher und ist durch die Beobachtung widerlegbar.

Der Unterschied zwischen diesen beiden Arten von Sätzen wird vielleicht besonders klar durch folgende Überlegung. Sagt mir jemand: „Ich habe dieses Stück Eisen auf 800° erhitzt und es ist dabei nicht rot geworden", so werde ich das nachprüfen; dabei wird sich vielleicht herausstellen, daß er gelogen hat, vielleicht wird sich herausstellen, daß er das Opfer einer Täuschung geworden ist, vielleicht wird sich aber auch herausstellen, daß — entgegen meinen bisherigen Ansichten — auch Fälle vorkommen, in denen ein Stück Eisen bei Erwärmung auf 800° nicht rotglühend wird, und dann werde ich eben meine Meinung über das Verhalten von Eisen bei Erhitzung abändern. Wenn mir aber jemand sagt: „Ich habe dieses Stück

Eisen auf 800° erwärmt und dabei ist es sowohl rot als nicht rot geworden" oder: „dabei ist es sowohl rot als weiß geworden", dann werde ich bestimmt gar nichts nachprüfen, ich werde auch nicht sagen: „der Mann hat gelogen" oder „der Mann ist Opfer einer Täuschung geworden" und ich werde ganz bestimmt meine Ansichten über das Verhalten von Eisen bei Erhitzung nicht ändern; sondern — man kann dies am besten durch ein jedem Kartenspieler geläufiges Wort ausdrücken — der Mann hat *Renonce gemacht:* er hat sich vergangen gegen die Regeln, nach denen wir sprechen wollen, und ich werde mich weigern, weiter mit ihm zu sprechen. Es ist ganz so, wie wenn jemand beim Tarockspielen versuchen wollte, mir den Skieß mit dem Mond zu stechen; auch da werde ich gar nichts nachprüfen, ich werde meine Ansichten über das Verhalten von Gegenständen nicht abändern, sondern ich werde mich weigern, mit ihm weiter Tarock zu spielen.

Fassen wir zusammen: Wir müssen unterscheiden zwischen zwei Arten von Sätzen: solchen, die etwas Tatsächliches aussagen, und solchen, die lediglich eine Abhängigkeit in der Zuweisung der Bezeichnungen an die Gegenstände ausdrücken; die Sätze dieser zweiten Art wollen wir *tautologisch* nennen;[10]) sie sagen nichts über Gegenstände aus und sind eben deshalb sicher, allgemein gültig, durch Beobachtung unwiderlegbar; die Sätze der ersten Art hingegen sind nicht sicher, können durch Beobachtung widerlegt werden. Die logischen Sätze vom Widerspruch und vom ausgeschlossenen Dritten sind tautologisch, ebenso z. B. der Satz: „Kein Gegenstand ist sowohl rot als blau".

Und nun behaupten wir, daß auch alle anderen Sätze der Logik tautologisch sind. Kommen wir also, um uns das wenigstens an einem Beispiel klarzumachen, noch einmal auf die Logik zurück! Wir sagten: gewissen Gegenständen wird die Bezeichnung „rot" beigelegt und es wird die Verabredung getroffen, jedem andern Gegenstand die Bezeichnung „nicht rot" beizulegen; diese Verabredung über den Gebrauch der Negation wird ausgedrückt durch die Sätze vom Widerspruch und vom ausgeschlossenen Dritten. Nun wird (um weiter an Gegenständen zu exemplifizieren, denen Farbe zukommt) noch die Verabredung getroffen, jedem Gegenstand, dem die Bezeichnung „rot" beigelegt wird, auch die Bezeichnungen „rot oder blau", „blau oder rot", „rot oder gelb", „gelb oder rot" etc. beizulegen, jedem Gegenstand, dem die Bezeichnung „blau" beigelegt wird, auch die Bezeichnungen „blau oder rot", „rot oder blau", „blau oder gelb", „gelb oder blau" etc. beizulegen u. s. f. Auf Grund dieser Verabredung können wir dann mit voller Sicherheit z. B. den Satz aussprechen: „Jeder rote Gegenstand ist rot oder blau"; das ist wieder ein tautologischer Satz; er handelt nicht von den Gegenständen, über die wir sprechen, sondern nur von der Art, wie wir über diese Gegenstände sprechen.

Vergegenwärtigen wir uns nochmals die Art, wie den Gegenständen die Bezeichnungen „rot", „nicht rot", „blau", „rot oder blau" etc. beigelegt werden, so können wir auch völlig sicher und unwiderleglich sagen: Jedem Gegenstand, dem die beiden Bezeichnungen „rot oder blau" und „nicht rot" beigelegt werden, wird auch die Bezeichnung „blau" beigelegt — gewöhnlich kurz so ausgesprochen: Ist ein Gegenstand rot oder blau und nicht rot, so ist er blau. Auch das ist ein tautologischer Satz; er enthält keinerlei Feststellung über das Verhalten der Gegenstände, er drückt nur aus, in welchem Sinne wir die logischen Worte „nicht" und „oder" verwenden.

Damit nun sind wir bei etwas ganz Grundlegendem angelangt: Die Verabredung über die Verwendung der Worte „nicht" und „oder" ist derart, daß, wenn ich die beiden Sätze ausspreche: „Der Gegenstand A ist rot oder blau" und „Der Gegenstand A ist nicht rot", ich dadurch schon mitgesagt habe: „Der Gegenstand A ist blau". Das ist das Wesen des sogenannten *logischen Schließens.* Es beruht also keineswegs darauf, daß zwischen Sachverhalten ein realer Zusammenhang besteht, den wir durch das Denken erfassen, es hat vielmehr mit dem Verhalten der Gegenstände überhaupt nichts zu tun, sondern fließt aus der Art, wie wir über die Gegenstände sprechen. Wer das logische Schließen nicht anerkennen wollte, hat nicht etwa eine andere Meinung über das Verhalten der Gegenstände als ich, sondern er weigert sich, über die Gegenstände nach denselben Regeln zu sprechen, wie ich; ich kann ihn nicht überzeugen, sondern ich muß mich weigern, mit ihm weiter zu sprechen, so wie ich mich weigern werde, weiter mit einem Partner Tarock zu spielen, der darauf beharrt, meinen Skieß mit dem Mond zu stechen.

Was das logische Schließen leistet, ist also dies: es macht uns bewußt, was alles wir — auf Grund der Verabredungen über den Gebrauch der Sprache — durch Behauptung eines Systemes von Sätzen (zwar nicht ausdrücklich, aber doch implizit und vielleicht unbewußt) mitbehauptet haben; so wie in unserem obigen Beispiel durch Behauptung der beiden Sätze: „Der Gegenstand A ist rot oder blau" und „Der Gegenstand A ist nicht rot" implizit mitbehauptet wird: „Der Gegenstand A ist blau".

Mit dieser Formulierung streifen wir auch schon die Antwort auf eine Frage, die sich wohl naturgemäß jedem Leser, der unseren Überlegungen folgte, aufgedrängt haben dürfte: Wenn es wirklich zutrifft, daß die Sätze der Logik tautologisch sind, daß sie nichts über die Gegenstände aussagen, wozu dient dann überhaupt die Logik?

Die logischen Sätze, an denen wir exemplifizierten, flossen aus den Verabredungen über den Gebrauch der Worte „nicht" und „oder" (und man kann zeigen, daß das gleiche von allen Sätzen der sogenannten Aussagenlogik gilt). Überlegen wir also zunächst, wozu man die Worte „nicht" und „oder" in die Sprache einführt. Das geschieht wohl deshalb, weil wir nicht allwissend sind. Fragt man mich nach der Farbe des Kleides, das Fräulein Erna gestern trug, so werde ich mich vielleicht an seine Farbe nicht erinnern können; ich kann nicht sagen, ob es rot oder blau oder grün war; aber vielleicht werde ich wenigstens sagen können: „gelb war es nicht"; wäre ich allwissend, so wüßte ich seine Farbe, ich hätte dann nicht nötig, zu sagen: „gelb war es nicht", sondern könnte etwa sagen: „es war rot". Oder: meine Tochter schrieb mir, sie habe einen Dackel geschenkt bekommen; da ich ihn noch nicht gesehen habe, kenne ich seine Farbe nicht; ich kann nicht sagen „er ist schwarz" und ich kann nicht sagen „er ist braun"; wohl aber kann ich sagen: „er ist schwarz oder braun"; wäre ich allwissend, so hätte ich dieses „oder" nicht nötig und könnte ohneweiters sagen: „er ist braun".

Und so haben auch die logischen Sätze, obwohl sie rein tautologisch sind, so hat auch das logische Schließen, obwohl es nichts anderes ist als tautologisches Umformen, für uns Bedeutung deshalb, weil wir nicht allwissend sind. Unsere Sprache ist so gemacht, daß wir bei Behauptung gewisser Sätze andre Sätze implizit mitbehaupten — aber wir sehen nicht sofort, was alles wir so implizit mitbehauptet haben, erst das logische Schließen macht uns das bewußt. Ich behaupte z. B. die Sätze: „Die Blume, die Herr Maier im Knopfloch trägt, ist entweder eine Rose oder eine Nelke". „Wenn Herr Maier eine Nelke im Knopfloch trägt, so ist sie weiß". „Die Blume, die Herr Maier im Knopfloch trägt, ist nicht weiß". Vielleicht bin ich mir gar nicht bewußt, daß ich dabei auch implizit behauptet habe: „Die Blume, die Herr Maier im Knopfloch trägt, ist eine Rose"; das logische Schließen aber bringt es mir zum Bewußtsein. Freilich weiß ich deswegen noch nicht, ob die Blume, die Herr Maier im Knopfloch trägt, wirklich eine Rose ist; bemerke ich, daß sie keine Rose ist, so darf ich nicht auf meinen früheren Behauptungen beharren — sonst vergehe ich mich gegen die Regeln des Sprechens, ich mache Renonce.

III.

Wenn ich hoffen darf, die Stellung der Logik nun einigermaßen klar gemacht zu haben, so kann ich mich nun über die Stellung der *Mathematik* ganz kurz fassen. Die Sätze der Mathematik sind von ganz derselben Art, wie die Sätze der Logik: sie sind tautologisch, sie sagen gar nichts über die Gegenstände aus, von denen wir sprechen wollen, sondern sie handeln nur von der Art, wie wir über diese Gegenstände sprechen wollen. Daß wir apodiktisch und allgemeingültig den Satz aussprechen können: $2+3=5$, daß wir schon vor Anstellen einer Beobachtung mit völliger Sicherheit sagen können, daß sich bei ihr nicht herausstellen wird, daß etwa $2+3=7$ ist, rührt daher, daß wir mit $2+3$ dasselbe meinen wie mit 5 — ähnlich wie wir mit „Helleborus niger" dasselbe meinen wie mit „Schneerose", und sich deshalb bei keiner noch so subtilen botanischen Untersuchung herausstellen kann, daß ein Exemplar der Schneerose kein Helleborus niger wäre. Daß wir aber mit $2+3$ dasselbe meinen wie mit 5, das bringen wir uns zum Bewußtsein, indem wir darauf zurückgehen, was wir mit 2, mit 3, mit 5, mit $+$ meinen, und solange tautologisch umformen, bis wir eben sehen, daß wir mit $2+3$ dasselbe meinen wie mit 5. Dieses sukzessive tautologische Umformen ist das, was man „Rechnen" nennt; das Addieren, das Multiplizieren, das man in der Schule lernt, sind Anweisungen zu solchen tautologischen Umformungen; jeder mathematische Beweis ist eine Aufeinanderfolge solcher tautologischer Umformungen. Der Nutzen beruht wieder darauf, daß wir z. B. keineswegs sofort sehen, daß wir mit 24×31 dasselbe meinen wie mit 744; rechnen wir aber das Produkt 24×31 aus, so formen wir es schrittweise um, so daß wir bei jeder einzelnen Umformung uns darüber klar sind, daß wir auf Grund der Verabredungen über die Verwendung der auftretenden Zeichen (hier Zahlzeichen und die Zeichen $+$ und $\times$) nach der Umformung noch dasselbe meinen wie vor der Umformung, bis uns am Schluß bewußt geworden ist, daß wir mit 744 dasselbe meinen wie mit 24×31.

Freilich ist der Nachweis des tautologischen Charakters der Mathematik noch nicht in allen Punkten erbracht; es handelt sich da um ein mühevolles und schwieriges Problem; doch zweifeln wir nicht daran, daß die Meinung vom tautologischen Charakter[11]) der Mathematik ihrem Wesen nach zutreffend ist.

Man hat sich lange dagegen gesträubt, in den Sätzen der Mathematik nichts als Tautologien zu sehen; *Kant* hat den tautologischen Charakter der Mathematik aufs lebhafteste bestritten,[12]) und der große Mathematiker *Henri Poincaré*, dem wir auch soviel an philosophischer Kritik verdanken, hat geradezu argumentiert: da die Mathematik unmöglich eine ungeheure Tautologie sein kann, so muß in ihr irgend ein a priorisches Prinzip stecken.[13]) Und in der Tat, es scheint auf den ersten Blick kaum glaublich, daß die ganze Mathematik mit ihren so schwer erkämpften Sätzen, mit ihren oft so überraschenden Resultaten, sich sollte in Tautologien auflösen lassen. Aber diese Argumentation übersieht nur eine Kleinigkeit: sie übersieht den Umstand, daß wir nicht allwissend sind. Ein allwissendes Wesen freilich wüßte unmittelbar, was alles bei Behauptung einiger Sätze mitbehauptet ist, es wüßte unmittelbar, daß auf Grund der Verabredungen über den Gebrauch der Zahlzeichen und des Zeichens $\times$ mit 24×31 und 744 dasselbe gemeint ist; ein allwissendes Wesen braucht keine Logik und keine Mathematik. Wir aber müssen uns dies erst durch sukzessive tautologische Umformung bewußt machen, und so kann es für uns sehr überraschend sein, daß wir durch Behaupten einiger Sätze einen von diesen anscheinend gänzlich verschiedenen Satz mitbehauptet haben, oder daß wir mit zwei äußerlich ganz verschiedenen Symbolkomplexen tatsächlich dasselbe meinen.

IV.

Und nun mache man sich klar, wie himmelweit unsere Auffassung entfernt ist von der alten — vielleicht darf man sagen: platonisierenden — Auffassung: die Welt sei konstruiert nach den Gesetzen der Logik und der Mathematik („immerzu treibt Gott Mathematik") und in unserem Denken, einem schwachen Abglanz von Gottes Allwissenheit, sei uns ein Mittel geschenkt, diese ewigen Gesetze der Welt zu erfassen. Nein! Keinerlei Realität kann unser Denken erfassen, von keiner Tatsache der Welt kann uns das Denken Kunde bringen, es bezieht sich nur auf die Art, wie wir über die Welt sprechen, es kann nur Gesagtes tautologisch umformen. Es besteht keine Möglichkeit durch das Denken hinter die durch die Beobachtung erfaßte Welt der Sinne zu einer „Welt des wahren Seins" vorzustoßen: jede Metaphysik ist unmöglich! Unmöglich, nicht weil die Aufgabe für unser menschliches Denken zu schwer wäre, sondern weil sie sinnlos ist, weil jeder Versuch, Metaphysik zu treiben, ein Versuch ist, in einer Weise zu sprechen, die den Vereinbarungen, wie wir sprechen wollen, zuwiderläuft, vergleichbar dem Versuche, mit dem Mond den Skieß zu stechen.

Kehren wir nun zurück zum Problem von dem wir ausgingen: welches ist die gegenseitige Stellung von Beobachtung und Theorie in der Physik?

Wir sagten, die übliche Auffassung sei etwa die: der Erfahrung entnehmen wir die Gültigkeit gewisser Naturgesetze, und da wir durch unser Denken die allgemeinsten Gesetze alles Seins erfassen, so wissen wir, daß auch alles. was aus diesen Naturgesetzen durch logisches oder mathematisches Denken folgt, sich realisiert finden muß. Wir sehen nun, daß diese Auffassung unhaltbar ist; denn unser Denken erfaßt keinerlei Gesetze des Seins. Nie und nirgends kann uns also das Denken ein Wissen über Tatsachen liefern. das über das Beobachtete hinausgeht. Wie aber sollen wir uns dann zu den auf theoretischem Wege gemachten Entdeckungen stellen, in denen — wie wir sagten — die übliche Auffassung scheinbar eine so starke Stütze findet? Überlegen wir uns z. B. um was es sich bei der Errechnung des Planeten Neptun durch *Leverrier* drehte!

Newton hat bemerkt, daß die bekannten Bewegungsvorgänge, himmlische wie irdische, sich sehr gut einheitlich beschreiben lassen durch die Annahme. daß zwischen je zwei Massenpunkten eine Anziehungskraft wirkt, die proportional ist ihren Massen, umgekehrt proportional dem Quadrat ihrer Entfernung. Und weil durch diese Annahme sich die bekannten Bewegungsvorgänge sehr gut beschreiben ließen, so *machte* er diese Annahme, d. h. er sprach versuchsweise, als Hypothese, das Gravitationsgesetz aus: „Zwischen je zwei Massenpunkten wirkt eine Anziehungskraft, die proportional ist ihren Massen, umgekehrt proportional dem Quadrat ihrer Entfernung." Als *Behauptung* konnte er dies Gesetz nicht aussprechen, sondern nur als Hypothese, denn niemand kann behaupten, daß sich je zwei Massenpunkte wirklich so verhalten, denn niemand kann alle Massenpunkte beobachten. Indem man aber das Gravitationsgesetz ausspricht, hat man implizit viele andere Sätze mit ausgesprochen, nämlich alle Sätze, die aus dem Gravitationsgesetze (zusammen mit unmittelbar der Beobachtung entnommenen Daten) durch Rechnen und durch logisches Schließen folgen: die theoretischen Physiker und Astronomen haben die Aufgabe, uns bewußt zu machen, was alles wir implizit mitsagen, wenn wir das Gravitationsgesetz aussprechen. Und Leverriers Rechnungen brachten zum Bewußtsein. daß durch Aussprechen des Gravitationsgesetzes mitgesagt ist, daß zu einer bestimmten Zeit an einer bestimmten Stelle des Himmels ein bis dahin unbekannter Planet zu sehen sein mußte. Man hat hingeschaut und hat tatsächlich diesen neuen Planeten gesehen — die Hypothese des Gravitationsgesetzes hatte sich bewährt. Aber nicht die Rechnung Leverriers hat ergeben, daß dieser Planet vorhanden ist, sondern das Hinschaun, die Beobachtung hat dies ergeben. Diese Beobachtung hätte ebensowohl anders ausgehen können, es hätte ebensogut geschehen können, daß an der betreffenden Stelle des Himmels nichts zu sehen war — dann hätte sich eben in diesem Falle das Gravitationsgesetz nicht bewährt und man hätte zu zweifeln begonnen, ob das Gravitationsgesetz wirklich eine geeignete Hypothese zur Beschreibung der beobachtbaren Bewegungsvorgänge sei. Und so ist es ja tatsächlich später gekommen: Indem man das Gravitationsgesetz ausspricht, ist implizit mitgesagt, daß zu einer bestimmten Zeit der Planet Merkur an einer bestimmten Stelle des Himmels zu sehen sein muß — ob er dann wirklich dort zu sehen war, konnte nur die Beobachtung lehren; die Beobachtungen aber ergaben, daß er *nicht* genau an der betreffenden Stelle des Himmels zu sehen war. Und was geschah? Man sagte sich: da wir bei Aussprechen des Gravitationsgesetzes implizit Sätze mitaussprechen, die nicht zutreffen, so können wir die Hypothese des Gravitationsgesetzes nicht aufrecht erhalten. *Newtons* Gravitationstheorie wurde durch die *Einsteins* ersetzt.

Es ist also nicht so, daß wir durch die Erfahrung wissen, daß gewisse Naturgesetze gelten, und — weil wir durch unser Denken die allgemeinsten Gesetze alles Seins erfassen — deshalb auch wissen, daß alles realisiert sein muß, was durch Denken aus diesen Naturgesetzen gefolgert werden kann. Vielmehr ist es so: wir wissen von keinem einzigen Naturgesetze, daß es gilt; die Naturgesetze sind *Hypothesen,* die wir versuchsweise aussprechen; durch Aussprechen solcher Naturgesetze werden aber implizit viele andere Sätze mitausgesprochen (und Aufgabe des Denkens ist es, uns bewußt zu machen, welche Sätze implizit mitausgesprochen wurden); insolange nun diese implizit mitausgesprochenen Sätze (soferne sie von unmittelbar Beobachtbarem handeln) durch die Beobachtung bestätigt werden, bewähren sich diese Naturgesetze und wir halten an ihnen fest; wenn aber diese implizit mitausgesprochenen Sätze durch die Beobachtung nicht bestätigt werden, bewähren sich die Naturgesetze nicht und werden durch andere ersetzt.

V.

Ich habe versucht, es klar zu machen, wie sich Logik und Mathematik in eine rein empiristische Philosophie einfügen, und ich glaube, daß erst durch diese, von der Auffassung früherer Empiristen fundamental abweichende Auffassung von Logik und Mathematik konsequenter Empirismus überhaupt möglich geworden ist. Damit ist aber noch keineswegs alles gesagt, was über die Beziehungen von Beobachtung und Theorie zu sagen ist, und so will ich mich nun einem Problem zuwenden, bei dessen Behandlung wir wieder stark von manchen früheren Empiristen abweichen.

Wir haben festgestellt, daß nur die Beobachtung uns ein Wissen über Tatsachen liefern kann; die Theorie ist dazu völlig außerstande. Manche Empiristen zogen daraus die Folgerung, es seien in den von Tatsachen handelnden Wissenschaften nur solche Sätze legitim, die (wenigstens prinzipiell) durch Beobachtung bestätigt oder widerlegt werden können, die also nur von (wenigstens prinzipiell) Beobachtbarem handeln; in einem legitimen Satze dürften also nur Terme auftreten, die aus Beobachtbarem zusammensetzbar, konstituierbar sind. Insbesondere *E. Mach*[14]) hat mit großer Schärfe diese Forderung vertreten und er hat nachdrücklichst darauf hingewiesen, daß die übliche Physik diese Forderung durchaus nicht erfüllt; insbesondere seien alle Sätze, die von Molekeln, Atomen (und wir können heute hinzufügen: Elektronen, Protonen, Quanten) handeln, nicht legitim, weil alle diese Terme prinzipiell unkonstituierbar (nicht aus Beobachtbarem zusammensetzbar) seien; alle solchen Sätze seien daher metaphysisch und hätten kein Bürgerrecht in der Wissenschaft. *L. Boltzmann* hat gegen diese Ansicht Machs aufs lebhafteste polemisiert,[15]) und ich glaube, wir müssen uns in diesem Streite auf Seite Boltzmanns stellen.

In der Tat, die ganze Wissenschaft ist voll von Sätzen, die prinzipiell nicht durch Beobachtung bestätigt werden können, weil sie unkonstituierbare Terme enthalten; nicht nur die Sätze über Molekeln, Atome, Elektronen etc. sind von dieser Art. In der theoretischen Physik wird (vermöge der Einführung von Koordinaten) die Stelle jedes Ereignisses in Raum und Zeit festgelegt gedacht durch Angabe von Zahlen; eine solche Festlegung aber geht prinzipiell über jede Beobachtungsmöglichkeit hinaus; wie sollte eine Beobachtung darüber entscheiden können, ob das Verhältnis zweier Längen exakt durch die Zahl $^1/_3$ angegeben wird, oder durch einen Dezimalbruch, der mit einer sehr großen Zahl von Stellen 3 beginnt und dann irgendwie anders weitergeht? Die Physik behauptet, daß sich im leeren Raume elektromagnetische Vorgänge abspielen; wie aber sollte das je durch Beobachtung festgestellt werden können? Dazu müßte man doch irgendwelche Apparate an die Stelle bringen, wo man das feststellen will, dann aber ist dort nicht mehr leerer Raum, denn dann sind ja diese Apparate dort! Vor vielen Jahren, auf einem mit einem Freunde unternommenen Spaziergange im Walde, machten wir, indem wir dem Treiben in einem Ameisenhaufen zusahen, die scherzhafte Bemerkung, die Zoologie könne doch gar nicht davon sprechen, wie sich Ameisen verhalten, sie könne nur davon sprechen, wie sich Ameisen verhalten, wenn Menschen ihnen zusehen; das war ein Scherz, aber es liegt viel Ernst in diesem Scherze: jeder Vorgang wird irgendwie dadurch gestört, daß man ihn beobachtet; die Physik aber spricht vom ungestörten Vorgange — daß das nicht eine zu vernachlässigende Spitzfindigkeit, sondern von prinzipieller Bedeutung ist, wird durch die neueste Entwicklung der Physik in klares Licht gerückt.

Aber wir brauchen gar nicht so weit zu gehen, um Beispiele von physikalischen Sätzen zu finden, die prinzipiell nicht durch Beobachtung bestätigt werden können, weil sie einen unkonstituierbaren Term enthalten. Das Wort „alle“, das doch irgendwie in jedem Naturgesetze auftritt, ist — abgesehen von dem Falle, wo ich die Individuen aufzähle, die unter „alle“ gemeint sind, in welchem Falle das Wort „alle“ nur eine an sich überflüssige Abkürzung darstellt — ein solcher unkonstituierbarer Term, dem nichts Beobachtbares entspricht. Wie sollte jemals durch Beobachtung festgestellt werden, daß wirklich *alle* Körper sich durch Erwärmung ausdehnen? Wie sollte durch Beobachtung auch nur festgestellt werden, daß *alle* Amseln schwarz sind? Denn wären durch Zufall selbst alle Amseln der Welt auf ihre Farbe beobachtet worden, so könnten wir doch nicht behaupten, alle Amseln seien schwarz, weil wir nie wissen könnten, daß die beobachteten Amseln alle Amseln sind. Der Term „alle“ ist also unkonstituierbar, jedes Naturgesetz somit — wie wir ja übrigens schon weiter oben festgestellt haben — ein prinzipiell nicht durch Beobachtung bestätigbarer Satz.

Wir sehen also, daß die Forderung *Machs,* es müßten alle Sätze aus der Wissenschaft entfernt werden, in denen unkonstituierbare Terme vorkommen, undurchführbar ist: es würden nicht nur, wie *Mach* das wollte, die Sätze über Moleküle, Atome etc. verschwinden, sondern die ganze Wissenschaft würde zusammenstürzen.

Wenn wir aber zugeben müssen, daß auch solche Sätze Bürgerrecht in der Wissenschaft genießen, die unkonstituierbare Terme enthalten, die prinzipiell nicht durch Beobachtung bestätigt werden können, ist damit nicht wieder der Metaphysik Tür und Tor geöffnet? Wir müssen uns also die Frage vorlegen: was ist der Sinn der legitimen wissenschaftlichen Sätze, die unkonstituierbare Terme enthalten? Wodurch unterscheiden sie sich von den illegitimen. metaphysischen Sätzen?

Jedesmal, wenn man unkonstituierbare Terme in die Wissenschaft einführt, muß man ihnen eine Gebrauchsanweisung mitgeben, man muß Regeln angeben, wie mit ihnen zu operieren ist, wie Sätze, in denen sie vorkommen, in andere Sätze transformiert werden sollen. Und diese Regeln müssen derart sein, daß wir schließlich auf Sätze kommen, in denen kein unkonstituierbarer Term mehr vorkommt, die durch Beobachtung unmittelbar bestätigt oder widerlegt werden können.[16] Mit den Operationsregeln für den Term „alle" beschäftigt sich ein eigenes Kapitel der Logik. Die wichtigste dieser Regeln lautet: „Was für alle gelten soll, soll auch für jedes einzelne gelten." Sage ich also „Alle Amseln sind schwarz" und „Der Vogel, der auf diesem Baume sitzt, ist eine Amsel", so habe ich — zufolge der Operationsregeln für das Wort „alle" — mitgesagt: „Der Vogel, der auf diesem Baume sitzt, ist schwarz", und damit bin ich auf einen Satz gekommen, der unmittelbar durch Beobachtung bestätigt oder widerlegt werden kann: wird er durch die Beobachtung bestätigt, so kann ich bei den zuerst ausgesprochenen beiden Sätzen verbleiben; wird er durch die Beobachtung widerlegt, so kann ich nicht — ohne Renonce zu machen — auf den beiden ersten Sätzen beharren; wird dann etwa auch noch der Satz „Der Vogel, der auf diesem Baume sitzt, ist eine Amsel" durch Beobachtung bestätigt, so muß ich den Satz „Alle Amseln sind schwarz" fallen lassen.

Die Einführung unkonstituierbarer Terme ist also an sich noch nicht Metaphysik; sie ist für die Wissenschaft unentbehrlich, und sie ist durchaus legitim, wenn diesen Termen Gebrauchsanweisungen mitgegeben werden, auf Grund derer aus Sätzen, in denen diese Terme vorkommen, Sätze gewonnen werden können, in denen sie nicht mehr vorkommen, und die durch Beobachtung kontrollierbar sind. Fehlt eine solche Gebrauchsanweisung, oder reicht sie nicht aus zur Wegschaffung der unkonstituierbaren Terme, dann freilich treibt man Metaphysik. Die legitimen Sätze der Wissenschaft mit unkonstituierbaren Termen sind wohlgedecktem Papiergeld vergleichbar, das bei der Nationalbank jederzeit in Gold umgewechselt werden kann — die metaphysischen Sätze gleichen ungedecktem Papiergeld, für das niemand Gold oder Waren gibt.

Obwohl nun die Terme „Molekel", „Atom", „Elektron" etc. unkonstituierbar sind, so ist ihre Verwendung in der Physik doch durchaus legitim: Die kinetische Gastheorie z. B. geht aus von Sätzen über das Verhalten von Molekeln und gewinnt aus ihnen durch geeignete Umformungen und Interpretationsregeln Aussagen über das Verhalten konkreter Gase, in denen der Term „Molekel" nicht mehr vorkommt, und die durch Beobachtung kontrollierbar sind. Die Elektronentheorie geht aus von Sätzen über das Verhalten von Elektronen und Protonen und gewinnt daraus in ähnlicher Weise Aussagen über das Aussehen von Spektren, in denen die Terme

„Elektron“ und „Proton“ nicht mehr vorkommen, und die durch Beobachtung kontrollierbar sind.

Wenn wir so, entgegen der Meinung von *Mach,* feststellen, daß die Verwendung von „Atomen“ etc. in der Physik durchaus legitim und nicht metaphysich ist, so gilt dies keineswegs von gewissen „philosophischen“ Erörterungen über Atome. Da findet man die Argumentation: qualitative Änderungen einer Substanz seien für uns unverstehbar, verstehen können wir nur Lageänderungen an sich unveränderlicher Substanzen; da die sinnlich wahrnehmbaren Körper qualitative Änderungen zeigen (aus Eis wird Wasser, aus Wasser wird Dampf), so müssen wir annehmen, die sinnlich wahrnehmbaren Körper bestehen aus sinnlich nicht wahrnehmbaren, völlig unveränderlichen Atomen, und was wir als qualitative Änderungen der Körper wahrnehmen, seien in Wirklichkeit nur Lageänderungen der Atome. Diese Argumentation scheint uns ganz sinnleer. Wir können überhaupt keine Tatsache verstehen, weder die Veränderung noch das Beharren einer Qualität, weder Bewegung noch Ruhe eines Körpers. Verstehen können wir eine tautologische Umformung, aber nie etwas Beobachtbares: „verstehen“ bezieht sich auf das Denken, sowie „sehen“ auf Farben, „hören“ auf Töne; da wir aber die Tatsachen nicht durch Denken, sondern immer nur durch Beobachtung erfassen, können wir keine Tatsache verstehen, so wie wir keine Farbe hören, keinen Ton sehen können; und das rührt nicht daher, daß unsere Ohren, unsere Augen zu schlecht sind — auch mit den feinsten Ohren könnten wir keine Farben hören, auch mit den schärfsten Augen könnten wir nicht Töne sehen; und ebenso können wir nicht Tatsachen verstehen — nicht weil unser Denken dazu zu schwach wäre, sondern weil Denken und Tatsachen nichts mit einander zu tun haben.

Die eben von uns als sinnleer erkannte Argumentation wird verwendet zur Begründung der These, die Welt, die uns unsere Sinne zeigen, sei bloßer *Schein,* wahres *Sein,* wahre *Realität* komme nur den Atomen und ihren Lageänderungen zu. Dies ist ein Musterbeispiel eines metaphysischen und darum sinnleeren Satzes. Er enthält die unkonstituierten Therme: „Schein“ und „Sein (Realität)“, ohne daß irgend eine Gebrauchsanweisung für diese Terme gegeben wird, auf Grund derer wir von der angeführten These zu Sätzen gelangen könnten, die durch die Beobachtung überprüfbar wären. Ich will damit nicht sagen, daß es nicht auch eine legitime Verwendung der Terme „Schein“, „Realität“ gibt; es hat seinen guten Sinn, wenn ich die gesehene Knickung eines in Wasser getauchten Stockes für bloßen Schein erkläre: es ist damit etwa gemeint, daß der Tastsinn diese Knickung nicht aufzeigt; es hat seinen guten Sinn, wenn ich eine von mir halluzinierte Gestalt für bloßen Schein erkläre: es ist damit gemeint, daß andere Leute, die sie meiner Meinung nach nomalerweise unbedingt hätten sehen müssen, versichern, sie haben sie nicht gesehen. Und immer, wenn das Wort „Schein" in legitimer Weise verwendet wird, wenn in legitimer Weise gewisse Wahrnehmungen als „scheinbar", als „irreal" bezeichnet werden, geschieht es vermöge eines Vergleiches dieser Wahrnehmungen mit anderen Wahrnehmungen. Metaphysik, illegitim, sinnleer aber ist es, *alle* Wahrnehmungen für bloßen Schein zu erklären — denn woran gemessen sollen sie bloßer Schein sein? Das Denken — wie dies die rationalistische Philosophie wollte — kann diesen Maßstab nicht abgeben; denn das Denken hat mit der Wahrnehmung nichts zu tun und kann deshalb über sie auch nichts ausmachen.

VI.

Ich möchte diese Ausführungen nicht schließen, ohne ein großes Problem wenigstens gestreift zu haben: das *Wahrheitsproblem.* Die alte, metaphysische Auffassung ist da etwa die: es gibt eine Realität, eine Welt wahren Seins, und eine Aussage ist wahr, wenn sie übereinstimmt mit dem, was in dieser Realität wirklich statt hat; die Aussage des Gravitationsgesetzes z. B. ist wahr, wenn sich in der Realität je zwei Körper wirklich so anziehen, wie dieses Gesetz es behauptet; leider ist uns aber diese Realität nicht ohneweiters zugänglich, sodaß wir nicht recht in die Lage kommen, festzustellen, daß ein Satz wahr ist; aber das ist unser menschliches Pech, am Wesen der Sache wird dadurch nichts geändert.

Entgegen dieser metaphysischen Auffassung, Wahrheit bestehe in der — doch nicht feststellbaren — Übereinstimmung mit der Realität, bekennen wir uns zur *pragmatistischen* Auffassung: Wahrheit eines Satzes besteht in seiner *Bewährung.*[17]) Freilich wird dadurch die Wahrheit ihres absoluten, ewigen Charakters entkleidet, sie wird relativiert, sie wird vermenschlicht, aber der Wahrheitsbegriff wird *anwendbar!* Und welchem Zwecke könnte ein Wahrheitsbegriff dienen, der nicht anwendbar ist?

Worin nun besteht die Bewährung eines Satzes? *H. Poincaré* hat gesagt — und er hat darin gewiß recht — das Wesen der Naturwissenschaft bestehe darin, daß sie Voraussagen macht, und zwar Voraussagen über unmittelbar Beobachtbares. Indem ich das Gravitationsgesetz ausspreche, mache ich implizit die Voraussage: der Stein, den ich jetzt schleudere, wird sich so und so bewegen, die Planeten werden morgen um neuen Uhr abends — falls der Himmel nicht bewölkt ist — an den und jenen Stellen des Himmels zu sehen sein.

Insolange nun die Voraussagen, die aus einem Satze der Naturwissenschaft fließen, zutreffen, oder wenigstens in der überwiegenden Mehrzahl der Fälle zutreffen, bewährt sich dieser Satz, er wird als wahr bezeichnet und es wird an ihm festgehalten (wie dies beim Gravitationsgesetze bis vor kurzem der Fall war); mehren sich aber Fälle, in denen diese Voraussagen nicht zutreffen, so bewährt sich der Satz nicht, er wird als falsch bezeichnet und fallen gelassen (wie dies mit dem Gravitationsgesetz neuerer Zeit geschah). Man wende hingegen nicht ein, dieser pragmatistische Wahrheitsbegriff sei nicht der wahre Wahrheitsbegriff; das Gravitationsgesetz sei eben immer falsch gewesen, und die Physiker hätten sich nur getäuscht, als sie es wegen seiner Bewährung lange Zeit für wahr hielten. Wer so argumentiert, verwendet den Term „wahr" in illegitimer, metaphysischer Weise, er ist nicht imstande zu sagen, unter welchen wirklich kontrollierbaren Umständen er bereit wäre zu behaupten, „Das Gravitationsgesetz ist wahr".

Poincaré war der Meinung, gerade dadurch, daß sie Voraussagen mache, unterscheide sich die Naturwissenschaft von der Geschichtswissenschaft; wenn der Historiker sage: „Hier hat Johann ohne Land geweilt; das ist eine Tatsache, dafür gebe ich alle Hypothesen der Welt hin", so antworte der Naturwissenschaftler: „Hier hat Johann ohne Land geweilt? Das ist mir ganz egal, er wird nie wieder hier weilen". So sehr Poincaré darin recht hat, daß das Wesen der Naturwissenschaft im Voraussagen besteht, so wenig hat er — glaube ich — darin recht, daß sie sich dadurch fundamental von der Geschichtswissenschaft unterscheidet. Auch die Aussage: „Hier hat Johann ohne Land geweilt" ist in letzter Linie eine Voraussage, genauer gesagt: eine Anweisung, Voraussagen zu machen, die sich bewähren kann oder nicht, ganz wie ein Satz der Naturwissenschaft; es dreht sich um Voraussagen etwa der Art: bei erneuter, genauerer Durchforschung der vorhandenen Quellen, bei Auffindung neuer Quellen, bei besserer Kenntnis der Gesetzmäßigkeiten im Ablauf historischer Ereignisse, werden die Menschen, die sich damit beschäftigen, immer wieder sagen: Hier hat Johann ohne Land geweilt. Tatsächlich ist die Bestätigung solcher Voraussagen durch die Beobachtung, ist diese Bewährung das einzige Kriterium für die Wahrheit eines historischen Satzes und daher auch der Sinn dieser Wahrheit. Denn ebensowenig wie bei einem naturwissenschaftlichen Satze kann bei einem historischen Satze das Kriterium der Wahrheit Übereinstimmung mit der Realität sein. Es ist doch nicht so, daß sich die historischen Tatsachen irgendwo — in der Welt der Ideen, im Reiche der Mütter — aufbewahrt finden, wie in einem Museum, und man brauchte dort nur nachzuschauen, um festzustellen ob ein historischer Satz wahr oder falsch ist, nur daß uns armen Menschen leider der Zutritt zu diesem Museum verwehrt ist! Es gibt also keinen prinzipiellen Unterschied zwischen historischer und naturwissenschaftlicher Wahrheit, es gibt ebenso wenig eine absolute historische, wie eine absolute naturwissenschaftliche Wahrheit, Kriterium für die eine wie für die andere ist die Bewährung; ein historischer Satz handelt ebenso viel oder ebenso wenig von Tatsachen, wie ein naturwissenschaftlicher, er ist ebenso sehr oder ebenso wenig Hypothese, wie ein naturwissenschaftlicher. Und ebenso wie jedes Naturgesetz enthält auch jeder historische Satz einen unkonstituierbaren Term: nämlich die grammatikalische Form der Vergangenheit.

Auch darin kann man nicht eine prinzipielle Trennungslinie zwischen historischen und physikalischen Wissenschaften finden wollen, daß die historischen Wissenschaften von menschlichem Verhalten handeln, während menschliches Verhalten in die Physik nicht eingehe. Ganz abgesehen davon, daß der Physiker vergangenen Beobachtungen, den eigenen wie denen anderer, ganz genau so gegenübertritt, wie der Historiker seinen Quellen — auch die Voraussagen, an denen sich ein physikalischer Satz zu bewähren hat, beziehen sich größtenteils auf menschliches Verhalten. Diese Voraussagen haben nicht nur die Form: „wenn du ins Spektroskop schaust, wirst du eine gelbe Linie sehen" und es ist nicht so, daß wenn ich hineinschaue und die gelbe Linie nicht sehe, schon die Nichtbewährung des physikalischen Satzes, aus dem die Vorhersage floß, festgestellt würde — auch ein Blinder kann schließlich Physik treiben und es wird ihm nicht einfallen, alle physikalischen Sätze zu leugnen, aus denen Voraussagen über Farbwahrnehmungen fließen. Zu den Voraussagen, an denen ein physikalischer Satz sich zu bewähren hat, gehören eben auch solche wie: „Wenn du einen andern Menschen veranlaßt, in das Spektroskop zu schauen, so wirst du hören, daß er sagt: ich sehe eine gelbe Linie" und viele ähnliche.

Wo also sollte man eine prinzipielle Trennungslinie zwischen Physik, Geschichte, Soziologie, Psychologie ziehen? Alle diese Disziplinen sind völlig miteinander verflochten, sie alle werden prinzipiell nach derselben Methode getrieben, in ihnen allen ist Kriterium der Wahrheit die Bewährung — es gibt, wie wir schon eingangs sagten, nur eine Wissenschaft: die *Einheitswissenschaft*.

Zweifrontenkrieg

Damit bin ich am Ende meiner Ausführungen. Ich weiß wohl, daß das, was ich vorbrachte, vielfach bestritten wird; denn wir stehen mit unseren Ansichten in einem schweren Zweifrontenkampf: gegen den sogenannten gesunden Menschenverstand einerseits, dem unsere Ansichten paradox erscheinen, weil sie ungewohnt sind (aber der „gesunde" Menschenverstand würde besser der träge Menschenverstand heißen, denn er ist nichts anderes als der Niederschlag alter, bequem und deshalb lieb gewordener Denkgewohnheiten und wehrt sich gegen alles Ungewohnte), und gegen den angeblichen Tiefsinn der Metaphysik, der in Wirklichkeit Un-Sinn ist. Aber wenn auch die Überzahl der Gegner groß ist und sie ihre Stellungen durch Jahrtausende ausgebaut haben, so wissen wir doch, daß unsere Gedanken siegreich vordringen, und unser Leitstern im Kampfe bleiben die Worte, die *Boltzmann* seiner Mechanik voranstellte:

Bring vor, was wahr ist;
schreib so, daß es klar ist
und verficht's, bis es mit dir gar ist!

ANMERKUNGEN

Diese Arbeit gibt zwei Vorträge wieder, die im Frühjahr 1932 in einem Vortragszyklus zugunsten der Errichtung eines Grabdenkmales für *Ludwig Boltzmann* und im Herbst 1932 im *Verein Ernst Mach* in Wien gehalten wurden.

[1]) Diese These von der „Einheitswissenschaft" steht in Gegensatz zur Auffassung, die Wissenschaften zerfielen in Naturwissenschaften und Geisteswissenschaften, die sich prinzipiell gänzlich verschiedener Methoden bedienen. Gibt man dieser These die Form, jeder sinnvolle Satz über Tatsachen lasse sich in der Sprache der Physik ausdrücken, so wird sie als „Physikalismus" bezeichnet. Näheres hierüber:

O. Neurath, „Physikalismus" in Scientia 1931. S. 297 ff.

O. Neurath. „Physikalism: The Philosophy of the Viennese Circle". The Monist. Oktober 1931, Seite 618.

O. Neurath. „Einheitswissenschaft und Psychologie", Einheitswissenschaft, Heft 1; Gerold & Co. Wien 1933.

R. Carnap. „Die physikalische Sprache als Universalsprache der Wissenschaft". Erkenntnis. Bd. II. S. 433. 1932.

Entsprechend der Tendenz der Sammlung „Einheitswissenschaft", der *ersten Orientierung* des Lesers zu dienen, weicht die vorliegende Schrift möglichst wenig von der üblichen „inhaltlichen" Sprechweise ab, will dadurch aber keineswegs einen Gegensatz zu *Carnaps* Lehre von den Vorteilen einer „formalen" Sprechweise betonen.

[2]) *Kant*. Kritik der reinen Vernunft. Herausgeber Th. Valentiner, 12. Auflage, Meiner, Phil. Bibl. Bd. 37, Seite 95. (Allgem. Anmerkungen zur transzendentalen Ästhetik.)

„Wir haben also sagen wollen: daß alle unsere Anschauung nichts als die Vorstellung von Erscheinung sei; daß die Dinge, die wir anschauen, nicht das an sich selbst sind, wofür wir sie anschauen, noch ihre Verhältnisse so an sich selbst beschaffen sind, als sie uns erscheinen; und daß, wenn wir unser Subjekt oder auch nur die subjektive Beschaffenheit der Sinne überhaupt aufheben, alle die Beschaffenheit, alle Verhältnisse der Objekte im Raum und Zeit, ja selbst Raum und Zeit verschwinden würden, und als Erscheinungen nicht an sich selbst, sondern nur in uns existieren können. Was es für eine Bewandtnis mit den Gegenständen an sich und abgesondert von all dieser Rezeptivität unserer Sinnlichkeit haben möge, bleibt uns gänzlich unbekannt... Wenn wir diese unsere Anschauung auch zum höchsten Grade der Deutlichkeit bringen könnten, so würden wir dadurch der Beschaffenheit der Gegenstände an sich selbst nicht näher kommen. Denn wir würden auf alle Fälle doch nur unsere Art der Anschauung, d. i. unsere Sinnlichkeit vollständig erkennen und diese immer nur unter den dem Subjekt ursprünglich anhängenden Bedingungen von Raum und Zeit; was die Gegenstände an sich selbst sein mögen, würde uns durch die aufgeklärteste Erkenntnis der Erscheinung derselben, die uns allein gegeben ist, doch niemals bekannt werden."

Kant. Aus „Anhang": „Von der Amphibolie der Reflexionsbegriffe ...", Kritik d. rein. Vern., Herausgeber Th. Valentiner, 12. Auflage, Meiner, Phil. Bibl. Bd. 37, Seite 296 ff.

„Was die Dinge an sich sein mögen, weiß ich nicht und brauche es nicht zu wissen, weil mir doch niemals ein Ding anders als in der Erscheinung vorkommen kann." Oder Kant: „Prolegomena" zitiert nach Ausgabe 1749, § 36: „Wie ist Natur in materieller Bedeutung, nämlich der Anschauung nach, als der Inbegriff der Erscheinung, wie ist Raum und Zeit und das, was beide erfüllt, der Gegenstand der Empfindung, überhaupt möglich? Die Antwort ist: vermittels der Beschaffenheit unserer Sinnlichkeit, nach welcher sie auf die ihr eigentümliche Art, von Gegenständen, die ihr an sich selbst unbekannt, und von jenen Erscheinungen ganz unterschieden sind, gerührt wird."

[3]) *René Descartes*. „Betrachtungen über die Grundlagen der Philosophie", Reklam-Ausg. Seite 26: „Alles nämlich, was ich bis heute für das Allerwahrste hingenommen habe, empfing ich unmittelbar oder mittelbar von den *Sinnen;* diese aber habe ich bisweilen auf Täuschungen ertappt, und es ist eine Klugheitsregel, niemals denen volles Vertrauen zu schenken, die uns auch nur ein einziges Mal getäuscht haben." Seite 44: „So erfasse ich also, das, was ich mit den Augen zu sehen meinte, in Wahrheit nur durch das *Urteilsvermögen,* welches meinem Geiste innewohnt." Seite 46: „Ich weiß jetzt, daß die Körper nicht eigentlich von den Sinnen oder von dem Vorstellungsvermögen, sondern von dem Ver-

stande erfaßt werden, und zwar nicht, weil wir sie berühren und sehen, sondern lediglich weil wir sie *denken* ...."
Leibniz. Nouveaux Essais IV, Kap. IV, § 5.

„Übrigens ruht die Grundlage unserer Sicherheit in betreff der universellen und ewigen Wahrheiten in den Ideen selbst, unabhängig von den Sinnen, wie denn auch die reinen und intelligiblen Ideen in keiner Weise von den Sinnen abhängen, z. B. die des Seins, des Einen, des Selben etc. Aber die Ideen der Sinnenqualitäten, wie der Farbe, des Geruches etc. (die in der Tat nur Phantome sind) kommen von den Sinnen, d. h. aus unseren verworrenen Wahrmungen."
Leibniz, Nouveaux Essais IV, Kap. XVII, § 3.

„Die Fähigkeit, die diese Verkettungen der Wahrheiten erfaßt, oder die Fähigkeit, zu denken, wird Vernunft genannt.... Diese Fähigkeit nun wurde hienieden einzig und allein dem Menschen zuteil, nicht aber den übrigen Geschöpfen; denn ich habe schon oben gezeigt, daß der Schatten von Vernunft, der sich bei den Tieren zeigt, lediglich das Erwarten eines ähnlichen Ereignisses in einem Falle ist, der einem vergangenen Falle ähnlich ist, ohne daß sie wissen, ob derselbe Grund vorlag. Und die Menschen selbst handeln nicht anders in den Fällen, wo sie nur empirisch sind. Aber sie erheben sich über die Tiere, insoferne sie die Verkettungen der Wahrheiten sehen; die Verkettungen, sage ich, die selbst wieder ewige und universelle Wahrheiten darstellen."

[4]) *John Locke.* Versuch über den menschlichen Verstand, I. Bd. Meiner 1913. II. Buch, Seite 101.

„Wir wollen also annehmen, der Geist sei, wie man sagt, ein unbeschriebenes Blatt ohne alle Eindrücke, frei von allen Ideen; wie werden ihm diese dann zugeführt? Wie gelangt er zu dem gewaltigen Vorrat von Ideen, womit ihn die geschäftige Phantasie des Menschen, die keine Schranken kennt, in nahezu unendlicher Mannigfaltigkeit beschrieben hat? Von wo hat er das gesamte *Material* für sein Denken und Erkennen? Ich antworte darauf mit einem einzigen Wort: aus der *Erfahrung.* Sie liegt unserem gesamten Wissen zu Grunde; aus ihr leitet es sich letzten Endes her. Unsere Beobachtung, die entweder auf äußere, sinnliche Objekte gerichtet ist, oder auf innere Bewußtseinsvorgänge, die wir wahrnehmen, und über die wir reflektieren, liefert unserem Verstand das gesamte *Material* des Denkens. Dies sind die beiden Quellen der Erkenntnis, aus denen alle Ideen entspringen, die wir haben oder naturgemäß haben können."

[5]) *J. St. Mill.* „System der deduktiven und induktiven Logik". 4. Auflage 1877, Vieweg, Braunschweig. Seite 318.

„Nichtsdestoweniger wird bei näherer Betrachtung erhellen, ... daß in einem jeden Schritt einer arithmetischen oder algebraischen Berechnung eine wirkliche Induktion, eine wirkliche Folgerung von Tatsachen aus Tatsachen enthalten ist; daß dies einfach nur durch die umfassende Natur der Induktion und die daraus folgende äußerste Allgemeinheit der Sprache verdeckt wird..... Die wissenschaftliche Sprache macht also keine Ausnahme von dem Schluß, zu dem wir früher gelangten, daß sogar *die Prozesse der deduktiven Wissenschaften ganz induktiv, und daß ihre ersten Prinzipien Generalisationen aus der Erfahrung sind."* (Gesperrt vom Herausgeber.) Kennzeichnend für Mills Anschauungen sind auch die Kapitelüberschriften: „Alle deduktiven Wissenschaften sind induktiv." „Die Sätze der Arithmetik sind nicht bloße wörtliche Urteile, sondern Generalisationen aus der Erfahrung."
J. St. Mill. Logik I. S. 294.

„Drei Steine in zwei getrennten Teilen und drei Steine in einem einzigen Haufen vereinigt, machen auf unsere Sinne nicht denselben Eindruck, und die Behauptung, daß dieselben Steine vermöge einer bloßen Änderung ihrer Anordnung und ihres Ortes bald den einen, bald den anderen Eindruck erzeugen können, ist nicht ein identischer Satz. Es ist eine Wahrheit, die durch lange und konstante Erfahrung erworben wurde, eine induktive Wahrheit, und auf solchen Wahrheiten beruht die Wissenschaft von den Zahlen. Die grundlegenden Wahrheiten dieser Wissenschaft beruhen alle auf dem Zeugnis der Sinne."

6) Vgl. z. B. *Kant.* Kritik der reinen Vernunft. Einleitung V, 2. Seite 62.
„*Naturwissenschaft* (Phisica) *enthält synthetische Urteile a priori als Prinzipien in sich.* Ich will nur ein paar Sätze zum Beispiel anführen, als den Satz: daß in allen Veränderungen der körperlichen Welt die Quantität der Materie unverändert bleibe, oder daß in aller Mitteilung der Bewegung Wirkung und Gegenwirkung jederzeit einander gleich sein müssen. An beiden ist nicht allein die Notwendigkeit, mithin ihr Ursprung a priori, sondern auch daß sie synthetische Sätze sind, klar."

7) Vgl. *M. Schlick* „Gibt es ein materiales Apriori?" Wissenschaftl. Jahresbericht der phil. Ges. a. d. Univ. Wien 1931/32. S. 55.

8) *Leibniz.* Nouveaux Essais IV, Kap I., § 3.
„Denn der Verstand bemerkt unmittelbar, daß eine Idee nicht die andere ist, daß das Weiße nicht das Schwarze ist."
Leibniz. Nouveaux Essais IV, Kap. II., § 1.
„Die Erkenntnis ist also intuitiv, wenn der Verstand die Übereinstimmung zweier Ideen unmittelbar durch sie selbst, ohne daß eine andere dazukäme, bemerkt. In diesem Falle hat der Verstand keinerlei Mühe, die Wahrheit zu beweisen oder zu prüfen. So, wie das Auge das Licht sieht, sieht der Verstand, daß das Weiße nicht das Schwarze ist, daß ein Kreis nicht ein Dreieck ist, daß drei zwei und eins ist. Diese Erkenntnis ist die klarste, die sicherste, deren die menschliche Schwäche fähig ist; sie wirkt in unwiderstehlicher Weise und gestattet dem Verstande kein Bedenken. Es ist die Erkenntnis, daß die Idee so in unserem Verstande ist, wie wir sie erfassen. Wer eine größere Sicherheit verlangt, weiß nicht, was er verlangt."

9) Daß es sich bei Sätzen wie „kein Gegenstand ist sowohl rot als blau" um Festsetzungen handelt, in welcher Weise die Farbwörter „rot", „blau" etc. verwendet werden sollen, kann man sich daran klar machen, daß an sich auch eine andere Art ihrer Verwendung durchaus denkbar wäre, und gelegentlich auch vorkommt: vielleicht wird mancher von einem gelbgrün gefärbten Gegenstande sagen, er sei sowohl gelb als grün. Bei Tönen ist (aus naheliegenden Gründen) sogar eine solche Sprechweise die allgemein übliche: wenn man einen c-dur-Dreiklang hört, so sagt man, man höre sowohl den Ton c, als den Ton e, als den Ton g. An sich wäre es durchaus denkbar, daß man für jeden Akkord eine eigene Bezeichnung hätte und analog der Festsetzung „kein Gegenstand ist sowohl rot als blau" die Festsetzung träfe: „Die Töne c und e können niemals gleichzeitig gehört weden". — *L. Wittgenstein* hat das so formuliert: Ein Satz wie „Kein Gegenstand ist sowohl rot als blau" gehört zur „Syntax" der Farbwörter.

10) Gewöhnlich wird nach dem Vorgange von *Wittgenstein* das Wort „tautologisch" im engeren Sinne verwendet; er nennt „tautologisch" einen Satz, der durch seine bloße Form wahr ist. Vgl. hiezu Wittgenstein, Tractatus Logico-Philosophicus, Paul Kegan, London 1922, Seite 98: „4.464: Die Wahrheit der Tautologie ist gewiß, des Satzes möglich, der Kontradiktion unmöglich" und Seite 156: „6.12: Daß die Sätze der Logik Tautologien sind, das zeigt die formalen — logischen — Eigenschaften der Sprache, der Welt."
Wittgenstein war es, der als erster klar die Bedeutung dieses Begriffes auseinandersetzte und dadurch entscheidend in die Entwicklung der hier vorgetragenen Gedankengänge eingriff.

11) Ein vollständiges System des Logikkalküls und des logischen Aufbaues der Mathematik:
A. Wittgenstein B. Russell. Principia Mathematica. Cambridge. 2. Aufl. 1925.
Die leitenden Grundgedanken dieses Werkes findet man dargelegt in:
B. Russell. Einführung in die mathemat. Philosopie... München, Drei Maskenverl. 1923.
R. Carnap. Abriß der Logistik. Schriften zur wissenschaftlichen Weltauffassung, Bd. 2, J. Springer 1929.

[12]) *Kant.* Kritik der reinen Vernunft. Einleitung V. 1. (Oben zitierte Ausgabe).
„*Mathematische Urteile sind insgesamt synthetisch.* Dieser Satz scheint den Bemerkungen der Zergliederer der menschlichen Vernunft bisher entgangen, ja allen ihren Vermutungen gerade entgegengesetzt zu sein, ob er gleich unwidersprechlich gewiß und in der Folge sehr wichtig ist. Denn weil man fand, daß die Schlüsse der Mathematiker alle nach dem Satze des Widerspruchs fortgehen (welches die Natur einer jeden apodiktischen Gewißheit erfordert), so überredete man sich, daß auch die Grundsätze aus dem Satze des Widerspruchs, anerkannt würden; worin sie sich irrten; denn ein synthetischer Satz kann allerdings nach dem Satze des Widerspruchs eingesehen werden, aber nur so, daß ein anderer synthetischer Satz vorausgesetzt wird, aus dem es gefolgert werden kann, niemals aber an sich selbst.... Man sollte anfänglich zwar denken: daß der Satz $7+5=12$ ein bloß analytischer Satz sei, der aus dem Begriffe einer Summe von Sieben und Fünf nach dem Satz des Widerspruchs erfolge. Allein, wenn man es näher betrachtet, so findet man, daß der Begriff der Summe von 7 und 5 nichts weiter enthalte, als die Vereinigung beider Zahlen in eine einzige, wodurch ganz und gar nicht gedacht wird, welche diese einzige Zahl sei, die beide zusammengefaßt. Der Begriff von Zwölf ist keineswegs dadurch schon gedacht, daß ich mir jene Vereinigung von Sieben und Fünf denke, und ich mag meinen Begriff von einer solchen möglichen Summe noch so lang zergliedern, so werde ich doch darin die Zwölf nicht antreffen. Man muß über diese Begriffe hinausgehen, indem man die Anschauung zu Hilfe nimmt, die einem von beiden korrespondiert, etwa seine fünf Finger, oder (wie Segner in seiner Arithmetik) fünf Punkte, und so nach und nach die Einheiten der in der Anschauung gegebenen Fünf zu dem Begriffe der Sieben hinzutun.... Der arithmetische Satz ist also jederzeit synthetisch; welches man desto deutlicher inne wird, wenn man etwas größere Zahlen nimmt, da es dann klar einleuchtet, daß, wir möchten unsere Begriffe drehen und wenden, wie wir wollen, wir, ohne die Anschauung zu Hilfe zu nehmen, vermittelst der bloßen Zergliederung unserer Begriffe die Summe niemals finden könnten."

[13]) *H. Poincaré.* Wissenschaft und Hypothese. I. Kapitel, I.
„Wenn im Gegenteil alle Behauptungen, welche die Mathematik aufstellt, sich auseinander durch die formale Logik ableiten lassen, wieso besteht die Mathematik dann nicht in einer ungeheuren Tautologie? Der logische Schluß kann uns nichts wesentlich Neues lehren, und wenn alles vom Prinzipe der Identität ausgehen soll, so müßte sich auch alles darauf zurückführen lassen. Wird man aber zugeben, daß alle die Lehrsätze, welche so viele Bände füllen, nichts anderes leisten, als auf Umwegen zu sagen, daß A gleich A ist!"

[14]) *Mach.* „Mechanik in ihrer Entwicklung." 8. Auflage. Seite 466.
„Nicht jede bestehende wissenschaftliche Theorie ergibt sich so natürlich und ungekünstelt. Wenn z. B. chemische, elektrische, optische Erscheinungen durch Atome erklärt werden, so hat sich die Hilfsvorstellung der Atome nicht nach dem Prinzip der Kontinuität ergeben, sie ist vielmehr für diesen Zweck eigens erfunden worden. Atome können wir nirgends wahrnehmen, sie sind wie alle Substanzen Gedankendinge. Ja, den Atomen werden zum Teil Eigenschaften zugeschrieben, welche allen bisher beobachteten widersprechen. Mögen die Atomtheorien immer geeignet sein, eine Reihe von Tatsachen darzustellen, die Naturforscher, welche Newtons Regeln des Philosophierens sich zu Herzen genommen haben, werden diese Theorien nur als provisorische Hilfsmittel gelten lassen und einen Ersatz durch eine natürliche Anschauung anstreben."

[15]) *Ludwig Boltzmann.* „Populäre Schriften." II. Auflage. Barth, 1911. Seite 142:
„Endlich wäre es nicht ein Schaden für die Wissenschaft, wenn man nicht noch heute die gegenwärtigen Anschauungen der Atomistik mit gleichem Eifer pflegte, wie die der Phänomenologie? Die Beantwortung dieser Fragen in dem der Atomistik günstigen Sinne bezeichne ich schon hier als das Resultat der folgenden Betrachtungen: Die Differentialgleichungen der mathematisch-physikalischen Phänomenologie sind offenbar nichts als Regeln für die Bildung und Verbindung von Zahlen und geometrischen

Begriffen, diese aber sind wieder nichts anderes als Gedankenbilder, aus denen die Erscheinungen vorhergesagt werden können. Genau dasselbe gilt auch von den Vorstellungen der Atomistik, so daß ich in dieser Beziehung nicht den mindesten Unterschied zu erkennen vermag. Überhaupt scheint mir von einem umfassenden Tatsachengebiete niemals eine direkte Beschreibung, stets nur ein Gedankenbild möglich. Man darf daher nicht mit *Ostwald* sagen, Du sollst Dir kein Bild machen, sondern nur, Du sollst in dasselbe möglichst wenig Willkürliches aufnehmen."

16) Bei dieser unmittelbaren Bestätigung, beziehungsweise Widerlegung spielen diejenigen Sätze eine wichtige Rolle, die Carnap und Neurath als Protokollsätze bezeichnet haben und auf die alle Systemsätze der Realwissenschaft zurückführbar sind.

17) In den Hauptwerken des Pragmatismus:
J. Dewey. Studies in Logical Theorie. 1903. Pg. 106 ff.
„Das was als hinreichend gesichert als Grundlage ferneren Handelns gelten kann, wird als wirklich und wahr betrachtet."
W. James. Der Pragmatismus. 1908. Seite 51.
„(Als wahr gilt) was uns am besten führt, was für jeden Teil des Lebens am besten paßt, was sich mit der Gesamtheit der Erfahrungen am besten vereinigen läßt."

Die Krise der Anschauung

von

HANS HAHN.

Unter allen führenden Philosophen war wohl I. Kant derjenige, der der Anschauung die weittragendste Bedeutung in unserer Erkenntnis zuschrieb. Er ging aus von der gewiß zutreffenden Bemerkung, daß in unserer Erkenntnis zwei entgegengesetzte Momente sich aufs innigste durchdringen: ein passives Moment bloßer Rezeptivität und ein aktives Moment der Spontaneität; wir lesen in der „Kritik der reinen Vernunft" (und zwar zu Beginn des Abschnittes, der überschrieben ist: „Der transcendentalen Elementarlehre zweiter Teil. Die transcendentale Logik"): „Unsere Erkenntniss entspringt aus zwei Grundquellen des Gemüths, deren die erste ist, die Vorstellungen zu empfangen (die Receptivität der Eindrücke), die zweite das Vermögen, durch jene Vorstellungen einen Gegenstand zu erkennen (Spontaneität der Begriffe); durch die erstere wird uns ein Gegenstand gegeben, durch die zweite wird dieser im Verhältniss auf diese Vorstellung (als blosse Bestimmung des Gemüths) *gedacht*. Anschauung und Begriffe machen also die Elemente aller unserer Erkenntniss aus...". Also: passiv verhalten wir uns, indem wir durch die Anschauung Vorstellungen in uns aufnehmen, aktiv, indem wir sie im Denken verarbeiten. Nach Kant sind nun in der Anschauung wieder zwei Bestandteile zu unterscheiden: ein der Erfahrung entstammender, empirischer, aposteriorischer Teil, der den *Inhalt* der Anschauung ausmacht: Farben, Töne, Gerüche, die Empfindungen des Tastsinnes, wie Härte, Weichheit, Rauhigkeit usw., und ein von aller Erfahrung unabhängiger, reiner, apriorischer Teil, der die *Form* der Anschauung ausmacht; und zwar haben wir zwei solche reine Anschauungsformen: den *Raum* als die Anschauungsform unseres *äußeren* Sinnes (vermittels dessen „wir uns Gegenstände als außer uns" vorstellen) und die *Zeit* als die Anschauungsform unseres *inneren* Sinnes, „vermittelst dessen das Gemüth sich selbst oder seinen inneren Zustand anschaut".

Diese reine Anschauung spielt nun nach Kants Auffassung eine äußerst wichtige Rolle in unserer Erkenntnis. Auf reine Anschauung (und nicht etwa auf das Denken) gründet sich seiner

Meinung nach die Mathematik: Die Geometrie, wie sie seit dem Altertum gelehrt wird, handelt von den Eigenschaften des uns in reiner Anschauung völlig exakt gegebenen Raumes; die Arithmetik (die Lehre von den reellen Zahlen) beruht auf der reinen, völlig exakten Anschauung der Zeit. Die reinen Anschauungsformen von Raum und Zeit bilden den apriorischen Rahmen, in den wir alle physikalischen Vorgänge, die uns die Erfahrung liefert, einordnen: jedes physikalische Ereignis hat seine ganz präzise, exakt feststehende Stelle in Raum und Zeit.

Wie plausibel diese Ansichten auch zunächst scheinen mögen und wie sehr sie auch dem Stande der Wissenschaft in den Tagen Kants entsprachen — durch den Weg, den die Wissenschaft seither genommen hat, wurden sie in ihren Grundfesten erschüttert.

Die physikalische Seite der Frage wurde schon in den beiden ersten Vorträgen behandelt, so daß ich mich da auf kurze Andeutungen beschränken kann. Die Ansichten Kants über die Stellung von Raum und Zeit in der Physik entsprechen der Newtonschen Physik, die ja in den Tagen Kants alleinherrschend war und bis in die neueste Zeit alleinherrschend geblieben ist. Einen ersten gewaltigen Stoß erhielt diese Auffassung durch Einsteins Relativitätstheorie: Nach der Kantschen Auffassung haben Raum und Zeit nichts miteinander zu tun; sie entstammen ja auch ganz verschiedenen Quellen: der Raum ist die Anschauungsform des äußeren, die Zeit die des inneren Sinnes; wir haben einen absolut ruhenden Raum und eine von ihm unabhängig dahinfließende absolute Zeit. Die Relativitätstheorie hingegen lehrt: es gibt keinen absoluten Raum und keine absolute Zeit; absolute physikalische Bedeutung hat nur eine Union von Raum und Zeit, die „Welt"[1]).

Einen viel schlimmeren Stoß aber erhielt Kants Auffassung von Raum und Zeit als apriorische Anschauungsformen durch die neueste Entwicklung der Physik. Wir sagten schon, daß nach jener Auffassung jedes physikalische Ereignis seine exakt feststehende Stelle in Raum und Zeit hat. Eine gewisse Schwierigkeit war da immer vorhanden: wir kennen die physikalischen Ereignisse nur durch die Erfahrung, und alle Erfahrung ist unpräzise, jede Beobachtung ist mit Beobachtungsfehlern behaftet; nach dieser alten Auffassung wäre es also so, daß zwar jedes physikalische Ereignis seine exakte Stelle in Raum und Zeit *hat*, daß es uns aber prinzipiell unmöglich ist, diese exakte Stelle *kennenzulernen*. Darin steckt zweifellos eine gewisse Unstimmigkeit. Betrachten wir etwa ein Kreidestück; sobald eine Längeneinheit gewählt ist, wird der Abstand zweier Punkte dieses Kreidestückes durch eine ganz präzise reelle Zahl gemessen; denken wir uns für je zwei Punkte des Kreidestückes den Abstand gebildet und nennen den größten aller dieser Abstände den „Durchmesser" des Kreidestückes; bei der Auffassung, daß dieses Kreidestück einen exakt feststehenden Teil des uns in präziser Anschauung gegebenen Raumes einnimmt, wäre nun die Frage: „Ist der Durchmesser dieses Kreidestückes rational oder irrational?" durchaus sinnvoll — aber sie könnte niemals beantwortet werden, denn der Unterschied zwischen rational und irrational ist viel zu fein, als daß er jemals durch Beobachtung festgestellt werden könnte; es gibt also bei dieser Auffassung sinnvolle Fragen, die prinzipiell unbeantwortbar sind, d. h. diese Auffassung ist *metaphysisch*.

Man hat diese Schwierigkeit früher nicht recht ernst genommen; man argumentierte etwa so: wenn auch jede einzelne Beobachtung ungenau, mit Beobachtungsfehlern behaftet ist, so werden doch unsere Beobachtungsmittel immer feiner und feiner; denken wir uns nun ein und dieselbe physikalische Größe immer erneut, durch immer feinere Beobachtungsmittel gemessen, so werden die so ermittelten Werte, deren jeder einzelne unexakt ist, sich unbeschränkt einem ganz bestimmten Grenzwerte annähern, und dieser Grenzwert ist dann der exakte Wert der betreffenden physikalischen Größe. So unbefriedigend diese Argumentation vom philosophischen Standpunkte ist — die neueste Entwicklung der Physik scheint darzutun, daß sie auch aus rein physikalischen Gründen unhaltbar ist: es scheint, daß aus rein physikalischen Gründen die Lokalisation eines Ereignisses in Raum und Zeit nicht mit unbeschränkter Annäherung erfolgen kann; es scheint, daß da aus rein physikalischen Gründen gewisse Genauigkeitsschranken nicht überschritten werden können[2]. Es bleibt also dabei: die Lehre von der exakten Lokalisation der physikalischen Ereignisse in Raum und Zeit ist metaphysisch und somit bedeutungsleer. So erschütternd nun die neueste revolutionäre Entwicklung der Physik auf die meisten auf dogmatisch-metaphysische Lehrmeinungen festgelegten Menschen — einschließlich der meisten Physiker — wirken mußte: für den an empiristischer Philosophie geschulten Denker hat sie nichts Paradoxes; sie erscheint ihm sofort vertraut und er heißt sie willkommen als einen gewaltigen Schritt nach vorwärts auf dem Wege der „Physikalisierung" der Physik, ihrer Säuberung von metaphysischen Elementen.

Nach diesen knappen Andeutungen über die physikalische Seite der Frage wenden wir uns nunmehr dem Gebiete der Mathematik zu, wo der Widerstand gegen Kants Lehre von der reinen Anschauung erheblich früher einsetzte als auf dem Gebiete der Physik: Ich will also von jetzt ab ausschließlich über das Thema „*Mathematik und Anschauung*" sprechen; und auch da will ich einen ebenso wichtigen als schwierigen Fragenkomplex, mit dem Herr Menger im letzten Vortrage dieses Zyklus sich noch beschäftigen wird, gänzlich beiseite lassen: ich werde nicht sprechen von der heftigen und erfolgreichen Opposition gegen Kants These, daß auch die Arithmetik, die Lehre von den Zahlen, auf reiner Anschauung beruht — eine Opposition, die unlösbar verknüpft ist mit dem Namen Bertrand Russell, und die es sich zum Ziel gesetzt hat, darzutun, daß ganz im Gegensatze zu Kants These die Arithmetik durchaus der Domäne des *Denkens*, der Logik angehört[3]. Ich enge also mein Thema weiter ein auf: „*Geometrie und Anschauung*", und will versuchen, zu zeigen, wie es dazu kam, daß auch auf dem Gebiete der Geometrie, die doch zunächst die ureigenste Domäne der Anschauung zu sein scheint, das Vertrauen zur Anschauung erschüttert wurde, so daß sie immer mehr in Mißkredit kam und schließlich auch aus der Geometrie völlig verbannt wurde.

Eines der erregenden Momente für diese Entwicklung war die Entdeckung, daß es, in offenbarem Gegensatz zu dem, was man anschauungsmäßig als sicher angenommen hatte, Kurven gibt, die in keinem Punkte eine Tangente besitzen, oder — was, wie wir sehen werden, auf dasselbe hinauskommt — daß Bewegungen eines Punktes denkbar sind, bei denen der bewegte Punkt in keinem Augenblick eine bestimmte Geschwindigkeit aufweist. Es erregte bei den Mathematikern gewaltigen Eindruck, als der große Berliner Mathematiker C. Weierstrass im Jahre 1861 diese Entdeckung bekanntmachte — heute wissen wir, aus Manuskripten, die in der Wiener Nationalbibliothek aufbewahrt werden, daß diese Tatsache dem österreichischen Philosophen, Theologen und Mathematiker B. Bolzano schon erheblich früher bekannt war. Da es sich dabei um Fragen dreht, die unmittelbar die Grundlagen der von Newton und Leibniz entwickelten *Differentialrechnung* betreffen, so muß ich zunächst einige Worte über die fundamentalen Begriffsbildungen dieser Disziplin vorausschicken[4].

Fig. 1.

Newton ging aus vom Begriffe der *Geschwindigkeit*. Man denke sich einen Punkt, der sich auf einer geraden Linie bewegt, etwa auf der in Fig. 1 gezeichneten Geraden; im Augenblicke t befinde sich der bewegte Punkt etwa an der Stelle q. Was hat man nun unter der Geschwindigkeit des bewegten Punktes in diesem Augenblicke t zu verstehen? Stellt man die Lage des bewegten Punktes in einem zweiten Augenblicke t' fest (er befinde sich in diesem zweiten Augenblicke etwa an der Stelle q'), so kennt man den Weg qq', den er in der zwischen den Augenblicken t und t' verflossenen Zeitspanne zurückgelegt hat. Dividiert man den zurückgelegten Weg qq' durch die Länge der zwischen den Augenblicken t und t' verflossenen Zeitspanne, so erhält man die sogenannte „mittlere Geschwindigkeit" des bewegten Punktes zwischen den Augenblicken t und t'. Diese „mittlere" Geschwindigkeit ist keineswegs die Geschwindigkeit im Augenblicke t selbst (sie kann z. B. sehr groß ausfallen, obwohl die Geschwindigkeit im Augenblicke t sehr gering war — wenn nur der Punkt sich während des größeren Teiles der betrachteten Zeitspanne sehr rasch bewegt); aber, wenn nur der zweite Augenblick t' entsprechend nahe am ersten Augenblicke t gewählt wurde, so wird doch diese mittlere Geschwindigkeit zwischen den Augenblicken t und t' eine gute Annäherung an die Geschwindigkeit im Augenblicke t selbst liefern, und zwar eine um so bessere, je näher der Augenblick t' am Augenblicke t gewählt war. Newtons Erwägung ist nun etwa die: denkt man sich den Augenblick t' immer näher und näher am Augenblicke t gewählt, so wird die mittlere Geschwindigkeit zwischen den Augenblicken t und t' sich unbeschränkt einem ganz bestimmten Werte annähern, sie wird — wie das in der Mathematik ausgedrückt wird — einem bestimmten Grenzwerte zustreben, und dieser Grenzwert ist das, was man die „Geschwindigkeit des bewegten Punktes im Augenblicke t" nennt. Also: Geschwindigkeit im Augenblicke t ist der Grenzwert, dem die mittlere Geschwindigkeit zwischen den Augenblicken t und t' zustrebt, wenn der Augenblick t' unbegrenzt dem Augenblicke t angenähert wird.

Leibniz ging vom sogenannten *Tangentenproblem* aus. Denken wir uns eine Kurve gegeben (Fig. 2) und fragen wir, welche Steigung sie in einem ihrer Punkte, etwa im Punkte p, gegen die Horizontale aufweist. Wir wählen auf der Kurve einen zweiten Punkt p' und bilden auch hier wieder zunächst die „mittlere Steigung" der Kurve zwischen den Punkten p und p', die man erhält, indem man die bei Durchlaufung des Kurvenstückes von p nach p' gewonnene Höhe (in Fig. 2 gegeben durch die Strecke $p'' p'$) dividiert durch die Horizontalprojektion des zurückgelegten Weges (in Fig 2 gegeben durch die Strecke $p p''$, die angibt, um wieviel man bei Durchlaufung des Kurven-

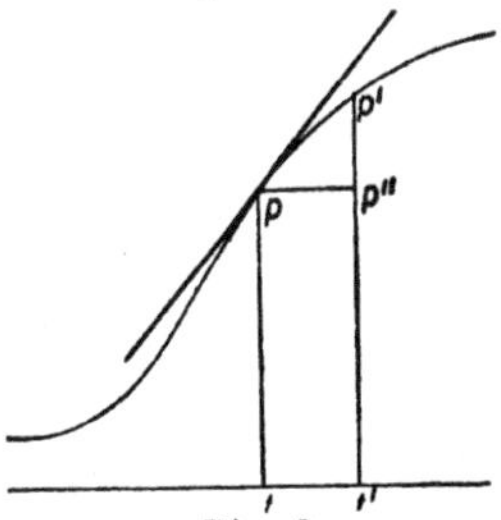

Fig. 2.

stückes von p nach p' in horizontaler Richtung weitergekommen ist). Diese mittlere Steigung der Kurve zwischen den Punkten p und p' ist nun zwar nicht identisch mit ihrer Steigung im Punkte p selbst (im Falle der Fig. 2 ist ersichtlich die Steigung im Punkte p größer als die mittlere Steigung zwischen p und p'), aber sie wird doch eine gute Annäherung an die Steigung der Kurve im Punkte p selbst liefern, wenn nur der Punkt p' entsprechend nahe am Punkte p gewählt war; und diese Annäherung wird um so besser sein, je näher p' an p gewählt wird. Und nun heißt es wieder: Nähert man insbesondere den Punkt p' unbegrenzt dem Punkte p an, so wird die mittlere Steigung der Kurve zwischen den Punkten p und p' einem bestimmten Grenzwerte zustreben, und dieser Grenzwert ist das, was man als die „Steigung der Kurve im Punkte p" bezeichnet. Also: Steigung im Punkte p ist der Grenzwert, dem die mittlere Steigung zwischen den Punkten p und p' zustrebt, wenn der Punkt p' unbegrenzt dem Punkte p angenähert wird. Als „Tangente unserer Kurve im Punkte p" bezeichnet man nun diejenige durch den Punkt p gehende Gerade, die (in ihrem ganzen Verlaufe) dieselbe Steigung aufweist, wie die Kurve im Punkte p.

Die Analogie dieses Verfahrens zur Ermittlung der Steigung einer Kurve mit dem oben auseinandergesetzten Verfahren zur Ermittlung der Geschwindigkeit eines bewegten Punktes springt in die Augen. Und in der Tat, die Aufgabe, die Geschwindigkeit des bewegten Punktes in einem bestimmten Augenblicke zu ermitteln, geht völlig über in die Aufgabe, die Steigung einer Kurve in einem gegebenen Punkte zu ermitteln, wenn man sich eines einfachen Verfahrens bedient, das von den graphischen Fahrplänen der Eisenbahnen her wohl ziemlich allgemein bekannt ist: man trage auf einer horizontalen Geraden (der „Zeitachse") die Werte der Zeit ein, so daß jeder Punkt dieser Geraden einen bestimmten Zeitpunkt repräsentiert, und auf der Geraden der Fig. 1, auf der der betrachtete Punkt sich bewegt fixiere man — ganz nach Belieben — irgendeinen Punkt o; befindet sich der bewegte Punkt im Augenblicke t an der Stelle q, so trage man (vgl. Fig. 2) in dem den Augenblick t repräsentierenden Punkte der Zeitachse senkrecht zu dieser Zeitachse die Strecke $o\,q$ ab; der Punkt p, zu dem man so gelangt, repräsentiert dann in Fig. 2 die Lage des bewegten Punktes im Augenblicke t; denkt man sich das für jeden einzelnen Augenblick durchgeführt, so erhält man als Darstellung der Bewegung unseres Punktes eine Kurve (die „Zeit-Weg-Kurve" des bewegten Punktes), aus der man alle Einzelheiten der Bewegung dieses Punktes ebenso entnehmen kann, wie bei einem Eisenbahnzuge aus seinem graphischen Fahrplane. Offenbar ist nun die mittlere Steigung der Zeit-Weg-Kurve zwischen den Punkten p und p' identisch mit der mittleren Geschwindigkeit des bewegten Punktes zwischen den Augenblicken t und t', und daher die Steigung der Zeit-Weg-Kurve im Punkte p identisch mit der Geschwindigkeit des bewegten Punktes im Augenblicke t. Das ist der einfache Zusammenhang zwischen Geschwindigkeitsproblem und Tangentenproblem; diese beiden Probleme sind also begrifflich nicht voneinander verschieden. Die *Grundaufgabe der Differentialrechnung* ist nun diese: Es sei die Bahn eines bewegten Punktes bekannt; daraus ist seine Geschwindigkeit in jedem Augenblicke zu berechnen — oder: es sei eine Kurve gegeben; in jedem ihrer Punkte ist ihre Steigung zu berechnen (in jedem ihrer Punkte ist ihre Tangente zu finden). Wir halten uns im folgenden an das Tangentenproblem. Alles, was wir über das Tangentenproblem auseinandersetzen werden, überträgt sich nach dem Gesagten ohne weiteres auf das Geschwindigkeitsproblem.

Wir sagten: Wird der Punkt p' auf der betrachteten Kurve unbeschränkt dem Punkte p angenähert, so wird die mittlere Steigung der Kurve zwischen p und p' unbeschränkt einem Grenzwerte zustreben, der dann die Steigung der Kurve im Punkte p selbst angibt. Ist es denn aber sicher wahr, daß die mittlere Steigung zwischen p und p' einem bestimmten Grenzwerte zustrebt, wenn der Punkt p' unbeschränkt an den Punkt p angenähert wird? Bei all den Kurven, mit denen man sich seit altersher üblicherweise beschäftigte, wie Kreisen, Ellipsen, Hyperbeln, Parabeln, Zykloiden usw., ist es, wie die Rechnung zeigt, tatsächlich der Fall; aber es ist nicht bei *jeder* Kurve der Fall, wie wir an einem verhältnismäßig einfachen Beispiele sehen können. Man betrachte die in Fig. 3 angedeutete Kurve. Sie ist eine Wellenlinie, die in der Nähe des Punktes p unendlich viele Wellen aufweist; Wellenlänge wie Amplitude der einzelnen Wellen nehmen bei Annäherung an den Punkt p unbeschränkt ab. Wir wollen versuchen, nach dem oben angegebenen Verfahren die Steigung dieser Kurve im Punkte p zu ermitteln. Wir nehmen also auf der Kurve einen zweiten Punkt p' an und bilden ihre mittlere Steigung zwischen p und p'; legen wir den Punkt p' in den Punkt p_1 (Fig. 3), so fällt die mittlere Steigung zwischen

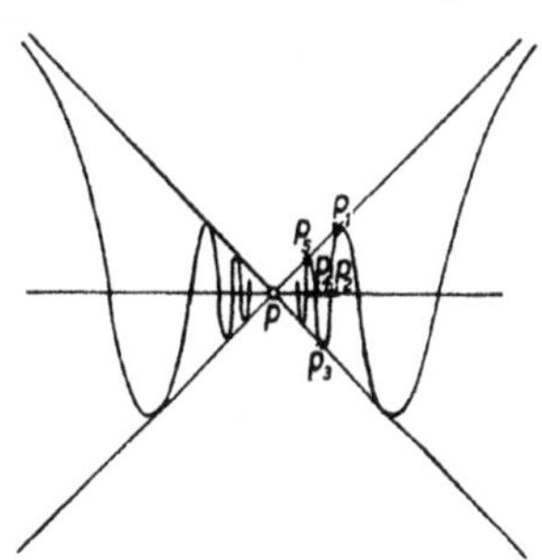

Fig. 3.

p und p_1 gleich 1 aus. Lassen wir den Punkt p' auf der Kurve gegen den Punkt p heranrücken, so nimmt, wie man sieht, die mittlere Steigung zunächst ab; wenn der Punkt p' in p_2 angekommen ist, so ist die mittlere Steigung (zwischen p und p_2) gleich 0 geworden; rückt der Punkt p' auf der Kurve weiter gegen p, so nimmt die mittlere Steigung zwischen p und p' weiter ab, sie wird negativ (wir haben ein mittleres „Gefälle") und sinkt bis zum Werte — 1, wenn der Punkt p' bis p_3: rückt; rückt p' weiter gegen p, so beginnt die mittlere Steigung wieder zu wachsen, sie erreicht wieder den Wert 0, wenn p' bis nach p_4 rückt, wächst durch positive Werte weiter und wird wieder gleich 1, wenn p' in p_5 angekommen ist. Rückt p' auf der Kurve weiter gegen p, so geht dasselbe Spiel nun wieder an: wenn der Punkt p' bei seiner Annäherung gegen p eine volle Welle unserer Wellenlinie durchläuft, so sinkt die mittlere Steigung zwischen p und p' vom Werte 1 bis zum Werte — 1, um dann wieder vom Werte — 1 bis zum Werte 1 anzusteigen. Nähert sich nun der Punkt p' unbeschränkt dem Punkte p, so muß er unendlich viele solcher Wellen durchlaufen, denn auf jeder Seite des Punktes p weist unsere Kurve unendlich viele Wellen auf; nähert sich also der Punkt p' unbeschränkt dem Punkte p, so schwankt die mittlere Steigung zwischen p und p' unablässig zwischen den Werten 1 und — 1 hin und her: es kann keine Rede davon sein, daß sie sich dabei unbeschränkt einem bestimmten Grenzwerte nähert; es kann also auch nicht von einer bestimmten Steigung unserer Kurve im Punkte p die Rede sein; diese Kurve hat daher auch im Punkte p keine bestimmte Tangente.

Dieses relativ einfache, der Anschauung gut zugängliche Beispiel zeigt also, daß eine Kurve nicht in jedem ihrer Punkte eine Tangente zu haben braucht; darüber kann kein Zweifel bestehen. Aber man war früher der Meinung, daß die Anschauung zwingend dartue, daß ein solcher Mangel doch nur in vereinzelten Ausnahmepunkten einer Kurve eintreten kann, keineswegs aber in *allen* Punkten einer Kurve; man war der Meinung, man könne aus der Anschauung mit voller Sicherheit entnehmen, daß eine Kurve — wenn schon nicht in allen — so doch in der überwiegenden Mehrzahl ihrer Punkte eine bestimmte Steigung aufweisen, eine bestimmte Tangente besitzen muß. Der Mathematiker und Physiker Ampère, dessen Verdienste in der Lehre von der Elektrizität allgemein bekannt sind, hat versucht, es zu beweisen, aber sein Beweis war falsch. Und groß war die Überraschung, als Weierstrass eine Kurve bekanntmachte, die in keinem einzigen Punkte eine bestimmte Steigung, eine bestimmte Tangente besitzt. Weierstrass kam zu einer solchen Kurve durch schwierige Rechnungen; ich kann nicht daran denken, diese Rechnungen hier vorzuführen. Wir können aber heute auf viel einfacherem Wege zu diesem Ziele gelangen, und ich will versuchen, einen solchen Weg hier wenigstens andeutungsweise vorzuführen[5]).

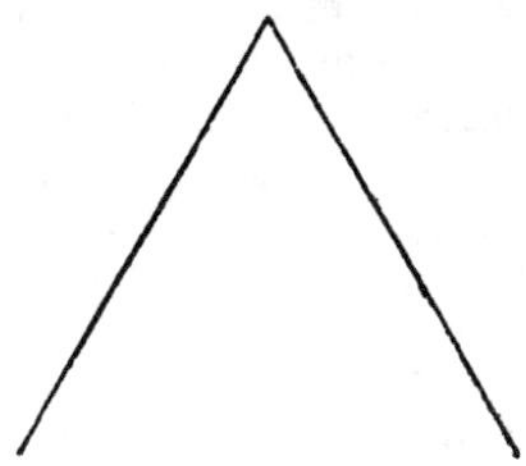

Fig. 4.

Wir gehen aus von der einfachen in Fig. 4 dargestellten Linie, die aus einer ansteigenden und einer abfallenden Strecke zusammengesetzt ist. Die ansteigende Strecke ersetzen wir, wie es Fig. 5 zeigt, durch einen aus sechs Strecken zusammengesetzten Streckenzug, der erst bis zur halben Höhe ansteigt, dann wieder ganz herabsinkt, dann neuerdings bis zur halben Höhe ansteigt, dann weiter steigt bis zur vollen Höhe, wieder bis zur halben Höhe zurücksinkt, schließlich wieder zur vollen Höhe ansteigt; ebenso ersetzen wir die abfallende Strecke von Fig. 4 durch einen aus sechs Strecken zusammengesetzten Streckenzug, der von der vollen Höhe bis zu halber Höhe sinkt, zur vollen Höhe zurücksteigt, wieder zur halben Höhe herabsinkt, weiter ganz herabsinkt, sich wieder zu halber Höhe erhebt, um schließlich wieder ganz herabzusinken. Von dem aus zwölf Strecken zusammengesetzten Streckenzug der Fig. 5, den wir so erhalten, gehen wir über zu dem aus 72 Strecken zusammengesetzten Streckenzug der Fig. 6, indem wir, analog wie beim Übergang von Fig. 4 zu Fig. 5, jede Strecke der Fig. 5 ersetzen durch einen aus sechs Strecken zusammengesetzten Streckenzug, und man sieht, wie dieses Verfahren immer weiter fortgesetzt werden kann und zu immer komplizierteren Streckenzügen führt. Es läßt sich nun (was wir natürlich hier nicht weiter ausführen wollen), in aller Schärfe zeigen, daß die so sukzessive konstruierten Streckenzüge sich unbeschränkt einer ganz bestimmten Kurve annähern, die die gewünschte Eigenschaft aufweist: sie hat in keinem Punkte eine bestimmte Steigung, besitzt daher in keinem Punkte

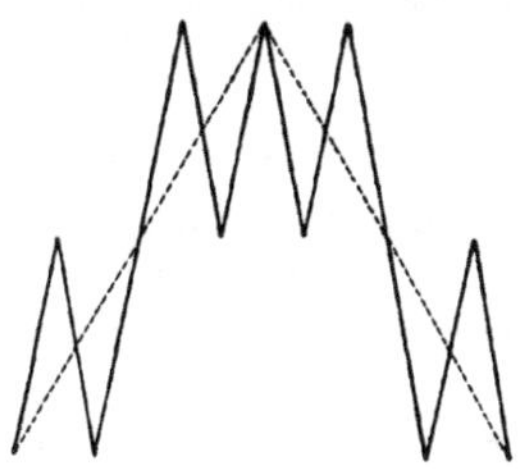

Fig. 5.

Fig. 6.

eine Tangente. Freilich entzieht sich der Verlauf dieser Kurve der Anschauung durchaus; und auch schon die sukzessive konstruierten Streckenzüge werden nach wenigen Schritten des Verfahrens so fein, daß die Anschauung nicht mehr recht folgen kann; bei der Kurve, der sich diese Streckenzüge unbeschränkt annähern, versagt sie jedenfalls gänzlich; nur das Denken, die logische Analyse kann bis zu dieser Kurve vorstoßen. Und wir sehen: hätte man sich in dieser Frage auf die Anschauung verlassen, so wäre man in Irrtum verharrt, denn die Anschauung schien zwingend darzutun, daß es Kurven, die in keinem Punkte eine Tangente haben, nicht geben kann.

Dieses erste Beispiel für das Versagen der Anschauung haben wir den Grundlagen der Differentialrechnung entnommen; ein zweites könnten wir den Grundlagen der *Integralrechnung* entnehmen. Die Grundaufgabe der Differentialrechnung war: bei gegebener Bahn eines bewegten Punktes seine Geschwindigkeit zu berechnen, bei gegebener Kurve ihre Steigung zu berechnen; die Grundaufgabe der Integralrechnung ist gerade die umgekehrte: von einem bewegten Punkte sei in jedem Augenblicke die Geschwindigkeit bekannt, es ist seine Bahn zu berechnen — oder: von einer Kurve sei überall die Steigung bekannt, es ist die Kurve selbst zu berechnen. Diese Aufgabe aber hat nur dann einen Sinn, wenn durch die Geschwindigkeit des bewegten Punktes seine Bahn, wenn durch die Steigung einer Kurve die Kurve selbst wirklich bestimmt ist. Wir stehen also vor der Frage, ob das der Fall ist oder nicht; präziser gesprochen: wir stehen vor der Frage: wenn zwei auf einer Geraden bewegliche Punkte sich im selben Augenblicke von derselben Stelle der Geraden aus in Bewegung setzen und in jedem Augenblicke übereinstimmende Geschwindigkeiten haben, müssen sie dann beisammen bleiben oder können sie auseinander geraten — bzw.: wenn zwei Kurven in einer Ebene vom selben Punkte ausgehen und immerzu übereinstimmende Steigung aufweisen, müssen sie sich dann in ihrem ganzen Verlaufe decken, oder kann sich die eine von beiden über die andere erheben? Die Anschauung scheint zwingend darzutun, daß die beiden bewegten Punkte immerzu beisammen bleiben müssen, daß die beiden Kurven sich in ihrem ganzen Verlaufe decken müssen; und doch lehrt die logische Analyse, daß es nicht notwendig so ist; für die üblicherweise in Betracht gezogenen Bewegungen; für die üblicherweise in Betracht gezogenen Kurven trifft es freilich zu; aber es sind gewisse recht komplizierte Bewegungen denkbar, es gibt gewisse recht komplizierte Kurven, für die es nicht zutrifft. Näher darauf einzugehen, fehlt uns der Raum[6]); wir müssen uns mit dem Hinweis begnügen, daß auch in dieser Frage die scheinbare Sicherheit der Anschauung sich als trügerisch erweist.

Die beiden bisherigen Beispiele für das Versagen der Anschauung waren den der Differential- und Integralrechnung zugrunde liegenden Erwägungen entnommen, also einem immerhin schwierigeren Gebiete, das man ja gemeinhin schon durch die Bezeichnung „höhere Mathematik" von den elementareren Teilen der Mathematik abhebt. Es wird also von Bedeutung sein, zu zeigen, wie auch schon in den elementaren Teilen der Mathematik sich ein Versagen der Anschauung feststellen läßt.

Ganz an der Schwelle der Geometrie steht der Begriff der *Kurve;* jedermann glaubt, eine anschaulich klare Vorstellung davon zu haben, was eine Kurve ist, und seit altersher glaubte man, diese Vorstellung durch die Definition einfangen zu können: Kurven sind diejenigen geometrischen Gebilde, die durch Bewegung eines Punktes[7]) erzeugt werden können. Aber siehe da! Im Jahre 1890 zeigte der (auch durch seine Forschungen über Logik hochverdiente) italienische Mathematiker Giuseppe Peano, daß zu den durch Bewegung eines Punktes erzeugbaren geometrischen Gebilden auch ganze Flächenstücke gehören: es ist z. B. eine Bewegung eines Punktes denkbar, bei der der bewegte Punkt in einer endlichen Zeitspanne sämtliche Punkte einer Quadratfläche durchläuft — und doch wird niemand eine volle Quadratfläche als eine Kurve ansehen wollen. Ich will versuchen, an der Hand einiger Figuren wenigstens eine angenäherte Vorstellung davon zu vermitteln, wie man zu einer solchen Bewegung eines Punktes gelangt[8]).

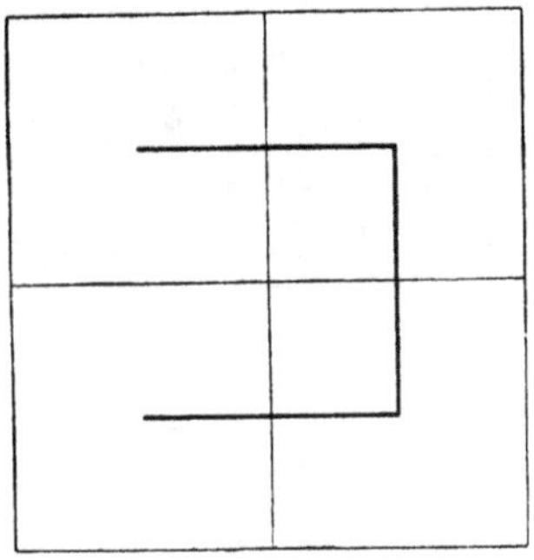

Abb. 7.

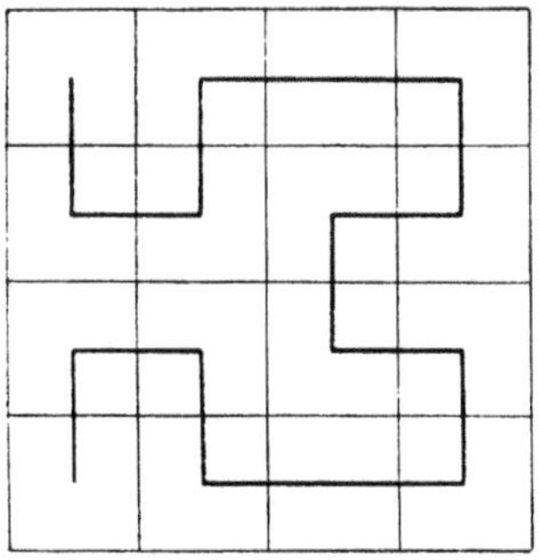

Abb. 8.

Man zerlege, wie Fig. 7 es zeigt, ein Quadrat in vier gleichgroße Teilquadrate, verbinde die Mittelpunkte dieser vier Teilquadrate durch einen Streckenzug und denke sich einen Punkt zunächst so bewegt, daß er in einer

4*

endlichen Zeitspanne — sagen wir: in der Zeiteinheit — mit gleichförmiger Geschwindigkeit diesen Streckenzug durchläuft. Sodann zerlege man (Fig. 8) jedes der vier Teilquadrate der Fig. 7 neuerdings in vier gleichgroße Teilquadrate, verbinde die Mittelpunkte dieser 16 Teilquadrate durch einen Streckenzug und denke sich den Punkt nunmehr so bewegt, daß er in der Zeiteinheit mit gleichförmiger Geschwindigkeit diesen neuen Streckenzug durchläuft. Sodann zerlege man (Fig. 9) jedes der 16 Teilquadrate der Fig. 8 neuerdings in vier gleichgroße Teilquadrate, verbinde die Mittelpunkte dieser 64 Teilquadrate durch einen Streckenzug und denke sich nunmehr den Punkt so bewegt, daß er in der Zeiteinheit mit gleichförmiger Geschwindigkeit diesen neuen Streckenzug durchläuft. Es ist ersichtlich, wie dieses Verfahren fortzusetzen ist; Fig. 10 zeigt einen der späteren Schritte, bei dem das Quadrat in 4096 Teilquadrate geteilt ist. Es läßt sich nun in aller Schärfe zeigen, daß die hier sukzessive in Betracht gezogenen Bewegungen — bei deren erster ein Punkt in der Zeiteinheit einen die Mittelpunkte der vier Teilquadrate der Fig. 7 verbindenden Streckenzug durchläuft, bei deren zweiter in derselben Zeit einen die Mittelpunkte der 16 Teilquadrate der Fig. 8, bei deren sechster in derselben Zeit einen die Mittelpunkte der 4096 Teilquadrate der Fig. 10 verbindenden Streckenzug — sich unbeschränkt einer ganz bestimmten Bewegung annähern, die den bewegten Punkt in der Zeiteinheit durch sämtliche Punkte der Quadratfläche hindurchführt. Freilich entzieht sich diese Bewegung jeder Möglichkeit der Anschauung, sie kann nur durch logische Analyse erfaßt werden.

Abb. 9.

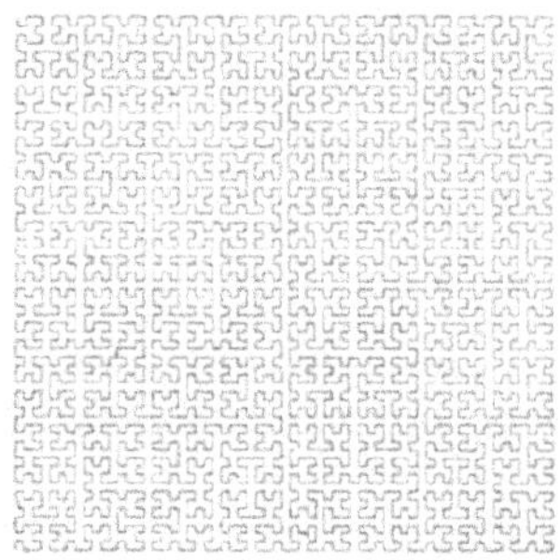

Abb. 10.

Während so, entgegen allem, was die Anschauung darzutun scheint, geometrische Gebilde, die niemand als Kurve ansehen wird, wie z. B. eine Quadratfläche, durch Bewegung eines Punktes erzeugt werden können, ist dies für andere geometrische Gebilde, die man viel eher geneigt sein wird, als Kurven anzusprechen, nicht der Fall. Man betrachte etwa das in Fig. 11 angedeutete geometrische Gebilde: eine Wellenlinie, die in der Nähe der (mit zu dem Gebilde zu rechnenden) Strecke *a b* unendlich viele Wellen mit unbeschränkt abnehmender Wellenlänge aufweist, deren Amplituden aber, im Gegensatz zu Fig. 3, nicht unbeschränkt abnehmen, sondern alle gleich groß sind; man zeigt unschwer, daß dieses geometrische Gebilde, trotz seines linienhaften Charakters, nicht durch Bewegung eines Punktes erzeugt werden kann: es ist keine Bewegung eines Punktes denkbar, die den bewegten Punkt in einer endlichen Zeitspanne durch alle Punkte dieses Gebildes hindurch führen würde.

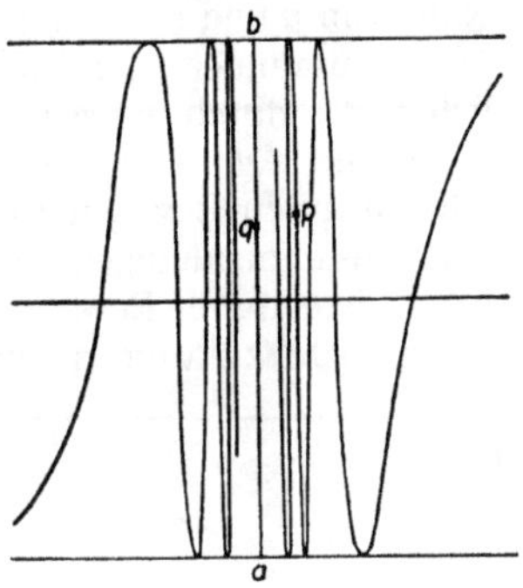

Fig. 11.

Hier erheben sich nun ganz naturgemäß zwei wichtige Fragen: 1. Da, wie wir sehen, die oben angeführte altehrwürdige Definition des Begriffes Kurve durchaus ungeeignet ist, unsere primitive Kurvenvorstellung einzufangen, durch welche andere, zweckdienlichere Definition ist sie zu ersetzen? 2. Da, wie wir sehen, die durch Bewegung eines Punktes erzeugbaren geometrischen Gebilde sich keineswegs mit den Kurven decken, welche geometrischen Gebilde sind es denn, die durch Bewegung eines Punktes erzeugt werden können? Beide Fragen sind heute befriedigend beantwortet; auf die Beantwortung der ersten kommen wir später zurück; über die zweite seien gleich einige Worte gesagt[9]): Ihre Beantwortung gelang mit Hilfe eines neuen geometrischen Begriffes, des „*Zusammenhanges im kleinen*" oder „lokalen Zusammenhanges". Betrachten wir einige durch Bewegung eines Punktes erzeugbare Gebilde, z. B. (Fig. 12) eine Strecke, eine Kreislinie, eine Quadratfläche; wir nehmen auf einem solchen Gebilde zwei recht nahe beieinander gelegene Punkte *p* und *q* an und sehen: wir können, ohne das betreffende Gebilde verlassen zu müssen, von *p* nach *q* auf einem Wege gelangen, der in großer Nähe von *p* und *q* verbleibt: diese Eigenschaft (in entsprechend präziserer Formulierung) bezeichnet man als „Zu-

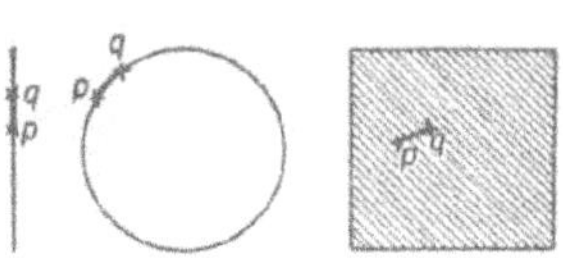

Fig. 12.

sammenhang im kleinen"; das in Fig. 11 angedeutete Gebilde hat diese Eigenschaft nicht: man betrachte auf ihm etwa die nahe beieinander gelegenen Punkte p und q; will man von p nach q gelangen, ohne das Gebilde zu verlassen, so müßte man die sämtlichen dazwischen gelegenen unendlich vielen Wellen der Wellenlinie durchlaufen; dieser Weg bleibt aber nicht in großer Nähe von p und q, da alle diese Wellen dieselbe Amplitude haben. Der „Zusammenhang im kleinen" nun ist es, der im wesentlichen die durch Bewegung eines Punktes erzeugbaren Gebilde charakterisiert: eine Strecke, eine Kreislinie, eine Quadratfläche können durch Bewegung eines Punktes erzeugt werden, denn sie sind zusammenhängend im kleinen; das Gebilde der Fig. 11 kann nicht durch Bewegung eines Punktes erzeugt werden, denn es ist nicht zusammenhängend im kleinen.

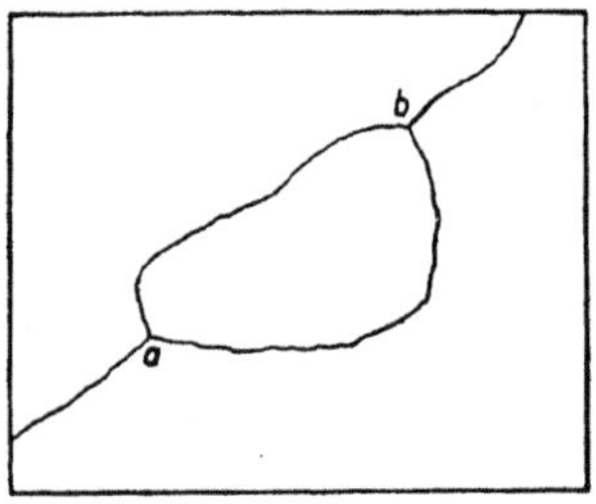

Fig. 13.

Wir wollen uns noch an einem zweiten Beispiele überzeugen, wie unzuverlässig sich die Anschauung schon bei geometrischen Fragen ganz elementarer Natur zeigt. Denken wir uns ein Landkartenblatt (Fig. 13), auf dem im ganzen drei verschiedene Länder vorkommen. Es werden dann Grenzpunke auftreten, in denen zwei Länder aneinander grenzen, es können aber auch Punkte auftreten, in denen alle drei Länder aneinander grenzen, sogenannte „Dreiländerecken", wie die Punkte a und b in Fig. 13. Die Anschauung scheint zwingend darzutun, daß solche Dreiländerecken nur vereinzelt auftreten können, daß in der weitaus überwiegenden Mehrzahl aller Grenzpunkte nur zwei Länder aneinander grenzen werden. Und doch ist, wie der holländische Mathematiker L. E. J. Brouwer im Jahre 1910 zeigte, eine Einteilung eines Kartenblattes in drei Länder möglich, bei der in *jedem* auftretenden Grenzpunkte *alle drei Länder aneinander grenzen*[10]). Es sei wieder versucht, das wenigstens andeutungsweise einigermaßen klarzumachen.

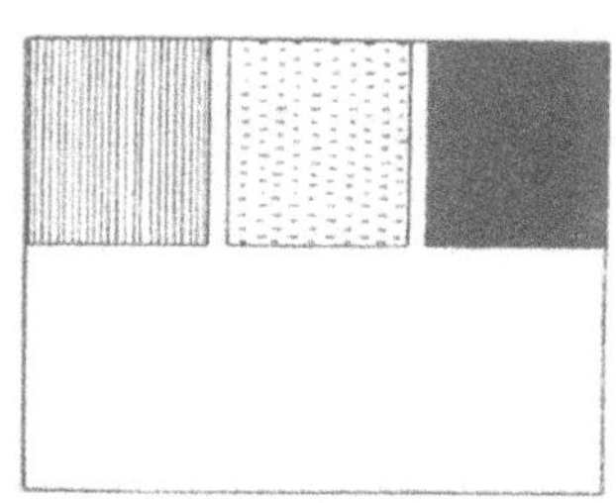
Fig. 14.

Wir gehen aus von dem in Fig. 14 gezeichneten Landkartenblatte, auf dem man drei verschiedene Länder, ein schraffiertes, ein punktiertes, ein schwarzes Land, findet; alles übrige sei herrenloses Gebiet. Nun beschließt das schraffierte Land, um das unbesetzte Gebiet in seine Einflußsphäre zu bringen, in dieses Gebiet einen Korridor vorzutreiben (Fig. 15), der jedem Punkte des unbesetzten Gebietes bis auf einen Kilometer nahekommt, aber — um jeden Konflikt zu vermeiden — an keines der beiden anderen Länder anstößt. Nachdem dies geschehen ist, denkt man sich im punktierten Land: das können wir auch! und nun treibt auch das punktierte Land einen Korridor in das noch unbesetzte Gebiet vor (Fig. 16), der jedem noch unbesetzten Punkte sogar bis auf einen halben Kilometer nahekommt, aber an keines der anderen Länder anstößt. Nachdem dies geschehen ist, denkt man sich im schwarzen Land: da können wir unmöglich zurückbleiben! und nun treibt auch das schwarze Land in das noch unbesetzte Gebiet einen Korridor vor (Fig. 17), der jedem Punkte dieses Gebietes sogar bis auf einen Drittel-Kilometer nahekommt, aber an keines der anderen Länder anstößt. Nachdem dies geschehen, sagt man sich im schraffierten Lande: wir sind übertrumpft worden und müssen neuerdings vorgehen! und das schraffierte Land treibt wieder einen Korridor in das noch unbesetzte Gebiet vor, der jedem Punkte dieses Gebietes sogar bis auf einen Viertel-Kilometer nahekommt, aber an keines der anderen Länder anstößt. Und so geht es immer weiter: als nächstes treibt dann das punktierte Land einen Korridor vor, der jedem noch unbesetzten Punkte bis auf einen Fünftel-Kilometer nahekommt, dann das schwarze Land einen Korridor, der jedem noch unbe-

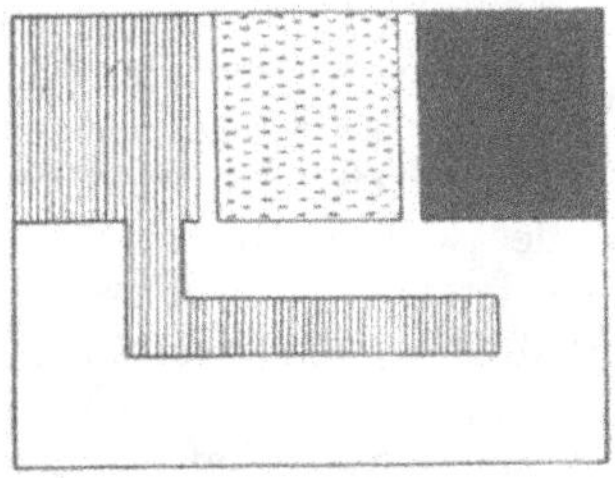

Fig. 15

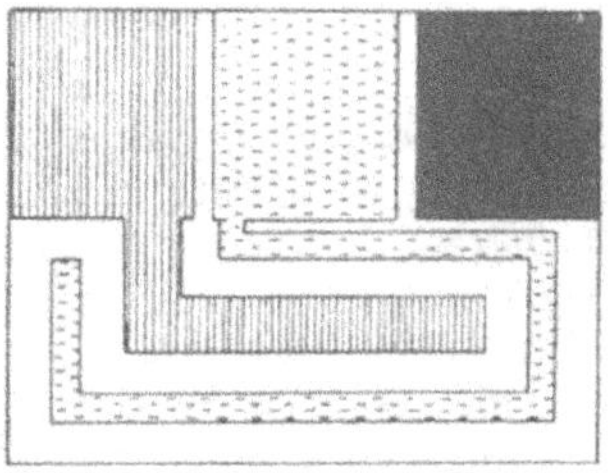

Fig. 16

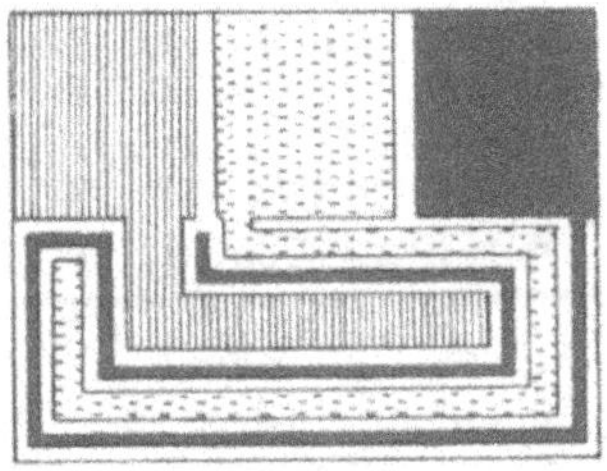

Fig. 17

setzten Punkte bis auf einen Sechstel-Kilometer nahekommt, dann rückt wieder das schraffierte Land vor usw. usw. Und da wir schon unserer Phantasie die Zügel schießen ließen, wollen wir noch annehmen, das schraffierte Land habe zum Vortreiben seines ersten Korridors ein Jahr gebraucht, das punktierte Land habe sodann seinen ersten Korridor im nächsten Halbjahre vorgetrieben, dann das schwarze Land seinen ersten Korridor im nächsten Vierteljahre, dann das schraffierte Land seinen zweiten Korridor im nächsten Achteljahre usw., so daß jeder weitere Korridor in der Hälfte der Zeit fertig wird, die der letztangelegte erfordert hatte. Man überzeugt sich unschwer, daß nach Ablauf von zwei Jahren kein unbesetztes Gebiet mehr da ist, und daß dann ein Zustand erreicht ist, bei dem das ganze Kartenblatt so auf die drei Länder aufgeteilt ist, daß nirgends nur zwei dieser Länder aneinander grenzen: in jedem auftretenden Grenzpunkte grenzen sie alle drei aneinander. Anschauungsmäßig freilich kann man diese Verteilung nicht erfassen, wieder ist es nur die logische Analyse, die uns so weit führt; auch hier sehen wir also wieder: hätte man sich bei dieser doch so einfachen Fragestellung auf die Anschauung verlassen, so wäre man in schweren Irrtum verfallen.

Und da die Anschauung sich in so vielen Fragen als trügerisch erwiesen hatte, da es immer wieder vorkam, daß Sätze, die der Anschauung als durchaus gesichert galten, sich bei logischer Analyse als falsch herausstellten, so wurde man in der Mathematik gegenüber der Anschauung immer skeptischer; es brach immer mehr die Überzeugung durch, daß es unzulänglich sei, irgendeinen mathematischen Satz der Anschauung zu entnehmen, daß es nicht anginge, irgendeine mathematische Disziplin auf Anschauung zu gründen; es entstand die Forderung nach völliger Eliminierung der Anschauung aus der Mathematik, die Forderung nach völliger *Logisierung der Mathematik:* Jeder neue mathematische Begriff muß durch rein logische Definition eingeführt werden, jeder mathematische Beweis muß mit rein logischen Mitteln geführt werden. Pioniere auf diesem Wege waren (um nur die berühmtesten zu nennen): Augustin de Cauchy (1789—1857), Bernard Bolzano (1781—1848), Carl Weierstrass (1815—1897), Georg Cantor (1845—1918), Richard Dedekind (1831—1916).

Die Aufgabe einer völligen Logisierung der Mathematik war eine mühevolle und schwierige — es war eine Reform an Haupt und Gliedern. Sätze, die man früher als anschaulich evident hingenommen hatte, mußten sorgsam bewiesen werden. Um nur ein Beispiel zu nennen: ein so einfacher geometrischer

Satz, wie: „Jedes geschlossene, sich selbst nicht durchsetzende Polygon zerlegt die Ebene in genau zwei getrennte Teile" erfordert einen recht langwierigen, recht kunstvollen Beweis[11]), und in noch höherem Maße gilt das von dem analogen Satz im Raume: „Jedes geschlossene, sich selbst nicht durchsetzende Polyeder zerlegt den Raum in genau zwei getrennte Teile"[12]).

Als Prototyp eines der reinen Anschauung entnommenen synthetischen Urteiles a priori führt Kant ausdrücklich den Satz an: *der Raum ist dreidimensional.* Aber auch dieser Satz erfordert nach unserer heutigen Auffassung eine eindringende logische Analyse: es muß zunächst rein logisch definiert werden, was unter der Dimensionszahl eines geometrischen Gebildes, einer „Punktmenge" zu verstehen ist, sodann muß rein logisch bewiesen werden, daß bei Zugrundelegung dieser Definition der Raum der üblichen Geometrie, der zugleich der Raum der klassischen Newtonschen Physik ist, wirklich dreidimensional ist. Das wurde erst in jüngster Zeit, in den Jahren 1921/22 geleistet, und zwar gleichzeitig durch den Wiener Mathematiker K. Menger und den russischen Mathematiker P. Urysohn, der mittlerweile, in der Blüte seines Schaffens, einem tragischen Unfalle zum Opfer fiel. Ich will wenigstens eine flüchtige Vorstellung davon geben, wie da die Dimensionszahl einer Punktmenge definiert wird[13]).

Eine Punktmenge wird als *nulldimensional* bezeichnet, wenn es zu jedem ihrer Punkte beliebig kleine Umgebungen gibt, deren Begrenzung keinen Punkt der Menge enthält; jede aus endlich vielen Punkten bestehende Menge z. B. ist nulldimensional (vgl. Fig. 18), aber es gibt auch sehr viele, sehr komplizierte nulldimensionale Punktmengen, die aus unendlich vielen Punkten bestehen. Eine nicht nulldimensionale Punktmenge heißt nun *eindimensional,* wenn es zu jedem ihrer Punkte beliebig kleine Umgebungen gibt, deren Begrenzung mit der Punkt menge nur eine nulldimensionale Menge gemein hat; jede Gerade, jede aus endlich vielen geradlinigen Strecken zusammengesetzte Figur, jede Kreislinie, jede Ellipse, kurz alle Gebilde, die man gemeinhin als Kurven bezeichnet, sind in diesem Sinne eindimensional (vgl. Fig. 19), aber auch das in Fig. 11 dargestellte geometrische Gebilde, das — wie wir sahen — nicht durch Bewegung eines Punktes erzeugbar ist. Eine weder nulldimensionale noch eindimensionale Punktmenge

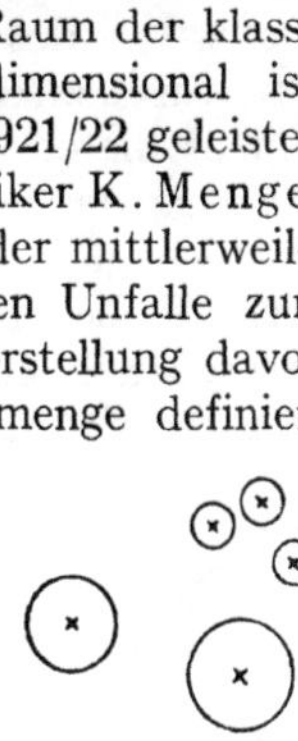
Fig. 18.

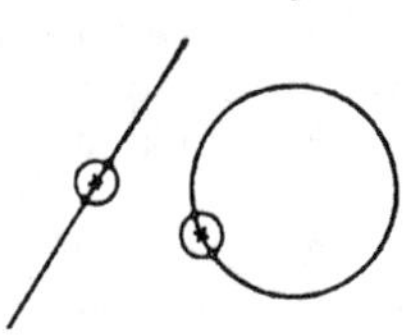
Fig. 19.

heißt sodann *zweidimensional*, wenn es zu jedem ihrer Punkte beliebig kleine Umgebungen gibt, deren Begrenzung mit der Punktmenge eine höchstens eindimensionale Menge gemein hat; jede Ebene, jede Polygon- oder Kreisfläche, jede Kugeloberfläche, kurz alle Gebilde, die man gemeinhin als Flächen bezeichnet, sind in diesem Sinne zweidimensional. Eine weder nulldimensionale, noch eindimensionale, noch zweidimensionale Punktmenge nun heißt *dreidimensional*, wenn es zu jedem ihrer Punkte beliebig kleine Umgebungen gibt, deren Begrenzung mit der Punktmenge eine höchstens zweidimensionale Menge gemein hat. Ein — freilich keineswegs einfacher — Beweis zeigt nun, daß der Raum der üblichen Geometrie tatsächlich in diesem Sinne dreidimensional ist.

Diese Theorie liefert nun auch eine wirklich befriedigende Definition des Begriffes *Kurve*[14]). Als wesentlichstes Merkmal der Kurve erscheint dabei ihre Eindimensionalität. Aber darüber hinaus liefert diese Theorie auch eine außerordentlich feine Analyse der Struktur von Kurven. Auch darüber möchte ich noch einige Worte sagen. Ein Punkt einer Kurve heißt *Endpunkt*, wenn es beliebig kleine Umgebungen dieses Punktes gibt, deren Begrenzung einen einzigen Punkt mit der Kurve gemein hat (vgl. in Fig. 20 die Punkte *a* und *b*); ein Punkt der Kurve, der nicht Endpunkt ist, heißt ein *gewöhnlicher Punkt*, wenn es beliebig kleine Umgebungen dieses Punktes gibt, deren Begrenzung genau zwei Punkte mit der Kurve gemein hat (vgl. in Fig. 20 den Punkt *c*); ein Punkt einer Kurve heißt ein *Verzweigungspunkt*, wenn die Begrenzung jeder hinlänglich kleinen Umgebung dieses Punktes mit der Kurve mehr als zwei Punkte gemein hat (vgl. in Fig. 20 den Punkt *d*).

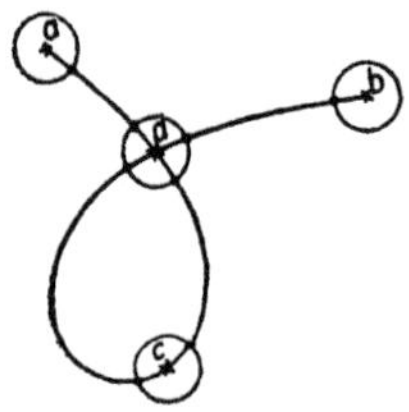

Fig. 20.

Die Anschauung scheint nun zu lehren, daß Endpunkte und Verzweigungspunkte auf einer Kurve eine Art Ausnahmestellung einnehmen, daß sie in gewissem Sinne nur vereinzelt auftreten können, daß eine Kurve unmöglich aus lauter Endpunkten bestehen kann, oder aus lauter Verzweigungspunkten. Diese Vermutung wird, was die Endpunkte anlangt, durch die logische Analyse präzisiert und bestätigt, was aber die Verzweigungspunkte anlangt, so wird die Vermutung durch die logische Analyse widerlegt. Es gibt, wie der polnische Mathematiker W. Sierpiński im Jahre 1915 zeigte, Kurven, *deren sämtliche Punkte Verzweigungspunkte sind*. Versuchen wir, uns dies durch einige Andeutungen näherzubringen.

Man denke sich in ein gleichseitiges Dreieck ein anderes gleichseitiges Dreieck eingeschrieben, so wie Fig. 21 es zeigt, und denke sich das (in Fig. 21 schraffierte) Innere des eingeschriebenen Dreieckes getilgt; es bleiben dann drei gleichseitige Dreiecke samt ihren Rändern übrig. In jedes dieser drei übriggebliebenen gleichseitigen Dreiecke schreibe man (Fig. 22) wieder ein gleichseitiges Dreieck ein und denke sich das Innere jedes der drei eingeschriebenen Dreiecke getilgt; es bleiben dann neun gleichseitige Dreiecke samt ihren Rändern übrig. In jedes dieser neun übriggebliebenen gleichseitigen Dreiecke schreibe man wieder ein gleichseitiges Dreieck ein und denke sich das Innere jedes dieser neun eingeschriebenen Dreiecke getilgt, so daß 27 gleichseitige Dreiecke übrigbleiben. Und dieses Verfahren denke man sich unbeschränkt fortgesetzt. (In Fig. 23 findet man den fünften Schritt dieses Verfahrens, bei dem 243 gleichseitige Dreiecke übrigbleiben, dargestellt.) Die Punkte des ursprünglichen gleichseitigen Dreieckes, die schließlich übrigbleiben (d. h. bei keinem der unendlich vielen Schritte unseres Verfahrens getilgt werden) bilden dann, wie man zeigen kann, eine Kurve, und zwar eine Kurve, deren sämtliche Punkte, mit Ausnahme der drei Eckpunkte *a*, *b*, *c* des ursprünglichen Dreieckes, Verzweigungspunkte sind. Es ist sehr leicht, daraus eine Kurve zu gewinnen, deren *sämtliche* Punkte Verzweigungspunkte sind, z. B. indem man die ganze Figur so verzerrt, daß die drei Eckpunkte *a*, *b*, *c* des ursprünglichen Dreieckes in einen Punkt zusammenrücken.

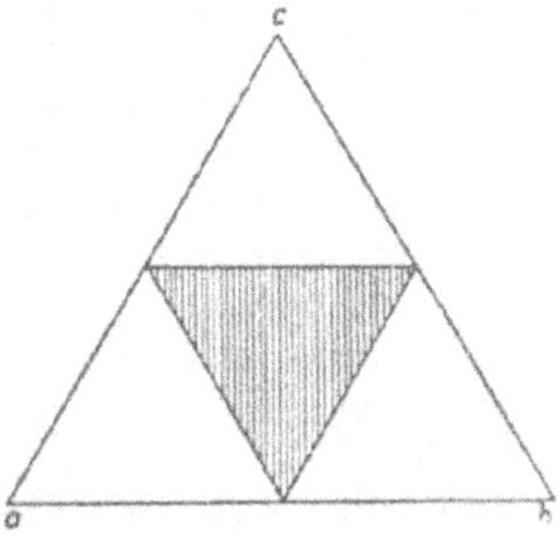

Fig. 21.

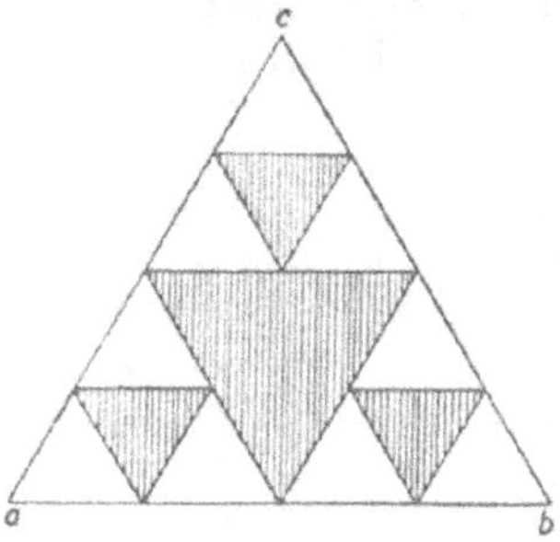

Fig. 22.

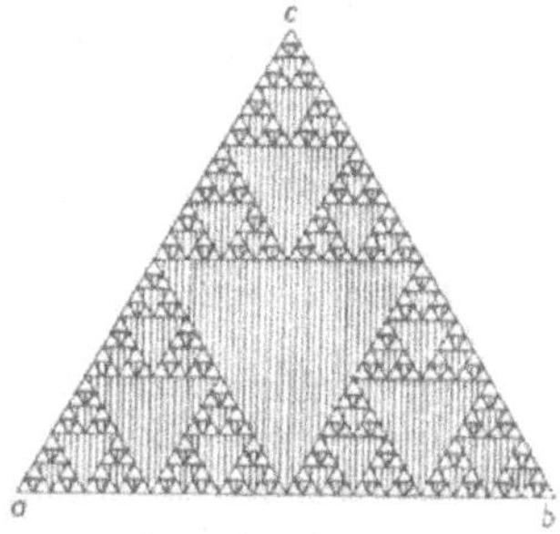

Fig. 23.

Doch nun genug der Beispiele, und fassen wir das Gesagte zusammen! Immer wieder haben wir gefunden, daß die Anschauung sich in Fragen der Geometrie, und zwar auch in prinzipiell sehr einfachen und elementaren Fragen, als durchaus unzuverlässige Führerin erweist. Ein so unzuverlässiges Hilfsmittel kann aber unmöglich den Ausgangspunkt, die Grundlage einer mathematischen Disziplin abgeben. Der Raum der Geometrie ist nicht eine Form reiner Anschauung, sondern eine logische Konstruktion.

Als widerspruchsfreie logische Konstruktionen aber sind auch andersgeartete Räume möglich, als der Raum der üblichen Geometrie: Räume z. B., in denen das sogenannte euklidische Parallelenpostulat durch ein gegenteiliges ersetzt wird („nichteuklidische" Räume), oder Räume, deren Dimensionszahl größer als drei ist, oder „nichtarchimedische" Räume, über die hier noch einige Worte gesagt seien, während den nichteuklidischen und mehrdimensionalen Räumen der ganze nächste Vortrag gewidmet ist.

Die Möglichkeit, die Länge einer Strecke durch eine reelle Zahl zu messen und die daraus fließende Möglichkeit, die Lage eines Punktes, wie es in der analytischen Geometrie geschieht, durch Angabe reeller Zahlen (seiner „Koordinaten") festzulegen, beruht auf dem sogenannten *Postulat des Archimedes*[15]), das besagt: Sind zwei Strecken gegeben, so gibt es stets ein Vielfaches der ersten, das größer als die zweite ist. Als logische Konstruktion aber sind durchaus auch Räume möglich[16]), in denen das archimedische Postulat durch das gegenteilige ersetzt ist, in denen es also Strecken gibt, die größer sind als jedes Vielfache einer gegebenen Strecke, in denen es demgemäß, nach Wahl einer Strecke als Längeneinheit, *unendlich große* und *unendlich kleine* Strecken gibt, während es im Raume der üblichen Geometrie unendlich große und unendlich kleine Strecken nicht gibt. Auch in einem solchen „nichtarchimedischen" Raume kann man Strecken messen und analytische Geometrie treiben, freilich nicht mit Hilfe der reellen Zahlen der üblichen Arithmetik, sondern mit Hilfe „nichtarchimedischer Zahlensysteme", die man aber ebenso zu überblicken vermag, mit denen man ebensogut rechnen kann, wie mit den reellen Zahlen der üblichen Arithmetik[17]).

Was sollen wir nun von dem oft gehörten Einwande halten: alle diese mehrdimensionalen, nichteuklidischen, nichtarchimedischen Geometrien — mögen sie auch als logische Konstruktionen widerspruchsfrei sein — zur Einordnung unserer Erlebnisse sind sie unbrauchbar, denn sie sind *unanschaulich*; zur Einordnung unserer Erlebnisse ist einzig brauchbar die übliche dreidimensionale, euklidische, archimedische Geometrie, denn sie ist die einzig anschauliche. Dazu wäre vor allem zu sagen — und mein ganzer Vortrag diente ja dazu, dies zu zeigen —, daß es auch mit der Anschaulichkeit der üblichen Geometrie nicht gar so weit her ist. Es ist eben *jede* Geometrie — dreidimensionale wie mehrdimensionale, eukli ische wie nichteuklidische, archimedische wie nichtarchimedische — eine logische Konstruktion. Die Entwicklung der Physik hat es mit sich gebracht, daß man bis in die jüngste Zeit sich zur Ordnung unserer Erlebnisse ausschließlich der logischen Konstruktion der dreidimensionalen, euklidischen, archimedischen Geometrie bediente; sie hat sich bis in die jüngste Zeit trefflich für diesen Zweck bewährt, und so hat man sich völlig an ihre Handhabung gewöhnt. Diese Gewöhnung an die Handhabung der üblichen Geometrie zur Ordnung unserer Erlebnisse ist es, was man als ihre Anschaulichkeit bezeichnet, jedes Abweichen davon gilt als unanschaulich, als anschauungswidrig, als anschauungsunmöglich. Solche „Anschauungsmöglichkeiten" aber finden sich, wie wir sehen, auch in der üblichen Geometrie: sie stellen sich ein, sobald man sich nicht mehr darauf beschränkt, nur Gebilde zu betrachten, an die man durch langen Gebrauch gewöhnt ist, sondern auch Gebilde in den Kreis unserer Betrachtungen zieht, an die man bisher nicht gedacht hatte.

Die neueste Physik läßt es nun als zweckmäßig erscheinen, zur Ordnung unserer Erlebnisse auch die logischen Konstruktionen mehrdimensionaler und nichteuklidischer Geometrien heranzuziehen (während wir bisher keinen Anhaltspunkt dafür haben, daß auch das Heranziehen nichtarchimedischer Geometrien sich als zweckmäßig erweisen könnte; aber ausgeschlossen ist dies keineswegs); da sich aber diese Entwicklung erst in jüngster Zeit vollzog, sind wir an die Handhabung dieser logischen Konstruktionen zur Ordnung unserer Erlebnisse noch nicht gewöhnt, darum gelten diese Geometrien als anschauungswidrig.

So ist es auch seinerzeit gegangen, als die Lehre von der Kugelgestalt der Erde aufkam; da wurde diese Lehre vielfach abgelehnt, weil die Existenz der Antipoden anschauungswidrig sei; mittlerweile aber hat man sich an diese Auffassung gewöhnt, und heute fällt es niemandem mehr ein, sie wegen angeblicher Anschauungswidrigkeit als unmöglich zu erklären.

Auch die spezifisch physikalischen Begriffe sind logische Konstruktionen, und man kann an ihnen deutlich sehen, wie diejenigen, an deren Handhabung man gewöhnt ist, anschaulichen Charakter bekommen, und die, an deren Handhabung wir nicht gewohnt sind, unanschaulich bleiben. Der Begriff „Gewicht" ist einer, dessen Handhabung ziemlich jedermann gewohnt ist, darum verbindet sich mit diesem Begriffe ziemlich für jedermann eine gewisse Anschaulichkeit; der Begriff „Trägheitsmoment", mit dessen Handhabung die meisten Menschen nichts zu tun haben, bleibt auch für die meisten Menschen unanschaulich, während er für manche Experimentalphysiker und Techniker, die ihn fortgesetzt handhaben müssen, ebenso anschaulichen Charakter gewinnt, wie der Begriff „Gewicht" für die meisten Menschen. Ähnlich wird für den Elektrotechniker der Begriff „Potentialdifferenz" anschaulichen Charakter haben, für die meisten Menschen aber nicht.

Wenn nun die Heranziehung der logischen Konstruktionen mehrdimensionaler und nichteuklidischer Geometrien zur Ordnung unserer Erlebnisse sich bewähren sollte, wenn man sich an ihre Handhabung immer mehr gewöhnt haben wird, wenn sie in den Schulunterricht eingedrungen sein wird, wenn man sie mit der Muttermilch einsaugen wird, wie es heute der Fall ist für die Handhabung der dreidimensionalen euklidischen Geometrie, dann wird es niemandem mehr einfallen, diese Geometrien als anschauungswidrig zu bezeichnen, dann werden sie ebenso als anschaulich gelten, wie heute die dreidimensionale euklidische Geometrie. Denn nicht, wie Kant dies wollte, ein reines Erkenntnismittel a priori ist die Anschauung, sondern auf psychischer Trägheit beruhende Macht der Gewöhnung!

Gibt es Unendliches?

Von
HANS HAHN.

Seit dem Altertum beschäftigt die Frage nach dem Unendlichen unablässig den Menschengeist. Immer wieder wurde es entschiedenst geleugnet, daß irgend etwas Unendliches existieren oder irgendeine Unendlichkeit vom Menschen erfaßt werden könne, und immer wieder fanden sich Denker, die dies für durchaus möglich hielten. Aristoteles hatte gelehrt, daß keinerlei vollendetes Unendliches möglich sei. Die Philosophie des Mittelalters — beherrscht einerseits von der Autorität des Aristoteles, der für sie „der Philosoph" schlechthin war, in mindestens gleichem Maße aber andrerseits beherrscht von der Autorität der Kirche — diskutiert unermüdlich die Frage, wie sich die von Aristoteles gelehrte Unmöglichkeit des Unendlichen mit der von der Kirche gelehrten Allmacht Gottes vertrage. Die These des Aristoteles verfeinernd und präzisierernd lehrt der heilige Thomas von Aquino, kein Unendliches könne *gegeben* sein, während nicht wenige aus der großen Reihe der scholastischen Philosophen, insbesondere solche aus der Nominalistenschule, die gegenteilige These verfechten[1]). Zum Teile wenigstens wird dabei eine bewundernswerte logische Schärfe entwickelt, die in den folgenden Jahrhunderten gänzlich verloren ging, und in manchen Punkten erst wieder in der kritischen Mathematik des 19. Jahrhunderts erreicht wurde. Um von den neuzeitlichen Denkern nur einige der ganz großen zu erwähnen, und nur solche, die mathematisch gerichtet waren: Descartes lehnt ausdrücklich jede Beschäftigung mit dem Unendlichen ab: „Wir werden uns — schreibt er in seinen Principia — nicht mit Streitigkeiten über das Unendliche ermüden, denn bei unserer eigenen Endlichkeit wäre es verkehrt, wenn wir versuchten, etwas darüber zu bestimmen und es so gleichsam endlich und begreiflich zu machen ... denn nur der, welcher seinen Geist für unendlich hält, kann glauben, hierüber nachdenken zu müssen." Leibniz hingegen schreibt in einem Briefe: „Je suis tellement pour l'infini actuel, qu'au lieu d'admettre que la nature l'abhorre, comme l'on dit vulgairement, je tiens qu'elle l'affecte partout, pour mieux marquer la perfection de son Auteur. Ainsi je crois qu'il n'y a aucune partie de la matière qui ne soit, je ne dis pas divisible, mais actuellement divisée; et par consequent la moindre particelle doit etre considerée comme un monde plein d'une infinité de créatures différentes." Aber er lehrt — in dieser Beziehung einig mit seinem sonstigen Antipoden Locke: „Nous n'avons pas l'idée d'un éspace infini, et rien n'est plus sensible que l'absurdité d'une idée actuelle d'un nombre infini." Im Jahre 1831 schrieb C. F. Gauß, der als der größte aller Mathematiker verehrt wird, die oft zitierten Worte: „So protestiere ich gegen den Gebrauch einer unendlichen Größe als einer vollendeten, welcher in der Mathematik niemals erlaubt ist"; und ebenso denkt A. Cauchy, der große Mathematiker, dessen Gedanken für die Entwicklung der Mathematik im 19. Jahrhundert vielfach bestimmend waren. Ganz anders sein Zeitgenosse, der Österreicher B. Bolzano, der sich mit Cauchy in den Ruhm teilt, die Grundsteine zu einer kritisch exakten Fundierung der Analysis gelegt zu haben, dessen Schicksal aber auch darin ein echt österreichisches war, daß seine Leistungen von den Zeitgenossen wenig beachtet wurden, so daß sein Einfluß auf die weitere Entwicklung nicht entfernt mit dem Cauchys verglichen werden kann. Er schrieb in den Jahren 1847/48 eine kleine Schrift „Die Paradoxien des Unendlichen", die erst nach seinem Tode gedruckt wurde[2]), in der er bestrebt ist, „den Schein des Widerspruches, der an diesen mathematischen Paradoxien haftet, als das, was er ist, als bloßen Schein zu erkennen", und so das Unendliche zum Objekte wissenschaftlicher Forschung zu machen. Der entscheidende Erfolg in diesem Sinne aber war Georg Cantor vorbehalten, der in den Jahren 1871—1884 eine ganz neue, ganz eigenartige mathematische Disziplin schuf, die Mengenlehre, in der zum ersten Male nach jahrtausendelangem Für und Wider mit aller Schärfe der modernen Mathematik eine Lehre vom Unendlichen begründet wurde[3]).

Wie so viele große Neuschöpfungen geht auch diese von einem ganz einfachen Gedanken aus. Cantor legt sich die Frage vor: Was meinen wir, wenn wir von zwei *endlichen* Mengen sagen, sie bestehen aus gleich viel Elementen, sie haben gleiche Anzahl, sie sind *gleichzahlig?* Offenbar nichts anderes als dies: Zwischen den Elementen der ersten und denen der zweiten Menge ist eine Zuordnung möglich, bei der jedem Elemente der ersten Menge genau eines der zweiten Menge zugeordnet ist, und auch jedem Elemente der zweiten Menge genau eines der ersten; eine solche Zuordnung heißt „umkehrbar eindeutig" oder kürzer „eineindeutig". Die Menge der Finger der rechten

Hand ist gleichzahlig der Menge der Finger der linken Hand, denn es ist zwischen den Fingern der rechten Hand und denen der linken Hand eine eineindeutige Zuordnung möglich; eine solche Zuordnung erhalten wir z. B. indem wir Daumen auf Daumen, Zeigefinger auf Zeigefinger legen, usw. Die Menge meiner Ohren aber und die Menge der Finger einer Hand sind nicht gleichzahlig, denn da ist offenbar keine eineindeutige Zuordnung möglich; versuche ich die Finger einer Hand den Ohren zuzuordnen, so bleiben, wie ich es auch anstellen mag, notwendig Finger über, die keinem Ohr zugeordnet sind. Die *Anzahl* (oder *Kardinalzahl*) einer Menge ist nun offenbar ein Merkmal, das sie mit allen gleichzahligen Mengen gemein hat und durch das sie sich von allen mit ihr nicht gleichzahligen Mengen unterscheidet. Die Anzahl „fünf" z. B. ist das Merkmal, in dem alle mit der Menge der Finger meiner rechten Hand gleichzahligen Mengen übereinstimmen und in dem sie sich von allen anderen Mengen unterscheiden.

Wir haben also die Definitionen: Zwei Mengen heißen *gleichzahlig*, wenn zwischen ihren Elementen eine eineindeutige Zuordnung möglich ist; das Merkmal, das eine Menge mit allen gleichzahligen Mengen gemein hat und durch das sie sich von jeder mit ihr nicht gleichzahligen Menge unterscheidet, heißt die *Anzahl* dieser Menge. Und nun machen wir die fundamentale Bemerkung: in diese Definitionen geht in keiner Weise die Endlichkeit der betrachteten Mengen ein; sie können ebensogut auf unendliche Mengen angewendet werden wie auf endliche Mengen. Und damit sind die Begriffe „gleichzahlig" und „Anzahl" auf Mengen aus unendlich vielen Elementen übertragen. Die Anzahlen endlicher Mengen, also die Zahlen 1, 2, 3, ... werden *natürliche Zahlen* genannt; die Anzahlen unendlicher Mengen nennt Cantor *transfinite Kardinalzahlen* oder auch *transfinite Mächtigkeiten.*

Aber gibt es überhaupt unendliche Mengen? Davon können wir uns sogleich an einem ganz einfachen Beispiel überzeugen: Es gibt offenbar unendlich viele verschiedene natürliche Zahlen; die Menge aller natürlichen Zahlen enthält also unendlich viele Elemente; sie ist eine unendliche Menge. Diejenigen Mengen nun, die gleichzahlig der Menge aller natürlichen Zahlen sind, deren Elemente also eineindeutig den natürlichen Zahlen zugeordnet werden können, heißen *abzählbar unendliche Mengen.* Diese Bezeichnung soll folgendes ausdrücken: Eine Menge der Anzahl „fünf" ist eine Menge, deren Elemente eineindeutig den *ersten fünf* natürlichen Zahlen zugeordnet werden können, also mit den Zahlen 1, 2, 3, 4, 5 numeriert werden können, die mit Hilfe der *ersten fünf* natürlichen Zahlen abgezählt werden kann; eine abzählbar unendliche Menge ist eine Menge, deren Elemente eineindeutig den *sämtlichen* natürlichen Zahlen zugeordnet werden können, also mit Hilfe der *sämtlichen* natürlichen Zahlen numeriert werden können, die mit Hilfe der *sämtlichen* natürlichen Zahlen abgezählt werden kann. Nach unseren Definitionen haben alle abzählbar unendlichen Mengen dieselbe Anzahl; diese Anzahl muß nun einen Namen bekommen, sowie in alten Zeiten die Anzahl der Menge der Finger einer Hand den Namen „fünf", geschrieben 5, erhalten hat. Cantor hat dieser Anzahl den Namen „Alef-Null", geschrieben $\aleph_0$, gegeben. (Warum er ihr gerade diesen etwas bizarren Namen gegeben hat, wird später verständlich werden.) Die Zahl $\aleph_0$ ist also ein erstes Beispiel einer transfiniten Kardinalzahl; sowie die Aussage: „Eine Menge hat die Anzahl 5" bedeutet: Ihre Elemente können eineindeutig den Fingern der rechten Hand, oder — was auf dasselbe herauskommt — den Zeichen 1, 2, 3, 4, 5 zugeordnet werden, so bedeutet die Aussage: „Eine Menge hat die Anzahl $\aleph_0$": Ihre Elemente können eineindeutig den *sämtlichen* natürlichen Zahlen zugeordnet werden.

Sehen wir uns nach Beispielen abzählbar unendlicher Mengen um, so kommen wir sofort zu höchst überraschenden Resultaten. Die Menge aller natürlichen Zahlen selbst ist abzählbar unendlich; das ist eine Trivialität, dies war ja geradezu die Definition des Begriffes „abzählbar unendlich". Aber auch die Menge aller *geraden* Zahlen ist abzählbar unendlich, hat also dieselbe Anzahl $\aleph_0$ wie die Menge *aller* natürlichen Zahlen, obwohl man doch geneigt wäre zu sagen, es gibt viel weniger gerade als natürliche Zahlen. Um unsere Behauptung zu beweisen, hat man lediglich die in Fig. 1 skizzierte Zuordnung heranzuziehen, d. h. jeder natürlichen Zahl ihr zweifaches zuzuordnen; das ist ersichtlich eine eindeutige Zuordnung zwischen *allen* natürlichen und allen *geraden* Zahlen, und damit ist alles gezeigt. Genau so sieht man, daß die Menge aller *ungeraden* Zahlen abzählbar unendlich

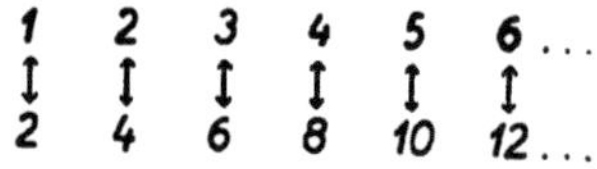

Fig. 1.

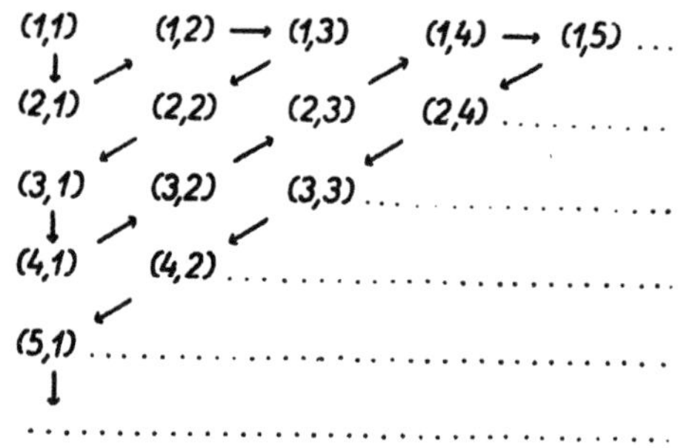

Fig. 2.

ist. Noch überraschender vielleicht ist die Tatsache, daß auch die Menge aller *Paare natürlicher Zahlen* abzählbar

1	2	3	4	5	6	7	8	9	10	11	12	13	14	15...
↕	↕	↕	↕	↕	↕	↕	↕	↕	↕	↕	↕	↕	↕	↕
(1,1)	(2,1)	(1,2)	(1,3)	(2,2)	(3,1)	(4,1)	(3,2)	(2,3)	(1,4)	(1,5)	(2,4)	(3,3)	(4,2)	(5,1)..

Fig. 3.

unendlich ist. Um dies einzusehen, hat man nur die Menge aller Paare natürlicher Zahlen, so wie dies in Fig. 2 angedeutet ist, „nach Diagonalen" zu ordnen, wodurch man sofort die in Fig. 3 angedeutete eineindeutige Zuordnung zwischen allen natürlichen Zahlen und allen Paaren natürlicher Zahlen erhält. Daraus fließt dann leicht die Tatsache, die Cantor schon als Student gefunden hat, daß auch die Menge aller rationalen Brüche (das sind die Quotienten zweier ganzen Zahlen, wie $\frac{1}{2}$, $\frac{2}{3}$ usw.) abzählbar unendlich, also gleichzahlig der Menge aller natürlichen Zahlen, ist, obwohl man doch geneigt wäre zu sagen, daß es ungeheuer viel mehr Brüche als natürliche Zahlen gibt. Aber noch mehr: Cantor konnte sogar beweisen, daß auch die Menge aller sogenannten algebraischen Zahlen, d. h. die Menge aller Zahlen, die einer algebraischen Gleichung $a_0 x^n + a_1 x^{n-1} + \ldots + a_{n-1} x + a_n = 0$ mit ganzzahligen Koeffizienten $a_0, a_1, \ldots, a_n$ genügen, abzählbar unendlich ist.

Nun aber wird sich bei manchem Leser wohl ein Verdacht geregt haben: Sind nicht am Ende *alle* unendlichen Mengen abzählbar unendlich, also untereinander gleichzahlig? Wäre dem so, so wären unsere Hoffnungen arg enttäuscht; es gäbe eben dann neben den endlichen Mengen die unendlichen, die alle untereinander gleichzahlig wären, und mehr wäre darüber nicht zu sagen. Aber im Jahre 1874 gelang Cantor der Beweis, daß es auch unendliche Mengen gibt, die *nicht* abzählbar sind, daß es also verschiedene unendliche Anzahlen, verschiedene transfinite Kardinalzahlen gibt. Insbesondere hat Cantor gezeigt: die Menge aller sogenannten reellen Zahlen (das sind alle ganzen Zahlen, alle Brüche und alle irrationalen Zahlen) ist *unabzählbar unendlich*. Der Beweis ist so einfach, daß ich seinen Gedankengang vorführen kann. Offenbar genügt es, zu zeigen, daß die Menge aller reellen Zahlen zwischen 0 und 1 nicht abzählbar unendlich ist; denn dann ist erst recht die Menge *aller* reellen Zahlen nicht abzählbar unendlich. Wir machen beim Beweise Gebrauch von der bekannten Tatsache, daß jede reelle Zahl zwischen 0 und 1 in einen unendlichen Dezimalbruch entwickelt werden kann. Der zu beweisende Satz: „Die Menge aller reellen Zahlen zwischen 0 und 1 ist nicht abzählbar unendlich" kann auch so ausgesprochen werden: Eine *abzählbar-unendliche* Menge reeller Zahlen zwischen 0 und 1 kann nicht die Menge *aller* reellen Zahlen zwischen 0 und 1 sein, oder noch anders: ist eine abzählbar unendliche Menge reeller Zahlen zwischen 0 und 1 gegeben, so gibt es immer eine reelle Zahl zwischen 0 und 1, die nicht zu der gegebenen Menge gehört. Denken wir uns, um das zu beweisen, eine abzählbar unendliche Menge reeller Zahlen zwischen 0 und 1 gegeben; da sie abzählbar unendlich ist, können die in ihr vorkommenden reellen Zahlen eineindeutig den natürlichen Zahlen zugeordnet werden. Nun schreiben wir die dabei der Zahl 1 zugeordnete reelle Zahl als unendlichen Dezimalbruch an, darunter die der natürlichen Zahl 2 zugeordnete reelle Zahl, darunter die der natürlichen Zahl 3 zugeordnete reelle Zahl, usw.; das wird z. B. so aussehen:

0·20745 . . .
0·16238 . . .
0·97126 . . .
. . . .
. . . .

Nun kann in der Tat sofort eine reelle Zahl zwischen 0 und 1 angegeben werden, die in der gegebenen abzählbar unendlichen Menge reeller Zahlen zwischen 0 und 1 nicht vorkommt: Man wähle ihre erste Dezimale verschieden von der ersten Dezimale der in der ersten Zeile stehenden Zahl, z. B. 3, ihre zweite Dezimale verschieden von der zweiten Dezimale der in der zweiten Zeile stehenden Zahl, z. B. 2, ihre dritte Dezimale verschieden von der dritten Dezimale der in der dritten Zeile stehenden Zahl, z. B. 5 usw. Es ist klar, daß man so eine reelle Zahl zwischen 0 und 1 erhält, die von allen den gegebenen abzählbar unendlich vielen reellen Zahlen verschieden ist, und das ist es, was zu zeigen war.

Damit ist also bewiesen, daß die Menge der natürlichen Zahlen und die Menge der reellen Zahlen nicht gleichzahlig sind; diese beiden Mengen haben verschiedene Kardinalzahlen. Die Kardinalzahl der Menge der reellen Zahlen nannte Cantor die „Mächtigkeit des Kontinuums", wir wollen sie mit $\mathfrak{c}$ bezeichnen. Da, wie früher besprochen, die Menge aller algebraischen Zahlen abzählbar unendlich ist, und, wie wir nun sahen, die Menge aller reellen Zahlen nicht abzählbar unendlich ist, muß es reelle Zahlen geben, die nicht algebraisch sind. Es sind die sogenannten *transzendenten* Zahlen, von denen schon Herr Menger in seinem Vortrag gesprochen hat, und deren Existenz durch diesen von Cantor herrührenden Gedankengang in der denkbar einfachsten Weise bewiesen ist.

Da bekanntlich die reellen Zahlen eineindeutig den Punkten einer Geraden zugeordnet werden können, ist $\mathfrak{c}$ auch die Kardinalzahl der Menge aller Punkte einer Geraden. Überraschender Weise konnte Cantor nachweisen, daß es auch eine eineindeutige Zuordnung zwischen der Menge aller Punkte einer Ebene und der Menge aller Punkte einer Geraden gibt. Diese beiden Mengen sind also gleichzahlig, d. h. $\mathfrak{c}$ ist auch die Kardinalzahl der Menge aller Punkte einer Ebene, obwohl man doch auch hier geneigt wäre zu sagen, daß eine Ebene außerordentlich viel mehr Punkte enthält, als eine Gerade; ja, wie Cantor gezeigt hat, ist $\mathfrak{c}$ auch die Kardinalzahl der Menge aller Punkte des dreidimensionalen Raumes, ja eines Raumes von beliebiger Dimensionszahl.

Wir kennen nun schon zwei verschiedene transfinite Kardinalzahlen: $\aleph_0$ und $\mathfrak{c}$, die Mächtigkeit der abzählbar unendlichen Mengen und die Mächtigkeit des Kontinuums. Gibt es noch andere? Ja, es gibt sicher unendlich viele verschiedene transfinite Kardinalzahlen; denn ist mir irgendeine Menge M gegeben, so kann sofort eine Menge mit größerer Kardinalzahl angegeben werden: die Menge aller möglichen Teilmengen von M hat nämlich größere Kardinalzahl als die Menge M selbst. Nehmen wir z. B. eine Menge aus drei Elementen, etwa die Menge der drei Zeichen 1, 2, 3; ihre sämtlichen Teilmengen sind dann die folgenden Mengen: 1; 2; 3; 1,2; 2,3; 1,3, und das sind mehr als drei; Cantor hat gezeigt, daß dies allgemein[4]), auch für unendliche Mengen gilt. Z. B. hat die Menge aller möglichen Punktmengen auf einer Geraden größere Kardinalzahl als die Menge aller Punkte der Geraden, d. h. ihre Kardinalzahl ist größer als $\mathfrak{c}$.

Nun handelt es sich noch darum, einen Überblick über alle möglichen transfiniten Kardinalzahlen zu gewinnen. Bei den Kardinalzahlen der endlichen Mengen, den natürlichen Zahlen, haben wir folgenden einfachen Sachverhalt: Es gibt unter ihnen eine kleinste, die 1; und: Ist eine endliche Menge M der Anzahl m gegeben, so erhält man eine Menge der nächstgrößeren Anzahl, indem man zur Menge M noch ein Element hinzufügt. Wie steht es diesbezüglich bei den unendlichen Mengen? Da zeigt man unschwer, daß es auch unter den transfiniten Kardinalzahlen eine kleinste gibt: es ist $\aleph_0$, die Mächtigkeit der abzählbar unendlichen Mengen (man glaube aber nicht, daß das selbstverständlich ist: unter allen positiven Brüchen gibt es keinen kleinsten). Nicht so leicht wie bei den endlichen Mengen ist es aber, zu einer transfiniten Kardinalzahl die nächst-

7*

größere zu bilden; denn indem man zu einer unendlichen Menge noch ein Element hinzufügt, bekommt man nicht eine Menge größerer, sondern eine Menge gleicher Anzahl. Aber Cantor hat auch dieses Problem gelöst, indem er zeigte, daß es zu jeder transfiniten Kardinalzahl eine nächstgrößere gibt (was wieder keineswegs selbstverständlich ist: zu einem Bruch gibt es keinen nächstgrößeren) und indem er zeigte, wie man sie erhält. Wir können darauf nicht eingehen, da es uns zu tief in rein mathematische Überlegungen hineinführen würde. Es genügt uns die Tatsache: Es gibt eine kleinste transfinite Kardinalzahl, nämlich $\aleph_0$; es gibt dazu eine nächstgrößere, sie heißt $\aleph_1$; zu dieser gibt es wieder eine nächstgrößere, sie heißt $\aleph_2$ usw. Aber damit sind noch nicht alle transfiniten Kardinalzahlen erschöpft; denkt man sich die Kardinalzahlen $\aleph_0, \aleph_1, \aleph_2, \ldots, \aleph_{10}, \ldots, \aleph_{100}, \ldots, \aleph_{1000}, \ldots$ gebildet, kurz alle Alefs $\aleph_n$, deren Index n eine natürliche Zahl ist, so gibt es wieder eine erste transfinite Kardinalzahl, die größer ist als sie alle, Cantor nannte sie $\aleph_\omega$, zu dieser wieder eine nächstgrößere $\aleph_{\omega+1}$, und so immer weiter.

Da die so sukzessive gebildeten Alefs alle möglichen transfiniten Kardinalzahlen sind, muß auch die Mächtigkeit $\mathfrak{c}$ des Kontinuums unter ihnen vorkommen. Es entsteht also die Frage: welches Alef ist die Mächtigkeit des Kontinuums? Das ist das berühmte Kontinuumproblem. Wir wissen schon, daß dies nicht $\aleph_0$ sein kann, denn, wie wir sahen, ist die Menge aller reellen Zahlen unabzählbar unendlich, d. h. nicht gleichzahlig der Menge der natürlichen Zahlen. Cantor hat vermutet, daß $\aleph_1$ die Mächtigkeit des Kontinuums ist. Doch ist die Frage bis heute offen, und wir sehen vorläufig auch nicht die Spur eines Weges zu ihrer Lösung.[5])

Ich möchte hier noch auf eine logische Feinheit aufmerksam machen, die erst einige Zeit nach dem eben skizzierten Aufbau der Mengenlehre durch G. Cantor von Zermelo bemerkt wurde: Beim Beweise, daß $\aleph_0$ die kleinste transfinite Kardinalzahl ist, sowie beim Beweise, daß jede transfinite Kardinalzahl in der Reihe der Alefs vorkommt, benützt man ein Schlußprinzip, das auch bei einigen anderen mathematischen Beweisen benötigt wird, ohne daß man sich davon explizit Rechenschaft gegeben hätte, und das — wie es scheint — nicht auf die anderen Prinzipe der Logik zurückführbar ist. Es handelt sich dabei um folgendes Prinzip, das als das *Auswahlpostulat* oder *Auswahlaxiom* bezeichnet wird[6]) (von dem schon im Vortrage des Herrn Menger die Rede war): *ist eine Menge von Mengen M gegeben, die zu je zweien kein Element gemein haben, so gibt es auch eine Menge, die mit jeder der Mengen M genau ein Element gemein hat.* Das macht keine Schwierigkeiten, wenn es sich um *endlich viele* Mengen handelt; dann kann man aus jeder dieser Mengen ein Element auswählen, das sind endlich viele Akte; hat man aus jeder der gegebenen Mengen ein Element ausgewählt, so ist man fertig: man hat dann eine Menge, die mit jeder der gegebenen Mengen genau ein Element gemein hat. Es macht auch keine Schwierigkeit, wenn *unendlich viele* Mengen gegeben sind, und zugleich eine Regel gegeben ist, die in jeder dieser Mengen ein Element auszeichnet: Die Menge der ausgezeichneten Elemente ist dann die gewünschte, die mit jeder der gegebenen Mengen genau ein Element gemein hat. Wenn aber unendlich viele Mengen gegeben sind, und keine Regel dazugegeben ist, die in jeder dieser Mengen ein Element auszeichnet, so kommt man nicht so durch, wie im Falle, wo nur endlich viele Mengen gegeben sind; denn wollte man, wie im Falle endlich vieler Mengen, aus jeder der gegebenen Mengen nach Belieben ein Element herausgreifen, würde man ja nie damit fertig, und käme also nie wirklich zur gewünschten Menge, die mit jeder der gegebenen ein Element gemein hat. Darum also stellt die Behauptung, daß es eine solche Menge in jedem Falle gibt, ein eigenes logisches Postulat dar. Der berühmte englische Logiker und Philosoph B. Russell hat das sehr hübsch durch folgenden Scherz anschaulich gemacht: In Kulturländern ist es üblich, daß bei einem Paar Schuhe der rechte sich vom linken unterscheidet, während bei einem Paar Strümpfe dieser Unterschied nicht gemacht wird; denken wir uns einen unendlich reichen Mann (ein Millionär oder Milliardär würde nicht ausreichen, er muß ein „Infinitär" sein) und der besitze aus unendlichem Spleen eine Kollektion von unendlich vielen Schuhpaaren und unendlich vielen Strumpfpaaren; dann verfügt er auch sofort über eine Menge von Schuhen, die aus jedem Paare genau einen enthält, z. B. die Menge der rechten Schuhe aus jedem Paar; aber wie soll man zu einer Menge gelangen, die aus jedem Strumpfpaar genau einen Strumpf enthält? Natürlich dreht es sich bei diesem Beispiel nur um eine scherzhafte Illustration; die Sache selbst aber ist eine ernste und schwerwiegende logische Entdeckung, an die sich manches Problem knüpft. Doch für den Augenblick genug hievon; wir kommen noch darauf zurück.

Durch den im vorstehenden freilich nur ganz skizzenhaft geschilderten Aufbau der Mengenlehre scheint die Frage „Gibt es Unendliches?" durchaus mit „ja" beantwortet: Es gibt nicht nur, wie dies schon Leibniz gesagt hatte, unendliche Mengen, sondern es gibt sogar, was Leibniz geleugnet hatte, unendliche Zahlen; und man kann auch noch zeigen, daß mit ihnen auch sehr wohl gerechnet werden kann, ähnlich, wenn auch nicht genau so, wie mit den endlichen natürlichen Zahlen.

Nun aber müssen wir das Erreichte mit kritischen Augen betrachten. Wenn in der Mathematik mit den Worten „es gibt" eine Existenz, mit den Worten „es gibt nicht" eine Nichtexistenz behauptet wird, so ist damit offenbar etwas ganz anderes gemeint, als wenn dies im täglichen Leben oder etwa in der Geographie oder in der Naturgeschichte geschieht. Machen wir uns das an Beispielen klar! Man weiß etwa seit den Tagen Platos: Es gibt regelmäßige Körper mit vier Seitenflächen (Tetraeder), mit sechs Seitenflächen (Hexaeder oder Würfel), mit acht Seitenflächen (Oktaeder), mit zwölf Seitenflächen (Dodekaeder) und mit zwanzig Seitenflächen (Ikosaeder). Andere regelmäßige Körper als die fünf eben genannten gibt es nicht. Offenbar ist doch mit diesem mathematischen „es gibt", „es gibt nicht" etwas ganz anderes gemeint, als wenn z. B. in der Geographie behauptet wird: es gibt Berge von über achttausend Meter Höhe, es gibt keine Berge von zehntausend Meter Höhe. Denn obwohl die Mathematik lehrt, daß es Würfel und Ikosaeder gibt, so gibt es in dem Sinne, in dem es Berge von über achttausend Meter Höhe gibt, nämlich in der physischen Welt, keine Würfel und keine Ikosaeder; denn das schönste Steinsalzkristall wird doch nicht genau ein Würfel im mathematischen Sinne sein, und ein noch so gut hergestelltes Ikosaedermodell wird doch nicht ein Ikosaeder im mathematischen Sinne sein. Während es nun ziemlich klar ist, was die Wissenschaften, die von der physischen Welt handeln, meinen, wenn sie „es gibt" sagen, ist zunächst gar nicht recht klar, was die Mathematik damit meint. Und in der Tat herrscht darüber unter den Fachleuten, Mathematikern und Philosophen keineswegs Einigkeit. Es wurden da sehr viele verschiedene Meinungen vertreten, fast möchte man sagen: quot capita — tot sententiae; aber geht man aufs Wesentliche, so kann man wohl sagen, daß es im ganzen drei verschiedene Standpunkte gibt, die ich nun kurz schildern will. Den ersten können wir als den *realistischen* oder den *Platonischen Standpunkt* bezeichnen. Er lehrt: Die Objekte der Mathematik haben reale Existenz in einer Welt der Ideen; diese Ideenwelt ist keineswegs identisch mit der physischen Welt, welche vielmehr nur ein unvollkommenes und entstelltes Abbild der Ideenwelt ist (darum gibt es in der physischen Welt keine genauen Würfel, wohl aber in der Welt der Ideen). Während wir mit den Sinnen nur die physische Welt erfassen können, erfassen wir im Denken die Welt der Ideen. Einem mathematischen

Begriffe kommt Existenz zu, wenn ihm ein reales Objekt in der Welt der Ideen entspricht; ein mathematischer Satz ist wahr, wenn er das reale Verhalten der entsprechenden Objekte in der Ideenwelt richtig wiedergibt. Der zweite Standpunkt, den wir den *intuitionistischen* oder *Kantschen Standpunkt* nennen können, lehrt: Wir haben eine reine Anschauung. Mathematik ist Konstruktion in reiner Anschauung, und einem mathematischen Begriffe kommt Existenz zu, wenn er in reiner Anschauung konstruierbar ist. Es handelt sich da um eine philosophische Formulierung einer Auffassung, die man populär etwa so formulieren möchte: Wenn es schon in der physischen Welt keinen genauen Würfel gibt, so kann ich mir einen genauen Würfel doch wenigstens vorstellen. Den dritten Standpunkt wird man am besten den *logistischen Standpunkt* nennen; wollte man ihn historisch auf seine Wurzeln zurück verfolgen, so könnte man ihn wohl an die Nominalistenschule der scholastischen Philosophie knüpfen und daher auch den *nominalistischen* Standpunkt nennen. Er lehrt: Die Mathematik ist eine rein logische Disziplin, spielt sich also wie die Logik durchaus innerhalb der Sprache ab; mit irgendeiner Realität, mit irgendeiner reinen Anschauung hat sie gar nichts zu tun; vielmehr handelt sie lediglich vom Gebrauch von Zeichen, von Symbolen. Zeichen, Symbole aber können wir verwenden wie wir wollen, nach Regeln, die wir selbst stellen, nur dürfen wir uns dabei niemals in Widerspruch zu diesen selbst gestellten Regeln setzen. Letztes Kriterium mathematischer Existenz ist also die *Widerspruchsfreiheit*: Mathematische Existenz kann jedem Begriffe zugeschrieben werden, dessen Gebrauch uns nicht in Widerspruch verwickelt.

Nehmen wir nun in aller Kürze zu diesen Standpunkten kritisch Stellung. Vor zwei Jahren habe ich mich an eben dieser Stelle[7]) bemüht, darzulegen, daß der erste, der realistische Standpunkt unhaltbar ist, weil er dem Denken Fähigkeiten zuschreibt, die es nicht besitzt: unser Denken besteht in tautologischem Umformen, zum Erfassen einer Realität ist es außerstande. Plato selbst hat denn auch eine mystische Rückerinnerung (Anamnesis) unserer Seele an einen Zustand angenommen, in dem sie die Ideen gewissermaßen von Angesicht zu Angesicht schaute. Jedenfalls: Dieser erste Standpunkt ist ein durchaus metaphysischer und erscheint völlig ungeeignet zur Begründung der Mathematik; nichtsdestoweniger richtet er auch heute noch, wenn auch oft unbewußt, in der mathematischen Grundlagenforschung manche Verwirrung an.

Was den zweiten, den intuitionistischen Standpunkt anlangt,

so habe ich mich im Vortragszyklus des vergangenen Jahres[5]) bemüht, darzulegen, daß es so etwas wie eine reine Anschauung überhaupt nicht gibt. Kant selbst hat ihr noch einen sehr weiten Umfang zuerkannt, der sich angesichts der Entwicklung der Mathematik seit jenen Tagen unmöglich aufrecht erhalten ließ; neuere Anhänger dieses zweiten Standpunktes sind denn auch in dieser Hinsicht viel bescheidener geworden. Aber was diese angebliche reine Anschauung nun eigentlich leisten kann und was nicht, was ihr gemäß ist und was nicht, darüber herrscht unter den Anhängern dieses Standpunktes keinerlei Einigkeit. Das zeigt sich sehr deutlich bei der uns hier interessierenden Frage: Gibt es unendliche Mengen, unendliche Zahlen? Einige Intuitionisten sagen da: Beliebig große natürliche Zahlen können wohl in reiner Anschauung konstruiert werden, die Menge *aller* natürlichen Zahlen aber nicht; sie leugnen daher die Existenz irgendeiner unendlichen Menge. Andere meinen, daß auch die Menge aller natürlichen Zahlen in reiner Anschauung konstruierbar sei, aber keine unabzählbar unendliche Menge; sie behaupten also die Existenz abzählbar unendlicher Mengen und leugnen die Existenz unabzählbar unendlicher Mengen (insbesondere die der Menge aller reellen Zahlen). Wieder andere schreiben auch gewissen unabzählbaren Mengen Konstruierbarkeit und damit Existenz zu. Man sieht, daß dieser zweite Standpunkt auf recht schwankender Grundlage ruht; in krassem Gegensatze dazu steht die Schroffheit, mit der die Anhänger dieses Standpunktes alles, was ihrer subjektiven Meinung nach nicht in reiner Anschauung konstruierbar ist, für sinnlos erklären.

Nach dieser Ablehnung der beiden ersten Standpunkte müssen wir uns also auf den dritten, den logistischen Standpunkt stellen.

Bevor wir aber nun die uns interessierende Frage: „Gibt es unendliche Mengen, unendliche Zahlen?" vom logistischen Standpunkt aus besprechen, wird es vielleicht zweckmäßig sein, den Unterschied der drei Standpunkte an einer anderen Fragestellung zu verdeutlichen, indem wir kurz besprechen, wie sich das schon oben erwähnte Auswahlaxiom von jedem der drei Standpunkte ausnimmt. Ein Vertreter des *realistischen* Standpunktes würde sagen: Ob wir in der Logik das Auswahlaxiom anzunehmen oder abzulehnen haben, hängt davon ab, wie die Realität eingerichtet ist; ist sie so eingerichtet, wie das Auswahlaxiom es sagt, so haben wir es anzuerkennen, ist sie nicht so eingerichtet, so haben wir es abzulehnen. Leider wissen wir nicht, welches von beiden der Fall ist, und bei der Mangelhaftigkeit unserer Erkenntnismittel werden wir es bedauerlicherweise auch nie wissen. Ein Vertreter des *intuitionistischen*

Standpunktes würde wohl so sagen: Wir müssen uns besinnen, ob eine Menge, wie sie das Auswahlaxiom verlangt, also eine Menge, die mit jeder Menge eines gegebenen Mengensystems genau ein Element gemein hat, durch Konstruktion in reiner Anschauung erzeugt werden kann; dazu müßte ich aus jeder Menge des gegebenen Mengensystems ein Element auswählen; besteht dieses Mengensystem aus unendlich vielen Mengen, so wären dazu unendlich viele einzelne Akte erforderlich, was nicht durchführbar ist; das Auswahlaxiom ist also abzulehnen. Ein konsequenter Vertreter des *logistischen* Standpunktes aber wird sagen: Wenn wirklich das Auswahlaxiom von den übrigen Prinzipien der Logik unabhängig ist, d. h. wenn sowohl die Aussage des Auswahlaxioms als auch die gegenteilige Aussage mit den übrigen logischen Prinzipien verträglich ist, dann können wir es nach Belieben anerkennen oder durch ein gegenteiliges ersetzen, d. h. wir können sowohl eine Mathematik treiben, in der das Auswahlaxiom als Schlußprinzip benützt wird, eine „Zermelosche Mathematik", als auch eine Mathematik, in der ein gegenteiliges Axiom zugrunde gelegt wird, eine „nicht-Zermelosche Mathematik". Mit dem Verhalten der Realität aber, wie die Realisten das meinen, oder mit reiner Anschauung, wie die Intuitionisten das meinen, hat die ganze Frage nichts zu tun. Es handelt sich vielmehr darum, in welchem Sinne wir uns entschließen wollen, das Wort „Menge" zu verwenden: Es handelt sich um eine Festlegung über die Syntax des Wortes „Menge".

Kehren wir zurück zu dem uns eigentlich beschäftigenden Problem und überlegen wir: Wie wird sich ein Vertreter des logistischen Standpunktes zur Frage stellen: Gibt es unendliche Mengen, unendliche Zahlen? Er wird etwa sagen: Unendliche Mengen, unendliche Zahlen gibt es, falls mit diesen Termen widerspruchsfrei operiert werden kann. Wie also steht es mit dieser Widerspruchsfreiheit?

Nun wurde tatsächlich wiederholt von philosophischer Seite gegen Cantors Schöpfung der Vorwurf erhoben, sie führe zu Widersprüchen. Ein solcher Einwand z. B. lautet: „Nach Cantor ist die Menge aller natürlichen Zahlen gleichzahlig mit der Menge der geraden Zahlen; das aber widerspricht dem Axiom: Das Ganze kann niemals gleich einem seiner Teile sein." Wir müssen uns, um diesen Einwand zu widerlegen, fragen: Was kann der Sinn dieses angeblichen Axioms sein? Gewiß nicht die Behauptung, daß die Realität so beschaffen ist, wie dieses Axiom es sagt; das wäre ein Rückfall in den metaphysischen realistischen Standpunkt. Sein Sinn könnte vielmehr nur der einer syntaktischen Festsetzung sein, in welcher Weise wir die Worte „Ganzes", „Teil", „gleich" verwenden wollen. Und da muß man feststellen, daß diese Festsetzung weder dem Sprachgebrauch in der Umgangssprache noch dem in der Sprache der Wissenschaft völlig entsprechen würde. Man kann doch z. B. nach allgemeinem Sprachgebrauch sicher sagen, daß ein Ganzes hinsichtlich der Farbe gleich einem Teile sein kann — warum soll es verboten sein, zu sagen, daß ein Ganzes hinsichtlich der Anzahl gleich einem seiner Teile sei? Tatsächlich wird das genannte Axiom auch weder beim Aufbau der Logik noch bei dem der Mathematik verwendet. Ein Widerspruch liegt hier also keinesfalls vor, sondern es liegt nur das vor, daß die unendlichen Mengen in der in Rede stehenden Hinsicht ein anderes Verhalten zeigen, als man es von den endlichen Mengen her gewohnt ist: Eine endliche Menge kann nicht gleichzahlig einem ihrer Teile sein, eine unendliche Menge aber kann das. Es läßt sich sogar zeigen, daß jede unendliche Menge Teile hat, mit denen sie gleichzahlig ist, so daß man — wie Dedekind das tat — dieses Verhalten geradezu zur Definition des Begriffes „unendliche Menge" verwenden kann.

Noch viele andere Widersprüche glaubte man in den Begriffen der unendlichen Menge, der unendlichen Zahlen finden zu können. Wie im eben besprochenen Falle laufen diese Einwände gewöhnlich darauf hinaus, zu zeigen, daß gewisse Eigenschaften, die den endlichen Mengen, den endlichen Zahlen notwendig anhaften, den unendlichen Mengen und Zahlen nicht zukommen (z. B.: Jede Anzahl muß entweder gerade oder ungerade sein, Cantors transfinite Kardinalzahlen aber sind weder gerade noch ungerade; oder: Jede Anzahl muß durch Addition der Zahl 1 vergrößert werden, für Cantors transfinite Kardinalzahlen aber trifft das nicht zu). Daß aber die transfiniten Anzahlen teilweise andere Eigenschaften haben als die endlichen, ist kein Widerspruch, es ist nicht einmal verwunderlich, ja es muß sogar so sein; denn würden sie sich in nichts von den endlichen Anzahlen unterscheiden, so wären sie eben endliche und nicht transfinite Anzahlen: auch wenn man eine neue Tierart entdeckt, wird sie sich irgendwie von den bekannten unterscheiden, denn sonst wäre sie eben keine neue Art. Was von allen diesen Argumentationen als allein sinnvoll zurückbleibt, ist nur die Frage: Unterscheiden sich etwa die Begriffe, die man in der Mengenlehre transfinite Zahlen nennt, so stark von dem, was man bis dahin Zahlen nannte, daß es unzweckmäßig ist, auch sie als Zahlen zu bezeichnen? Das kann man natürlich halten wie man will, das ist — wie so viele sogenannte philosophische Probleme — eine bloße Frage der Terminologie, und gerade um solchen nur terminologischen Streitigkeiten auszuweichen, hat Cantor seinen transfiniten Kardinalzahlen auch den neutraleren, weniger belasteten Namen „Mächtigkeiten" gegeben; uns aber mutet dieser Streit ebenso an, wie der in früheren Zeiten ausgefochtene Streit, ob „eins" eine Anzahl sei, oder ob die Anzahlen erst bei „zwei" beginnen. Stellen wir nur fest: Die Terminologie „transfinite Zahlen" hat sich als durchaus zweckentsprechend erwiesen.

Mit Einwendungen dieser Art konnte denn Cantor leicht fertig werden, sie vermochten nicht, sein Gebäude zu erschüttern. Aber dadurch, daß man einige unzulängliche Beweise vermeintlicher Widersprüche widerlegt, ist noch nicht die Widerspruchsfreiheit erwiesen. Und es haben sich tatsächlich in Cantors Gebäude auch ernste Widersprüche ergeben, freilich nicht von so simpler Art wie die vermeintlichen Widersprüche, über die wir uns eben unterhielten. Es zeigten sich Widersprüche bei gewissen allzu umfassenden Mengenbildungen, wie der Menge aller Gegenstände, der Menge aller Mengen, der Menge aller unendlichen Anzahlen. Als Quelle dieser Widersprüche aber erwies sich nicht der Unendlichkeitsbegriff als solcher, diese Quelle lag vielmehr in gewissen Mängeln der klassischen Logik: Es wurde eine Reform der Logik nötig, über die Herr Menger im Vortragszyklus des vergangenen Jahres[9]) sprach. Sie bestand hauptsächlich in einer vorsichtigeren Verwendung des Wortes „alle", wie sie in Russells Theorie der logischen Typen[10]) gelehrt wird. So gelang es, alle bekannten Widersprüche, die sich in der Mengenlehre gezeigt hatten, aufzuklären und zu vermeiden. Im heutigen Aufbau der Mengenlehre ist ein Widerspruch nicht mehr bekannt.

Aber daraus, daß kein Widerspruch bekannt ist, folgt nicht, daß keiner vorhanden ist, ebensowenig wie daraus, daß im Jahre 1900 noch kein Okapi bekannt war, folgen konnte, daß es kein Okapi gibt. Und hier erhebt sich nun die Frage: wie kann überhaupt ein Widerspruchsfreiheitsbeweis geführt werden? Auch über diese Frage hat Herr Menger im vorjährigen Vortragszyklus gesprochen. Allem Anscheine nach ist dazu zu sagen: *Absolute* Widerspruchsfreiheitsbeweise gibt es nicht; jeder Widerspruchsfreiheitsbeweis ist *relativ*, er kann nur die Widerspruchsfreiheit eines Systemes auf die eines anderen zurückführen. Ist aber diese Erkenntnis nicht deletär für den logistischen Standpunkt, der mathematische Existenz auf Widerspruchsfreiheit gründen will? Ich glaube nein! Es zeigt sich nur eben auch hier wie überall, daß die Forderung nach einem absolut gesicherten Wissen eine überspannte Forderung ist — es gibt auf gar keinem Gebiete ein absolut gesichertes Wissen! Auch die von manchen Philosophen herangezogene sogenannte Evidenz der inneren Wahrnehmung bei einer Aussage wie „ich sehe jetzt etwas Weißes" liefert kein absolut sicheres Wissen. Indem ich den Satz „ich sehe etwas Weißes" formuliere und statuiere, bezieht er sich schon auf Vergangenes und ich kann nie wissen, ob in der auch noch so kurzen verflossenen Zeit mich mein Gedächtnis nicht getrogen hat.

Einen absoluten Widerspruchsfreiheitsbeweis für die Mengenlehre und damit einen Beweis für die mathematische Existenz unendlicher Mengen und unendlicher Zahlen gibt es also nicht. Einen solchen gibt es aber auch nicht für die Arithmetik der endlichen Zahlen, ja nicht einmal für die einfachsten Teile der Logik. Tatsache aber ist, daß ein Widerspruch im System der Mengenlehre nicht bekannt ist, und daß auch nicht die Spur eines Indizes zu sehen ist, daß ein solcher Widerspruch vorhanden sein könnte. Wir können mit angenähert derselben Sicherheit, mit der wir den endlichen Zahlen mathematische Existenz zubilligen, sie auch den unendlichen Mengen und Cantors transfiniten Zahlen zubilligen.

Wir haben bisher nur die Frage behandelt, ob es unendliche Mengen, unendliche Anzahlen gibt; nicht minder wichtig scheint aber die Frage, ob es unendliche *Ausdehnungen* gibt. Gewöhnlich wird sie so gestellt: *Ist der Raum unendlich?* Behandeln wir auch diese Frage zunächst vom rein mathematischen Standpunkt!

Da müssen wir sogleich sagen, daß in der Mathematik sehr verschiedene Arten von Räumen betrachtet werden. Uns interessieren hier einzig die sogenannten Riemannschen Räume, über die Herr Nöbeling im Vortragszyklus des Vorjahres ausführlich gesprochen hat[11]), und zwar interessieren uns insbesondere die dreidimensionalen Riemannschen Räume. Auf ihre genaue Definition kommt es uns hier nicht an; es genügt uns, folgendes festzuhalten: Ein solcher Riemannscher Raum ist eine Menge von Elementen, Punkte genannt, in der gewisse Teilmengen, Linien genannt[12]), betrachtet werden; jeder solchen Linie kann durch ein Rechenverfahren eine positive Zahl, die Länge dieser Linie genannt, zugeordnet werden und unter diesen Linien gibt es solche, von denen jedes hinlänglich kleine Stück AB kleinere Länge hat als jede andere dieselben Punkte A, B verbindende Linie; diese Linien heißen die *geodätischen* oder auch *geraden* Linien des betreffenden Raumes. Es kann nun sein, daß es im betrachteten Riemannschen Raume gerade Linien von beliebig großer Länge gibt; dann werden wir sagen, der betreffende Raum hat *unendliche Ausdehnung*; es kann aber auch sein, daß im betrachteten Riemannschen Raume die Länge aller geraden Linien unter einer festen Zahl bleibt; dann werden wir sagen, der betreffende Raum hat *endliche Ausdehnung*. Bis Ende des 18. Jahrhunderts war nur ein einziger mathematischer Raum bekannt, der deshalb „der Raum" schlechthin genannt wurde; es ist der Raum, dessen Geometrie in allen Schulen gelehrt wird, und der heute als der *euklidische* Raum bezeichnet wird, nach dem griechischen Mathematiker Euklid, der als erster die Geometrie dieses Raumes systematisch entwickelt hat; er ist im Sinne unserer Definition von *unendlicher* Ausdehnung.

Es gibt aber sehr wohl auch dreidimensionale Riemannsche Räume *endlicher* Ausdehnung; die bekanntesten sind die sogenannten *sphärischen* (und die mit ihnen nahe verwandten elliptischen) Räume, die dreidimensionale Analoga einer Kugeloberfläche sind. Eine Kugeloberfläche kann als zweidimensionaler Riemannscher Raum aufgefaßt werden, dessen geodätische oder „gerade" Linien die Bögen von Großkreisen sind (ein Großkreis einer Kugel ist ein Kreis, der aus der Kugel durch eine Ebene herausgeschnitten wird, die durch den Kugelmittelpunkt hindurchgeht, wie z. B. der Äquator und die Meridiane auf der Erdkugel). Ist r der Radius der Kugel, so ist der volle Umfang eines Großkreises gegeben durch $2r\pi$; kein Großkreisbogen kann also größere Länge haben als $2r\pi$, d. h. die Kugel, als zweidimensionaler Riemannscher Raum betrachtet, ist ein Raum endlicher Ausdehnung. Ganz analog geht es im dreidimensionalen sphärischen Raume zu; auch er ist ein Raum endlicher Ausdehnung; trotzdem aber hat er keinerlei Grenzen, er ist unbegrenzt, so wie eine Kugeloberfläche unbegrenzt ist; man kann auf einer seiner Geraden immer weiter wandern, ohne je durch eine Grenze des Raumes aufgehalten zu werden; nur kommt man nach endlicher Zeit an den Ausgangspunkt zurück, ganz ebenso wie auf einer Kugeloberfläche, wenn man auf einem Großkreise immer weiter und weiter wandert. So wie auf der Erdoberfläche eine Erdumseglung, so kann also in einem sphärischen Raume eine Raumumseglung durchgeführt werden.

Wir sehen also: Im mathematischen Sinne gibt es Räume unendlicher Ausdehnung (z. B. den euklidischen Raum) und Räume endlicher Ausdehnung (z. B. die sphärischen und elliptischen Räume). Das ist aber gar nicht das, was die meisten Menschen interessiert, wenn sie fragen: „Ist der Raum unendlich?" Das, wonach sie da fragen ist vielmehr dies: Ist der Raum, in dem sich unsere Erfahrung, in dem sich das physische

Geschehen abspielt, ist der Raum der physischen Welt von endlicher oder unendlicher Ausdehnung?

Solange man keinen anderen mathematischen Raum als den euklidischen kannte, war man selbstverständlich der Ansicht, der Raum der physischen Welt sei der unendlich ausgedehnte euklidische Raum, und Kant hat diese Auffassung ganz explizit formuliert; seiner Meinung nach ist die Einordnung unserer Beobachtungen in den euklidischen Raum eine Anschauungsnotwendigkeit, die Ausgangssätze der euklidischen Geometrie sind synthetische Urteile a priori.

Aber als man darauf kam, daß rein mathematisch auch andere Räume als der euklidische existieren, d. h. widerspruchsfrei sind, erhoben sich naturgemäß Zweifel, ob es wirklich so apodiktisch sicher sei, daß der Raum der physischen Welt der euklidische Raum sei, und es entstand die Meinung, es sei eine Frage der Erfahrung, also eine Frage, die durch das Experiment entschieden werden müsse, ob der Raum der physischen Welt der euklidische ist oder nicht. Gauß hat denn auch tatsächlich solche Experimente angestellt. Seit H. Poincaré, dem größten Mathematiker am Ende des 19. Jahrhunderts, wissen wir, daß die Frage, so gestellt, keinen Sinn hat[13]). Es steht in weitem Maße in unserer Willkür, in was für einen mathematischen Raum wir unsere Beobachtungen einordnen wollen. Die Frage gewinnt erst einen Sinn, wenn etwas darüber festgesetzt wird, *wie* diese Einordnung vorgenommen werden soll. Das wesentliche an einem Riemannschen Raume ist nun die Art und Weise, wie in ihm jeder Linie eine Länge zugewiesen wird, d. h. wie in ihm Längen gemessen werden. Setzen wir fest, daß die Längenmessungen im Raume des physischen Geschehens so vorgenommen werden sollen, wie es seit Urzeiten tatsächlich geschieht, nämlich durch Aneinanderlegen sogenannter „starrer" Maßstäbe, dann bekommt die Frage, ob der in dieser Weise zu einem Riemannschen Raum gemachte Raum des physischen Geschehens ein euklidischer oder nichteuklidischer Raum ist, einen Sinn; und dasselbe gilt von der Frage, ob er von unendlicher oder endlicher Ausdehnung ist.

Die vielleicht manchem auf der Zunge liegende Antwort: „Selbstverständlich wird durch diese Art des Messens der physische Raum zu einem mathematischen Raum von unendlicher Ausdehnung" wäre etwas voreilig. Um ein paar Worte zu dieser Problemstellung sagen zu können, müssen wir eine kurze, ganz simple mathematische Erörterung vorausschicken. Der euklidische Raum ist dadurch charakterisiert, daß in ihm die Summe der drei Winkel eines Dreieckes 180 Grad beträgt; in einem sphärischen Raume ist die Winkelsumme in jedem Dreiecke größer als 180 Grad, und der Überschuß über 180 Grad ist um so größer, je größer das Dreieck ist. Beim zweidimensionalen Analogon eines sphärischen Raumes, der Kugeloberfläche, sieht man das anschaulich vor sich; die Rolle der geradlinigen Dreiecke des sphärischen Raumes spielen, wie schon besprochen, auf der Kugeloberfläche die Dreiecke aus Großkreisbogen, und es ist eine sehr bekannte Tatsache der Elementargeometrie, daß die Winkelsumme in einem solchen sphärischen Dreiecke größer als 180 Grad ist, und daß der Überschuß um so größer ist, je größer der Flächeninhalt des Dreieckes ist. Vergleicht man ferner auf verschieden großen Kugeln sphärische Dreiecke gleichen Flächeninhaltes, so sieht man sofort, daß der Überschuß der Winkelsumme über 180 Grad um so größer ist, je kleiner der Durchmesser der Kugel, d. h. je größer die Krümmung der Kugel ist. Das gab Anlaß zur Einführung folgender Terminologie (und zwar handelt es sich dabei lediglich um eine *Terminologie*, hinter der sich gar nichts Geheimnisvolles verbirgt): Man nennt einen mathematischen Raum *gekrümmt*, wenn es in ihm Dreiecke gibt, deren Winkelsumme von 180 Grad abweicht, und zwar positiv gekrümmt, wenn in ihm — wie in den elliptischen und sphärischen Räumen — die Winkelsumme in allen Dreiecken größer als 180 Grad ist, negativ gekrümmt, wenn sie kleiner als 180 Grad ist — wie dies in den von Bolyai und Lobatschefsky entdeckten „hyperbolischen" Räumen der Fall ist.

Aus den mathematischen Ansätzen von Einsteins *allgemeiner Relativitätstheorie* ergibt sich nun, daß bei Zugrundelegung der vorhin besprochenen Art des Messens der Raum in der Umgebung gravitierender Massen, in einem „Gravitationsfelde" gekrümmt sein muß[14]). Das einzige uns unmittelbar zugängliche Gravitationsfeld, das der Erde, ist viel zu schwach, um das direkt nachprüfen zu können; auf indirektem Wege aber, nämlich durch die Ablenkung der Lichtstrahlen im viel stärkeren Gravitationsfelde der Sonne, die bei totalen Sonnenfinsternissen festgestellt wurde, konnte dies bestätigt werden. Soweit unsere heutigen Erfahrungen reichen, können wir also sagen: Machen wir durch die vorhin besprochene Art des Messens den Raum des physischen Geschehens zu einem mathematischen, Riemannschen Raum, so wird dieser mathematische Raum gekrümmt sein, und zwar wird seine Krümmung von Ort zu Ort verschieden sein, größer in der Nähe gravitierender Massen und kleiner ferne von ihnen.

Nun zurück zu der uns hier beschäftigenden Frage! Können wir nunmehr auch etwas darüber aussagen, ob dieser Raum endliche oder unendliche Ausdehnung haben wird? Das bisher Besprochene reicht dazu noch nicht aus. Wir müssen vielmehr noch gewisse einigermaßen plausible Annahmen machen. Eine solche plausible Annahme ist die, die Massen seien im ganzen Weltraume im groben Durchschnitte ziemlich gleichmäßig verteilt, es herrsche im Weltraume im Durchschnitt räumlich konstante Massendichte. Die bisherigen Beobachtungen der Astronomen können damit wenigstens bei einigem guten Willen einigermaßen in Einklang gebracht werden. Natürlich kann es nur im groben Durchschnitte stimmen, etwa so, wie wir recht gut sagen können, ein Stück Eisen habe im Durchschnitt überall dieselbe Dichte, obwohl doch seine Masse in sehr vielen ganz kleinen Korpuskeln konzentriert ist, die durch Zwischenräume getrennt sind, die im Verhältnisse zur Größe dieser Korpuskeln riesenhaft sind, so wie die Sterne im Weltraum durch Zwischenräume getrennt sind, die im Verhältnis zu ihrer Größe riesenhaft sind. Machen wir weiter die auch recht plausible Annahme, daß die Welt, im groben Durchschnitt genommen, stationär sei in dem Sinne, daß sich diese durchschnittlich konstante Massendichte unverändert erhält — so wie wir ein Stück Eisen doch als stationär ansehen, obwohl wir der Ansicht sind, daß die es zusammensetzenden Korpuskeln sich in lebhafter Bewegung befinden, so wie auch die Sterne im Weltraum in lebhafter Bewegung begriffen sind. Dann aber ergibt sich aus den Ansätzen der allgemeinen Relativitätstheorie, daß der mathematische Raum, in dem wir das physische Geschehen deuten wollen, im Durchschnitte überall dieselbe positive Krümmung aufweisen muß; ein solcher Raum ist aber, ebenso wie im zweidimensionalen die Kugeloberfläche, notwendig von endlicher Ausdehnung. Also: Wollen wir das physische Geschehen unter Zugrundelegung der bekannten Art der Längenmessung in einen mathematischen Raum einordnen, und machen wir die beiden besprochenen plausiblen Annahmen, dann kommen wir zum Schlusse, daß dieser Raum von *endlicher* Ausdehnung sein muß.

Ich sagte, daß die erste unserer Annahmen, die von der räumlich konstanten Massendichte, einigermaßen zu den Beobachtungen stimmt. Ist das auch mit der zweiten Annahme, der von der zeitlich konstanten Massendichte der Fall? Bis vor kurzem konnte man dieser Meinung sein. Nun scheinen aber gewisse astronomische Beobachtungen darauf hinzudeuten, daß — wieder im groben Durchschnitte gesprochen — alle Himmelskörper (Fixsterne und Nebelflecke) sich von uns entfernen, mit um so größerer Geschwindigkeit, je weiter sie von uns abstehen, die am weitesten von uns entfernten, die noch daraufhin untersucht werden konnten, mit geradezu phantastischen Geschwindigkeiten. Dann aber kann die durchschnittliche Massendichte der Welt unmöglich zeitlich konstant sein, sie muß vielmehr mit der Zeit immer geringer werden. Das würde bei Festhalten an den übrigen Zügen unseres Weltbildes besagen, daß wir den mathematischen Raum, in dem wir das physische Geschehen deuten, als zeitlich veränderlich annehmen müssen: In jedem Augenblicke wäre er ein Raum von im Durchschnitte konstanter positiver Krümmung, also von endlicher Ausdehnung, aber diese Krümmung nähme fortwährend ab, seine Ausdehnung also fortwährend zu. Eine solche Einordnung des physischen Geschehens in einen sich expandierenden Raum erweist sich als mathematisch durchaus möglich und in vollem Einklange mit der allgemeinen Relativitätstheorie.

Aber ist diese Deutung die einzig mögliche, die mit unserer bisherigen Erfahrung im Einklang steht, oder zeigen sich auch andere Deutungsmöglichkeiten? Ich sagte oben, die Annahme, der Weltraum sei im Durchschnitt überall gleich dicht mit Masse erfüllt, ließe sich leidlich gut mit den bisherigen astronomischen Beobachtungen in Einklang bringen; aber mit den astronomischen Beobachtungen steht auch die ganz andere Annahme nicht im Widerspruch, wir befänden uns mit unserem Fixsternsystem in einem Teile des Raumes, in dem eine stärkere Massenkonzentration statthat, während in größerer Entfernung von dieser Gegend des Raumes die Massenverteilung immer dünner wird; dann würden wir — immer bei Festhalten an der üblichen Art der Längenmessung — zur Einordnung des physischen Geschehens auf einen Raum geführt, der in der Gegend unseres Fixsternsystems eine gewisse Krümmung aufweist, die aber weiter draußen immer geringer und geringer wird[15]). Ein solcher Raum kann nun sehr wohl von unendlicher Ausdehnung sein. Und auch das Phänomen, daß die Sterne sich im Durchschnitt von uns wegbewegen, um so schneller, je weiter sie von uns entfernt sind, läßt sich so ganz einfach deuten[16]): nehmen wir an, daß irgendeinmal viele, mit ganz verschiedenen Geschwindigkeiten begabte Massen sich in einem relativ kleinen Bezirk des Raumes, sagen wir in einer Kugel *K*, konzentriert fanden; sie werden sich dann im Verlaufe der Zeit, jede mit der ihr eigenen Geschwindigkeit, aus diesem Bezirk des Raumes herausbewegen, und nach Verlauf einer genügend langen Zeit werden diejenigen dieser Massen, die die größte Geschwindigkeit haben, sich schon am weitesten von der Kugel *K* entfernt haben, die mit geringeren Geschwindigkeiten begabten werden noch näher an *K* sein, und die mit den geringsten Geschwindigkeiten begabten werden noch ganz nahe an K oder noch innerhalb K sein; dann aber wird sich für einen Beobachter, der sich innerhalb K befindet oder wenigstens nicht allzuweit von K entfernt ist, gerade das Bild zeigen, das uns, wie oben gesagt, die Sternenwelt darbietet: es werden sich im Durchschnitte die Massen von ihm entfernen, und zwar gerade die am weitesten von ihm abstehenden mit der größten Geschwindigkeit. Und damit hätten wir eine Deutung in einem ganz anders gearteten mathematischen Raume wie früher, und zwar in einem *unendlich* ausgedehnten Raume.

Zusammenfassend können wir wohl sagen: Die Frage, „ist der Raum des physischen Geschehens unendlich oder endlich ausgedehnt?" hat an sich keinen Sinn. Sie bekommt erst einen Sinn, wenn gewisse Forderungen an die Art der Einordnung des physischen Geschehens in einen mathematischen Raum gestellt werden. Und das läuft hinaus auf die Frage: „Eignet sich ein endlich oder ein unendlich ausgedehnter mathematischer Raum besser zur Einordnung des physischen Geschehens?" Auf diese Frage aber können wir beim heutigen Stande unseres Wissens keine einigermaßen solid begründete Antwort geben. Es scheint, daß sich mathematische Räume von endlicher und von unendlicher Ausdehnung etwa gleich gut zur Einordnung unseres bisherigen Beobachtungswissens eignen.

Vielleicht werden unentwegte Finitisten bei dieser Lage der Dinge sagen: „Dann ziehen wir die Deutung in einem Raume endlicher Ausdehnung vor, denn alles Unendliche ist uns nun einmal zuwider". Das mögen sie halten, wie sie wollen, aber sie dürfen nicht glauben, daß sie damit allem Unendlichen entronnen sind. Denn auch die endlich ausgedehnten Riemannschen Räume enthalten unendlich viele Punkte und auch der mathematische Ansatz für die Zeit ist so, daß jedes noch so kleine Zeitintervall unendlich viele Zeitpunkte enthält.

Muß das so sein? Zwingt uns irgend etwas dazu, unsere Erfahrung in einem mathematischen Raume, in einer mathematischen Zeit zu deuten, die aus unendlich vielen Punkten bestehen? Ich glaube: Nein! Prinzipiell wäre sehr wohl eine Physik denkbar, in der es nur endlich viele Raumpunkte und endlich viele Zeitpunkte, also in der Sprache der Relativitätstheorie nur endlich viele Weltpunkte gibt. Ich glaube: keine Logik, keine Anschauung, keine Erfahrung kann je die Unmöglichkeit einer solchen wahrhaft finiten Physik dartun. Ich weiß nicht, ob die verschiedenen Lehren von einer atomistischen Struktur der Materie, ob die heutige Quantenphysik die ersten Schatten sind, die eine künftige finite Physik vorauswirft. Wenn sie aber je käme, so wären wir nach einem ungeheuren Kreislaufe zu einem der Ausgangspunkte des abendländischen Denkens zurückgekehrt: zur pythagoräischen Lehre, daß alles in der Welt beherrscht wird durch die natürlichen Zahlen. Wenn der berühmte Lehrsatz vom rechtwinkligen Dreiecke den Namen des Pythagoras mit Recht trägt, so war es Pythagoras selbst, der seine Lehre, daß alles durch die natürlichen Zahlen beherrscht sei, in ihren Grundfesten erschütterte. Denn aus dem Lehrsatz vom rechtwinkligen Dreiecke folgt die Existenz von Strecken, die in irrationalem Verhältnis stehen, deren Verhältnis also nicht durch natürliche Zahlen ausdrückbar ist. Und da man nicht zwischen mathematischer Existenz und physischer Existenz unterschied, schien jede finite Physik unmöglich. Wenn man sich aber darüber klar ist, daß mathematische Existenz und physische Existenz Grundverschiedenes meinen, daß aus einer mathematischen Existenz niemals eine physische Existenz folgen kann, daß physische Existenz in letzter Linie nur durch Beobachtung dargetan werden kann, daß aber der mathematische Unterschied zwischen rational und irrational jede Beobachtungsmöglichkeit für immer transzendiert, dann wird man die prinzipielle Möglichkeit einer finiten Physik kaum leugnen können. Sei dem aber wie immer, mögen künftige Zeiten eine finite Physik bringen oder nicht, an der Möglichkeit, an der großartigen Schönheit einer *Logik* und einer *Mathematik* des Unendlichen wird dadurch nicht gerührt.

Literatur.

[1]) Vgl. hiezu: P. Duhem, Études sur Léonard de Vinci, ceux qu'il a lus et ceux qui l'ont lu. Seconde série (Paris, Hermann, 1909). IX. Léonard de Vinci et les deux infinis.

[2]) Neue Ausgabe (mit Anmerkungen versehen von H. Hahn) bei Felix Meiner, Leipzig 1920. (Der Philosophischen Bibliothek Band 99.)

[3]) Man findet die grundlegenden Abhandlungen von G. Cantor in: Georg Cantor, Gesammelte Abhandlungen mathematischen und philosophischen Inhalts, herausgegeben von E. Zermelo (Berlin, J. Springer, 1932). Dieser Band enthält auch eine von A. Fraenkel verfaßte Biographie Cantors und ein Bildnis Cantors. Zum eingehenderen Studium der Mengenlehre sei empfohlen: A. Fraenkel, Einleitung in die Mengenlehre. Dritte Auflage (Berlin, J. Springer, 1928).

[4]) Für Mengen M, die aus einem einzigen oder aus zwei Elementen bestehen, gilt es nur, wenn man auch die „leere Menge“, die gar kein Element enthält, und die Menge M selbst mit zu den Teilmengen von M rechnet.

[5]) Näheres über das Kontinuumproblem: W. Sierpiński, Hypothèse du continu (Monografje Matematyczne, Tom IV, Warszawa-Lwow 1934).

8*

[6]) Näheres über das Auswahlpostulat: W. Sierpiński, L'axiome de M. Zermelo et son rôle dans la théorie des ensembles et l'analyse. Bull. de l'Acad. des Sciences de Cracovie, Classes des sciences math. et nat., Série A, 1918, S. 97—152; W. Sierpiński, Leçons sur les nombres transfinis (Paris, Gauthier-Villars, 1928). Chap. VI.

[7]) Im Vortragszyklus zugunsten der Errichtung eines Grabdenkmales für Ludwig Boltzmann. Dieser Vortrag ist erschienen unter dem Titel: H. Hahn, Logik, Mathematik und Naturerkennen (Einheitswissenschaft, Heft 2, Wien, Gerold, 1933).

[8]) Krise und Neuaufbau in den exakten Wissenschaften (Leipzig u. Wien, Deuticke, 1933): H. Hahn, Die Krise der Anschauung.

[9]) Krise und Neuaufbau in den exakten Wissenschaften. K. Menger: Die neue Logik.

[10]) Vgl. A. N. Whitehead und B. Russell, Principia Mathemati Vol. I (Sec. ed., Cambridge 1925) S. 37 ff. (Deutsche Übersetzung unter dem Titel: Einführung in die mathematische Logik, München—Berlin, Drei Masken-Verlag, 1932).

[11]) Krise und Neuaufbau in den exakten Wissenschaften. G. Nöbeling, Die vierte Dimension und der krumme Raum.

[12]) Der Einfachheit halber verwende ich hier und im Folgenden das Wort „Linie" statt der in der Mathematik üblichen Bezeichnung „einfacher Kurvenbogen".

[13]) H. Poincaré, La science et l'hypothèse (Paris, Flammarion. Deuxième partie: L'espace). Deutsche Ausgabe (mit erläuternden Anmerkungen von F. und L. Lindemann): Wissenschaft und Hypothese (Leipzig, Teubner, 1906).

[14]) Man vergleiche zum Folgenden die leichtfaßliche Darstellung von A. Haas, Kosmologische Probleme der Physik (Leipzig, Akademische Verlagsgesellschaft, 1934), sowie die ausführliche, aber viel schwierigere Darstellung von H. Weyl, Raum, Zeit, Materie (5. Aufl., Berlin, Springer, 1923).

[15]) Man wird so auf mathematische Räume geführt, die dreidimensionale Analoga eines Rotationsparaboloides sind. Vgl. H. Weyl, a. a. O., S. 257.

[16]) Die folgende Deutung rührt her von E. A. Milne. Vgl. A. Haas, a. a. O., S. 59, und E. Freundlich, Die Naturwissenschaften 21 (1933), S. 54.

Testimonials – Lebenszeugnisse

Aus einem Brief von Hans Hahn an Paul Ehrenfest vom 26. 12. 1909

Im vergangenen Jahr bin ich innerlich der Mathematik nahezu untreu geworden, umgarnt von den Reizen der – Philosophie. Bei Poincaré, Mach, Hertz fängt es herrlich an, dann kommt Kant, und unaufhaltsam weiter geht es bis zu Aristoteles und Konsorten. Heute kommt es mir ganz absurd vor, die Geringschätzung mit der unsere Fachgenossen über diese Leute sprechen; die glauben vielfach ernstlich, daß ein Mann, dessen Name nach fast 2000 Jahren noch wirksam war wie am ersten Tag, nur läppischen Unsinn geschrieben habe. Ich bin Gemütserregungen wenig zugänglich; einem Freund, der so weit ist wie Du, gestehe ich es: ich habe hie und da Ehrfurcht gespürt bei meinen flüchtigen Leseversuchen in der aristotelischen Metaphysik – und viel Bedauern, daß mir jede Möglichkeit fehlt, diese Sachen durchzugrübeln, wie ich etwa die Variationsrechnung durchdacht habe. Ich habe wohl manche Komplimente gehört über meine Anlage zur Philosophie, erforsche ich mein Gewissen genau, so kann ich nicht leugnen, daß auch ich glaube, in dieser Richtung veranlagt zu sein, und ich kann nur eins sagen: ich bin überzeugt, daß *Probleme* diesen ganzen Fragen zugrunde liegen und daß die so vielfach zu hörende gegenläufige Behauptung ein läppisches, teils auf Ignoranz, teils auf Unfähigkeit (speziell in dieser einen Richtung) beruhendes Geschwätz ist.

Hans Hahns Lebenslauf, geschrieben anläßlich seiner Wahl zum korrespondierenden Mitglied der Österreichischen Akademie der Wissenschaften, im Frühjahr 1921

Ich wurde geboren zu Wien, am 27. IX. 1879, besuchte da das Gymnasium, wurde an der Universität Wien immatrikuliert im Herbst 1898, um daselbst, dem Wunsche meines Vaters folgend, Jus zu studieren. Da mir aber dieses Studium in keine Weise zusagte, beschloß ich alsbald, mich der Mathematik zuzuwenden. Ernsthafte mathematische Studien begann ich an der Universität Straßburg, wo ich im Jahre 1899/1900 H. Weber und Th. Reige hörte, sodann verbrachte ich das Wintersemester 1900/01 in München, wo mir durch Vorlesungen von A. Pringsheim der Sinn für arithmetische Strenge der Beweisführung erschlossen wurde. Sodann kehrte ich nach Wien zurück, wo ich, durch Vorlesungen von G. v. Escherich angeregt, mich eingehend mit Variationsrechnung zu befassen begann, was für eine Reihe von Jahren für meine wissenschaftliche Tätigkeit ausschlaggebend wurde. Mit vielem Interesse hörte ich damals auch die eigenartigen Vorlesungen von F. Mertens über Algebra und Zahlentheorie. Im Sommer 1902 wurde ich in Wien zum Dr. phil. promoviert. Im Wintersemester 1903/04 war ich in Göttingen, wo ich insbesondere durch Vorlesungen und Seminare von D. Hilbert gefördert und beeinflußt wurde. Im Jahre 1905 habilitierte ich mich an der Universität Wien, wo ich bis zu meiner Ernennung zum a. o. Professor a. d. Universität Czernowitz (1909) als Privatdozent Vorlesungen hielt, mit einer einsemestrigen Unterbrechung: im Wintersemester 1905/06 suppliierte ich an der Universität Innsbruck das durch den Rücktritt von O. Stolz erledigte Ordinariat. Vom 1. Oktober 1909 bis 1. Oktober 1916 war ich a. o. Professor a. d. Universität Czernowitz, beschäftigte ich mich intensiv mit dem Aufarbeiten der Literatur über Mengenlehre und die Theorie der reellen Funktionen. Ich hatte nämlich über Aufforderung von A. Schoenflies die Aufgabe übernommen, gemeinsam mit ihm die zweite Auflage seines Berichtes über die Entwicklung der Lehre von den Punktmannigfaltigkeiten herauszugeben, wobei ich insbesondere die Anwendung der Mengenlehre auf die Theorie der reellen Funktionen bearbeiten sollte. Ich sah alsbald, daß hier eine völlig neue und selbständige Darstellung erforderlich sei; diese war im ersten Entwurf fast fertiggestellt, als der Krieg ausbrach. Im Jänner 1915 wurde ich einberufen, kam im Herbste desselben Jahres ins Feld, wurde schwer verwundet und nach meiner Wiederherstellung als Lehrer an der Kadettenschule in Breitensee und am k. u. k. Lehrervorbereitungskurse verwendet, bis ich im Oktober 1916 einem Rufe als a. o. Professor an die Universität Bonn folgte, wo ich bereits ein halbes Jahr später, als Nachfolger von F. London, ein Ordinariat erhielt. In Bonn konnte ich endlich meine Darstellung der Theorie der reellen Funktionen weiter führen, die zu einem umfassenden Werke gediehen war, von dem zu Ende des Jahres 1920 der erste Band erschien, weiters der zweite Band im Manuskripte fertig gestellt ist. Die Jahre in Bonn waren für mich vor allem wertvoll durch meinen freundschaftlichen Verkehr mit G. Study, durch daß ich viele Einblicke in die mir ferner liegenden Gebiete der Geometrie gewann. Im April 1921 erfolge meine Ernennung zum ord. Prof. a. d. Universität Wien als Nachfolger von G. Escherich.

Brief von Wilhelm Wirtinger, Vorstand des Instituts für Mathematik der Universität Wien, an Hans Hahn (23. Jänner 1923)

Vielen Dank für Ihren lieben Brief vom 1. 1. 1920. Sie haben also den Mut gehabt, trotz all der trüben und schlechten Aussichten nach Wien zu kommen und ich möchte mir aufrichtig wünschen, daß Sie es nicht bereuen. Ich bewundere Ihren Optimismus und wünsche von Herzen, daß er Recht behalten möge gegenüber meinen Vorstellungen, daß alle die momentanen Mittel, die über augenblickliche Schwierigkeiten hinweghelfen, doch einmal versagen werden und eine Stabilisierung und ruhigere Entwicklung erst in einer Zeit eintreten wird, die wir nicht mehr erleben werden. Aber bei dieser Auffassung werde ich es als eine besondere Wohltat empfinden, wenn Sie mir wieder von Ihren Arbeitsgebieten erzählen werden und über Ihre Pläne sprechen.

Eine andere wichtige Sache für das Seminar ist die Fortführung der Monatshefte. Ich möchte Sie jetzt schon bitten in die Redaction einzutreten und diese auch wirklich zu führen. Ich habe das nun 17 Jahre gethan. Es ist bei den heutigen Verhältnissen wegen des Schriftenaustausches besonders wichtig, der von allen Ländern ausgenommen Frankreich wieder aufgenommen wurde. In Rußland sind wohl die inneren Zustände daran schuld, daß es nicht geschehen ist. Sie können nicht mehr jahrgangsweise erscheinen, sondern in „Bänden", das heißt wenn ich Geld und Material habe, drucke ich wieder einen Band. Der letzte Band kam so zu Stande, daß die Regierung 20.000 Kronen und Holländer 20.000 Kronen gaben. Für den nächsten Band stehen 40.000 aus einer Spende des Herrn Dr. Stonborough, die Einnahmen für den Verkauf dieses Jahrgangs und ein Rest aus den früheren Beträgen zur Verfügung. Ich hoffe im ganzen auf 50.000, auch hat für die nächsten Jahre Dr. Stonborough und Frau 40.000 pro Jahr in Aussicht gestellt.

An die Regierung bin ich noch nicht herangetreten, weil ich erst die Abrechnung des Verkaufs des letzten Jahrganges abwarten will. Vielleicht setzt Ihre frische Energie da auch noch mehr durch – wenn noch etwas da ist. Die Möglichkeit fortzuwursteln ist also vorhanden und irgendwie wird es noch eine Weile weitergehen.

Aus einem Brief von Leopold Vietoris an Leopold Schmetterer vom 27. 1. 1994

Ich war in den Studienjahren 1922/23 bis 1926/27 Assistent am Mathematischen Seminar (so hieß es damals) der Universität Wien, zugleich mit Josef Lense; zwei Assistenten für die drei Professoren Wirtinger, Furtwängler und Hahn! Trotzdem ging es sehr gut. Lense und ich hatten neben unserer Assistententätigkeit noch reichlich Zeit für eigene wissenschaftliche Arbeit.

Dabei hatte ich auch Gelegenheit, Vorlesungen von Hans Hahn zu hören. Sie zeichneten sich durch wohldurchdachten und ansprechenden Vortrag aus. Besonders in den von ihm abgehaltenen Seminaren, an denen ich auch als Vortragender mitwirken durfte, habe ich sehr viel von Hans Hahn gelernt.

Besonders schätze ich seine von weltanschaulichen oder politischen Gesichtspunkten nicht beeinflußte Beurteilung wissenschaftlicher und dienstlicher Leistungen.

Aus einem Brief von Georg Nöbeling an Leopold Schmetterer vom 7. 3. 1993

Während meiner vier Wiener Jahre (1929–1933) hat mich Hahn (neben Menger natürlich) sehr gefördert. Noch heute steht mir seine hohe Gestalt und seine eindringliche Redeweise lebhaft vor Augen.

Hahn war sehr vielseitig interessiert. So war er ein sehr engagiertes Mitglied des „Wiener Kreises“ um Schlick mit Carnap, Kaufmann, Neurath, Frau Neurath (Hahns blinder Schwester), Menger, Gödel. Der Kreis tagte im Mathematischen Institut in der Strudlhofgasse. Während der Sitzungen rauchte Hahn sehr gern eine Virginia, die ihm aber während jeder seiner Diskussionsbeiträge regelmäßig erlosch. – Hahn interessierte sich auch für den Spiritismus. Er soll auch an Séancen teilgenommen haben. Zu welchen Erkenntnissen er dabei gekommen ist, weiß ich nicht. Daß ihn seine Skepsis nicht von vornherein zu einer Ablehnung veranlaßte, habe ich nicht verstanden.

Eine Sportlernatur war Hahn nicht. Immerhin brach er sich beim Skifahren einmal ein Bein.

Im Institut erzählte man sich einmal eine für Hahn typische Szene. Zur Winterszeit wurde Hahn Zeuge, wie auf der glatten Gasse ein Pferd (das gab es damals noch, wenn auch selten) stürzte. Der Fuhrwerkslenker versuchte, das arme Tier wieder hochzuprügeln. Hahn warf sich dazwischen und rief nach der Polizei. Wie die Sache ausgegangen ist, weiß ich nicht.

Kurt Gödel: Besprechung des Buches von Hans Hahn: Theorie der reellen Funktionen Bd. I [erschienen in den Monatsheften für Mathematik und Physik, ***40*** *(1933)]*

H. Hahn, Reelle Funktionen. (Mathematik und ihre Anwendung, Bd. 13.) Akademische Verlagsgesellschaft, Leipzig 1932. Preis geb. RM 30,–.

Die seit langem vergriffene „Theorie der reellen Funktionen" desselben Autors erscheint hier in einer sowohl inhaltlich als in der Darstellungsform völlig neuen Bearbeitung. Die im vorliegenden ersten Band enthaltenen Kapitel (die Theorie der Mengenfunktionen ist einem zweiten Band vorbehalten) sind entsprechend den seither erzielten neuen Ergebnissen weit eingehender behandelt als in dem früheren Buche und ein neues Kapitel über analytische Mengen ist dazugekommen. Die Beweise sind mit einer Exaktheit und Ausführlichkeit gegeben, die in der mathematischen Lehrbuchliteratur wohl kaum ihresgleichen hat und die nicht mehr weit von der völligen „Formalisierung" (im Sinne etwa der Principia Mathematica) entfernt ist. Dabei ist die Anzahl der explizit formulierten Lehrsätze so groß, daß jeder einzelne unter Berufung auf die vorgehenden verhältnismäßig kurz bewiesen werden kann (ein Beweis überschreitet nur selten eine halbe Seite und ist meistens noch wesentlich kürzer). Ein weiterer charakteristischer Zug des Buches ist das überall durchgeführte Prinzip, mit der geringsten Zahl von Voraussetzungen auszukommen und daher auch alles möglichst abstrakt zu formulieren.

Trotz der detaillierten Angabe der Beweise ist das Werk auch inhaltlich außerordentlich reichhaltig. Viele Resultate aus der modernsten Zeit, die in den verschiedensten Zeitschriften und Büchern verstreut sind, wurden hier zum ersten Male systematisch dargestellt, mit zahlreichen Verbesserungen, Ergänzungen und neuen Beweisen, die vielfach (insbesondere in dem Kapitel über B a i r e sche Funktionen) auch für den Kenner Neues bringen werden. Jedem Abschnitt ist ein ausführliches Verzeichnis der Originalarbeiten beigegeben.

Vorkenntnisse werden außer den elementarsten Dingen keine vorausgesetzt und im I. Kapitel werden daher die wichtigsten Sätze aus der abstrakten Mengenlehre über Kardinalzahlen, Ordinalzahlen und Mengensysteme abgeleitet.

Das II. Kapitel behandelt die Theorie der Punktmengen (topologische und metrische Räume) und beschränkt sich dabei durchaus nicht auf das im folgenden Verwendete; z. B. wird die Metrisierbarkeit der regulären separablen Räume und ihre Einbettbarkeit in den Hilbert schen Raum bewiesen. Besonders ausführlich sind die Begriffe „von I. und von II. Kategorie", „residual" etc. sowie die entsprechenden lokalen Begriffe „von I. Kategorie in einem Punkt" etc. behandelt.

Es folgt ein Kapitel (III) über den Begriff der Stetigkeit. Zunächst werden beliebige (auch mehr-mehrdeutige) Abbildungen zweier metrischer Räume aufeinander betrachtet, wobei der Begriff der Stetigkeit aufgespalten wird in „oberhalb stetig" (d. h. die Urbilder der abgeschlossenen Mengen sind abgeschlossen) und „unterhalb stetig" (d. h. die Urbilder der offenen Mengen sind offen), wodurch auch die Lehrsätze entsprechend in Paare zerfallen. Auch die Sätze über die Erweiterbarkeit von in Teilräumen definierten stetigen Funktionen auf den ganzen Raum werden auf mehr-mehrdeutige Abbildungen verallgemeinert. Im Zusammenhang mit den stetigen Abbildungen werden die oberhalb stetigen Zerlegungssysteme behandelt und wird ferner die Homöomorphie jedes absoluten G_δ mit einem vollständigen Raum (nach Hausdorff), sowie die stetige Durchlaufbarkeit der im kleinen zusammenhängenden Kontinua nach einem neuen Verfahren von Whyburn bewiesen. Es folgt die Theorie der Schrankenfunktionen und der Schwankung, weiter ein Abschnitt über die verschiedenen Arten der gleichmäßigen Konvergenz, über Ungleichmäßigkeitsgrad, gleichgradige Stetigkeit u. ä. Der (aus dem Jahre 1922 stammende) Satz, daß es zu jeder in einem separablen vollständigen Raum R definierten Funktion eine in R dichte Menge gibt, auf der sie stetig ist, wird bewiesen, samt der Ergänzung, daß es nicht immer eine abzählbare Menge dieser Art gibt; ferner auch die Hurewicz sche Verallgemeinerung des Satzes von der Beschränktheit jeder auf einer kompakten Menge stetigen Funktion auf Folgen stetiger Funktionen.

Kapitel IV behandelt die Borel schen Mengen und Baire schen Funktionen. Die dieser Theorie zugrunde liegenden Sätze werden ganz abstrakt für beliebige, gewissen einfachen Bedingungen genügende Funktionssysteme mit einer beliebigen Menge (nicht notwendig metrischem

Raum) als Definitionsbereich ausgesprochen. Zwischen den Begriffen und Sätzen über Borel sche Mengen und Baire schen Funktionen wird ein vollkommener Parallelismus hergestellt. Zum Beispiel erscheinen die Lusin schen Trennungssätze (deren einfachster Fall ist die Trennbarkeit zweier fremder G_δ durch Mengen, die sowohl F_σ als G_δ sind) als Analoga zu den Einschiebungssätzen (deren einfachster Fall die Existenz einer stetigen Funktion zwischen einer oberhalb stetigen und einer größeren unterhalb stetigen Funktion ist). Neu gegenüber der früheren „Theorie der reellen Funktionen" sind ferner u. a. Bedingungen dafür, daß der Limes einer Folge von Funktionen α-ter Klasse *zur* selben Klasse gehört und dafür, daß zwei Funktionen f und g $\underline{\lim}$ und $\overline{\lim}$ derselben Funktionsfolge eines Funktionssystems $\mathfrak{S}$ sind. Das Kapitel schließt mit Untersuchungen über partiell stetige Funktionen, wobei u. a. bewiesen wird, daß, wenn $f(x_1 x_2 \ldots x_n)$ eine partiell (d. h. nach jeder Koordinate) stetige Funktion im R_n ist, $f(x\, x \ldots x)$ eine Funktion höchstens $n-1$-ter Klasse im R_1 ist, und daß umgekehrt jede Funktion $n-1$-ter Klasse im R_1 sich in dieser Form durch eine partiell stetige Funktion darstellen läßt.

Das letzte (V.) Kapitel bringt eine ausführliche Theorie der analytischen Mengen, u. a. ihre Darstellbarkeit als Wertmenge stetiger und halbstetiger Funktionen und als Menge der Werte k–facher (abzählbarer, unabzählbarer) Vielfachheit stetiger Funktionen im Baire schen Nullraum, ferner eine Reihe von Sätzen über das Verhältnis der analytischen zu den Borel schen Mengen. Es folgt eine Anwendung der analytischen Mengen auf die durch Baire sche Funktionen implizit definierten Funktionen und den Schluß bildet die Theorie der Lusin schen Siebe.

Im Rahmen einer Besprechung ist es leider völlig unmöglich, den behandelten Stoff auch nur annähernd zu erschöpfen, doch dürfte auch schon aus dem Gesagten zur Genüge die außerordentliche Reichhaltigkeit des Werkes insbesondere an modernen Ergebnissen hervorgehen, welche es vor allem auch als Nachschlagewerk für selbständig Arbeitende vorzüglich geeignet macht.

K. Gödel

Aus unveröffentlichten Erinnerungen von Prof. Olga Taussky-Todd, University of Pasadena

Recollections of Hans Hahn

– *1925–1931:* When I entered the University of Vienna in 1925, determined to study number theory, the three Full professors were Ph. Furtwängler (aged 56), W. Wirtinger (aged 60) and Hans Hahn (aged 46) (from 1927 on Karl Menger [1902–1985] was an Extraordinarius). The basic courses were given by them in turn und in my case Wirtinger was in charge. It was not until my third year that I had any significant contact with Hahn. This came about in the following way. Hahn, by far the most active of the three, ran various seminars, but had a rule that students presenting talks should be rehearsed, so that time would not be wasted. Walter Mayer, then a Privat-Dozent, later associated with Einstein, recruited and rehearsed me – I do not now remember what subject I spoke in, but it must have gone down well for after that, Hahn took an interest in my welfare. As far as I remember I did not attend any of his regular courses. In due course (7 March 1930) I received my Ph. D. – my supervisor was Furtwängler.

In the year after graduation I attend various seminars and I had my first really personal contact with Hahn. He suggested that we wrote together reviews of the van der Waerden volumes on Modern Algebra which had just appeared. The review of Part I (1930) appeared under both our signatures in Monatshefte f. Math. u. Physik 30 (1931) (Literaturberichte, 11–12). Apparently Hahn was satisfied with my work for the review of Part II (1931) appeared under my own signature in the next volume of the Monatshefte 40 (1932) (Literaturberichte, 3–4). Thus I owe also to Hahn personal instruction in the writing of reviews, which I continued to do for 60 years.

– *1931–1932:* I owe my first proper job to Hans Hahn. It happened like this. Courant was organizing the editing of Hilbert's „Werke“ and talked to Hahn about some young people who could do the work. Hahn recommended me and I, together with W. Magnus and H. Ulm, spent the academic year 1931/32 in Göttingen on the job, which was expected to be completed for Hilbert's 70th birthday on 23 January 1932. Vol. 1, with which I was mainly concerned, actually had a 1932 imprint while Vol. 2 appeared in 1933 and Vol. 3 in 1935.

– *1932–1934:* I returned to Vienna in the summer of 1932 and spent the next two years there – this was the time of my closest contact with

Hahn. Hahn and Menger were more sympathetic to the younger mathematician and e. g. organized public lectures which were published and which provided some loose money to support them.

At this time Hahn asked me to supervise the publication of a thesis by J. Pfleger. I do not know how I got competent in the subject – probably just on-the-job training. After this was done, Hahn, who was ailing, handed over another student, August Fröhlich to me.

I remember visiting Hahn, when he was in a „nursing-home" for treatment, to report progress to him and how he always introduced me to the staff. Fröhlich's thesis was completed in May 1934. Unfortunately Hahn died on 24 July 1934 and I went to Bryn Mawr College for the academic year 1934/35 and to Girton College for 1935–1937.

Next let me mention another thing for which I am indebted to Hahn, my long time friendship with Dr. Auguste Dick. As I recall it happened this way: Hahn assigned students of his course to Teaching Assistants and she was on the list assigned to me. Incidentally Hahn told his students that if they wanted to change their TA (for whatever reason), it could be arranged. Auguste stuck with me and completed her doctorate (in differential geometry) and became a well recognized teacher. On her retirement she began writing the definitive biography of Emmy Noether and got into contact with me again in this connection. In addition she has contributed greatly by her historical work on Schrödinger and on members of the Vienna Mathematical Community, in particular by her Erinnerungen an Hans Hahn.

Finally let me give some general impressions of Hahn, more from the point of view of an assistant, than as a student. There is no doubt that Hahn was a powerful mathematician of great breadth and great depth – witness, e. g. his students Hurewicz, Menger, Gödel. Apart from his interest in mathematics (including logic), he was politically active for the socialist party and an ardent follower of ESP (Extra Sensory Perception). In connection with ESP I recall attending a lecture of his on this subject and observing how he handled questions from doubting Thomases and what he thought of fakers who harmed the relevant research.

The Hahn – Banach theorem is one of the staples of Functional Analysis. There has been some controversy recently that E. Helly did not get sufficient credit for his contribution to this theorem. I would like to record that Hahn, when I was telling him about a lecture I was due to give, was insistent that I mention Helly's contribution (this in the early thirties).

On some occasion there was a discussion among the senior mathematicians about what they considered their best work. Hahn was quite sure that his was concerned with the „Streckenbild“, the topological characterization of a continuous map of a line segment (as compact, connected and locally connected). There was quite a lot of activity in this area at one time, but now it seems of little general interest.

No matter which of Hahn's contributions is regarded as the most memorable, there is no doubt that his overall contribution to mathematics will remain outstanding among those of the Austrian mathematicians of his time.

Schriftenverzeichnis / List of Publications
Hans Hahn

[1] (1903) Zur Theorie der zweiten Variation einfacher Integrale, *Monatshefte f. Mathematik u. Physik,* **14**, 3–57.

[2] (1903) Über die Lagrangesche Multiplikationsmethode in der Variationsrechnung, *Monatshefte f. Mathematik u. Physik,* **14**, 325–342.

[3] (1904) Bemerkungen zur Variationsrechnung, *Mathematische Annalen,* **58**, 148–168.

[4] (1904) Über das Strömen des Wassers in Röhren und Kanälen (gemeinsam mit G. Herglotz und K. Schwarzschild), *Zeitschr. f. Mathem. u. Physik,* **51**, 411–426.

[5] (1904) Über den Fundamentalsatz der Integralrechnung, *Monatshefte f. Mathematik u. Physik,* **16**, 161–166.

[6] (1904) Über punktweise unstetige Funktionen, *Monatshefte f. Mathematik u. Physik,* **16**, 312–320.

[7] (1904) Weiterentwicklung der Variationsrechnung in den letzten Jahren (gemeinsam mit E. Zermelo), *Enzyklopädie der mathemat. Wissensch.,* Teubner, Leipzig, **II A**, 8a, 627–641.

[8] (1905) Über Funktionen zweier komplexer Veränderlichen, *Monatshefte f. Mathematik u. Physik,* **16**, 29–44.

[9] (1906) Über einen Satz von Osgood in der Variationsrechnung, *Monatshefte f. Mathematik u. Physik,* **17**, 63–77.

[10] (1906) Über das allgemeine Problem der Variationsrechnung, *Monatshefte f. Mathematik u. Physik,* **17**, 295–304.

[11] (1907) Über die nicht-archimedischen Größensysteme, *Sitzungsber. d. Akademie d. Wiss. Wien, math.-naturw. Klasse,* **116**, 601–655.

[12] (1907) Über die Herleitung der Differentialgleichungen der Variationsrechnung, *Math. Annalen,* **63**, 253–272.

[13] (1908) Bemerkungen zu den Untersuchungen des Herrn M. Fréchet: Sur quelques points du calcul fonctionnel, *Monatshefte f. Mathematik u. Physik,* **19**, 247–257.

[14] (1908) Über die Anordnungssätze der Geometrie, *Monatshefte f. Mathematik u. Physik,* **19**, 289–303.

[15] (1909) Über Bolzas fünfte notwendige Bedingung in der Variationsrechnung, *Monatshefte f. Mathematik u. Physik,* **20**, 279–284.

[16] (1909) Über Extremalenbogen, deren Endpunkt zum Anfangspunkt konjugiert ist, *Sitzungsber. d. Akademie d. Wissenschaften Wien, math.-naturw. Klasse,* **118**, 99–116.

[17] (1910) Über den Zusammenhang zwischen den Theorien der zweiten Variation und der Weierstraßschen Theorie der Variationsrechnung, *Rendiconti del Circolo Matematico di Palermo,* **29**, 49–78.

*[18] (1910) Arithmetik, Mengenlehre, Grundbegriffe der Funktionenlehre, in E. Pascal, *Repertorium der höheren Mathematik,* Teubner, Leipzig, **Bd. I, 1,** Kap. I, 1–42.

[19] (1911) Bericht über die Theorie der linearen Integralgleichungen, *Jahresbericht der D. M. V.,* **20**, 69–117.

[20] (1911) Über räumliche Variationsprobleme, *Math. Annalen,* **70**, 110–142.

[21] (1911) Über Variationsprobleme mit variablen Endpunkten, *Monatshefte f. Mathematik u. Physik,* **22**, 127–136.

[22] (1912) Über die Integrale des Herrn Hellinger und die Orthogonalinvarianten der quadratischen Formen von unendlich vielen Veränderlichen, *Monatshefte f. Mathematik u. Physik,* **23**, 161–224.

[23] (1912) Allgemeiner Beweis des Osgoodschen Satzes der Variationsrechnung für einfache Integrale, *II. Weber-Festschrift,* 95–110.

[24] (1913) Ergänzende Bemerkungen zu meiner Arbeit über den Osgoodschen Satz in Band 17 dieser Zeitschrift, *Monatshefte f. Mathematik u. Physik,* **24**, 27–33.

[25] (1913) Über einfach geordnete Mengen, *Sitzungsber. d. Akademie d. Wissenschaften Wien, math.-naturw. Klasse,* **122**, 945–967.

[26] (1913) Über die Abbildung einer Strecke auf ein Quadrat, *Annali di Matematica,* **21**, 33–55.

[27] (1913) Über die hinreichenden Bedingungen für ein starkes Extremum beim einfachsten Probleme der Variationsrechnung, *Rendiconti del Circolo Matematica di Palermo,* **36**, 379–385.

[28] (1914) Über die allgemeinste ebene Punktmenge, die stetiges Bild einer Strecke ist, *Jahresbericht der D. M. V.,* **23**, 318–322.

[29] (1914) Über Annäherung an Lebesguesche Integrale durch Riemannsche Summen, *Sitzungsber. d. Akademie d. Wiss. Wien, math.-naturw. Klasse,* **123**, 713–743.

[30] (1914) Mengentheoretische Charakterisierung der stetigen Kurve, *Sitzungsber. d. Akademie d. Wiss. Wien, math.-naturw. Klasse,* **123**, 2433–2490.

[31] (1915) Über eine Verallgemeinerung der Riemannschen Integraldefinition, *Monatshefte f. Mathematik u. Physik,* **26**, 3–18.

[32] (1916) Über die Darstellung gegebener Funktionen durch singuläre Integrale I und II, *Denkschriften d. Akademie d. Wiss. Wien, math.-naturw. Kl.,* **93**, 585–692.

[33] (1916) Über Fejérs Summierung der Fourierschen Reihe, *Jahresbericht der D. M. V.,* **25**, 359–366.

[34] (1917) Über halbstetige und unstetige Funktionen, *Sitzungsber. d. Akademie d. Wiss. Wien, math.-naturw. Klasse,* **126**, 91–110.

[35] (1918) Über stetige Funktionen ohne Ableitung, *Jahresbericht der D. M. V.,* **26**, 281–284.

[36] (1918) Über das Interpolationsproblem, *Mathem. Zeitschrift,* **1**, 115–142.

[37] (1918) Einige Anwendungen der Theorie der singulären Integrale, *Sitzungsber. d. Akademie d. Wiss. Wien, math.-naturw. Klasse,* **127**, 1763–1785.

[38] (1919) Über die Menge der Konvergenzpunkte einer Funktionenfolge, *Archiv d. Math. u. Physik,* **28**, 34–45.

[39] (1919) Über die Vertauschbarkeit der Differentiationsfolge, *Jahresbericht der D. M. V.,* **27**, 184–188.

[40] (1919) Über Funktionen mehrerer Veränderlicher, die nach jeder einzelnen Veränderlichen stetig sind, *Mathem. Zeitschrift,* **4**, 307–313.

[41] (1919) Besprechung von Alfred Pringsheim: Vorlesungen über Zahlen- und Funktionenlehre, *Göttingische gelehrte Anzeigen,* **9–10**, 321–347.

*[42] (1920) Bernhard Bolzano, Paradoxien des Unendlichen (mit Anmerkungen versehen von H. Hahn), Leipzig.

[43] (1921) Über die Komponenten offener Mengen, *Fundamenta mathematicae,* **2**, 189–192.

[44] (1921) Arithmetische Bemerkungen (Entgegnung auf Bemerkungen des

Herrn J. A. Gmeiner), *Jahresbericht der D. M. V.*, **30**, 170–175.

[45] (1921) Schlußbemerkungen hiezu, *Jahresbericht der D. M. V.*, **30**, 178–179.

[46] (1921) Über die stetigen Kurven der Ebene, *Mathem. Zeitschrift*, **9**, 66–73.

[47] (1921) Über irreduzible Kontinua, *Sitzungsber. d. Akademie d. Wiss. Wien, math.-naturw. Klasse*, **130**, 217–250.

[48] (1921) Über die Darstellung willkürlicher Funktionen durch bestimmte Integrale (Bericht), *Jahresbericht der D. M. V.*, **30**, 94–97.

*[49] (1921) Theorie der reellen Funktionen I, Berlin, Springer Verlag.

[50] (1922) Über Folgen linearer Operationen, *Monatshefte f. Mathematik u. Physik*, **32**, 3–88.

[51] (1922) Über die Lagrangesche Multiplikatorenmethode, *Sitzungsber. d. Akademie d. Wiss. Wien, math.-naturw. Klasse*, **131**, 531–550.

[52] (1923) Über Reihen mit monoton abnehmenden Gliedern, *Monatshefte f. Mathematik u. Physik*, **33**, 121–134.

[53] (1923) Die Äquivalenz der Cesàroschen und Hölderschen Mittel, *Monatshefte f. Mathematik u. Physik*, **33**, 135–143.

[54] (1924) Über Fouriersche Reihen und Integrale, *Jahresbericht der D. M. V.*, **33**, 107.

[55] (1925) Über ein Existenztheorem der Variationsrechnung, *Anzeiger d. Akad. d. W. in Wien*, **62**, 233.

[56] (1925) Über die Methode der arithmetischen Mittel, *Anzeiger d. Akad. d. W. in Wien*, **62**, 233–234.

[57] (1925) Über ein Existenztheorem der Variationsrechnung, *Sitzungsber. d. Akademie d. Wiss. Wien, math.-naturw. Klasse*, **134**, 437–447.

[58] (1925) Über die Methode der arithmetischen Mittel in der Theorie der verallgemeinerten Fourierschen Integrale, *Sitzungsber. d. Akademie d. Wiss. Wien, math.-naturw. Klasse*, **134**, 449–470.

*[59] (1925) Einführung in die Elemente der höheren Mathematik, (gemeinsam mit H. Tietze), Leipzig, 12 + 330 S.

[60] (1926) Über eine Verallgemeinerung der Fourierschen Integralformel, *Acta mathematica*, **49**, 301–353.

[61] (1927) Über lineare Gleichungssysteme in linearen Räumen, *Journal f. d. reine u. angew. Mathematik*, **157**, 214–229.

*[62] (1927) Variationsrechnung, in *Repertorium der höheren Mathematik*, E. Pascal, Teubner, Leipzig, **Bd. I, 2**, Kap. XIV, 626–684.

[63] (1928) Über additive Mengenfunktionen, *Anzeiger d. Akad. d. W. in Wien*, **65**, 65–66.

[64] (1928) Über unendliche Reihen und totaladditive Mengenfunktionen, *Anzeiger d. Akad. d. W. in Wien*, **65**, 161–163.

[65] (1928) Über stetige Streckenbilder, *Anzeiger d. Akad. d. W. in Wien*, **65**, 281–282.

[66] (1928) Über stetige Streckenbilder, *Atti del Congresso Internazionale dei Matematici*, Bologna, Band 2, 217–220.

[67] (1929) Über den Integralbegriff, *Anzeiger d. Akad. d. W. in Wien*, **66**, 19–23.

[68] (1929) Über den Integralbegriff, *Festschrift der 57. Versammlung Deutscher Philologen und Schulmänner in Salzburg vom 25. bis 29. September 1929*, 193–202.

[69] (1929) Empirismus, Mathematik, Logik, *Forschungen und Fortschritte*, **5**, 409–410.

[70] (1929) Mengentheoretische Geometrie, *Die Naturwissenschaften*, **17**, 916–919.

*[71] (1929) Die Theorie der Integralgleichungen und Funktionen unendlich vieler Variablen und ihre Anwendung auf die Randwertaufgaben bei gewöhnlichen und partiellen Differentialgleichungen (gemeinsam mit L. Lichtenstein und J. Lense), in E. Pas-

cal, *Repertorium der höheren Mathematik,* Teubner, Leipzig, **Bd. I, 3**, Kap. XXIV, 1250–1324.

[72] (1930) Über unendliche Reihen und absolut-additive Mengenfunktionen, *Bulletin of the Calcutta Mathem. Society,* **20**, 227–238.

[73] (1930) Die Bedeutung der wissenschaftlichen Weltauffassung, insbesondere für Mathematik und Physik, *Erkenntnis,* **1**, 96–105.

[74] (1930) Überflüssige Wesenheiten (Occams Rasiermesser) (Veröff. Ver. Ernst Mach), Wien, 24 S.

[75] (1931) Diskussion zur Grundlegung der Mathematik, *Erkenntnis,* **2**, 135–141.

*[76] (1932) Reelle Funktionen, Teil 1: Punktfunktion, *Math. und ihre Anwendung in Monogr. u. Lehrb.,* **13**, Leipzig, 11 + 415 S.

[77] (1933) Über separable Mengen, *Anzeiger d. Akad. d. W. in Wien,* **70**, 58–59.

[78] (1933) Über die Multiplikation total additiver Mengenfunktionen, *Annali di Pisa,* **2**, 429–452.

[79] (1933) Logik, Mathematik und Naturerkennen, *Einheitswissenschaft,* Heft **2**, Wien, 33 S.

[80] (1933) Die Krise der Anschauung, in Krise und Neuaufbau in den exakten Wissenschaften, 5 Wiener Vorträge, 1. Zyklus, Leipzig, Wien, 41–64.

[81] (1934) Gibt es Unendliches?, in Alte Probleme – Neue Lösungen in den exakten Wissenschaften, 5 Wiener Vorträge, 2. Zyklus, Leipzig, Wien, 93–116.

*[82] (1948) Set functions, hrsg. v. A. Rosenthal, University of New Mexico Press, 9 + 324 S.

Inhaltsverzeichnis, Band 1
Table of Contents, Volume 1

Hahn's Work on Curve Theory / Hahns Arbeiten zur Kurventheorie

Hahn's Work in Algebra / Hahns Arbeit zur Algebra

Inhaltsverzeichnis, Band 2
Table of Contents, Volume 2

Hahn's Work in Real Analysis /
Hahns Arbeiten zur reellen Analysis

Hahn's Work in Hydrodynamics /
Hahns Arbeit zur Hydrodynamik

Springer-Verlag
and the Environment

WE AT SPRINGER-VERLAG FIRMLY BELIEVE THAT AN international science publisher has a special obligation to the environment, and our corporate policies consistently reflect this conviction.

WE ALSO EXPECT OUR BUSINESS PARTNERS – PRINTERS, paper mills, packaging manufacturers, etc. – to commit themselves to using environmentally friendly materials and production processes.

THE PAPER IN THIS BOOK IS MADE FROM NO-CHLORINE pulp and is acid free, in conformance with international standards for paper permanency.

Springer-Verlag
und Umwelt

Als internationaler wissenschaftlicher Verlag sind wir uns unserer besonderen Verpflichtung der Umwelt gegenüber bewußt und beziehen umweltorientierte Grundsätze in Unternehmensentscheidungen mit ein.

Von unseren Geschäftspartnern (Druckereien, Papierfabriken, Verpackungsherstellern usw.) verlangen wir, daß sie sowohl beim Herstellungsprozeß selbst als auch beim Einsatz der zur Verwendung kommenden Materialien ökologische Gesichtspunkte berücksichtigen.

Das für dieses Buch verwendete Papier ist aus chlorfrei hergestelltem Zellstoff gefertigt und im pH-Wert neutral.

www.ingramcontent.com/pod-product-compliance
Lightning Source LLC
LaVergne TN
LVHW062004041026
849558LV00006B/183

* 9 7 8 3 2 1 1 8 2 7 8 1 9 *